T0300840

Tuatara

Tuatara

Biology and conservation of a venerable survivor

Alison Cree

Canterbury University Press gratefully acknowledges generous financial support from the Society for Research on Amphibians and Reptiles in New Zealand (SRARNZ) and the Department of Conservation.

First published in 2014 by
CANTERBURY UNIVERSITY PRESS
University of Canterbury
Private Bag 4800, Christchurch
NEW ZEALAND
www.cup.canterbury.ac.nz

ISBN 978-1-927145-44-9

A catalogue record for this book is available from the National Library of New Zealand.

Cover design concept: Ken Miller
Book design and layout: Quentin Wilson & Associates, Christchurch, New Zealand
Typeset in Caslon Pro, 11pt/14pt

Cover images: Main cover image of adult tuatara, modified from the photograph on page 513, by Alison Cree
Back cover image of juvenile tuatara by Ken Miller
Author photo by Ken Miller

Printed in China through Asia Pacific Offset Limited

Contents

PART THREE: FUTURE SURVIVAL

Acknowledgements

Tuatara fascinate people from many walks of life. As a result, I have been blessed with assistance from a remarkably diverse range of contacts.

Among my academic colleagues, I am deeply indebted to Charles Daugherty and Michael Thompson for introducing me to wild tuatara. In 1986 Charlie invited me to join his research group at Victoria University of Wellington Te Whare Wānanga o te Ūpoko o te Ika a Māui as a postdoctoral fellow. For several years I worked alongside Charlie and Mike, the latter also a postdoctoral fellow in Charlie's group. During our adventures I spent over a year in total on Stephens Island (also known as Takapourewa, the premier island for studying tuatara in abundance), and observed tuatara on at least 13 other islands. The enthusiasm of Charlie and Mike for this book, including feedback supplied on several chapter drafts, is greatly appreciated. Charlie also hosted me during a period of study leave at Victoria University of Wellington, and allowed me to use photographs and unpublished information on body sizes from our island surveys of tuatara.

Since moving to the University of Otago Te Whare Wānanga o Otāgo in 1991, I have been inspired and assisted by members of my own research group, most of whom have worked with New Zealand's tuatara, lizards and frogs. I am especially grateful to Anne Besson, Kelly Hare and Claudine Tyrrell for their thoughtful appraisals of many chapters, and for assistance in numerous other ways. Claudine's dissections led to new diagrams herein of internal anatomy, and her cheerful assistance with the bibliography was immensely welcome. Others who provided research assistance at various stages include Joanne Connolly, Karina Holmes, Scott Jarvie, Leigh Marshall, Sophie Penniket and Jennifer Rock. Permission to reproduce data from a thesis was granted by Elaine Whitworth. Comments on chapter drafts from Amanda Caldwell (née Chamberlain), Jignasu Dolia, Jennifer Germano, Jonathon Hare, Scott Jarvie, Carey Knox, Marieke Lettink, Jolene Oldman, Sophie Penniket, Ricardo de Sá Rocha Mello and Aude Thierry were also appreciated. Thank you all for tolerating my distraction from our other research activities.

In a book with this diversity of topics, it is not possible for one author to be a specialist in all. For providing detailed guidance in their areas of expertise, including feedback on chapter drafts or sections thereof, I am indebted to the following. For information and comments on the palaeontology of reptiles and evolution of rhynchocephalians, I am deeply indebted to Marc Jones, and also extend thanks to Susan Evans. For information and comments on the geological history and palaeoecology of the Gondwanan landmass of Zealandia (and its later subsidiary, New Zealand), I thank Trevor Worthy, and, for his molecular perspective, Graham Wallis. Jennifer Hay provided insights from her ongoing work on the genetic relationships among tuatara past and present. For discussion and comments on the ecological impacts of human arrival in New Zealand, I thank Trevor Worthy, David Towns and Ian Smith. For comments on the study of tuatara by early taxonomists, I thank Kraig Adler. Valuable feedback on the history of study of tuatara within New Zealand was provided by Donald (Don) Newman, Ross Galbreath and Tony Whitaker. Tony's sustained interest in this project was greatly appreciated and his sudden death during the final stages of production is an enormous loss to New Zealand herpetology.

Until very recently, tuatara have been managed in New Zealand with advice from a recovery group. For the opportunity to participate as a member of this group from 1991, I am grateful to the Department of Conservation Te Papa Atawhai. For comments on chapter drafts involving populations in their area of care, I thank the most recent recovery group leader, Peter Gaze, and other group members, including Andrea Booth, Rob Chappell, John Heaphy, Shane McInnes, Leigh Marshall, Donald Newman, Oliver Overdyck and Richard Parrish. Another recent member, Barbara Blanchard (the captive-management coordinator for tuatara), supplied valuable information on tuatara in captivity in New Zealand and elsewhere.

For additional feedback on drafts concerning the ecology, island populations and/or conservation of tuatara, I thank Lindsay Anderson, Richard Griffith, Susan Keall, Kim Miller, Nicola Nelson, David Towns, Graham Ussher and Tony Whitaker. Mike Bull provided helpful commentary on parasites. For feedback on a chapter on reproduction and life history, I thank Nicola Nelson, and for comments on a chapter on environmental relationships, I am grateful to Marion Preest and Stephen Adolph.

The relationship between New Zealand reptiles (including tuatara) and te ao Māori (the Māori world) was an important and challenging topic to include. For their guidance and comments on draft material, I am indebted to the following: Mark Brunton, Khyla Russell and Stephen Scott. Valuable comments were also supplied by Dougal Austin, Mere Roberts and Paul Moon. I wish to acknowledge the generous help provided by members of Te Arawa, especially James (Jim) Schuster (also of the Historic Places Trust Pou Here Taonga and the Rangitiaria Schuster Whānau Trust). Jim hosted me on a visit to Rotorua, allowing me to absorb and photograph the place of reptiles in the carved history of his people – rarely have my eyes been opened so much in one day. Additionally, my thanks to Toby Curtis (Rakeiao Marae Komiti), the Reverend Joe Huta, John Merito (Houmaitawhiti Marae Komiti), Piki Thomas (Taheke Marae Komiti), Napi Waaka (Kahumatamomoe Marae Komiti), Te Ohu Wi Kingi (Hinemihi Marae Komiti) and others who graciously humoured my enquiries and supported my requests to reproduce photographs of carved reptiles.

For other information and access to carved reptiles in museum collections, I thank Dougal Austin (Museum of New Zealand Te Papa Tongarewa) and Dimitri Anson (Otago Museum). The assistance of Kipa Rangiheuea and Janine Love (Auckland Museum Tamaki Paenga Hira) is also appreciated. For further information on Māori perspectives, I thank the late James Elkington, Marlin Elkington, Wikuki Kingi Snr, Whetu McGregor, Te Rangi Martrell, John Moorfield, Glenice Paine, Hori Porata and Jim Williams. These and others who provided valuable perspectives affiliate to iwi including Ngāti Koata, Ngāti Rehua, Ngāti Whātua, Ngātiwai, Ngāi Tahu, Ngāi Tai and Te Ati Awa.

Others who answered enquiries and/or provided information and references on various topics include Sebastián Apesteguía, Matt Baber, Christine Baines, Ian Barber, Barbara Barratt, Andy Bassett, Robert Brassey, the late Peter Bristow, Derek Brown, Reva Calvert, Linda Campbell, Kenneth Cree, Janet Davidson, Charles Dawson, Raewyn Empson, Euan Fordyce, George Gibbs, Stephanie Godfrey, Margie Grant-Caplan, Louis Guillette, Jill Hamel, Anthony Harris, Richard Holdaway, Alan Hollows, Chris Jacomb, Phil Latham, Dave King, Matt McGlone, Whetu McGregor, Phil Millener, Hilary Miller, Craig Miller, Jennifer Moore, Les Moran, Kereti Rautangata, Kate Richardson, Neville Ritchie, Jonathan Ruffell, Steve Sawyer, Paul Scofield, Mary Simons, Bruce Thomas, Richard Walle, Tamsin Ward-Smith and Dylan van Winkel.

For information on or access to collections of tuatara held at their institutions, or on activities of former researchers of tuatara, I thank the following: Sandy Bartle and Raymond Coory (Museum of New Zealand Te Papa Tongarewa), Brian Gill (Auckland Museum), Geoff Tunnicliffe and Paul Schofield (Canterbury Museum), Erena Barker, Emma Burns, Simon Wylie and Otto Hyink (Otago Museum), Malcolm Mannering, Marilyn Duxson and Fieke Newman (Department of Anatomy, University of Otago) and Lindsay Hazley, Maurice Watson and Karl Gillies (Southland Museum and Art Gallery). For equivalent information from overseas, I thank Peter-Rene Becker (Übersee-Museum, Bremen), Simon Chaplin (Museums of the Royal College of Surgeons of England, London), Jakob Hallermann (Zoological Institute and Zoological Museum, University of Hamburg), Anne Heimhold (Museum für Völkerkunde, Hamburg), Colin McCarthy (Natural History Museum, London), Jiri Moravec (National Museum, Prague), Jill Sales (King's College Natural History Collection, London) and Franz Tiedemann (Natural History Museum, Vienna).

The services of librarians and archivists were invaluable, sometimes leading to the discovery of material presented for the first time in this book. I thank past librarians of the University of Otago (especially Helen Edwards and Inger Gledhill), as well as those of the Alexander Turnbull Library, Archives New Zealand Te Rua Mahara o te Kāwanatanga, the Museum of New Zealand (especially Jennifer Twist) and Victoria University of Wellington. I note with gratitude the permission of the chief archivists at Archives New Zealand and the Museum of New Zealand to quote from their institutions' files in this book. Tiena Jordan and Fiona Hall of the Whakatane Museum and Gallery, and Stephanie Smith of the Tauranga City Council, also provided valuable leads. I thank all those whose efforts led to the creation of the online resources Papers Past (http://paperspast.natlib.govt.nz) and Dictionary of New Zealand Biography (http://www.teara.govt.nz/en/biographies), resources that added so much to the historical content of this narrative.

Much of the early literature on tuatara was published in German. For assistance with forwarding or making translations of selected material, I thank Smita Apte, Bernard Goetz, Mario Grünreif, Barbara Kirschoff, Emmanuelle Martinez, Christoph Matthaei, Donald Newman, Ilka Söhle, Doris Zuur and Bob Zuur. I also appreciated assistance from Anne Besson with material in French and Mariano Rodríquez Recio with material in Spanish.

Illustrations add immensely to any account of biology. For his help in producing illustrations (many new), and for photographic advice and assistance on countless occasions, I cannot overstate my gratitude to my department's scientific illustrator, Kenneth Miller. This book would not be what it is (and producing it wouldn't have been nearly as much fun) without Ken's sterling contributions. I am also grateful to the Society for Research on Amphibians and Reptiles in New Zealand (SRARNZ) and the Department of Conservation for helping fund the costs of colour reproduction in this book.

Many individuals generously allowed reproduction of photographs or illustrations from their collections, photography of items in their collections, or otherwise assisted with access to illustrations. These include the following: Clare Allen, Pip Aplin, Anne Besson, Stu Bisset, Craig Briggs, Michael Bunce, Philip Carthew, Linda Cartland-Shaw, Mary Coughlan, Kenneth Cree, Len Doel, Rudolph Dührkoop (provided free on the internet), Alex Eagles, Dean Evans, James Gillingham, Jorge González, Richard Griffith, Heather Gunn, Jenny Halstead, Lindsay Hazley, John Heaphy, Klaus Henle, Martin Hill, Colin Hitchcock, Peter Kamp, Susan Keall, Colleen King, Peter King, Peter Lawless, Dallen Lee, David Lupton (courtesy Horizons Regional Council), Tom Lynch, Mary McIntyre, Rod Morris, Nicola Nelson, Donald Newman, Daniel Ormsby, Lee Pagni, Joanne Peace, Warren Pohatu, Brett Robertson, Jennifer Rock, Michael Schneider, Marcus Simons, the late James Taylor, Alan Tennyson, Aude Thierry, Claudine Tyrrell, Graham Ussher, Geoff Walls, Rufus Wells, Tony Whitaker, Trevor Worthy and Patricia J. Wynne.

I am also grateful for the use of photographs from the collections of, or with the permission or assistance of: Alexander Turnbull Library; Archives New Zealand; Auckland Museum; Auckland Radiology Group; Christchurch City Libraries Ngā Kete Wānanga-o-Ōtautahi; Department of Anthropology,

University of Otago; Otago Conservancy, Department of Conservation; GNS Science Te Pū Ao; Hocken Collections Uare Taoka o Hākena, University of Otago; Landcare Research Manaaki Whenua; Mt Olympus Wines; Museum of New Zealand; Natural History Museum, London; National Library of Australia; National Institute of Water and Atmospheric Research Taihoro Nukurangi; New Zealand Post; Ngāi Te Rangi; Otago Museum; Rotorua Museum of Art and History Te Whare Taonga o Te Arawa; Southland Museum and Art Gallery; Te Arawa Lakes Trust; Te Wānanga o Aotearoa; The Star of Dunedin; Tuatara Brewing Company Ltd; Übersee-Museum, Bremen (photographer Gabriele Warnke); Weta Workshop; and Zealandia™ (formerly the Karori Wildlife Sanctuary).

For permission to reproduce or adapt material or information from published sources, I thank: Company of Biologists; Department of Conservation; GNS Science; John Wiley and Sons; Ministry for the Environment – Manatū Mō Te Taiao; National Academy of Sciences; National Institute of Water and Atmospheric Research; Natural History Magazine Inc.; Prensa Cientifica SA; Springer Science+Business Media; Taylor & Francis Group; and Warner Music. To anyone I have overlooked who has helped in material ways, my heartfelt thanks.

The development to 'maturity' of this book has essentially matched in its duration the life-history characteristic of its subject. For their long-term faith, and support at critical moments, I am extremely grateful to some special people. I thank everyone at Canterbury University Press, especially publishers Catherine Montgomery and Rachel Scott, and former publishing assistant Kaye Godfrey, for seeing this project to completion amid the aftermath of the Canterbury earthquakes of 2010 and 2011. I would like to acknowledge their contractors: editor Sue Hallas, book designer Quentin Wilson, cartographer Tim Nolan, proofreader Ray Prebble, and indexer Diane Lowther. I am also grateful to the late Richard King, former managing editor of Canterbury University Press, for early support. Among my departmental colleagues, I thank in particular my former Heads of Department Professors Carolyn Burns, Alison Mercer, Hamish Spencer and Colin Townsend for their wise counsel. Finally, for their sustained interest in this project, and exceptional support in so many ways, I thank my husband Marcus Simons and members of the wider Cree, Simons and Cary families. I am indebted to you all.

Preface

It has been said that in every New Zealand animal, no matter how humble, there is a story. This book tells the story of one of the most famous: the reptile named by Māori (the first people of New Zealand) as tuatara.[1] Known formally to scientists the world over as the genus *Sphenodon*, tuatara are represented today by the sole living species *Sphenodon punctatus*.

Humans and tuatara have been in contact for almost eight centuries. Once viewed as food, then as fearsome guardians, tuatara are today considered within New Zealand as cultural treasures (in the language of Māori, taonga). Internationally, they hold distinctive value as subjects for scientific research. Formal study began with the first published description in 1831, but advanced greatly from 1867 when tuatara became recognised as the only surviving example of rhynchocephalian reptiles. This once-diverse group is related to lizards, but has a separate evolutionary history spanning nearly a quarter of a billion years.

From 1867 onwards tuatara have grown in fascination for scientists, collectors and lovers of the biologically curious from around the globe.[2] The focus of scientific interest has changed over the decades, from initial studies on unusual anatomical features, through research into ecology and environmental relationships, to today's concern with conservation. However, in the absence of a modern and comprehensive text summarising what has been learnt over nearly 200 years of published research, and almost 800 years of human contact, tuatara remain something of an enigma to many.

Several factors contribute to the intrigue. The first involves an evolutionary conundrum: the tuatara is clearly living today, and is therefore a modern reptile. Yet it is often described as a 'living fossil', as if it were a direct transplant from Mesozoic times. Another contributing factor is geographic isolation. Wild tuatara exist only in New Zealand, a small country in the South Pacific. Within New Zealand natural populations are now restricted to offshore islands that are often cold, windswept and uninhabited by humans. Strict legal protection for over 100 years has played a role in keeping the species from public view. Finally, those tuatara that have been visible in captivity have often frustrated their observers with an apparent disinclination for activity. According to a lighthouse keeper of 1913, one tuatara held indoors, and around which a pencilled outline was drawn on the floor, did not shift its position in over six weeks![3]

Despite these challenges, those of us fortunate to have worked with wild tuatara know that these stoical reptiles are in many ways amenable to research. They are small enough to be manageable by hand, yet large enough to attach radio-transmitters to, or to take blood or other tissues from for study. They are not poisonous or seriously dangerous, and they are impressively numerous in some populations. They are sedentary as adults and very long-lived, making long-term studies of the same individuals possible. And, with patience, much can be learned of their behaviour.

This book has been motivated by several aims. My first is to provide a comprehensive resource that will be of value to university students and professional herpetologists (biologists who study reptiles and amphibians) worldwide. Accurate information is essential to understanding the biology of tuatara and to making evolutionary comparisons with other reptiles. In addition, and especially within New Zealand, this book is intended to benefit those directly involved with conservation management. In the past this audience has favoured those from professional backgrounds in government departments, universities, museums, zoos and wildlife parks. However, as the wider human community engages more deeply with conservation, many groups have become involved in restoration of habitats for tuatara. Among these are various iwi (Māori tribes), who justifiably seek closer involvement with tuatara, including renewed recognition as kaitiaki (guardians), and who bring new understanding of what tuatara signify to human culture. Some of the ways in which this book is structured to help meet the needs of both specialist and non-specialist audiences are explained below.

A second aim in writing this book is to challenge – and ultimately, I hope, to lay to rest – the idea of tuatara as a 'living fossil'. The view of tuatara as a Mesozoic relic, a species unchanged for hundreds of millions of years, perhaps even ancestral to other living reptiles, is now viewed as quaint by the majority of scientists. Yet it still emerges in some publications and continues to hold tremendous popular appeal. Instead I seek to portray tuatara as a modern rhynchocephalian, undoubtedly somewhat conservative in the evolution of some characters compared with other living reptiles, but derived or specialised in others. This view does not diminish the importance of what remains a highly intriguing animal, although it does create more challenges for scientists in identifying what are truly primitive (ancestral) characters relative to those of lizards.

A third aim, which developed only as I began writing this book, is a more local one. Could knowledge of tuatara be used as a way to explore the story of New Zealand – its geological origins, settlement by people, development of a community of scientists, and growing awareness of the fragility of its biota?

This subsidiary theme has been a personal voyage of discovery, but one that I hope holds resonance for others interested in the way that people relate to nature.

This book is organised into three parts. Part 1 (Origins) begins with a chapter clarifying the evolutionary relationships between tuatara and both ancient and living reptiles. In particular, it summarises extensive new knowledge from palaeontology, which confirms tuatara as the only surviving, but not the most 'primitive', of the rhynchocephalian reptiles. In Chapter 2 I discuss how the lineage leading to tuatara came to be isolated in New Zealand, as well as the enormous geological and climatic upheavals that must have been experienced by this lineage over the past 80 million years. This chapter acknowledges the incredible biological diversity, of which tuatara were a part, that existed in New Zealand at the time that people arrived.

The appearance of humans in New Zealand, and their overall effects on the distribution of tuatara, are the subjects of Chapter 3. New Zealand was one of the last landmasses of the world to experience the human footprint, and the impacts, from both Polynesian and European waves of settlement, have been severe. Fortunately, though much reduced in abundance and distribution, tuatara were not among the many endemic species that became extinct. In Chapter 4 I explore interactions between New Zealand reptiles (tuatara and lizards collectively, as these reptiles cannot always be separated in traditional accounts) and the first human settlers, the Polynesians who became Māori. Chapter 5 covers the history of scientific study of tuatara, beginning with early collections sent to European museums, through to the flourishing present-day research on live tuatara centred in New Zealand.

In Part 2 (Biology of tuatara today) I describe current biology, with an emphasis on tuatara in their wild habitat. An evolutionary theme continues, with comparisons made where appropriate with other reptiles. Successive chapters cover current distribution, habitats and population histories, including geographic and genetic variation (Chapter 6); ecology, feeding and behaviour (Chapter 7); reproduction and life history (Chapter 8); and relationships with environmental factors, including temperature, oxygen, water and light (Chapter 9).

Part 3 (Future survival) is once again concerned directly with interactions between humans and tuatara. Chapter 10 covers the history of conservation, picking up from Chapter 5 with early conservation efforts, continuing through to the present-day recovery programme and issues that lie ahead, including the deeply troubling issue of climate change. Chapter 11 covers developments during 2012/13 in research and conservation of tuatara that emerged as this book went into production. It thus provides a brief update of preceding chapters.

To make this book as accessible to as wide an audience as possible, yet still valuable to the professional herpetologist, a number of features are incorporated. Major technical terms, defined at first use in the main text, are also explained in a scientific glossary near the book's end. A second glossary includes Māori words used in the text, many of which are in common use among speakers of New Zealand English.[4] The chapter texts are supplemented by numerous illustrations, as well as by boxes that develop topics of general interest in a more informal style, including anecdotes from my and others' experience. Scientific references, which are essential to allow ideas and information to be traced to source, are cited using numbered chapter endnotes rather than by surname and year in the body of the text. Some previously unpublished research findings are included in the text, but information from tests of statistical significance is saved for chapter endnotes.

Further explanation of the location of references will be helpful. This book includes a scientific bibliography of tuatara; thus, references that appear in full in the bibliography are cited in the chapter endnotes only by surname and year. Other sources, less specifically about tuatara, are cited in full in the chapter endnotes but not in the bibliography. Inclusion in the bibliography was ultimately somewhat arbitrary, but articles generally include those published in peer-reviewed sources and that contain significant new information or syntheses of the biology of tuatara. Thus, some chapters from the seminal series *Biology of the Reptilia* are included in the bibliography, whereas others with only passing mention of tuatara are not. Masters' and doctoral theses available by inter-loan through New Zealand university libraries are included, whereas undergraduate reports such as honours dissertations and items in unpublished government files (neither of which are generally available to most readers) are not – although the latter may be cited within chapter endnotes. The bibliography also includes some popular articles and newspaper accounts, especially where I judged the article to have new biological information or a historical perspective of particular interest. However, there are numerous popular articles on tuatara that did not meet these criteria, and these are generally not included. Additionally, articles containing evidence of the past natural distribution of tuatara are included, whereas those making passing mention of bones from tuatara in human-associated deposits are generally not. The bibliography, which contains over 850 references spanning the years from 1817 until 2013, will be made available as a pdf and an Endnote® file at http://www.tuatarabook.info.

Endnotes

1 The Māori word is the same in both singular and plural.
2 The phrase 'lovers of the curious' was first used by Buller 1877 about those who kept tuatara in captivity.
3 Thomson 1915.
4 The text follows modern Māori practice of using macrons to indicate long vowels. Quotations, of course, are reproduced in their original format.

Conventions, units and abbreviations

ANZ	Archives New Zealand Te Rua Mahara o te Kāwanatanga
asterisks	Asterisks are used in graphs to indicate levels of statistical significance. One asterisk indicates that a difference between the means (averages) of two groups has a statistical probability (*p*) of less than 5 in 100, two asterisks a probability of less than 1 in 100, and three asterisks a probability of less than 1 in 1000 (in scientific parlance, * = $p < 0.05$, ** = $p < 0.01$ and *** = $p < 0.001$). In short, the more asterisks, the less likely a given result is by chance alone, and the more likely it is to have some underlying biological meaning.
bird names	Names of New Zealand birds (scientific, common and Māori) are generally from the following: Checklist Committee (B. G. Gill, convenor). 2010. *Checklist of the birds of New Zealand, Norfolk and Macquarie Islands, and the Ross Dependency, Antarctica*. 4th edn. Te Papa Press, in association with the Ornithological Society of New Zealand.
DOC	Department of Conservation Te Papa Atawhai (New Zealand)
et al.	abbreviation of the Latin *et alia* or *et alii* (meaning 'and others'), used when citing references that have at least two authors in addition to the first-named author
g	gram(s). One gram equals about 0.0353 ounces.
island areas	Areas for offshore islands with tuatara follow those given in the two Tuatara Recovery Plans, which rely primarily on the following source: Taylor, G. A. S. 1989. *A register of northern offshore islands and a management strategy for island resources*. Northern Region Technical Report Series No. 13. Department of Conservation, Auckland. Other sources may give slightly different values.
island names	At first mention in each chapter, islands are named as recorded by the New Zealand Place Names Database (Archived) (http://www.linz.govt.nz/placenames/find-names/topographic-names-db/index.aspx; not maintained since 31 October 2008). These names often include Māori and English versions, but without consistent order or punctuation. At second mention within a chapter, only the first-listed name is used unless the context requires otherwise. Island names not listed in the database are noted here as informal (inf.) at first mention.
m	metre(s). One metre equals about 39.4 inches or 3.28 feet.
NMNZ	item in the collection of the Museum of New Zealand Te Papa Tongarewa (the national museum of New Zealand)
op. cit.	abbreviation of the Latin *opere citato*, used in chapter endnotes to indicate a reference that was cited in full earlier in the endnotes of the same chapter
p	statistical probability
pers. comm.	personal communication from the person named to the author
pers. obs.	personal observation by the author
s	second(s)
SVL	snout–vent length, a standard way of expressing the body length (excluding the tail) of a lizard-like reptile such as a tuatara
terms of orientation	Common terms for describing specific orientations of the body of a lizard-like reptile such as a tuatara are shown in the figure on the following page.

Figure illustrating terms of orientation in a young male tuatara on Stephens Island (Takapourewa).

Photo: Marcus Simons.

PART ONE

ORIGINS

CHAPTER ONE

Evolution: last of the rhynchocephalians

Biological curiosities challenge our understanding of the natural world, and among living reptiles there are no better examples than tuatara. Known to scientists as members of the genus *Sphenodon*, tuatara superficially resemble some of the dragon lizards of the Family Agamidae, found in Australia, Africa and Asia. Yet tuatara are not lizards: instead, features of their skeleton, and their skull in particular, identify them as the last of the rhynchocephalian reptiles.

Introduction

For nearly a quarter of a billion years, rhynchocephalians have existed as an evolutionary lineage separate from other reptiles. Tuatara, the only surviving members of that lineage, thus form an isolated but fascinating limb in the evolutionary history of vertebrates. This chapter introduces the classification and physical appearance of the sole living species of tuatara, *Sphenodon punctatus*. Features of anatomy and genetics that reveal evolutionary relationships between tuatara and other living reptiles – the turtles (Testudines or Chelonia); the lizards and their specialised offshoot, the snakes (Squamata);[1] and the crocodiles, alligators and gharials (Crocodylia) – are then explored. Of course in a strict sense birds are reptiles too, having evolved from the same ancestors as crocodilians. In this book the term 'reptiles' is used in its popular sense, with birds referred to only by specific mention.

The second part of this chapter steps back in time to the Mesozoic Era (250 to 65 million years ago), a period often known as the Age of Reptiles. It was during this era that rhynchocephalians appeared within a wave of new reptiles known as diapsids. Rhynchocephalians, and especially their more specialised forms (sphenodontians) were, for at least some tens of millions of years, a diverse and widespread group. Recent discoveries by palaeontologists have vastly expanded our understanding of these Mesozoic species, clarifying their relationships to each other, to other extinct reptiles and to their modern relatives, tuatara.

Fig. 1.1 Tuatara (*Sphenodon punctatus*) from Stephens Island (Takapourewa) in Cook Strait. This posed photo illustrates typical differences in appearance and size between an adult male (above) and a female (below).

Photo: Alison Cree.

There is a certain romantic appeal about the concept of 'last survivors'. Indeed, a closer link with Mesozoic than with modern reptiles has shrouded tuatara with misconceptions about evolutionary stasis. Biologists once considered rhynchocephalians in general, or tuatara specifically, as the most ancient of reptiles, even ancestral to many living forms. Such views are no longer supported. A popular view that tuatara have existed unchanged as a 'living fossil' since the Mesozoic is also untenable, as this chapter will argue.

Scientific classification of tuatara

Along with other organisms, tuatara are given scientific names in a nested hierarchy. The classification that follows tells us that tuatara belong to a major group (in classical Linnean taxonomy, a superorder) of reptiles known as Lepidosauria, a group that also includes the lizards and snakes. Among the lepidosaurs tuatara are placed within an order named Rhynchocephalia, and then within a subgroup (suborder) known as Sphenodontia (as an alternative, some researchers have used the name 'Sphenodontia' or 'Sphenodontida' for the order). Tuatara and the fossil sphenodontians most closely related to them are placed in the family Sphenodontidae, and then at a more precise level within the sub-family Sphenodontinae. More specifically still, the living tuatara are placed in the genus *Sphenodon* (*S.* for short). All living populations are here considered as members of the species *S. punctatus* (Figure 1.1).

Reptilia
Lepidosauria
Rhynchocephalia
Sphenodontia
Family Sphenodontidae
Sub-family Sphenodontinae
Genus *Sphenodon*
Sphenodon punctatus

The name Rhynchocephalia (from the Greek rhynchos, 'snout', + kephalē, 'head') can be translated as the 'snout heads', although 'beak heads' is a more commonly used term. The names of the suborder, family and genus recognise the wedge- or chisel-like nature of the teeth of tuatara (Greek sphen, 'wedge', + odous, 'tooth'). Hence, sphenodontians can informally be described as the 'wedge-toothed reptiles'. The specific name *punctatus* means 'spotted'.

The genus for tuatara was first recognised scientifically in 1831,[2] with the species known today as *S. punctatus* being described in 1842.[3] In 1877 a second species of tuatara, *S. guntheri*, was proposed, based on subtle differences in appearance and behaviour (Figure 1.2; the specific name honoured Albert Günther, the scientist who first recognised the anatomical features that distinguish tuatara from lizards).[4] This distinction soon fell into disuse, but, in 1990, a study examining variation in blood proteins among tuatara on different islands revived recognition of *S. guntheri*.[5] This taxon became known as the Brothers tuatara because the only natural population was found on the northernmost of The Brothers, a pair of small islands in Cook Strait.

Fig. 1.2 Tuatara on North Brother Island were recognised briefly in the late 1880s, and again during the 1990s, as a second species (*Sphenodon guntheri*). Some tuatara on this island, such as this young male, have blotches as well as spots on the dorsal surface. However, genetic differences are now considered insufficient to recognise this population as a distinct species. Instead, tuatara on North Brother Island appear to form an inbred population within the species *S. punctatus*.
Photo: Alison Cree.

BOX 1.1

Rhynchocephalia or Sphenodontia?

Tuatara are widely accepted as the last living members of a major clade or evolutionary group of reptiles, but the past few decades have seen debate over what that group (in classical taxonomy, an order) should be called: Rhynchocephalia or Sphenodontia? Rhynchocephalia, the term in longest usage, was coined in 1867 by Albert Günther of the British Museum, who was the first to recognise the unique features that distinguished tuatara from squamate reptiles (lizards and snakes).

By choosing the name Rhynchocephalia for the group to which tuatara belonged, Günther was ensuring continued recognition for his mentor and superior, Richard Owen, Superintendent of the Natural History Departments at the British Museum.[1] Owen had suggested the name *Rhynchocephalus* for the tuatara genus in 1845, apparently unaware that the name *Hatteria* had already been given to tuatara in 1842 by John Edward Gray, also of the British Museum.[2] Günther realised in 1867 that *Hatteria* had priority over *Rhynchocephalus* for the name of the genus, but chose to use a modification of Owen's name for the overall group. In fact, an even earlier name, *Sphænodon*, assigned by Gray in 1831, had ultimate priority for the genus (the original spelling used by Gray was formally emended to *Sphenodon* in 1957).[3] However, by the time that this sequence of events was realised in 1869,[4] the name Rhynchocephalia for the group as a whole was already in use.

The Order Rhynchocephalia soon became a repository for a variety of Mesozoic fossils, including a group of heavily built reptiles from the Triassic known as rhynchosaurs ('snout reptiles'). Rhynchosaurs had an overhanging bony beak, as well as rows of enlarged tooth plates on the maxillary bone of their upper jaws. In 1845 Owen had suggested an affinity between rhynchosaurs and tuatara, although in 1867 Günther expressed some reservation. By 1985 palaeontologists including Michael Benton concluded that the Rhynchocephalia had become a 'collection bag' for a variety of fossil forms, not all of which were closely related to each other or to tuatara.[5] Rhynchosaurs, for instance, lack the specialised palatine tooth row and acrodont teeth of tuatara, and their premaxillary 'beak' is made of bone, not teeth. Today, rhynchosaurs are considered as herbivorous archosauromorphs (relatives of the lineage that includes dinosaurs, birds and crocodilians).[6] In other words, rhynchosaurs and tuatara are no longer considered close relatives.

To avoid confusion between tuatara and rhynchosaurs, Benton recommended that the name Rhynchocephalia be dropped from use, and that the name Sphenodontia (until then, the name of a subcategory, or suborder) replace Rhynchocephalia for the order.[7] An alternative name, Sphenodontida, was independently proposed for the order in the same year.[8] One or other of these recommendations was initially followed, but most palaeontologists[9] now use a redefined Rhynchocephalia[10] for the major group to which tuatara belongs; under this scheme, the name Sphenodontia is sometimes used for the subgroup (suborder) of more advanced rhynchocephalians that includes tuatara, but which excludes *Gephyrosaurus* and perhaps one or two other basal genera. This usage, which recognises the historical precedence of the name Rhynchocephalia, is followed in this book. Thus, tuatara are here considered as rhynchocephalians, as sphenodontians and as sphenodontids, reflecting membership of the order, suborder and family, respectively.

The two proposed species of tuatara occurred on different islands, and thus the potential for natural interbreeding could not be tested.[6] Moreover, no reliable skeletal,[7] other morphological or chromosomal[8] differences could be detected, leading some to question whether *S. guntheri* should be recognised as a separate species.[9] Recent studies exploring variation in genetic information (as deoxyribonucleic acid or DNA) among populations have clarified the issue: no convincing support for recognising two species has been found.[10] So, from 2009, the living tuatara have been considered as one species, *S. punctatus*,[11] with tuatara from North Brother Island being a (possibly inbred) population of low genetic diversity within the wider group.[12]

On the other hand, recent studies do recognise two genetic stocks within *S. punctatus*. One of these, formally named as the subspecies *S. p. punctatus*,[13] is known collectively as northern tuatara. The other group, not formally named as a subspecies, is referred to as Cook Strait tuatara. Given the wide geographic and climatic separation between the 27[14] natural populations of northern tuatara, which are found on islands between latitudes 35°S and 38°S, and the five natural populations of Cook Strait tuatara, which lie at least 300 kilometres south at 40–41°S, this distinction is useful for ecological and political as well as genetic reasons.

Although all living populations of tuatara are now considered as a single species, it is worth noting that most aspects of biology have been described in detail for populations from only a few islands. Furthermore, genetic variation in fossil specimens from mainland sites has yet to be fully explored (such specimens from the North and South Islands of New Zealand are simply referred to as *Sphenodon* sp., meaning that their species identity is unclear). Early studies of living tuatara, including some of the anatomical studies referred to in this chapter, involved specimens harvested from Karewa Island (with a population of northern tuatara) and from North Brother Island (a population of Cook Strait tuatara). Since the late 1800s most studies have focused on Cook Strait tuatara from Stephens Island (an island known also by the Māori name Takapourewa). Stephens Island is home to an estimated 30,000 to 50,000 or more tuatara,[15] the most abundant of all populations.

Tuatara – first discovered and named by Māori

The first humans to settle in New Zealand were of East Polynesian origin and, over time, evolved into the people known today as Māori. At the time of European settlement the Māori culture had no written language, but tuatara were well known in oral tradition. European writers in the 1800s recorded several Māori words for New Zealand's spiny-backed reptile,[16] of which 'tuatara' is the one that remains in wide use. The name (Māori tua, 'back' or 'far side', + tara, 'spike' or 'spine') is often translated as 'peaks on the back', in reference to the spiny crest that runs along the nape of the neck and down the back of adults.

The word 'tuatara' is pronounced roughly as 'too-uh-tar-uh', with all vowels being short and the stress occurring on the first syllable. Like other nouns in the Māori language, it is the same in both singular and plural (in Māori, number is distinguished by the article: te tuatara refers to a single tuatara and ngā tuatara to several). Following Māori practice, contemporary New Zealand English tends not to make tuatara plural by adding an 's'; instead, just as with the English words sheep and species, context explains whether one or more tuatara is meant. In this book, the term 'tuatara' without a preceding article ('a' or 'the') usually means 'tuatara in the plural', whereas 'the tuatara' means, depending on context, 'the genus known as *Sphenodon*' or 'the living species *S. punctatus*'. Quotations from the literature, of course, preserve the authors' original usage.

Tuatara at a glance

Tuatara have a body shape typical of many terrestrial lizards – an elongate trunk with four short, splayed legs and a tail (the tail, if intact, accounts for slightly over half the total length). Body length in tuatara and other reptiles that drop and regrow their tails is conveniently expressed both as total length (ideally with the tail intact) and as snout–vent length, which is the length from the tip of the snout to the vent (the vent being the opening of the cloaca at the base of the tail, from which urine, faeces and eggs or sperm exit). Total lengths of up to 609 millimetres have been observed in tuatara,[17] although as this largest value was from an animal with a regrown tail, it probably slightly underestimates maximum length. Even though tuatara are longer than most living lizards, they are not exceptional in size: for example, green iguanas (*Iguana iguana*) reach 2 metres in total length, and Komodo dragons (*Varanus komodoensis*), 3 metres.

The maximum size reached by tuatara does, however, vary greatly among island populations. The largest animals have consistently been found on Tawhiti Rahi Island in the Poor Knights Islands, at the northern (warmest) end of the distributional range. In a survey during 1989, males on Tawhiti Rahi reached up to 311 millimetres in snout–vent length and 1.1 kilograms in mass (females, which are smaller on average, attained up to 260 millimetres and 540 grams).[18] Maximum size of tuatara on small islands is much less. For instance, on 4-hectare North Brother Island during the same time period, males reached only up to 256 millimetres in snout-vent length and 655 grams (females up to 213 millimetres and 350 grams).[19] Similar body sizes have been observed in populations of northern tuatara on small islands.[20] Tuatara on at least two islands have shown declines in size (mass, snout–vent length or the relationship between these) since the 1950s as habitats have changed.[21]

The colour of the uppermost (dorsal) surface of most adult tuatara is olive brown, with scattered small white or yellowish-white spots. On North Brother Island the spots sometimes take the form of larger blotches. Some

Fig. 1.3 An adult female tuatara from Whatupuke Island showing what I refer to as 'tiger banding' colouration.

Photo: Alison Cree.

Fig. 1.4 Colour variation among adult tuatara from Lady Alice Island (male, left and female, right). The identifying pen marks on the flanks will be lost at the next moult.

Photo: Alison Cree.

tuatara, especially on northern islands, have vivid colours of rust, red or brown, sometimes with darker banding (Figures 1.3, 1.4). A black throat stripe, present in many tuatara, is most obvious in males. The underside of the throat is grey or charcoal (darkest in males), with rows of small white spots running rearwards from the snout (Figure 1.5). The belly and underside of the tail are white, speckled with grey or charcoal. Tuatara show little capacity for colour change, but the skin is markedly brighter in colour immediately after moulting (typically in late summer).

The crest of white spines that runs along the back of the neck (nuchal spines) and back (dorsal spines) is best developed in adult males. It is eye-catching when held erect during courtship, but surprisingly soft to the touch. The skin of wild tuatara is rather loosely fitting, especially around the

throat, sides of the abdomen and upper hind limbs (in contrast, tuatara in captivity have a tendency to be overweight, with prominent nodules of fat under the skin).[22] The scales of the dorsal surface are small and granular, whereas the belly scales are larger and almost hexagonal.

Tuatara have a robust head, with a blunt snout extending slightly forward of the front margins of the jaws (Figure 1.6). Paired external nostrils (nares) are located near the front of the snout. Internal ears are present, but reveal no external sign.[23] The large, round eyes are encircled by a ring of small scales, often yellow. A vertical pupil within the gold-flecked iris expands at night, giving the eyes a black appearance.

The limbs of tuatara are short and sturdy, and the belly ground-hugging except during rare bursts of rapid locomotion. On each limb, the five toes are armed with strong claws, used when digging burrows and capable of inflicting deep, raking scratches in naïve handlers. As noted above, the tail can break off and regenerate. This important feature helps align tuatara with squamates, their closest relatives.

Fig. 1.5 Underside of two tuatara (female, left and male, right) on Stephens Island. Males tend to have a darker throat and chest than females. Note also the transverse vent or cloacal slit, a feature shared with squamate reptiles.

Photo: Marcus Simons.

Fig. 1.6 Head and shoulders of an adult male tuatara from Lady Alice Island (with distinctive red colouration on the body). Note the strong claws, the external nostrils near the front of the snout, and the absence of external ears.

Photo: Alison Cree.

BOX 1.2

Caudal autotomy: tails that drop

Tuatara and many lizards have the disconcerting ability to drop their tails at predetermined zones of weakness. This phenomenon, known as caudal autotomy, is an effective anti-predator strategy: the would-be predator is distracted by a wriggling piece of tail, while the tail's owner escapes to live another day (and often to grow another tail). Fossil rhynchocephalians from the Mesozoic also had this ability, whereas crocodilians and turtles do not. The existence of caudal autotomy in rhynchocephalians and squamates is a clue to the shared evolutionary history of these groups.

The anatomical basis for autotomy in tuatara is similar to that in most lizards.[1] Most of the bones that support the tail (the caudal vertebrae, of which there are up to 36 in tuatara) have fracture planes passing through them. In other words, the split occurs within a vertebra, rather than between vertebrae as one might at first think. In tuatara fracture planes are found in all caudal vertebrae apart from the five to seven vertebrae closest to the body. In cross-section, the vertebrae and major blood vessels are surrounded by fatty tissue, which is itself surrounded by muscle blocks (myomeres). The myomeres in each autotomy segment have processes that extend deeply between segments, rather like interlocking fingers. On the surface of the skin, each autotomy segment contains an enlarged crest scale known as a dorsal tubercle, and grooves between the scale rows mark the potential lines of breakage.

When the tail breaks, the skin splits, the vertebra fractures[2] and the muscle segments separate. The spinal cord and blood vessels are severed, but blood loss quickly stops and the event is not obviously painful. The regenerating tail grows very slowly in tuatara (often only a few millimetres a year)[3] and is never identical to the original. The regrown tail is shorter and unsegmented, and its spinal cord is protected by a tube of cartilage rather than by bony vertebrae.[4] The regenerating surface is initially covered by a dark, pliable epidermis, and eventually by an irregular arrangement of small scales.[5] There are no fracture planes in the regrown tail, although it can be forcibly broken. It can also break again at the junction with the original tail, or even closer to the tail base. Partial (incomplete) breaks followed by regeneration can result in forked tails with two tips (referred to as bifid tails) or even three (trifid tails).

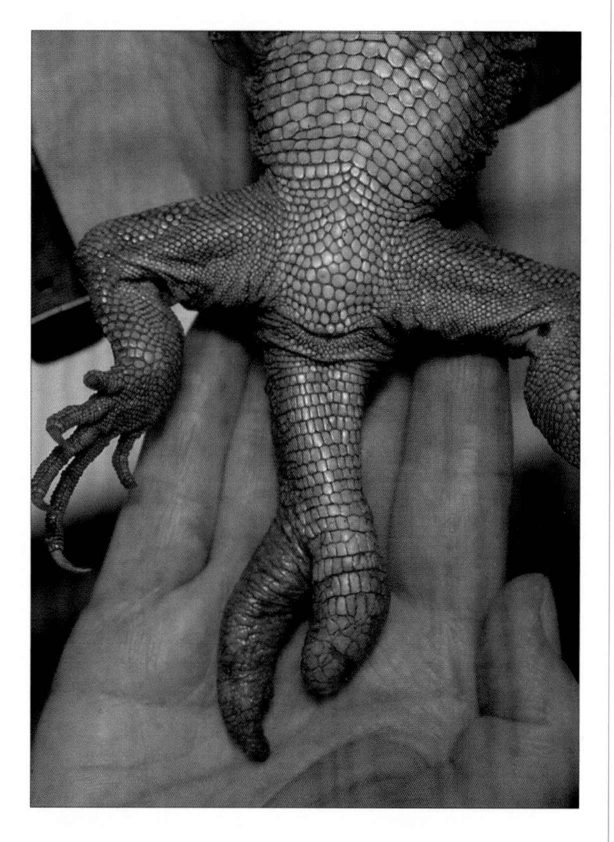

Box 1.2 Fig. 1 The underside of a tuatara with a forked (bifid) tail. The original tail (right) has a regenerated tip, as well as regrowth (left) from an incomplete break.

Photo: Brett Robertson/Victoria University of Wellington.

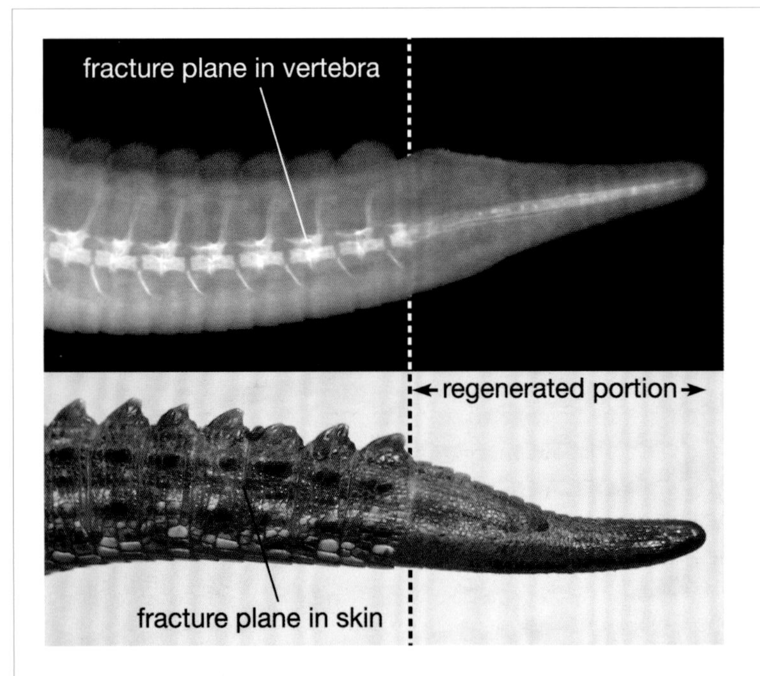

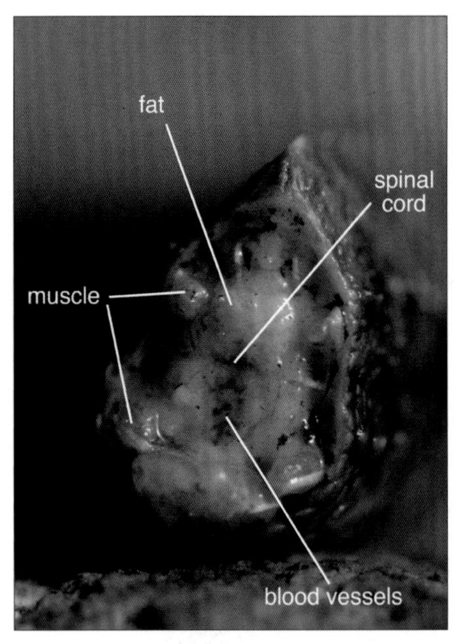

Box 1.2 Fig. 2 Lateral (side) views of regenerating tails in tuatara, showing original tail to left and regenerated tail to right. Top: X-ray of tail in a live adult. Bottom: Tail of a different individual, with a similar amount of regeneration. Both tuatara are of Stephens Island origin.

Photos: top, Elles Road Veterinary Centre, courtesy of Lindsay Hazley, Collection: Southland Museum & Art Gallery; bottom, Alison Cree.

Box 1.2 Fig. 3 The broken end of a freshly dropped piece of tail. Labelling based on Ali (1941).

Photo: Lindsay Hazley, Collection: Southland Museum & Art Gallery.

Caudal autotomy may help a tuatara escape from a predator or an aggressive fellow tuatara. However, it is not without cost. Studies on lizards have shown that tail loss can lead to at least temporary reductions in growth, locomotion, egg production, home range, social status, mating opportunities and survival.[6] Many of these costs reflect the roles of the tail as a fat store and a structure for signalling between individual lizards. In some species, tails are used as burrow plugs – which seems likely for tuatara also.

Tuatara seem to autotomise their tails less readily than many lizards. Although some scientists have interpreted this as indicating a more primitive process of autotomy,[7] the costs of tail loss have not been investigated. In tuatara – a large-bodied species lacking abdominal fat organs as alternative stores of fat – tail loss may be relatively costly, so autotomy is probably avoided in all but extreme situations. In addition, the benefits of tail loss to tuatara may be low, given that tuatara may not be able to run fast enough to allow reliable escape once the tail is shed, and that teeth and claws offer other options for defence.

The circumstances that lead to tail loss in tuatara are not well understood. Undoubtedly humans have sometimes contributed. On my first field trip to study tuatara, and eager to impress the trip's leaders with a high capture rate, I held too tightly onto an adult tuatara escaping down a burrow – only to find that the tail had come away in my hand (a mistake never repeated).[8] Cattle and sheep were, until recently, farmed on Stephens Island and probably also contributed to tail loss there. Natural instances of autotomy have been witnessed infrequently. During a study of social behaviour on Stephens Island, James Gillingham observed a fight between adult male tuatara. The resident male bit the tail of an intruder to his territory, the intruder ran off, and the dropped, squirming piece of tail was eaten by a nearby female![9] Tail loss is also suspected to occur during fights among females on nesting rookeries.[10]

Box 1.2 Fig. 4 Lighthouse keepers on Stephens Island (Takapourewa) hold tuatara destined for the San Francisco Zoo in about 1915. Although tuatara seem not to drop their tails readily, this method of holding is still not recommended.

Photo: Collection of Emily Scott (née Smith), courtesy of Heather Gunn.

Populations of tuatara on various islands differ in the frequency of tail loss, although the reasons are unclear. In a study led by Claudine Tyrrell of five northern populations, the proportion of adults with tail loss varied significantly among islands in both sexes. In males, the proportion varied from 69% (of 13 males) on Aorangi Island to 94% (of 50 males) on Coppermine Island.[11] The proportion among females was similar to that among males on some islands, but on Ruamahuaiti Island there was a significant difference between the sexes (75% of 36 males and 95% of 38 females had lost their tails).[12] More research is needed to explore the influence of individual characteristics including age and sex, as well as environmental features including activity of humans, density of tuatara and density of seabirds, on tail loss in different populations.

Unusual features of the dentition and skull

Tuatara are notorious for a bite sufficiently vice-like to bring tears to the eyes of grown men and women alike.[24] Nonetheless, a look inside the mouth is worth the risk, for it is here that the distinctiveness of tuatara starts to become apparent (Figure 1.7). Instead of having at most just one main row of teeth in the upper jaw, tuatara have two rows in parallel – an outer row on the jaw margin on a bone named the maxilla, and an inner, shorter row on the lateral edge of the

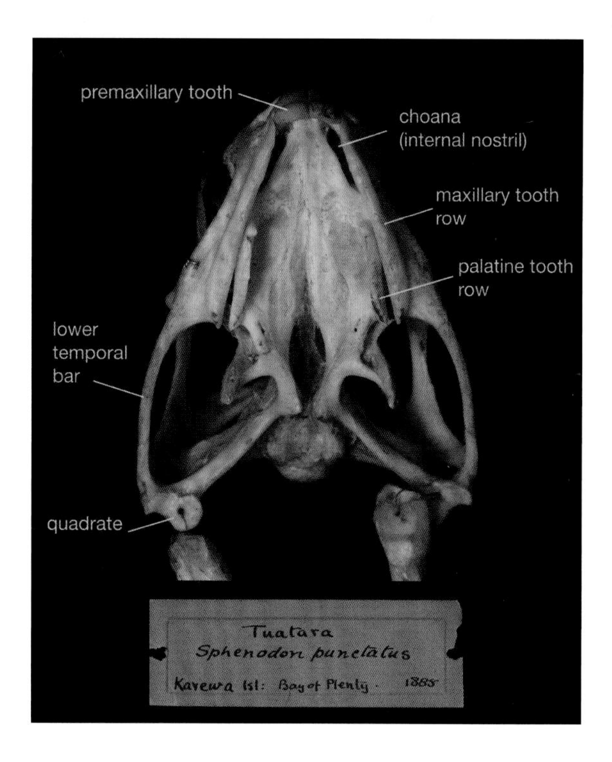

Fig. 1.7 Distinctive features in the skull of tuatara, including a complete lower temporal bar and two rows of teeth in the upper jaw, have attracted the attention of biologists since the 19th century. This photograph shows a ventral view of the palate, with the lower jaw reflected. The specimen, held in the Department of Anthropology and Archaeology at the University of Otago, was probably collected from Karewa Island by Professor T. Jeffery Parker of the University of Otago and Otago Museum. Skull length 60.3 mm.

Photo: Alison Cree.

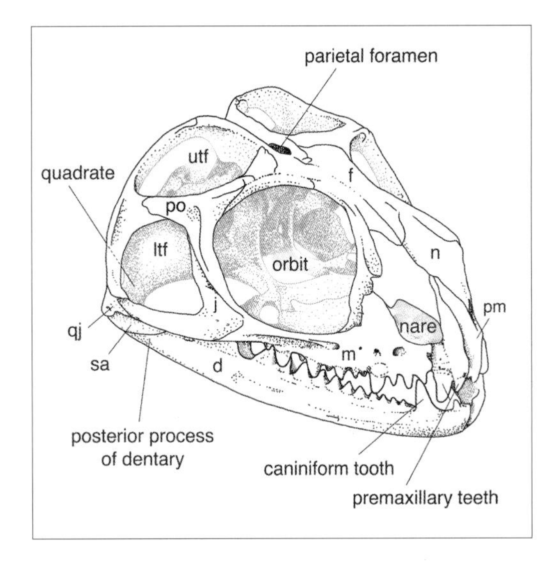

Fig. 1.8 (above) Line diagram showing significant features of the head skeleton of a tuatara. The long posterior process of the dentary helps distinguish tuatara and other rhynchocephalians from squamates; note also that the quadrate is firmly fixed to the palate, skull and upper jaw. On the other hand, the complete diapsid condition (in which upper and lower temporal fenestrae, the large openings at the rear of the skull, are both present) is no longer considered as a feature distinguishing rhynchocephalians from squamates. Abbreviations: d = dentary; f = frontal; j = jugal; ltf = lower temporal fenestra; m = maxilla; n = nasal; pm = premaxilla; po = postorbital; qj = quadratojugal; sa = surangular; utf = upper temporal fenestra. This specimen (c. 180 mm snout–vent length) is about the size of a small adult. The two teeth on each premaxilla have not yet developed the chisel-like appearance of the compound premaxillary teeth of large adults.

Line drawing by Ken Miller of specimen VT 669, Otago Museum, Dunedin, New Zealand.

Fig. 1.9 (top right) Head skeleton of a tuatara with relatively intact teeth. The long posterior process of the dentary, the tooth-bearing bone of the lower jaw, is characteristic of rhynchocephalians (see Figure 1.8 for labels). The acrodont teeth on the jaw margins form a serrated series that is often worn down later in life. The caniniform tooth or 'fang' near the front of the lower jaw is present only in tuatara and its closest relatives in the Family Sphenodontidae. At about 180 mm snout–vent length, this specimen is about the size of a small adult.

Photo: Jennifer Rock, specimen VT 669, Otago Museum, Dunedin, New Zealand

palatine bone. Between these rows is a gap or groove, into which fits the tooth row of the lower jaw (Figures 1.8, 1.9). The presence of teeth on various bones of the palate, as well as on the margin of the upper jaws, was a common feature of early reptiles, and palatal teeth remain present today in some lizards as well as in a highly modified, recurved form in snakes. However, the presence of a row of enlarged palatine teeth running essentially parallel with the maxillary teeth is considered a diagnostic character of rhynchocephalians.[25] Another rhynchocephalian characteristic is the long posterior process of the dentary, the tooth-bearing bone of the lower jaw.[26] This bony extension helps brace the lower jaw.

Tuatara have other unusual features of dentition shared with at least some ancient rhynchocephalians.[27] There is a pair of overhanging, incisor-like teeth at the front of the upper jaw, on a bone called the premaxilla. The marginal teeth have a form of implantation known as acrodont. Teeth of this type are not set in sockets. Instead, they are located on the crest of the jawbone and generally fused and ankylosed (firmly attached by surrounding bone) to it. Although often mistakenly described as mere 'serrations of the jawbone',[28] the acrodont teeth of tuatara are genuine teeth, with components of dentine, cementum and enamel.[29] During juvenile life, the remnants of the hatchling dentition near the front of the mouth are supplemented by teeth added towards the rear of the mouth.[30] Only a few of the hatchling teeth can be replaced. A caniniform (canine-like) tooth or 'fang' develops in juveniles near the front of the dentary as one of the successional (replacement) teeth. When the jaw is closed, this tooth slips into a notch just behind the premaxilla.

Upon capturing food, a tuatara closes its mouth and moves the lower jaw forward relative to the upper jaw. This forward-moving propalinal jaw action (more specifically known as proal)[31] produces a sawing-chew similar to that of a steak knife.[32] It enables large, hard prey items such as chitin-clad insects, and seabird chicks, to be crushed and severed (Figure 1.10). Old tuatara sometimes have their teeth worn down to the margins of their jaws, but the toothless ridges still function as cutting edges.

The cheek region of the skull reveals another significant feature.[33] Behind the orbit (eye socket) are two large openings. These are known as the upper (or superior) and lower (or inferior) temporal fenestrae (Latin fenestra, 'window'). Defining the lower margins of the fenestrae are two arches of bone, forming a skull type known as diapsid ('two-arched'). The presence of these arches, or bridges, explains a common name for tuatara in German, der Brūckenechse ('bridge-lizard', [34] although of course tuatara are not lizards). A complete diapsid skull with two arches was once thought to be a primitive (ancestral) feature of tuatara relative to lizards. However, the earliest rhynchocephalians lacked a complete lower temporal bar so, compared with these ancestral forms, tuatara are now considered to have a derived condition, as explained later in this chapter.

Fig. 1.10 With two rows of teeth in the upper jaw, and one in the lower, tuatara are able to apply three-point bending to large prey items, such as this well-armoured tree wētā (*Hemideina crassidens*, an orthopteran insect) on Stephens Island, as well as small vertebrates including seabird chicks. The lower jaw, with its serrated tooth row, is moved forwards relative to the upper jaw (a form of propalinal jaw action). In this way, large food items are sawn apart into smaller chunks – although, as here, pieces sometimes fall outside the mouth. The orange patches on the skin are clusters of ectoparasitic mites.

Photo: Alison Cree.

Features retained from pre-rhynchocephalian ancestors

The complete diapsid skull of tuatara is no longer viewed as a primitive feature relative to lizards. However, some other characteristics of the skeleton appear to be so. Note that the word 'primitive' is used here in the evolutionary sense of being ancestral, not in the functional sense of necessarily being inferior or maladaptive. Features of tuatara described as primitive are those that have been retained in a more ancestral condition than in a specified group or groups of relatives. Although there was once a belief that tuatara – the entire animal – could be considered as the most primitive of living reptiles,[35] biologists today restrict use of the term 'primitive' to specific features that have been modified to a lesser extent than in an organism's relatives – or avoid it altogether by using the technical term 'plesiomorphy'.

One example of a plesiomorphic or ancestral feature in tuatara is the presence of a parietal foramen (Figure 1.8). This opening in the top of the skull

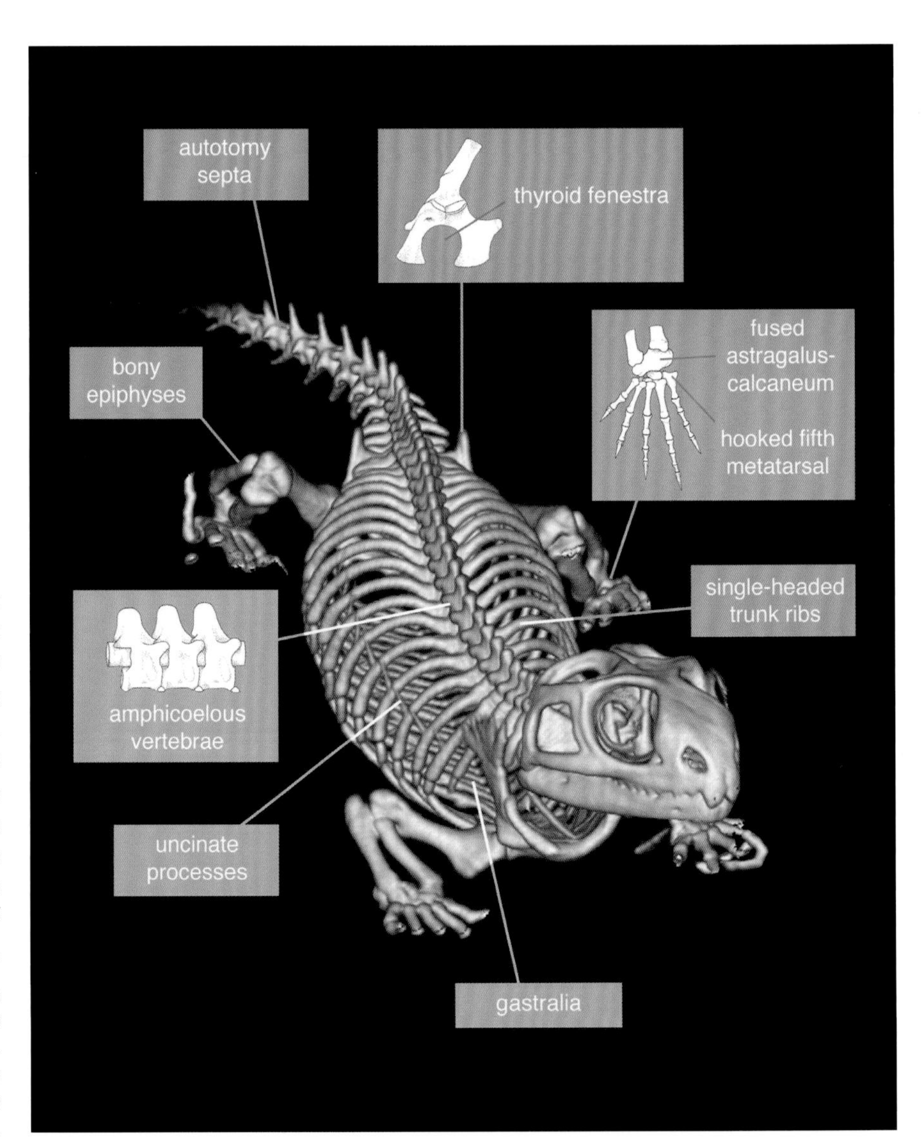

Fig. 1.11 CT scan of an adult female tuatara. Features that are plesiomorphic (ancestral) with respect to squamates are identified with brown boxes. Features that are shared with squamates, suggesting a common evolutionary origin, are identified with green boxes. Note that pointers to the main figure identify a single example of features that exist in the plural.

CT scan of tuatara from Auckland Zoo by Dallen Lee, Auckland Radiology Group (see 'healthy female' of Lee 2010). Inset line drawings by Ken Miller from specimens VT 001 and VT 669, Otago Museum, Dunedin, New Zealand.

lies above the parietal eye (the so-called 'third eye'), which is part of a sensory, hormone-secreting structure known as the pineal complex.[36] A skull opening for the pineal complex was present in many early reptiles. It remains today in tuatara and some squamates, but has been lost from turtles and crocodilians.

Another plesiomorphy of tuatara is the presence of amphicoelous vertebrae (Figure 1.11). These are vertebrae in which the centrum (the central bony cylinder) is bi-concave (hollowed on both ends). The centrum is perforated by a flexible, rod-like structure described in adult tuatara as a persistent, though much constricted, notochord (in other vertebrates, a notochord is present in embryos but typically reduced or absent in adults). Amphicoelous vertebrae are seen today in only a few squamates, including some geckos, in which the structure may have been reacquired. In other living reptiles, the centra are not perforated, and adjacent vertebrae typically interlock in a concave–convex arrangement (depending on whether the surface towards the head or towards the tail is concave, these conditions are described as procoelous or opisthocoelous, respectively). In tuatara, cartilaginous material between adjacent vertebrae results in a condition resembling opisthocoely.[37]

Uncinate processes and gastralia are other plesiomorphies of the skeleton of tuatara relative to squamates. Uncinate processes are small, hook-like structures connected to the ribs that provide attachment for muscles in the chest wall.[38] They also remain present in modern crocodilians and birds, but not in squamates. Gastralia or 'abdominal ribs' are not true ribs but bones that develop within the dermis (inner layer of the skin) of the animal's under-surface; they probably provide support and protection for internal organs.[39] Gastralia persist today also in crocodilians and possibly in modified form in the plastron (the shell on the under-surface) of turtles, but are not found in modern squamates.[40]

Derived features shared between rhynchocephalians and squamate reptiles

Many features of the skeleton of tuatara and extinct rhynchocephalians are also present in squamates but not in other modern reptiles.[41] These shared derived characters (synapomorphies) reflect the common ancestry of the two groups. Rhynchocephalians and squamates are each other's sister group, or closest relatives, and collectively form the lepidosaurs (Greek lepis, 'scale' + sauros, 'lizard', hence reptile).

One derived feature of the skeleton of Lepidosauria is a pleurodont form of tooth attachment, in which the teeth are loosely attached to the outside edge of the jaw. This form of attachment is common in modern lizards, although a few species have evolved acrodonty. Basal rhynchocephalians also exhibited a pleurodont-like condition, or a mixture of pleurodonty and acrodonty,[42] whereas in later rhynchocephalians, including tuatara, acrodont implantation has evolved. (In contrast, crocodilians show a thecodont (socketed) condition, and turtles have replaced teeth with a horny bill.)

Turning to other parts of the skeleton, rhynchocephalians and squamates share a common form of the sternum and have single-headed trunk ribs. They also have bony epiphyses (secondary ossification centres at the ends of the long bones), resulting in growth being determinate, or ceasing well before maximum age is reached.[43] Both groups have a large opening in the pelvic girdle known as the thyroid fenestra. There are many shared features of the limbs, including a fused astragalus-calcaneum in the ankle and a hooked fifth metatarsal in the foot. The fifth metatarsal is the bone at the base of the outermost toe; its hooked (bent or L-shaped) form allows it to work in opposition to the first digit like a pair of tongs, improving the foot's grip on uneven terrain.[44] Rhynchocephalians and squamates also both have autotomy septa (fracture zones in the tail vertebrae that enable the tail to be dropped and later regrown).

In addition, surviving lepidosaurs have some common features of their soft parts. For example, both tuatara and living squamates have a transverse vent (cloacal slit). In other words, the external opening of the cloaca has a side-to-side orientation, an alignment that contrasts with the anteroposterior (front-to-back) opening in other reptiles.[45] Tuatara and squamates also both produce an 'epidermal generation', an outer portion of skin that is shed in cycles.[46] Biochemical similarities have also been noted. For instance, tuatara and lizards, but not other reptiles, produce varanic acid as a common bile acid,[47] and both produce ornithuric acid as the major metabolite of benzoic acid in faeces and urine.[48] These comparisons suggest similarities in liver function between the two groups.

More differences between rhynchocephalians and squamates

Several derived (apomorphic) features of rhynchocephalians that distinguish tuatara and their relatives from squamate reptiles have been noted above. In the converse way, squamates have evolved many features not present in tuatara or other rhynchocephalians. Many of these relate to the structure of the skull and jaw, and thus to the detection and capture of food (Figure 1.12). In particular, the quadrate bone of the squamate skull evolved greater freedom at its lower end, enabling it to rotate freely where it articulates with the skull.[49] Evolution of this condition, known as streptostyly, probably allowed force to be applied more effectively when the jaws closed,[50] with a reduced risk of food falling out from the front of the mouth.[51] Early squamates were thus equipped to exploit a wider range of prey and to develop new feeding strategies, trends that have been carried further in later squamates such as snakes with the evolution of various forms of cranial kinesis, or flexibility between bones of the skull.[52] Rhynchocephalians, on the other hand, with a quadrate firmly fixed to the palate, skull and upper jaw, evolved in a different direction – one in which they could bite powerfully but slowly, and shear, cut or chew their food before swallowing.[53]

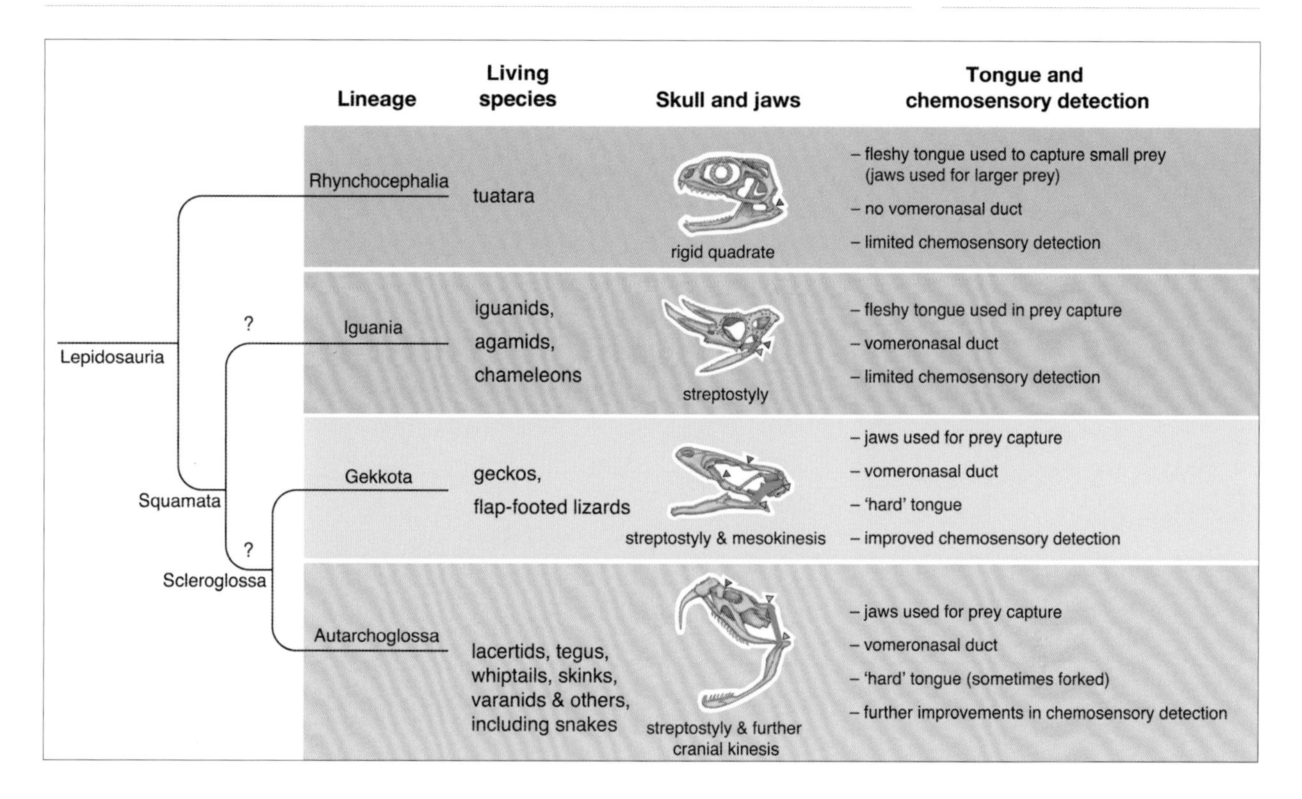

Fig. 1.12 Rhynchocephalians and squamates are each other's sister groups (closest relatives), but evolved in different directions in terms of skull and jaw structure. Importantly, tuatara and other rhynchocephalians have a quadrate bone firmly attached to the palate, skull and upper jaw, whereas in squamates, the jaw is able to rotate more freely with the skull, a form of jaw suspension known as streptostyly (quadrate shown in green; points of movement shown by arrowheads). In addition, the more derived, scleroglossan squamates, including geckos, skinks and snakes, have evolved further points of movement in the bones of the skull (conditions known as mesokinesis or prokinesis). These fundamental changes have contributed to a greater diversity of feeding methods in squamates, especially in scleroglossans, including the primary use of the jaw rather than the tongue for prey capture, and a greater reliance on the tongue and a well-developed chemosensory structure in the roof of the mouth (vomeronasal or Jacobson's organ) for prey detection. In contrast, tuatara are able to bite powerfully, and shear their food before swallowing, but are more sedentary than actively foraging scleroglossans, and rely less on chemosensory detection for prey capture and communication. Note that the hypothesised relationships within living squamates as shown here are well supported by morphological evidence but have recently been challenged by molecular data. This especially affects the basal placement of Iguania as a lineage distinct from other squamates, as indicated by question marks.

Skull diagrams by Patricia J. Wynne, from Vitt, L. J. and Pianka, E. R. 2006. The scaly ones. *Natural History* 116 (July/August): 28–35.

Another related difference between squamates and tuatara involves the development and complexity of the vomeronasal organ (also known as Jacobson's organ; Figure 1.13).[54] These paired organs in the roof of the mouth provide a chemical sense akin to smell. They develop in embryos as part of the nasal system, and remain as distinct organs in adult tuatara and squamates, although they differ in complexity. (In turtles, what is probably the equivalent structure lies within the nasal cavity, and in adult crocodilians the organ is absent.[55]) The vomeronasal organ of adult tuatara has a simple, tubular structure with a relatively low density of chemoreceptor cells.[56] It opens into the nasal system, and thus is only indirectly connected with the oral cavity. Tuatara

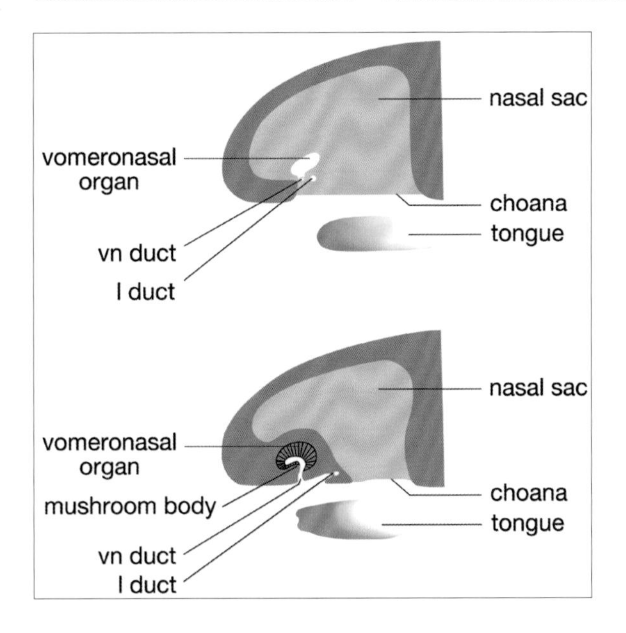

Fig. 1.13 Diagrams comparing the nasal organ, vomeronasal (Jacobson's) organ and tongue tip of a tuatara (top) with a lizard possessing a slightly notched tongue tip (gecko, bottom). In tuatara, the duct from the vomeronasal organ opens into the choanal tube (part of the nasal system) before the latter opens to the mouth. The vomeronasal organ is thus connected with the mouth only indirectly (see Figure 1.7 for the location of the choanae or internal nares, which open to the mouth). In squamates (bottom), the vomeronasal organ opens directly to the mouth via a duct separate from and anterior to the choana. This direct connection to the mouth, combined with the evolution of tongue-flicking behaviour and a forked tongue in advanced scleroglossans such as snakes, provides many squamates with a more effective chemosensory sense. l duct = lachrymal duct; vn duct = vomeronasal duct.

Modified from Bellairs A. d'A. 1984. Closing address: with comments on the organ of Jacobson and the evolution of Squamata, and on the intermandibular connection in Squamata. In: Ferguson M. W. J., ed. *The structure, development and evolution of reptiles. Symposia of the Zoological Society of London No. 52*. Academic Press, London. 665–683.

also have a broad, fleshy tongue (used when capturing small, invertebrate prey) and do not tongue-flick, so odorants probably reach the vomeronasal organ mainly by inhalation[57] and perhaps gaping.[58]

In adult squamates, the vomeronasal organ has a more complex structure than in tuatara. Hemispherical in shape, it contains a cartilaginous structure, the mushroom (fungiform) body, allowing for an enlarged layer of sensory cells. The vomeronasal organ of squamates is also separate from the nasal system, opening directly into the mouth via a vomeronasal duct. Although squamates in the iguanian lineage (which includes iguanas, chameleons and agamids) also use their tongues for capturing small prey, species in the more numerous and diverse scleroglossan lineage (including geckos, skinks and snakes) use their jaws for prey capture. Scleroglossans ('hard-tongued' squamates) also often flick their tongues, and in some species, including snakes, the tongue has become narrow and forked. The tongue has thus evolved an important role in transporting odorant chemicals from the outside environment to the vomeronasal organs,[59] in some cases providing directional information about the location of prey.[60] The evolution of a greater chemosensory sense is widely viewed as important to the success and diversity of squamates (especially scleroglossan squamates) today.[61]

Other differences between squamates and tuatara involve aspects of reproduction. For instance, male squamates have paired hemipenes whereas male tuatara lack an intromittent organ, although paired out-pouchings of the cloaca might be hemipenial precursors.[62] (Other living reptiles (crocodilians and turtles) have a single penis.) Male squamates have evolved a renal sex segment, a secretory region of the kidney not seen in male tuatara[63] or other reptiles. And the spermatozoa of squamates are unlike those of tuatara, which produce spermatozoa resembling those of crocodilians and turtles.[64] In female squamates the follicle surrounding developing eggs in the ovary has a multilayered granulosa, and contains a distinct cell type known as the pyriform cell. In tuatara the granulosa remains essentially single layered, without pyriform cells[65] (as in crocodilians and turtles).[66] Many lineages of squamates have evolved viviparity (live-bearing), whereas female tuatara (again like crocodilians and turtles) always lay eggs (oviparity) and have uterine tissues specialised for eggshell production. Hatchling squamates have a true egg tooth to aid hatching from the egg, whereas hatchling tuatara (like crocodilians and turtles) have a horny shell-breaker on the tip of the snout for this purpose.[67] However, whether any or all of these features were present in the same condition in ancestral rhynchocephalians as in tuatara is hard to know.

Do molecular studies support a close relationship between rhynchocephalians and squamates?

So far we have seen that tuatara present a composite of morphological features: some unique to rhynchocephalians, some shared with squamates and others shared with crocodilians and/or turtles. However, the weight of anatomical evidence clearly allies tuatara and other rhynchocephalians most closely with squamates. A recent hypothesis of evolutionary relationships among major groups of surviving reptiles and other amniote vertebrates (birds and mammals) is shown in Figure 1.14. This branching diagram indicates that mammals, reptiles and birds all evolved from a common amniote ancestor. A lineage leading to mammals diverged first. This may have been followed by one leading to turtles, although the exact origin of turtles remains controversial.[68] The remaining reptiles (diapsids) split into two main lineages: the lepidosaurs (represented today by tuatara, lizards and snakes) and the archosaurs (crocodilians and birds).

Morphological differences between species are, at least in part, a product of genetic differences. Therefore, as lineages become distinct morphologically, genetic differences might be expected to accrue at a similar rate. To some biologists' surprise, early molecular studies did not always yield the expected

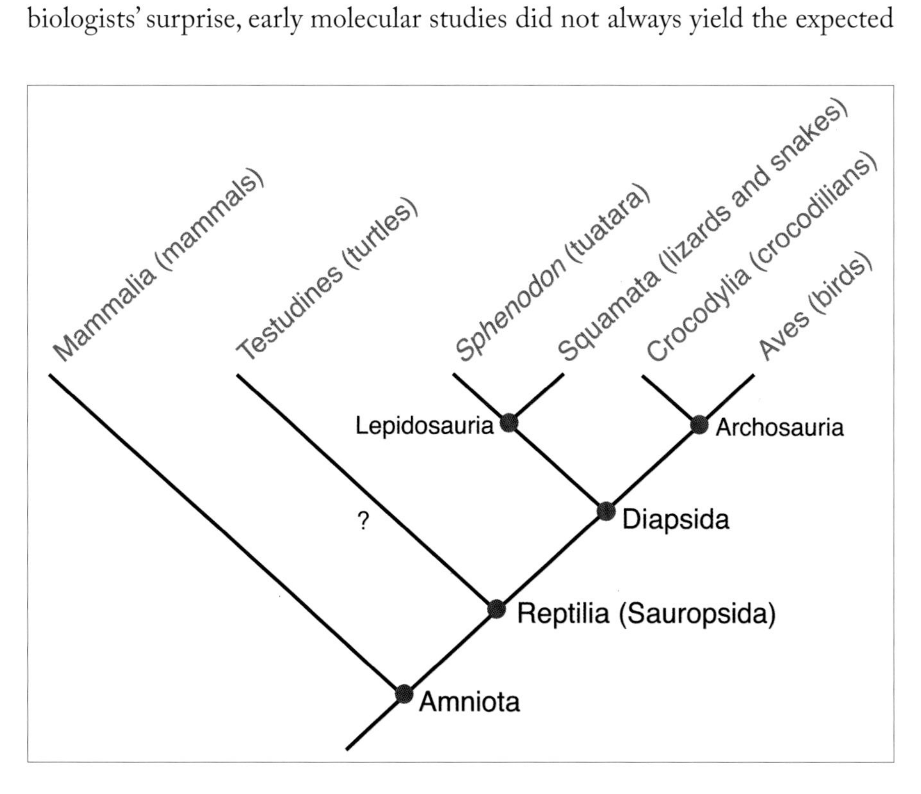

Fig. 1.14 A diagram summarising recent views about the phylogenetic relationships of living amniotes (Amniota). The branching indicates a relative, not absolute, time scale. Rhynchocephalians (represented today by tuatara) and squamates (lizards and snakes) are collectively known as lepidosaurs. They are each other's sister group (closest relatives) and probably diverged about 240–250 million years ago. The placement of turtles is debated, with some recent molecular analyses suggesting a placement within Diapsida.

relationships among reptiles, including tuatara. For example, sequences of haemoglobin proteins[69] and some nuclear and mitochondrial genes[70] group tuatara more closely with crocodilians and birds than with squamates. On the other hand, analyses of the nuclear genes C-mos[71] and RAG-1[72], of the entire mitochondrial genome,[73] and of SINES (short, interspersed, retro-transposable elements of the genome)[74] support the conventional view that tuatara are the sister group of living squamates.

In 2001 a comprehensive study using data then available from both sources – morphological and molecular studies – supported the traditional view of reptilian relationships, including the sister-group relationship of tuatara with squamates.[75] It seems that interpretations from early molecular studies were limited in part by minimal sampling of living forms. Typically, early analyses examined only a modest number of living species[76] as well as a small number of genes. Although rapid advances are being made,[77] the sequence of the full genome (the complete genetic make-up or hereditary information of an organism) has not yet been established for tuatara. Another limitation of early molecular studies – the absence of fossil forms[78] – is more problematic. Given the morphological diversity that we know once existed among fossil rhynchocephalians (see below), it is clear that data derived from sampling even all living populations of tuatara will not be genetically representative of all rhynchocephalians.

A further issue for molecular studies is that the rate of molecular evolution in reptiles, as in vertebrates generally, is inversely related to body size (in other words, larger-bodied species tend to accrue genetic changes more slowly than smaller-bodied ones).[79] Thus, if differences in body size were sustained over geological time, one might expect molecular divergence of tuatara from a common ancestor to be less than that of a small-bodied lizard. It was therefore surprising that tuatara were reported in 2008 to have a rapid rate of evolution in a specific region of mitochondrial DNA (the hypervariable regions, or HVRs, of the control region), in comparison with other vertebrates then studied (mammals and birds).[80] This study has already been contested on the grounds that it did not have an adequate dataset from ancient samples, and that it confounded geographic variation (between ancient samples, from mainland tuatara, and modern samples, from populations on offshore islands) with genuine differences over time.[81] Genetic studies thus continue to raise challenges for interpretation, but the living tuatara remains an important species for exploring relationships between morphological and molecular evolution in reptiles.

Origins of rhynchocephalians within the amniote lineage

We now turn back the clock several hundred million years to examine the origins of, and evolutionary development within, the rhynchocephalian lineage. About 320 million years ago, during the Middle Carboniferous, a major new group of vertebrates known as amniotes appeared (Figure 1.15).[82] Reptiles,

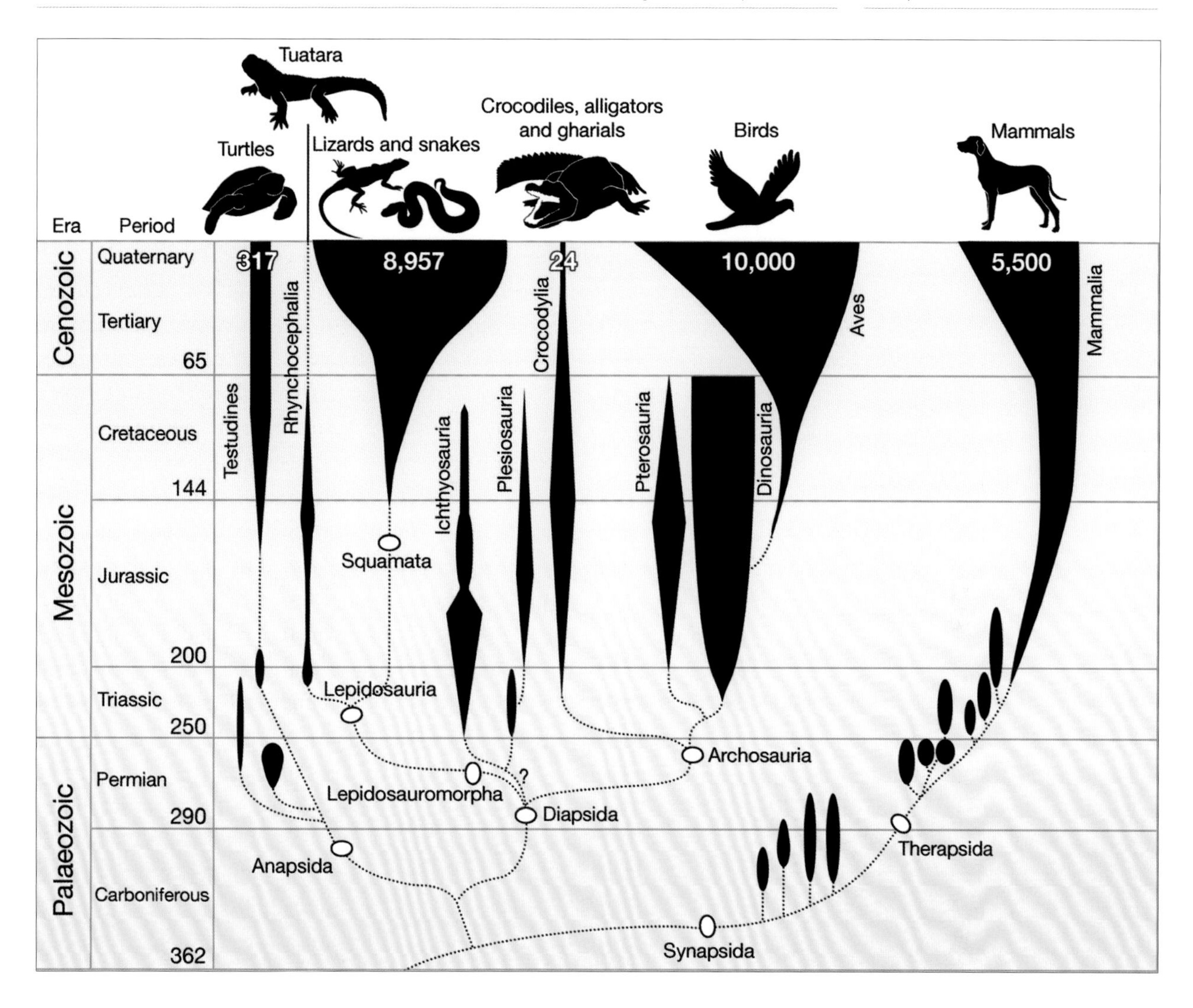

together with mammals and birds, constitute the surviving amniotes. Major lineages of amniotes are traditionally distinguished by the presence and nature of the bony arches that surround openings in the temporal region of the skull. Early in amniote evolution, a group known as synapsids diverged (Greek syn, 'together', + apsis, 'arch'). Synapsids have a single opening on the lower cheek and are represented today by mammals. Turtles have an anapsid skull, meaning one without arches or fenestrae (Greek an, 'not'). Turtles have usually been considered as another early branch of amniotes, although an alternative view, supported by some but not all molecular and morphological evidence, is that they arose from within the diapsids.[83]

The remaining reptiles and birds belong to a major new group, the diapsid lineage. The complete diapsid condition involves the presence of two arches. Modified conditions include the absence of some, or all, temporal bars: the lower temporal bar (as in many lizards) or both bars (snakes and some lizards). Diapsids were present by the Late

Fig. 1.15 A phylogeny of reptiles, emphasising the fossil history of major groups and relationships with other amniotes. The vertical extent of each group (solid colour) indicates its known fossil record through geological time, and the width of the group indicates its relative abundance through time. (The vertical axis is in a scale of millions of years ago, and species numbers as of 2012 are indicated at the top of each group's record.) Dotted lines indicate proposed phylogenetic relationships. The placement of turtles is controversial.

Modified from: Benton. M. J. 2005b. Reptilia (Reptiles). In: *Encyclopedia of Life Sciences*. John Wiley and Sons, Chichester. http://www.els.net/ [doi: 10.1038/npg.els.0004126], with species numbers updated using various sources.

Carboniferous[84] and diverged early on into two main branches. One, the lepidosauromorphs, includes the surviving lepidosaurs and some related fossils. The other, the archosauromorphs, includes some early fossils such as rhynchosaurs, once thought closely related to tuatara. This branch also includes the archosaurs (Greek arch, 'first' or 'chief') – the crocodilians, pterosaurs, dinosaurs and the descendants of dinosaurs, the birds. Archosauromorphs are characterised among other things by a limb structure that allows a more upright posture and erect gait than in lepidosauromorphs. The ichthyosaurs and plesiosaurs are two fossil groups of marine diapsids whose affinities are uncertain.

There is a common misconception – at least, within New Zealand – that tuatara are more closely related to dinosaurs than to lizards. Even university students are prone to make this error (73% of first-year ecology students did so in a survey that I conducted in 2005).[85] It should by now be clear that tuatara are most closely related to lizards (and, from Figure 1.15, that the closest living relatives of dinosaurs are their descendants, the birds, followed by crocodilians). The significance of tuatara thus lies not in being closely related to dinosaurs, but in representing a once-diverse lineage (rhynchocephalians) that had already split from other major groups of reptiles (including dinosaurs) at about the time when all the main groups were diversifying. We turn now to examining the evolutionary radiation within that rhynchocephalian lineage.

Diversity within the rhynchocephalians

The first rhynchocephalians appear in the fossil record in the Late Triassic, about 225 million years ago.[86] However, the actual divergence between rhynchocephalians and squamates may have occurred somewhat earlier, about 240–250 million years ago during the Early Triassic.[87] The Triassic is the first of the three major periods making up the Mesozoic Era (250 to 65 million years ago). The Triassic was a time of great radiation within the reptiles; many extinct forms as well as all surviving major lineages of reptiles seem to have differentiated within this period.

Although fossils of a few species of Mesozoic rhynchocephalians had been discovered by the late 19th century, detailed knowledge of this group has been achieved only during the last three decades. At least eight new species were described during the 1980s, and a further 20 or more since 1990. By 2011, about 34 genera of Mesozoic rhynchocephalians had been recognised, encompassing about 40 species (Table 1.1). Most of these (perhaps all but the most basal genus, *Gephyrosaurus*)[88] belong to the more derived sphenodontian suborder (Figure 1.16). These totals exclude *Sphenodon*, a modern genus.

The Mesozoic rhynchocephalians formed a widespread and abundant group. Their fossils have been found in most of the major continents, including Europe, Asia, North America, South America and Africa, though not yet Antarctica or Australia (Figure 1.17). Some Mesozoic genera occurred on several continents; for instance, species of *Clevosaurus* are known from

TABLE 1.1

Rhynchocephalian genera from the Mesozoic Era[1]

Genus	Period	Country
Ankylosphenodon	Early Cretaceous	Mexico
Brachyrhinodon	Late Triassic	Scotland
Clevosaurus (several species)	Late Triassic–Early Jurassic	Belgium, Brazil, Canada, China, England, South Africa, US
Cynosphenodon	Middle Jurassic	Mexico
Derasmosaurus	Early Cretaceous	Italy
Diphydontosaurus	Late Triassic	Belgium (?), England, Italy (?), US (?)
Eilenodon	Late Jurassic	US
Gephyrosaurus	Early Jurassic	Wales
Godavarisaurus	Early Jurassic	India
Homoeosaurus	Late Jurassic	England, France, Germany
Kaikaifilusaurus	Late Cretaceous	Argentina
Kallimodon (two species)	Late Jurassic	France, Germany
Kawasphenodon	Late Cretaceous	Argentina
Lamarquesaurus	Late Cretaceous	Argentina
Leptosaurus	Late Jurassic	France, Germany
Opisthias	Late Jurassic–Early Cretaceous (?)	England, US, South Africa (?)
Palaeopleurosaurus	Early Jurassic	Germany
Paleollanosaurus	Late Triassic	US
Pamizinsaurus	Early Cretaceous	Mexico
Pelecymala	Late Triassic	England
'*Piocormus*'[2]	Late Jurassic	France, Germany
Planocephalosaurus	Late Triassic	England
Pleurosaurus	Late Jurassic	France, Germany
Polysphenodon	Late Triassic	Germany
Priosphenodon	Late Cretaceous	Argentina
Rebbanasaurus	Early Jurassic	India
Sapheosaurus	Late Jurassic	France
Sigmala	Late Triassic	England
Sphenovipera	Middle Jurassic	Mexico
Theretairus	Late Jurassic–Early Cretaceous (?)	South Africa (?), US
Tingitana	Early Cretaceous	Morocco
Toxolophosaurus	Early Cretaceous	US
Whitakersaurus	Late Triassic	US
Zapatodon	Early Jurassic	Mexico
Undescribed or indeterminate taxa	Late Triassic–Late Cretaceous	Argentina, Brazil, England, Italy, Morocco, Mexico, Niger, Poland, Portugal, South Africa, US, Wales, Zimbabwe

Notes

1 Based on Evans et al. (2001, 2004), with additional data from Reynoso and Clark (1998), Reynoso (2000, 2003, 2005), Apesteguía and Novas (2003), Jones (2006a, 2008), Bonaparte and Sues (2006), Apesteguía and Rougier (2007), Jones et al. (2009b), Malafaia et al. (2010) and Marc Jones (pers. comm. 30 October 2008). *Priosphenodon* is possibly synonymous with *Kaikaifilusaurus* (Albino 2007). See chapter endnotes or the bibliography for full citations.

2 This and perhaps some other genera require re-examination of early material.

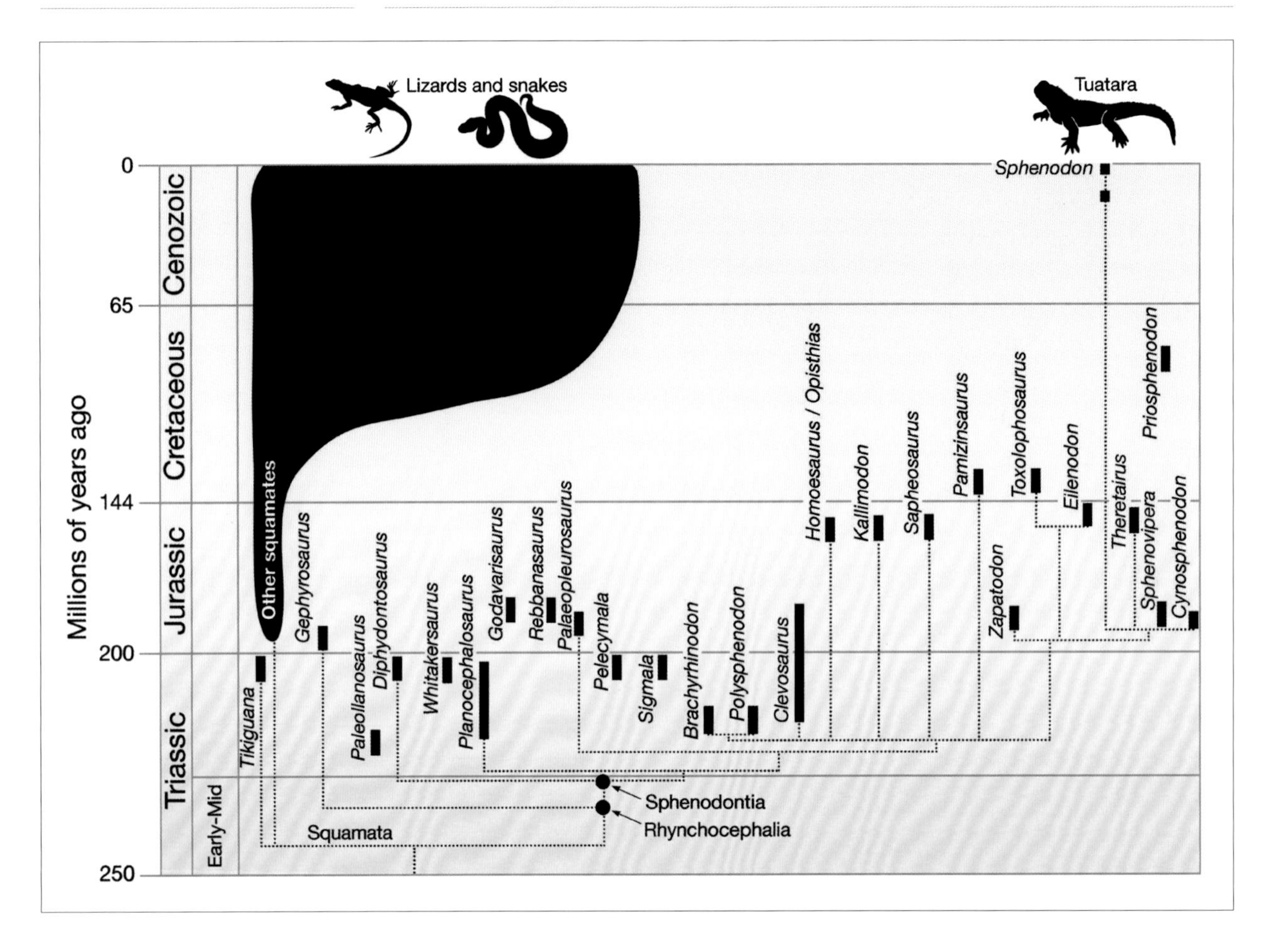

Fig. 1.16 A phylogeny of 25 genera of rhynchocephalians, including *Sphenodon*. The fossil history of each group (and squamates, for comparison) is shown in solid colour.

Modified from Heckert, A. B., Lucas, S. G., Rinehart, L. F. and Hunt, A. P. 2008. A new genus and species of sphenodontian from the Ghost Ranch *Coelophysis* Quarry (Upper Triassic: Apachean), Rock Point, Formation, New Mexico, USA. *Palaeontology* 51: 827–845. John Wiley and Sons. The fossil history of the New Zealand lineage that includes *Sphenodon* is updated following Jones et al. (2009b).

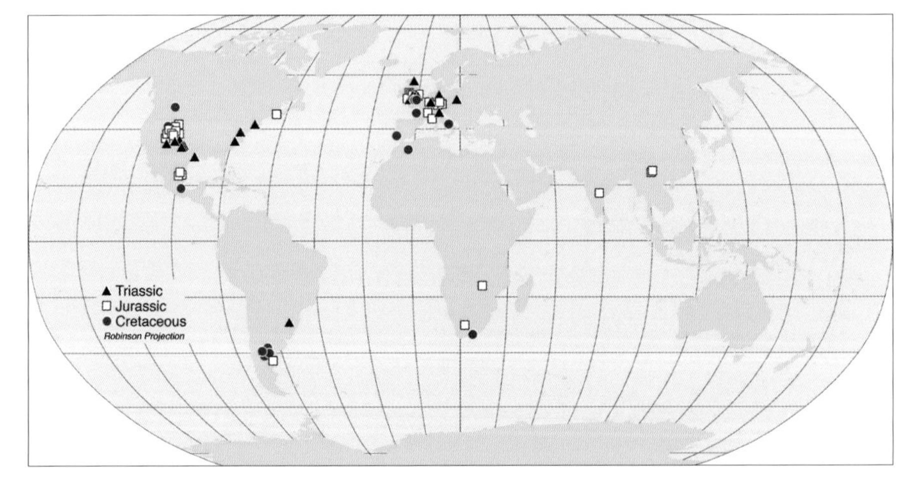

Fig. 1.17 Localities of rhynchocephalian fossils from the Mesozoic, superimposed on a current-day global map. Fossils from the Triassic are shown with black triangles, from the Jurassic with white squares, and from the Cretaceous with red circles.

Based, with one addition, on data from SM Table 1 of Jones, M. E. H., Tennyson, A. J. D., Worthy, J. P., Evans, S. E. and Worthy, T. H. 2009. A sphenodontine (Rhynchocephalia) from the Miocene of New Zealand and palaeobiogeography of the tuatara (*Sphenodon*). *Proceedings of the Royal Society B* 276: 1385–1390.

Europe, Asia, Africa, and North and South America.[89] No fossils of Mesozoic rhynchocephalians have been discovered in New Zealand, the only place that *Sphenodon* occurs. However, since 2002 significant finds have been made in New Zealand of three partial dentaries from relatively recent (Cenozoic) deposits.[90] These pieces of jawbone date from the Early Miocene about 19–16 million years ago. They have teeth typical of advanced rhynchocephalians such as *Sphenodon*, but the fragments provide too little information to be considered as those of tuatara. For now, they have been placed in the subfamily Sphenodontinae,[91] which includes *Sphenodon* and some Mesozoic rhynchocephalians from Mexico.

Mesozoic rhynchocephalians were, for the most part, terrestrial and lizard-like in general form (Figure 1.18). At around 150–300 millimetres in total length, many were smaller than today's tuatara, and some, such as *Homoeosaurus*, were relatively long-legged and probably more agile (Figure 1.19). The Mexican species *Zapatodon ejidoensis* is the smallest known; skull length of a juvenile was estimated at 11.3 millimetres.[92] A Mexican species, *Ankylosphenodon pachyostosus*, had a skull length of about 8 centimetres,[93] similar to that of an adult tuatara. At the upper extreme, adults of the recently described species *Priosphenodon avelasi* from Argentina had skulls reaching 15 centimetres in length, and a total body length approaching 1 metre. This

Fig. 1.18 An artist's impression of some well-known genera or groups of rhynchocephalians. Mesozoic forms ranged in size from the tiny *Zapatodon* to the substantial *Priosphenodon*, and included aquatic or semi-aquatic lineages (pleurosaurs and sapheosaurs) as well as terrestrial species. Unlike tuatara (*Sphenodon*), early rhynchocephalians such as *Gephyrosaurus* lacked a lower temporal bar and acrodont teeth (inset skull). Colourings and crests of Mesozoic forms are, of course, speculative.

Artwork by Jorge A. Gonzalez (gonzalezaurus.yahoo.com.ar) in: Apesteguia, S. 2007. La evolucíon de los lepidosaurios. *Investigacíon y Ciencia* April: 54–63, with minor modifications to text.

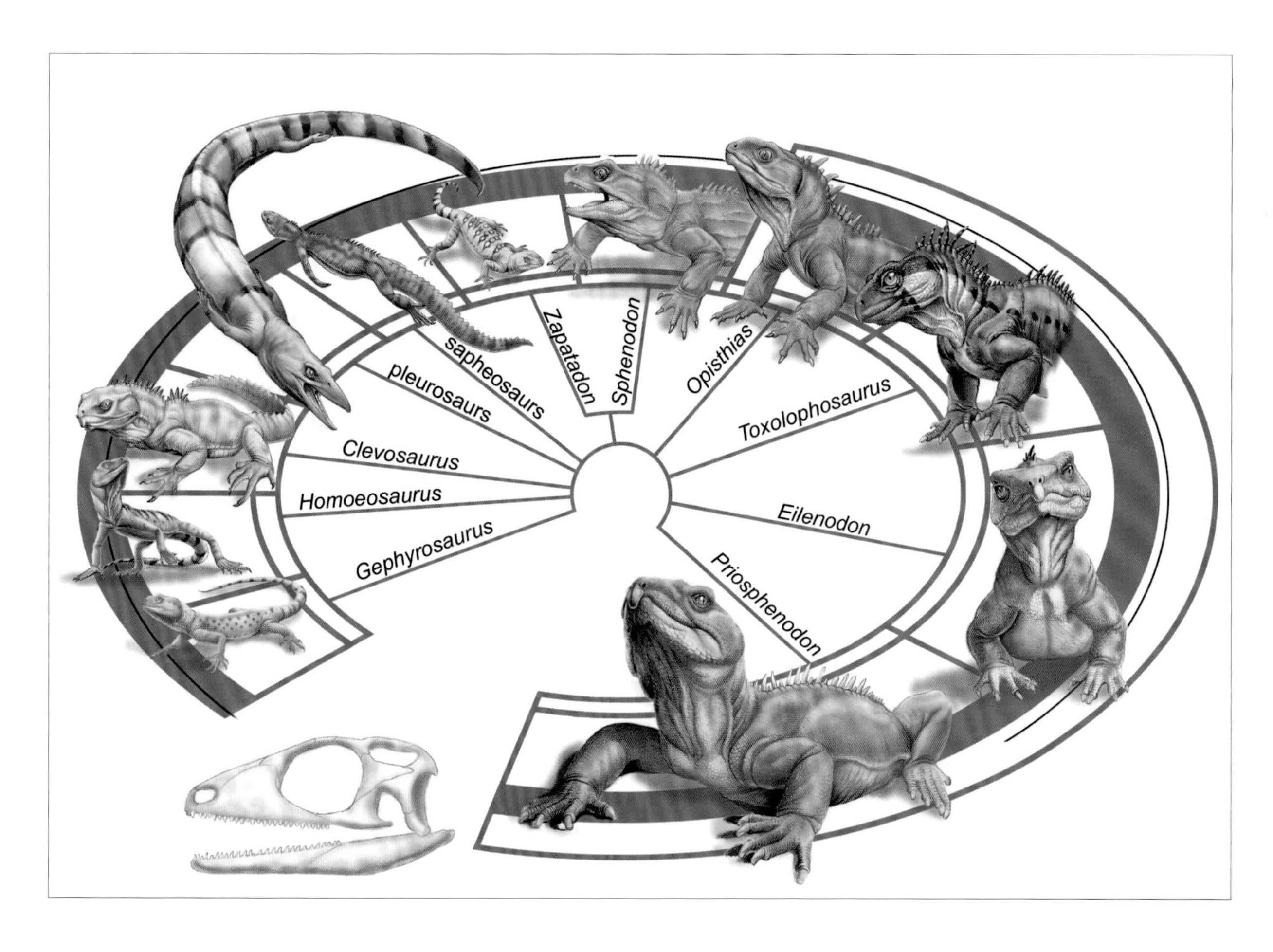

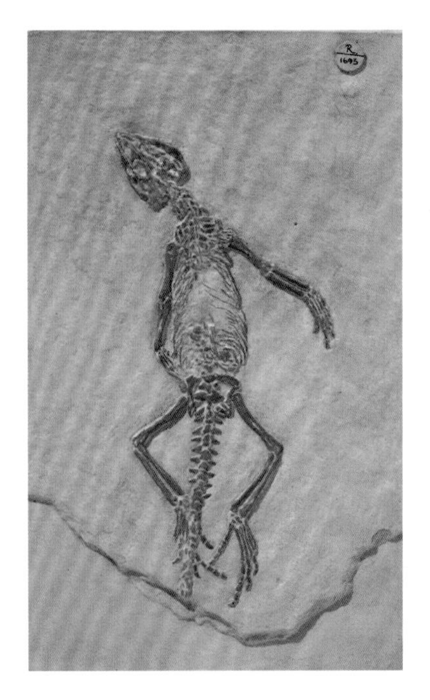

Fig 1.19 Tuatara were often described in the older literature as being more or less identical with the Mesozoic rhynchocephalian *Homoeosaurus*. Although similarities of jaw and tooth structure help define both genera as sphenodontians, *Homoeosaurus* is no longer considered a close relative of tuatara. It was a smaller, longer-legged and probably more agile genus.

Photo: *Homoeosaurus maximiliani*. © Natural History Museum, London, ref. 4894.

species, with its massive, sharply beaked skull and robust skeleton, is the largest known of the terrestrial sphenodontians.[94]

Two lineages of sphenodontians were partly or fully aquatic. One of these, the sapheosaurs, includes the Mexican species *Ankylosphenodon pachyostosus*, mentioned above. *Ankylosphenodon* had a robust skeleton and may have been partly aquatic, perhaps swimming with a pattern similar to a modern lizard, the marine iguana (*Amblyrhynchus cristatus*).[95] Members of a second lineage, the pleurosaurs, were fully aquatic.[96] The marine *Pleurosaurus* was up to 1 metre long. Although its skull reveals its sphenodontian affinities, the rest of its skeleton was modified for aquatic life. The trunk and tail were long (the tail for propulsion), and the limbs were relatively short. Another member, *Palaeopleurosaurus*, has some features intermediate between *Pleurosaurus* and the terrestrial sphenodontians.[97]

Evolutionary trends within the rhynchocephalians

When we combine knowledge of Mesozoic rhynchocephalians with that about tuatara, it is clear that rhynchocephalians were not the static lineage once thought. As a group, they exhibit considerable diversity, with several evolutionary trends.

One recent and important conclusion is that some of the distinctive characteristics of tuatara, previously considered plesiomorphic (ancestral) relative to those of lizards, are in fact derived. In particular, a complete lower temporal bar in the skull has evolved secondarily in tuatara and other advanced sphenodontians. The most basal rhynchocephalians, including *Gephyrosaurus*, *Diphydontosaurus* and perhaps *Planocephalosaurus*, had an incomplete lower temporal bar.[98] In tuatara, the way the bar forms during embryonic development, and the sutures (joints between the bones) of the bar, also suggest that the bar does not have the same evolutionary origin as that in early non-rhynchocephalian diapsids.[99] Recent computer simulations suggest that redevelopment of a lower temporal bar helped reduce the otherwise-high stresses on the quadrate and skull during hard biting.[100]

The absence of certain ear structures in tuatara is also now considered a derived condition within rhynchocephalians. Tuatara lack an external ear opening, a functional tympanic membrane and an air-filled tympanic cavity.[101] These absences suggest that tuatara are not as good at detecting high-frequency airborne sounds as lizards that have these structures.[102] However, in early rhynchocephalians the skull includes a feature known as the quadrate-quadratojugal conch, which indicates that a tympanic membrane was present. Some genera, such as *Clevosaurus*, may have had an impedance-matching middle ear comparable in function with that of most lizards.[103] The condition of the middle ear in tuatara is therefore now considered degenerate rather than primitive.

BOX 1.3

Are tuatara 'living fossils'?

Scientific views change as new evidence accumulates and prevailing ideas are challenged – a situation well illustrated by beliefs about the antiquity and evolutionary significance of tuatara.

When Albert Günther proposed in 1867 that tuatara should be recognised as a rhynchocephalian reptile, only 8 years had elapsed since publication of Charles Darwin's *On the origin of species*. Specialists in reptile classification soon embraced the new theory of natural selection, and perhaps not surprisingly, given Victorian certainties about 'progress', some came to believe that ancestor–descendant relationships could be found among living species. The tuatara genus, geographically isolated in the South Pacific and with no close living relatives, was a prime candidate for an ancestral reptile.

In 1889, for instance, the British Museum's herpetologist, George Boulenger, described tuatara as being unquestionably the 'oldest existing Reptilian type'[1] – a view taken to an extreme by Walter Buller, a leading New Zealand naturalist, who wrote of *Sphenodon* in 1895 that: 'It is perhaps generically the oldest inhabitant of the earth'.[2] In 1904 Samuel Williston, Professor of Paleontology at the University of Chicago, proposed a classification in which rhynchocephalians were ancestral to both squamates (on one branch) and to crocodilians, dinosaurs and pterosaurs (on another).[3] And in 1912 William Benham, professor of biology and curator of the Otago University Museum, described tuatara as 'the most ancient Reptile on earth', and suggested that crocodiles and turtles had 'descended from ancient Tuataras'.[4]

Although new discoveries gradually made views of tuatara as an 'ancestor' or 'oldest reptile' untenable, tuatara remained widely perceived as a living fossil, unchanged since Mesozoic times. During the mid-20th century, tuatara were often described using phrases such as 'the zoological equivalent of Methuselah'.[5] In the 1967 edition of his book *The meaning of evolution*, George Gaylord Simpson referred to tuatara as one of the 'immortals' of the animal world, saying that it represented 'one of the most remarkable examples of evolutionary stagnation ... the evolutionary rate of this line has been virtually nil for the last 140,000,000 years or more.'[6] When asked in 1952 whether tuatara existed in the world 200 million years ago, tuatara biologist Bill Dawbin reportedly replied, 'Yes, it was about that time, as nearly as we can place it'.[7] And in 1966 Richard Sharell wrote enthusiastically of tuatara in his popular book *The tuatara, lizards and frogs of New Zealand* as follows: 'Imagine the excitement of having the chance to see, to study, to observe a true saurian of Mesozoic times in the flesh, still living'.[8]

In fact, the oldest described fossils of tuatara – the genus *Sphenodon* – are no more than about 34,000 to 100,000 years old.[9] Although the discovery of Miocene fossil jawbones resembling those of tuatara could push this back further (by up to 19 million years for the subfamily),[10] it is apparent that the fossil record for tuatara is not an especially ancient one. Compared with its Mesozoic relatives, *Sphenodon* must be viewed as an essentially modern genus.

When talking about antiquity, it is important to make a distinction between tuatara in particular and rhynchocephalians in general. The fossil record for rhynchocephalians dates from at least 220 million years ago in the Triassic, and tuatara clearly share some characteristics with the earliest forms. But tuatara are not identical to Mesozoic species, having their own unique combination of specialised characters.[11] Furthermore, other members of the group have evolved features such as herbivory, protective armour or modifications for aquatic life.[12] These variations show that, contrary to early impressions, rhynchocephalians did not genetically or developmentally lack the capacity for change.[13]

Although biologists no longer think of rhynchocephalians as a static group, is it still correct to talk of these animals as an 'ancient' lineage? This depends on your reference point. Compared with humans, rhynchocephalians are undoubtedly ancient: our own genus, *Homo* (which includes several fossil species) has been in existence for only about 2.5 million years, and our own species, *Homo sapiens*, for probably no more than 250,000.[14] However, relative to the other main lineages of extant reptiles, the point is moot. Turtles (Testudines) arose in the Late Triassic and crocodilians by the Early Jurassic. Fossil squamates have been recorded from the Late Triassic,[15] and their sister-group relationship with rhynchocephalians means that they must have evolved by the time that rhynchocephalians appeared. In other words, it seems likely that each of the four major lineages of surviving reptiles had appeared by about 200 million years ago, making them all of similar antiquity.

Returning to the relationships between tuatara and Mesozoic rhynchocephalians, it is worth noting that although skeletons can tell us much, they do not reveal everything about the similarities between fossil and modern forms. Indeed, the concept of morphological stasis is misleading if it leads us to assume that the physiology, ecology and behaviour of tuatara were identical with those of their Mesozoic relatives.[16] Clearly, Mesozoic environments – both physical and biological – were very different from those of today; climates, for instance, were often much warmer. During subsequent isolation on the New Zealand landmass, the lineage leading to tuatara would have experienced climates ranging from subtropical to near glacial[17] – yet tuatara are currently adapted to a cool-temperate climate and die quickly in warm conditions.[18]

Mesozoic rhynchocephalians also experienced a different suite of predators, competitors, prey, parasites and vegetation than today. For instance, the seabirds that tuatara include in their diet[19] would not have been available to rhynchocephalians of the Early Mesozoic because they had not yet evolved.[20] Similarly, ticks, which now parasitise tuatara on some islands, may not have bothered early Mesozoic rhynchocephalians because ticks are not known to have existed prior to the Cretaceous.[21] Ecologically and behaviourally, as well as physiologically and genetically, tuatara cannot have been identical with Mesozoic rhynchocephalians.

The concept of morphological stasis can be downright dangerous to conservation. In the early 1900s when tuatara were considered the most ancient of reptiles and an example of evolutionary stagnation, it was argued (at least within New Zealand's popular literature) that extinction of tuatara was not only imminent and inevitable, but also morally desirable![22] However, as scientists have increasingly recognised, especially following an influential article by Carl Gans

Box 1.3 Fig. 1 Unlike their Mesozoic relatives, tuatara have evolved a unique relationship with seabirds, such as this fairy prion (tītī-wainui, *Pachyptila turtur*) on Stephens Island. Tuatara make use of burrows dug by seabirds and also prey on seabird eggs and chicks – food items that were not available for much if any of the Mesozoic Era.

Photo: Brett Robertson/Victoria University of Wellington

Box 1.3 Fig. 2 This early 20th-century postcard from New Zealand reveals a then-popular but incorrect evolutionary perspective: tuatara were seen as not only the oldest reptile, but the oldest animal, on the face of the earth.

Photo: Muir & Moodie postcard No. 48R, courtesy of James Taylor.

in 1983, tuatara are not 'maladapted' for modern life and, in the absence of introduced mammals, have been very successful.[23] For example, despite the fact that males lack intromittent organs (penises or their equivalent), there is no evidence to support a 19th-century view that reproduction is impaired as a result.[24] Indeed, tuatara were extremely widespread in the pre-human New Zealand environment, and appear no less adapted for survival than New Zealand lizards, with which their lineage has co-existed for tens of millions of years.

So, where does this leave us with respect to the notion of tuatara as 'living fossils'? Although this term has great popular appeal, it may come as a surprise to learn that it has no clear meaning for scientists – at least seven possible definitions have been recognised.[25] In the case of tuatara, it cannot mean that the genus has survived unchanged, morphologically or otherwise, since the Mesozoic. Nor does it mean that the lineage to which tuatara belong is substantially more ancient than other surviving reptilian lineages. The concept simply seems to boil down to one of relative numbers. Tuatara – a single genus represented by just one species – are the sole survivors of the rhynchocephalians, a once widespread and moderately diverse group. The status of tuatara as a phylogenetic relict thus contrasts with that of the crocodilians, turtles and squamates, which are today represented by tens, hundreds or thousands of species, respectively. Crocodilians and turtles, with their distinctive body plans, are other reptiles that are occasionally referred to as living fossils, and there seems little doubt that they would be described this way more often if the number of surviving species was lower.

Finally, it is possible to acknowledge the imprecision of the term 'living fossil' for tuatara without undermining the extraordinary significance of these animals. The genus remains the only living representative of a lineage that has been on its own 'evolutionary trajectory' for 240 million years or more. Tuatara thus represent a unique connection with the Mesozoic, even though they are not transplanted unchanged from it. With their particular combination of derived and ancestral features, they illustrate a different approach to reptilian life from that in other living species. Tuatara will undoubtedly continue to play an important role in helping biologists to understand how the various characteristics of reptiles have evolved in relation to each other. Such research is based on the modern understanding that tuatara represent a 'branch tip' of the rhynchocephalian lineage, rather than a Mesozoic genus or one ancestral to any other living reptile.

BOX 1.4

Skull structure and feeding in rhynchocephalians, past and present

Rhynchocephalians vary in their skull and tooth structure, suggesting that they also varied in the ways in which they captured and processed food. Four broad groups, represented by well-preserved skulls, have recently been described by palaeontologist Marc Jones.[1] *Gephyrosaurus bridensis* (see A in figure), representing basal, non-sphenodontian taxa, had an incomplete lower temporal bar, a relatively slender snout, and numerous small, relatively weakly attached teeth. It probably fed on small and soft insects, swallowing without much chewing. More derived taxa (sphenodontians: B, C, D and E) have or had wedge-shaped incisor-like teeth at the front of the mouth, and fewer but larger, more rigidly attached (acrodont) teeth on the maxilla. Clevosaurs, represented by *Clevosaurus hudsoni* (B), had strong, rounded snouts and maxillary teeth that formed a blade-like flange, perhaps cutting through large insect, small vertebrate and plant food like a pair of strong scissors. Pleurosaurs, represented by *Pleurosaurus goldfussi* (C), were a fully aquatic group in which the skull was long and flattened, and lacked a lower temporal bar. They may have fed on fish. Eilenodontines, represented by *Priosphenodon avelasi* (D), had robust skulls with a large postorbital area (the region posterior to the eye socket), and no lower temporal bar; they also possessed transversely expanded teeth on the dentary (the main bone of the lower jaw; not shown in figure). Eilenodontines were probably herbivorous. Sphenodontines, represented by an adult tuatara (*Sphenodon*, E), have a short snout with a large postorbital area (related to the jaw muscle volume), and a substantial lower temporal bar (secondarily derived). As tuatara grow from hatching (compare F, near-hatching, with E), they exhibit some of the same trends seen among adult rhynchocephalians of other genera: they develop a larger postorbital area, a more ventrally positioned jaw joint, and a parietal crest (a bony attachment site for muscles). These developmental changes may be related to a change from feeding exclusively on small invertebrates as hatchlings to feeding on larger and tougher food (including small vertebrates) as adults.

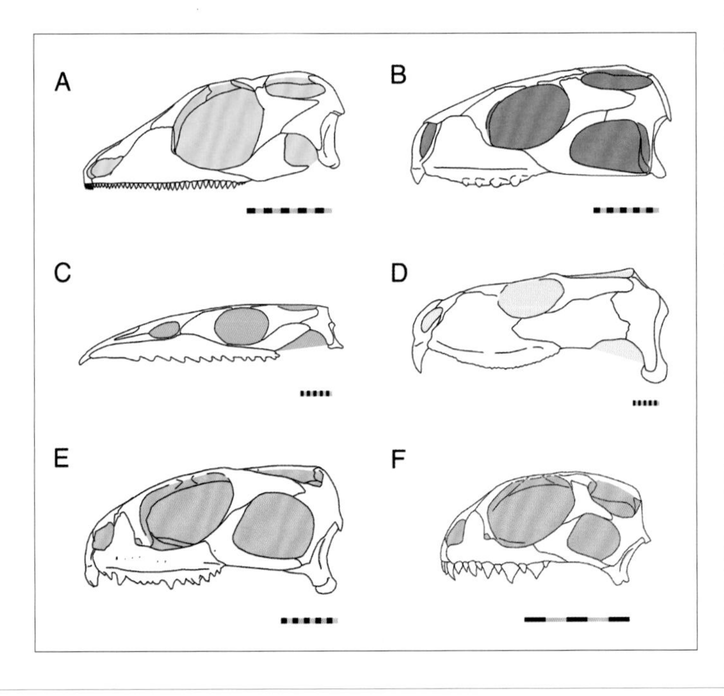

Box 1.4 Fig. 1 Skulls of representative rhynchocephalians. A: *Gephyrosaurus bridensis*; B: *Clevosaurus hudsoni*; C: *Pleurosaurus goldfussi*; D: *Priosphenodon avelasi*: E: adult tuatara (*Sphenodon*); F: tuatara at about hatching. Scale bars = 10 mm, except for F = 5 mm.

From Jones, M. H. E. 2008. Skull shape and feeding strategy in *Sphenodon* and other Rhynchocephalia (Diapsida: Lepidosauria). *Journal of Morphology* 269 (8): 945–966. Copyright 2008 Wiley-Liss, Inc. This material is reproduced with permission of John Wiley & Sons, Inc.

The secondary development in tuatara and other advanced rhynchocephalians of the lower temporal bar, and the loss of an impedance-matching middle ear, may be related to changes in feeding.[104] Basal rhynchocephalians had teeth with pleurodont-like attachment, resembling the implantation in many lizards today. In addition to the maxillary and palatine tooth rows, there were numerous small teeth on other bones in the upper palate (the inner palatines, pterygoids and vomers). *Gephyrosaurus* and other early rhynchocephalians may have fed by impaling relatively small, soft insects with a fast but weak bite, probably swallowing without much chewing.[105] Trends in later rhynchocephalians (the sphenodontians) include a reduced number of palatal teeth, more strongly attached (acrodont) teeth, a more robust skull, and bony crests on the parietal bone that imply more powerful jaw musculature. Sphenodontians probably had a more powerful but slower bite than did early forms, enabling more effective shearing of larger prey into bite-sized pieces before swallowing.[106]

Some advanced Mesozoic sphenodontians had well-developed propalinal jaw action, and several were apparently herbivorous. Two, *Eilenodon* from the Late Jurassic and *Toxolophosaurus* from the Early Cretaceous, had broadened cheek teeth that were probably used to grind and shred vegetation.[107] *Ankylosphenodon*, from the Early Cretaceous, had deeply ankylosed teeth that were apparently long-lasting and constantly growing rather than broad.[108] Among the Mesozoic genera, *Cynosphenodon* and one or two other genera with anterior caniniform teeth on the dentary and maxilla are considered the closest relatives of tuatara.[109] One of these, a recently described Mexican species with the delightful name '*Sphenovipera jimmysjoyi*', had grooves on the surface of the two fang-like teeth at the front of each lower jaw. The presence of these grooves, combined with a widely gaping mouth, suggested to the discoverers that this species was able to deliver venom to its prey;[110] however, alternative uses for the grooves, such as redirecting food and fluid from the tooth tip, have since been proposed.[111]

Rhynchocephalians in the Age of Reptiles

The Mesozoic Era, particularly in its later stages, was a time of great radiation of diapsid reptiles. Diapsids dominated the air (pterosaurs) and the land (dinosaurs and other archosaurs in particular), and were conspicuous in the sea (ichthyosaurs, plesiosaurs and mosasaurs – a group of marine squamates). Dinosaurs attract popular attention, unsurprisingly given the enormous size, ferocity and bizarre appearance of some forms, but what was life like for the terrestrial rhynchocephalians that lived, both literally and figuratively, in their shadows?[112]

During the Triassic, when the first rhynchocephalians appeared, the globe was a fundamentally different place. All of the continents were connected at times to form the great super-continent Pangaea, allowing many terrestrial species, including some rhynchocephalians, to have widespread distributions.

Fig. 1.20 The small rhynchocephalian *Brachyrhinodon* (about 150–200 mm long, shown in brown at top of the rock in foreground) lived in the Late Triassic. This reconstruction, based on fossils found at Elgin in Scotland, includes various archosaurs. Some, such as the early dinosaur at bottom left, may have been predators of *Brachyrhinodon*. The large, beaked rhynchosaurs, shown in the right foreground feeding on seed-ferns, were once thought closely related to rhynchocephalians, but are now classed as archosaurs.

Artwork by Jenny Halstead.

Climates were becoming drier and more seasonal, and were warmer than today, with no polar ice caps. Emergent plants were changing from a flora dominated by seed-ferns to one dominated by conifers, cycadophytes and ginkgoes. Angiosperms, the flowering plants that dominate the world's vegetation today, had yet to evolve. Shrub and ground cover was provided by ferns, lycopods, horsetails, mosses and liverworts.[113]

A valuable 'window' into the world of early rhynchocephalians comes from excavations of Late Triassic and Early Jurassic deposits in the United Kingdom. In parts of England and Wales small limestone islands were surrounded by a shallow sea. Flash floods swept many small animals into fissures in the limestone, trapping them until modern quarries uncovered their bones. One such fossil assemblage includes the early rhynchocephalian *Gephyrosaurus*, three other rhynchocephalians, archosaurs, mammal-like synapsids, and several mammals, including one large enough to have been a carnivore.[114] *Gephyrosaurus* was small (100–120 millimetres snout–vent length, about the size of a half-grown juvenile tuatara) and probably fed on beetles and snails found at the site, as well as on flies, grasshoppers and other invertebrates that had evolved by this time.[115] At another site nearby, six species of rhynchocephalians have been found in an assemblage that also contains vertebrate herbivores (procolophonid reptiles, mammal-like synapsids, as well as an early prosauropod dinosaur up to 3 metres long) and carnivores (a gracile, terrestrial crocodile the size of a large cat, and a theropod dinosaur).[116] Of the better-known rhynchocephalians, *Planocephalosaurus* probably fed on small insects and soft grubs. *Clevosaurus hudsoni* probably ate larger insects as well as small vertebrates (possibly including rhynchocephalians) and perhaps some plant material, whereas *Sigmala* was probably herbivorous.[117] A reconstruction of another Late Triassic community from Scotland shows the small rhynchocephalian *Brachyrhinodon* as part of an assemblage including herbivorous rhynchosaurs, predatory dinosaurs and other archosaurs (Figure 1.20).

These and other recent excavations around the world confirm that terrestrial rhynchocephalians were widespread and abundant in early Mesozoic assemblages. What defences did they have against the carnivores and large herbivores that shared their world? At least some were lightly built and agile, with good hearing and a fast turn of speed.[118] Many, perhaps all, had caudal fracture planes, dropping their tails when pursued by predators. Perhaps their colouration provided camouflage – we cannot say for sure. A recent discovery in Mexico suggests that one later form had protective armour in its skin. This Early Cretaceous genus has been named *Pamizinsaurus* (meaning, with reference to the Náhuatl language, 'lizard covered with corn').[119] The bead-like dermal osteoscutes that covered the body may have offered protection from other animals, in much the same way as similar structures do today in the lizards of the genus *Heloderma*, including the Gila monster, *Heloderma suspectum*.

Decline of the rhynchocephalians

From their first appearance in the Early Mesozoic, rhynchocephalians remained in the fossil record for at least 140 million years. They diversified widely in the Late Triassic and survived a mass extinction event at the end of this period, during which many diapsid reptiles such as rhynchosaurs died out. Rhynchocephalians disappeared from the Asian record in the Early Jurassic, but remained present in Europe and North America until the Early Cretaceous.[120] In South America they were a diverse group until the Late Cretaceous, as shown by the recent discovery of several species (some as large or larger than tuatara) from Patagonian Argentina (Figure 1.21).[121]

Reasons for the demise of Mesozoic rhynchocephalians are unclear.

Fig. 1.21 The large rhynchocephalian *Priosphenodon* (up to 1 metre long, bottom foreground) survived until at least the Late Cretaceous in Patagonian Argentina. This species was probably herbivorous. Mammals and other terrestrial reptiles have been found at the same site, the latter including turtles, crocodiles, dinosaurs, and snakes with legs.

Artwork by Jorge A. Gonzalez in: Apesteguia, S. 2007. La evolucíon de los lepidosaurios. *Investigacíon y Ciencia* April: 54–63.

During the early Mesozoic, rhynchocephalians seem to have rarely co-existed with squamates, perhaps indicating that the two groups competed with each other or that they had preferences for different environments.[122] Rhynchocephalians may have eventually been out-competed by squamates, but the relative importance of this and other factors, such as competition with insectivorous mammals, predation and environmental change, remains unclear.

Whatever the causes, most rhynchocephalians seem to have disappeared before the mass extinction that occurred at the end of the Cretaceous. This extinction event is widely attributed to the effects of an asteroid about 10 kilometres in diameter hitting Earth at what today is the Yucatán Peninsula of Mexico.[123] The massive impact site, the Chicxulub crater, measures nearly 200 kilometres across. Immediate and delayed effects caused catastrophic environmental change, and about half of all plant and animal species became extinct. Dinosaurs, pterosaurs and the last of the South American sphenodontians disappeared at about this time, but most rhynchocephalians had already died out.

Today, living rhynchocephalians are represented by just the species *S. punctatus* in New Zealand. Other surviving lineages of reptiles have been more successful, in terms of both species diversity and geographic range. Crocodilians are represented by 24 species spread through many tropical and subtropical regions.[124] Turtles, numbering around 317 species, have distributions encompassing most tropical to temperate areas, including oceanic waters. But it is the squamates, whose diversity is an order of magnitude greater again (at least 8957 species), and whose geographic ranges include all but the highest latitudes, that have been the major success story of the (non-avian) reptiles.

Although it might seem that rhynchocephalians are an evolutionary 'dead-end', that conclusion would be premature. At least one lineage survived the mass extinction at the end of the Cretaceous, and then a further 65 million years until the present. Tuatara have been much reduced in numbers in New Zealand since the arrival of humans and their biological entourage, but under appropriate conditions tuatara continue to thrive. The next chapter traces what is known of the evolutionary, geological and climatic history of the lineage leading to tuatara, from the Mesozoic Era until the time that humans set foot in New Zealand.

Endnotes to Chapter 1

See the bibliography for full references to articles on tuatara, which are cited here at first mention by surname and year only. References on other topics are cited here in full at first mention, and then using 'op. cit.'. Except where required for clarity, the convention of parentheses around years has been dispensed with for economy of space.

1 Amphisbaenians (worm lizards) are here included within 'lizards'. Snakes have also originated in an evolutionary sense from within 'lizards', but because of their distinctive body plan they are often treated separately, as here.
2 Gray 1831; see Chapter 5 for more on taxonomic history.
3 Gray 1842.
4 Buller 1877.
5 Daugherty et al. 1990b.
6 The potential for interbreeding between the Brothers and other populations has also not been tested in captivity.
7 Worthy 1997; Worthy and Grant Mackie 2003. A possible skeletal difference is under study: see Jones 2007a.
8 Norris et al. 2004; Norris 2007.
9 Worthy and Holdaway 2002, see p. 460.
10 Hay et al. 2003, 2004, 2010; Bell et al. 2004; MacAvoy et al. 2007; Miller et al. 2006; Norris 2007.
11 This article was first published on-line in 2009. For the print version, see Hay et al. 2010.
12 MacAvoy et al. 2007.
13 von Wettstein 1931. For further discussion, see Daugherty et al. 1990b.
14 This total includes tuatara on Mauitaha Island (in The Hen and Chickens Islands), which are arguably not a 'population' in that only one tuatara has been seen in the last few decades (Tennyson and Pierce 1995). The total also treats the two halves of Motunau (Plate Island) as separate populations, as these landmasses are separated by about 2 metres of sea.
15 Newman 1982c.
16 See Chapter 4.
17 On Tawhiti Rahi Island, pers. obs.
18 Pers. obs. for Tawhiti Rahi during 1989. On Stephens Island, longer and heavier females (up to 267 mm snout–vent length and 690 grams, respectively) were recorded during the 1950s, but maximum sizes have since declined (Moore et al. 2007) to lower values than on Tawhiti Rahi.
19 Thompson et al. 1992 (for 1988); pers. obs. during 1989.
20 Tyrrell et al. 2000. See Chapter 6 for more on geographic variation in body size.
21 Hoare et al. 2006 for North Brother Island; Moore et al. 2007 for Stephens Island.
22 Pers. obs. See Chapter 8 for more on obesity in captivity.
23 See Chapter 7 for more on ears and hearing.
24 For descriptions of being bitten by tuatara, see Anon 1882b; Reischek 1882; Hutton and Drummond 1904, p. 358; Schmidt 1953; Wojtusiak 1973; Chorlton 1977; Daugherty et al. 1990a.
25 Evans, S. E. 1984. The classification of the Lepidosauria. *Zoological Journal of the Linnean Society* 82: 87–100; Fraser, N. C. 1988. The osteology and relationships of *Clevosaurus* (Reptilia: Sphenodontida). *Philosophical Transactions of the Royal Society of London B* 321: 125–78; Rieppel, O. 1994. The Lepidosauromorpha: an overview with special emphasis on the Squamata. In: Fraser, N. C. and Sues, H.-D. (eds). *In the shadow of the dinosaurs.* Cambridge University Press, Cambridge, pp. 23–37; Evans 2003.
26 Gauthier, J., Estes, R. and de Queiroz, K. 1988. A phylogenetic analysis of Lepidosauromorpha. In: Estes, R. and Pregill, G. (eds). *Phylogenetic relationships of the lizard families.* Stanford University Press, Stanford, pp. 15–98; Evans 2003, op. cit.
27 Fraser 1988, op. cit.; Reynoso, V.-H. 2003. Growth patterns and ontogenetic variation of the teeth and jaws of the Middle Jurassic sphenodontian *Cynosphenodon huizachalensis* (Reptilia: Rhynchocephalia). *Canadian Journal of Earth Science* 40: 609–619.
28 For example, Sharell 1966, see p. 26. As noted by Kieser et al. 2009, the misinterpretation apparently dates back to Colenso 1886.
29 Kieser et al. 2009. See also Howes and Swinnerton 1901.
30 Robinson 1976.
31 Sometimes referred to as prooral (e.g. Jones et al. 2009a, 2009b in supplementary material).
32 Robinson 1976.
33 For a detailed description of the skull of tuatara and brief history of study, see Evans 2008. For three-dimensional images of skulls prepared using X-ray computed tomography, see http://digimorph.org/.
34 Mertens 1958.
35 For example, Carroll, R. L. 1969. Origin of reptiles. In: Gans, C., Bellairs, A. d'A. and Parsons, T. S. (eds). *Biology of the Reptilia. Vol. 1. Morphology A.* Academic Press, London, see p. 3; Robb 1977, see p. 1.
36 See Chapter 9, Box 9.2 for more on the pineal complex.
37 Hoffstetter and Gasc 1969.
38 Bellairs, 1969, Vol.1, p. 51.
39 In dinosaurs, gastralia may have evolved a role in lung ventilation. See Claessens, L. P. A. 2004. Dinosaur gastralia; origin, morphology, and function. *Journal of Vertebrate Palaeontology* 24: 89–106.
40 Romer, A. S. 1970. *The vertebrate body.* 4th edn. W. B. Saunders, Philadelphia, see pp. 152–153.
41 Evans 1984, op. cit.; Evans, S. E. 1988. The early history and relationships of the Diapsida. In: Benton, M. J. (ed.), *The phylogeny and classification of the tetrapods. Vol. 1. Amphibians, reptiles, birds.* Systematics Association Special Volume No. 35A, Clarendon Press, Oxford, pp. 221–260; Benton, M. J. 1985. Classification and phylogeny of the diapsid reptiles. *Zoological Journal of the Linnean Society* 84: 97–164; Gauthier et al. 1988, op. cit.; Carroll, R. L. and Currie, P. J. 1991. The early radiation of diapsid reptiles. In: Schultze, H.-P. and Trueb, L. (eds). *Origins of the higher groups of tetrapods: controversy and consensus.* Cornell University Press, Ithaca, pp. 354–424.
42 Jones 2008.
43 Growth in tuatara may, however, be more flexible with age than once thought. Adult tuatara from North Brother Island that were translocated to Titi Island, where there was little competition with other tuatara for food, showed an unexpected resumption of growth in snout–vent length as well as in mass (Nelson et al. 2002b). It remains possible that growth ceases before maximum age is reached.

44 Robinson, P. L. 1975. The functions of the hooked fifth metatarsal in lepidosaurian reptiles. *Colloques International du Centre National de la Recherché Scientifique* 218: 461–483.
45 Gauthier et al. 1988, op. cit.
46 Alibardi and Gill 2007. See Chapter 9.
47 Schaffner, F. 1998. The liver. In: Gans, C. and Gaunt, A. S. (eds). *Biology of the Reptilia. Vol. 19. Morphology G. Visceral organs*. Society for the Study of Amphibians and Reptiles, Ithaca, New York, pp. 485–531; Hagey, L. R., Vidal, N., Hofmann, A. F. and Krasowski, M. D. 2010. Evolutionary diversity of bile salts in reptiles and mammals, including analysis of ancient human and extinct giant ground sloth coprolites. *BMC Evolutionary Biology* 10 (133): 1–23.
48 Jordan et al. 1980.
49 Intriguingly, a lizard with a fixed quadrate and a complete lower temporal bar has recently been described. Lu, J-C., Ji, S.-A., Dong, Z.-M and Wu, X-C. 2008. An Upper Cretaceous lizard with a lower temporal arcade. *Naturwissenschaften* doi: 10.1007/s00114-008-0364-1; Mo, J.-Y., Xu, X. and Evans, S. E. 2010. The evolution of the lepidosaurian lower temporal bar: new perspectives from the Late Cretaceous of South China. *Proceedings of the Royal Society B* 277: 331–336.
50 It has sometimes been suggested that the bite of squamates also became faster and more powerful (e.g. Vitt, L. J. and Pianka, E. R. 2006. The scaly ones. *Natural History* 115(6): 1–7). See also Schaerlaeken et al. 2008 for the contribution of the loss of the lower temporal bar to bite force in lizards compared with tuatara.
51 See diagrams in Benton, M. J. 2005a. *Vertebrate palaeontology*. 3rd edn. Blackwell, Oxford, p. 240.
52 Vitt, L. J., Pianka, E. R., Cooper Jr, W. E. and Schwenk, K. 2003. History and the global ecology of squamate reptiles. *American Naturalist* 162: 44–60; Pianka, E. and Vitt, L. J. 2003. *Lizards: windows to the evolution of diversity*. University of California Press, Berkeley; Vitt and Pianka 2006, op. cit.
53 Robinson, P. L. 1973. A problematic reptile from the British Upper Trias. *Journal of the Geological Society of London* 129: 457–479; Jones 2008.
54 Parsons 1959a, b, 1970.
55 Halpern, M. 1992. Nasal chemical senses in reptiles: structure and function. In: Gans, C. and Crews, D. (eds). *Biology of the Reptilia. Vol. 18. Hormones, brain, and behavior*. University of Chicago Press, Chicago, pp. 423–523.
56 Pratt 1948; Gabe and Saint Girons 1976; Bellairs 1984.
57 Schwenk 1986; Cooper et al. 2001.
58 Besson et al. 2009.
59 Bellairs 1984; Halpern 1992, op. cit.
60 Schwenk 2000.
61 Bellairs 1984; Cooper, W. E. 1994. Chemical discrimination by tongue-flicking in lizards: a review with hypotheses on its origin and its ecological and phylogenetic relationships. *Journal of Chemical Ecology* 20: 439–487; Vitt et al. 2003, op. cit.; Pianka and Vitt 2003, op. cit.; Vitt and Pianka 2006, op. cit.
62 Arnold 1984.
63 Saint Girons and Newman 1987.
64 Healy and Jamieson 1992; Jamieson and Healy 1992.
65 Osawa 1898b; Gabe and Saint Girons 1964a.
66 Calderón M. L., De Pérez G. R. and Ramírez Pinilla M. P. 2004. Morphology of the ovary of *Caiman crocodilus* (Crocodylia: Alligatoridae). *Annals of Anatomy* 186: 13–24.
67 De Beer 1949.
68 Lyson, T. R., Bever, G. S., Bhullar, B.-A. S., Joyce, W. G. and Gauthier, J. A. 2010. Transitional fossils and the origin of turtles. *Biology Letters* doi: 10.1098/rsbl.2010.0371.
69 Gorr et al. 1998.
70 Seutin, G., Lang, B. F., Mindell, D. P. and Morais, R. 1994. Evolution of the WANCY region in amniote mitochondrial DNA. *Molecular Biology and Evolution* 11: 329–340; Hedges, S. B. and Poling, L. L. 1999. A molecular phylogeny of reptiles. *Science* 283: 998–1001.
71 Saint, K. M., Austin, C. C., Donnellan, S. C. and Hutchinson, M. N. 1998. C-*mos*, a nuclear marker useful for squamate phylogenetic analysis. *Molecular Phylogenetics and Evolution* 10: 259–263. See also Vidal and Hedges 2005.
72 Hugall, A. F., Foster, R. and Lee, M. S. Y. 2007. Calibration choice, rate smoothing, and the pattern of tetrapod diversification according to the long nuclear gene RAG-1. *Systematic Biology* 56: 543–563.
73 Rest et al. 2003.
74 Piskurek, O., Austin, C. C. and Okada, N. 2006. Sauria SINEs: novel short interspersed retroposable elements that are widespread in reptile genomes. *Journal of Molecular Evolution* 62: 630–644.
75 Lee, M. S. Y. 2001. Molecules, morphology, and the monophyly of diapsid reptiles. *Contributions to Zoology* 70: http://dpc.uba.uva.nl/ctz/vol70/nr01/a01.
76 For example, genes of the major histocompatibility complex (MHC) in tuatara do not group strongly with those of other reptile species examined, a pattern attributed to limited sampling of taxa (Miller et al. 2005a) and high levels of divergence in these genes (Miller et al. 2006).
77 A large-insert bacterial artificial chromosome (BAC) library has been constructed from a male tuatara (Wang et al. 2006; see also O'Meally et al. 2009). Recent advances include sequences for CR1 LINE (Shedlock 2006), HoxD (Di-Poï et al. 2010) and EDGR-LINE (Lowe et al. 2010).
78 Lee 2001, op. cit.
79 Bromham, L. 2002. Molecular clocks in reptiles: life history influences rate of molecular evolution. *Molecular Biology and Evolution* 19: 302–309.
80 Hay et al. 2008.
81 Miller et al. 2008b. For a response, see Subramanian et al. 2008. For further comment, see Lanfear and Ho 2009.
82 Benton, M. J. 2005b. Reptilia (Reptiles). In: *Encyclopedia of life sciences*. John Wiley and Sons, Chichester. http://www.els.net/ [doi: 10.1038/npg.els.0004126].
83 Benton 2005b, op. cit.; Mallatt, J. and Winchell, C. J. 2007. Ribosomal RNA genes and deuterostome phylogeny revisited: more cyclostomes, elasmobranchs, reptiles, and a brittle star. *Molecular Phylogenetics and Evolution* 43: 1005–1022.
84 Benton 2005a, op. cit., see p. 113.
85 I surveyed first-year university students in ecology with the multiple-choice question: 'The closest relatives of tuatara are: turtles OR lizards OR dinosaurs OR crocodiles'. Of the 85 responses received from students living in New Zealand for the previous 3 years or more, 73% thought that the closest relatives of tuatara were dinosaurs. Only 20% correctly chose lizards.
86 Evans, S. E., Prasad, G. V. R. and Manhas, B. K. 2001. Rhynchocephalians (Diapsida: Lepidosauria) from the Jurassic Kota Formation of India. *Zoological Journal of the Linnean Society* 133: 309–334; Evans 2003.
87 Evans 2003; Jones et al. 2009b; Evans, S. E. and Jones, M. E. H. 2010. The origin, early history and diversification of lepidosauromorph reptiles. In: Bandyopadhyay, S. (ed.), *New aspects of Mesozoic diversity*. Springer-Verlag, Berlin, pp. 27–44.

88 Evans 2003.
89 Sues, H.-D. and Reisz, R. R. 1995. First record of the Early Mesozoic sphenodontian *Clevosaurus* (Lepidosauria: Rhynchocephalia) from the southern hemisphere. *Journal of Paleontology* 69: 123–126; Bonaparte, J. F. and Sues, H.-D. 2006. A new species of *Clevosaurus* (Lepidosauria: Rhynchocephalia) from the Upper Triassic of Rio Grande do Sul, Brazil. *Palaeontology* 49: 917–923; Jones, M. E. H. 2006. The Early Jurassic clevosaurs from China (Diapsida: Lepidosauria). In: Harris, J. D., Lucas, G. G., Spielmann, J. A., Lockley, M. G., Milner, A. R. C. and Kirkland, J. I. (eds.), The Triassic–Jurassic terrestrial transition. *New Mexico Museum of Natural History and Science Bulletin* 37: 548–562.
90 Worthy et al. 2002b; Jones 2007a; Jones et al. 2009b.
91 Jones et al. 2009b.
92 Reynoso, V.-H. and Clark, J. M. 1998. A dwarf sphenodontian from the Jurassic La Boca Formation of Tamaulipas, México. *Journal of Vertebrate Paleontology* 18: 333–339.
93 Reynoso 2000, op. cit.
94 Apesteguía, S. and Novas, F. E. 2003. Large Cretaceous sphenodontian from Patagonia provides insight into lepidosaur evolution in Gondwana. *Nature* 425: 609–612. Some authors consider *Priosphenodon* a synonym of *Kaikaifilusaurus*, a genus described in less detail in the same year. For discussion, see Albino, A. 2007. Lepidosauromorpha. In: Gasparini, Z., Salgado, L. and Coria, R.A. (eds). *Patagonian Mesozoic reptiles.* Indiana University Press, Bloomington, pp. 87–115.
95 Reynoso, V. H. 2000. An unusual aquatic sphenodontian (Reptilia: Diapsida) from the Tlayua Formation (Albian), Central Mexico. *Journal of Paleontology* 74: 133–148.
96 Carroll, R. L. 1985. A pleurosaur from the Lower Jurassic and the taxonomic position of the Sphenodontida. *Palaeontographica A* 189: 1–28; Carroll, R. L. and Wild, R. 1994. Marine members of the Sphenodontia. In: Fraser, N. C. and Sues, H.-D. (eds). *In the shadow of the dinosaurs.* Cambridge University Press, Cambridge, pp. 70–83.
97 Carroll 1985, op. cit.
98 Fraser, N. C. 1982. A new rhynchocephalian from the British Upper Trias. *Palaeontology* 25: 709–725; Whiteside 1986; Evans 1988, op. cit.
99 Whiteside 1986; Rieppel 1994, op. cit.
100 Moazen et al. 2009. See also Jones and Lappin 2009.
101 Baird 1970; Gans and Wever 1976.
102 For more on hearing in tuatara, see Chapter 7.
103 Wu, X.-C. 1994. Late Triassic–Early Jurassic sphenodontians from China and the phylogeny of the Sphenodontia. In: Fraser, N. C. and Sues, H.-D. (eds). *In the shadow of the dinosaurs.* Cambridge University Press, Cambridge, pp. 38–69.
104 Robinson 1973, op. cit.; Whiteside 1986.
105 Fraser 1982, op. cit.; Fraser 1988, op. cit.; Fraser, N. C. and Walkden, G. M. 1983. The ecology of a Late Triassic reptile assemblage from Gloucestershire, England. *Palaeogeography, Palaeoclimatology, Palaeoecology* 42: 341--365.
106 Robinson 1973, op. cit.; Fraser, N. C. and Walkden, G. M. 1984. The postcranial skeleton of the Upper Triassic sphenodontid *Planocephalosaurus robinsonae. Palaeontology* 27: 575–595; Jones 2008.
107 Rasmussen, T. E. and Callison, G. 1981. A new herbivorous sphenodontid (Rhynchocephalia: Reptilia) from the Jurassic of Colorado. *Journal of Paleontology* 55: 1109–1116; Throckmorton, G. S., Hopson, J. A. and Parks, P. 1981. A redescription of *Toxolophosaurus cloudi* Olson, a Lower Cretaceous herbivorous sphenodontid reptile. *Journal of Paleontology* 55: 586–597.
108 Reynoso 2000, op. cit.
109 Reynoso, V.-H. 2003. Growth patterns and ontogenetic variation of the teeth and jaws of the Middle Jurassic sphenodontian *Cynosphenodon huizachalensis* (Reptilia: Rhynchocephalia*). Canadian Journal of Earth Science* 40: 609–619; Jones 2008.
110 Reynoso, V. H. 2005. Possible evidence of a venom apparatus in a Middle Jurassic sphenodontian from the Huizachal red beds of Tamaulipas, México. *Journal of Vertebrate Paleontology* 25: 646–654.
111 Jones et al. 2009b (supplementary material).
112 Fraser, N. C. and Sues, H.-D. (eds). 1994. *In the shadow of the dinosaurs.* Cambridge University Press, Cambridge.
113 Stevens, G., McGlone, M. and McCullogh, B. 1995. *Prehistoric New Zealand.* Reed Books, Auckland; Benton 2005a, op. cit., see pp. 137–138.
114 Evans, S. E. and Kermack, K. A. 1994. Assemblages of small tetrapods from the Early Jurassic of Britain. In: Fraser, N. C. and Sues, H.-D. (eds). *In the shadow of the dinosaurs.* Cambridge University Press, Cambridge, pp. 271–283.
115 Evans, S. E. 1980. The skull of a new eosuchian reptile from the Lower Jurassic of South Wales. *Zoological Journal of the Linnean Society* 70: 203–264; Evans, S. E. 1981. The postcranial skeleton of the Lower Jurassic eosuchian *Gephyrosaurus bridensis. Zoological Journal of the Linnean Society* 73: 81–116.
116 Fraser 1988, op. cit.; Fraser, N. C. 1994. Assemblages of small tetrapods from British Late Triassic fissure deposits. In: Fraser, N. C. and Sues, H.-D. (eds). *In the shadow of the dinosaurs.* Cambridge University Press, Cambridge, pp. 214–226.
117 Fraser, N. C. and Walkden, G. M. 1983. The ecology of a Late Triassic reptile assemblage from Gloucestershire, England. *Palaeogeography, Palaeoclimatology, Palaeoecology* 42: 341–365; Fraser 1988, op. cit.
118 Evans 1980, op. cit.; Evans 1981, op. cit.
119 Reynoso, V.-H. 1997. A 'beaded' sphenodontian (Diapsida: Lepidosauria) from the Early Cretaceous of Central Mexico. *Journal of Vertebrate Paleontology* 17: 52–59.
120 Evans et al. 2001, op. cit.; Evans, S. E., Raia, P. and Berbera, C. 2004. New lizards and rhynchocephalians from the Lower Cretaceous of southern Italy. *Acta Palaeontologica Polonica* 49: 393–408. See also Malafaia, E., Ortega, F., Escaso, F., Dantas, P., Pimentel, N., Gasulla, J. M., Ribeiro, B., Barriga, F. and Sanz, J. L. 2010. Vertebrate fauna at the Allosaurus fossil-site of Andrés (Upper Jurassic), Pombal, Portugal. *Journal of Iberian Geology* 36: 193–204.
121 Apesteguía and Novas 2003, op. cit.; Apesteguía, S. 2005. A Late Campanian sphenodontid (Reptilia, Diapsida) from northern Patagonia. *Comptes Rendus Palevol* 4: 663–669; Apesteguía, S. and Rougier, G. W. 2007. A Late Campanian sphenodontid maxilla from northern Patagonia. *American Museum Novitates* No. 3581: 1–11. See also Albino 2007, op. cit.
122 Jones 2006, op. cit.; Evans and Jones 2010, op. cit.
123 Benton 2005a, op. cit., see pp. 253–255.
124 Species numbers in this paragraph (as of 1 January 2011) are from Uetz, P. et al., *The Reptile Database*, http://www.reptile-database.org, accessed 1 April 2011.

Endnotes to boxed text

For explanation of citation format in box endnotes, see endnotes to the main text.

Box 1.1

1. For more on the contributions of Gray, Owen and Günther to the taxonomy of tuatara, see Chapter 5.
2. The name *Rhynchocephalus* was unavailable in any case, having been assigned to another species in 1806. See Hemming 1957.
3. Hemming 1957.
4. Gray 1869.
5. Benton, M. J. 1985. Classification and phylogeny of the diapsid reptiles. *Zoological Journal of the Linnean Society* 84: 97–164.
6. Benton, M. J. 2005. *Vertebrate palaeontology*. 3rd edn. Blackwell, Oxford. See pp. 146–147.
7. Benton 1985, op. cit.
8. Carroll, R. L. 1985. A pleurosaur from the Lower Jurassic and the taxonomic position of the Sphenodontida. *Palaeontographica* A 189: 1–28. The name Sphenodontida has also sometimes been used for the suborder. In either context (order or suborder) the term introduces ambiguity in what is meant by the vernacular 'sphenodontid': does it mean a member of a higher-level category or a member of the family Sphenodontidae? See Jones et al. 2009b (supplementary material).
9. Jones et al. 2009b (supplementary material).
10. Following: Gauthier, J., Estes, R. and de Queiroz, K. 1988. A phylogenetic analysis of Lepidosauromorpha. In: Estes, R. and Pregill, G. (eds), *Phylogenetic relationships of the lizard families*. Stanford University Press, Stanford, pp. 15–98.

Box 1.2

1. Sources for this paragraph: Howes and Swinnerton 1901; Ali 1941; Hoffstetter and Gasc 1969; Bellairs and Bryant 1985; Seligmann et al. 2008.
2. An additional portion of the vertebra may be ablated after the initial breakage in some cases – see Seligmann et al. 2008.
3. Regrowth was about 5 mm over 10 months in animals held at 22–24°C (Alibardi 2010).
4. Alibardi and Meyer-Rochow 1989, 1990a.
5. Alibardi and Maderson 2003b.
6. Reviewed by Clause, A. R. and Capaldi, E. A. 2006. Caudal autotomy and regeneration in lizards. *Journal of Experimental Zoology* 305A: 965-973.
7. Ali 1941; Seligmann et al. 2008.
8. For other examples of tail loss during collection, see Ward 1882 and Schmidt 1953.
9. Gillingham et al. 1995.
10. See discussion on p. 738 of Seligmann et al. 2008.
11. Tyrrell et al. 2000 ($p < 0.05$ for differences among islands in both sexes).
12. $x^2 = 5.691$, df = 1, $p = 0.017$.

Box 1.3

1. Boulenger, G. A. 1889. *Catalogue of the chelonians, rhynchocephalians, and -emydosaurians in the British Museum (Natural History)*. British Museum of Natural History, London. 311 pp.
2. Buller 1895, p. 93.
3. Williston, S. W. 1904. The temporal arches of the Reptilia. *Biological Bulletin 7*: 175–192.
4. Benham to Hamilton, May 1912, ANZ IA 1 46/18/4 pt 1; the same letter was published in an unidentified New Zealand newspaper on 11 June 1913 (copy in ANZ ABWN 6095 W5021 Box 688 23/856 pt 1).
5. Anon. 1966. George Gaylord Simpson also described tuatara as among the 'Methuselahs': Simpson, G. G. 1967. *The meaning of evolution: a study of the history of life and of its significance for man. Revised edn.* Yale University Press, New Haven, p. 191.
6. Simpson op. cit., p. 49.
7. Anon. 1952.
8. Sharell 1966.
9. Worthy and Grant-Mackie 2003; Jones et al. 2009b.
10. Jones et al. 2009b.
11. Benton 1986; Whiteside 1986.
12. Reynoso, V.-H. 1997. A 'beaded' sphenodontian (Diapsida: Lepidosauria) from the Early Cretaceous of Central Mexico. *Journal of Vertebrate Paleontology* 17: 52–59; Reynoso, V. H. 2000. An unusual aquatic sphenodontian (Reptilia: Diapsida) from the Tlayua Formation (Albian), Central Mexico. *Journal of Paleontology* 74: 133–148.
13. Carroll, R. L. and Wild, R. 1994. Marine members of the Sphenodontia. In: Fraser, N. C. and Sues, H.-D. (eds). *In the shadow of the dinosaurs*. Cambridge University Press, Cambridge, pp. 70–83.
14. Groves, C. P. 2006. Humans among primates. In: Merrick, J. R., Archer, M., Hickey, G. M. and Lee, M. S. Y. *Evolution and biogeography of Australasian vertebrates*. AUSCIPUB (Australian Scientific Publishing), Oatlands, NSW, pp. 779–831.
15. See p. 347 of Evans 2008.
16. Dawbin 1962a.
17. For more on the climatic history of New Zealand through geological time, see Chapter 2.
18. For more on temperature tolerance in tuatara, see Chapter 9. Castanet (1983) was among those who recognised that ecophysiological or life history features such as slow growth, low metabolism and long lifespan might be derived and not representative of the ancestral condition of Triassic rhynchocephalians.
19. For more on the relationships between tuatara and seabirds, see Chapter 7.
20. Dawbin 1962a; Crook 1975. Recent evidence shows that the Procellariidae (the family to which prions and petrels belong) is represented by fossils dating from the Early Eocene, although possible procellariiforms of uncertain affinity date from the latest Cretaceous: Warheit, K. I. 2002. The seabird fossil record and the role of paleontology in understanding seabird community structure. In: Schreiber, E. A. and Burger, J. (eds). *Biology of marine birds*. CRC Marine Biology series, CRC Press, Boca Raton, pp. 17–55.
21. De La Fuente, J. 2003. The fossil record and the origin of ticks (Acari: Parasitiformes: Ixodida). *Experimental and Applied Acarology* 29: 331–344.
22. Drummond 1911. For more on the concept of evolutionary stagnation in relation to conservation, see Chapter 10.
23. Gans 1983. See also Thompson and Daugherty 1992.
24. Cree and Daugherty 1990.
25. Schopf, T. J. M. 1984. Rates of evolution and the notion of 'living fossils'. *Annual Review of Earth and Planetary Science* 12: 245–292.

Box 1.4

1. Jones 2006a, 2008, 2009.

CHAPTER TWO

Isolation: Zealandia adrift

Modern New Zealand forms an emergent part of Zealandia, sometimes described as 'the largest drowned continent in the world'.[1] But has the tuatara lineage survived here from Gondwanan times? Or did the ancestors of tuatara arrive more recently, following re-emergence of all land above the sea, as some scientists have lately suggested?

Introduction

When rhynchocephalians first evolved, the land that was to become New Zealand was little more than a few volcanic islands near the South Pole. Over the following 200 million years, this landmass changed radically as it uplifted, became isolated and drifted north. Just how did tuatara come to be present in this South Pacific location? And what environmental and ecological changes has the tuatara lineage had to adjust to over such an extraordinary length of time?

This chapter explores the geological origins of New Zealand to help answer these questions. Our story begins during the Mesozoic Era, continuing

Fig. 2.1 Continental reconstruction of the Late Jurassic, about 150 million years ago. The mega-continent of Pangaea consisted of the northern super-continent, Laurasia, and the southern super-continent, Gondwana (with Zealandia developing at the bottom of this view, below the eastern margin of Australia and Antarctica). Rhynchocephalians were present on both super-continents during the Mesozoic Era. Green indicates inferred areas of vegetated land; brown/grey, un-vegetated land; palest blue, very shallow sea (possibly emergent at times when sea level dropped slightly); and dark blue, deep ocean.

Base image from Ron Blakey, Colorado Plateau Geosystems, Inc.

BOX 2.1

New Zealand today

Geologically, New Zealand is part of the now largely submerged continent of Zealandia. About 82 million years ago Zealandia separated from the Antarctic region of Gondwana and began to move north. The landmasses making up Zealandia can be seen from the continental crust, which connects modern New Zealand with the Campbell Plateau, Chatham Rise, Lord Howe Rise and Norfolk Ridge. The shape of Zealandia (and New Zealand) has changed over time as landmasses on the Pacific and Australian plates have moved relative to each other.

In a political sense the New Zealand archipelago consists of two main islands, North and South (here collectively referred to as the mainland), plus over 700 smaller islands including Stewart Island/Rakiura. These extend in latitude from the Kermadec Islands at about 29°S to Campbell Island at about 52°S. Longitudinally, New Zealand extends from the southern South Island in the west to the Chatham Islands in the east. Islands surrounding the mainland can be separated into two main groups.[1] Those that lie within about 50 kilometres of the North, South and Stewart

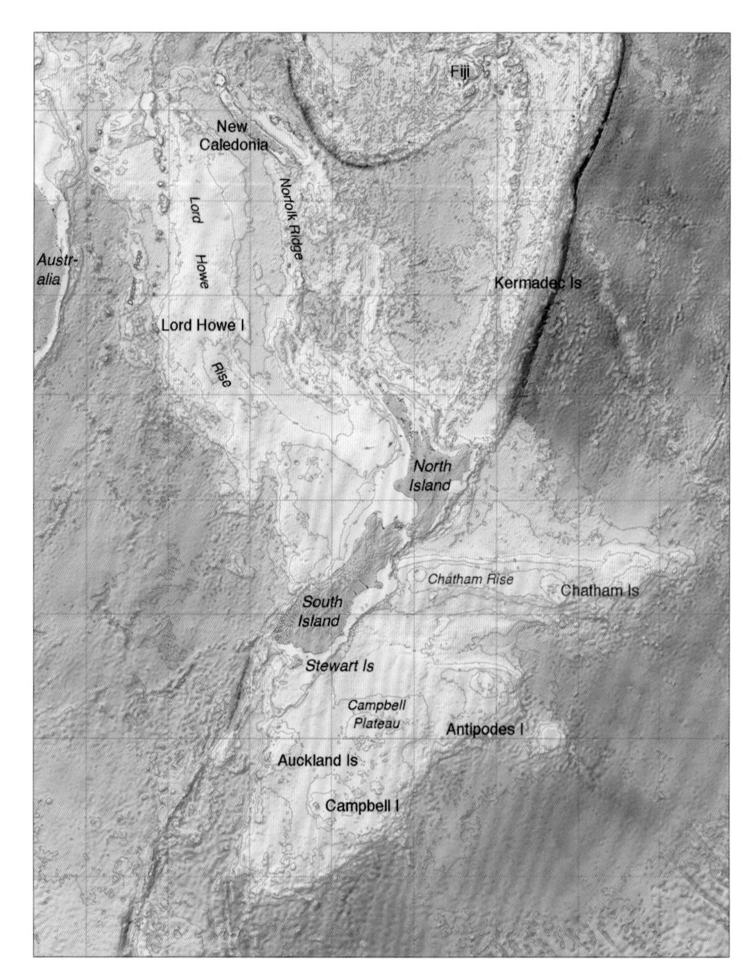

Box 2.1 Fig. 1 Modern New Zealand forms part of Zealandia, a continent now largely drowned. Emergent landmasses (in green), including New Caledonia in the north, Campbell Island in the south, and the Chatham Islands in the east, are connected by shallowly submerged continental crust (yellow-beige).

Source: base image from GNS Science.

islands are offshore or continental-shelf islands, and would have been connected to the mainland during the last glacial (70,000–14,000 years ago), when sea level dropped about 120–135 metres below today's level. Outlying or oceanic islands are found more than 50 kilometres off the mainland coast and have effectively always been isolated. The total land area of New Zealand (270,500 square kilometres) is similar to that of the British Isles or Japan. The nearest large landmass, Australia, lies about 1600 kilometres to the west.

The New Zealand mainland spans more than 1600 kilometres in length, with a width never greater than about 450 kilometres. The terrain is often mountainous, especially in the Southern Alps of the South Island (highest peak, 3754 metres). The long coastline surrounded by large ocean masses delivers a largely maritime climate. Mean annual temperatures range from about 16°C in the subtropical north to below 10°C in the cool-temperate south.[2] Inland parts of the southern South Island have a climate approaching continental conditions (dry hot summers and cold winters with snow).

New Zealand was first settled by East Polynesians in about AD 1230–1300.[3] European settlement began in the late 1700s. The human population surpassed four million in 2003 and there are three official languages: English, Māori and New Zealand sign language. One possible translation of the modern Māori name for New Zealand, Aotearoa, is 'land of the long clear day'.[4]

through the Cenozoic until the time of human arrival. The ways in which various organisms, including the ancestors of tuatara, came to be present in New Zealand, the interactions that might have occurred between tuatara and other animals (including some now extinct), and the effects of past changes in climate on the lineage leading to tuatara are discussed. Despite many gaps in our knowledge about the history of the lineage, some inferences are possible for early time periods, and detailed information has recently become available for the last 10,000 years.

New Zealand during the Mesozoic: formation and the break from Gondwana

During the Triassic, about 220 million years ago,[2] when early rhynchocephalians evolved, the world's continents were closely associated to form the mega-continent Pangaea (Chapter 1). By the Late Jurassic, about 150 million years ago, Pangaea had largely separated again into the northern super-continent of Laurasia and the southern super-continent of Gondwana, surrounded by ancient seas (Figure 2.1). The origins of New Zealand, along with the other southern hemisphere landmasses of Australia–New Guinea, Antarctica, South America and Africa–Madagascar, as well as India and some northern hemisphere terranes, lie in this ancient super-continent of Gondwana.[3]

The history of New Zealand's formation has been well described by geologists.[4] During Triassic times what was to become New Zealand lay largely underwater off the coast of Antarctica. Although a few small volcanic islands were probably all that showed above the ocean surface, the submerged land was being enlarged by volcanic debris and eroding sediments. In Jurassic times this area slowly uplifted, forming by the Late Jurassic a new landmass ten

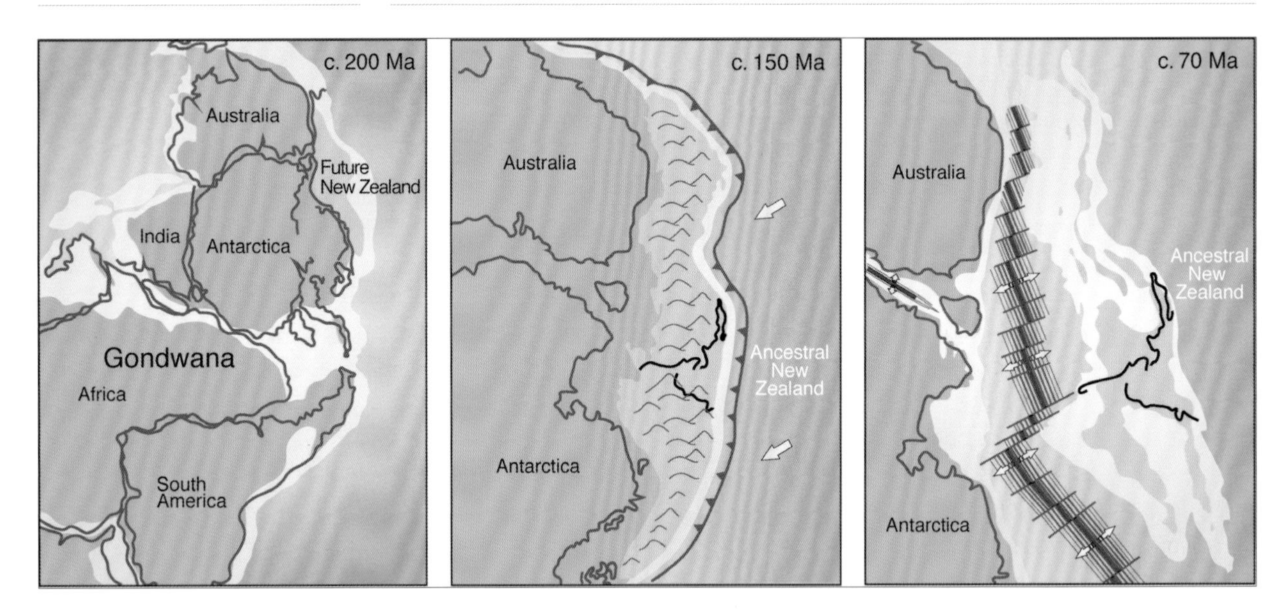

Fig. 2.2 Development of the ancestral New Zealand landmass of Zealandia from the Early Jurassic, about 200 million years ago (Ma), until the Late Cretaceous, about 70 million years ago. Zealandia initially stretched to encompass what are now New Caledonia and the Chatham Islands. Close connections with Australia and Antarctica allowed Zealandia to be colonised by plants and animals that were part of wider Gondwana, including almost certainly the rhynchocephalian ancestors of tuatara. By about 70 million years ago, the once-mountainous landmass was reduced in size and land connections with Australia and Antarctica had been severed. Dark green indicates major landmasses of Gondwana; pale green, the developing Zealandia; pale blue indicates marine regions to c. 2000 m depth; dark blue, deep ocean; red indicates oceanic spreading ridge, with arrows indicating direction of spreading; black-toothed line indicates subduction zone.

Palaeogeographic reconstructions modified with permission from: McSaveney, E. and Sutherland, R. 2005. *New Zealand Adrift*. 2nd edn. Information Series No. 69. Institute of Geological and Nuclear Sciences, Lower Hutt.

times the size of modern New Zealand (Figure 2.2). The new land, now known as Zealandia, stretched westwards to abut the edges of Australia and southwards to Antarctica. To the north it extended to include what is now New Caledonia, and to the east it encompassed today's Chatham Islands.

During the Jurassic, the part of Zealandia that would become modern New Zealand lay at about 60–78°S, about 20–40° further south than now. The climate was humid and warmer than at present. By mid-Cretaceous times New Zealand had moved to about 85°S. Although this was virtually on the South Pole, polar ice was unlikely (the isolating effects of a circumpolar current were not yet established) and the climate was probably merely cool to cold temperate.[5] Winter darkness, however, would have lasted for several months each year. The possible effects on dinosaurs known from these latitudes have aroused much speculation, but we can wonder also about the effects of long winter nights on the rhynchocephalian ancestors of tuatara.

At the same time as New Zealand was developing during the Jurassic, Gondwana was itself starting to fragment. The first separations were evident by about 110 million years ago in the Early Cretaceous, when Africa and the Indian sub-continent broke away.[6] Zealandia began separating from Australia at about this time, although separation from the region of Gondwana that became Antarctica did not occur until somewhat later, about 82 million years ago. By the end of the Cretaceous at 65 million years ago, Zealandia may have been isolated from other southern landmasses for over 15 million years.[7] Now eroded to half its former size, Zealandia was moving northwards, towards New Zealand's present-day location.

Zealandia would then have been populated with plants and animals that had colonised it during Triassic, Jurassic and Cretaceous times.

BOX 2.2

Polar dinosaurs – and rhynchocephalians?

Several groups of terrestrial reptiles had widespread distributions across Gondwana during Jurassic and Cretaceous times. These include not only rhynchocephalians but also dinosaurs, and probably some early types of lizards. Dinosaurs, which tend to arouse the most public attention, have been discovered as fossils from the Gondwanan lands of Antarctica, Australia, mainland New Zealand and the Chatham Islands, suggesting that they co-existed with the rhynchocephalian ancestors of tuatara in Zealandia.

An intriguing point about these discoveries is that, at the time, Zealandia was close to or even within the polar circle. Climates were much warmer than at the same latitudes today, reflecting the immense Gondwanan landmass and the absence of circumpolar currents. Forest clearly grew in what is now New Zealand. One analysis, based on leaf structure, suggests a mean annual temperature of about 10°C (about the same as the southern South Island coast today) at a time when latitude was between 70° to 90°S.[1] Additional methods have given estimates within a broad range of –6° to +8°C for similar latitudes in southern Australia.[2] In other words, estimated air temperatures may have been similar to, or only a few degrees lower than, those currently experienced in parts of the South Island where lizards still survive and where tuatara occurred until a few hundred years ago. It seems that cool temperature may not have seriously constrained reptilian life in Zealandia during the Late Cretaceous.

Daily light cycles, on the other hand, are determined by the sun. Assuming a similar tilt on the Earth's rotational axis as now, Cretaceous New

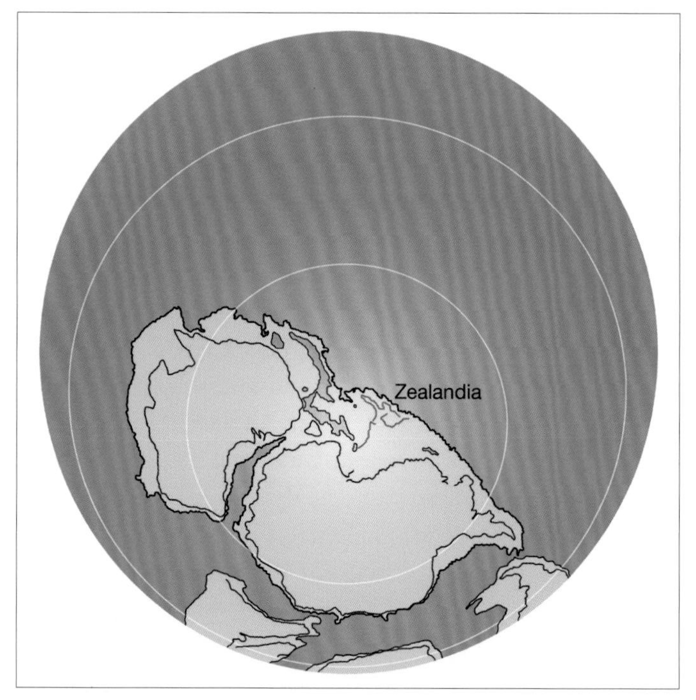

Box 2.2 Fig. 1 During Jurassic and Cretaceous times, Zealandia was at times within the polar circle. Although temperatures were warm enough for forest growth, the ancestors of tuatara would have experienced much longer winter nights than in modern New Zealand. The figure shows a palaeogeographic reconstruction of parts of Gondwana at c. 105 Ma, when the developing region of Zealandia was close to the south pole. Pale blue indicates continental regions that are shallower than 2000 m at the present day; red lines indicate the present-day coastline of New Zealand.

Based on an animation by Rupert Sutherland, GNS Science. In: Brockie, B. Native plants and animals – overview – species unique to New Zealand. Te Ara – the Encyclopedia of New Zealand, updated 14 November 2012. http://www.TeAra.govt.nz/en/interactive/10592/split-from-gondwana

Zealand would have experienced a winter darkness of 3 months or more. What did reptiles do under such a regime? Perhaps the ancestors of tuatara sought crevices where they could wait out the winter in a torpid condition, going without food for several months, as tuatara are capable of doing today. It is also possible, given the nocturnal habits of modern tuatara, that their polar ancestors emerged periodically during the long winter night if temperatures ameliorated and food was available.

As for dinosaurs, the possibility that some might have remained active during the winter months by being endothermic (generating their own body heat by metabolic processes) remains an ongoing debate. Modern reptiles, including tuatara, are ectothermic: their body temperature can be raised above that of their surroundings only by absorbing heat from an external source (usually sunlight). Ectothermy is an energy-efficient way of life, but sets limits on activity when sunlight is not available. Discoveries by palaeontologists Pat Vickers-Rich, Tom Rich and their colleagues raise the possibility that some Australian dinosaurs resembling hypsilophodonts (a group of fast, two-legged dinosaurs) were endothermic. These dinosaurs had large eyes and optic lobes, suggesting that they remained active during the dark winter months,[3] possibly making use of burrows.[4] At least one species shows an apparently continuous rate of bone deposition with no lines of arrested growth, implying that growth was continuous, not seasonal.[5] In contrast, in tuatara[6] and other modern reptiles of temperate zones, lines of arrested growth are produced on a seasonal basis.

The debate over whether some polar dinosaurs were endothermic will no doubt continue. However, the conclusion from the presence of tuatara in New Zealand today is that some small reptiles, including rhynchocephalians, almost certainly survived the same polar conditions while remaining ectothermic.

Forests of gymnosperms (conifers and their relatives) and ferns were widespread across Gondwana during the Jurassic. Some fossil deposits of forests from Jurassic times remain (Figure 2.3); these reveal a layer of emergent conifers (araucarian pines and podocarps), a canopy of tree ferns and cycads, and a forest floor cover including ferns, lycopods (club mosses and their relatives) and bryophytes (mosses and liverworts).[8] By mid-Cretaceous times angiosperms (flowering plants) had evolved, and these rapidly diversified.[9]

Rhynchocephalians were widespread during Mesozoic times throughout Pangaea, including the Gondwanan lands of Africa, Madagascar, India and South America (Chapter 1). In other words, rhynchocephalians almost certainly had the opportunity to establish in Zealandia before its separation from Gondwana. However, no Mesozoic fossils of rhynchocephalians have been discovered from the New Zealand mainland, the Chatham Islands or indeed New Caledonia, so we cannot say for sure how many types were present, or how similar they were to modern tuatara. Rhynchocephalians might have arrived in Zealandia via Antarctica, Australia or both, but these lands have not yet revealed any evidence. It is probably only a matter of time and effort before fossils of Mesozoic rhynchocephalians are discovered from Australia, a discovery that would greatly expand our understanding of the evolution of this group. No doubt, the news media will sensationalise such a discovery as the embarrassing presence there of a New Zealand icon 'absent without leave', as it were. The reality, however, is that if rhynchocephalian fossils are found in Australia, they are unlikely to be tuatara, but rather, their relatives.

The rhynchocephalian ancestors of tuatara that we suspect were present in Zealandia would have shared their landmass with other terrestrial reptiles. Among these were several types of dinosaurs, although only a few fragmentary remains are known from mainland New Zealand. Theropods, whose remains occur in Late Jurassic and Late Cretaceous sediments, may have reached about 1–10 metres in length.[10] Juveniles and small adults of these two-legged, meat-eating dinosaurs might have preyed on rhynchocephalians. Herbivorous dinosaurs from the Late Cretaceous of New Zealand included a four-legged sauropod about 12 metres long, a four-legged ankylosaur about 3 metres long, and a hypsilophodont (fast and two-legged) about 3 metres long.[11] Although these species probably had few direct interactions with terrestrial rhynchocephalians, the heavy-bodied sauropods and ankylosaurs might have flattened a few individuals, just as cattle squashed the occasional tuatara while present on Stephens Island (Takapourewa) during the 1900s.[12] Other reptiles (not dinosaurs) were present during the Cretaceous: in the air were pterosaurs, and in the seas ichthyosaurs, plesiosaurs, mosasaurs and a species of marine turtle. However, interactions with terrestrial rhynchocephalians are unlikely.

In fact, the fossil record for terrestrial animals in New Zealand during the late Mesozoic is frustratingly meagre. Terrestrial sediments suited to fossil deposition are rare, so scientists rely on indirect lines of evidence for the presence of major animal types. The evidence includes the current and fossil distribution of related types among southern landmasses, as well as analyses of genetic differences among modern forms, with estimated times of divergence. Such evidence suggests that the ancestors of tuatara may have shared ancient Zealandia with the ancestors of today's leiopelmatid frogs (family Leiopelmatidae), acanthisittid wrens (family Acanthisittidae) and a

Fig. 2.3 At Curio Bay, on the Catlins coast, the remains of a Jurassic forest are exposed at low tide. Here, fossilised logs and stumps (from trees possibly felled and buried after a volcanic eruption about 180 million years ago) are examined by the author. Although direct evidence is lacking, the ancestors of tuatara might once have roamed these forests.

Photo: Marcus Simons.

BOX 2.3

On board the Zealandian ark?

When Zealandia began its major break from the remnants of Gondwana about 82 million years ago, it had been sharing in the biological diversity of the southern super-continent for some tens of millions of years. Until about a decade ago, it was thought that many of the distinctive plants and animals of modern New Zealand traced their presence here back to this time. Increasingly, however, such assumptions have been challenged by molecular studies, which imply closer relationships with southern relatives than allowed by vicariance (the physical separation of populations following development of ocean barriers). If molecular dating is correct, many plants and animals actually arrived in New Zealand more recently, via oceanic dispersal.[1]

Tuatara, however, are one of several lineages whose origins in New Zealand remain consistent with a vicariant origin. Other groups with plausible archaic origins include the ancestors of the forest giant, kauri (*Agathis australis*). Molecular distances between New Zealand kauri and other members of the family Araucariaceae, including the monkey-puzzle tree of Chile, the Norfolk Island pine and the Woomera pine of Australia, are large enough that the ancestors of kauri may still have been present on the New Zealand landmass since the time that New Zealand was part of Gondwana.[2] Similarly, several groups of invertebrates, including mite harvestmen, giant wētā and some of the onychophorans (velvet worms) present molecular evidence consistent with a vicariant origin.[3] New Zealand's endemic leiopelmatid frogs also form a lineage that appears to have a vicariant origin from a Gondwanan source,[4] although this currently seems less likely on molecular grounds for the endemic geckos within the family Diplodactylidae.[5]

New Zealand's distinctive ratite birds – the moa and kiwi – conceivably remain part of the archaic fauna. Ratites once occurred on all the southern continents originating from Gondwana; in addition to the kiwi species, today's surviving lineages include the rheas of South America, the ostrich of Africa, the emu of Australia and the cassowaries of Australia and New Guinea. Molecular analyses suggest that the ancestors of moa and kiwi arrived separately about 76–53 million years ago, perhaps only subsequently and independently losing the ability to fly.[6] Although these proposed times of arrival are after Zealandia began unzipping at its southern margin from what became Antarctica, continental crust may still have remained connected between eastern Australia and the northwestern extremity of Zealandia at the Lord Howe Rise.[7] Among remaining lineages of New Zealand birds, wrens of the family Acanthisittidae are also suggested to have a vicariant origin in New Zealand,[8] although many other lineages seem likely to have arrived through dispersal.

Box 2.3 Fig. 1 Representatives of several biological lineages appear to have co-existed for tens of millions of years in Zealandia with the lineage leading to tuatara. All of the modern representatives shown here co-existed with tuatara, at least until the arrival of humans. Top left: Cook Strait giant wētā (*Deinacrida rugosa*). Top right: Hamilton's frog (*Leiopelma hamiltoni*). Bottom left: New Zealand kauri (*Agathis australis*). Bottom right: An artist's impression of the stout-legged moa *Euryapteryx curtus*, which overlapped extensively with tuatara in coastal sites and lowland open forest until human arrival.

Photos: wētā and frog: Alison Cree; kauri: Rod Morris; moa: watercolour by Paul Martinson, Museum of New Zealand Te Papa Tongarewa, 2006-0010-1/24.

New Zealand lineage of diplodactylid geckos (family Diplodactylidae),[13] as well as with numerous invertebrates.[14] Many of these small animals (and no doubt others no longer present today) could have been prey for the rhynchocephalian ancestors of tuatara.

Also inferred to be present at or soon after the separation of Zealandia were members of a distinctive lineage of birds – the ratites.[15] In the absence of large terrestrial mammals, flightless moa developed from one early lineage, evolving important roles as large (20–300 kilogram) browsers in pre-human New Zealand (nine species are known, but all are now extinct[16]). Based on the behaviour of large and primarily herbivorous birds surviving in New Zealand today (Chapter 7), juveniles of early Zealandian ratites might have preyed occasionally on small rhynchocephalians.

An outstanding biological feature of Zealandia was the virtual, if not complete, lack of terrestrial mammals. This absence, which makes modern New Zealand very unusual by world standards, was once explained by inferring that mammals evolved too recently to disperse to New Zealand before its separation from Gondwana. However, we now know that monotremes and other small mammals existed in Australia from at least the Early Cretaceous.[17] Thus, it was not completely unforeseen when fossil fragments diagnosed as those of a small mammal were recently uncovered from New Zealand. The fragments date from about 19–16 million years ago and suggest a mouse-sized species, seemingly not closely related to any modern mammals.[18] Clearly, we have much to learn about the ability of rhynchocephalians to survive in the presence of land mammals.

Land snakes remain a group whose apparent absence from the New Zealand fossil record is puzzling.[19] Snakes worldwide date from the Early Cretaceous[20] and terrestrial snakes, which were conspicuous in the South American fauna by the early Late Cretaceous, are thought to have shown their earliest radiation in Gondwana.[21] The Zealandian ancestors of tuatara may have co-existed with land snakes, but direct evidence is lacking.

New Zealand at the end of the Cretaceous: effects of extraterrestrial impact

At the close of the Cretaceous, some 65 million years ago, the world experienced a catastrophic reduction in biodiversity. This mass-extinction event is widely attributed to the effects of an asteroid, 10 kilometres in diameter, hitting the Earth in shallow seas off the Yucatán Peninsula in Mexico.[22] Known as the K-T (Cretaceous–Tertiary) boundary, the event is marked in geological strata by an accumulation of iridium, a metal rare on Earth but abundant in asteroids. New Zealand was among the first landmasses in which the iridium anomaly was detected – clear evidence that the effects of the collision extended far from the site of impact.[23]

Initially, the collision would have produced a massive earthquake (with an estimated magnitude one thousand times greater than any recorded earthquake), as well as shock waves, hurricane-force winds, fireballs and a giant tidal wave that perhaps reached as far as New Zealand.[24] An 'impact winter' would have followed – a period of several months of total darkness, with a severe drop in temperature resulting from accumulated dust and aerosols in the atmosphere. Within a few years sulphate aerosols would have fallen as acid rain.[25] These effects alone would have caused the virtual collapse of many ecosystems, and further impacts were to follow. Sufficient carbon dioxide may have vaporised from limestone rocks at the impact site to cause a prolonged period of global warming, with temperatures possibly increasing by up to 7.5°C over a few thousand years.[26] The climate may have continued to oscillate markedly for the first million years of the early Cenozoic.[27]

About half of all animal and plant species alive at the time of the impact became extinct,[28] including about 43% of families of tetrapod vertebrates.[29] There were clear effects on sea life around New Zealand: plesiosaurs and mosasaurs, plus various types of plankton and shellfish,[30] all disappeared from New Zealand seawaters in the changed regime.[31] Impacts on land are not as well understood. Chemicals known as fullerenes appear in K-T boundary clays, suggesting that wildfires occurred; later changes in pollen and spore counts imply a sudden shift from a flora dominated by gymnosperms to one dominated initially by ferns.[32] Gymnosperms later recovered, but angiosperms, which disappeared at the time of impact, did not recover to near Cretaceous levels until the late Early Palaeocene. Dinosaurs present on Zealandia during the Late Cretaceous, including on the land that became modern New Zealand and the Chatham Islands, appear not to have survived beyond the K-T boundary.[33]

The effects of the asteroid's impact on the distribution and diversity of rhynchocephalians in Zealandia are completely unknown. If, like modern

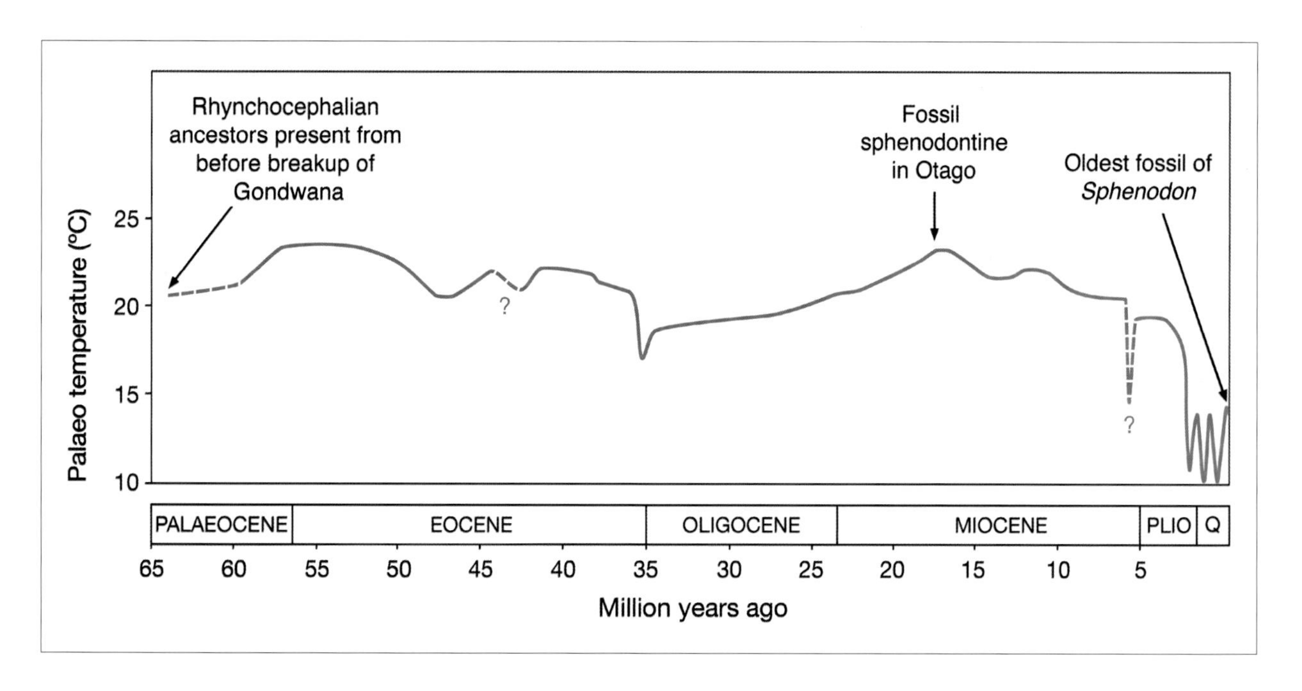

Fig. 2.4 The lineage leading to tuatara has experienced substantial changes in temperature during its long history of isolation in New Zealand. The inferred presence of rhynchocephalian ancestors and dates of oldest fossils are indicated. Plio = Pliocene Epoch; Q = Quaternary Period, comprising the Pleistocene and Holocene epochs.

Sea surface palaeo-temperature curve (normalised to the latitude of Wellington, with only four of at least ten glacial cycles shown) from Cooper, R. A. and Millener, P. R. 1993. The New Zealand biota: historical background and new research. *Trends in Ecology and Evolution* 8: 429–433 (after Hornibrook 1992). Minor changes to epoch boundaries have been made for consistency with the text.

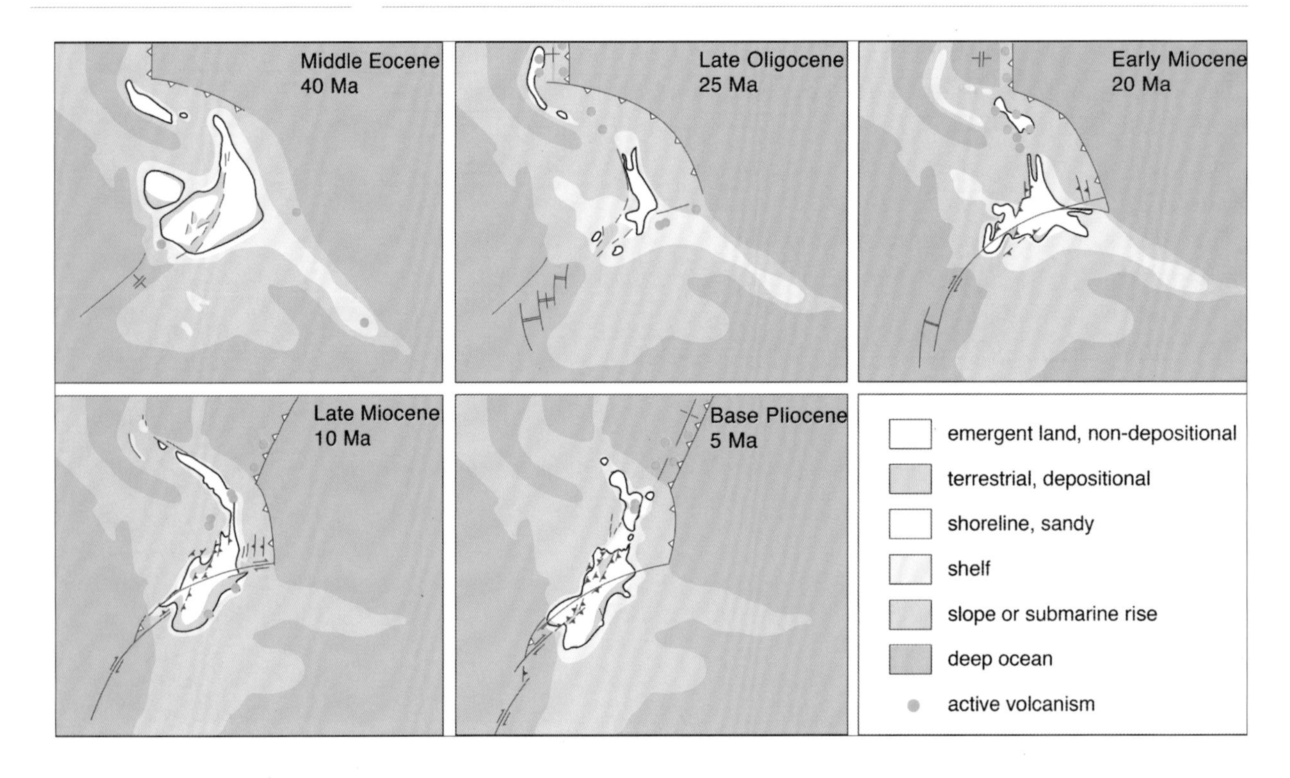

Fig. 2.5 Zealandia has undergone substantial changes in shape and area since separating from Gondwana. The most dramatic reduction and fragmentation occurred during the Oligocene Drowning, about 25 million years ago (Ma). By the Early Miocene (20 million years ago), there is evidence for two main landmasses that may have remained separated for a further 18 million years. The southern landmass (eventually reforming into today's South Island) may have been an important refuge for animals that survived the Oligocene Drowning, including the ancestors of tuatara. The black line indicates palaeo-coastline; red lines indicate faults, subduction zones or seafloor spreading centres.

Palaeogeographic reconstructions modified with permission from: King, P. R. 2000. New Zealand's changing configuration in the last 100 million years: plate tectonics, basin development, and depositional setting. In: *2000 New Zealand Petroleum Conference Proceedings, 19–22 March*, Christchurch Convention Centre. Ministry of Economic Development, Wellington, pp. 131–145. http://www.nzpam.govt.nz/cms/pdf-library/petroleum-conferences-1/2000-conference-proceedings/king-1-9-mb-pdf

tuatara, ancestral forms were nocturnal, burrowing species with a relatively small body size for vertebrates, they would have been well placed to tolerate the disruptions that undoubtedly occurred to Zealandian ecosystems.

New Zealand during the Cenozoic until the Pleistocene ice ages

The end of the Cretaceous marks the end of the Mesozoic Era and the beginning of the Cenozoic, which extends until the present time. The Cenozoic Era is divided into seven epochs. During the first two of these, the Palaeocene (65–57 million years ago) and the Eocene (56–35 million years ago), Zealandia continued to drift north. By the Late Eocene a subtropical environment existed and sea was developing between New Zealand and New Caledonia (Figures 2.4 and 2.5).[34] Antarctica, now separated from Australia, had moved to the South Pole, where it began to receive ice and snow, cooling the waters of the Southern Hemisphere.[35] Wind-assisted arrivals in Zealandia during the early Cenozoic probably included various species of plants, birds, invertebrates and possibly mystacinid bats (family Mystacinidae); conversely, there may also have been dispersal from Zealandia.[36] Procellariiform seabirds (order Procellariiformes: the 'tubenoses', or petrels, prions and shearwaters, a group with which tuatara have close associations today) had evolved by the Eocene and would have had the opportunity to establish along New Zealand's extensive coastline, as early penguins had done by the Palaeocene.

The Oligocene (34–24 million years ago) was a dramatic period in New Zealand's history. Initially, conditions cooled quite rapidly as the Circum-Antarctic Current and westerly winds established around Antarctica. The winds encouraged more birds to establish from other lands. Of particular importance, high sea levels produced a marine transgression known as the Oligocene Drowning. However, the extent to which the land was covered by sea remains controversial. Some scientists have inferred that emergent land, in the form of a series of low-lying islands, was reduced in area to perhaps 18% of today's amount.[37] Maximum altitudes may have been no more than a few hundred metres. The combinations of geographic isolation and low relief, plus sustained leaching of soil nutrients, have been described as making New Zealand an 'island prison' for animals and plants during this time.[38] Genetic bottlenecks and extinctions became more likely,[39] but there was also opportunity for speciation in those forms that became separated among the fragmented lands.

A recent and more provocative idea is that New Zealand was completely submerged during the Oligocene epoch. This proposal was first made in 2001 based on geological evidence for marine conditions across substantial areas of what is now land.[40] According to this proposal, the burden of proof should be shifted from assuming that there was emergent land to proving it. If the hypothesis of complete submergence is correct, all terrestrial and freshwater plants and animals must have repopulated Zealandia (now proto-New Zealand) by subsequent dispersal across ocean, from lands such as Australia or South America.

With respect to tuatara, the suggestion of total submergence creates a number of difficulties. First, no known source population existed (if it did, it must have subsequently disappeared without trace, despite extensive fossil excavations of other vertebrates from Australia, at least). Second, further dispersal must have been unsuccessful (for instance, there is no evidence that tuatara reached the Chatham Islands following the islands' re-emergence above the sea about 1–4 million years ago). In fact, today's tuatara appear unlikely to be good dispersers (they are not known to enter seawater voluntarily,[41] they have moderately high rates of water loss through the skin,[42] and their eggs are not hard-shelled and must absorb water to hatch[43]). Possibly their ancestors possessed different characteristics, but this is uncertain. Third, dispersal of other terrestrial reptiles (such as agamid lizards and snakes) from Australia to New Zealand has never been recorded, despite ample time since the Oligocene for such events to have occurred.

Furthermore, recent fossil discoveries show that rhynchocephalians,[44] as well as geckos and skinks,[45] leiopelmatid frogs,[46] and several distinctive types of birds including acanthisittid wrens[47] and moa,[48] were already present in New Zealand during the Early Miocene, 19–16 million years ago. While there is molecular support for an arrival of skinks at about this time (via long-distance dispersal from New Caledonia),[49] such an explanation seems highly unlikely for all of the remaining lineages. Additional evidence includes the presence of plant fossils (including araucarians and *Banksia*) in deposits from the Late Oligocene–Early Miocene,[50] as well as fossils of orchids,[51] of a diverse and distinctive fauna of birds,[52] and (perhaps most stunningly, for herpetologists) of a turtle that may have been terrestrial by the Early Miocene.[53] These discoveries lead many palaeontologists to believe that land is more likely to have remained emergent during the Oligocene than not. In defence of the idea of a continuous presence of rhynchocephalians in New Zealand, Marc Jones and colleagues have pointed out that tuatara have survived for about 10,000 years (since the last glaciation ended) on small and sometimes tiny islands.[54] For instance, Stephens Island (150 hectares, supporting 30,000 tuatara or more) represents less than one hundred-thousandth the area of modern New Zealand. It would seem extremely difficult to prove that islands of this size did not exist during the Oligocene. In view of all of the assumptions involved on either side of the argument, it seems most likely that the ancestors of tuatara did survive on emergent land during the Oligocene transgression.

During the Early Miocene land areas enlarged and reconnected. A recent (2009) geological model suggests that by about 20 million years ago, New Zealand existed largely as two main islands: a smaller, northern island, roughly equivalent to today's Northland, widely separated from a larger, southern island (with a more compact shape than today's South Island).[55] Sea temperatures became tropical – about 5°C to 7°C warmer than today – and subtropical plants flourished on land, with palms growing in the southernmost regions.[56] An enormous lake came to dominate the Central Otago region. With an area of at least 5600 square kilometres, Lake Manuherikia was about nine times the area of Lake Taupo, the largest lake in New Zealand today. The habitat around the palaeolake appears initially to have been largely subtropical rainforest, including cycads. Over time it became drier (with palms and then woodland), eventually becoming a mosaic of rainforest and open, possibly frosty herb-fields.[57] It is from the St Bathans Fauna of the Manuherikia sediments that the oldest fossils of rhynchocephalians in New Zealand (estimated age: 19–16 million years) have been found (Figure 2.6). The jawbones resemble those of the modern species *Sphenodon punctatus* closely enough to be placed in the same subfamily (the sphenodontines), but the remains are too fragmentary to be more specific.[58]

Other vertebrate elements of the lake and its environs included those of fish, ducks, bats[59] and a crocodilian,[60] as well as New Zealand's only reported terrestrial mammal, a creature of unknown affinities about the size of a mouse.[61] By the Middle–Late Miocene (15–6 million years ago), ice sheets were building up in Antarctica, and the resultant cooling in New Zealand may have contributed to the disappearance of groups such as crocodilians and land mammals.[62]

The Miocene was also a time when New Zealand came to straddle a zone of subduction in the Earth's crust. Thus began a history of land movement above the boundary of the Australian and Pacific plates. The Pliocene (5–2 million years ago) was marked by mountain building, including the uplifting of the Southern Alps, which now form the jagged backbone of the South Island. On the North Island volcanic activity began and continues to this day. The two main islands present in New Zealand by the early Miocene may have remained separated by a major seaway (the Manawatu Strait) until about 1.5–2 million years ago.[63] Genetic and fossil evidence suggests that reconnection of these landmasses may have allowed distinctive taxa such as moa, which clearly survived on the southern landmass, to repopulate the enlarging northern landmass.[64] An absence of deep genetic splits within surviving populations[65] supports a similar idea for tuatara also.[66]

Fig. 2.6 The oldest fossil evidence for rhynchocephalians in New Zealand: two views of a jaw fragment from a sphenodontine living in the Manuherikia region (St Bathans Fauna) of Central Otago during the Middle to Late Miocene, about 19–16 million years ago.

Photo: Museum of New Zealand Te Papa Tongarewa, S.43075.

The Late Pliocene was marked by a rapid drop in temperature,[67] a consequence of earlier continental movements and changes in ocean circulation.[68] During the Pleistocene (from 2 million to 10,000 years ago), the planet entered what is popularly known as the Ice Age. Although the term suggests just one extended period of cooling, the Pleistocene actually consisted of a series of at least 20 cold periods (glacials) alternating with brief warmer periods (interglacials). Climates in much of New Zealand during glacials were probably similar to the conditions of today's subalpine or alpine regions, but during the warmest periods of interglacials climates were temperate, and perhaps even subtropical.[69]

With so much water locked up in ice and snow, sea levels dropped markedly during glacials. At the peak of the last (Otiran) glacial around 18,000–20,000 years ago, sea levels fell to about 120–135 metres below those of today,[70] so that the New Zealand landmass was much enlarged (Figure 2.7). Land areas that today are offshore islands with isolated populations of tuatara were then connected to the mainland. The South Island was also connected to Stewart Island/Rakiura across Foveaux Strait. Another connection, between the North and South Islands across western Cook Strait, may also have existed,[71] though the fossil record provides no evidence that populations of vertebrate animals mixed.[72]

Vegetation patterns were certainly affected by these climatic swings. During the last glacial the conditions were windy and average temperatures may have fallen to about 5°C below those of today.[73] Continuous forest was probably largely limited to northern New Zealand. Although the southwestern South Island was extensively ice-covered and glaciated, pollen evidence suggests that much of the remaining land in central and eastern New Zealand was

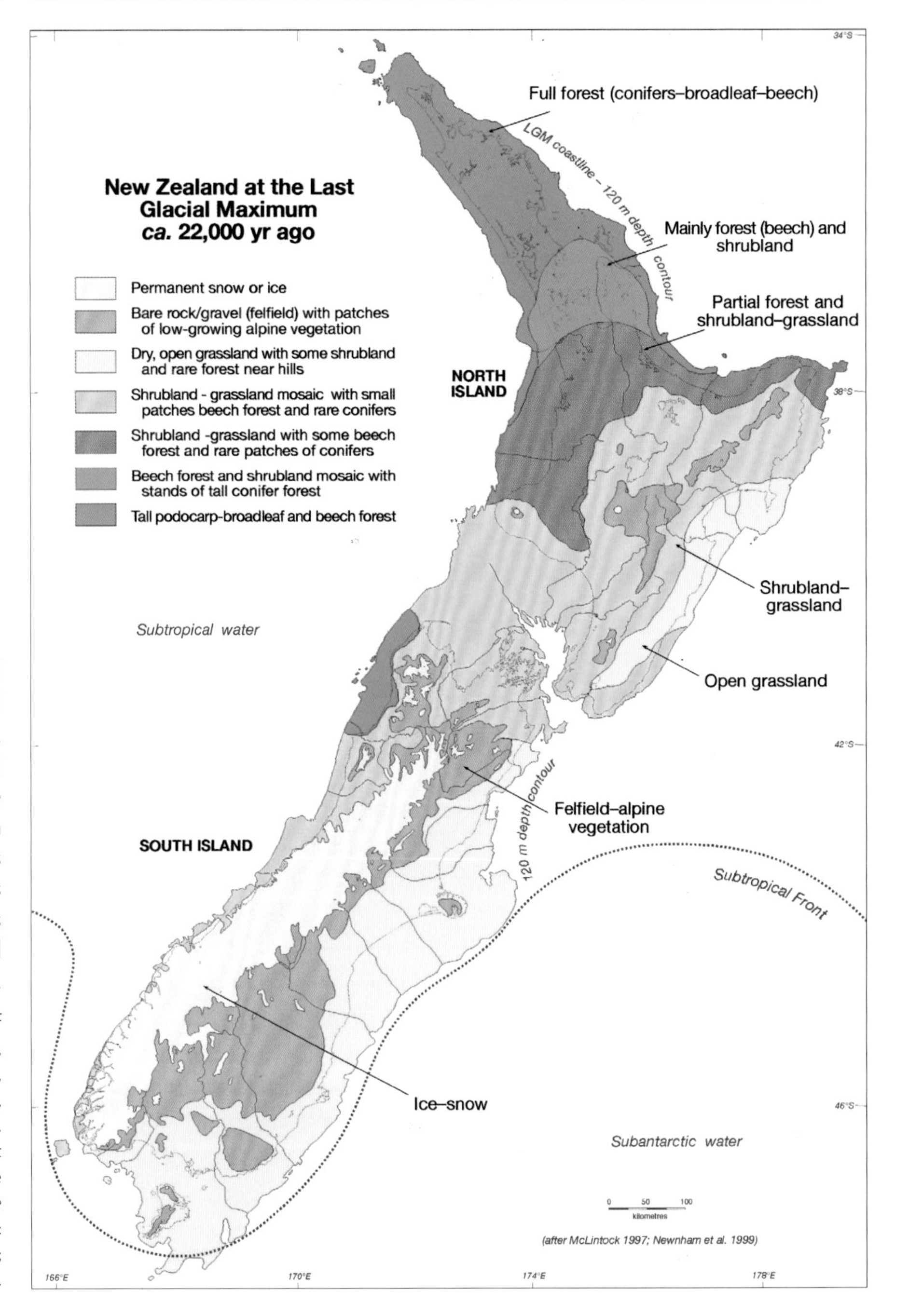

Fig. 2.7 The New Zealand coastline and vegetation inferred for the Last Glacial Maximum, the peak of the last glacial period about 22,000 years ago when ice sheets reached their maximum extent. Fossil evidence implies that tuatara survived in patches of lowland forest or mosaics of forest/grassland/shrubland during this time.

From supplementary information file of Bunce, M., Worthy, T. H., Phillips, M. J., Holdaway, R. N., Willerslev, E., Haile, J., Shapiro, B., Scofield, R. P., Drummond, A., Kamp, P. J. J. and Cooper, A. 2009. The evolutionary history of the extinct ratite moa and New Zealand Neogene paleogeography. *Proceedings of the National Academy of Sciences* 106: 20646–20651 (after McLintock 1997; Newnham et al. 1999).

vegetated with hardy grasslands and shrublands. Even in the South Island, patches of forest clearly survived in sheltered locations and at low altitudes,[74] and recent discoveries of fossilised, forest-dependent beetles imply that forest existed even further inland than indicated by pollen.[75] It is worth noting that at latitudes with similar climates today (near glaciers in Patagonian South America), the flora and fauna remain diverse, suggesting to ecologist George Gibbs that there may have been 'a tendency to overstate the impact of the Pleistocene in New Zealand'.[76]

Tuatara must have survived in southern New Zealand during the last glacial (judging by their widespread distribution in the South Island soon afterwards; see below), but the effects of the cooler climate on their activity and reproduction are unknown. By 14,000 years ago the glaciers were in retreat and the vegetation recovered quickly. By about 10,000 years ago, as the Pleistocene ended and the Holocene began, forest cover had re-established over most of New Zealand, accounting for about 81% of the land area around the time of human arrival.[77]

Tuatara during the Cenozoic until the arrival of humans

The oldest known bones attributed to the genus *Sphenodon* come from Cape Wanbrow, near Oamaru (Figure 2.8; see Appendices 1 and 2 for lists of sites reported to contain tuatara bones). The most recent estimate is that the bones

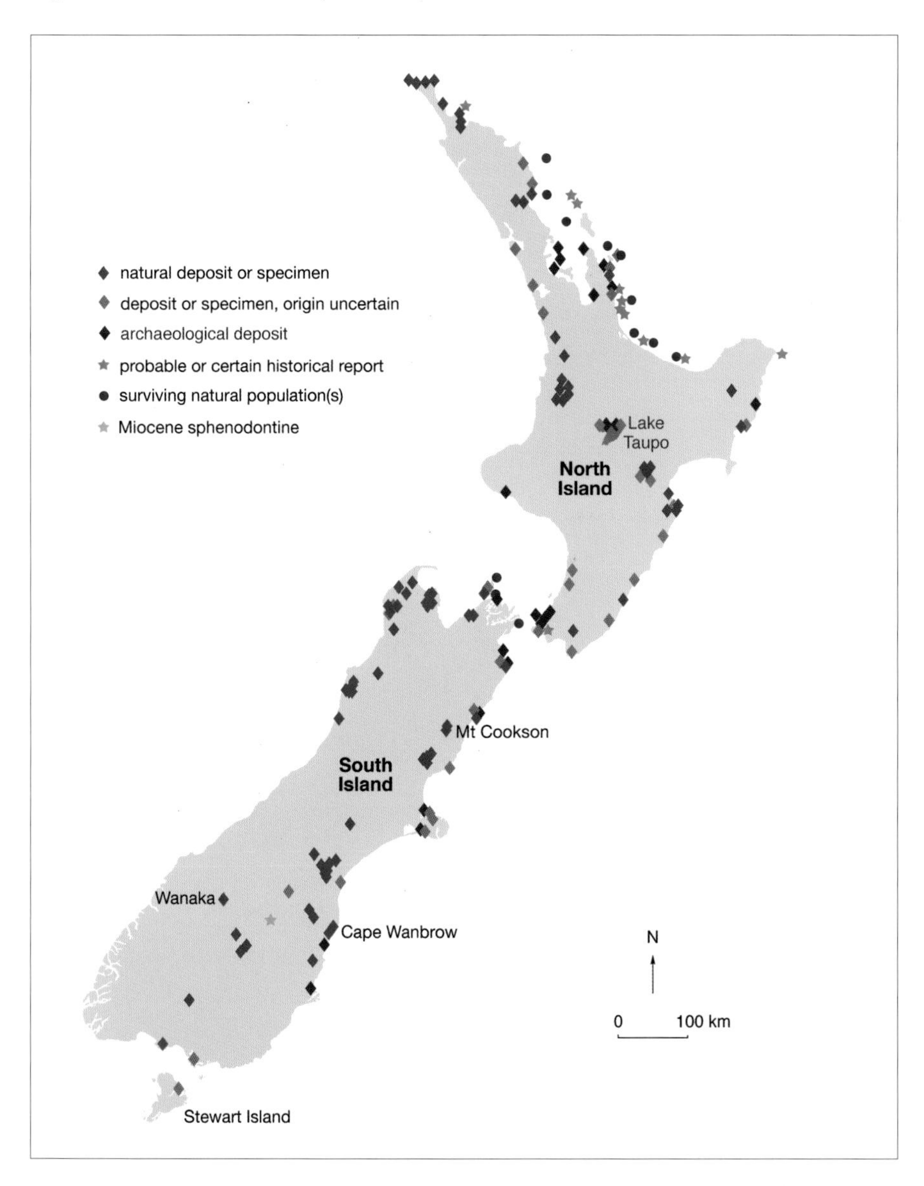

Fig. 2.8 Tuatara (*Sphenodon* sp.) were widely distributed in pre-human New Zealand. In this map of inferred distribution, sites with bones or specimens are coloured according to whether the deposit is likely to be natural (red diamond), of uncertain origin (natural or archaeological, brown diamond), or archaeological (black diamond). Archaeological sites are less certain indicators of natural distribution because of the possibility of human transport (see Chapter 3 for discussion). Overlapping diamonds of different colours may indicate differing interpretations of the same deposit by different authors. Mt Cookson and Wanaka (labelled) are examples of sites with substantially colder winters (because of higher altitude and/or latitude) than tuatara now experience. Although most mainland deposits date from within the past 20,000 years, a few sites are of Pleistocene age (e.g. those at Cape Wanbrow may be up to about 100,000 years old). For reference, the site near St Bathans where jaw fragments of a Miocene fossil sphenodontine, not necessarily attributable to *Sphenodon*, have been found is indicated (orange star). Also shown are islands where natural populations of tuatara survive today (dark blue circle), plus islands where tuatara no longer exist but have been reported historically with a high probability of occurrence (pale blue star). Each symbol may indicate more than one island or deposit in close proximity. Most if not all islands with tuatara became isolated from the mainland following sea-level rises at the end of the last glacial about 10,000 years ago. For sources, see Appendices 1, 2 and 3.

are between 35,000 and 100,000 years old.[78] However, most of our knowledge of tuatara is limited to the current interglacial epoch, the Holocene, from 10,000 years ago until the present. This is a mere blink of an eye, geologically speaking, in the history of the lineage (about one ten-thousandth of the time since Zealandia broke from Gondwana, for instance).

Within the relatively short Holocene, the evidence for the past presence of tuatara is extensive. Numerous tuatara bones from natural sources have been found throughout much of the North and South Islands (in the past these relatively young and un-mineralised bones were often described in New Zealand as 'sub-fossils'). The sources include natural pitfalls, such as vertical shafts or tomo that lead to subterranean caves in limestone. Bones have also been found in coastal dunes, and occasionally in swamps, as well as in the regurgitated remains left by predatory birds (Figures 2.9–2.11).[79] Tuatara may have experienced periodic local extinctions following volcanic activity in parts

Fig. 2.9 Bones of tuatara are frequently found in Northland dunes, including this site at Tokerau Beach.

Photo: Trevor Worthy.

Fig. 2.10 Tuatara once lived at the aptly named site of Earthquakes, in the limestone country of North Otago. Predatory birds (thought to be now-extinct laughing owls, (*Sceloglaux albifacies*) roosted on the cliffs, and left bones of tuatara in the remains of their meals.

Photo: Marcus Simons.

of the North Island,[80] but the vegetation and fauna seem to have recovered quickly from such natural disturbance. Collectively, the available material indicates that tuatara were widespread on both main islands of New Zealand prior to human settlement about 730 years ago.

As sea levels rose during the Late Pleistocene and Early Holocene, tuatara also became marooned on dozens of offshore islands, surviving today on about 32 of these islands,[81] and judging from fossil evidence and historical records they were present until recently on about 25 others (Chapter 3). Stewart Island is an interesting case for which the evidence of past presence is equivocal. As mentioned in the preceding section, this large, southerly island was connected to the South Island during the last glacial. Tuatara certainly occurred nearby on the Southland coast during the Holocene,[82] but searching in the limited range of suitable sites for fossils on Stewart Island has so far not revealed any tuatara material.[83] Perhaps sea levels rose sufficiently quickly after the last glacial that tuatara could not reinvade from the South Island. On the other hand a newspaper account from 1883 refers to 'one of the large lizards, peculiar to the North Island' and of '15 or 16 inches [c. 38 or 41 centimetres] in length' being collected from 'the Brothers Islands (near Stewart's Island) recently' by Captain Fairchild (Master of the government vessel servicing lighthouses),[84] and a tuatara specimen is recorded in the catalogue of the Southland Museum and Art Gallery 'from Native Island, Stewart Island'.[85] However, whether the latter specimen originated from a natural population on Native Island (located near the mouth of Paterson Inlet on Stewart Island) is unclear.

Fig. 2.11 A member of Trevor Worthy's research team sieves the food remains left by predatory birds (probably laughing owls) on a ledge at Gordons Valley in South Canterbury. The remains confirm that tuatara once lived in the vicinity.

Photo: Trevor Worthy.

Most of the tuatara material from natural deposits is in the form of disarticulated bones rather than complete skeletons. Thus, we cannot be certain how different, if at all, mainland tuatara were in appearance from island populations today. Although it was suggested in 1885 that the bones of a mainland specimen might represent a species (*Sphenodon diversum*) different from that surviving today,[86] this has not received contemporary support. A recent preliminary examination of dentaries from several mainland sites suggests that those from the northernmost part of the North Island were slightly smaller on average than those from more southern and cooler locations,[87] and further study of shape variation in the dentary and maxilla is underway.[88] In addition, at least one complete mainland specimen is known: an air-dried carcass from an Otago cave (Figure 2.12).[89] The specimen looks superficially similar to tuatara living today on islands, but closer study is needed. Genetic analysis of mainland fossils has begun,[90] and promises to enlighten us about the diversity that once existed within the genus *Sphenodon*.

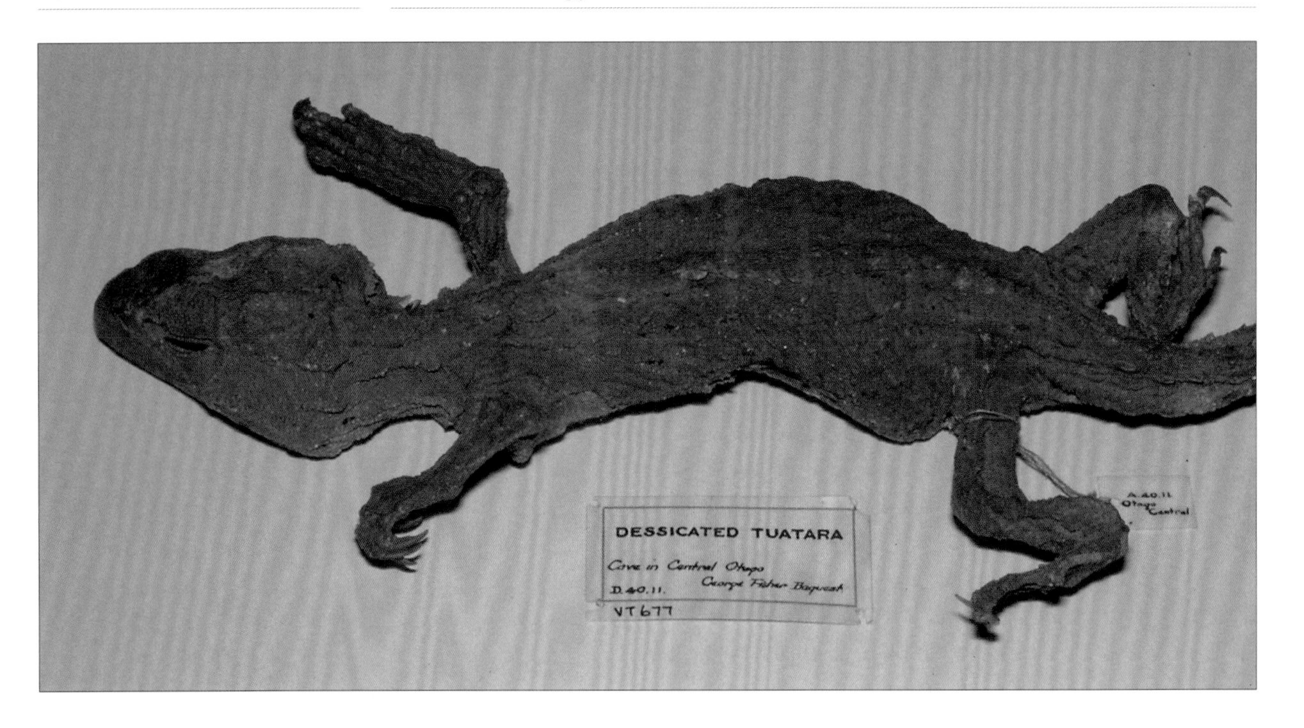

Fig. 2.12 Most evidence for the past presence of tuatara on the New Zealand mainland consists of un-mineralised bone fragments (usually the dentary – the major bone from the lower jaw – along with its characteristic serrated margin of acrodont teeth). On rare occasions complete, air-dried specimens have been found, such as this example from a cave in Central Otago. The specimen, now in the Otago Museum, appears to be that of an adult male, with a snout–vent length of about 205 millimetres.

Photo: Alison Cree.

One point already clear from palaeoecological studies is that the mainland sites where tuatara existed in the Holocene provided a wider range of habitats than occupied by tuatara today. The distribution of mainland tuatara extended slightly further north (to 34°S) and considerably further south (to 46°S) than that of today's natural populations on islands (35–41°S), and included sites that were coastal dune forests as well as inland forests and forest/shrubland mosaics. Dominant forest trees in these locations over this timeframe included kauri (*Agathis australis*) in the north, rimu (*Dacrydium cupressinum*) and northern rātā (*Metrosideros robusta*) in the west, mataī (*Prumnopitys taxifolia*) and kahikatea (*Dacrycarpus dacrydioides*) in the east, and beech (*Nothofagus* spp.) in cooler, upland locations.[91] Natural populations of tuatara are rare or absent from such forest types today.

Recent reconstructions for the South Island by palaeoecologists Trevor Worthy, Richard Holdaway and others have been particularly revealing.[92] Their excavations show that tuatara existed on the western South Island in dense podocarp–hardwood (rimu–rātā) forest where rainfall today exceeds 2800 millimetres per year, but appear to have been more abundant in drier eastern locations, where rainfall today is sometimes as low as 400 millimetres per year. Eastern habitats included lowland areas that were then covered by tall podocarp–broadleaf forest dominated by mataī, as well as higher altitude areas in beech forest or grassland/shrubland mosaics. The highest South Island sites where tuatara occurred are around 620–800 metres above sea level in northwest Nelson,[93] and around 500 metres above sea level on Mt Cookson in North Canterbury[94] and near Wanaka in Otago.[95] Both of the latter two regions have extreme climates today, with hot, dry conditions in summer and frosts with snow lying on the ground for days or even weeks in winter.

BOX 2.4

The lost world of Holocene New Zealand

Time-travel is a popular fantasy. Were it possible, a trip back to see mainland tuatara in the pre-human forests of Holocene New Zealand would be top of my list. Along with tuatara, there would have been an array of creatures – birds, lizards, amphibians, invertebrates and a few mammals – with which mainland tuatara would have lived in noisy co-existence. Many of these species are now globally extinct, others no longer overlap in distribution with tuatara, and some, such as the giant Delcourt's gecko (*Hoplodactylus delcourti*) have a presence in New Zealand that remains unconfirmed. Today we can only guess at the kind of interactions that tuatara would have had with such creatures.

Although free of mammalian predators, the pre-human forests and shrublands were undoubtedly not free of all predators, as sometimes portrayed. Birds that were probable or known predators of mainland tuatara include the now extinct adzebills (*Aptornis* spp.), the New Zealand raven (*Corvus antipodum*), the

Box 2.4 Fig. 1 Animals of Holocene New Zealand include extinct species that had known or suspected interactions with tuatara. Left: A North Island adzebill (*Aptornis otidiformis*) dwarfs a tuatara, and was a probable predator of it. Right: Lyall's wren (*Traversia lyalli*), a fast-moving but flightless wren, was probably caught and eaten occasionally by tuatara. Also shown are the fairy prion (*Pachyptila turtur*) and a green gecko of the genus *Naultinus*, species with which tuatara survive today on Stephens Island.

Source: watercolours by Paul Martinson, Museum of New Zealand Te Papa Tongarewa (reg. 2006-0010-1/40; 2006-0010-1/3).

New Zealand owlet-nightjar (*Aegotheles novaezealandiae*) and the laughing owl (*Sceloglaux albifacies*). The predatory New Zealand falcon (kārearea, *Falco novaeseelandiae*) and omnivorous, ground-dwelling birds such as the weka (*Gallirallus australis*), which still survive today although rarely in tuatara habitat, would also have preyed on tuatara. Hunting strategies, as well as the sizes of tuatara taken, no doubt varied among species. The large, robust and flightless adzebills, for instance, had fearsome bills and were probably significant predators of adult tuatara by day, possibly even digging tuatara from their burrows. In contrast, the flighted laughing owls probably took only smaller tuatara that emerged at night. Juveniles of several species of moa (now extinct) may also have preyed on small tuatara. With a wingspan reaching up to about 2.6 metres,[1] the enormous Haast's eagle (*Aquila moorei*, now extinct) preyed on moa but was unlikely to bother with any but perhaps the largest tuatara. This eagle would certainly have added drama for a time-travelling herpetologist, however, just as it is believed to have done for early Polynesian settlers.

Small vertebrates, especially slow-moving, ground-dwelling species, would have been vulnerable to capture by tuatara. Novel prey species may have included tiny, flightless birds such as the now-extinct bush wren (*Xenicus longipes*) and Lyall's wren (*Traversia lyalli*), as well as mystacinid bats (pekapeka). Unusually for bats, mystacinids forage on the forest floor, even (like tuatara) entering seabird burrows.[2] One species (*Mystacina robusta*) is now extinct and the other (the lesser short-tailed bat, *M. tuberculata*) now co-occurs with a (relict) population of tuatara only on Hauturu/Little Barrier Island, where direct interaction is probably minimal. At least six species of terrestrial or semi-aquatic frogs (*Leiopelma* spp.) may have been preyed on by tuatara; three are now extinct and only one still overlaps in distribution with tuatara, on Stephens Island (Takapourewa; where it is occasionally eaten[3]). Many species of large invertebrates that no longer exist today, and some lizards that are either extinct or currently non-overlapping in distribution with tuatara, are also likely to have been part of the diet of tuatara on the mainland.

Thus, today's island populations give us incomplete information about the habitats of tuatara and the interactions that were possible with other species. Behaviours that still occur in tuatara today could reflect ecological opportunities and pressures that once existed on the New Zealand mainland and about which we have at best only limited awareness.

The same research team has greatly expanded our understanding of the types of animals with which South Island tuatara must have shared their Holocene forests. Of particular value has been the discovery of deposits left by two species of predatory birds over the last few thousand years.[96] These birds, the nocturnal laughing owl (whēkau, *Sceloglaux albifacies*; Figure 2.13) and the diurnal New Zealand falcon (kārearea, *Falco novaeseelandiae*; Figure 2.14), regurgitated indigestible material and left it on or below cliffs and ledges (sometimes in cave entrances[97]), where it was later found by intrepid searchers. Most deposits with tuatara bones were apparently produced by laughing owls, which weighed about 500–600 grams. Laughing owls were probably major predators of small-to-medium-sized vertebrates in pre-human Holocene ecosystems[98] and became extinct only about a century ago. New Zealand falcons still exist.[99]

Holocene forests and shrublands had a much more diverse fauna than that co-existing with today's tuatara populations. The South Island, for instance, had about 113 breeding species of birds and the North Island nearly as many.[100] In addition to the laughing owl, several now-extinct birds were probably major predators of tuatara. These include the adzebills (*Aptornis* spp.), which were large (10–13 kilograms), flightless and probably omnivorous; the New Zealand raven (*Corvus antipodum*), a medium-sized (0.9–1-kilogram), flighted omnivore; and the New Zealand owlet-nightjar (*Aegotheles novaezealandiae*), a small (c. 200-gram) weakly flying, nocturnal predator.[101] Two flightless species that barely survive on mainland New Zealand, the weka (*Gallirallus australis*, 0.7–1 kilograms, omnivorous) and the South Island takahē (*Porphyrio hochstetteri*, 2–3 kilograms, primarily herbivorous), both occasionally consume lizards.[102] They, the extinct North Island takahē (*P. mantelli*) and juveniles of the extinct moa might all have included small tuatara in their diet.

Tuatara are opportunistic predators, and in mainland forests would have had access to a wider range of prey species (many now extinct) than on islands today. The forest floor probably teemed with invertebrates, including large orthopterans (wētā), carabid beetles and large weevils.[103] Small (less than 35-gram) vertebrates that were widespread and might have been captured by tuatara included flightless wrens, mystacinid bats and leiopelmatid frogs. Procellariiform seabirds were present around the coastline and in some inland parts of the mainland; chicks of the smaller species would have fallen prey to tuatara, as they do today on islands.

Geckos and skinks were diverse and probably abundant on mainland New Zealand during the pre-human Holocene. The smaller, ground-dwelling forms of these lizards would have occasionally been captured and eaten by tuatara, as surviving species are today.[104] There has also been much speculation

Fig. 2.13 The laughing owl was a widespread predator of small tuatara on mainland New Zealand. The last known specimens date from about 1914.

Photo (mounted specimen): Rod Morris.

Fig. 2.14 The New Zealand falcon (*Falco novaeseelandiae*), which still survives today, was one of several predators of mainland tuatara during the pre-human Holocene.

Photo: Rod Morris.

Fig. 2.15 Tuatara are the largest living reptiles in New Zealand today, but were they once dwarfed by an enormous gecko? The only known specimen of Delcourt's gecko (*Hoplodactylus delcourti*, top) was found in a French museum, where it was lodged from an unknown source prior to the early 1900s. Morphologically, it resembles living geckos in the genus *Hoplodactylus*, known only from New Zealand. At 370 millimetres in snout–vent length, this giant gecko is substantially longer than an adult female tuatara from North Brother Island (bottom). Geckos are more arboreal than tuatara, so interactions may have been few.

Photo: Brett Robertson/ Victoria University of Wellington.

that a giant gecko, *Hoplodactylus delcourti* – too large as an adult to be preyed on by tuatara – might once have been present in New Zealand.[105] The species is known only from a single stuffed specimen, in a French museum, and is considerably larger than most adult tuatara (Figure 2.15). Although the specimen's origin is unknown, its morphology leads it to be placed in a genus otherwise known only from New Zealand. The specimen could correspond with tantalising reports of a kawekaweau, a large lizard reported by Māori to early Europeans in New Zealand.[106] However, there is no fossil record for such a large lizard here, despite extensive excavations.[107]

To conclude, just a few thousand years after the last glaciation ended, the two main islands of New Zealand had become largely forested. Tuatara were widespread and sometimes locally abundant predators of terrestrial invertebrates and small vertebrates. They were themselves eaten by a variety of flighted and flightless birds, many of which are extinct today. As best we can imagine them, the forests were noisy and active places. All this was to change in the short space of about 730 years, as the following chapter relates.

Endnotes to Chapter 2

See the bibliography for full references to articles on tuatara, which are cited here at first mention by surname and year only. References on other topics are cited here in full at first mention, and then using 'op. cit.'. Except where required for clarity, the convention of parentheses around years has been dispensed with for economy of space.

1 Stevens, G., McGlone, M. and McCulloch, B. 1995. *Prehistoric New Zealand.* Reed Books, Auckland. See pp. 30, 60.

2 The geological time scale used here follows that of the Geological Society of America (http://www.geosociety.org/science/timescale/, accessed 10 January 2009).

3 For an animated illustration of the separation of Gondwana, see http://www.scotese.com/Default.htm (accessed 3 July 2012).

4 Sources include: Brazier, R., Keyes, I. and Stevens, G. 1990. *The great New Zealand fossil book.* New Zealand Department of Scientific and Industrial Research, Lower Hutt, New Zealand; Stevens et al. 1995, op. cit.; Schellart, W. P., Lister, G. S. and Toy, V. G. 2006. A Late Cretaceous and Cenozoic reconstruction of the Southwest Pacific region: tectonics controlled by subduction and slab rollback processes. *Earth-Science Reviews* 76: 191–233; Campbell, H. and Hutching, G. 2007. *In search of ancient New Zealand.* Penguin Group, North Shore and GNS Science, New Zealand; Graham, I. J. (ed.). 2008. *A continent on the move: New Zealand geoscience into the 21st century.* Geological Society of New Zealand in association with GNS Science, Wellington.

5 Stevens et al. 1995, op. cit.

6 McLoughlin, S. 2001. The breakup history of Gondwana and its impact on pre-Cenozoic floristic provincialism. *Australian Journal of Botany* 49: 271–300.

7 However, some uncertainty remains about the time of final separation of emergent land. In the region of the Lord Howe Rise, Zealandia's continental crust continued to abut eastern Australia until about 55 million years ago (Schellart et al. 2006, op. cit.), although whether the connection included land above sea level is unclear.

8 Thorn, V. 2001. Vegetation communities of a high palaeolatitude Middle Jurassic forest in New Zealand. *Palaeogeography, Palaeoclimatology, Palaeoecology* 168: 273–289.

9 McLoughlin 2001, op. cit.

10 Molnar, R. E. and Wiffen, J. 1994. A Late Cretaceous polar dinosaur fauna from New Zealand. *Cretaceous Research* 15: 689–706; Molnar, R. E., Wiffen, J. and Hayes, B. 1998. A probable theropod bone from the latest Jurassic of New Zealand. *New Zealand Journal of Geology and Geophysics* 41: 145–148.

11 Molnar and Wiffen 1994, op. cit.; Cox, G. and Wiffen, J. 2002. *Dinosaur New Zealand.* HarperCollins Publishers New Zealand Ltd, Auckland.

12 Pers. obs. Cattle have since been removed from Stephens Island (Takapourewa).

13 Wallis, G. P. and Trewick, S. A. 2009. New Zealand phylogeography: evolution on a small continent. *Molecular Ecology* 18: 3548–3580.

14 Gibbs, G. 2006. *Ghosts of Gondwana: the history of life in New Zealand.* Craig Potton Publishing, Nelson, New Zealand. See pp. 95, 104.

15 Cooper, A., Lalueza-Fox, C., Anderson, S., Rambaut, A., Austin, J. and Ward, R. 2001. Complete mitochondrial genome sequences of two extinct moas clarify ratite evolution. *Nature* 409: 704–707.

16 Bunce, M., Worthy, T. H., Phillips, M. J., Holdaway, R. N., Willerslev, E., Haile, J., Shapiro, B., Scofield, R. P., Drummond, A., Kamp, P. J. J. and Cooper, A. 2009. The evolutionary history of the extinct ratite moa and New Zealand Neogene paleogeography. *Proceedings of the National Academy of Sciences* 106: 20646–20651.

17 Long, J. A. 1998. *The dinosaurs of Australia and New Zealand and other animals of the Mesozoic era.* University of New South Wales Press Ltd, Sydney.

18 Worthy, T. H., Tennyson, A. J. D., Archer, M., Musser, A. M., Hand, S. J., Jones, C., Douglas, P. J., McNamara, J. A. and Beck, R. M. D. 2006. Miocene mammal reveals a Mesozoic ghost lineage on insular New Zealand, southwest Pacific. *Proceedings of the National Academy of Sciences* 103: 19419–19423.

19 Bones in Miocene material from the Manuherikia Group sediments in Otago that were initially thought to be those of a snake (Worthy et al. 2002b) were later reported as from fish (Gibbs 2006, op. cit.; see pp. 65, 109).

20 Benton, M. J. 2005. *Vertebrate palaeontology.* 3rd edn. Blackwell Publishing, Malden, MA. See p. 244.

21 Albino, A. 2007. Lepidosauromorpha. In: Gasparini, Z., Salgado, L. and Coria, R.A. (eds). *Patagonian Mesozoic reptiles.* Indiana University Press, Bloomington, pp. 87–115.

22 For discussion, see Campbell and Hutching 2007, op. cit.

23 Vajda, V., Raine, J. I. and Hollis, C. J. 2001. Indication of global deforestation at the Cretaceous-Tertiary boundary by New Zealand fern spike. *Science* 294: 1700–1702; Hollis, C. J. 2003a. The Cretaceous/Tertiary boundary event in New Zealand: profiling mass extinction. *New Zealand Journal of Geology and Geophysics* 46: 307–321.

24 Hollis, C. J. 2003b. Fatal impact: the asteroid that wiped out the dinosaurs. *Alpha* 116. Royal Society of New Zealand, Wellington.

25 Pope, K. O., Baines, K. H, Ocampo, A. C. and Ivanov, B. A. 1997. Energy, volatile production, and climatic effects of the Chicxulub Cretaceous/Tertiary impact. *Journal of Geophysical Research* 102: 21645–21664.

26 Beerling, D. J., Lomax, B. H., Royer, D. L., Upchurch, J., G.R. and Kump, L. R. 2002. An atmospheric pCO_2 reconstruction across the Cretaceous–Tertiary boundary from leaf megafossils. *Proceedings of the National Academy of Sciences* 99: 7836–7840.

27 Vajda et al. 2001, op. cit.

28 Hollis 2003b, op. cit.

29 Benton 2005, op. cit.; see p. 250.

30 Stevens et al. 1995, op. cit.

31 Strong, C. P. 2000. Cretaceous–Tertiary foraminiferal succession at Flaxbourne River, Marlborough, New Zealand. *New Zealand Journal of Geology and Geophysics* 43: 1–20.

32 Vajda et al. 2001, op. cit. See also Hollis 2003a, op. cit.

33 Stilwell, J. D., Consoli, C. P., Sutherland, R., Salisbury, S., Rich, T. H., Vickers-Rich, P. A., Currie, P. J. and Wilson, G. J. 2006. Dinosaur sanctuary on the Chatham Islands, Southwest Pacific: first record of theropods from the K-T boundary Takatika Grit. *Palaeogeography, Palaeoclimatology, Palaeoecology* 230: 243–250.

34 Stevens et al. 1995, op. cit.

35 Nelson, C. S. and Cooke, P. J. 2001. History of oceanic front development in the New Zealand sector of the Southern Ocean during the Cenozoic – a synthesis. *New Zealand Journal of Geology and Geophysics* 44: 535–553.
36 Wallis and Trewick 2009, op. cit.
37 Cooper, R. A. and Millener, P. R. 1993. The New Zealand biota: historical background and new research. *Trends in Ecology and Evolution* 8: 429–433.
38 Stevens et al. 1995, op. cit.; see p. 69.
39 Cooper and Millener 1993, op. cit.; Cooper, A. and Cooper, R. A. 1995. The Oligocene bottleneck and New Zealand biota: genetic record of a past environmental crisis. *Proceedings of the Royal Society of London B* 261: 293–302.
40 Campbell, H. and Landis, C. 2001. New Zealand awash. *New Zealand Geographic* 51 (May–June): 6–7; Campbell and Hutching 2007, op. cit. See also Waters and Craw 2006.
41 Dead tuatara have occasionally been found in seawater close to islands, where they were probably washed by waves (Chapter 9) or dropped by predatory birds (Chapter 7).
42 Cree, A. and Daugherty, C. H. 1991. *High rates of cutaneous water loss in nocturnal New Zealand reptiles.* Unpublished report to the Department of Conservation, Wellington. Without access to burrows to reduce evaporative water loss, tuatara become dependent on freshwater for drinking, at least in the laboratory (Hill 1982).
43 Thompson 1990.
44 Jones et al. 2009b.
45 Lee, M. S. Y., Hutchinson, M. N., Worthy, T. H., Archer, M., Tennyson, A. J. D., Worthy, J. P. and Scofield, R. P. 2009. Miocene skinks and geckos reveal long-term conservatism of New Zealand's lizard fauna. *Biology Letters* 5: 833–837.
46 Tennyson, A. J. D. 2010. The origin and history of New Zealand's terrestrial vertebrates. *New Zealand Journal of Ecology* 34: 6–27.
47 Worthy, T. H., Hand, S. J., Nguyen, J. M. T., Tennyson, A. J. D., Worthy, J. P., Scofield, R. P., Boles, W. E. and Archer, M. 2010. Biogeographical and phylogenetic implications of an early Miocene wren (Aves: Passeriformes: Acathisittidae) from New Zealand. *Journal of Vertebrate Paleontology* 30: 479–498.
48 Tennyson, A. J. D., Worthy, T. H., Jones, C. M., Scofield, R. P. and Hand, S. J. 2010. Moa's Ark: Miocene fossils reveal the great antiquity of moa (Aves: Dinornithiformes) in Zealandia. *Records of the Australian Museum* 62: 105–114. See also Worthy, T. H., Tennyson, A. J. D., Jones, C., McNamara, J. A. and Douglas, B. J. 2007. Miocene waterfowl and other birds from Central Otago, New Zealand. *Journal of Systematic Palaeontology* 5: 1–39; Tennyson 2010, op. cit.
49 Chapple, D. G., Ritchie, P. A. and Daugherty, C. H. 2009. Origin, diversification and systematics of the New Zealand skink fauna (Reptilia: Scincidae). *Molecular Phylogenetics and Evolution* 52: 470–487.
50 Carpenter, R. J., Jordan, G. J., Lee, D. E. and Hill, R. S. 2010. Leaf fossils of *Banksia* (Proteacaea) from New Zealand: an Australian abroad. *American Journal of Botany* 97: 288–297.
51 Conran, J. G., Bannister, J. M. and Lee, D. E. 2009. Earliest orchid macrofossils: early Miocene *Dendrobium* and *Earina* (Orchidaceae: Epidendroideae) from New Zealand. *American Journal of Botany* 96: 466–474.
52 Worthy et al. 2007, op. cit.
53 Worthy, T. H., Tennyson, A. J. D., Hand, S., Godthelp, H. and Scofield, R. P. 2011. Terrestrial turtle fossils from New Zealand refloat Moa's Ark. *Copeia* 2011: 72–76.
54 Jones et al. 2009b.
55 Bunce et al. 2009, op. cit.
56 Brazier et al. 1990, op. cit.
57 Pole, M., Douglas, B. and Mason, G. 2003. The terrestrial Miocene biota of southern New Zealand. *Journal of the Royal Society of New Zealand* 33: 415–426.
58 Jones et al. 2009b.
59 Molnar, R. E. and Pole, M. 1997. A Miocene crocodilian from New Zealand. *Alcheringa* 21: 65–70; Pole et al. 2003, op. cit.
60 Molnar and Pole 1997, op. cit.
61 Worthy et al. 2006, op. cit.
62 Worthy et al. 2006, op. cit.
63 Bunce et al. 2009, op. cit.
64 Bunce et al. 2009, op. cit.
65 Hay et al. 2003.
66 Bunce et al. 2009, op. cit.
67 Cooper and Millener 1993, op. cit.
68 Stevens et al. 1995, op. cit.
69 Stevens et al. 1995, op. cit.
70 Shackleton, N. J. 1987. Oxygen isotopes, ice volume and sea level. *Quaternary Science Reviews* 6: 183–190; Gibbs 2006, op. cit., p. 82.
71 Stevens et al. 1995, op. cit.
72 Worthy and Holdaway 1994, 2002. See also Bunce et al. 2009, op. cit.
73 McGlone, M. S. 1988. New Zealand. In: Huntley, B. and Webb, I. T. (eds). *Vegetation history. Vol. 7.* Kluwer Academic Publishers, Dordrecht, pp. 557–599.
74 Worthy and Holdaway 1993.
75 Marra, M. 2008. Quaternary fossil beetles from New Zealand. *New Zealand Entomologist* 31: 5–16.
76 Gibbs 2006, op. cit., p. 86.
77 Leathwick, J., Wilson, G., Rutledge, D., Wardle, P., Morgan, F., Johnston, K., McLeod, M. and Kirkpatrick, R. 2003. *Land environments of New Zealand: Nga taiao o Aotearoa.* David Bateman, Auckland.
78 Worthy and Grant-Mackie 2003. Earlier estimates dated the fossils simply as over 36,000 years old, reflecting the effective limit for radiocarbon dating at the time.
79 Millener 1981; Worthy 1984, 1997, 1998a, b, c; 2000, 2001; Worthy and Holdaway 1993, 1994, 1995, 1996, 2000, 2002; Clark et al. 1996; Holdaway and Worthy 1997; Worthy et al. 2002a; Worthy and Roscoe 2003; Wood 2006.
80 Worthy and Holdaway 2000.
81 Following the second Tuatara Recovery Plan (Gaze 2001), this total treats the two parts of Motunau Island (Plate Island) as separate populations (the tuatara there are separated by about 2 m of sea). The total also includes rat-inhabited Mauitaha Island, on which a single tuatara was seen in 1994 (Tennyson and Pierce 1995), although this hardly constitutes a viable population.
82 Worthy 1998b.
83 Worthy and Holdaway 2002, p. 610.
84 Anon. 1883. See Chapter 3 for further discussion of this report, including the possibility that the Brothers Islands near Stewart Island were confused with the Brothers Islands in Cook Strait, not far from Stephens Island.
85 The entry was written in 1951. The specimen was not locatable in 2003 (Maurice Watson, Southland Museum and Art Gallery, pers. comm. 23 July 2003), although a preserved tuatara thought to be this specimen has since been returned to the museum (L. Hazley, Southland Museum and Art Gallery, pers. comm. 13 January 2010).

86 Colenso 1886.
87 Dickison, M. 1998. *Are island tuatara representative?* Unpublished BSc Hons dissertation. Victoria University of Wellington, Wellington.
88 Jones 2007a.
89 An additional specimen is recorded in the catalogue of the Southland Museum (L. Hazley, pers. comm.).
90 Hay et al. 2008; Jennifer Hay, Massey University, pers. comm. 2 March 2008.
91 Stevens et al. 1995, op. cit.; McGlone 1988, op. cit.; McGlone, M. S. 1989. The Polynesian settlement of New Zealand in relation to environmental and biotic changes. In: Rudge, M. R. (ed.). Moas, mammals and climate in the ecological history of New Zealand. *New Zealand Journal of Ecology* 12 (Supplement), pp. 115–129.
92 Worthy 1997, 1998a, b, c, 2001; Worthy and Holdaway 1993, 1994, 1995, 1996, 2002; Clark et al. 1996; Holdaway and Worthy 1997; Worthy and Roscoe 2003; Wood 2006, 2009.
93 Worthy 2001; Worthy and Roscoe 2003.
94 Worthy and Holdaway 1995.
95 Worthy 1998a.
96 Worthy and Holdaway 1995, 1996; Worthy 1997, 1998a.
97 See supplementary material of: Wilmshurst, J. M., Anderson, A. J., Higham, T. F. G. and Worthy, T. H. 2008. Dating the late prehistoric dispersal of Polynesians to New Zealand using the commensal Pacific rat. *Proceedings of the National Academy of Sciences* 105: 7676–7680.
98 Worthy, T. H. and Holdaway, R. N. 1994. Scraps from an owl's table – predator activity as a significant taphonomic process newly recognised from New Zealand Quaternary deposits. *Alcheringa* 18: 229–245.
99 Worthy 1997. See also Worthy et al. 2002a for a North Island site.
100 Holdaway, R. N., Worthy, T. H. and Tennyson, A. J. D. 2001. A working list of breeding bird species of the New Zealand region at first human contact. *New Zealand Journal of Zoology* 28: 119–187.
101 Weight estimates from: Atkinson, I. A. E. and Millener, P. R. 1991. An ornithological glimpse into New Zealand's prehuman past. *Acta XX Congressus Internationalis Ornithologici* 1: 129–192. Note that two of today's predators of tuatara, Australasian harriers (*Circus approximans*) and sacred kingfishers (*Todiramphus sanctus*), appear to have been rare or absent from New Zealand prior to habitat modifications by humans (see Chapter 7).
102 Atkinson and Millener 1991, op. cit.; Whitaker, A. H. 1991. *Research on the striped gecko (*Hoplodactylus stephensi*) on Maud Island, Pelorus Sound, Marlborough: 6–12 March 1991.* New Zealand Department of Conservation, Nelson.
103 Worthy and Holdaway 1993.
104 Walls 1981, 1982; Moller 1985; Fraser 1993; Southey 1985. See Chapter 7 for more on the contemporary diets of tuatara.
105 Whitaker, A. H. 1992. Was the kawekaweau the world's largest gecko? *Forest and Bird* May No. 264: 44–46.
106 Whitaker 1992, op. cit. See Chapter 4 for more on Māori traditions about reptiles.
107 Worthy and Holdaway 2002, pp. 463–464.

Endnotes to boxed text

For explanation of citation format in box endnotes, see endnotes to the main text.

Box 2.1

1 Daugherty, C. H., Towns, D. R., Atkinson, I. A. E. and Gibbs, G. W. 1990. The significance of the biological resources of New Zealand islands for ecological restoration. In: Towns, D. R., Daugherty, C. H. and Atkinson, I. A. E. (eds). *Ecological restoration of New Zealand islands*. Conservation Sciences Publication No. 2, Department of Conservation, Wellington, pp. 9–21.
2 http://www.niwa.co.nz/education-and-training/schools/resources/climate/overview (accessed 4 January 2010).
3 Lowe, D. J., Newnham, R. M., McFadgen, B. G. and Higham, T. F. G. 2000. Tephras and New Zealand archaeology. *Journal of Archaeological Science* 27: 859–870; Wilmshurst, J. M., Anderson, A. J., Higham, T. F. G. and Worthy, T. H. 2008. Dating the late prehistoric dispersal of Polynesians to New Zealand using the commensal Pacific rat. *Proceedings of the National Academy of Sciences* 105: 7676–7680; Wilmshurst, J. M., Hunt, T. L., Lipo, C. P. and Anderson, A. J. 2011. High-precision radiocarbon dating shows recent and rapid initial human colonization of East Polynesia. *Proceedings of the National Academy of Sciences* 108: 1815–1820.
4 King, M. 2003. *The Penguin history of New Zealand*. Penguin Books, Auckland. See p. 41.

Box 2.2

1 Parrish, J. T., Daniel, I. L., Kennedy, E. M. and Spicer, R. A. 1998. Paleoclimatic significance of mid-Cretaceous floras from the Middle Clarence Valley, New Zealand. *Palaios* 13: 149–159.
2 Discussed in: Vickers-Rich, P. and Rich, T. H. Australia's polar dinosaurs. *Scientific American* July 1993: 40–45; Martin, A. J. 2009. Dinosaur burrows in the Otway Group (Albian) of Victoria, Australia, and their relation to Cretaceous polar environments. *Cretaceous Research* 30: 1223–1237.
3 Vickers-Rich and Rich 1993, op. cit.
4 Martin 2009, op. cit.
5 Chinsamy, A., Rich, T. and Vickers-Rich, P. 1998. Polar dinosaur bone histology. *Journal of Vertebrate Palaeontology* 18: 385–390.
6 Castanet et al. 1988.

Box 2.3

1 Goldberg, J., Trewick, S. A. and Paterson, A. M. 2008. Evolution of New Zealand's terrestrial fauna: a review of molecular evidence. *Philosophical Transactions of the Royal Society B* 363: 3319–3334; Wallis, G. P. and Trewick, S. A. 2009. New Zealand phylogeography:

evolution on a small continent. *Molecular Ecology* 18: 3548–3580.
2 Knapp, M., Mudaliar, R., Havell, D., Wagstaff, S. J. and Pockhart, P. J. 2007. The drowning of New Zealand and the problem of *Agathis*. *Systematic Biology* 56: 862–870.
3 Pratt, R. C., Morgan-Richards, M. and Trewick, S. A. 2008. Diversification of New Zealand weta (Orthoptera: Ensifera: Anostostomatidae) and their relationships in Australasia. *Philosophical Transactions of the Royal Society B* 363: 3427–3437; Boyer, S. and Giribet, G. 2009. Welcome back New Zealand: regional biogeography and Gondwanan origin of three endemic genera of mite harvestmen (Arachnida, Opiliones, Cyphophthalmi). *Journal of Biogeography* 36: 1084–1099; Allwood, J., Gleeson, D., Mayer, G., Daniels, S., Beggs, J. R. and Buckley, T. R. 2010. Support for vicariant origins of the New Zealand Onychophora. *Journal of Biogeography* 37: 669–681. See also Gibbs, G. 2006. *Ghosts of Gondwana: the history of life in New Zealand*. Craig Potton Publishing, Nelson, New Zealand, pp. 45, 90, 104, 128.
4 Gibbs 2006, op. cit., p. 45.
5 Nielsen, S.V., Bauer, A. M., Jackman, T. R., Hitchmough, R. A. and Daugherty, C. H. 2011. New Zealand geckos (Diplodactylidae): cryptic diversity in a post-Gondwanan lineage with trans-Tasman affinities. *Molecular Phylogenetics and Evolution* 59: 1–22.
6 Phillips, M. J., Gibb, G., Crimp, E. A. and Penny, D. 2010. Tinamous and moa flock together: mitochondrial genome sequence analysis reveals independent losses of flight among ratites. *Systematic Biology* 59: 90–107.
7 Schellart, W. P., Lister, G. S. and Toy, V. G. 2006. A Late Cretaceous and Cenozoic reconstruction of the southwest Pacific region: tectonics controlled by subduction and slab rollback processes. *Earth-Science Reviews* 76: 191–233.
8 Wallis and Trewick 2009, op. cit.; Worthy, T. H., Hand, S. J., Nguyen, J. M. T., Tennyson, A. J. D., Worthy, J. P., Scofield, R. P., Boles, W. E. and Archer, M. 2010. Biogeographical and phylogenetic implications of an early Miocene wren (Aves: Passeriformes: Acathisittidae) from New Zealand. *Journal of Vertebrate Paleontology* 30: 479–498.

Box 2.4

1 Worthy and Holdaway 2002, p. 312. The same authors (p. 332) considered terrestrial birds of 1 kg or larger as the primary prey of the eagle. The largest tuatara reach just over 1 kg in weight.
2 Lloyd, B. L. 2005. Lesser short-tailed bat (pp. 110–126) and Greater short-tailed bat (pp. 127–129). Both chapters in: King, C. M. (ed.). *The handbook of New Zealand mammals*. 2nd edn. Oxford University Press, Melbourne.
3 Newman 1977.

CHAPTER THREE

Turmoil: the arrival of humans and other mammals

New Zealand was among the last major landmasses of the world to feel the human footprint. The resulting wave of extinctions over the seven centuries that followed is now considered the 'dominating feature' of the country's recent faunal history.[1] Although tuatara were among the survivors, the species nonetheless experienced a dramatic reduction in distribution and abundance during this period.

Introduction

No one knows for certain why the first humans came, or even precisely when. But come they did and in increasing numbers and variety, leading to the nation of over 4 million people that exists in New Zealand today. Human settlement has undoubtedly led to new cultural diversity and creative opportunities. The dilemma for biologists, however, is that humans and their accompanying mammals have also brought near or total annihilation for many of the country's unique life forms, including the subject of this story.

The human impact on New Zealand ecosystems has been described as nothing short of an 'ecological revolution'.[2] Two main waves of species extinctions are recognised: the first centres around the impacts that followed the arrival of rats and the hunting and burning that began with Polynesian settlement, and the second around the further introductions of mammals and changes in habitat that followed European settlement.[3] Whether or not the first, Polynesian phase was subdivided in time between rat-introduction and human-settlement phases has until recently been a matter of vigorous debate.

This chapter explores these two waves, focusing on the consequences of each for tuatara, with an emphasis on the period up until the early 1900s. Related chapters include Chapter 4, which explores Māori traditions regarding tuatara; Chapter 5, which documents the history of scientific study; and Chapter 10, which covers the history of conservation efforts for tuatara since the late 1800s.

The first wave: Polynesian arrival and settlement

The first humans to colonise New Zealand came from an area of East Polynesia including the Society Islands (Figure 3.1).[4] Settlement of New Zealand, 3000 kilometres to the southwest, was among the final steps in a great wave of expansion throughout Polynesia by people with large, fast, sea-going canoes and well-developed navigational skills. From the perspective of ecologists Matt McGlone and colleagues, settlement through the Pacific islands may have arisen from 'an ingrained tradition of leaving in search of better prospects before resources were totally exhausted.'[5] Birds and marine resources were typically abundant on unexploited islands, and, as seabirds funnelled in vast clouds towards land, they may themselves have provided signals of new resources to be harvested.[6]

Once New Zealand was settled, in about AD 1230 to 1300,[7] the bounty proved substantial. Along with the petrels and shearwaters that burrowed the hillsides, the coastline supported gulls, shags, terns, seals, penguins, fish and shellfish. Inland were more petrels and their relatives, as well as geese,

Fig. 3.1 Final stages in the dispersal of the first humans in the South Pacific. Dispersal from West Polynesia (blue arrow) to central East Polynesia including the Society Islands (aubergine oval) occurred about AD c. 1025–1121. Within one or two centuries, East Polynesians had dispersed to the north, east and southwest (beige 'triangle'). The Kermadec Islands may have been used as a 'stepping stone' to reach New Zealand.

Modified from Wilmshurst, J. M. et al. 2011. High-precision radiocarbon dating shows recent and rapid initial human colonization of East Polynesia. *Proceedings of the National Academy of Sciences* 108: 1815–1820.

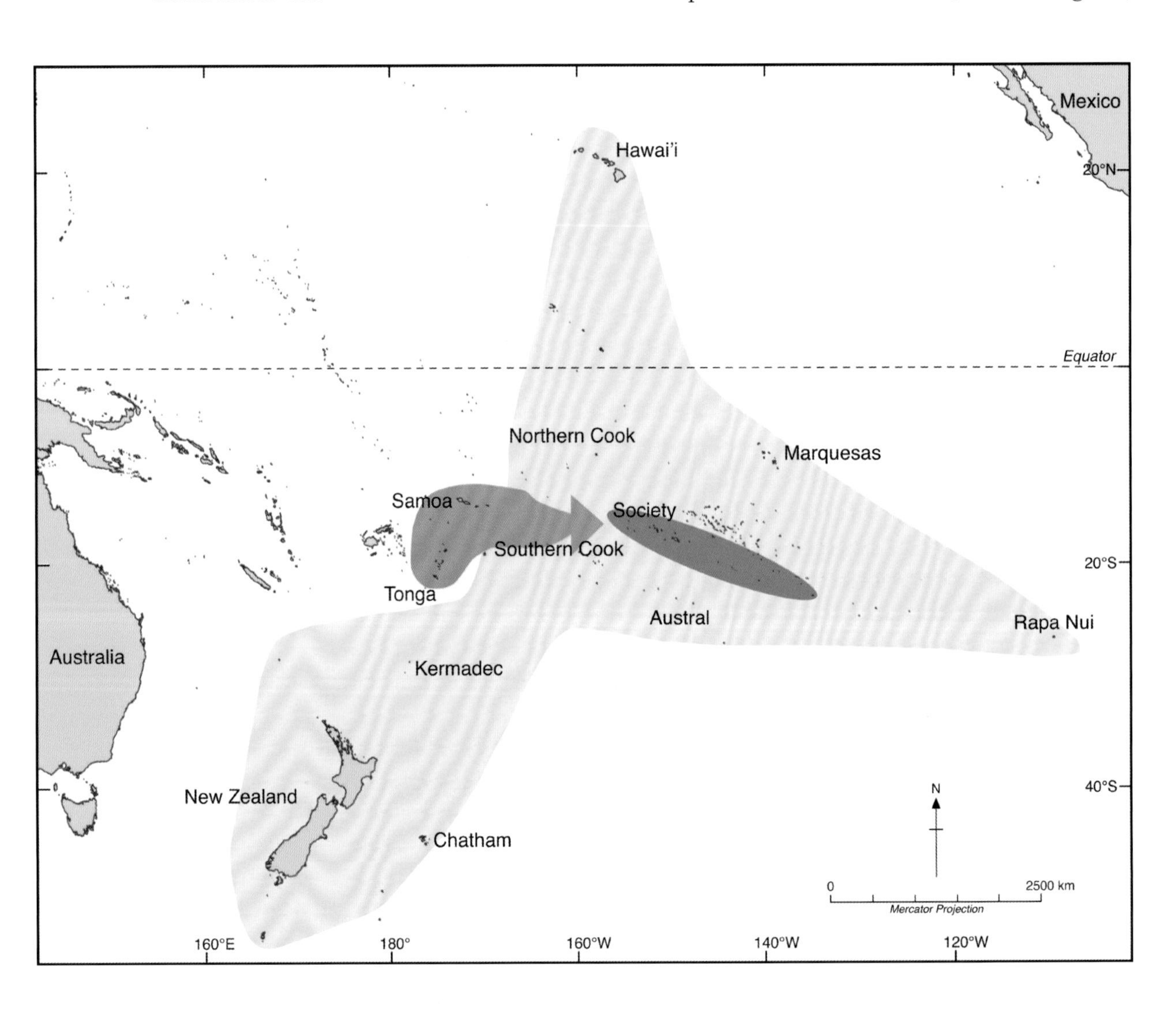

Fig. 3.2 A reconstruction by Chris Gaskin of an East Polynesian campsite on the banks of the Hawks Burn, near Cromwell in the southern South Island. Moa were butchered at the site in their hundreds, and the surrounding forest was burned. Tuatara once lived in the vicinity and could have been harvested, as suggested by evidence from middens elsewhere.

Source: Crown copyright: Department of Conservation Te Papa Atawhai.

swans and other waterfowl, plus numerous forest birds including the large and flightless adzebills (*Aptornis* spp.), takahē (*Porphyrio* spp.), kākāpō (*Strigops habroptilus*), rails, and about nine species of moa in six genera.[8] In East Polynesia the word 'moa' refers to domestic fowl, but these avian giants were walking communal feasts of 20 to 300 kilograms each. This early period of Polynesian settlement, often described as the Archaic Phase, was once referred to as the Moa-hunter Period because of the conspicuous presence of moa in human diets; however, other big game birds, and marine mammals such as the New Zealand fur seal (*Arctocephalus forsteri*), were also heavily hunted.[9] The resources were so plentiful, and the fauna so näive, that obtaining food has been likened by palaeoecologists Trevor Worthy and Richard Holdaway as more akin to 'supermarket shopping' than hunting.[10]

Evidence from dated artefacts tells us that the early settlers ranged quickly about the country, moving along the coasts and inland in both the North and South Islands (Figure 3.2).[11] Evidence from another source – dated bones – reveals that a rodent commensal of those first humans also spread rapidly throughout the main islands. Pacific rats (*Rattus exulans*, Figure 3.3) originated in Southeast Asia and have been distributed widely by people throughout the Pacific (in New Zealand, whether arrival was by accident or

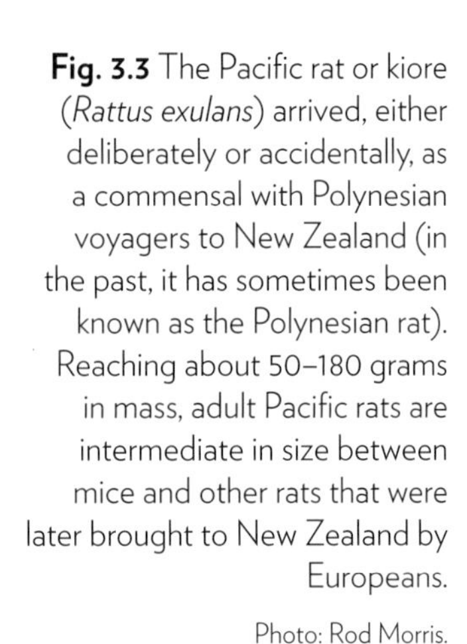

Fig. 3.3 The Pacific rat or kiore (*Rattus exulans*) arrived, either deliberately or accidentally, as a commensal with Polynesian voyagers to New Zealand (in the past, it has sometimes been known as the Polynesian rat). Reaching about 50–180 grams in mass, adult Pacific rats are intermediate in size between mice and other rats that were later brought to New Zealand by Europeans.

Photo: Rod Morris.

design is unclear). Genetic evidence suggests repeated introductions of Pacific rats (and by inference, humans) from East Polynesia, and possibly from more westerly locations as well, via the Kermadec Islands as a 'stepping stone'.[12] Following establishment in New Zealand, these small, nocturnal rats became known locally by the Māori name kiore.

The opportunity that Pacific rats found in New Zealand matched that for humans. Richard Holdaway has envisaged a rat 'blitzkrieg' advancing across the landscape, penetrating through dense forest ahead of human spread, and with exponential reproduction in the presence of unlimited food.[13] Clearly, the rats penetrated into some of the wettest areas avoided until later by Māori.[14] Pacific rats are omnivores, and would have consumed not only seeds and plants but also invertebrates and small vertebrates – eggs, chicks and even adults of small birds, as well as frogs, bats and reptiles, including tuatara. Unlike most predatory birds of New Zealand, the new mammalian predator hunted on the ground at night by smell,[15] thus proving a peerless foe for flightless and nocturnal species.

Inferences about the impacts of Pacific rats on mainland tuatara come in part from changes in species composition within natural deposits of bones. Once rat bones appear in the layers, the abundance of tuatara and other small endemic vertebrates declines.[16] Many of these deposits are the regurgitated remains left by the laughing owl (*Sceloglaux albifacies*). It seems that as the native prey of laughing owls disappeared, the owls switched to preying on the cause – Pacific rats – until the owl itself went extinct.

Other, more circumstantial evidence for past impacts of Pacific rats on mainland tuatara comes from studies of the effects of the presence and removal of rats on offshore islands. These impacts are discussed in more detail in Chapter 7. In brief, at least ten populations of tuatara on offshore islands appear to have become extinct in the probable or known presence of Pacific rats, and reduced density and recruitment failure have been noted for at least seven other populations.[17] Although direct evidence is lacking (and would be hard to come by), it seems that Pacific rats prey on the eggs and juveniles of tuatara, and compete with juveniles and adults for food. Crucially, where Pacific rats have been eradicated from islands with relict populations of tuatara, juvenile tuatara have once again appeared in meaningful numbers.[18]

The date when Pacific rats arrived in New Zealand has until recently been a vexed issue. Evidence presented in 1996 suggested unexpectedly that some rat bones from both the North and South Islands dated from about AD 50,[19] more than a thousand years older than the oldest evidence of human artefacts or activities. As Pacific rats are very poor swimmers, these dates implied the early arrival of humans who either left or died out without trace. However, further analyses now suggest that these early dates resulted from technical inaccuracy.[20] Biological evidence also supports a first arrival of rats at about AD 1250–1300, around the time of human settlement. For example, rat predation on the shells of large land snails in Northland appears not to begin until about that time,[21] and plant seed cases show the distinct gnaw marks of rodents only from about the same period.[22]

Pacific rats were not the only predators of mainland tuatara. Despite the avian plenty, tuatara were apparently also harvested by humans during the Archaic Phase. The evidence (usually in the form of the distinctive dentaries, the tooth-bearing bones of the lower jaw)[23] comes from archaeological deposits described as middens, or rubbish heaps containing domestic waste (Figure 3.4;

Fig. 3.4 The distinctive dentaries of tuatara, with their jagged margin of teeth, have often been found in archaeological deposits left by early Polynesian settlers. These charcoal-stained examples are from the Sunde site on Motutapu Island (collection of the Department of Anthropology and Archaeology, University of Otago).

Photo: Ken Miller and Alison Cree.

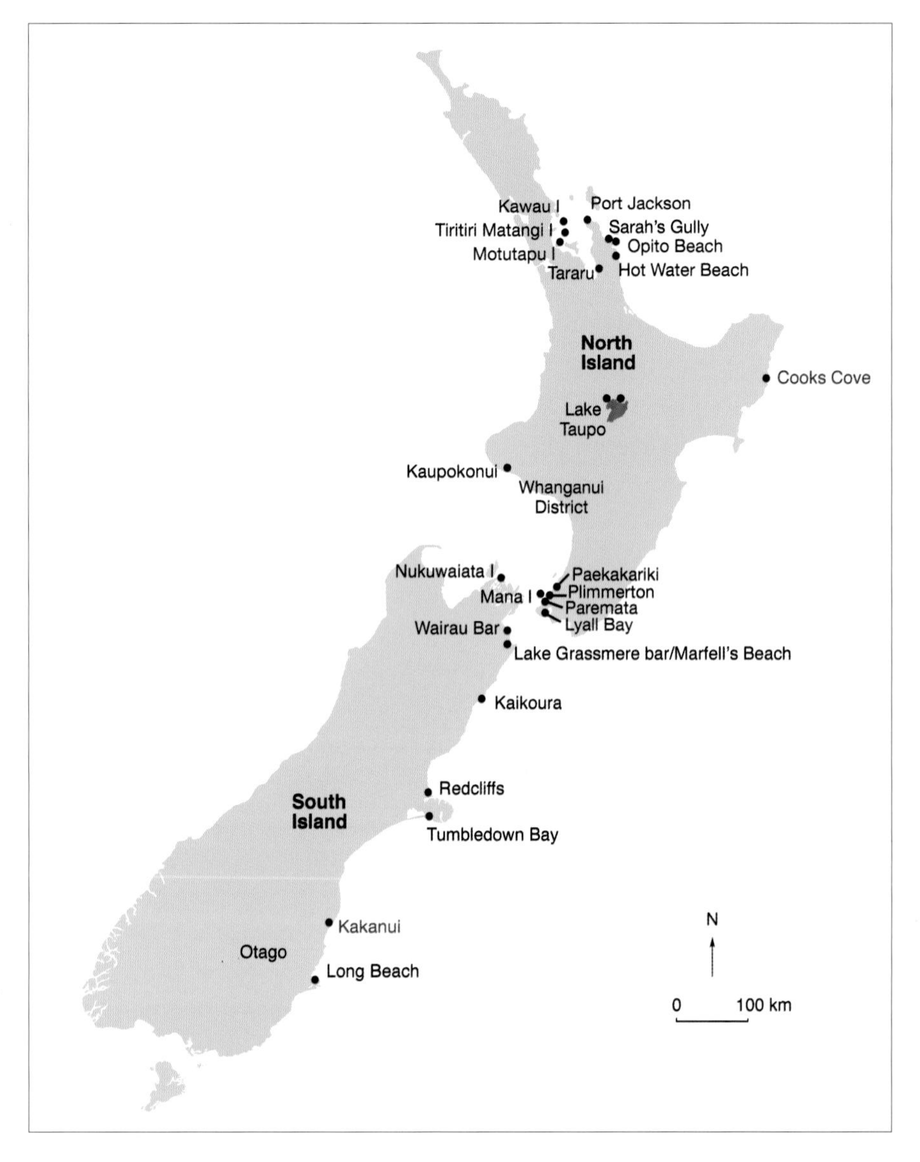

Fig. 3.5 Distribution of tuatara bones in sites described as archaeological deposits or middens. For sources, see Appendix 3. For additional sites that may include midden material, see Chapter 2.

Appendix 3). Assuming that such sites were indeed middens,[24] then it appears that humans were harvesting tuatara at a wide range of locations (Figure 3.5). In the North Island, these include the coastlines of the Coromandel, East Cape, Taranaki, Whanganui and Wellington regions, as well as inland near Lake Taupo; in the South Island, sites are scattered along the east coast from Wairau Bar in Marlborough to Long Beach in Otago. Additional mainland sites with tuatara bones may include midden material (see Chapter 2), but disturbance by wind, livestock, rabbits and people sometimes makes it difficult to distinguish material in middens from underlying natural deposits.[25] Bones of tuatara have also been reported from middens on several islands in the Hauraki Gulf, including Kawau, Tiritiri Matangi and Motutapu islands, as well as on Mana Island near Wellington and on Nukuwaiata Island (the inner of the Chetwode Islands) in the Marlborough Sounds.

The evidence of tuatara from archaeological deposits is too fragmentary to estimate the proportional contribution to human diets (most remains of tuatara are likely to have passed undetected, especially in older digs, which focused on human artefacts and large, easily identified bird bones). Although tuatara bones from middens have themselves not been dated, radiocarbon dates for associated material indicate that deposition typically occurred between about AD 1200 and AD 1500. This fits with the frequent presence in the same deposits of bones from birds such as moa, crows and swans (Appendix 3), whose populations became extinct in New Zealand within the first few centuries of human colonisation.

Natural populations of tuatara sometimes existed near the locations of midden deposits (Chapter 2), so tuatara whose remains appear in middens could have been locally captured. However, the Polynesian colonists are known to have carried moa haunches and marine shellfish to new sites, so perhaps tuatara were also transported at times.[26] As Ian Crook and Les Moran once suggested, tuatara appear 'an almost ideal source of meat' for itinerant humans because they can survive extended periods (several weeks or more) without food in small containers.[27] Perhaps, like giant tortoises on board 18th-century sailing ships,[28] tuatara were a useful, low-maintenance and transportable supplement to other food supplies. The possibility that tuatara were transported as companion animals also cannot be discounted.[29]

Harvesting of tuatara by the first colonists could have been assisted by kurī (*Canis familiaris*), a companion dog and food source introduced by Polynesian settlers. Kurī were low-set, short-legged animals that later interbred with dogs of European origin, eventually disappearing as a distinct breed.[30] In more recent times European dogs have proved capable of killing tuatara,[31] and one lighthouse keeper reported having 'a dog that would hunt [tuatara] out without touching them' by which means he 'came across many [that] would have passed unnoticed.'[32] In the last few years conservation biologists have pursued the possibility of training dogs as odour-detectors for tuatara and lizards, with promising results.[33] Thus, a role for kurī in the demise of mainland tuatara is possible, though speculative.

At about the same time as human settlement began, pollen deposits on the North and South Islands begin to reveal widespread evidence of vegetation change (Figure 3.6). Fires became conspicuous (as shown by an increase in charcoal),[34] and pollen changed in composition from plants of the forest towards those of grassland and fernland.[35] Burning of the damp and often impenetrable forests would have brought a variety of benefits for Polynesian settlers, including easier passage when travelling, open areas for dwelling and the cultivation of crops, and increased growth of bracken (*Pteridium esculentum*), a fern whose rhizomes provided a valuable source of carbohydrate.[36] Tuatara could not have escaped from fires, and although some may have survived in burrows, others (especially young animals living under surface cover) were probably burnt. Densities may subsequently have declined following changes in vegetation, which probably altered the availability of food,

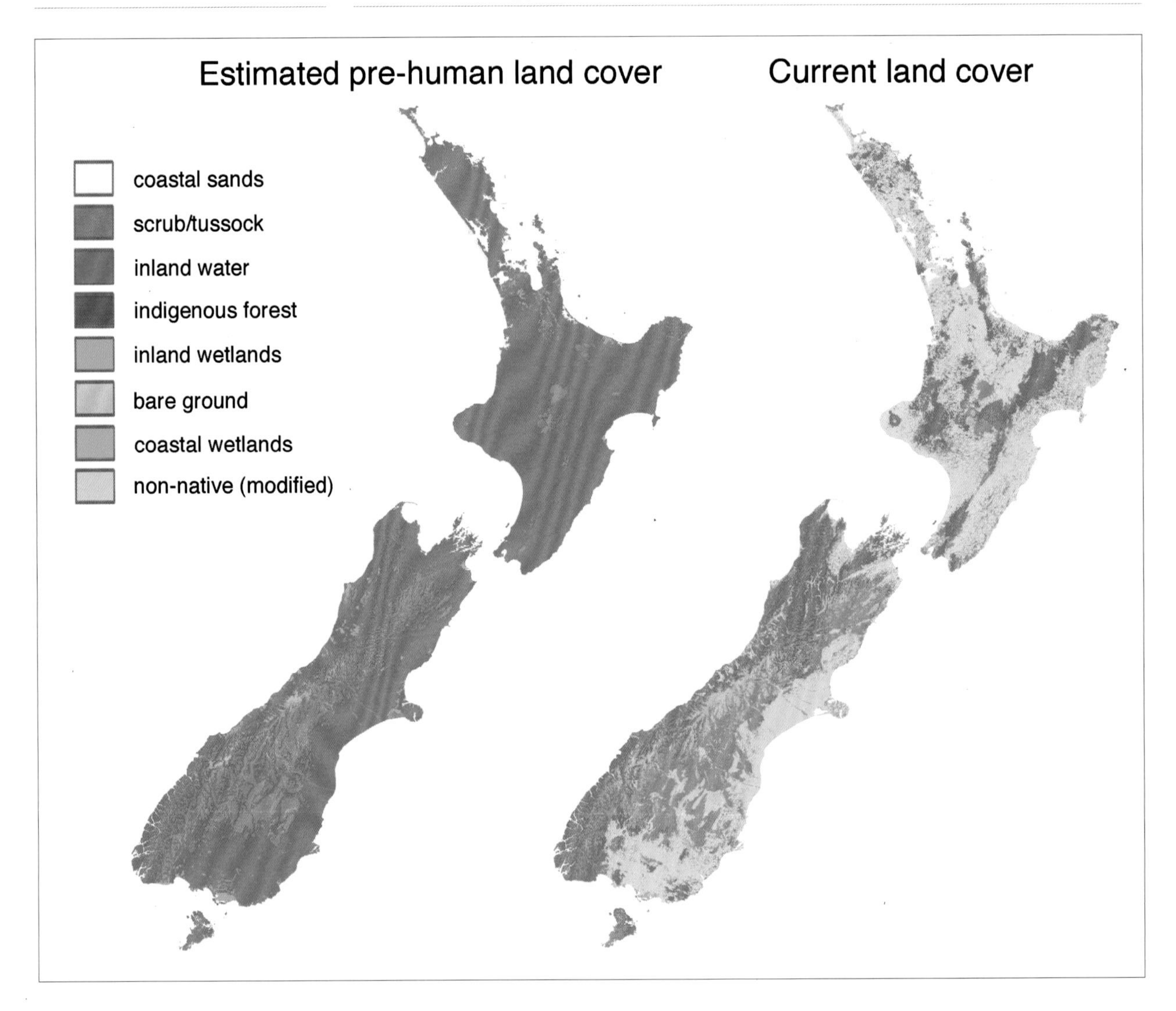

Fig. 3.6 Loss of indigenous forest cover in New Zealand following human arrival. In recent pre-human times, when tuatara were widely distributed on the mainland, indigenous (native) forests covered about 81% of the land area (green area of map on left). Settlement by Polynesians led to substantial losses, especially in the eastern South Island, where tuatara were once common. Native vegetation was further reduced following European settlement. By the end of the 20th century, tuatara were long extinct on the two main islands, and native forest had been reduced to about 23% of the land area (map on right). Based on Leathwick, J., Wilson, G., Rutledge, D., Wardle, P., Morgan, F., Johnston, K., McLeod, M. and Kirkpatrick, R. 2003. *Land Environments of New Zealand Nga Taiao o Aotearoa*. David Bateman, Auckland, in association with Manaaki Whenua – Landcare Research and Ministry for the Environment.

Source: Manaaki Whenua – Landcare Research.

including invertebrates, petrels and other small vertebrates. In addition, loss of cover would have increased the vulnerability of tuatara to birds of prey, as discussed below in relation to habitat change during European times.

Overall, the omens from human settlement were dire for many species, most obviously flightless birds. Butchery sites with moa carcasses estimated to number in the thousands are known.[37] Models of the population collapse developed by Richard Holdaway and Chris Jacomb indicate that the reproductive rate (one to two eggs per clutch) of these long-lived birds was so low that virtually any harvesting of adults would have been unsustainable. Within about 100–200 years of human arrival, moa were effectively extinct.[38] Breeding colonies of fur seals had also disappeared from the northern North Island by about AD 1500 and were dwindling in the south.[39] As the co-evolutionary relationships between plants and avian herbivores were decoupled, the structure and composition of the vegetation would have changed,[40] and flow-on extinctions were inevitable. The immense Haast's eagle (*Aquila moorei*), for example, which was dependent on forest for habitat and on moa for prey, soon disappeared. Other birds that succumbed during this period include the two species of adzebills, the two species of flightless geese (*Cnemiornis* spp.), the New Zealand raven (*Corvus antipodum*), the New Zealand coot (*Fulica prisca*) and the New Zealand population of black swans (*Cygnus atratus*).[41]

Initially the colonists may have moved their settlements frequently as local resources became depleted.[42] They also shifted to eating larger amounts of smaller items of food, including small penguins, fish and shellfish, as well as plants.[43] The warmer North Island was more conducive to growth in the human population, which there became increasingly dependent on crops. As a probable consequence of intensified conflict over resources,[44] inter-tribal warfare and fortified villages (pā) developed (Figure 3.7), and cannibalism became known. Cultural changes in the South Island appear less rapid and extreme, with fewer pā sites and a greater emphasis on movements to harvest tī kōuka (cabbage tree, *Cordyline australis*) and other food supplies on a seasonal basis.[45] These overall changes in lifestyle, many of which were evident in the North Island by about AD 1500, mark the transition to the Classic Phase of Māori culture.

The environmental impact of Polynesian settlement was considerable. By the late 1700s, when Europeans first arrived, nearly half the mainland forests had gone. More than 30% of breeding bird species had become extinct from the North, South and Stewart/Rakiura Islands,[46] and some were globally extinct. Procellariiform seabirds suffered heavy losses, depleting the transfer of marine nutrients to mainland ecosystems.[47] Small vertebrates, while possibly not subject to human harvesting, were vulnerable to Pacific rats. The three largest species of endemic frogs (*Leiopelma* spp.) became globally extinct during this period. Also to essentially disappear from the mainland were the robust short-tailed bat (*Mystacina robusta*, now globally extinct) and several lizards, including Duvaucel's gecko (*Hoplodactylus duvaucelii*) and several species of skinks (*Oligosoma* spp.), some of which survive today on offshore islands.[48]

As for tuatara, by the time of European settlement, New Zealand's largest surviving reptile was effectively extinct on the North and South Islands. Although a few sightings from the North Island were reported in European times (discussed further in Chapter 5), no mainland specimens were preserved for conclusive identification and none obviously involved populations with juveniles. The most recent radiocarbon dates for bones from mainland tuatara

Fig. 3.7 By the time of European settlement, Māori society exhibited a high level of latent, if not overt, hostility between neighbouring tribes and sub-tribes. Fortified villages, or pā, were commonplace on elevated sites, especially in the North Island. Storehouses were raised on legs to protect food from Pacific rats. This view shows part of the outer palisade of Rangihaeata's pā at present-day Plimmerton on the southwest coast of the North Island, with Mana Island in the distance. Bones of tuatara have been found in middens on the coast near Plimmerton and on Mana Island, implying that tuatara were once harvested here, but the species no longer survives at either place.

Source: Plate 57 in G. F. Angas's *The New Zealanders Illustrated* (Thomas McLean, London, 1846), Alexander Turnbull Library, Wellington, New Zealand, PUBL-0014-57.

are from about AD 1300–1450.[49] The effects for populations on offshore islands were in general less severe, although not non-existent. On some, the Poor Knights Islands and Hauturu/Little Barrier Island, for example, tuatara survived alongside human settlement. Tuatara also remained on many other offshore islands that were visited seasonally to harvest crops or seabirds. However, numbers were dwindling on many of those islands on which, like Hauturu, Pacific rats had become established. The practice of sometimes burning the vegetation to aid harvesting of procellariiform chicks (known from European times as mutton-birding) would have contributed to lower densities. On at least seven offshore islands for which there is evidence for human harvesting, or the presence of Pacific rats, tuatara were already at or close to extinction by late Polynesian or early European times (Table 3.1).

Tuatara retained a strong presence in the oral traditions of Māori (discussed further in Chapter 4). Traditional knowledge gathered by Europeans from Māori soon after settlement included reports that tuatara had lived, or still lived, on the mainland. Members of some iwi (tribes) also retained a tradition of having eaten tuatara, whereas others expressed a sense of revulsion at the idea. Tuatara-like reptiles also featured in Māori carvings from around this period, and the living animal was well known to those who settled on or visited northern or Cook Strait islands. Thus, tuatara retained an active presence in the life and lore of many Māori of this period.

The second wave: European arrival and settlement

Like the Polynesians before them, the first Europeans arrived on oceanic voyages of discovery. Abel Tasman's party, from the Dutch East India Company, charted parts of the western coastlines of both main islands in 1642 but the crew did not venture ashore. Four members were killed in a skirmish with Māori in Golden Bay and one body was taken, an event interpreted as evidence of cannibalism in a time of resource crisis.[50]

The three voyages led by James Cook (in 1769, 1772–1773 and 1777) generated a fundamental shift in the level of European interest in New Zealand (including the scientific study of tuatara: Chapter 5). Subsequent European visits were at first intermittent, mainly involving sealers, whalers and ships seeking timber. From the early 1800s, however, missionary activity began and trading and exploration increased. In 1840 the British government negotiated the Treaty of Waitangi with Māori chiefs, and New Zealand was annexed as a British Colony. Today the treaty is regarded as New Zealand's founding document,[51] although the exact nature of the agreement and the extent to which it has been upheld remain hotly debated (including with respect to cultural treasures such as tuatara; see Chapter 10).

Europeans rapidly settled in their tens of thousands. Suffering from diseases and warfare, and increasingly alienated from tribal lands, Māori accounted by 1896 for only 5.7% of the total population[52] (from the 1940s, the proportion increased[53]). With the development of new towns, industries and agriculture, modification of New Zealand's natural habitats continued at a rapid rate. Indigenous forest cover on the mainland declined overall from an estimated 81% of land area in pre-human times to about 23% today.[54]

European settlement also brought a new and devastating wave of introduced mammals. First to arrive was the Norway rat (*Rattus norvegicus*), probably from the time of Cook's first voyage in 1769 but undoubtedly supplemented by whaling and sealing ships (fortunately, a catastrophic shipwreck of the *Endeavour* off the Brothers Islands,[55] the northern island of which had a vulnerable population of tuatara, was narrowly averted).

TABLE 3.1

Islands surrounding the North and South Islands on which tuatara have been reported, and are presumed or known to have become extinct

Presumed time of extinction: during Polynesian or early European settlement (no confirmed sightings since European arrival)

Island[1]	Nature of evidence for past presence of tuatara[2]	Sources for past presence of tuatara[3]	Introduced mammals present[4]
Motuopao Island	Bone(s) in dunes	Millener 1981	Pacific rat (†)
Kawau Island	Bone(s) in midden c. AD 1600s	Robert Brassey (Department of Conservation, pers. comm. 18 September 2003)	Ship rat, cat, brushtail possum, brush-tailed rock wallaby, numerous other exotic species (†), post-European arrival
Tiritiri Matangi Island	Bone(s) in midden c. AD 1400s; possible sighting c. 1902 but no source reference provided	Robert Brassey (pers. comm. 13 October 1999); Rimmer 2004	Pacific rat (†); plus rabbit (†), cat (†), pig (†), goat (†), sheep (†) post-European arrival
Motutapu Island	Bone(s) in midden c. AD 1400	Scott 1970; Davidson 1972	Norway rat (†), ship rat (§), house mouse (§), European hedgehog (§), pig (†), cat (§), rabbit (§), stoat (§), fallow deer (†), brush-tailed rock wallaby (†) and brushtail possum (†) post-European arrival
Korapuki Island	Bones from at least two individuals, one in burrow	Hicks et al. 1975; Mike Meads (pers.comm. in Cree et al. 1995a); David Towns (Department of Conservation, pers. comm. 3 September 2003)	Pacific rat (†); plus European rabbit (†) post-European arrival
Mana Island	Bones in midden	Miskelly 1999	House mouse (†), sheep (†), rabbit (captive only?) post-European arrival
Motunau Island [Canterbury]	Bone (museum specimen CM Rep 340)		European rabbit (†) post-European arrival

Presumed time of extinction: survival into European times reported or known

Island[1]	Nature of evidence for past presence of tuatara[2]	Sources for past presence of tuatara[3]	Introduced mammals present[4]
Manawatawhi/ Three Kings Islands	Present, according to information supplied by Cheeseman and/or Bollons in 1922 to Department of Lands and Survey	R. P. Greville to Under Secretary of Department of Lands and Survey 19 May 1922 (ANZ ABWN 6095 W5021 Box 688 23/856 pt 1); T. W. Brodrick to Under Secretary of Internal Affairs 25 May 1922 (NMNZ MU000001 Box 21 Item 2 File 14/2/10)	Dog (†), pig (†), goat (†) post-European arrival
Moturoa Islands	Individuals reported 'from … small islands off Cape Karakara, Mangonui County' by Cheeseman in 1913; named as Moturoa Islands in information supplied by Cheeseman and/or Bollons in 1922	T. F. Cheeseman to A. Hamilton 22 March 1913 (NMNZ MU 000207 Box 4 Item 1); R. P. Greville to Under Secretary of Department of Lands and Survey 19 May 1922 (ANZ ABWN 6095 W5021 Box 688 23/856 pt 1)	Pigs (†) on Whale Island post-European arrival; otherwise mammal-free
Atihau Island [previously known also as Trig Island], Mokohinau Islands	Reported by lighthouse keepers on 'the island known as Goat Island – the most westerly of the group', and that 'on the outside island (Goat) they are plentiful'	A. Hamilton to Minister of Internal Affairs 15 May 1913 (ANZ IA 1 46/18/4 pt 1); see also McCallum 1980	Pacific rat (†); plus goat post-European arrival (†)
Burgess Island (Pokohinu), in Mokohinau Islands	Repeated sightings by lighthouse keepers c. 1880s–1920s, e.g. 'The tuatara lizard is found in abundance on two of the [Mokohinau] islands' (Sandager 1890)	Sandager (1890); see also J. Duthie to Dominion Museum 12 April 1913 (refers specifically to Burgess Island; in NMNZ MU 000207 Box 4 Item 1); A. Hamilton to Minister of Internal Affairs 15 May 1913 (ANZ IA 1 46/18/4 pt 1); R. S. Wilson to Secretary of Marine 1 January 1931 (ANZ M1 25/611 pt 3)	Pacific rat (†); plus cattle (†), goat (†), sheep (†), pig (†) post-European arrival

Presumed time of extinction: Survival into European times reported or known

Island[1]	Nature of evidence for past presence of tuatara[2]	Sources for past presence of tuatara[3]	Introduced mammals present[4]
Burgess Island (Pokohinu), in Mokohinau Islands (cont.)	Individuals (8 or 9) said to have been collected in 1890 from the 'principal island' Present, according to information supplied by Cheeseman and/or Bollons in 1922 to Department of Lands and Survey	Anon. 1931 R. P. Greville to Under Secretary of Department of Lands and Survey 19 May 1922 (ANZ ABWN 6095 W5021 Box 688 23/856 pt 1); T. W. Brodrick to Under Secretary of Internal Affairs 25 May 1922 (NMNZ MU000001 Box 21 Item 2 File 14/2/10)	
Fanal Island (Motukino), Mokohinau Islands	Reported 1922, observer not stated	Crook 1970	Pacific rat (†)
Great Barrier Island (Aotea Island)	'It is common on ... the Barrier Islands'	Buller 1895	Pacific rat; plus ship rat, house mouse, European rabbit, cattle, pig, goat († underway), cat post-European arrival
Shoe Island (Motuhoa)	'I have actually seen specimens from ... Shoe Island'	T. F. Cheeseman to A. Hamilton, 22 March 1913 (NMNZ MU 000207 Box 4 Item 1). Also reported by Falla (1935) but unclear whether this involved separate sightings.	Norway rat, European rabbit post-European arrival
Slipper Island (Whakahau)	Present, according to information supplied by Cheeseman and/or Bollons to Department of Lands and Survey	R. P. Greville to Under Secretary of Department of Lands and Survey 19 May 1922 (ANZ ABWN 6095 W5021 Box 688 23/856 pt 1); T. W. Brodrick to Under Secretary of Internal Affairs 25 May 1922 (NMNZ MU000001 Box 21 Item 2 File 14/2/10). Also reported by Falla (1935) but unclear whether this involved separate sightings.	Pacific rat (†), dog (†); plus Norway rat (and European rabbit?) post-European arrival
Hauturu Island	Five, probably large males, sighted 1978	P. J. Burstall (pers. comm. to D. G. Newman in Newman 1986)	House mouse (†), Norway rat (†, but probably reinvaded) post-European arrival
Whenuakura Island	Tuatara (n = 110) sighted in 1981, but extinct by 1985	Newman 1986	House mouse (†), Norway rat († but re-invaded) post-European arrival
Mayor Island (Tuhua)	Sightings and a photograph of an individual living near a sacred pōhutukawa tree in late 1800s; teacher in Native School on Motiti Island told by Māori that tuatara were present on Mayor Island, apparently in numbers	Prebble 1971; C. M. Clench to Rev. J. Crews, Wellington Zoological Society 18 August 1913 (ANZ IA 1 46/18/4 pt 1)	Pacific rat (†); plus Norway rat (†), pig (†), cat (†) post-European arrival
Motiti Island [Cook's 'Flat Island']	Individual collected in 1867 from deserted kūmara pit	G. Mair (pers. comm. to W. L. Buller in Buller 1877); G. Mair to H. D. Bell 8 June 1913 (ANZ IA 1 46/18/4 pt 1); see also Box 5.3	(Pacific rat?); Norway rat, cattle, pig, horse, sheep, dog post-European arrival
Moutuhora Island [previously known as Whale Island]	'Tradition says that they were plentiful on Whale Island' (W. Mair); 'it is abundant ... on Whale Island' (Buller); members of Ngāti Awa reported that tuatara were felt in burrows by mutton-birders in 1930s (Owen 1998)	Mair 1873; Buller 1895[5]; Owen 1998	Norway rat (†), rabbit (†), sheep (†), goat (†), cat (†) post-European arrival
East Island (Whangaokeno Island)	Many sightings and some preserved specimens from late 1800s–early 1900s	See Box 3.1	Pacific rat (†); goat (†), cat (†) post-European arrival
Matiu/Somes Island	Individual pulled from burrow of a pet rabbit in 1842 and sent to British Museum; sighting in 1877 reported	Knox 1870; Acting Director of Dominion Museum to Secretary of Marine, 19 March 1919 (ANZ M 1 25/611 pt 1); see also Box 10.6	Rabbit (pets, at least †), ship rat (†) post-European arrival; various other exotic species in animal quarantine station after disappearance of tuatara (†)

Presumed time of extinction: survival into European times reported or known

Island[1]	Nature of evidence for past presence of tuatara[2]	Sources for past presence of tuatara[3]	Introduced mammals present[4]
Rangitoto ke te Tonga (D'Urville Island)	Bone (museum specimen NMNZ RE 112); live tuatara found occasionally at Port Hardy Peninsula, c. 1890s? (Webber); individuals reported at southwest end, c. 1950s (Wright)	Webber 1953; Wright undated	Pacific rat; plus house mouse, European rabbit (†), cat, stoat, pig, goat (†), red deer post-European arrival
At least one of the Chetwode Islands	Bones in midden on Nukuwaiata Island; 'It inhabits islands in Cook Straits ... [including] the Chetwynd Islands' (Buller)	Richard Walle (Department of Conservation, pers. comm. 15 April 2010); Buller 1895	Pacific rat (†); plus pig (†), white rabbit (†) post-European arrival
Native Island (near Stewart Island/ Rakiura)	Specimen recorded in catalogue of Southland Museum and Art Gallery in 1951 (A51.18)	Maurice Watson (Southland Museum and Art Gallery, pers. comm. 23 July 2003); Lindsay Hazley (Southland Museum and Art Gallery, pers. comm. 13 January 2010)	Ship rat, pig (†), goat (†) post-European arrival
Brothers Islands (near Stewart Island)	'Captain Fairchild ... caught one of the large lizards peculiar to the North Island ... It measures 15 or 16 inches in length'	Anon. 1883	?

Notes

1 Within each time period, islands are listed approximately from north to south, following the eastern coastline. The North and South Islands, not listed here, also have extensive evidence for the past presence of tuatara (Chapter 2).

2 ANZ = Archives New Zealand; CM = Canterbury Museum; NMNZ = Museum of New Zealand.

3 References not in the bibliography are cited here in full.

Davidson, J. M. 1972. Archaeological investigations on Motutapu Island, New Zealand. *Records of the Auckland Institute and Museum* 9: 1–14.

Hicks, G. R. F., McColl, H. P., Meads, M. J., Hardy, G. S. and Roser, R. J. 1975. An ecological reconnaissance of Korapuki Island, Mercury Islands. *Notornis* 22: 195–220.

Miskelly, C. 1999. *Mana Island ecological restoration plan*. Department of Conservation, Wellington, p. 72.

Rimmer, A. 2004. *Tiritiri Matangi: a model of conservation*. Tandem Press, Auckland, p. 107.

Scott, S. D. 1970. Excavations at the 'Sunde Site', N38/24, Motutapu Island, New Zealand. *Records of the Auckland Institute and Museum* 7: 13–30.

Wright, A. [undated, c. 1950–1955]. *Report on bird life around French Pass*. Unpublished report received and extracted by Wildlife Service 2/25 May. ANZ IA 1 46/18/4 pt 2.

4 Present as wild, feral or farmed populations around the presumed time of tuatara extinction. '†' indicates animals that have since been eradicated, removed or have died out; § indicates that eradication is underway. Various sources including the tuatara references cited, DOC staff and chapters in: King, C. M. (ed.) 2005. *The handbook of New Zealand mammals*. 2nd edn. Oxford University Press, Auckland. Scientific names as follows: brushtail possum (*Trichosurus vulpecula*), brushtailed rock wallaby (*Petrogale penicillata*), cat (*Felis catus*), cattle (*Bos taurus*), dog (*Canis familiaris*), European hedgehog (*Erinaceus europaeus*), European rabbit (*Oryctolagus cuniculus*), fallow deer (*Dama dama*), goat (*Capra hircus*), horse (*Equus caballus*), house mouse (*Mus musculus*), Norway rat (*Rattus norvegicus*), Pacific rat (*Rattus exulans*), pig (*Sus scrofa*), red deer (*Cervus elaphus scoticus*), sheep (*Ovis aries*), ship rat (*Rattus rattus*) and stoat (*Mustela erminea*). White rabbits are a domestic breed of the European rabbit.

5 Note that an implication that tuatara were collected from Moutohora Island by Gilbert Mair and Walter Buller in 1886 (Park, G. 2006. *Theatre country: essays on landscape and whenua*. Victoria University Press, Wellington, p. 79) is misleading. Mair's diary makes clear that the animals were collected from Moutoki Island (Mair, G. *Diary and notes Nov. 1885–Mar. 1886* [Diary 36]. Alexander Turnbull Library, Wellington, New Zealand).

Numerous introductions followed.[56] Goats (*Capra hircus*), pigs (*Sus scrofa*) and rabbits (*Oryctolagus cuniculus*) came with Cook, and house cats (*Felis catus*) probably about the same time. Cattle (*Bos taurus*), sheep (*Ovis aries*) and horses (*Equus caballus*) arrived in the early 1800s. House mice (*Mus musculus*) were present from about the 1830s, and ship rats (*Rattus rattus*, also known as black rats) were widespread by the 1870s. During the 1860s acclimatisation societies began to promote further introductions of animals deemed as 'useful', and the pace of introductions increased.[57] By the 1990s, 34 species of land mammals, 33 species of birds, one species of lizard, three species of frogs, 20 species of freshwater fish, about 1500 species of insects and about 1600 species of plants had established in New Zealand as alien species (deliberately or accidentally).[58] These figures include species introduced by Polynesians, but the vast majority have established only since European arrival.

From a modern biological perspective, it is tempting to view many of the deliberate introductions (of predatory mammals in particular) as ecological sabotage. Yet, the settlers' motivations (for food, gifts, industry, agriculture, recreational sports and remembrance of 'Home' in the 'Britain of the South', as New Zealand was for a time known) were hardly malicious, and even biologists supported some importations. Fortunately for tuatara, attempts to establish mongooses and foxes, two known predators of reptiles, were unsuccessful.[59]

Fig. 3.8 Lighthouses on islands enhanced human safety at sea, but at the cost of ecological losses on land. This view from 1990, looking to the east, shows the lighthouse on Cuvier Island (Repanga Island), which was staffed from 1889. Pacific rats were already present and feral cats (*Felis catus*) (known predators of tuatara) soon became numerous. By 1914 the keepers (by then appointed as Special Protectors of the tuatara) were attempting to control both, but tuatara were already scarce, and cats and rats remained present for decades.

Photo: Alison Cree.

Many of the introduced mammals probably had little initial impact on tuatara because mainland populations of the reptiles were already extinct, or close to it. However, European mammals eventually reached some of the offshore islands that still housed tuatara and contributed to further losses of populations, as described below.

Tuatara and lighthouses on offshore islands

One group having a particular association with tuatara from the late 1800s were the keepers of the now-automated Marine Department's lighthouses. Safety at sea was paramount for the new nation, with its long and dangerous coastline. Lighthouse stations were constructed on three islands on which tuatara survive today. With dates of first operation of the light in parentheses,[60] these were North Brother Island (1877), Cuvier Island (Repanga Island, 1889; Figures 3.8 and 3.9) and Stephens Island (Takapourewa, 1894; Figure 3.10). Lighthouse stations were also established on four islands on which tuatara are now extinct but are known or suspected to have been present at the time (though sometimes as relict populations),[61] namely Tiritiri Matangi Island (1865), Matiu/Somes Island (1866), Burgess Island (Pokohinu, in the Mokohinau Islands, 1883) and East Island (Whangaokeno Island, 1900). An un-staffed gas-fired beacon was established on Coppermine Island, which still has tuatara, from 1913.

Although the construction teams and keepers were remarkably successful in preventing establishment of European rodents on islands with tuatara, cats were another story. Ships carrying construction crews perhaps carried cats to help keep rodents under control (dogs were also sometimes present on islands during lighthouse construction, which could take up to 4 years). Cats became established on several islands with lighthouses,[62] and were noted in keepers' records, as well as newspaper reports of the time, as predators of tuatara. In

Fig. 3.9 While Cuvier Island was a staffed lighthouse station, about a quarter of the island was managed as a farm. The remaining forest was seriously depleted by feral goats (*Capra hircus*) plus wandering cattle (*Bos taurus*) and sheep (*Ovis aries*). By the 1960s goats and cats had been eradicated and roaming stock had been fenced. Since automation of the lighthouse in 1982, Cuvier Island has been managed as a nature reserve. Some houses were removed, while others have been retained for their historic value and use by conservation workers. Since this image was taken in 1990 (looking to the west from near the lighthouse), Pacific rats have been eradicated, shrublands have regenerated and forest cover has enlarged. Cuvier's relict population of tuatara, reduced in the 1990s to about a dozen known adults, is now experiencing a slow recovery, in part with the assistance of captive breeding in zoos (see also Chapters 6 and 10). Photo: Alison Cree.

Fig. 3.10 Like that of Cuvier Island, the ecology of Stephens Island (Takapourewa) was altered radically by the construction of a lighthouse station. Up to three keepers and their families operated the station from 1894 until automation of the light in 1990. The northern slopes of the once forest-clad island were transformed into sheep pasture, while feral cats and unrestrained stock roamed widely elsewhere. About 14 species of native land birds became locally extinct (some globally), but with the elimination of cats by 1925, and in the absence of rodents, tuatara survived. The remnant patches of forest (shown here in 1990) have since been enlarged by re-plantings following the removal of stock.

Photo: Alison Cree.

1904, for example, the botanist Leonard Cockayne described at a gathering of the New Zealand Institute in Wellington how wild cats were 'playing havoc' with tuatara on Stephens Island; it was said that 'wherever a visitor went on the island he would find the heads of these lizards [*sic*]'.[63] On Cuvier and Stephens Islands, cats were eventually eliminated (see below and Chapter 10) and tuatara survive today on both (although only with intensive management on Cuvier, which also had Pacific rats and goats). Cats contributed to the rapid extinction of tuatara on East Island.

Keepers and their families had to be self-sufficient for long periods, sometimes many months. Not surprisingly, natural vegetation was modified substantially, not only to accommodate the lighthouse and dwellings, but also on larger islands to create gardens and pasture for livestock.[64] Pigs (rooting omnivores) were brought to at least two of the lighthouse islands with tuatara,[65] including Burgess Island,[66] where tuatara are now extinct. A pen-housed piglet escaped for about 6 months on Stephens Island; by the time it was re-captured in about 1979 by the Wildlife Service, it had grown to a very large size. At recapture, 'numerous tuatara skins and remains of prion/petrel eggs' were found in its gut: clear evidence of predation on tuatara. Unfortunately for those who caught the pig, the flesh (considered to have been tainted by the tuatara) was inedible.[67]

Goats, sheep and sometimes cattle often roamed freely on lighthouse islands. The damage to natural vegetation contributed to reduced densities of tuatara[68] and greater vulnerability of the remaining tuatara to cats, Pacific rats and predatory native birds such as harriers (*Circus approximans*). Goats probably hastened the extinction of tuatara from Burgess and East Islands, as well as the virtual extinction on Cuvier Island. On Stephens Island in 1904, only 10 years after the light was first exhibited, the bush was disappearing so rapidly that Cockayne reported that 'the island would soon be fairly bare' (Figure 3.11).[69]

BOX 3.1

Story of an extinction: the disappearance of tuatara from East Island

East Island is a small (13-hectare) island lying about 2 kilometres off the North Island at East Cape. Known also as Whangaokeno, the island is composed of a distinctive but unstable white papa clay, and is surrounded by strong tidal rips. The island's connections with tuatara date from pre-history.

In the traditions of Ngāti Porou, tribal ancestors travelled in the canoe *Mangarara* from Hawaiki and passed by Whangaokeno on their arrival. There Wheketoro, the ship's captain, offloaded his cargo of tuatara, geckos, insects and birds and placed a tapu (restriction or taboo) on the island to protect them.[1] Tradition also records that the island was visited at times to harvest seabirds and tutu (*Coriaria* spp.) berries and to till the soil for crops.[2] At some point Pacific rats (*Rattus exulans*) became established.

In 1769 East Island received its European name from Lieutenant James Cook, for whom the island formed an eastern-most marker when the North Island's coastline was charted. By 1898 a string of shipwrecks nearby led the Marine Department to begin constructing a lighthouse. From the start the location was fraught with difficulties: four men drowned in 1899 while bringing material ashore, and one keeper who became 'extremely dangerous and violent' in 1905 had to be lashed up by the other keepers, for fear of them losing their lives.[3] Staffed by three keepers and their families, the lighthouse beamed from the island from 1900 until 1922, when erosion forced its reconstruction on the mainland.

By then the arrival of a lighthouse had brought the final disaster for the island's most famous reptile. International interest in tuatara had expanded dramatically following recognition in 1867, by Albert Günther of the British Museum, that tuatara were sufficiently different from lizards to be placed in their own order of reptiles, the Rhynchocephalia. Soon afterwards, in 1871, the existence of a population on East Island was brought to European attention by Walter Buller. A leading naturalist in New Zealand at the time, Buller drew (in his publication) on correspondence from his brother-in-law Captain Gilbert Mair, a resident of the Bay of Plenty who was familiar with tuatara elsewhere. According to Mair the East Island tuatara were of a 'bright green colour'.[4] Six years later, in another publication, Buller quoted a newspaper report in which it was stated that 'a dozen of large East Cape Island green lizards' had been caught and stuffed by Māori, who received payment of four shillings per specimen. The reptiles ('the ugliest of all creeping things, with the exception of frogs', the newspaper was said to have written) were apparently exported to the Sydney Museum.[5]

Buller commented at the time that the 'green lizard' of East Island was probably the same species of tuatara that he had just described from North Brother Island, *Sphenodon guntheri* (a species no longer recognised as distinct from *S. punctatus*; see Chapter 1 for the taxonomic history).[6] Buller's interest must have been roused, because within 2 years he had described a captive specimen from East Island that had been given to a Mr John White as the gift of a Māori chief. Based on the specimen's short and thickset form, as well as its colour and markings, Buller concluded that the population should be considered a new 'variety' of *S. guntheri*.[7]

As interest in the population grew, its status became more precarious. Legal protection had been introduced for tuatara everywhere in 1895, but this did not guarantee security. In 1898 an anonymous correspondent wrote ominously in the local newspaper, the *Poverty Bay Herald*, that European and Māori visitors had 'destroyed many' and that with 'the advent of lighthousekeepers and their cat and dog pets to East Island, there will probably soon disappear the few tuatara lizards that remain'. One

Box 3.1 Fig. 1 East Island from the mainland (February 1998). The white papa clay contributed to the instability of the lighthouse, which was eventually removed for reassembly on the mainland. This unusual substrate is not seen on other tuatara islands, and its effects on burrowing and nesting by tuatara are unknown. Perhaps it contributed to tuatara there having the thickset form suggested by Buller. The date of first operation of the light was 1900, not 1906 as stated in the sign.

Photo: Alison Cree.

account from 1901 referred to tuatara being collected, as many as 40 at one time, 'for people all over New Zealand.'[8]

By 1913 public concern about the state of tuatara populations had grown, prompting the Minister of Internal Affairs to solicit a report from the Director of the Dominion Museum. The Director, in turn, sought the opinion of lighthouse keepers and others, and collated these for the Minister.[9] One keeper reported that tuatara had been plentiful on East Island in 1900 and that 'a great many were sent away during building operations'. Another reported seeing cats killing them. Keeper Ansin wrote that there was 'hardly a Tuatara to be seen when I left the Island', and attributed the decrease to cats and hawks.[10] A keeper on the island in 1913 reported seeing cats but no tuatara.

Despite public recognition that extinction of the East Island population was imminent,[11] nothing seems to have been done to address the immediate causes. In particular, there was no known attempt to stop cats being taken to the island, nor to appoint the keepers as Special Protectors of the tuatara (which would have required them to kill cats), as was done in 1913 for several other islands with tuatara. Cats had been recognised as a serious predator of tuatara on Stephens Island since the late 1890s, and by 1913 a concerted effort by the keepers had seen them virtually eradicated from that island (which was more than 10 times the area of East Island). Perhaps it was all considered too late for the tuatara of East Island; in any case, there are no known sightings of tuatara there since 1914.[12]

Box 3.1 Fig. 2 (right)Two adult tuatara from East Island, collected by keeper F. S. Edmonds in the early 1900s and now held at the Otago Museum (OM A33: 56). The 'bright green' colour that adults are said to have had in life will have been leached by the preservative.

Photo: Alison Cree.

Box 3.1 Fig. 3 (far right) Tuatara on East Island were still producing eggs at the time of lighthouse construction. Two keepers reported finding nests in the early 1900s, one with a dozen eggs and one with the surprisingly large number of 17 (Thomson 1915). The collector of these eggs, which are now in the Museum of New Zealand Te Papa Tongarewa, is unknown (NMNZ R1115).

Photo: Alison Cree.

It seems that several factors contributed to extinction of tuatara. We know that Pacific rats were present (and would almost certainly have preyed on eggs and hatchlings), that tuatara were collected in their dozens and that, with construction of the lighthouse, cats were introduced and killed tuatara. Goats arrived in 1906 and became feral, causing damage to the vegetation,[13] which probably increased the vulnerability of tuatara to hawks.

Sometime after the disappearance of tuatara, the disturbances were reversed: the lighthouse was removed in 1922, goats were eradicated in the early 1960s and Pacific rats in 1997 (cats had already died out or been removed). Now, the possibility of reintroducing tuatara to East Island is under discussion between local Māori and the Department of Conservation. The ecosystem may thus be restored, but the tuatara will have to come from elsewhere and may not be the same genetically as those that were lost.[14] Perhaps we should not let this trouble us, because the island is slowly eroding into the sea in any case[15] and will not provide a secure habitat indefinitely, especially in the light of global climate change.

Fencing was eventually erected on Stephens to protect some of the remaining bush.[70] Nonetheless, the largest forest remnant was still unfenced and being damaged by stock in 1922.[71] Further criticism surfaced in 1935, this time in an illustrated article in the American *National Geographic Magazine* by Frieda Cobb Blanchard, who had visited Stephens in 1927 with her husband Frank Blanchard, an American herpetologist. According to Frieda Blanchard, 'The lighthouse keepers' sheep wander at large, destroying the undergrowth and in time killing off the bush',[72] a situation that caught government officials by surprise, given that fencing was said to be already in place.[73]

At least partly offsetting the challenges of remoteness, lighthouse islands were exhilarating places for those with interests in natural history. Some of

Fig. 3.11 On Stephens Island the penetrating wind aggravated the damage to forest from free-roaming stock, making tuatara more vulnerable to birds of prey. Remnant native forest was eventually fenced, and has subsequently regenerated considerably.

Source: View of Stephens Island, 1930s, Alexander Turnbull Library, Wellington, New Zealand, Royal Forest and Bird Protection Society of NZ Collection, PAColl-3295-1-27.

the keepers (aided by their children) supplied early information on behaviour of tuatara (Figure 3.12). On Stephens Island during the late 1800s, the keepers collected tuatara or their eggs for others to study, and hosted visits by early scientists (Chapter 5). Captains of the vessels servicing the lighthouse islands assisted with the delivery of tuatara for study and display. Captain John Fairchild (Figure 3.13), master of the government steamers the *Luna* from 1871–1877, then variously the *Hinemoa* and the *Stella* until 1895 and the *Tutanekai* from 1896 to 1898,[74] is often mentioned in connection with tuatara.[75] On 9 February 1882 the *New Zealand Herald* wrote that 'Captain Fairchild rarely visits this port [Auckland] without having some natural curiosities on board' – on that occasion, he had 19 tuatara from Karewa Island.[76] Fairchild's modus operandi on the Hen and Chickens Islands (probably Coppermine Island, more specifically) was to dig a deep pit near where tuatara were found; the next time he visited, he would collect whatever specimens were unlucky enough to have fallen in.[77] Fairchild delivered 20 tuatara to the Sydney Museum in 1892,[78] and supplied at least six to the Auckland Museum, according to a register of 1909.[79] Buller reported in 1877 that Fairchild had collected 'a batch' of tuatara from the Brothers Islands during construction of the lighthouse (these were later forwarded to England by Sir Julius Vogel, the Premier of New Zealand).[80] Buller, a keen local naturalist, himself received tuatara from Fairchild on at least one occasion.[81]

Fig. 3.12 Two keepers and their wives on Stephens Island in the 1930s. Prolonged stays on a lighthouse station could pose challenges to mental health. The presence of tuatara was an asset for those keepers and family members with an interest in natural history.

Source: Alexander Turnbull Library, Wellington, New Zealand, Royal Forest and Bird Protection Society of NZ Collection, PAColl-3295-1-04.

Fig. 3.13 Captain John Fairchild, Master of four government steamers from 1871 until 1898, had an intimate association with lighthouses. Through them he developed a connection with tuatara, often transporting individuals from islands to the mainland, prior to legal protection in 1895.

Source: Portrait by John Nicol Crombie, c. 1860, Alexander Turnbull Library, Wellington, New Zealand, PA2-0567.

The chief engineer on the *Hinemoa* in 1890 was said to have kept a pet tuatara on board.[82]

Captain John Bollons (Figure 3.14) was Fairchild's successor as master of the *Hinemoa,* from 1898 to 1922; he then captained the *Tutanekai* until around the time of his death in 1929. He is said to have followed Fairchild's pit-fall technique for collecting tuatara from Coppermine Island.[83] In 1913, in response to calls from the public and some lighthouse keepers for better protection of tuatara, Bollons was requested by the Minister of Internal Affairs to transfer eggs from Stephens Island to several other islands. If implemented, these early translocations for conservation were, by modern standards, flimsily conceived (Chapter 10). Bollons later supplied information to officials about the existence of tuatara on other islands, including some without lighthouses (Table 3.1).

Also in 1913 several of the keepers were appointed as Special Protectors of tuatara.[84] Among their duties (discussed further in Chapter 10) was to report on whether the numbers of tuatara appeared to be increasing or decreasing. Some reports provided useful information about predation by cats and about the past presence of tuatara on other lighthouse islands.[85] However, the quality varied widely, especially concerning estimates of abundance. Although later publicity material from the Marine Department claimed that '[p]ractically all the early information [on tuatara] was supplied by these men',[86] this was an exaggeration in light of various studies that had been underway on offshore islands since the late 1890s (Chapter 5). Augustus Hamilton, the Director of the Dominion Museum, collated the keepers' reports for the Minister of Internal Affairs in 1913, concluding with some frustration that 'The letters from these persons do not give very much information and it is extraordinary how little they appear to have noticed things.'[87]

Although the keepers may not have fulfilled the expectations of professional scientists, many showed affection for these 'truly wonderful reptiles',[88] as keeper D. J. Grindlay on Cuvier Island described them. Some keepers put out water dishes for tuatara to bathe in, or fed the tuatara by hand. One keeper on the Brothers in 1928 reported that 'on one late occasion a Tutara [*sic*] came inside the dwelling, just after dark ... and by its instinct made a course for the fire ... When requested to leave, it was evidently half asleep in the enjoyment of the heat.'[89] Grindlay wrote in 1919 of observing a tuatara ('Jimmie') in his garden, which could 'be seen every evening at sunset going into the creek for his well loved and well earned bath ... I have stood talking to this reptile while he has been bathing for fully half an hour.'[90]

Some of the keepers also assisted with early initiatives to reduce the impacts of predators. In a sustained and well-organised effort beginning from 1897,[91] the keepers on Stephens Island had eradicated cats by about 1925.[92] Harriers, a native hawk, were killed in the hundreds on Stephens Island during c. 1914–1915[93] and on North Brother Island during 1936–1943.[94] A warrant was also issued to the keepers on Stephens Island to kill kingfishers (kōtare, *Todiramphus sanctus*) for a 3-month period in 1924.[95] Although harriers and kingfishers were known predators of tuatara,[96] their impacts were exacerbated by clearance of native vegetation and they are no longer considered pests (the bounty for shooting harriers was discontinued in 1929, in part a response to better fencing that was considered to give tuatara the benefit of remnant forest cover[97]). Some of the keepers on Cuvier Island trapped and poisoned rats during the 1920s with the intention of assisting tuatara,[98] but this attempt appears to have been localised and short lived. Ammunition was supplied for killing goats on Cuvier; Grindlay wrote in 1919 that he had 'lost count' of the number he had killed (over 50).[99] In later decades keepers on Stephens and North Brother Islands assisted in various ways with government and university-based research on tuatara. The government tenders the *Enterprise* and the *Enterprise II*, and in later years the helicopters that assisted with deliveries, became valued means of transportation for scientists.

By 1990 all of the lighthouse stations on islands with tuatara had been automated, ending a unique period of human–tuatara association. Today most lighthouse islands are uninhabited by humans. Many have had their reserve status upgraded, allowing more-intensive management, including revegetation and the removal of the remaining introduced mammals, by the Department of Conservation in consultation with local Māori.

Fig. 3.14 Captain John Bollons, Master of two government steamers from 1898 until about 1929, continued the connection begun by Fairchild between the Lighthouse Service and tuatara. Highly regarded for his seamanship and crew management, Bollons also had interests in natural history. He supplied information to government officials about islands on which tuatara then survived; he may also have assisted with early translocations of eggs between islands.

Source: Captain John Peter Bollons, and an unidentified woman, on board the ship *Tutanekai* at Akaroa, November 1923, Alexander Turnbull Library, Wellington, New Zealand, ref. 1/2-112289-F, PAColl-5479.

Net declines in distribution of tuatara following human settlement

By the late 1900s evidence for the past presence of tuatara had been reported for at least 26 offshore islands on which tuatara apparently no longer exist (Table 3.1; Figure 3.15; Appendix 3). Nonetheless, the number and timing of extinctions are imprecise for various reasons. As mentioned above for bones in middens, we cannot rule out the possibility that animals were sometimes transported from elsewhere. In the late 1900s two large tuatara were reported to be living under the house of a kaumātua (Māori elder) on Motiti Island, but these specimens are known to have been brought from Motunau Island (Plate Island) to live as pets.[100] Transfers from nearby Stephens Island could account for 19th- and 20th-century reports of tuatara on Rangitoto ke te Tonga (D'Urville Island), just as transfers are known to have been made from Stephens to the nearby mainland at French Pass (see Chapter 5, Box 5.2). In the late 1800s a lone tuatara known as Nga uri apo was photographed living under a sacred pōhutukawa tree of the same name on Mayor Island (Tuhua) (Figure 3.16). While the animal is unquestionably a tuatara, there is

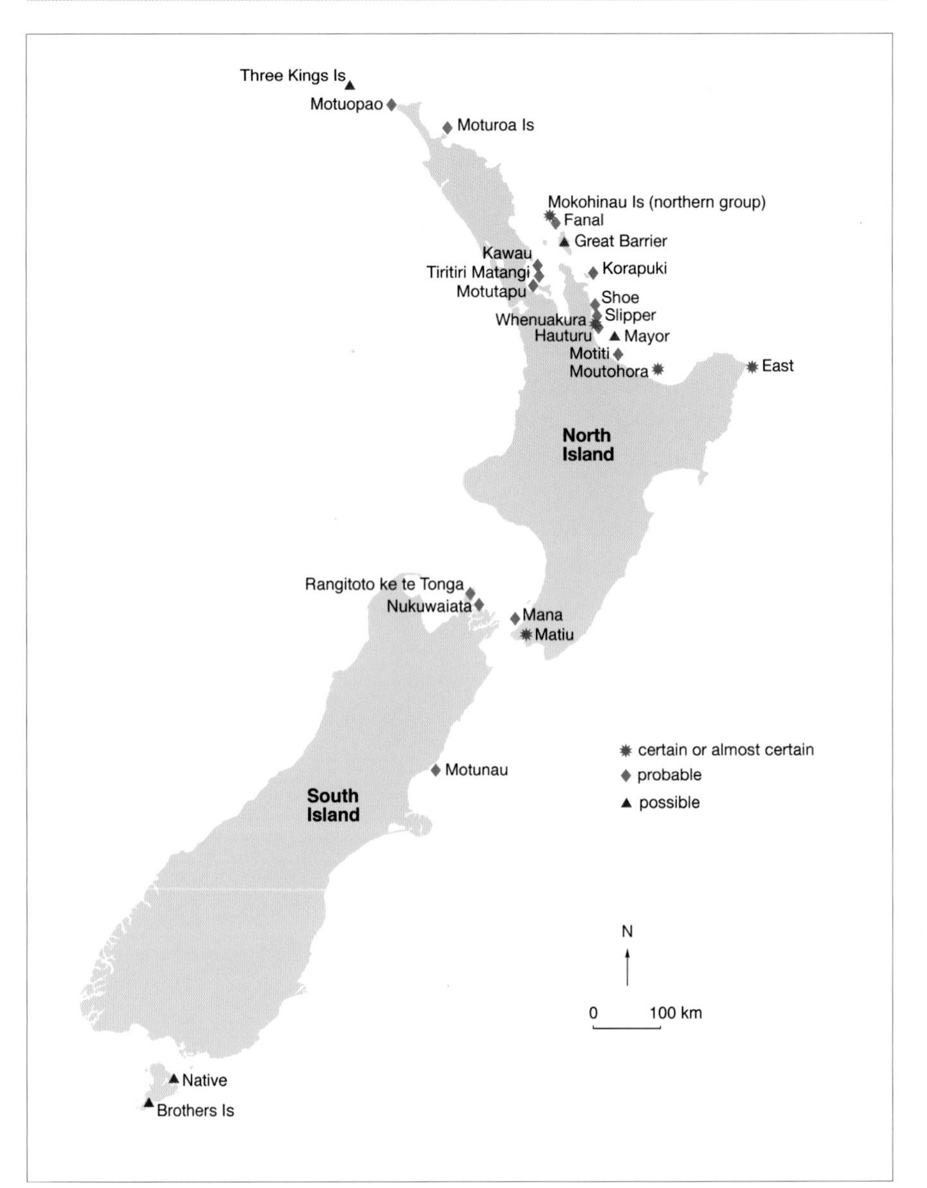

Fig. 3.15 Map showing New Zealand islands (in addition to the North and South Islands) on which tuatara may have become extinct since human arrival, based on reported sightings or evidence from biological remains. The likelihood of past presence and extinction is rated in terms of the quality of evidence available. 'Is' (for islands) indicates that tuatara were implied or known to be present on more than one island in the group. Tuatara probably occurred on many additional islands for which no direct evidence of a population remains. For details, see text and Table 3.1.

some doubt about whether a population could have survived on this volcanic island following a major eruption that occurred about 6340 years ago.[101] Thus, the photographed animal possibly represents a specimen shifted from elsewhere.[102]

In other cases, island identities may have become confused. As a potential example, although two sets of islands known as the Brothers Islands are reported sources of tuatara, one near Stephens Island (in Cook Strait) and one near 'Stewart's Island' (in Southland), only the former is a confirmed source.[103] Other islands, such as Atihau in the Mokohinau Islands, have had multiple English names, formal and informal. There are also islands on which tuatara have been reported since the late 1800s but where extinction is unconfirmed (access is difficult and tuatara may yet survive, albeit in low numbers). These are not included in Table 3.1, but are mentioned in Chapter 6.

Given that evidence for the past presence of tuatara on islands is of varying quality (Table 3.1), I have estimated the likelihood of a past population using categories of 'certain or almost certain', 'probable' and 'possible' (Figure 3.15).[104] These assessments take into account the number and nature of observations, whether specimens exist, whether the island is within the known latitudinal range for island populations, and whether the island would have been connected to the mainland during the last ice age and provided viable habitat since. Excluding the North and South Islands (for which extinction is certain: see Chapter 2), at least six island populations are here considered as certain or almost certain extinctions, at least fifteen as probable extinctions and at least five as possible extinctions.

As Ian Crook pointed out in 1973, there is a high probability that tuatara have become extinct on other islands during the human period but for which no definite evidence is recorded. One example is Middle Island in the Aldermen Islands.[105] This island is centred within a cluster of seven other islands, all of which have tuatara but no rats. Middle Island, by contrast, had a population of Pacific rats until recently (they were eradicated by 1993[106]). Thus, the circumstantial evidence suggests that tuatara became extinct on Middle Island because of the presence of rats. Double Island (Moturehu) in the Mercury Islands and Rurima Island have been considered by Crook and others as further examples.[107] Several islands (Victory (Moutiti), Titi, Motuara, Motungarara, Long, Allports and Mabel) were recently reported by Māori elders as past or present habitat of tuatara in the Cook Strait–Marlborough Sounds region, based on traditional ecological knowledge;[108] although surviving natural populations of tuatara are unlikely, all are plausible sites for past populations.[109]

Of the predators introduced following European settlement, Norway rats and cats were almost certainly the most destructive to tuatara (fortunately, mustelids never established on islands with tuatara). Norway rats are strongly implicated in about five probable or known extinctions, and cats in at least two (cats also temporarily threatened two populations that survive today). By preying on small seabirds and invertebrates, Norway rats and cats would also have reduced the food supply for tuatara. There are, however, at least three apparent extinctions for which there is no evidence to implicate rodents.[110] Possible explanations are that tuatara were never actually present, that rodents

Fig. 3.16 A solitary tuatara from Mayor Island (Tuhua). This tuatara was known as Nga uri apo ('descendant of the past') after the sacred pōhutukawa tree of the same name near where it lived. It was photographed by E. C. Gold-Smith, the District Surveyor from Tauranga (Gold-Smith conducted a survey of Mayor Island in 1884, although his published account did not mention this tuatara). Such isolated sightings create uncertainties for inferring the existence of a natural population, as tuatara have sometimes been shifted between islands as companion animals.

Source: Prebble, G. K. 1971. *Tuhua – Mayor Island*. Ashford-Kent, Tauranga. Hocken Collections, Uare Taoka o Hākena, University of Otago, S10–550a.

were present but died out, or that other factors (such as the introduction of pigs and/or rabbits, habitat alteration from human activities, or the genetic and demographic consequences of small population size) contributed to the extinction of tuatara.

The most recent extinction of an entire population of tuatara occurred in the early 1980s, when Norway rats reached Whenuakura Island, a tiny (2-hectare) island near Whangamata at the base of the Coromandel Peninsula (Figures 3.17, 3.18). The island's tuatara were already under study by Donald Newman of the Wildlife Service following the arrival of mice, possibly within the previous five years. During a survey in 1980, a total of 110 tuatara, including juveniles, were observed.[111] But Norway rats arrived about two years later, and by 1985, no tuatara could be found. Whenuakura Island is only about 800 metres offshore, and connected by water gaps of no more than about 400 metres from other islands even closer to shore. The closest, Hauturu Island (not to be confused with Hauturu/Little Barrier Island), is connected by a causeway to the mainland at low tide. Norway rats were present on Hauturu Island by 1980,[112] and probably swam from there to Whenuakura Island, perhaps via intervening islets (Norway rats can cross water distances of 400 metres or more[113]). Norway rats and mice were eradicated from Whenuakura Island during 1984–1985, and from Hauturu Island in 1994.[114] However, tuatara have not been seen on Whenuakura or Hauturu Islands since, and

Fig. 3.17 Whenuakura Island, site of the most recent extinction of a population of tuatara. Despite lying only about 800 metres from Otahu Beach at Whangamata (and connected to it by gaps of no more than about 400 metres across water), Whenuakura Island maintained a small population of tuatara until the early 1980s. Following the arrival of Norway rats (*Rattus norvegicus*), the tuatara disappeared.

Photo: Donald Newman.

Norway rats have reinvaded Whenuakura Island, at least.[115] Although the extinction event happened too rapidly for interactions between tuatara and Norway rats to be observed, the loss of tuatara from Whenuakura (and probably from Hauturu Island as well: see Table 3.1) provides circumstantial evidence for vulnerability. No population of tuatara survives today in the presence of Norway rats.

Although direct evidence for the effects of rats on wild tuatara is scant, circumstantial evidence for vulnerability comes from recent observations in a captive colony. In spring of 2007 a search for 10 tuatara in an outdoor enclosure on the mainland revealed only one tuatara alive. The animals had been left undisturbed over winter, during which time it appears that one or more rats discovered a gap in the roof left by builders. Although rats were not sighted, the elevated location of the numerous droppings left around the gap suggests that one or more ship rats had entered. Skeletons of tuatara (one from an adult female, and others from juveniles aged 3–4 years) were found in their burrows, the adult with some skin still left on the head.[116] This sad event emphasises the vulnerability of ectothermic tuatara to rodents. Rats, being endotherms, maintain a high demand for food energy while active during the winter, and seemingly killed their prey while the tuatara were torpid, and least able to defend themselves.

Summary: Humans as 'future eaters'

Isolated in the South Pacific, New Zealand was one of the last major landmasses to be discovered by humans. Yet, late settlement – the span of only about 15 human lifetimes – has been no guarantee of wise stewardship. The collapse of the megafauna that followed human settlement is well known, and parallels that documented for many other landmasses of the world. The propensity for humans to cause the widespread extinction of large vertebrates is one example of what renowned ecologist Tim Flannery has termed 'future eating' – the consumption and depletion of resources that will be required in the future.[117] Less widely appreciated are the effects of humans and their commensal mammals on small or medium-sized vertebrates such as tuatara, and on invertebrates and vegetation.

Prior to human arrival tuatara were distributed over much of the North and South Islands as well as on about 57 or more offshore islands. Following human settlement the land occupied by tuatara declined over about 700 years to about 32 offshore islands, an area estimated as probably less than 0.5% of that occupied before human arrival.[118] About 18 of the reported extinctions from offshore islands have occurred within the past 150 years. The current relict distribution of tuatara is that of an 'island pseudo-endemic' – an organism restricted today to offshore islands not because it evolved there, but because of mainland extinction.[119] Protected offshore islands continue to be an essential refuge against further decline, and a reservoir from which translocations to areas of past occupancy can be attempted.

Fig. 3.18 A rare photograph of a tuatara from Whenuakura Island. The population became extinct in the 1980s.

Photo: Klaus Henle.

Although this chapter has separated the waves of human impact on a chronological basis, this is not an attempt to weigh blame among different cultural groups. The different waves of settlement have, in a geological timeframe, been essentially simultaneous, and would have been devastating whichever group of humans arrived first.[120] Much could still be learnt about the sequence of events. For example, research on the dates, quantity and features of the bones of tuatara in middens could help refine the estimated time of disappearance from the North and South Islands. Stable isotope analyses might give clues as to the diets of tuatara from different locations, including changes over time in the contribution from seabirds.[121]

Fortunately for tuatara there is evidence that, over time, humans become 'ecologically attuned', or reconciled to the ecological limitations of their land.[122] The extent to which conservation in New Zealand has come to ensure a more certain future for tuatara, including reintroduction as part of ecological restoration of island and mainland habitats, is the subject of Chapter 10.

Endnotes to Chapter 3

See the bibliography for full references to articles on tuatara, which are cited here at first mention by surname and year only. References on other topics are cited here in full at first mention, and then using 'op. cit.'. Except where required for clarity, the convention of parentheses around years has been dispensed with for economy of space.

1 Worthy and Holdaway 2002, p. 529.

2 Caughley, G. 1989. New Zealand plant–herbivore systems: past and present. In: Rudge, M. R. (ed.). Moas, mammals and climate in the ecological history of New Zealand. *New Zealand Journal of Ecology* 12 (supplement): 3–10.

3 Holdaway, R. N. 1989. New Zealand's pre-human avifauna and its vulnerability. In: Rudge, M. R. (ed.). Moas, mammals and climate in the ecological history of New Zealand. *New Zealand Journal of Ecology* 12 (supplement): 11–25. At the time of publication, Holdaway recognised these as three waves, the arrival of rats then being seen as greatly preceding settlement by humans.

4 Sutton, D. G. 1994. Preface. In: Sutton, D. G. (ed.). *The origins of the first New Zealanders.* Auckland University Press, Auckland, pp. 1–18; Wilmshurst, J. M., Hunt, T. L., Lipo, C. P. and Anderson, A. J. 2011. High-precision radiocarbon dating shows recent and rapid initial human colonization of East Polynesia. *Proceedings of the National Academy of Sciences* 108: 1815–1820.

5 McGlone, M. S., Anderson, A. J. and Holdaway, R. N. 1994. An ecological approach to the Polynesian settlement of New Zealand. In: Sutton, D. G. (ed.). *The origins of the first New Zealanders.* Auckland University Press, Auckland, pp. 136–163. See p. 139 for quote.

6 McGlone et al. 1994, op. cit.

7 Lowe, D. J., Newnham, R. M., McFadgen, B. G. and Higham, T. F. G. 2000. Tephras and New Zealand archaeology. *Journal of Archaeological Science* 27: 859–870; Wilmshurst, J. M., Anderson, A. J., Higham, T. F. G. and Worthy, T. H. 2008. Dating the late prehistoric dispersal of Polynesians to New Zealand using the commensal Pacific rat. *Proceedings of the National Academy of Sciences* 105: 7676–7680; Wilmshurst et al. 2011, op. cit.

8 Bunce, M., Worthy, T. H., Phillips, M. J., Holdaway, R. N., Willerslev, E., Haile, J., Shapiro, B., Scofield, R. P., Drummond, A., Kamp, P. J. J. and Cooper, A. 2009. The evolutionary history of the extinct ratite moa and New Zealand Neogene paleogeography. *Proceedings of the National Academy of Sciences* 106: 20646–20651.

9 Smith, I. 2005. Retreat and resilience: fur seals and human settlement in New Zealand. In: Monks, G. G. (ed.). *The exploitation and cultural importance of sea mammals. Proceedings of the 9th Conference of the International Council of Archaeozoology, Durham, August 2002.* Oxbow Books, Oxford, pp. 6–18.

10 Worthy and Holdaway 2002, p. 540.

11 Higham, T., Anderson, A. and Jacomb, C. 1999. Dating the first New Zealanders: the chronology of Wairau Bar. *Antiquity* 73: 420–427; see also discussion in McGlone et al. 1994, op. cit.

12 Matisoo-Smith, E., Roberts, R. M., Irwin, G. J., Allen, J. S., Penny, D. and Lambert, D. M. 1998. Patterns of prehistoric human mobility in Polynesia indicated by mtDNA from the Pacific rat. *Proceedings of the National Academy of Science USA* 95: 15145–15150.

13 Holdaway 1989, op. cit.

14 Wilmshurst, J. M., Higham, T. F. G., Allen, H., Johns, D. and Phillips, C. 2004. Early Maori settlement impacts on northern coastal Taranaki, New Zealand. *New Zealand Journal of Ecology* 28: 167–179.

15 Holdaway, R. N. 1999. Introduced predators and avifaunal extinction in New Zealand. In: McPhee, R. D. E. (ed.). *Extinctions in near time: causes, contexts, and consequences.* Kluwer Academic/Plenum Publishers, New York, pp. 189–239.

16 Worthy and Holdaway 1995, 1996; Worthy 1997, 1998a.

17 See, for example, Crook 1973a; Cree et al. 1995a.

18 Towns et al. 2007.

19 Holdaway, R. N. 1996. Arrival of rats in New Zealand. *Nature* 384: 225–226.

20 Wilmshurst et al. 2008, op. cit.

21 Brook, F. J. 2000. Prehistoric predation of the landsnail *Placostylus ambagiosus* Suter (Stylommatophora: Bulimulidae), and evidence for the timing of establishment of rats in northernmost New Zealand. *Journal of the Royal Society of New Zealand* 30: 227–241.

22 Wilmshurst, J. M. and Higham, T. F. G. 2004. Using rat-gnawed seed to independently date the arrival of Pacific rats and humans in New Zealand. *The Holocene* 14: 801–806; Wilmshurst et al. 2008, op. cit.

23 However, if older reports used less distinctive bones, species identity may sometimes have been confused with that of other small vertebrates (Paul Scofield, Canterbury Museum, pers. comm. 12 August 2002).

24 Material in dunes, originally reported as coming from a midden, must be considered as perhaps from a natural deposit. For example, bones from dunes at Wainui, near Gisborne, may not be from a midden as reported by Hutchinson (1898) (Trevor Worthy, Palaeofaunal Surveys, Nelson, pers. comm. 4 October 1999). Given this and the point made in the preceding footnote, a detailed reassessment of material reported as the bones of tuatara from middens would be valuable.

25 Millener 1981; Worthy and Holdaway 2000.

26 Crook 1975.

27 Crook, I. and Moran L. Unpublished manuscript made available by Les Moran to the author (13 October 1999).

28 Chambers, P. 2004. *A sheltered life: the unexpected history of the giant tortoise.* John Murray, London, p. 94.

29 Leahy, A. 1976. Whakamoenga Cave, Taupo, N94/7: a report on the ecology, economy and stratigraphy. *Records of the Auckland Institute and Museum* 13: 29–75.

30 Clark, G. 2005. Kuri. In: King, C. M. (ed.). *The handbook of New Zealand mammals.* Oxford University Press, Auckland, pp. 255–260.

31 Donne 1942; B. Sladden to Director of the Dominion Museum, 9 April 1929, ANZ IA 1 52/30 pt 1.

32 A. Duncan, 30 April 1913, in Thomson 1915.

33 Browne 2005.

34 McGlone et al. 1994, op. cit.; McWethy, D B., Whitlock, C., Wilmshurst, J. M., McGlone, M. S., Fromont, M., Li, X., Dieffenbacher-Krall, A., Hobbs, W. O., Fritz, S. C. and Cook, E. R. 2010. Rapid landscape transformation in South Island, New Zealand, following initial Polynesian

settlement. *Proceedings of the National Academy of Sciences* 107: 21343–21348.

35 McGlone et al. 1994, op. cit.; McGlone, M. S. and Wilmshurst, J. M. 1999. Dating initial Maori environmental impact in New Zealand. *Quaternary International* 59: 5–16.

36 McGlone and Wilmshurst 1999, op. cit.

37 Anderson, A. 1989. *Prodigious birds: moas and moa-hunting in prehistoric New Zealand*. Cambridge University Press, Cambridge.

38 Anderson, A. 2000. Defining the period of moa extinction. *Archaeology in New Zealand* 43: 195–201; Holdaway, R. N. and Jacomb, C. 2000. Rapid extinction of the moas (Aves: Dinornithiformes): model, test, and implications. *Science* 287: 2250–2254.

39 Smith 2005, op. cit.; Nagaoka, L. 2001. Using diversity indices to measure changes in prey choice at the Shag River mouth site, southern New Zealand. *International Journal of Osteoarchaeology* 11: 101–111.

40 Caughley 1989, op. cit.

41 Holdaway 1989, op. cit.; Holdaway 1999, op. cit.

42 Higham et al. 1999, op. cit.

43 Nagaoka 2001, op. cit.

44 McGlone et al. 1994, op. cit.

45 Hamel, G., Allen, R., Davis, L., McGovern-Wilson, R., Petchey, P. and Smith, I. 2003. The human factor. In: Darby, J., Fordyce, R. E., Mark, A., Probert, K. and Townsend, C. (eds). *The natural history of southern New Zealand*. University of Otago Press, Dunedin, pp. 129–151.

46 Holdaway 1999, op. cit.

47 Hawke, D. J., Holdaway, R. N., Causer, J. E. and Ogden, S. 1999. Soil indicators of pre-European seabird breeding in New Zealand at sites identified by predator deposits. *Australian Journal of Soil Research* 37: 103–113.

48 Towns and Daugherty 1994.

49 A bone from Earthquakes in the South Island is dated at 500 years before present (BP, where present is by convention 1950), with a standard error of ± 100 years, in supplementary material of Wilmshurst et al. 2008, op. cit. Slightly older most-recent dates (649–659 years BP, ± 30–50 years) for bones from other sites in the North and South Islands are presented in the supplementary material of Hay et al. 2008.

50 Flannery, T. F. 1994. *The future eaters*. Reed Books, Chatswood, New South Wales, p. 248.

51 Jacobs, W. and Wilson, J. 1995. *The birth of New Zealand: a nation's heritage*. Kowhai Publishing, Auckland.

52 Statistics New Zealand. 1997. *New Zealand official yearbook. 100th edition*. GP Publications, Wellington.

53 In 2010 those who identified as Māori accounted for 16% of the population.

54 Leathwick, J., Wilson, G., Rutledge, D., Wardle, P., Morgan, F., Johnston, K., McLeod, M. and Kirkpatrick, R. 2003. *Land environments of New Zealand: Nga taiao o Aotearoa*. David Bateman, Auckland.

55 Beaglehole, J. C. 1955. *The voyage of the Endeavour 1768–1771*. Vol. 1. Published for the Hakluyt Society at the University Press, Cambridge. See p. 248.

56 For dates of arrival, see Atkinson, I. A. E. 2001. Introduced mammals and models for restoration. *Biological Conservation* 99: 81–96.

57 Atkinson, I. A. E. and Cameron, E. K. 1993. Human influence on the terrestrial biota and biotic communities of New Zealand. *Trends in Ecology and Evolution* 8: 447–451; McDowall, R. M. 1994. *Gamekeepers for the nation*. Canterbury University Press, Christchurch.

58 Atkinson and Cameron 1993, op. cit.; Atkinson 2001, op. cit.

59 McDowall 1994, op. cit.

60 Dates for first operation (except for Matiu/Somes) from Beaglehole, H. 2006. *Lighting the coast: a history of New Zealand's coastal lighthouse system*. Canterbury University Press, Christchurch.

61 Tuatara have since been reintroduced to Tiritiri Matangi Island and Matiu/Somes Island; see Chapter 10.

62 Exactly who introduced the cats has sometimes been a thorny issue, especially for Stephens Island where cats famously contributed to the global extinction of a flightless wren (*Traversia lyalli*). According to one retrospective account, cats were introduced to Stephens by 'one of the men employed in building the lighthouse', apparently a factor in the Marine Department's later agreement to contribute to a bounty for every cat shot (see R. McNab to The Hon. Mr Russell, 14 July 1916, ANZ M1 25/611 pt 1). However, a detailed consideration of additional sources led to the conclusion that the feral population may have established from a single pregnant cat landed in early 1894, perhaps after the lighthouse was completed. Whether or not the keepers were responsible, the existence of the lighthouse station was clearly the initiating factor that led to the arrival of cats. See Galbreath, R. and Brown, D. 2004. The tale of the lighthouse-keeper's cat: discovery and extinction of the Stephens Island wren (*Traversia lyalli*). *Notornis* 51: 193–200. See also Medway, D. G. 2004a. The land bird fauna of Stephens Island, New Zealand in the early 1890s, and the cause of its demise. *Notornis* 51: 201–211.

63 Anon. 1904, p. 4.

64 North Brother Island was an exception: its small, rocky nature made gardening and farming impractical, so food supplies were shipped in and it operated as a station for 'single men' only.

65 Tuatara were, however, able to survive the presence of pigs from about the 1820s until 1936 on Aorangi Island in the Poor Knights Islands (see Chapter 6); Aorangi had no lighthouse.

66 Cameron, E. K. 1990. Vascular plants of the main northern Mokohinau Islands, north-east New Zealand. *Tane* 32: 113–130.

67 Primary information and quote supplied by John Cheyne (pers. comm. 10 August 2004), based on a conversation with Kerry Horgan, who captured the pig (Cheyne and Horgan worked for the Wildlife Service). Supporting information supplied by Donald Newman, Department of Conservation (pers. comm. 27 October 2004). See also Brown [2000].

68 For example, on Stephens Island by the 1980s–2000s, density of tuatara in sheep pasture was less than half that in forest (see Chapter 7 for further discussion, and Appendix 4 for references).

69 Anon. 1904.

70 For example, 'this Department [Internal Affairs] took initiatives to have some of the bush fenced off to prevent destruction by cattle'; P. J. Kelleher to the Secretary, Public Service Commissioner, 29 March 1932. ANZ M1 25/611 pt 3.

71 Medway 2004a, op. cit.

72 Blanchard 1935, p. 657.

73 J. Heenan to the Secretary for Marine, 25 January 1937. ANZ M 1 25/611 pt 3.

74 Beaglehole 2006, op. cit. See also Martin, E. R. 1969. *Marine Department centennial history 1866-1966*. Government Printer, Wellington.

75 For example, Anon. 1883.

76 Anon. 1882b.

77 Cowan 1908.
78 Anon. 1892.
79 According to the museum's first register of land vertebrates (copy supplied by Brian Gill, Auckland Museum).
80 Buller 1877.
81 Buller 1893.
82 Anon. 1931.
83 Skegg, P. D. G. 1965. Historical notes on the offshore islands recently visited by A. U. Field Club. *Tane* 11: 93–98.
84 The caretaker on Hauturu/Little Barrier Island, which had no lighthouse but had been declared a sanctuary for flora and fauna, was also recognised in this way.
85 See Thomson 1915 for a compilation. See also Sandager 1890.
86 Marine Department. [c. 1966]. Marine Department's work in protection of tuataras. *Press* Feature, p. 3.
87 A. Hamilton to Minister of Internal Affairs, 15 May 1913. ANZ IA 1 46/18/4 pt 1.
88 D. J. Grindlay to Under Secretary for Internal Affairs, 17 March 1919. ANZ IA 1 46/18/8 pt 1.
89 W. Marshall to Secretary of Internal Affairs, 1 July 1928. ANZ M 1 25/611 pt 2.
90 D. J. Grindlay to Under Secretary for Internal Affairs, 17 March 1919. ANZ IA 1 46/18/8 pt 1.
91 Feral cats were observed on Stephens Island by 1895 and the keepers began to shoot them from 1897. Further concern was raised after a visit by the Governor in c. 1901, and (as recommended by the Principal Keeper) visiting mailmen were thereafter paid to shoot cats. The keepers had also been supplied with guns and ammunition and later received payments for each cat killed, in response to further pressure on the Marine Department (see Chapter 10). From 1905 a supplement for each cat killed was added from the Tourist Department. Files in ANZ M 1 25/611 pt 1 Tuatara lizards 1901–1922; see also Medway 2004a, op. cit.
92 By 1912 over 700 cats were reported to have been destroyed (H. G. Ell to F. H. Bell, 1 November 1912. ANZ IA 1 46/18/4 pt 1). In his reply (6 November 1912), Bell noted that by this time 'there are scarcely any cats to be found on the island', though keepers were still being paid for each one killed. See also Medway 2004a, op. cit.
93 Thomson 1915.
94 Keeper to S. F. Marriott, Secretary of Marine Department 7 September 1944. ANZ IA 46/18/4 pt 1.
95 Secretary of Marine Department to Principal Keeper, 15 March 1924. ANZ M 1 25/611 pt 2 1923–1928.
96 For harriers see, for example, Buller 1879; Thomson 1915 (and Chapter 7 for further discussion). For kingfishers see Medway, D. G. 2004b. The land bird fauna of Stephens Island, New Zealand 1915–1933. *Notornis* 51: 229–230.
97 For example, 'this Department took initiatives to have some of the bush fenced off to prevent destruction by cattle'; P. J. Kelleher to the Secretary, Public Service Commissioner, 29 March 1932. ANZ M1 25/611 pt 3.
98 G. W. Brown to the Secretary of Marine, 15 February 1924. ANZ IA 1 46/18/8 pt 1.
99 D. J. Grindlay to Under Secretary for Internal Affairs, 17 March 1919. ANZ IA 1 46/18/8 pt 1.
100 John Heaphy, Department of Conservation, pers. comm. 6 August 2003 and 17 August 2005.
101 Buck, M. D. 1985. An assessment of volcanic risk on and from Mayor Island, New Zealand. *New Zealand Journal of Geology and Geophysics* 28: 283–298.
102 One possibility includes small islands surrounding Mayor Island and reported to have tuatara, such as Motuoneone (see Chapter 6).
103 To complicate the matter further, there appear to be two sets of islands known as the Brothers Islands near Stewart Island/Rakiura, one officially recognised by the New Zealand Geographic Board in Port Pegasus and another informally recognised as part of the Titi or Muttonbird Islands on the southwestern coast (http://en.wikipedia.org/wiki/Titi/Muttonbird_Islands, accessed 13 May 2010).
104 Not listed in Table 3.1 is a dubious report for Green Island on the Otago coast, apparently from about 1870 but recorded much later in a family history (Robinson, L. 1974. *The Sinclair story*. Family Reunion Committee, Gore, p. 45): 'On the Island were dozens of tuataras, brown, green and spotted. They would bask in the sun, on their beds … the men teased them with sticks to start them fighting.' The lack of other reports of tuatara is surprising for a location so close to Dunedin, and there is nothing to account for a subsequent extinction on this rat-free island; probably, the reptiles were lizards.
105 Note that Middle Island in the Aldermen Islands was then known as Middle Chain Island (Crook 1973), and should not be confused with Atiu or Middle Island in the Mercury Islands.
106 Atkinson, I. A. E. and Towns, D. T. 2005. Kiore. In: King, C. M. (ed.). *The handbook of New Zealand mammals*. 2nd edn. Oxford University Press, Auckland, pp. 159–174.
107 Crook 1973a; Whitaker 1978; Cree et al. 1995a.
108 Ramstad et al. 2007a.
109 All except Motungarara (referred to as Cooper's Island in Ramstad et al. 2007) have a history of introduced mammals (Peter Gaze, Department of Conservation, pers. comm. 21 May 2010).
110 Motunau Island in Canterbury, and the Moturoa Islands. The present-day absence of tuatara from the Moturoa Islands is considered surprising (Hitchmough, R. A. 1977. The lizards of the Moturoa Island group. *Tane* 23: 37–46.)
111 Newman 1986.
112 Newman 1986.
113 Russell, J. C., Towns, D. R., Anderson, S. H. and Clout, M. N. 2005. Intercepting the first rat ashore. *Nature* 437: 1107.
114 Innes 2005 (Norway rat, pp. 174–187); Ruscoe and Murphy 2005 (House mouse, pp. 203–221). Both chapters in: King, C. M. (ed.). *The handbook of New Zealand mammals*. Oxford University Press, Auckland.
115 Innes 2005, op. cit.
116 Barbara Blanchard, Wellington Zoo, pers. comm. 20 May 2010 (based on information supplied verbally by Anne Richardson of Peacock Springs).
117 Flannery 1994, op. cit.
118 Cree and Butler 1993.
119 Towns, D. R. and Robb, J. 1986. The importance of northern offshore islands as refugia for endangered lizard and frog species. In: Wright, A. E. and Beever, R. E. (eds). *The offshore islands of northern New Zealand*. New Zealand Department of Lands and Survey Information Series No. 16, Wellington, pp. 197–210.
120 Worthy and Holdaway 2002, p. 529.
121 See Cree et al. 1999 for an example of the use of this technique for surviving populations.
122 Flannery 1994, op. cit.; see pp. 344 and 389.

Endnotes to boxed text

For explanation of citation format in box endnotes, see endnotes to the main text.

Box 3.1

1 This tradition suggests that the tuatara were of special significance, as objects described by Māori as having originated from Hawaiki were often those that were greatly valued. See Orbell 1996, pp. 61–63.

2 Hovell, H. 11 February 2004. *Cultural history relating to tuatara on Whangaokena* [sic]. Unpublished report, East Coast Hawke's Bay Conservancy, Department of Conservation. NHE 03-30.

3 ANZ ML-East Cape 1 Daily journals 1. 1900–1906.

4 Buller 1871.

5 Buller 1877. Both quotes from p. 325.

6 Buller 1877.

7 Buller 1878.

8 Evidence of Hoani te Kahaki (10-5-01), *Maori Land Court Minutes*, Waipiro Bay 1901, p. 159.

9 A. Hamilton to Minister for Internal Affairs 15 May 1913; A. Hamilton to the Under-Secretary for Internal Affairs 24 August 1913. ANZ IA 1 46/18/4 pt 1. See also the published summary of Thomson 1915.

10 J. V. E. Ansin to the Director of the Dominion Museum, 16 July 1913. NMNZ MU000207 Box 4, Item 1.

11 Anon. 1913a; Anon. 1913c (copy in ANZ IA 1 46/18/4 pt 1).

12 Bell, B. D. and Blackburn, A. 1960. The islands of Gisborne, East Coast. *Notornis* 8: 244–254.

13 Moors, P. J. 1980. East Island. *Wildlife: A Review* 11: 48–52.

14 Analysis of preserved material is underway by Jennifer Hay, Massey University (pers. comm. 11 December 2006).

15 One prediction, based on recent rates of erosion, is that the island will be reduced to a rock platform in about 1700 years. See note 159, p. 313 in Beaglehole, H. 2006. *Lighting the coast: a history of New Zealand's coastal lighthouse system*. Canterbury University Press, Christchurch.

CHAPTER FOUR

Ngārara: tuatara and other reptiles in Māori tradition

Tuatara – the reptiles with peaks on their backs – were named by the East Polynesian people who settled New Zealand over 700 years ago and became Māori. By the 19th century, when written records became widely available, Māori culture had integrated ngārara (tuatara and other reptiles) into unique traditions involving the natural and spiritual world.

Introduction

The first people of New Zealand, or tangata whenua (people of the land), settled in about AD 1230–1300 from islands in central East Polynesia.[1] Facing the challenges of survival in a large, cool and forested landscape, they learnt to harvest new natural resources, and sustained themselves spiritually with modified beliefs and traditions. By the time of European arrival in the late 1700s, Māori[2] numbered about 100,000, had strong social ties with others related by descent, including whānau (extended family), hapū (sub-tribe) and iwi (tribe), and identified closely with particular areas of land. Connections between tribes were established through intermarriage and migration, but competition for resources, or insults to mana (status, prestige), led to sporadic inter-tribal warfare.

This chapter explores the traditions and beliefs developed by Māori about reptiles, based largely on published or publicly available sources. Some of these traditions involved tuatara, but because tuatara and lizards were not always distinguished from each other as biologists do today, both are included here. Traditions, of course, are not necessarily static in space or time. This chapter takes a broadly chronological approach, beginning with inferences about relationships with reptiles during the period prior to European arrival. The core of the chapter draws attention to the rich oral traditions that were subsequently recorded in writing, and illustrated in whakairo rākau (wood carving), with an emphasis on material from the 19th century. The possibility of geographic or tribal variation is acknowledged by identifying the source region or iwi whenever possible. Some contemporary beliefs are mentioned in passing, but these are not the prime focus of this chapter (for viewpoints expressed by Māori about conservation management of tuatara, see Chapter 10).

A few words about context and interpretation are appropriate. First, this chapter does not attempt to speak on behalf of contemporary Māori, who may prefer to give their own interpretations of the material presented here. Rather, it is an attempt to draw attention to and explore, from the perspective of a Pākehā herpetologist (a New Zealander of European descent who studies reptiles), the rich evidence for contact between humans and reptiles, including tuatara, in a South Pacific landscape. Second, most published records concerning the traditions of 19th- and early 20th-century Māori were made by European males, including explorers, missionaries, ethnologists and others. Valuable though these accounts are, their understandings may be seen in part as a product of the beliefs and customs, religious and otherwise, of those who wrote them. Finally, oral traditions and artwork from the past are not of mere historical interest: they are highly valued as taonga (treasured material) by Māori today, are imbued with spiritual dimensions, and must be treated with respect. Depths of meaning and additional, more spiritually powerful material may exist that have not been shared with Pākehā. Conversely, perhaps new interpretations are possible from an understanding of herpetology; this is the perspective taken in this chapter.

Reptiles in middens and early artwork

Bones of tuatara are widely distributed in middens throughout much of New Zealand, implying that tuatara were eaten by East Polynesian settlers (see Chapter 3). At one early site (at Wairau Bar, in Marlborough; Figure 4.1), a tuatara mandible worked as an awl has been found. This may have been used, as with similar awls made from bird bones, to puncture holes in animal skins before these were sewn into garments.[3]

Reptiles also appear fleetingly in artwork from this period. An evocative piece carved in whalebone confirms that Polynesian settlers were aware of reptiles (Figure 4.2), but the significance of the carving is unknown. The bone was found in sand hills at Wainui, on the East Coast of the North Island (Te Ika a Māui); it was described in 1903 by Augustus Hamilton, who soon afterwards became Director of the Colonial (later Dominion) Museum.[4] A hole under the base of the tail suggests that it was intended to hang as a pendant, with the head facing down. Although the carving has not been dated, it is Archaic in style. While the notches down the midline might seem to indicate a tuatara's crest, notches are common on other Archaic pieces and may be a stylistic rather than a biologically informative feature. Another pendant from this early period, which has similar notches down the midline, is thought to represent a turtle. Such pendants may have been worn by people of authority and buried with them.[5]

Fig. 4.1 Some locations, regions (in capitals) and iwi areas (in italics) mentioned in the text. Modern spellings used here may differ from historical sources quoted in the text. For locations of islands with existing populations of tuatara, see Figure 6.1.

Fig. 4.2 This small whalebone pendant (about 111 x 32 mm) in the form of a lizard-like reptile has a Polynesian style. The hole near the base of the tail is what suggests that the object was hung as a pendant. The tail is stumpy – perhaps it was autotomised. Ngā Kakano 1100–1300 or Te Tipunga 1300–1500, iwi unknown, from Gisborne region.

Museum of New Zealand Te Papa Tongarewa, ME000643.

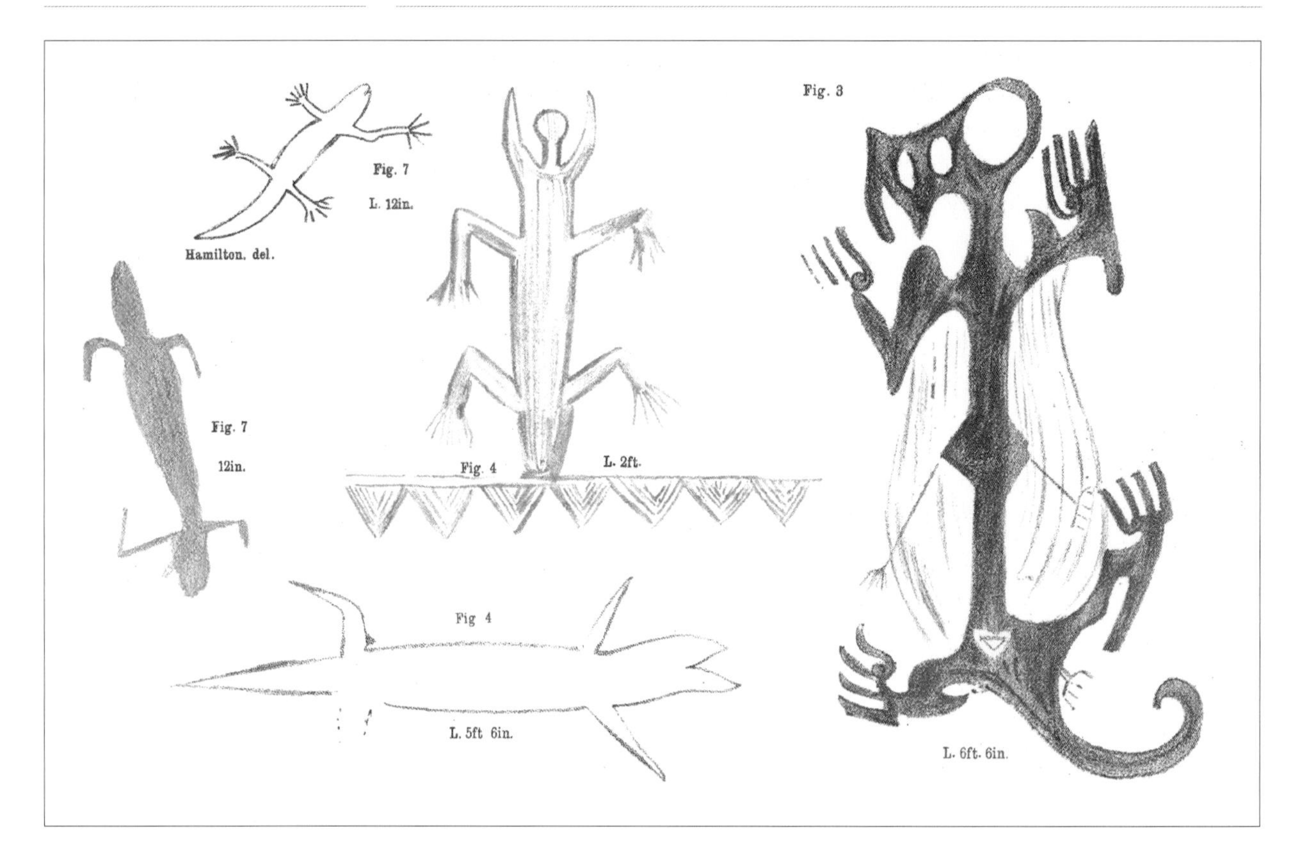

Fig. 4.3 Reptile-like creatures appear in rock art, probably dating from the first few hundred years of Polynesian settlement. These drawings were copied from caves in the Opihi River area by Augustus Hamilton in 1898. The large figure on the right, interpreted as a taniwha (reptilian water-monster), is the inspiration for the logo of New Zealand's professional herpetological society, the Society for Research on Reptiles and Amphibians in New Zealand (SRARNZ).

Composite of drawings, with original scales, from Hamilton, A. 1898. On rock pictographs in South Canterbury. *Transactions and Proceedings of the New Zealand Institute* 30: 24–29 + plates 1–10.

Other examples of reptiles in artwork from this early period are on a very different scale. In inland parts of the South Island (Te Wai Pounamu), especially in eastern parts of South Canterbury and North Otago, are limestone rock shelters in which human visitors have created long-lasting reminders of their presence. The drawings, in charcoal and ochre, date from the early period of settlement through to European arrival, as revealed by topics spanning moa to European sailing ships. They are attributed to the Rapuwai and Waitaha cultures, but fall today within the rohe (tribal area, domain) of the iwi Ngāi Tahu.

These rock drawings are often heavily stylised, but representations of lizard-like reptiles (Figure 4.3) as well as humans, birds and dogs are recognisable. Although tuatara appear not to have been specifically represented,[6] our knowledge may be incomplete. In 1923 the rock paintings were said to include 'a very realistic lizard rendered in black. This interesting and important piece was cut from the rock by a visitor, and is now, if it has survived at all, in private hands in America.'[7] Ngāi Tahu now manages a South Island Māori Rock Art Project,[8] which aims to record the surviving art as a permanent photographic record and to help guide further preservation.

Reptiles in creation narratives

By the time of European landfall in 1769, what is recognised today as Classic or Tribal Māori society was well established.[9] Life in the North Island, where the population was largest, was relatively sedentary, being based on crops that could be cultivated and on villages that could be defended. Carving, weaving and oratory were highly valued and practised widely (Figure 4.4). In the more challenging climate of the South Island, where crop production was more difficult, population density was lower, fortifications were fewer, and life was more nomadic.

Within Classic Māori culture (and contemporary culture today) a strong emphasis was placed on whakapapa (genealogy, lineage, family history). Creatures of the natural and spiritual worlds, including reptiles, were incorporated within wider traditions about whanaungatanga (kinship). Living things were seen as descending from the primal parents Papa-tūā-nuku (Papa, earth mother) and Rangi-nui-atea (Rangi, sky father),[10] who in some accounts were themselves descended from Io-matua (a supreme being). Papa and Rangi had many children, from whom all things known to Māori descend (Figure 4.5). In some accounts, Tangaroa (god and father of the ocean) was the progenitor of reptiles, as well as of whales, sharks and other fish. In another version published by Governor Grey in 1855, reptiles are the great-grandchildren of Tangaroa, through his offspring Punga (parent of ugly things) and then Tū-te-wehiwehi, or Tū-te-wanawana (father of reptiles).[11] According to Tainui iwi, there is a distinction between tuatara (seen as direct offspring of Tangaroa) and other reptiles (seen as offspring of Tū-te-wehiwehi/Tū-te-wanawana).[12] In southern accounts, Tāne may be seen as the father of lizards.[13] Other variations on creation narratives, as they affect reptiles, are noted later in this chapter.

Fig 4.4 A fortified Māori village or pā. The original caption reads 'The ariki or paramount Chief, is seen standing among subsidiary chiefs, priests and other important men; behind him is the Council House, next to it is a carved store for food, sharks are hanging on the food frames, in the foreground a woman slave is placing food near the earth oven.'

From Donne, T. E. 1927. *The Maori past and present: an account of a highly attractive, intelligent people, their doubtful origin, their customs & ways of living, art, methods of warfare, hunting and other characteristics, mental & physical*. Seeley Service, London. Hocken Collections, Uare Taoka o Hākena, University of Otago, S06-545.

An Old Time *Pa* Surrounded by Palisading.

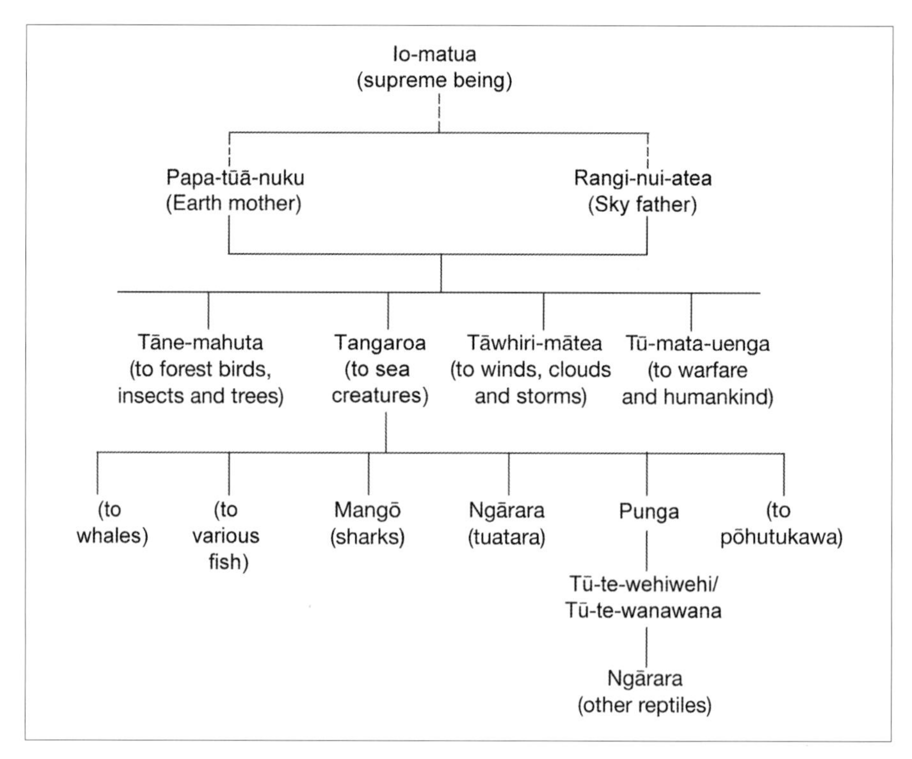

Fig. 4.5 An abbreviated whakapapa depicting the origins of tuatara and other reptiles.
Based on various sources, including Auckland Museum 2001. *Te Ao Turoa education kit* (available at http://www.aucklandmuseum.com/68/teacher-resources, accessed 11 July 2012), and Mere Roberts, Auckland University of Technology (pers. comm. 24 January 2008).

Reptiles as dangerous spirits

Acknowledgement of common ancestry with humans does not mean that reptiles were always viewed benignly. In a cultural system in which elements of life were classified either as tapu (under religious restriction, forbidden, feared or sacred) or noa (safe, unrestricted, not tapu), reptiles were clearly tapu. And for humans living in close contact with the natural world, and seeking, as humans do everywhere, to find explanations for events that befell them, reptiles (especially lizards) filled an important role as symbols of life or death, or, as Simmons put it in 1987, the 'mortal principle'.[14] Among many accounts (although not all, as we shall see), lizards were believed to represent a dangerous atua (god or spirit, sometimes referred to as Whiro).

Such a belief posed a difficulty for Europeans enquiring about reptiles. An early account was published by Richard Cruise, who spent nearly 10 months in the Bay of Islands in 1820. As commanding officer of a military detachment on the store ship *Dromedary*, he seems a relatively impartial observer. Cruise reported that 'one of the gentlemen, having found a lizard, carried it to a native woman to ask her the name of it. She shrunk from him in a state of terror that exceeded description, and conjured him not to approach her, as it was in the shape of the animal he held in his hand that the Atua was wont to take possession of the dying, and to devour their bowels.'[15] The situation was evidently still similar on the South Island in 1920, when the ethnologist James Herries Beattie observed: '[The topic of lizards] is also a difficult subject to collect information about as the average Māori shuns these harmless creatures through superstitious fears.'[16]

BOX 4.1

Tuatara, ngārara, kārara: what's in a name – or carving?

Written accounts from the 19th and early 20th centuries offer several Māori words for reptiles or reptile-like creatures. The most commonly used is 'ngārara' (or 'kārara' in southern dialect, in which a 'k' often replaces 'ng'). Both were general terms for reptiles[1] that, depending on context, could refer to tuatara; however, unless there is a specific reference to features such as the distinctive crest and/or teeth of tuatara, lizards cannot be eliminated. Indeed, although ngārara or kārara (a crawling thing) usually refers to a living reptile, it could also refer to insects.[2] The same word was sometimes also used in reference to fabulous monsters of caves and waterways,[3] as in the tale of the taniwha Ngārara-huarau related in the main text.

The Māori word 'tuatara', on the other hand (tua meaning 'back', tara meaning 'spine'), seems specific to the animal that scientists know today as *Sphenodon punctatus*: fortunately for our interpretation of older texts, no New Zealand lizard has a crest. Tuatara were also sometimes known as ruatara, duatara, tuatete and tuakeke.[4] The first two of these may be variant spellings arising from slight differences in pronunciation; the term 'tuatete', however, is more ambiguous. In some contexts a tuatete seems to have been considered a legendary or extinct animal, 'something like a lizard, but many times larger'.[5] Additional terms that seem specific to tuatara were 'ika tuatara' (meaning the 'fish that is a tuatara') and 'ngārara-tuatara' ('kārara-tuatara').[6] A common English word for a large reptile in 19th-century accounts is 'guana'[7] (a corruption of 'iguana') – this word is likely to refer only to tuatara, as no New Zealand lizard has a crest, dewlap or body size resembling those of an iguana.

The Māori words 'kaweau' or 'kawekaweau' could occasionally indicate tuatara.[8] More often, however, these seem to refer to a large tree-lizard,[9] which some biologists consider a description of an

Box 4.1 Fig. 1 The fierce jaws of this reptile may look tuatara-like, but the curling tail is reminiscent of that of a gecko. It may have been the guardian lizard Kataore that the carver (possibly Tene Waitere of Ngāti Tarawhai) had in mind. This poutokoihi or central porch-post is part of a building known at times as the Te Arawa church, English church or Nuku-Te-Apiapi portico house, at Whakarewarewa in Rotorua. Vapours from nearby mud-pools have dissolved the outer coat of green paint, exposing a layer of red underneath.

Photo: Alison Cree.

Box 4.1 Fig. 2 A stylised, heavy-jawed reptile in a carving from the East Cape region. The provenance of this tekoteko is unknown.

Photo: Auckland Museum, OA018745.

extinct giant gecko, *Hoplodactylus delcourti*, which may once have existed in New Zealand.[10] 'Moko' is a general term that seems to refer to lizards rather than tuatara (it also refers to body carving or tattoos, but the relationship, if any, between these and lizards is unclear). 'Moko' could be qualified (as in moko-kākāriki for green gecko, moko-pāpā or moko-tāpiri for dark-coloured or grey-brown geckos living in holes in trees, and mokomoko for at least some types of skink).[11] The toropakihi or tupakihi was a common grass-dwelling lizard in the South Island, probably a skink, although in some contexts it could be an insect.[12] 'Kaurehe', although listed in one source as a synonym for tuatara,[13] was usually a big, amphibious, lizard-like creature of mythical nature,[14] which some European naturalists took optimistically to be a beaver or otter (no such mammal has been discovered in New Zealand).

A final difficulty with early records arises from imprecision in use of the English word 'lizard'. Until 1867 tuatara were not recognised by scientists as rhynchocephalians, a group distinct from lizards. For many decades afterwards tuatara were known in New Zealand as the 'tuatara lizard', and even today tuatara are not always clearly distinguished from lizards in non-specialist accounts. This means that, even if the word 'tuatara' was used by Māori, some interpreters might have translated this ambiguously as 'lizard'. Such imprecision could account for variations in tales, such as that of Ngārara-nui related in the main text, which sometimes involve a large lizard and sometimes a tuatara instead.[15]

Turning to carvings, which in the absence of a written language were an important way of recording tribal history, similar issues arise. Many specialists of carving or culture have concluded that reptiles are among the very few animals faithfully or realistically depicted in Māori carving.[16] Does this mean that herpetologists can reliably infer the identity of tuatara, geckos or skinks (or even different species of geckos or skinks) in 19th-century carvings? The answer seems to be 'no'.

One reason is that rarely is the intention of the carver publicly known. In a few cases where it is, such as in some of the carvings of Wero and Tene Waitere

Box 4.1 Fig. 3 Although lacking crests, the reptiles in this carved door panel are among the most tuatara-like of all Māori carvings. Tradition records that they depict the pet lizards of the bird-woman Kurangaituku, between whose legs they are placed. The door panel, by Tene Waitere of Ngāti Tarawhai, comes from the house Nuku-Te-Apiapi, and remains on display in the Rotorua Museum.

Reproduced with consent of the Te Arawa Lakes Trust. Photo: Rotorua Museum of Art and History Te Whare Taonga o Te Arawa, New Zealand, ref. X.518.

Box 4.1 Fig. 4 Detail of the pet lizards of Kurangaituku, from the previous figure.

of Ngāti Tarawhai, it is evident from written or oral tradition that the carving represents the artist's concept of a kaitiaki or taniwha rather than a living species of reptile.

Another reason for cautious interpretation is that reptiles in older carvings rarely show combinations of features that would unequivocally identify them to a herpetologist as one of the three major groups. For example, some have the jaws of a tuatara but the crestless body and curling tail of a gecko. Some are more elongate than any living New Zealand reptile, with a closer resemblance to reptiles elsewhere such as varanids (monitor lizards or goannas), or even crocodiles. Yet others have a creative combination of features not possible in any living reptile, such as an otherwise tuatara-like wood carving from Taranaki in which the head has been rotated 180° ventrally (the jaw points upwards and the eye downwards relative to the body).[17]

In short, carvings, as a medium for telling a story, may never have been intended as an accurate depiction of a living animal. Herpetologists must be careful to avoid making assumptions about the 'species' of reptiles depicted.

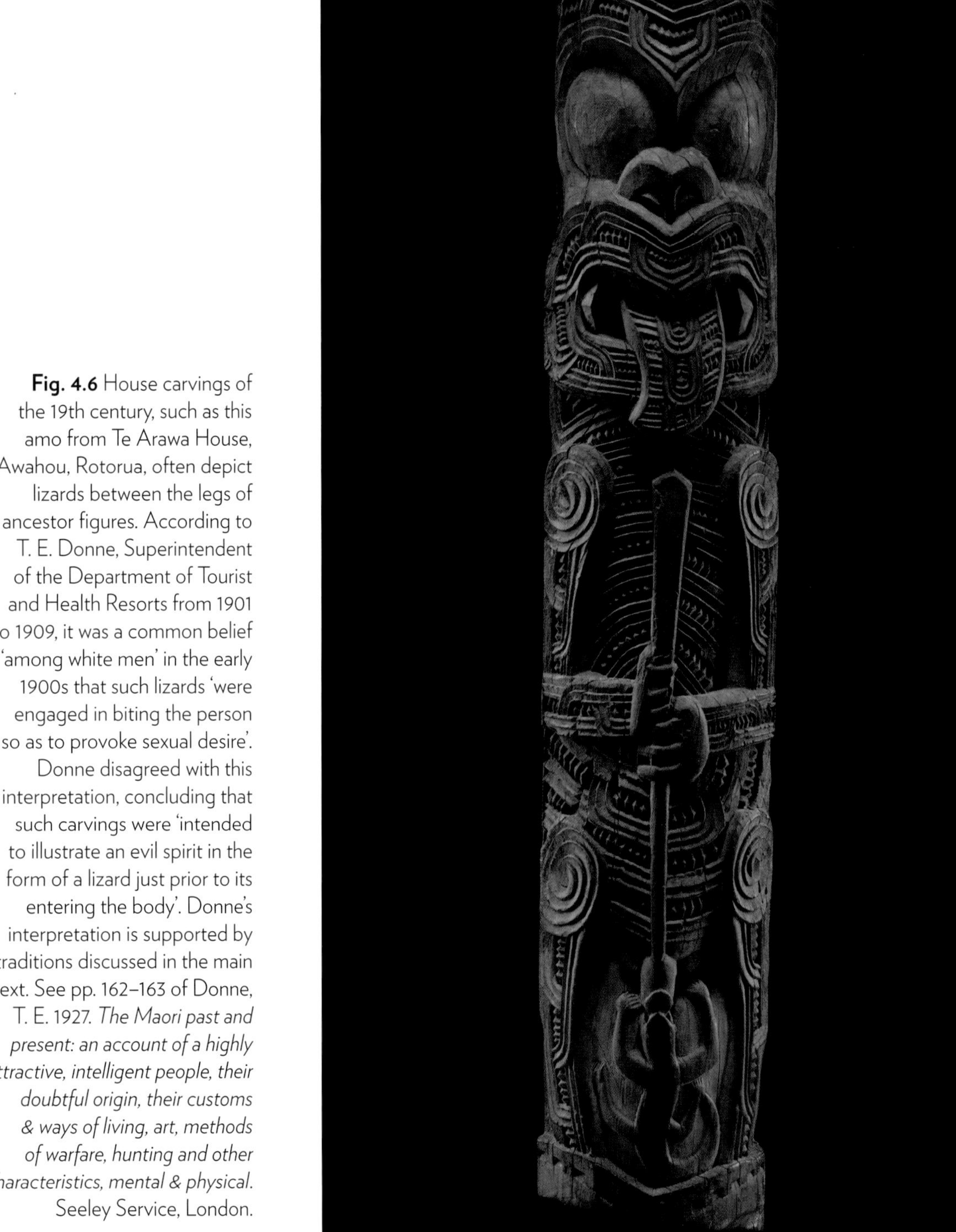

Fig. 4.6 House carvings of the 19th century, such as this amo from Te Arawa House, Awahou, Rotorua, often depict lizards between the legs of ancestor figures. According to T. E. Donne, Superintendent of the Department of Tourist and Health Resorts from 1901 to 1909, it was a common belief 'among white men' in the early 1900s that such lizards 'were engaged in biting the person so as to provoke sexual desire'. Donne disagreed with this interpretation, concluding that such carvings were 'intended to illustrate an evil spirit in the form of a lizard just prior to its entering the body'. Donne's interpretation is supported by traditions discussed in the main text. See pp. 162–163 of Donne, T. E. 1927. *The Maori past and present: an account of a highly attractive, intelligent people, their doubtful origin, their customs & ways of living, art, methods of warfare, hunting and other characteristics, mental & physical.* Seeley Service, London.

Museum of New Zealand Te Papa Tongarewa, ME014417.

A characteristically colourful account of such beliefs was given by Joel Samuel Polack in 1840 (Polack was an English trader based in the Bay of Islands).

> The favourite form of the Gods is supposed to be that of a lizard or a bird … Lizards are accounted as virulent deities, that enter the orifices of the human body, devouring the inside of the hapless sufferer, until a priest intercedes, and by means foul and fair, drives the intruder from his quarters … [After payment of a fee the priest commences an exorcism] and will not hesitate to assert that he discovered the rascal leaving the mouth, nostrils, or ears, of the sick man in the shape of a lizard, (*ruatára*) or some such animal.[17]

This belief that reptiles could enter the body's orifices may be an explanation for the many carvings that depict a reptile between an ancestor figure's legs (Figure 4.6).[18] It was also said that to snore was to risk inviting lizards – suggesting that snoring was as unpopular in the past as today.[19]

Fear was exhibited towards lizards, both live and dead, and by those with the strongest of reputations for bravery. George Cooper, who made an expedition during 1849–1850 from Auckland to Taranaki with Governor Grey, wrote of seeing 'twenty or thirty strong able-bodied men … take to the water for refuse [*sic*]' when shown the dead body of a 'common green lizard',[20] and H. C. Field in 1882 witnessed 'a leading chief who had distinguished himself in many fights, bleed at the nose, scream and become quite convulsed at the sight of a little brown lizard.'[21] Lizards preserved in spirits were also considered life threatening, according to experiences in the Waikato district by the artist George Angas in 1844[22] and the mission-school teacher James Stack in the 1850s.[23]

Ways of avoiding the threat posed by lizards seem to have varied. Experiences in the Bay of Islands in 1827 led the artist Augustus Earle to write that 'the lizard is sacred, and never injured',[24] and the missionary William Yates added in 1835 that the older and more superstitious among the Māori were, on account of their beliefs, 'particularly careful not to do it [a lizard] an injury.'[25] A later author, Thomas Moser, described a seemingly different approach.[26] Following some illness and deaths in a village in 1862, a poropiti (prophet) led a search for lizards among the fern; eventually several lizards were caught and thrown on a fire while incantations were made, and potatoes were later roasted in the ashes. This procedure was considered to remove a tapu that had inadvertently been violated, an event seen as the cause of the original illness. T. W. Downes, Supervisor of Works on the Whanganui River, gave an essentially similar account in 1937.

> On the river it was formerly, and to some extent is still, the idea that lizards would crawl into the bodies of the natives and thus cause all of their diseases. I myself have often wondered why, when they thought the lizard was the personification of an *atua*, they would kill it. On enquiry, I

> was told, 'You say in the bible we are to resist the devil, we therefore resist him by killing him.' If after some searching a lizard was found, the luckless creature was burnt with considerable ceremony and its ashes sprinkled over potatoes which were eaten. In this way was the power of the evil spirit destroyed and diseases warded off from the participators and their children ... I do not think that eating the creature is considered necessary now, but killing it certainly is.[27]

In the South Island during the early 1900s, Beattie was told that to destroy tapu on a place, a lizard was trampled on and then burnt by a tohuka (southern dialect for tohunga; a priestly expert).[28] A Māori informant told Beattie that, having seen a lizard following a night during which he had had a bad dream, he killed the lizard with a big stone, lit a fire and burnt it. The informant commented: 'A wise old man told me afterwards that it was a good job I had killed and burnt the lizard and so stopped any evil coming to me because of my dream.'[29] Other reports, not specific as to region or date, add that an alternative method for removing the tapu of a lizard, and thus rendering it harmless, was to have a woman, who was noa, step over the lizard.[30]

These accounts suggest some differences in responses to lizards. Whether this variation reflects tribal variation in traditions, a change over time (including the influence of Christianity), the context in which lizards were encountered (for example, whether a person was healthy, or already ill), or simply the personal views of the individuals concerned is not clear.

Above all other reptiles, it was the small, velvety geckos that were considered the most dangerous. Their function in the Māori world-view can be seen as explaining otherwise inexplicable bad fortune such as death and disease.[31] The powers of geckos extended to both major types, moko-pāpā (interpreted today as one or more of the so-called 'brown geckos', all known until recently as members of the genus *Hoplodactylus*,[32] which tend to be darker and more terrestrial),[33] and moko-kākāriki (the various species of green geckos in the genus *Naultinus*, which are largely arboreal). Downes related in 1937 the following account from a man who was lying with his brother on the ground when a moko-pāpā appeared between them.

> They both rolled away at first and then Titi chased it, when it reared its head and laughed. He killed it with a stick, and then came the problem what to do with it. They were both frightened to cut it up and throw the pieces to the four points of the compass, as was the old custom among the Tawata people to prevent death or disaster or war parties coming in, as they did not know the correct *karakia* [prayer, incantation] to use for so doing, so they tied it to a tree with the legs extended but this proved to be useless, for a week later come the message that a cousin had died at Parinui, some seventy-five miles above Whanganui.[34]

The same man related a subsequent capture of a lizard and, being too frightened to kill it after his previous experience, he released it; 'then almost

BOX 4.2

Why were geckos and other reptiles so feared?

Records from the 19th century attest to a deep fear among Māori of reptiles, especially geckos, that were essentially harmless. Is there an underlying explanation? An early view, expressed by Shortland in 1856 and later developed by ethnologists and anthropologists, was that reptiles represented an ancestral memory or 'former knowledge' of potentially dangerous species such as crocodiles[1] or snakes.[2] For example Elsdon Best observed that 'moko', a name for lizard among Māori and other Polynesians, is similar to 'mokotolo', a name for crocodile in Melanesia. He concluded that 'memories of crocodiles of other lands' influenced the carving of reptilian creatures and the development of taniwha myths among Māori.[3]

More recently anthropologist Margaret Orbell proposed in the 1990s what can be termed the 'anomalous ika' idea. Orbell argued that reptiles were feared because they were thought to be 'anomalous creatures, closely related to fish (ika) yet living on the land and even, in some cases, climbing trees'.[4] In support she pointed out a visual similarity between gurnard (the fish with which tuatara quarrelled before leaving the sea, according to one version of a popular tale related in the main text) and tuatara themselves.[5] The sinusoidal movement of some reptiles (including tuatara and geckos, which also climb), and the presence of overlapping scales on others (skinks), could help to reinforce such a belief.

One additional feature of geckos that clearly

Box 4.2 Fig. 1 Geckos, such as the common gecko shown here (an undescribed species of *Woodworthia*, formerly placed in the genus *Hoplodactylus*), were interpreted by Māori as spirits capable of entering a person's orifices to cause disease. Perhaps a tendency of nocturnal geckos to seek warm surfaces at night, and cling when disturbed, contributed to such a belief.

Photo: Ken Miller.

Box 4.2 Fig. 2 Green geckos, such as this jewelled gecko (*Naultinus gemmeus*), have vividly coloured mouths and can make a startling, chattering call. To hear this was considered by Māori as an omen of death or disaster.

Photo: Rod Morris.

contributed to a particular fear (over and above that of tuatara and skinks) was vocal ability. Sound production is not dramatic in the so-called 'brown geckos' (a group of 11-plus species that until recently were all placed in the genus *Hoplodactylus*), although some of these species chatter when disturbed by other lizards in their daytime retreat sites. On the other hand, green geckos (several species of *Naultinus*, known to Māori as moko-kākāriki and visible by day on foliage), sometimes make a dramatic noise with their mouths wide open. I heard such a call only once during several years of fieldwork on jewelled geckos (*N. gemmeus*), but the effect (like a mini-machine gun going off in my hand) was so startling that I almost dropped the lizard. Māori termed this chattering sound 'kata', meaning 'to laugh', but it was a laughter that brought terror to the observer.[6]

It is not difficult to see how, if geckos in general were already feared, such an encounter might be taken as a message or warning. Best reported in 1923 that if a moko-kākāriki was seen by a war party on its way to attack an enemy, the party would immediately return home, presumably to avert disaster. He also quoted an earlier source to the effect that, during the period of European settlement when battles for land between Māori and European became rife, 'the native belief was that the European settlers wanted the lizard to destroy the Māori folk so that they might obtain possession of their land'.[7] It was said that local people resorted in one case to destroying groves of karaka trees to prevent the lizards in them (presumably a species of green gecko) from achieving this result.[8]

Features of other geckos that might have contributed in combination to fear (especially the belief that geckos could enter a person's body) are nocturnal thermophily (a preference for warmth at night) and adhesive ability. Nocturnal geckos emerge to forage when air temperatures are low, but become heat-seeking when warmer temperatures become available. For example, common geckos from a population in southern New Zealand (an undescribed species within the new genus *Woodworthia*[9]) emerge with body temperatures as low as 12°C at night, but choose warm surfaces up to about 28°C when these are offered in the laboratory.[10] Conceivably, like some snakes elsewhere today, nocturnal geckos might have sought the warmth of human bodies sleeping on the forest floor at night.[11] They could then have used their adhesive toepads (not present in tuatara or skinks, and less prominent in the diurnal green geckos) to cling disconcertingly when disturbed.

In addition, both geckos and skinks in New Zealand (but not, so far as is known, tuatara[12]) can carry *Salmonella,* a bacterium present in reptiles overseas that can be transmitted to humans.[13] If lizards were handled and a person experienced stomach cramps or vomiting soon afterwards, perhaps that experience might come to be associated with lizards (although the experience of New Zealand herpetologists provides little to indicate a high risk of transmission). Finally, we must also allow the possibility that there is no explanation for the particular fear among 19th-century Māori of geckos, or that the origin is too ancient to be fruitfully considered.

Fig. 4.7 'A tabood store house in New Zealand' by Augustus Earle, c. 1827. The support post of this pātaka or storehouse, which is believed to be from the Bay of Islands or Hokianga Harbour, has a lizard extending downwards toward the head of a human figure. Earle wrote that 'a lizard taking hold of the top of a man's head' was a favourite subject in carvings, the 'tradition being, that that was the origin of man'. Later Māori carvings and written traditions do not seem to have presented lizards in the same way.

See p. 142 of Earle, A. 1832. *A narrative of a nine months' residence in New Zealand, in 1827; together with a journal of residence in Tristan d'Acunha, an island situated between South America and the Cape of Good Hope*. Longman, Rees, Orme, Brown, Green and Longman, London. National Library of Australia, nla.pic–an2822475.

immediately came the influenza epidemic and Ngaranga's whole family (relatives), eight persons in all, died. This could certainly have been averted, had the proper incantation been used and the creature cut up and thrown round as was the former custom.'[35]

An interesting corollary is that such powerful creatures could be put to work on one's behalf. Reptiles were carved on items such as waka huia (treasure boxes, feather boxes) or waka tūpāpaku (bone caskets)[36] to guard the tapu contents, and on pātaka (storehouses) to protect valuable resources (Figure 4.7). A slender lizard is carved on a kōauau (flute) made from human bone, and may have been played during funerary ceremonies and other rites of passage.[37] In an account from Tai Tokerau (Northland), a waka tūpāpaku with a carved lizard was a kaitiaki (guard or guardian) of a burial cave, and a person who inadvertently walked over rather than around it when bringing bones to inter was bitten by the 'spirit of the lizard', fell ill and died.[38] Lizards were sometimes tattooed on human genitals, although in one traditional story this seemingly contributed not a barrier but an attraction that led to incest.[39]

There are many accounts of geckos being used to guard sacred places or objects. During construction of an important building, green geckos were sometimes buried under a post supporting the ridgepole to act as the occupants' guardian.[40] A lizard known as a moko-tāpiri could be used to guard significant stones, concealed at the base of a tree, which marked the mauri (life force) of the forest.[41] Skinks seem not to have been feared to the same extent as geckos, though they too were attributed with special powers. In one account, an East Coast Māori related that to bite a mokomoko or skink on the head 'so as to crush it and cause blood to flow' was considered a certain cure for toothache.[42]

Complex feelings about tuatara

Tuatara, although not as deeply feared as geckos,[43] were still treated warily by many. In a detailed report by 'A Party of Officers of the 58th Regiment' in 1852, the authors relate their attempt to get assistance from Bay of Plenty Māori to collect tuatara from the Rurima Rocks (presumably Moutoki Island). With much persuasion and recompense, paddlers were obtained for the canoe. When the party returned to Whakatane, the sight of the captured tuatara in a box was said to be enough to disperse a crowd of at least 100 people, and no Māori craft would transport the tuatara to Auckland.[44]

An account from a Māori informant from Nelson, recorded by Beattie in 1920, indicates a reluctance to visit islands with tuatara in that region also. This fear was said to be waning with European influence. 'In regard to titi on Stephen's [*sic*] Island he [the informant] added later that he thought the reason the old Maoris had not gone there for muttonbirds was because of their horror of the tuatara, but when the whalers came and half-castes appeared these latter went to the island and had done so for the last sixty years.'[45]

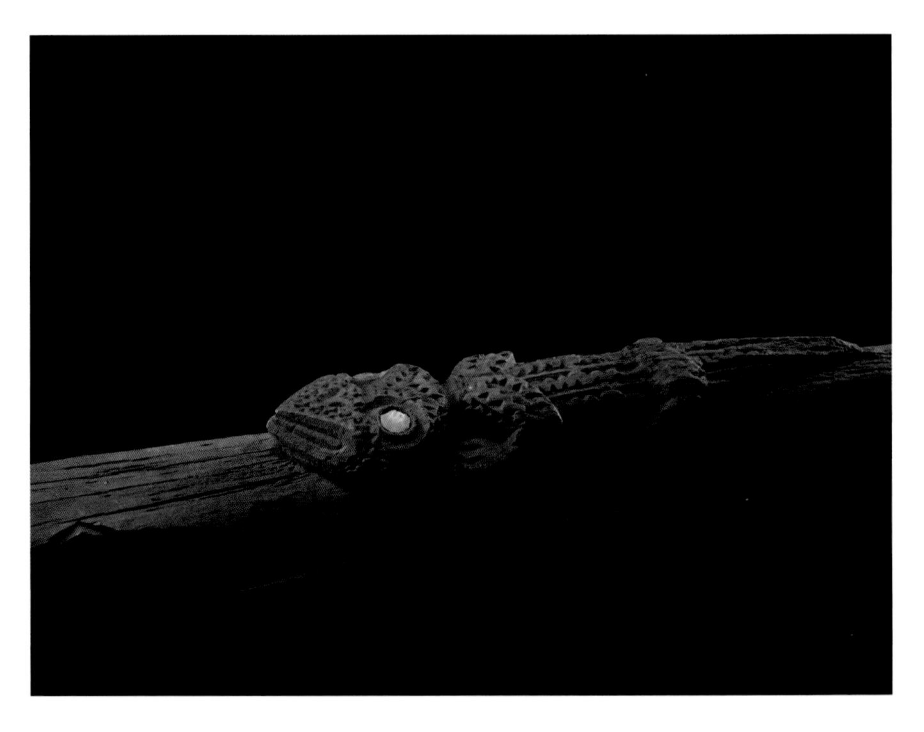

Fig. 4.8 Reptiles sometimes appear on taumanu (thwarts) of waka taua (war canoes). This example features two inward-facing reptiles, one with paua-shell eyes. The decorative style, a series of alternating zig-zags and ridges known as taratara-o-kai, is often seen on pātaka (storehouses). (Taratara-o-kai literally means 'peaks of food', but the term may also be a reference to the jagged teeth of Kai (Kae), who stole and ate a pet whale.) Although the geographic origins of this taumanu are unknown, note the similarities (including the heavy jaw and taratara-o-kai) to the reptile on the pātaka Te Puāwai o Te Arawa at Auckland Museum (Box 4.4).

Museum of New Zealand Te Papa Tongarewa, ME001968.

BOX 4.3

The significance of being named 'Tuatara'

Although tuatara were sometimes seen as powerful guardians and defenders of eggs, to refer to a person as a tuatara could be less than flattering. In 1840 J. S. Polack recorded that among the nicknames given by Māori to each other and to visitors were those that referred to animals, adding that 'A lazy man or slow walker is termed Tuatāra or lizard.'[1] Thomas Moser wrote in 1863 of 'a small [Māori] boy of a most precocious nature, who was termed "Tua Tara", from a horrid sort of lizard that the natives abhor.'[2]

Other people received the nickname for less negative reasons. Captain James Reddy Clendon, a kauri trader in the Hokianga district during the 1860s, was nicknamed 'Tuatara'; apparently, his habit of hunting for good wood in various harbours reminded local Māori of a tuatara hunting for food in nooks and crannies.[3] Polack had himself noted in 1838 that the name 'Tuatara' was given to an acquaintance of his for the same reason.[4] Interestingly, one of three waka taua (war canoes) that greeted the Duke of Edinburgh during a royal visit to Auckland in 1869 was called Te Tuatara,[5] although the reason was not recorded.[6]

Those individuals who did venture to offshore islands to collect tītī (muttonbirds, the chicks of the burrow-nesting grey-faced petrel *Pterodroma macroptera* and the sooty shearwater *Puffinus griseus*) had the opportunity to observe tuatara, reinforcing knowledge of their natural history, including movements and feeding behaviour. Mutton-bird harvesters also learnt to cope with close encounters. One such episode, with the tuatara coming off worse, was related by William Goodlet of the Otago Museum in the 1920s. 'I was told by a halfcaste Maori that when he was mutton-birding he put his hand into a burrow to get a small mutton bird, but made a mistake, and got the tuatara instead. As a matter of fact the reptile got him, and hung on until he got a stick and killed it.'[46]

In recent years a contrasting practice (and a brave one, considering the claws and teeth of tuatara) has been related by elders of several northern iwi, including Ngāti Rehua and Ngāti Manuhiri (part of the wider Ngātiwai iwi) and Ngā Puhi, as well as by others of the Hauraki region. Tuatara were placed inside one's shirt (especially by women) to keep cool while mutton-birding. At the end of the day, the tuatara were returned to their burrows with a prayer of thanks.[47] (In an interesting parallel, the procedure of placing tuatara inside one's shirt as a convenient place for storage while collecting them was reported, apparently by a European, in 1908. Whether this approach was modified from Māori practices or developed independently is unknown.[48])

Awareness that tuatara were feared, by at least some Māori, has been used to interpret carvings of tuatara-like reptiles on taumanu (thwarts) of waka taua (war canoes; Figure 4.8).[49] In 1925 Best suggested that such carvings might indicate a tapu position in the canoe that was reserved for a tohunga or chief.[50] However, a more complex basis for the tradition comes from a Bay of Plenty resident, in 1966, who explained that the 'ancient Maori seafarers of old' carried live tuatara in their canoes to serve as compasses.[51] Two such tuatara were said to have been kept in the prow of Mātaatua (an ancestral canoe bringing settlers from Hawaiki, the mythical homeland, to the Bay of Plenty), where they were closely watched by Taneatua, the tohunga. If the tuatara turned their heads to right or left, that indicated the direction to be followed by the canoe. In a related tradition from the Opotiki region, two boys in ancient times were said to have been in the habit of taking live tuatara with them as pets or compass-bearers in a small canoe while fishing. One day the boys drowned in a storm, but the tuatara swam to the Rurima Rocks and established a population there. Their grieving mother, a revered woman, placed a ban on fishing in that region of the Bay of Plenty, a restriction that broke down only after the arrival of Europeans.[52]

Reactions to pet tuatara on the mainland were mixed. T. E. Donne, the Superintendent of the Department of Tourist and Health Resorts, kept pet tuatara in his office in the early 1900s; some Māori visitors were curious to see the tuatara but departed fearfully once shown.[53] Beattie reported, 'an old Maori refuse to go near or touch a pet tuatara' in Murihiku (southern South Island).[54] On the other hand, another informant of Beattie's talked of a 'lizard' being kept as a pet at Kaitangata,[55] and Stack was told that 'A ngarara (reptile) known as Te iha was kept a long time at Kaiapoi. It was fed on small birds and prepared fern-root. It was very gentle and liked being stroked, uttering at the time a guttural sound expressive of pleasure.'[56]

Tuatara as guardians of the dead and defenders of eggs

Tuatara, like lizards, could be employed as guardians of special places. In 1838 the missionary and naturalist-explorer William Colenso was told by East Cape Māori of a monstrous animal called a moa, which resembled an immense domestic cock but had a face like a man. It was to be found in a cavern on a mountain named Wakapunake, where 'it lived on air, and was attended or guarded by two immense *Tuataras*'.[57] Reverend Richard Taylor was told a similar tale in 1839, although in his version the immense bird was said to live on the mountain named Hikurangi and to be guarded by a large lizard. Māori also told Taylor that a possible cause of the demise of the moa was the extinction of the tuatara, on which (along with lizards) moa were said to have fed.[58]

Other examples relate more explicitly to defending the dead. Downes was told by an informant of an ancestor named Piki,

> who was buried between Whakatane and Rotorua in a native ceremony at Ngaroto. His grave is now closely guarded by a ngarara shaped like a

> tuatara, only about 18 inches [45 centimetres] long, but with such powerful mana that nearly 200 people have already been killed by venturing too near. Piki's spirit still visits his old whare [house, dwelling] near by, and daily food is pushed into the whare with a forked stick either to feed the spirit or appease the Ngarara. All the people, said Kauae-o-rangi, know of this.'[59]

In a tangi (lament) for a chief who died a natural death, a tuatara was sometimes referred to during the eulogies.[60] When the eminent Māori doctor, politician and anthropologist Sir Peter Buck (Te Rangi Hiroa) died in 1954, his funeral bier had carved prominently on it a very large 'tuatara (lizard), the almost universal symbol of death.'[61]

Māori of the 19th century were well aware that tuatara were egg-laying reptiles,[62] and related tales of females that guarded or reclaimed their eggs from humans. In a story from about 1875, a man travelling found a tuatara colony and, among the burrows, a nest. He took an egg away with him, carefully placing it beside him at night when he went to bed in a deserted hut. In the morning the egg was gone and he was perplexed as to how it could have been taken. An elder who heard the report then observed: 'that there was but little doubt that the mother tuatara had followed him and the scent of the egg, and had got into the hut in the night and carried it off, for, concluded the elder, it is a well-known fact that the tuatara will use every imaginable means for securing its eggs and young.'[63] Similar tales were told to Beattie by South Island Māori. In one such account, 'It was said that if you took the eggs and the mother tuatara returned to the rua [burrow] and found them gone, she could scent them and would follow you for miles.'[64]

While these accounts may seem implausible at face value, they could have deeper meanings containing an element of natural history. Wild female tuatara often return to their nests for several nights after egg-laying; during this time they defend their nests from other females that might otherwise dig up the same site.[65] In one instance at Wellington Zoo, a captive female that had laid eggs five days previously made repeated biting lunges at the feet of three human keepers who entered the enclosure to collect the eggs.[66] This suggests that observations of nest defence directed at other female tuatara, or at people's feet, could have been enlarged by Māori over time into stories about nest defence directed more generally at humans.

Harvesting of tuatara on the mainland

Part of the reason that tuatara were not as deeply feared as lizards may have been that traditions of harvesting tuatara still existed. Consider the following well-known tale. In this translation, by Colenso in 1879, the reptile is referred to as a 'large lizard'; however, it is clear from the spiny crest and description as 'guana' that a tuatara is meant. In some other versions, the shark is replaced by a gurnard.[67]

The Fable of the Shark and the Large Lizard – (Guana).

> In the days of yore the large lizard and the shark lived together in the sea, for they were brothers, both being of the children of Punga. The lizard was the elder and the shark the younger. After some time they fell out, and as the quarrel was great and protracted, the lizard, vexed at the conduct of his younger brother, determined to leave off dwelling in the sea, and to reside on the dry land, so he left the water. But just as he had got on the shore, his brother the shark swam up to where he was on a rock, and wished him to return, saying – 'Let you and I go out to sea, to the deep water.' The lizard replied, with a bitter curse, saying – 'Go thou to the sea, that thou mayst become a relish of fish for the basket of cooked roots. On this, the shark retorted with another curse, saying – 'Go thou on shore that thou mayst be smothered with the smoke of the fire of green fern.' Then the lizard replied, with a laugh, 'Indeed, I will go on shore, away up to the dry land, where I shall be looked upon as the personification of the demon-god Tu, with my spines and ridgy crest causing fear and affright, so that all will gladly get out of my way, hurrah![68]

The identity of tuatara as food is alluded to in this traditional story by the reference to smothering by smoke.[69] Edward Shortland, who as Private Secretary to Governor Hobson and an interpreter of land claims lived in close contact with Māori during the 1840s, was told that ngārara of the 'iguana' type had been caught by lighting a fire at the entrance to the burrow. He added: 'The tribe of Rangitani hunt it in the way above described for food; and it is by them esteemed very good eating.'[70] 'A Party of Officers of the 58th Regiment' were told in 1852 by Māori of the Bay of Plenty that 'A fire lighted at night attracts them from their holes, and they crowd round it', adding that 'Nararas [*sic*], we were told, were formerly used as food by some tribes.'[71]

In a similar vein, Colenso recorded in 1879 that 'I had often heard of the old mode of capturing this (the edible) lizard, which lived in holes (burrows) at the foot of trees, and was made to appear by smoking them out.'[72] He went on to say that: 'forty years ago this animal was still being eaten by an inland tribe named Rangitane'.[73] In an earlier account, from 1844, he indicated that this was not a universal practice: 'The flesh of the *Tuatara* alone [of all the reptiles] is made use of by the natives as an article of food; only however by one or two tribes inhabiting the interior of the island, for which they have been often spoken contemptuously of by their countrymen.'[74]

Another report indicates that to eat a tuatara provided a certain mana or even notoriety to the person involved. A Kaiapoi chief, Hakopa te Ata o Tu, told the missionary James Stack of eating ngārara during the time that he had been held prisoner in Otaki by Te Rauparaha, apparently between about 1828 and 1849.[75]

> During his residence at Otaki, Hakopa often joined in ngarara hunts, when as many as forty were caught and eaten. It would appear from his

> account, and from what other natives say, that the large lizard was formerly eaten, but not as a common article of diet, only by those who had acquired a taste for what was generally regarded with abhorrence. By rising superior to popular prejudice in this matter individuals obtained a certain notoriety in the tribe, which they turned to their own advantage in other ways.[76]

Elsewhere, tuatara were said to have been eaten in the Taranaki district,[77] and ngārara[78] or 'lizards'[79] were eaten in the South Island.[80] Ernst Dieffenbach was told that 'guana' (and kiore, the Pacific rat, *Rattus exulans*) were 'once favourite dishes' when they still occurred in the North Island.[81] Sources referred to in the section above also hint at a possible harvesting of eggs. A sound-producing device known as a pūrerehua (said to have been used among other things to lure lizards) might have been connected with harvesting tuatara.[82] The word 'tuatara' is also connected with different types of soil (for example, tuatara wawata is a brown, crumbly soil suitable for kūmara, a sweet potato),[83] but whether such names are connected with tuatara and their burrowing is not known.

The most detailed accounts of harvesting are those provided by the Tūhoe people of the Urewera district, published by Best. Tuatara were the only reptiles eaten, and had no superstitious feeling attached to them. They were said to have been 'numerous on the mainland in pre-European days, and certain places were famed for the number of these lizards [*sic*] they produced. Such places were Wai-o-hau, Tawhiu-au and Mount Edgecumbe … *Tuatara* were formerly collected for food, placed in baskets, and taken alive to the village, where they were cooked and eaten'[84] (Figure 4.9).

Fig 4.9 'A feast at Mata-ta' by George French Angas. Putauaki or Mt Edgecumbe, visible in the background of this lithograph, was a site famed for harvesting of tuatara, according to traditions recorded by Elsdon Best.

Plate 36 (upper panel) from Angas, G. F. 1847. *The New Zealanders illustrated*. T. McLean, London. Alexander Turnbull Library, Wellington, New Zealand, ref. PUBL–0014–36.

BOX 4.4

Ngāraranui and other reptile-eaters

Ngāraranui (also written Ngārara-nui or Ngārara Nui, meaning 'great reptile') was a son of Rangi-Uru and the chief Whakaue-Kaipapa, who lived on the island of Mokoia in Lake Rotorua and from whom Ngāti Whakaue descend.[1] Ngāraranui acquired his name through the bravery of eating a reptile. In an account recorded by Captain Gilbert Mair of the Bay of Plenty in 1885, 'Ngararanui or Taua went to Kawhia and killed the lizard on top of Pirongia ... [illegible] was the name of the lizard'.[2] A later account adds detail.

> The chief, Ngararanui, was a descendant of Whakaue of Te Arawa. He is always depicted in carving as having a lizard going into his mouth. When he was young he travelled to Waikato and eloped with a beautiful girl of that tribe. Together they arrived at Rotorua; but the girl was not long contented with her new home, and back the couple returned to Waikato. In claiming forgiveness, she exhibited her husband to Waikato and boasted that there was no feat he could not accomplish. After consultation, Waikato chiefs placed a large lizard in front of Ngararanui, daring him to eat it. This the chief reluctantly did, so he was named Ngararanui, and his fame has not lessened to this day.[3]

Another context for the ritualistic eating of a reptile was to prove that one's word could be trusted. T. E. Donne, a former Superintendent of the Department of Tourist and Health Resorts, explained it thus in 1942: 'When a tribe of Maoris contemplated making war, the important men sent delegates to a friendly tribe to solicit their aid in the fighting. If the solicited party agreed to an alliance, the delegates required that their ariki, or paramount chiefs, eat a small lizard in a raw state, in ratification of the pledge. This was the most fearsome and binding oath of integrity that could be administered.'[4] Eating a reptile could also be connected with 'acquiring the art of magic', according to ethnologist Elsdon Best.[5]

These traditions reveal that having to swallow a reptile, especially certain lizards, could be considered as 'the supreme test of a person's resolution and powers', in the words of anthropologist Margaret Orbell.[6] Many carvings of the Rotorua district (Te Arawa), several from the East Coast (Ngāti Porou) and two from the Urewera district (Tūhoe),[7] dating from the late 1870s into the early 1900s, seem to be illustrating the bravery of ancestors in this way.[8]

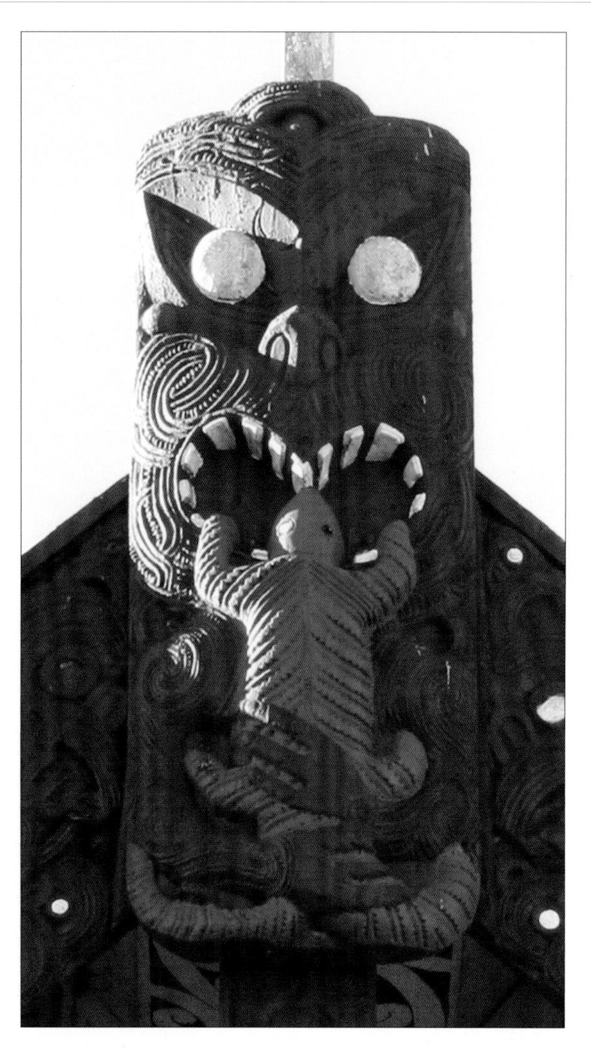

Box 4.4 Fig. 1 (opposite) On each amo of the meeting house Rangitihi by Lake Rotoiti, a large green lizard touches the tongue of an ancestor figure. Intriguingly, the tail of each lizard is itself engulfed by an ancestor's head between the legs of the main figure.

Box 4.4 Fig. 2 (above, left) An amo of the meeting house Kahumatamomoe near Mourea has a green lizard with its head inside the mouth of an ancestor figure.

Box 4.4 Fig. 3 (above, right) The koruru of a house known at times as the Arawa Church, English Church or Nuku-Te-Apiapi portico house at Whakarewarewa in Rotorua.

Photos: Alison Cree.

Box 4.4 Fig. 4 (left) A reptile enters the mouth of an ancestor figure in this Ngāti Porou carving from the late 1870s, part of the unfinished meeting house of a Hawke's Bay chief.

Photo: Alison Cree, © Otago Museum, Dunedin, New Zealand.

Box 4.4 Fig. 5 The koruru on the rear of Te Puāwai o Te Arawa, a pātaka of a chief of Ngāti Pikiao. This ornate storehouse was carved between about 1878 and 1880 by the Ngāti Tarawhai carver Wero, assisted by Tene Waitere, and is now on display in the Auckland Museum.

Photo: Alison Cree.

Later, Best expanded that, according to Māori of the Bay of Plenty district,

> the practice [of eating tuatara] was a common one with their ancestors, and [they] quote the following sayings in corroboration: *Ko Putauaki te maunga, he ngarara tona kai* (Putauaki is the mountain; its food product is the lizard). Putauaki is the native name of Mount Edgecumbe. Another is: *Ko te kekerewai, ko te tuatara nga kai o Wai-o-hau* (The green beetle and the *tuatara* are the food products of Wai-o-hau) ...The two places named were famed haunts of the *tuatara*. Women were not allowed to eat these reptiles. It was said that if they did so then they would be attacked or surrounded by numbers of such reptiles.[85]

Elsewhere, Best was told that women who ate tuatara would probably perish.[86]

Another account of harvesting was given in Māori by Takaanui Tarakawa in the *Journal of the Polynesian Society* in 1911, and translated in the same volume. A tohunga and historian of Ngāti Rangiwewehi and Ngāi Te Rangi, Tarakawa was not universally popular for his efforts to record Māori knowledge in such a public fashion.[87]

> There is a certain *hapu* of Ngati-Hine-kuia who live to the north of Rotorua. At Te Rua-hakoakoa is the dwelling place of the Tuatara, which are found in the holes among the rocks. When those people go to catch them, they start at early daylight, whilst all the others in the village are asleep. The reason for this is, that it is not right that the people of the village should eat whilst the others are absent, because the Tuataras will be angry and they will squirm, and bite those who seek to draw them out of their holes, and will writhe about in the baskets in which they are carried. Should those who go to catch the Tuatara know that the people of the village have eaten, then they return home and are angry on account of the fruitlessness of their journey.[88]

The translator added Tarakawa's view that tuatara were 'still to be found at Te Rua-hakoakoa', a name meaning the hole or burrow of the hākoakoa, a type of petrel.

Traditions of harvesting reptiles could, it seems, be used in psychological warfare. During his travels in Hawke's Bay, Colenso came across an earthwork in the form of a massive reptile on the banks of the River Waitio. Some years later (in 1879) he described it as

> a huge earthwork representation of a *ngarara*, or *ika*, *i.e.*, a lizard, or crocodile, which, several generations back, had been cut and dug and formed in the ground by a chief of that time named Rangitauira ... It had the rude appearance of a huge Saurian extended, with its four legs and claws and tail, but crooked, not straight, as if to represent it wriggling or living, and not dead. It was many yards in length, and of corresponding width and thickness, and by no means badly executed.[89]

The earthwork was known as Te-Ika-a-Rangitauira and had been constructed while Rangitauira's war party had been engaged in a dispute with another pā (fortified village). Unable to complete the attack on that occasion, his party instead created the sculpture (which originally consisted of three such creatures) 'to indicate to the people of the *pa* they had set out to attack, how they had intended to serve (*i.e.* devour) them'. Sadly, by the time of writing in 1879, Colenso reported that the sculpture was 'no longer to be seen as I saw it'.[90]

Teeth of tuatara were apparently also put to use. James Stack, who became an interpreter and an inspector of native schools in Canterbury, was told by locals of a large burrowing ngārara 2–3 feet (60–90 centimetres) long, with a serrated crest from the nape of the neck to the tail. According to Stack's informants, the 'mouth was full of teeth, some grew large and caused the upper lip to project. These when taken from the jaw were three or four inches long, and half an inch at the base; when split in two and polished they were prized as mat pins.'[91] Aurei (mat or cloak pins) were used to pin cloaks together at the shoulder, and sometimes were attached in small bunches to make rattling sounds that drew attention to the wearer.[92] Apart from the size of the teeth, which even with the adjacent skull bones (premaxillae) attached is surely exaggerated,[93] the description recorded by Stack fits well with that of a tuatara.

Past distribution of tuatara on the mainland

From the accounts above, it is clear that some Māori of the 19th and early 20th century retained not only traditions of having harvested tuatara: some also believed that tuatara still existed in parts of the North and South Islands. Beattie and Stack collected many additional references to the recent presence of tuatara in the South Island. On the eastern coast tuatara were said to have lived at Motunau, Takitu (Belfast), Whakamoa (Island Bay, on Banks Peninsula) and in the Catlins district,[94] as well as on the banks of the Waimakariri River westwards and also at Waitui.[95] Most of these places are near to sites where naturally deposited bones of tuatara have been found (see Chapter 2, Figure 2.8). Other locations reported to Beattie included an island near Okarito on the West Coast,[96] and Mason Bay on Stewart Island/Rakiura.[97] Although evidence that tuatara once occurred in these two places is lacking, the presence of tuatara bones on the nearby mainland makes these reasonable possibilities. A reference to tuatara being 'down on Auckland Island'[98] is implausible in a literal sense (the subantarctic Auckland Islands have never been connected to the mainland); an alternative interpretation is that it was meant in the context of a spiritual boundary.[99]

Despite many accounts, no solid evidence exists that tuatara survived into European times as breeding populations on either the North Island or South Island. Natural historians of the 19th century reported, at best, only occasional sightings of tuatara on the mainland (with no specimens preserved to confirm their identity; see Chapter 5, Box 5.1). Most likely, mainland populations were effectively already extinct. An oral tradition of mainland populations could have been reinforced by collections of tuatara from offshore islands as pets or curiosities, with some animals later escaping or being released (see Chapter 5, Box 5.2).

Taniwha: human-eating reptiles of fabulous dimensions

One very real difficulty for those interested in herpetological history is that Māori tradition also included a belief in taniwha, or fabulous monsters (Figure 4.10). These creatures often had reptilian characteristics and were sometimes described using the same word (ngārara) as living reptiles. Many

Fig. 4.10 Moko-hiku-waru, the 'lizard with eight tails', is a well-known taniwha from North Island traditions. Moko-hiku-waru is sometimes attributed to the Taranaki area, but was also said by Tainui to have accompanied their waka to New Zealand and thereafter inhabited the Tamaki River. This contemporary interpretation attributes Moko-hiku-waru with a crest and other tuatara-like features.

of the reptiles featuring in carvings from the 19th century depict taniwha; if one was treated well it could also be a kaitiaki or guardian of local people.

In many accounts, such reptilian monsters would be eventually overcome by a heroic ancestor. One well-known story relates the slaying of Ngārara-huarau, a cave-dwelling creature who took a Māori woman to be his wife. With the help of her relatives, she devised a plan to end his power. The monster was invited to be the guest of her people in a large house built to accommodate him; there, while he slumbered, fires were lit outside and the monster was burnt and speared to death. In one version, the woman later gives birth to a child that is part human and part reptile, which dies.[100] In another version, the monster is female and pursues a human male before being burnt to death.[101]

Other variations on this story were recorded by Beattie in the South Island. In one, Kārara-huarau is said to have had two ure (penises), one for lizards and one for women[102] – which is consistent with the two hemipenes found in living lizards (but lacking in tuatara). Another version included an explanation of the origin of lizard-like reptiles: 'When the great giant Kararahuarau was burnt two inohi (scales) flew out from his funeral pyre and one formed the karara in the sea (a kind of shark) and the other landing on the earth formed the karara on the land (lizards)'.[103]

To add to the complexity, sometimes taniwha were described as tuatara. The following account was related in 1922 by Hare Hongi (born H. M.

Stowell, of Ngā Puhi and European descent). Hongi received his information in 1887 from the old chief Nokora Te Manukarioi:

> [Nokora] assured me that in earlier years the tuatara abounded on Rangitoto off D'Urville Island. He said that all Maoris had an instinctive dread of that reptile, and a cave on Rangitoto Island was reputed to be the home of a most ferocious type, a type that could lacerate and devour a living man. Because of that evil repute the cave was severely avoided. This tradition came to the ears of a young and very daring chief ... despite the warnings of his fellows, and the entreaties of the women, he went off alone on his quest. Naturally the villagers remained uneasy and under a grave anxiety as day after day passed and the young chief failed to return. A party was organised, which proceeded direct to the tuatara cave, and there, to their horror, they saw the fresh skeleton of the young man, stripped of its flesh to the bare bones. They had no difficulty in recognising the remains, and having made a stretcher, they reverently placed them upon it and bore them home ... Nakora [*sic*] personally assured me that of his knowledge he knew that this incident was true; a little later old Wi Hapi Pakau ... also told me of his own knowledge the same story, without variation. As for Nakora, he concluded his story with these rather remarkable words: 'Ki a au nei he ika kino rawa tena, te tuatara; he ika haehae tena i te ao katoa' (literally I am fully convinced that that fish, the tuatara, is a very bad one; the knowledge of its habit of lacerating and devouring mankind is world-wide).[104]

Taniwha were generally creatures of dangerous places, such as caves, or rivers with strong currents, that travellers should avoid. To ensure safe passage, it was necessary to appease the taniwha with an offering and a prayer. Taniwha can thus be viewed as a 'metaphorical way of saying that [a] place is either dangerous or has special spiritual or historical significance'; in other words, that the place should be respected.[105]

Softer regional views?

Māori of the 19th century were not a single and unified people but a collection of interrelated and sometimes hostile tribes. It is unrealistic to expect a single 'Māori view' on reptiles, and indeed, accounts from certain regions, especially the Bay of Plenty and East Coast, seem distinctly more sympathetic than those recounted above.

Ngāti Porou, for instance, have a tradition that tuatara and geckos, among other creatures, were brought to New Zealand from Hawaiki, then released onto Whangaokeno (East Island), on which a tapu was placed to protect them.[106] Generally, items that are said to have come from Hawaiki, such as kūmara and greenstone, were ones that were greatly treasured and valued,[107] so this suggests that reptiles were too. Another East Coast tradition holds that tuatara were created by an ancestor or deity named Peketua, a brother of Tāne (as compared with descent from Punga, as related earlier). Peketua fashioned an egg from clay and gave it life; when the egg cracked open, tuatara emerged. Lizards also seemed to get more gentle treatment, and sometimes were spoken of, with tuatara and insects, as Te Whānau a Peketua (the family of Peketua).[108] An early 19th-century Māori named Uhia is said to have placed a gecko in his hand as a medium for interpreting battle omens;[109] similarly, the famous Tūhoe chief Tu Purewa consulted his pet lizard Peketahi for advice.[110]

For people of the central North Island also, reptiles could be omens of good fortune – even overseas. During World War II, members of the B Company of 28 (Māori) Battalion were drawn from Rotorua, Bay of Plenty, Taupo and Thames-Coromandel. Stationed in Crete during 1941, they saw some of the fiercest fighting of the war. At a critical moment when under threat from German soldiers, Captain Royal had a vision of 'a tuatara with its head pointing directly over a cliff'. Seizing the inspiration, he led his men down the steep cliff, whence they escaped.[111] A similar premonition of good fortune, narrated by Sonny Sewell, concerned a lizard found on his boot: 'I let it play around in my hand a bit and one of my uncles came along and asked me what I had … and he said, "We're saved … we'll survive this war." When I asked why, he said: "That's our guardian. That's our taniwha, that one there. He'll look after us."'[112]

In a southern account, the Ngāti Māmoe (Kāti Māmoe) chief Te Raki-Tauneke is said to have had a guardian spirit in the form of a giant lizard named Matamata. On occasions, Matamata saved or restored his master from certain death in battle by hypnotising his enemies, carrying him on his back or licking him all over to reassemble his severed body parts. When Te Raki-Tauneke left his dwelling near Dunedin to travel and neglected to inform his guardian lizard, the distraught Matamata set out to look for him, forming various features of the landscape such as lagoons and river bends as he passed.[113]

Reptiles in the carvings of Te Arawa and Ngāti Porou

Few if any examples of reptiles in Māori carvings from pre-European times are known, perhaps because of the association between reptiles and death.[114] Yet reptiles later featured prominently in carvings from the Bay of Plenty and East Coast regions, where carving traditions continued to develop following European arrival. Carving within the Te Arawa tribes of Ngāti Tarawhai and Ngāti Pikiao, especially, flowered in the late 1800s. In part, this was a response to wealthy tourists drawn to the thermal wonderland near Rotorua, and to government officials and other patrons who began to appreciate the attractions of Māori art for the young colony.

However, Māori were themselves driving demand for a new type of carving – large whare whakairo (carved houses), which served as tribal assembly halls, symbols of status and repositories of ancestral history. Meeting

BOX 4.5

Whare whakairo: carved houses

By the mid-1800s war canoes, once a premier vehicle for carving prowess, were in declining demand among Māori. Instead carved houses in the form of large meeting houses (sometimes known as whare rūnanga or wharenui) became increasingly popular. Meeting houses provided a venue for large tribal gatherings and served as symbol of hapū pride and identity, incorporating aspects of the earlier chief's house, guesthouse and European-style church under one roof.[1] The arrival of metal tools meant that large carved surfaces could be created more quickly, and polychrome paints allowed an expanded colour palette. Often, however, 'new' houses incorporated carvings from earlier work, making it sometimes difficult today to know the age or significance of features such as reptiles.

Box 4.5 Fig. 1 Rangitihi, a whare whakairo of various hapū of Ngāti Pikiao, including Ngāti Hinerangi, lies on the shore of Lake Rotoiti. Carved by George Emery, Kaka Niao and others, Rangitihi was opened in 1951. Here the restoration is examined by James Schuster of Te Arawa and the Historic Places Trust Pouhere Taonga. Each amo features an ancestor figure with a large, green, lizard-like reptile on its tongue. Lizards on exterior surfaces are often painted green in carved houses of the Rotorua region today.

Photo: Alison Cree.

Meeting houses are often named for a significant ancestor. The physical form of the house symbolises the ancestor's body: the tāhuhu or ridgepole represents the backbone; the poupou or side wall posts, the ribs; the koruru or carved face, the head; the maihi or bargeboards, the embracing arms; and so on.[2] Below is a list of main components mentioned in the text, including those on which reptiles might appear.[3]

tekoteko – freestanding ancestor figure at the apex of the maihi
amo – post at the side front, supporting the maihi
koruru – carved face at front apex of house; gable head
maihi – barge board
pou – post (in general)
poupou – side wall post
poutāhu(hu) – central post of front wall inside house, supporting ridgepole
poutokoihi – central post at front of porch, supporting ridgepole
poutokomanawa – central free-standing post inside house, supporting ridgepole
poutuarongo – central post of rear wall inside house, supporting ridgepole
tāhuhu – ridgepole

houses provided new surfaces on which to carve, and with a tradition of reptiles as kaitiaki in the central and eastern North Island, there were obvious reasons for reptiles to be featured. Surviving examples from the late 19th and early 20th century show lizard-like reptiles crawling up and down pou, as accompanying figures on tekoteko and koruru, on amo, and on poupou, inside or out (see Box 4.5 for explanation of terms). They also lead in file, mouth to tail, down the exterior tāhuhu of the famous pātaka (storehouse) known as Te Puāwai o Te Arawa (Flower of Te Arawa; Figure 4.11). Several examples of reptiles on carved meeting houses of the Rotorua area are described below.

Fig. 4.11 Lizards lead mouth to tail along the tāhuhu or ridgepole (shown here prior to assembly) of the pātaka Te Puāwai o Te Arawa. This large and richly detailed storehouse was carved between 1878 and 1880 by the master carver Wero, assisted by Tene Waitere (Ngāti Tarawhai). The pātaka, which remains on display in the Auckland Museum, was also known as Tuhua Kataore (the pit or lair of Kataore, a guardian lizard for local people).

Photo: G. Archey, Auckland Museum, B10392.

BOX 4.6

Kataore: the taniwha-lizard of Tikitapu

The tale of Kataore illustrates well the intermingled identities of taniwha and living species of reptiles in Māori tradition. Kataore was one of several famous taniwha of the Rotorua area. To the local people (including those who became Ngāti Tarawhai and Ngāti Hinemihi), she was the harmless pet lizard of the ancestress Hinemihi, and was viewed benignly by the chief Tangaroa Mihi.[1] But to strangers who came too close to the creature's lair, in and around the shores of Tikitapu (Blue Lake), Kataore was a dangerous monster, or taniwha.

As one group after another of travellers disappeared, the people of neighbouring Ngāti Tama grew suspicious. A party of experienced taniwha-hunters was dispatched to capture and destroy the monstrous reptile. Kataore was snared with ropes, and eventually killed and eaten. This was the source of warfare between the tribes.

Box 4.6 Fig. 1 The warriors of Ngāti Tama who eventually overcame Kataore also killed several other taniwha of the central North Island. In this modern interpretation, warriors are shown overcoming the taniwha Hotupuku of Kaingaroa Plains near Kapenga (about 30 kilometres from Tikitapu), using a similar method to that which they would later use on Kataore. The insides of the two taniwha, when opened, were full of the bodies and possessions of the people that they had eaten. Both taniwha were eaten by their captors.

Kataore the taniwha was vanquished, but belief in Kataore the guardian lizard has remained strong. For Wero, Tene Waitere and other master carvers of Ngāti Tarawhai, Kataore was an important and frequently used motif. There are examples (some with the name Kataore written alongside) in the whare whakairo Hinemihi, famously buried in the Tarawera Eruption of 1886 and now located in England, in another Hinemihi in Whakarewarewa, and on a short post outside a third Hinemihi near Rotorua. Kataore was probably the basis for many other carvings of reptiles by Ngāti Tarawhai carvers, including some illustrated elsewhere in this chapter.[2]

Jim Schuster, a descendent of Tene Waitere's family, explains that Kataore has remained a kaitiaki of Ngāti Tarawhai to this day. A modern version of Kataore guards the grave of Emily Schuster, Jim's mother, and the sight of a live lizard, whether in New Zealand or overseas, can provide an important validation to family members of the right course of action in moments of uncertainty or tension.[3]

Box 4.6 Fig. 2 The whare Hinemihi at Whakarewarewa was carved by Tene Waitere (1854–1931) of Ngāti Tarawhai for his granddaughter Guide Rangi, and opened in 1928. This house of treasures is lovingly maintained today by Tene and Rangi's descendants. Kataore, the guardian lizard of Ngāti Tarawhai, features on the poutokomanawa (an interior support post). Although this Kataore has a tuatara-like crest, its head is perhaps more like that of a gecko, illustrating the stylised nature of taniwha carvings.

Photo: Alison Cree.

Box 4.6 Fig. 3 (left) A depiction of Kataore (with the name carved below) on a post outside the meeting house Hinemihi of Ngāti Hinemihi near Rotorua. The pou originated as a fence post at Whakarewarewa Village.

Photo: Alison Cree.

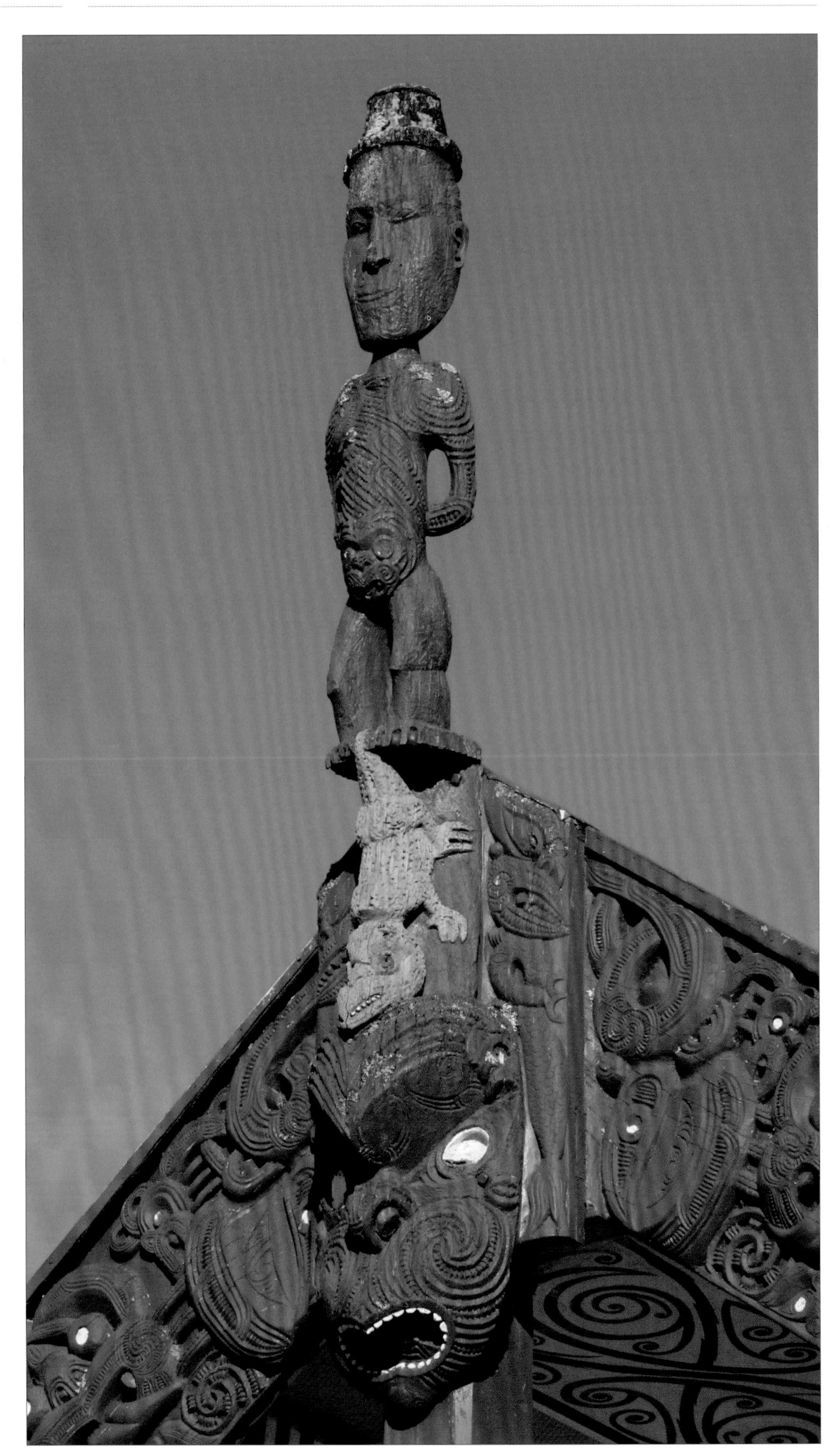

Fig 4.12 The tekoteko (ancestor figure, with top hat) and koruru of Houmaitawhiti, at Otaramarae, Lake Rotoiti. Between the two is the guardian lizard Wharetoroa, in green. This carving, by Wero of Ngāti Tarawhai, dates from the 1870s. Wero is noted for his quirky inclusion of European-style garments.

Photo: Alison Cree.

For convenience, these are described as 'lizards' following popular usage; readers may consider for themselves whether they have the characteristics of tuatara or not (but see Box 4.1 for cautions).

Houmaitawhiti, a fully carved house of Ngāti Hinekura (a hapū of Ngāti Pikiao), is regarded as 'one of the oldest completely carved meeting houses still standing on a marae in New Zealand'.[115] It was constructed at Maketu in the 1870s but was later shifted and re-erected in its current location on Pounamunui Marae at Otaramarae, on the shore of Lake Rotoiti, during the early 1900s; it was renovated in the 1960s. The house is named after the ancestral chief of Te Arawa who encouraged his people to voyage from Hawaiki. Below Houmaitawhiti on the tekoteko is the guardian lizard Wharetoroa (Figure 4.12), which also appears (along with another lizard, Pounamunui), on the poutuarongo.

Several well-known examples of carved reptiles come from **Nuku-Te-Apiapi** ('Abode of the multitudes'), no longer standing. Erected in 1906 as a show house for European tourists alongside the Geyser Hotel at Whakarewarewa,[116] Nuku-Te-Apiapi featured two striking lizards about 1.3 metres long on the poutokoihi (Figure 4.13). Carved by Tene Waitere of Ngāti Tarawhai, these represent the kaitiaki or taniwha Kataore. Unlike many carvings of reptiles on exterior surfaces today, they are painted red-brown rather than green. Also carved by Tene, the door panel (in black) depicted the notorious bird-woman Kurangaituku, with her pet lizards between her legs (the window alongside illustrated Hatupatu, a cultural hero of Te Arawa, who escaped from and finally killed Kurangaituku). Kurangaituku's lizards are rather tuatara-like, albeit without crests. Both the post and the door panel remain on display today in the Rotorua Museum of Art and History Te Whare Taonga o Te Arawa. Also in Nuku-Te-Apiapi, the end of the tāhuhu above the porch had a lizard crawling up the body of a human-like figure.[117]

Fig. 4.13 An elegant pair of lizards, representing Kataore, descends the poutokoihi of Nuku-te-Apiapi. The post was carved during 1904–1905 by Tene Waitere and survives today, painted red-brown, in the Rotorua Museum (the meeting house was dismantled in 1946).

Photo: James McDonald, 1920 (cropped image). Museum of New Zealand Te Papa Tongarewa MA_B.000096.

An intriguingly different way of depicting reptiles is found on the whare **Rakeiao** of Ngāti Rongomai, a hapū of Ngāti Pikiao. Rakeiao was built not long after the Tarawera eruption of 1886[118] and erected in its current location, on the shore of Lake Rotoiti, in 1912. On both amo, the ancestor figures (which probably represent descendants of Rakeiao) have lizards as eyes (Figure 4.14). A kaumātua of Rakeiao Marae recalls being told as a child that

Fig. 4.14 Unusually, both amo of the whare Rakeiao (of Ngāti Rongomai) incorporate lizards as part of the ancestor-figures' eyes. The carvings, which are believed to date from 1886, have recently been renovated.
Photo: Alison Cree.

the lizards were the guardian protectors of Rakeiao, and always on the lookout for pending danger.[119]

A final Te Arawa example is that of **Kahumatamomoe**, near Mourea on the northern shore of Lake Rotorua. This Ngāti Pikiao house was reconstructed in its current location in 1914, using timbers from an older house.[120] Here the poutokoihi fairly crawls with stunning green lizards – all nine of which are individually named (Figures 4.15, 4.16). The right amo of the same whare features an ancestor figure with a green lizard entering its mouth. Inside, a large lizard – said to represent an atua named Awhipapa – is carved on the poutāhuhu.[121]

As well as on houses, reptiles featured on other carvings of Ngāti Tarawhai and Ngāti Pikiao during the late 19th and early 20th century, including those produced for the tourist market. Examples include ceremonial paddles and ornate 'zoomorphic' bowls with legs.[122]

Reptiles were also depicted on the carved houses of Ngāti Porou of Te Tai Rāwhiti (East Cape to Gisborne). Three lizards appear on the doorjamb of the right side of **Rauru nui a Toi**, built about 1850 but with some reconstruction since.[123] Two lizards flank a human head on the windowsill of **Tuauau**, built about 1880.[124] A pair of lizards appears, unusually in profile, between human figures on a baseboard from an unknown house, now in a Swiss museum.[125] Spectacular carvings from the unfinished house of the Hawke's Bay chief Karaitiana Tumoana (Takamoana), attributed to itinerant carvers of Ngāti Porou working in the late 1870s and now on display in the Otago Museum, feature reptiles in several places (Figures 4.17–4.19).[126] Some are small and lizard-like, placed discreetly between the legs of human figures; others, more tuatara-like, are gripped firmly by the hand, one entering the mouth. The naturalistic style of animals in these carvings is considered a relatively modern development.[127]

Memories of reptiles in the landscape

Given the vivid role that reptiles, both natural and spiritual, once played in the Māori world, it is no surprise to learn that many became incorporated in placenames or features of the wider landscape. Examples in official use today include the Maungataniwha Range (mountain of the taniwha, in Northland), near which one may find Tuataranui Stream (stream of the great tuatara); Ngararahuarau Stream (near Taumarunui); Ngararanui Stream (near Lake Taupo); the localities Ruataniwha (lair of the taniwha, in Hawke's Bay) and Motukarara (island of reptiles, on Banks Peninsula); Karara Stream (in Otago, inland of Oamaru); and four islands named Motungarara (island of reptiles: in the Bay of Islands, in Lake Waikareiti in Urewera National Park, near Kapiti Island, and in Queen Charlotte Sound).

Yet other places were described in the 1800s and early 1900s as having connections with reptiles. Sacred red rocks on the hillside at Port Levy on Banks Peninsula were known as Te Ngarara, and great offence was caused

Fig. 4.15 Lizards adorn the poutokoihi of Kahumatamomoe, a whare of Ngāti Pikiao. A lizard also appears on the right amo.

Photo: Alison Cree.

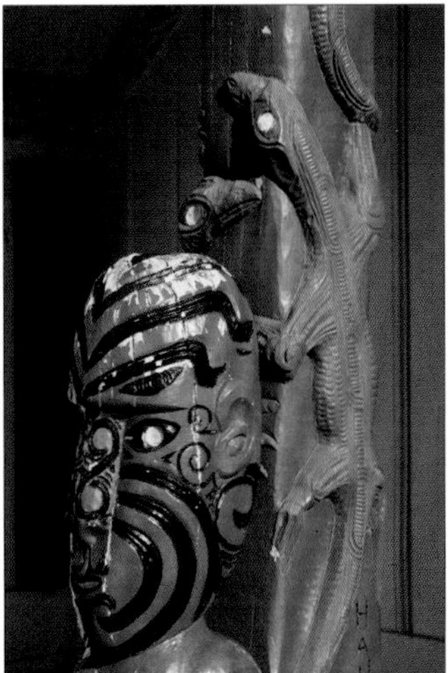

Fig. 4.16 Detail of a lizard and ancestor figure at the base of the poutokoihi of the whare Kahumatamomoe.

Photo: Alison Cree.

when a European surveyor stood on them.[128] A winding part of the 'Taiari' (i.e. Taieri) River known as Te Rua-taniwha was haunted by the ghost of a kārara. Rocks in a field near Orawia, in Southland, were described as the petrified remains of a man-eating kārara, and caves nearby at Clifden were known as Te-ana-o-te-karara[129] (cave of the reptile; a patch of forest nearby is known officially as Te Karara Bush). Members of Ngāi Tahu have told me

that rock formations in the southern landscape can still speak of the past presence of large reptiles.[130]

Concluding remarks

The discovery and settlement of New Zealand by East Polynesian people who became Māori provides a unique South Pacific example of contact between people and reptiles. Written records from the 19th century suggest that Māori distinguished tuatara from at least three major kinds of lizards, the latter corresponding at least broadly to 'brown' geckos (moko-pāpā), green geckos (moko-kākāriki) and skinks (mokomoko). Views about reptiles were complex, however, and seem to have varied both regionally (among iwi) and with the species concerned, and perhaps among individual people too. Reptiles could be tapu (and therefore feared or even dreaded), but also kaitiaki (guardians). Traditions about living reptiles merged with those of taniwha, which were human-eating creatures of supernatural status, often reptilian in form.

Traditional knowledge among Māori must surely have been evolving from the time of first settlement of New Zealand. Following European arrival, oral traditions would have been exposed to new influences, including Christianity, the introduction of a moneyed economy and scientific study. Carving traditions in general have clearly continued to evolve over the last 200 years, but specific influences on the ways that reptiles were viewed or illustrated in various media have yet to be closely examined.

Today there is increasing recognition of the role that traditional ecological knowledge, transmitted orally by indigenous cultures, can contribute to ecological management and conservation as well as to cultural diversity. Traditional ecological knowledge about tuatara from both 19th- and early 20th-century records until the present day[131] seems limited, at least in part because dwindling abundance restricted tuatara to offshore islands that often were declared tapu; by comparison, more detailed information is available about species that continued to be harvested, such as tuna or eels (*Anguilla* spp.), and harakeke or flax (*Phormium tenax*).[132] However, information about tuatara was not non-existent, and reported or inferred methods for harvesting might, with appropriate modification, contribute to new survey techniques. Another possibility is that traditional records about harvesting might provide clues as to the past abundance and distribution of tuatara, not only on offshore islands in recent decades,[133] but also on the mainland in the 19th century. In other words, perhaps tuatara survived most recently in the central and lower North Island, where traditions of harvesting seem stronger than further north or south.

Modern comparisons of 19th-century traditions between Māori and indigenous cultures elsewhere, including but not limited to Polynesia, would be a fascinating topic; such comparisons are, however, beyond the scope of this chapter. Māori were by no means unique in viewing reptiles as the holders of supernatural powers, or as food; such has been – and sometimes continues to be – the fate of reptiles in many cultures.[134] An important message for herpetologists is that reptiles were not ignored by Māori: they held (and continue to hold) a special and distinctive place, and there is much opportunity for further dialogue about our shared interest. Future interpretations of traditions in early written records,[135] and in current times (ideally using carefully chosen control groups), including by herpetologists who are also Māori, will aid our understanding of this intriguing topic.

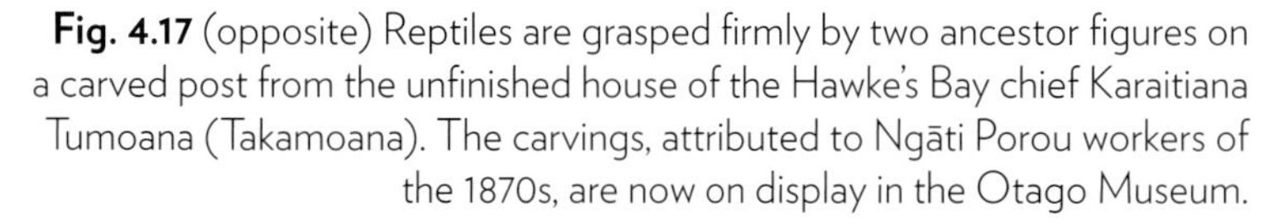
Fig. 4.17 (opposite) Reptiles are grasped firmly by two ancestor figures on a carved post from the unfinished house of the Hawke's Bay chief Karaitiana Tumoana (Takamoana). The carvings, attributed to Ngāti Porou workers of the 1870s, are now on display in the Otago Museum.

Photo composite: Alison Cree and Ken Miller, © Otago Museum, Dunedin, New Zealand.

Fig. 4.18 (top right) Detail from upper part of Figure 4.17.

Photo: Alison Cree, © Otago Museum, Dunedin, New Zealand.

Fig. 4.19 (right) A tuatara-like reptile held firmly by an ancestor figure, on a post from the unfinished house of Karaitiana Tumoana (Takamoana). The naturalistic style of animals in these Ngāti Porou carvings is considered a relatively modern development.

Photo: Alison Cree, © Otago Museum, Dunedin, New Zealand.

Endnotes to Chapter 4

See the bibliography for full references to articles on tuatara, which are cited here at first mention by surname and year only. References on other topics are cited here in full at first mention, and then using 'op. cit.'. Except where required for clarity, the convention of parentheses around years has been dispensed with for economy of space.

1 Lowe, D. J., Newnham, R. M., McFadgen, B. G. and Higham, T. F. G. 2000. Tephras and New Zealand archaeology. *Journal of Archaeological Science* 27: 859–870; Wilmshurst, J. M., Anderson, A. J., Higham, T. F. G. and Worthy, T. H. 2008. Dating the late prehistoric dispersal of Polynesians to New Zealand using the commensal Pacific rat. *Proceedings of the National Academy of Sciences* 105: 7676–7680; Wilmshurst, J. M., Hunt, T. L., Lipo, C. P. and Anderson, A. J. 2011. High-precision radiocarbon dating shows recent and rapid initial human colonization of East Polynesia. *Proceedings of the National Academy of Sciences* 108: 1815–1820.

2 The term 'Māori' (usual or ordinary), as a uniting term for all tangata whenua, and the corresponding term 'Pākehā' (New Zealander of European descent), developed after first contact as people perceived the need for a binary distinction.

3 Duff, R. 1977. *The moa-hunter period of Maori culture.* 3rd edn. E. C. Keating, Government Printer, Wellington. See pp. 29, 222.

4 Hamilton, A. 1903. Notes on a bone pendant in the form of a lizard(?), found on the sandhills at Wainui; and on some other bone objects. *Transactions of the New Zealand Institute* 35: 111–113 + plates VIII–IX.

5 Prickett, N. 1999. *Nga tohu tawhito: early Maori ornaments.* David Bateman and Auckland Museum, Auckland. See p. 29 for illustration of turtle pendant.

6 Gerard O'Regan, Project Director, South Island Māori Rock Art Project, pers. comm. 2003.

7 Skinner, H. D. 1923. The lizard in Maori art and belief. *Journal of the Polynesian Society* 32: 184–185. Some examples, including faint lizard-like creatures, are on display in the Otago Museum, November 2012.

8 http://www.ngaitahu.iwi.nz/rockart/, accessed 11 July 2012.

9 For a very readable introduction to te ao Māori (the Māori world), including background to this paragraph, see Chapter 4 (pp. 76–91) of King, M. 2003. *The Penguin history of New Zealand.* Penguin, Auckland.

10 However, in a tradition from the far south, Tangaroa (Takaroa) is seen as the first husband of Papa and the father of at least one of her children; Rangi (Raki), the nephew of Tangaroa, is her second husband. See p. 6 of Tremewan, C. 2002. *Traditional stories from southern New Zealand: He kōrero nō Te Wai Pounamu.* Macmillan Brown Centre for Pacific Studies, University of Canterbury, Christchurch.

11 Grey, G. 1855. *Polynesian mythology & ancient traditional history of the New Zealanders, as furnished by their priests and chiefs.* John Murray, London. See pp. 1–15. In a similar version from Ngāti Hau of the Wanganui region, Tangaroa was 'the ancestor of all the fish, and of the tuatara and other kinds of lizards' [sic] (from John White's *Ancient World of the Maori*, as reproduced in Anon. 1965. The gods of the Ancient Maori World. *Te Ao Hou* 52: 16–20).

12 Auckland Museum. 2001. *Te ao turoa education kit.* Auckland Museum, Auckland. Available at http://www.aucklandmuseum.com/68/teacher-resources (accessed 11 July 2012).

13 In an account recorded by Shortland, Tāne, son of Papa and Rangi, fathers lizards as well as trees, plants, water and stones. See Tremewan 2002, op. cit., p. 48.

14 Simmons, D. R. 1987. *Whakairo: Māori tribal art.* Oxford University Press, Auckland. See p. 49.

15 Cruise, R. A. 1957. *Journal of a ten months' residence in New Zealand [1820].* (Bagnall, A. G. (ed.)). Pegasus Press, Christchurch. See pp. 184, 217. See also Colenso 1879, p. 84.

16 Beattie 1994. See p. 187.

17 Polack 1840. Vol. I, excerpts from pp. 241–264.

18 For other examples of reptiles between an ancestor figure's legs, see Best 1923.

19 Beattie, H. 1990. *Tikao talks (Ko taoka tapu o te ao kohatu: Treasures from the ancient world of the Maori, as told by Teone Taare Tikao).* Penguin Books, Auckland. See pp. 92, 94.

20 Cooper, G. S. 1851. *Journal of an expedition overland from Auckland to Taranaki by way of Rotorua, Taupo, and the West Coast, undertaken in the summer of 1849–50, by His Excellency the Governor-in-Chief of New Zealand.* Williamson and Wilson, Auckland. See p. 138.

21 Field, H. C. 1882. *Naultinus sylvestris, Buller. Journal of Science* [old series] 1: 177–178 (see p. 178 for quote).

22 Angas, G. F. 1847. *Savage life and scenes in Australia and New Zealand: being an artist's impressions of countries and people at the Antipodes, with numerous illustrations. Vol II.* 2nd edn. Smith, Elder and Co., London. See p. 67.

23 Stack, J. W. 1936. *More Maoriland adventures of J. W. Stack* (Reed, A. H. (ed.)). A. H. and A. W. Reed, Dunedin. See pp. 195–196.

24 Earle, A. 1832. *A narrative of a nine months' residence in New Zealand, in 1827; together with a journal of residence in Tristan d'Acunha, an island situated between South America and the Cape of Good Hope.* Longman, Rees, Orme, Brown, Green and Longman, London. See p. 142.

25 Yate, W. 1835. *An account of New Zealand; and of the Church Missionary Society's mission in the northern island.* 2nd edn. Seeley and Burnside, London. See p. 71.

26 Moser, T. 1863. *Mahoe leaves: being a selection of sketches of New Zealand and its inhabitants.* William Lyon, Wellington. Reprint published 1974 by Capper Press, Christchurch. See pp. 76–81.

27 Downes 1937, pp. 217–218. Elsdon Best, New Zealand's first professional ethnographer, also related from the Wanganui district the tradition of killing a lizard, then burning it while reciting chants; see Best 1923.

28 Beattie 1994. See p. 382.

29 Beattie, H. 1920a. Nature-lore of the southern Maori. *Transactions and Proceedings of the New Zealand Institute* 52: 53–77 (see pp. 55–56 for quote) For another account of probably the same incident, see Beattie, H. 1919. Traditions and legends collected from the natives of Murihiku (Southland, New Zealand). Part XI. *Journal of the Polynesian Society* 28: 212–225.

30 Reed, A. W. 2002. *Taonga tuku iho: Illustrated encyclopedia of traditional Māori life.* (Revised by Mikaere, B. (ed.)). New Holland, Auckland. See also Orbell 1996, pp. 89–90.

31 Orbell 1996, p. 89.
32 Some species have recently been placed in new genera. See Nielsen, S. V., Bauer, A. M., Jackman, T. R., Hitchmough, R. A. and Daugherty, C. H. 2011. New Zealand geckos (Diplodactylidae): cryptic diversity in a post-Gondwanan lineage with trans-Tasman affinities. *Molecular Phylogenetics and Evolution* 59: 1–22.
33 A very dark-skinned child might sometimes be jokingly called a mokopāpā. See pp. 148 and 157 of Beattie 1990, op. cit.
34 Downes 1937, p. 218.
35 Downes 1937, p. 218.
36 For illustration, see p. 160 of Brown, D. 2003. *Tai Tokerau whakairo rākau: Northland Māori wood carving*. Reed Publishing, Auckland.
37 This kōauau, now in the collection of the British Museum, is illustrated on p. 48 of Davidson, J., Te Awekotuku, N., Hakiwai, A. T., Neich, R., Pendergrast, M., and Starzecka, D. C. (eds). 1996. *Maori art and culture*. David Bateman, Auckland.
38 Brown 2003, op. cit. See p. 152.
39 In a traditional story, Whiro had a lizard carved on his penis that was seen by his daughter-in-law. One suggestion is that the daughter-in-law may have been attracted in part by the powerful status that the lizard implied; in any case, the pair later committed incest. This account was recorded on Ruapuke Island in Foveaux Strait in the 1840s or early 1850s by the German missionary J. F. H. Wohlers. See pp. 314 and 320 of Tremewan (2002, op. cit.). On the same theme, Ramstad et al. (2007a) interviewed elders of Te Atiawa and Ngāti Koata as well as a member of Ngāti Wai/Ngā Puhi in 2004, who reported that women would sometimes have reptiles tattooed near their genitals to signify a powerful boundary.
40 Best 1923.
41 Hamilton, A. 1901. *Maori art*. New Zealand Institute, Wellington. See p. 205.
42 Best 1923, p. 331.
43 Orbell, M. 2003. *The illustrated encyclopedia of Maori myth and legend*. Canterbury University Press, Christchurch. See p. 155.
44 A Party of Officers of the 58th Regiment 1852. By the 1870s, however, Māori were assisting Gilbert Mair with his observations and collections of tuatara from islands in the Bay of Plenty (see Chapter 5).
45 Beattie 1994. See p. 507. See also p. 509.
46 Goodlet c. 1922, p. 17.
47 Christine Baines (Ngāti Manuhiri) and Whetu McGregor (Ngāti Rehua), pers. comm. 12 November 2006. See also Whetu McGregor's account in Ramstad et al. 2007a; Levine 2010; Lyver, P. O'B. and Moller, H. 2010. An alternative reality: Maori spiritual guardianship of New Zealand's native birds. In: Tideman, S. and Gosler, A. (eds). *Ethno-ornithology: birds, indigenous peoples, culture and society*. Earthscan, London, pp. 240–264.
48 As recorded in a newspaper account of the activities of a Mr G. G. Norris (probably brother of Colin Norris of Tauranga): 'As both hands are needed [when moving about Karewa Island], it is impossible to carry anything to put the lizards [sic] in, and the method consequently followed is to wear a loose shirt and place them inside it, where the warmth of the body keeps them quiet and harmless' (Anon. 1908a).
49 This thwart, one half of which is shown in Figure 4.8, is illustrated in its entirety on p. 89 of Best, E. 1925. *The Maori canoe: an account of various types of vessels used by the Maori of New Zealand in former times, with some description of those of the Isles of the Pacific, and a brief account of the peopling of New Zealand*. Dominion Museum Bulletin No. 7, Dominion Museum, Wellington. Best considered that the thwart was probably carved with metal tools, implying that it post-dates European arrival. Several other examples of taumanu with reptiles exist. One, with two heavy-jawed, tuatara-like reptiles facing outwards, is from Waiuku (Ngāti Te Ata; AM 46039; see Brown 2003, op. cit., p. 94 for photo). Another, in simpler style, is from Kawhia (AM 6178), and one with a distinct crest is from the East Coast (for illustration, see p. 83 of Neich, R. 1996. *Wood carving*. In Davidson et al. (Starzecka, (ed.) op cit.), pp. 69–113. A taumanu that appears to have been reconstructed in or after the 1930s, and which may have been based on the example shown in Figure 4.8, is found on the canoe *Teremoe*, recently on display in the Museum of New Zealand Te Papa Tongarewa.
50 Best 1925, op. cit. See p. 89. This interpretation was elaborated upon by Barrow, T. 1984. *An illustrated guide to Maori art*. Heinemann Reed, Auckland. See p. 62: 'Lizards were sometimes carved on a thwart [of a waka] near the stern to reserve it for the use of the commanding chief and others of authority who stood to chant and spur on the paddlers'.
51 Kingsley-Smith 1966, p. 124.
52 Kingsley-Smith 1966.
53 Donne 1942.
54 Beattie 1994. See p. 187.
55 Beattie 1920a, op. cit.
56 Stack 1875, p. 296.
57 Colenso 1844, p. 81; see also Colenso 1846.
58 Taylor, R. 1873. An account of the first discovery of moa remains. *Transactions and Proceedings of the New Zealand Institute* 5: 97–101.
59 Downes 1937, p. 216.
60 Bird, W. W. 1955. Songs of the Maori. *Te Ao Hou* No. 12: 20–24.
61 Anon. 1954. Te Rangihiroa: his burial marks the end of an epoch. *Te Ao Hou* No. 9: 34–40, 43 (see p. 34 for quote).
62 A Party of Officers of the 58th Regiment 1852.
63 Hongi 1922, p. 15.
64 Beattie 1994, see p. 509. See also Beattie 1920a, op. cit.
65 Cree and Tyrrell 2001; Refsnider et al. 2009. See Chapter 8 for more on nest-guarding behaviour.
66 Blanchard 2002, p. 88.
67 Best 1909.
68 Colenso 1879, p. 101.
69 Shortland, E. 1856. *Traditions and superstitions of the New Zealanders: with illustrations of their manners and customs*. Longman, Brown, Green, Longmans and Roberts, London. See p. 57.
70 Shortland 1856, op. cit. See p. 57.
71 A Party of Officers of the 58th Regiment 1852, p. 3.
72 Colenso 1879, p. 101.
73 Colenso 1879, p. 101.
74 Colenso 1844, p. 95; see also Colenso 1846. In an early herpetological discussion between Māori and Europeans, during James Cook's third voyage in 1777, reference was also made to reptiles that were killed by making fires at the mouths of burrows. However, in that case the reptiles were said to be about 8 feet in length and man-eating, and thus clearly of the taniwha variety (see Chapter 5).
75 Based on the dates of Te Rauparaha-led attacks by Ngāti Toa on Ngai Tahu at Kaiapoi pā during c. 1828–1831, and of Te Rauparaha's death at Otaki in 1849. See Oliver, S. 2003. Te Rauparaha ? – 1849. *Dictionary of New Zealand Biography*,

updated 31 July 2003 (http://www.teara.govt.nz/en/biographies, accessed 11 July 2012).
76 Stack 1875, p. 297.
77 Downes 1937, op. cit.; Phillips, W. J. 1997. *Maori carving illustrated*. 4th edn. (Revised by Simmons, D. R.) Reed, Auckland. See p. 31.
78 Stack 1875.
79 Beattie 1920a, op. cit.
80 Elders among Te Atiawa, Ngāti Koata and Ngāti Wai/Ngā Puhi interviewed in 2004 also acknowledged a possible consumption of tuatara as ritual, for example, to demonstrate bravery or to acquire knowledge, but harvesting for food was more controversial. See Ramstad et al. 2007a, b.
81 Dieffenbach 1843, Vol. 1. See p. 45.
82 A pūrerehua (also known as a 'bull roarer') consisted of a blade made of wood, stone or bone, swung round overhead on a string to make a whirring sound. Although it could be used to summon rain and farewell the dead, it was also used to lure lizards (Mo Tātou – Ngāi Tahu display, Museum of New Zealand Te Papa Tongarewa, seen March 2007). Tuatara are easily drawn from their burrows by strange sounds (see Chapter 7 of this book): could sound production have been a way of attracting them for harvesting?
83 Williams, H. W. 1957. *A dictionary of the Maori language*. 6th edn. Government Printer, Wellington.
84 Best 1909, pp. 235–237.
85 Best 1923, pp. 331–332.
86 Best 1909.
87 Ballara, A. and Tapsell, P. 2010. Tarakawa, Takaanui Hohoia 1952–1919. In: *Dictionary of New Zealand Biography*. http://www.teara.govt.nz/en/biographies/3t4/1 (accessed 11 July 2012).
88 Tarakawa 1911, pp. 40–41.
89 Colenso 1879, pp. 85–86.
90 Colenso 1879, p. 85.
91 Stack 1875, p. 296.
92 Auckland Museum display, seen November 2006.
93 Perhaps the entire jaw rather than just the teeth was used (with thanks to Marc Jones, University College London, for discussion on the size of tuatara teeth and premaxillae).
94 Beattie 1994. See pp. 187, 349, 350, 354.
95 Stack 1875.
96 Beattie 1994. See p. 530.
97 Beattie 1920a, op. cit.
98 Beattie 1920a, op. cit.
99 Stephen Scott, University of Otago, pers. comm. 2007.
100 Best, E. 1893. Te patunga o Ngarara-Huarau. *Journal of the Polynesian Society* 2: 211–219.
101 Tremewan 2002, op. cit., pp. 337–345. In this account, 'Te Ngārara-hua-rau' is translated as 'Lizard-of-numerous-progeny'.
102 Beattie, H. 1920b. Traditions and legends collected from the natives of Murihiku (Southland, New Zealand) Part XII. *Journal of the Polynesian Society* 29: 128–138.
103 Beattie 1994. See p. 434.
104 Hongi 1922, p. 15. Hongi is said to have been 'often gleefully theatrical' when recounting his life experiences, a trait that may have entered this account. See Gibbons, P. J. Stowell, Henry Matthew 1859–1944. *Dictionary of New Zealand Biography* http://www.teara.govt.nz/en/biographies/1h32/1, (accessed 11 July 2012).
105 Michael King, quoted in Hubbard, A. 2003. Hunting for the taniwha – and Maori. *Sunday Star Times*, 2 February, p. 5.
106 Orbell 2003, op. cit. See p. 104.
107 Orbell 1996. See pp. 61–63.
108 Orbell 2003, op. cit. See p. 137. See also Best 1923.
109 Orbell 1996. See p. 90.
110 Stafford, D. M. 2002. *Te Arawa: a history of the Arawa people*. Reed, Auckland. See p. 247.
111 Cody, J. F. 1956. 28 Maori Battalion. In: Kippenberger, H. K. (ed.). *The official history of New Zealand in the Second World War 1939–1945*. Historical Publications Branch, Wellington, p. 127 (http://nzetc.victoria.ac.nz/tm/scholarly/tei-WH2Maor-c5.html, accessed 11 July 2012).
112 From the papers of Brigadier George Dittmer, first Commanding Officer of 28 Māori Battalion, as presented, courtesy of George Rehu, in a display at the Rotorua Museum of Art and History Te Whare Taonga o Te Arawa (seen 27 June 2007).
113 Beattie, H. 1915. Traditions and legends collected from the natives of Murihiku (Southland, New Zealand). Part II. Te Mano-o-Te-Rapuwai. *Journal of the Polynesian Society* 24: 130–139; Beattie 1920b, op. cit.
114 Barrow 1984, op. cit. See p. 40.
115 Neich 2001, op. cit., pp. 326–327. The carvings are by Wero of Ngāti Tarawhai and a Ngāti Pikiao associate.
116 For the history of Nuku-Te-Apiapi see Neich (2001: Neich, R. 2001. *Carved histories: Rotorua Ngāti Tarawhai woodcarving*. Auckland University Press, Auckland, pp. 326–327. The carvings are by Wero of Ngāti Tarawhai and a Ngāti Pikiao associate.), pp. 210–211. See also Phillips, W. J. 1970. Historical notes on the carved house Nuku Te Apiapi. *Journal of the Polynesian Society* 79: 71–85. Note that the photograph of the two lizards representing Kataore on the poutokoihi in Best (1923) is reproduced upside down.
117 Phillipps 1970, op. cit. The location of this carving is unknown to me.
118 Phillips, W. J. and McEwen, J. M. 1948. Carved houses of Te Arawa. *Dominion Museum Records in Ethnology*. Vol. 1, No. 2. See p. 94. The carvers were Te Ngaru Whakapuka of Ngāti Pikiao and Te Ipu of Mourea. See also Cresswell, J. C. M. 1977. *Maori meeting houses of the North Island*. PCS Publications, Auckland. See pp. 55–56.
119 Toby Curtis, Ngāti Rongomai, pers. comm. 30 September 2013.
120 Phillips and McEwen 1948, op. cit. See pp. 85–86. The principal carver was Te Ipu Whakatara of Ngāti Pikiao. See also Cresswell 1977, op. cit., p. 53.
121 Phillips and McEwen 1948, op. cit. See pp. 85–86.
122 Neich 2001, op. cit. See pp. 59, 239–241 and 248 for discussion and examples.
123 Simmons, D. 2006. *Meeting-houses of Ngāti Porou o Te Tai Rāwhiti: an illustrated guide*. Reed Books, Auckland. See p. 92.
124 Simmons 2006, op. cit. See p. 106.
125 Simmons 2006, op. cit. See p. 159.
126 For details of the carvings' confused history, see Anson, D. 2004. What's in a name? The house carvings that Dr Hocken gave to the Otago Museum. *Journal of the Polynesian Society* 113: 73–90.
127 Neich 1996, op. cit. See pp. 108–109. See also Anson 2004, op. cit.
128 Evison, H. 1993. *Te Wai Pounamu, the greenstone island: a history of the southern Maori during the European colonisation of New Zealand*. Aoraki Press in association with the Ngai Tahu Maori Trust Board and Te Runanganui o Tahu, Christchurch. See p. 324.

129 Beattie 1919, op. cit.
130 For example, Kelly Davis, Ngāi Tahu, pers. comm. June 2005.
131 See Ramstad et al. 2007a, b for contemporary views among elders of some iwi.
132 For example, eels are recognised by a multitude of names, and flax by at least 27 varieties. See Auckland Museum 2001, op. cit., pp. 12 and 21.
133 Ramstad et al. 2007a, b.
134 For discussion, see pp. 281–298 of Pianka, E. R. and Vitt, L. J. 2003. *Lizards: windows to the evolution of diversity*. University of California Press, Berkeley. For example, in Morocco I was told of the tradition of burning a chameleon to avert a perceived misfortune, such as the failure of a female relative to marry (also reported by Hardy, P., Vorhees, M. and Edsall, H. 2005. *Morocco*. Lonely Planet, Australia, p. 39).
135 Including less accessible records not examined here, such as land court records, Māori language periodicals and personal accounts. See Orbell 2003, op. cit., pp. 17–21.

Endnotes to boxed text

For explanation of citation format in box endnotes, see endnotes to the main text.

Box 4.1

1 Dieffenbach, quoted by Gray 1843, see p. 204; Best 1909; Beattie 1994, see pp. 348–349, 508.
2 Taylor 1848, see p. 4; Taylor, R. 1870. *Maori and English dictionary* [revision of *A leaf from the natural history of New Zealand*, 1848]. George T. Chapman, Auckland; Beattie 1994, see p. 508.
3 Beattie, H. 1919. Traditions and legends collected from the natives of Murihiku (Southland, New Zealand). Part XI. *Journal of the Polynesian Society* 28: 212–225; Dieffenbach 1843, Vol. 1, see p. 140.
4 Taylor 1848; Best 1923.
5 Cooper, G. S. 1851. *Journal of an expedition overland from Auckland to Taranaki, by way of Rotorua, Taupo, and the west coast, undertaken in the summer of 1849–50, by His Excellency the Governor-in-Chief of New Zealand*. Williamson and Wilson, Auckland, p. 138; see also Colenso 1879.
6 Beattie 1994, see p. 349.
7 Polack 1838, p. 317; Colenso 1879.
8 Williams, H. W. 1957. *A dictionary of the Maori language*. 6th edn. Government Printer, Wellington.
9 Buller 1871; Best 1909.
10 Whitaker, A. H. 1992. Was the kawekaweau the world's largest gecko? *Forest and Bird* No. 264: 44–46.
11 Dieffenbach, quoted by Gray 1843, see p. 202; Colenso 1844, 1846; Best 1909; Beattie 1994, see pp. 349–350.
12 Beattie 1994, see pp. 349–350, 508.
13 Williams 1957, op. cit.
14 Beattie 1994, see p. 354.
15 For example, Sharell 1966, see p. 58; cf. Colenso 1879.
16 Best 1923; Mead, S. M. 1967. *The art of Maori carving*. Reed, Wellington, see p. 24; Phillips, W. J. 1997. *Maori carving illustrated*. 4th edn. (Revised by Simmons, D. R.) Reed, Auckland, see p. 31. See also p. 57 of Sharell 1966 (a popular book about the New Zealand herpetofauna) for a similar comment.
17 This carving, from a swamp near Waitara, Taranaki, is of unknown age and has broken off from a larger piece. It has tuatara-like jaws and body but no crest. Its current location is unknown, but for a photograph see p. 37 of Day, K. 2001. *Māori wood carving of the Taranaki region*. Reed Publishing, Auckland.

Box 4.2

1 Best 1923; see also Sharell 1966, p. 57; Phillips, W. J. 1997. *Māori carving illustrated*. 4th edn. (Revised by Simmons, D. R.). Reed, Auckland, p. 32.
2 JHS [Skinner, J. H., c. 1920s–1930s]. Maori nature notes written for the *Otago Daily Times*, XXXI and XXXVII. In *Teviotdale Clippings* Vol. 3, p. 21, 38. Hocken Collections, University of Otago, Dunedin.
3 Best 1923, p. 325. Skinner likewise drew parallels between Māori artwork and Indonesian traditions of crocodiles, although some of his parallels seem unconvincing (Skinner, H. D. 1964. Crocodile and lizard in New Zealand myth and material culture. *Records of the Otago Museum Anthropology No. 1*. Otago Museum, Dunedin). More recently Simon Best has speculated that a taniwha tradition might have been reinforced by vagrant crocodiles reaching New Zealand (Best, S. 1988. Here be dragons. *Journal of the Polynesian Society* 97: 239–259).
4 Orbell, M. 2003. *The illustrated encyclopedia of Māori myth and legend*. Canterbury University Press, Christchurch. See p. 154.
5 Orbell 1996, see p. 92.
6 See p. 409 of Taylor 1855; Best 1923.
7 Best 1923, p. 329.
8 Best 1923.
9 Until recently named as a member of the *Hoplodactylus maculatus* species complex. See Nielsen, S. V., Bauer, A. M., Jackman, T. R., Hitchmough, R. A. and Daugherty, C. H. 2011. New Zealand geckos (Diplodactylidae): cryptic diversity in a post-Gondwanan lineage with trans-Tasman affinities. *Molecular Phylogenetics and Evolution* 59: 1–22.
10 Data for Macraes Flat population in Rock, J., Andrews, R. M. and Cree, A. 2000. Effects of reproductive condition, season, and site on selected temperatures of a viviparous gecko. *Physiological and Biochemical Zoology* 73: 344–355.
11 This idea seems consistent with a story related in the main text, in which T. W. Downes, Supervisor of Works on the Whanganui River, was told of a moko-pāpā that appeared between two men lying on the ground.
12 Gartrell et al. 2006, 2007.
13 For a suspected link between salmonellosis in humans and New Zealand lizards, see De Hamel, F. A. and McInnes, H. M. 1971. Lizards as vectors of human salmonellosis. *Journal of Hygiene* 69: 247–53.

Box 4.3

1 Polack 1840, Vol. I, p. 125.
2 Moser, T. 1863. *Mahoe leaves: being a selection of sketches of New Zealand and its inhabitants*. William Lyon, Wellington. Reprinted 1974 by Capper Press, Christchurch. See p. 47.

3 White, M. 2006. All care taken. *New Zealand Heritage* No. 102 (Spring): 12–15.
4 Polack 1838, Vol. II, p. 81.
5 Anon. 1953. Old newspapers give a fascinating glimpses of a royal visitor in troubled times. *Te Ao Hou* No. 6 (Royal Tour): 13–14.
6 Presumably, it was not an indication of a lack of speed.

Box 4.4

1 Grey, G. 1855. *Polynesian mythology and ancient traditional history of the New Zealanders, as furnished by their priests and chiefs*. John Murray, London, p. 235; Stafford, D. M. 2002. Te Arawa: a history of the Arawa people. Reed, Auckland. See p. 84.
2 Mair, G. Entry for 12 May 1885. In *Diary and notes 1885*. MS-Papers-0092 (MS-Copy-Micro-0688-4), Alexander Turnbull Library, Wellington.
3 Phillips, W. J. 1997. *Māori carving illustrated*. 4th edn. (Revised by Simmons, D. R.) Reed, Auckland. See p. 32.
4 Donne 1942, p. 101.
5 Best 1923, p. 329.
6 Orbell 1996. See p. 90.
7 Best 1909.
8 Sharell 1966. See p. 58. Sharell speculated that lizards crawling out of the mouth were also a phallic symbol and part of a fertility rite, but support for this interpretation is lacking.

Box 4.5

1 Neich, R. 2001. Carved histories: *Rotorua Ngāti Tarawhai woodcarving*. Auckland University Press, Auckland. See pp. 49, 131 and 174. See also p. 12 of Simmons, D. 2006. *Meeting-houses of Ngāti Porou o Te Tai Rāwhiti: an illustrated guide*. Reed Books, Auckland.
2 Brown, D. 2003. *Tai Tokerau whakairo rākau: Northland Māori wood carving*. Reed Publishing, Auckland. See p. 12.
3 Neich 2001, op. cit., pp. 399–402; Brown 2003, op. cit., pp. 233–234.

Box 4.6

1 For the stories of Kataore and Hotupuku, see Colenso 1879. See also Stafford, D. M. 2002. *Te Arawa: a history of the Arawa people*. Reed, Auckland, pp. 67–74.
2 This includes the famous pātaka Te Puāwai o Te Arawa (Flower of the Arawa), which was also sometimes known as Tuhua Kataore (the pit or lair of Kataore). See Neich, R. 2001. *Carved histories: Rotorua Ngāti Tarawhai woodcarving*. Auckland University Press, Auckland, pp. 314–315.
3 James Schuster, Historic Places Trust, pers. comm. 27 June 2007.

CHAPTER FIVE

Discovered by science: the first 200 years

The first written reference to 'tuatara' as a reptile dates from 1820, in an early missionary dictionary. By 1844 the missionary-printer Colenso knew enough about local natural history to record that 'the productions of New Zealand are, generally speaking, peculiar to herself, and highly curious in structure.'[1] This understanding, which remains true of tuatara today, has yielded nearly two centuries' worth of scientific study.

Introduction

Terra australis incognita, the 'unknown southern landmass', was the lure that brought the first European explorers to New Zealand. This hypothetical landmass was considered, by geographers of the 17th and 18th centuries, to be present south of 60° latitude in order to balance lands of the Northern Hemisphere. Of course, a great southern land did not exist as such – but New Zealand did, and once located, offered tremendous opportunities for European exploration and discovery.

Yet for more than 150 years from first European arrival, and despite knowledge of reptiles among indigenous Māori, the existence of tuatara went unnoticed by Europeans. It was only during the early 1800s, when William Colenso and other missionaries explored widely in and around the North Island, that tuatara finally made their entrance into the scientific literature. And to establish their affinities required that they be examined halfway around the globe in the leading scientific institutions of the day.

In many ways, the history of the scientific study of tuatara tells in microcosm much of the history of European settlement of New Zealand. This chapter traces that study through first explorations of local natural history, later taxonomic study in England, and finally the development of formal scientific study within New Zealand. It is a story rich in diverse personalities and nationalities, all united by contact with an outwardly unremarkable but surprisingly significant reptile.

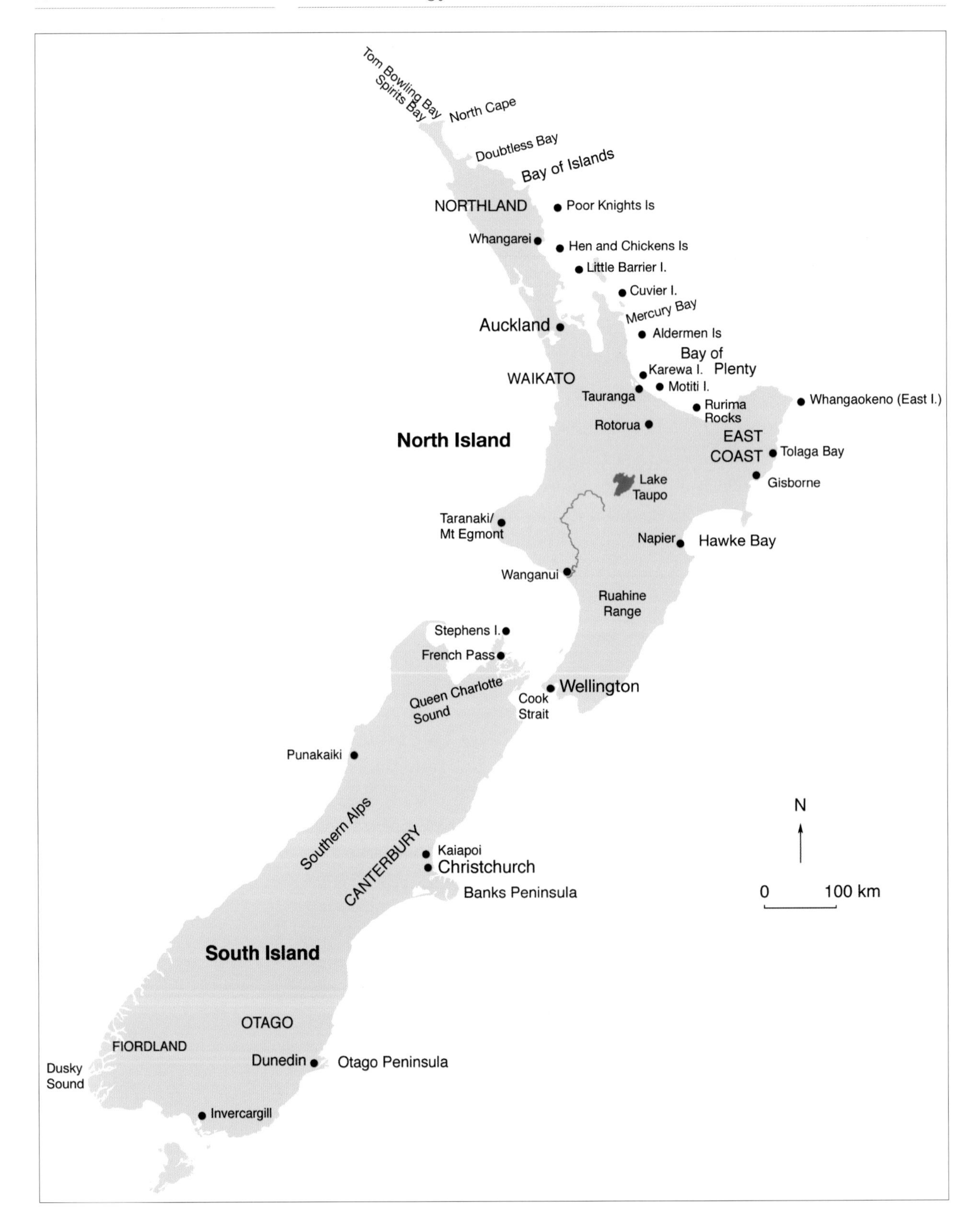

Fig. 5.1 Locations of key places mentioned in the text.

European discovery of New Zealand: 1642–1800

The first Europeans to visit New Zealand worked for the Dutch East India Company. Led by Abel Tasman, the crews of the *Heemskerck* and the *Zeehaen* searched for the hypothesised southern continent and for shipping routes useful for trade. In December 1642 Tasman recorded sighting 'a large land, uplifted high' – the mountains of the South Island, viewed from the West Coast near Punakaiki.[2] Tasman proceeded north, entering Cook Strait and anchoring within sight of Stephens Island (Takapourewa; Figure 5.1).[3] After a hostile reception from Māori in what is now known as Golden Bay, Tasman and his men continued northwards, mapping the western coast of the North Island. Never landing, they remained unaware of the existence of tuatara and other unique New Zealand animals.

It was the momentous expeditions of the English naval officer **James Cook** that brought the full extent of New Zealand to wide attention. Cook, one of the greatest maritime explorers and cartographers of all time, was appointed to command a scientific expedition to the South Pacific in 1768. Organised by the British Admiralty and the Royal Society, the expedition had the public objective of observing, from Tahiti, the transit of the planet Venus across the sun's disc. The secondary, secret objective was to search for the hypothesised southern continent. Cook's ship, the *Endeavour*, carried a retinue of natural historians and artists, led by the energetic and independently wealthy Joseph Banks.

The crew of the *Endeavour* sighted New Zealand in October 1769 from near present-day Gisborne. Between then and April 1770, Cook circumnavigated and mapped both main islands (Figure 5.2). Members of the party went ashore at several places, making strenuous efforts towards peaceable interactions with Māori. Banks and his team worked incessantly, collecting numerous specimens of natural history that were eventually delivered back to England in triumph. Further expeditions led by Cook followed in 1772 (on the *Resolution* and *Adventure)* and in 1777 (*Resolution* and *Discovery*).

All told, Cook spent a total of 328 days exploring the coast of New Zealand.[4] He and his natural historians had numerous encounters with Māori (aided on the first voyage by Tupaia, a Tahitian chief) and spent many weeks on shore. Extensive periods were spent at Ship Cove in Queen Charlotte Sound and in Dusky Sound in Fiordland, as well as briefer periods in Poverty Bay and Tolaga Bay on the East Coast, at Mercury Bay and in the Bay of Islands. Cook sailed within sight of all but one island or island group that would then have had tuatara populations, and gave European names to virtually all of these.[5] Yet the existence of tuatara remained undiscovered by Cook's parties, the only reptiles found being 'two or three sorts of small harmless Lizards'.[6] This result strongly suggests that tuatara were already rare, if not extinct, on the coasts of mainland New Zealand.

The only hint in Cook's expeditions of the possible existence of tuatara came during the third voyage. In 1777 Cook was told by a Māori man that:

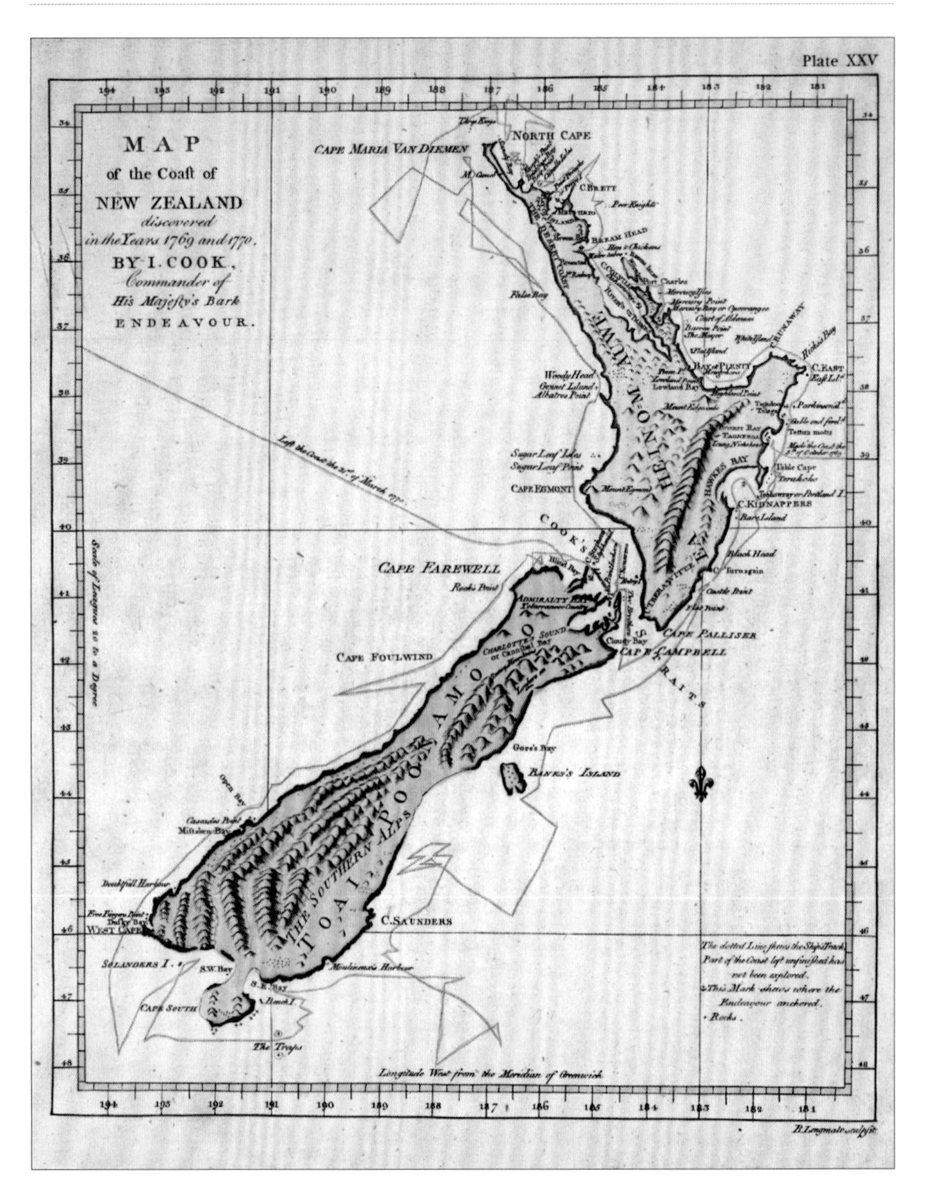

Fig. 5.2 During his first voyage to New Zealand on the *Endeavour* in 1769–1770, Cook mapped the coastline in detail. European names were given to many of the islands or island groups on which tuatara occur naturally today, including the Poor Knights Islands, Hen and Chickens Islands, Barrier Islands, Mercury Islands, Aldermen Islands, Stephens Island (Takapourewa), the Brothers Islands and the Admiralty Islands, which include the Trio Islands (Kuru Pongi). Cook also mapped several islands or island groups that still have tuatara today, including Karewa Island, Motunau Island and Cuvier Island (Repanga Island), and charted several others on which tuatara were probably present at the time, such as East Island (Whangaokeno), Moutuhora Island and Motiti Island. Despite sailing within sight of island habitats on many occasions, Cook's parties never discovered the existence of tuatara.

Engraving from Parkinson, S. 1773. *Journal of a voyage to the South Sea, in his Majesty's ship, the Endeavour*. Longmate sculpsit., London. Plate 25, opp. p. 124. Alexander Turnbull Library, Wellington, New Zealand, ref. PUBL–0037–25.

> there are Lizards and Snakes in New Zealand of an enormous size, he [his Māori informant] described the first as being Eight feet [c. 2.4 metres] in length and as big round as a mans body, he said they would sieze on and devour men, that they burrow in the ground and that they are killed by making fires at the mouth of the holes. We could not be misstaken in the Animal, as he, with his own hand drew a very good representation of a Lizard on a piece of paper as also a Snake in order to shew what kind of an animal he meant.[7]

This account is often interpreted as a reference to a giant, reptile-like taniwha, a powerful and sometimes dangerous creature of the Māori spiritual world. However, it may also be imbued with characteristics of living tuatara, as later Europeans were told by Māori that tuatara were smoked from their burrows before being eaten.[8]

Overlapping with Cook's first visit, a French expedition likewise failed to detect tuatara. Led by Jean De Surville, the *St Jean Baptiste* voyaged for commercial reasons, without natural historians. The vessel entered Doubtless Bay, Northland, in December 1769, and repeated shore visits were made to obtain fresh food and water for the scurvy-affected crew. Contact with local Māori was made, and sightings of numerous birds, as well as fish, shellfish, and a small black lizard were reported.[9] However, there was no mention of tuatara alive or dead, despite the subsequent discovery in Doubtless Bay of numerous bones of tuatara in the dunes.[10] Again, it seems likely that tuatara must already have been locally extinct.

Another French visit to Northland was made soon afterwards, between April and July 1772. Commanded by Marion Du Fresne, the crew of the *Mascarin* and *Marquis de Castries* included educated and thoughtful observers, although no scientists or artists.[11] There were opportunities to interact with local Māori and to observe the northern flora and fauna during brief visits ashore in Spirits Bay and Tom Bowling Bay (again, sites where tuatara bones have since been found), followed by more than a month anchored in the Bay of Islands. During the latter period members of the crew went inland to cut trees for masts, and birds, many remarkably tame, were hunted. The only reptiles mentioned, however, were 'little lizards'.[12]

Missionaries, travellers and traders: 1800–1880

The existence of tuatara was first brought to scientific attention by the church missionaries, traders and their associates active in the Bay of Islands, in Northland, during the early decades of the 1800s (Figure 5.3). The first missions, established by the Church Missionary Society, were under the oversight of Samuel Marsden.[13] An English-born chaplain, Marsden felt compelled to bring Christianity, as well as European skills and trades, to the Māori people. From 1814 to 1823 he established several missions in the Bay of Islands, which he supervised from Sydney, Australia. He travelled widely in the upper North Island during many visits until 1837, but did not directly provide information about tuatara.

The traveller **John Nicholas**, who accompanied Marsden on the latter's first visit during 1814–1815,[14] contributed what is almost certainly a reference to the existence of tuatara. After cutting trees in the Bay of Islands with Marsden and a Māori guide, Nicholas related the following:

> Observing a hole at the foot of one of these trees, which evidently appeared to have been burrowed by some quadruped, we inquired of Gunnah [the Māori guide] what animal he supposed it was; and from his description of it, we had reason to believe that it must be the Guana [Nicholas went on to note that such an animal was not mentioned in the Forster's account from Cook's second voyage] ... Wishing to know how far our surmise was correct, we desired our friend to thrust a stick into the hole, and endeavour to worry the animal out of it; but this he tried with

Fig. 5.3 The arrival of Samuel Marsden (1765–1838) and the Church Missionary Society in the Bay of Islands in 1814, and resulting contact with Māori, led ultimately to the supply of specimens of tuatara to British taxonomists.

Engraving by Samuel Williams, Landing of the Rev. S. Marsden in New Zealand, Dec. 19, 1814; frontispiece in Selwyn, J. C. ed. 1847. *Annals of the diocese of New Zealand.* Society for the Promotion of Christian Knowledge, London. Alexander Turnbull Library, Wellington, New Zealand, ref. F–542–1/2–MNZ.

> no effect, for either it was not in the hole at the time, or, if there, not to be dislodged by such means. Gunnah, however, was rather well pleased than otherwise at not meeting with this animal; for his dread of it was so great, that he shrunk back with terror at the time he thought it would come out, nor did he examine the hole but with great reluctance. This we thought very strange, for the Guana (the animal we took it for) is perfectly harmless.[15]

The term 'guana', a corruption of iguana, was subsequently often used by Europeans when writing about tuatara. It clearly meant a species distinct from small lizards. However, there is no indication that Nicholas actually saw a specimen.

By 1820 the missionaries were apparently aware of the existence of tuatara; the word (as Tūa tāra) is listed in Kendall's first Māori–English dictionary as 'a species of lizard', as well as 'name of a person'.[16] Nonetheless, specimens, alive or dead, continued to elude scientific expeditions. One such expedition, in 1824, was led by French explorer Duperrey of the *Coquille*. About 2 weeks were spent in the Bay of Islands, but only one reptile, an unidentified species of skink, was reported.[17] The same vessel, renamed the *Astrolabe* and now under the command of **Dumont D'Urville** (second in command on the previous voyage), explored New Zealand waters for a longer period during December 1826 to March 1827. D'Urville charted Cook Strait and all of the North Island's east coast to North Cape to improve on Cook's records, and once again many islands on which tuatara existed were sighted – but not visited. Naturalists went ashore at several places (including Tasman Bay in the South Island, and Tolaga Bay, the Waitemata Harbour and Whangarei in the North). They discussed the country's fauna with **William Williams**, a missionary in the Bay of Islands. Williams reported the existence of 'a few fairly large lizards'[18] – it is possible that he had seen tuatara from offshore islands during a recent visit to the Bay of Plenty. On his final visit in 1840, D'Urville (commanding the *Astrolabe* and the *Zelée*) came ashore on the Otago Peninsula, as well as at Akaroa on Banks Peninsula, at Poverty Bay and the Bay of Islands, where birds and invertebrates were collected. As with earlier explorers, D'Urville's failure to record a single specimen of tuatara during three visits strongly indicates that tuatara were already extinct from the eastern coasts of the North and South Islands.

Among the early traders near the mission station in the Bay of Islands was the colourful figure of **Joel Polack**. This widely travelled Englishman settled first in the Hokianga in 1831, and about a year later established what would become a prosperous trading base in the Bay of Islands (Figure 5.4).[19] Polack was an articulate observer, writing two lively accounts soon after his return to England in 1837. The following passage, from his first book, *New Zealand: being a narrative of travels and adventures during a residence in that country between the years 1831 and 1837*, implies that Polack was aware of the existence of tuatara:

> The gigantic lizard, or *guana*, exists principally in the Island of Victoria [South Island]. Some are found in the isles of the Bay of Plenty. The natives relate ogre-killing stories of this reptile, but doubtless it is harmless.[20]

Earlier in the same book, Polack related catching 'a small lizard, or tuātārā, beautifully striated with bright green lines',[21] but it seems that this was a

Fig. 5.4 Kororareka (now Russell) in the Bay of Islands was an early centre for European activities in New Zealand. The illustration, from about 1835 or 1836, shows European yachts, rowboats and houses (including Joel Polack's), as well as Māori dwellings (centre) and canoes. Polack (1807–1882), a trader, was one of the first Europeans to report the existence of tuatara, although it is unclear whether he saw one himself.

Frontispiece from Polack, J. S. 1838. *New Zealand: being a narrative of travels and adventures during a residence in that country between the years 1831 and 1837*. Vol 1. Richard Bentley, London. Alexander Turnbull Library, Wellington, New Zealand, ref. PUBL-0115-1-front.

Fig. 5.5 The missionary-printer and naturalist William Colenso (1811–1899) was among the first in New Zealand to publish observations on captive tuatara and endemic lizards. He also collected traditional stories from Māori about tuatara, and reported the existence of bones of tuatara from the North Island. During his years in the Bay of Islands from 1834 to 1844, Colenso met visiting explorers and probably assisted with the supply of tuatara to British museums. Well known for his botanical contributions, Colenso may also be considered the father of New Zealand herpetology.

Photo: William Colenso, c. 1868, photographer unidentified. Alexander Turnbull Library, Wellington, New Zealand, ref. F–3097–1/4.

Fig. 5.6 (top right) The Reverend Richard Taylor (1805–1873) was a missionary with an amateur interest in natural history. He briefly described tuatara in natural history books between 1848 and the 1870s, though most of his information seems derivative.

Photo: Reverend Richard Taylor, c. 1860–1873, photographer unidentified. Alexander Turnbull Library, Wellington, New Zealand, ref. C–14302–1/2.

small lizard, as distinct from his 'guana', the real tuatara. In his second book, *Manners and customs of the New Zealanders*, Polack made further reference to the 'guana' – 'a gigantic species of … frigid animal, exist[ing] in recesses of the forest and mountain gorges of the heights of the Island of Victoria';[22] he also recorded Māori beliefs about the powers of reptiles in general.

The sources of Polack's information are not stated. Probably they included the Bay of Islands missionaries, whom Polack knew, as well as Māori and settlers with whom he traded. However, it is unclear whether Polack saw a tuatara himself. In 1842 Polack returned from England to New Zealand and shifted his business from the Bay of Islands to Auckland. He later purchased, for their mineral deposits, several islands or island groups that would then have had populations of tuatara (the Poor Knights Islands, Hen and Chickens Islands and Fanal Island in the Mokohinau Islands).[23] However, Polack contributed no further information on tuatara, and eventually settled in California.

Two English-born missionary-naturalists who arrived in the Bay of Islands at a similar time to Polack and who did see and write about tuatara were William Colenso and the Reverend Richard Taylor. **William Colenso**, who was initially appointed by the Church Missionary Society as a printer, arrived at the Bay of Islands in late 1834 (Figure 5.5). There he expanded his interest in natural history, boosted by visits from passing scientists, including Charles Darwin on the *Beagle* in 1835 (Darwin spent a weary 9 days in the Bay of Islands towards the end of his voyage, was unimpressed with many aspects of the place, and made no reference to tuatara). Colenso, however, garnered a near-lifetime of observations as he evangelised his way around remote parts of the North Island, including the East Coast, the Urewera region, Rotorua, the Waikato and the Wairarapa coast. In 1844 Colenso took over a new mission station in Hawke's Bay, and expanded his travels into Palliser Bay, Wellington, through the Ruahine Range and to Lake Taupo.

BOX 5.1

When did the last mainland tuatara disappear?

The date when tuatara populations last survived on the North and South Islands is of biological and cultural interest. Several types of information are available, including radiocarbon dating of natural deposits containing tuatara bones, dating of midden deposits and reports of actual sightings. Each of these has its limitations. Natural deposits with reliable dating are the most valuable but, until recently, few precise records have been available. New evidence reveals that fossils from the North and South Islands date from as recently as about 550 years ago,[1] thus overlapping in time with fossils of rats introduced by humans. Midden deposits also suggest that tuatara survived at North and South Island sites until after human settlement, perhaps as recently as about 500 years ago (Appendix 3). However, midden deposits are less reliable indicators of natural distribution because, at least in some cases, specimens could have been collected from nearby offshore islands.

Reports of live tuatara on the mainland are the most controversial. The more recent sightings are of tuatara known or suspected to have been taken from offshore islands. Examples include those living under buildings at French Pass from the 1890s until the 1950s, as well as those seen for years about the cliffs at Tauranga in the early 1900s, evidently having escaped from the Norris–Thilenius colony.[2] Other likely escapees include isolated specimens that have turned up in improbable habitats, such as the tuatara found at the entrance to the Domain tunnel in Auckland in 1879,[3] another discovered in a drain at the back of a house in Auckland in 1915,[4] and one found under a house in Nelson in the 1970s.[5] Several other reports of mainland 'tuatara' exist from the last century;[6] some were probably indeed tuatara (perhaps released after people had second thoughts about the wisdom of illegal collection), whereas others may have been simple cases of species misidentification.[7]

Among older written accounts, however, a few seem plausible in terms of species identity and habitats. In 1900 a Gisborne resident wrote to James Hector, Director of the Colonial Museum as follows:

> Yesterday a gentleman working in his garden [in a Gisborne suburb] unearthed a tuatara. The situation was in a gully which has been cleared for a considerable time – but contains puriri stumps and debris – it was in a burrow about 6 ft [1.8 metres] long at a depth varying from 6 to 12 inches [15–30 centimetres] below the ground. Am I not right in supposing it to be a very unusual circumstance to find the tuatara on the mainland? The specimen in question is [illegible] 15 inches [38 centimetres] long.[8]

Even more plausibly, since he himself had kept tuatara for years and would have been certain of their identity, Hector wrote in reply:

> About 1865 we caught several alive at the back of Wtn [Wellington] in Makara Valley. Pigs have now destroyed most except on detached rocky islands and even there they are being rapidly exterminated by commercial collectors – especially Germans.[9]

Hector's observation was probably the source for A. K. Newman's published report in 1877 that tuatara had been captured at Makara; Newman also reported the Hutt Valley and Mt Victoria as sites of captures from the Wellington region. Walter Buller wrote of obtaining a preserved specimen that had been captured 'in Evans Bay about the year 1842' by an early Wellington settler.[10] Further reports from the Evans Bay–Miramar area date from around 1900 to 1910.[11] Hector McLeod, who was involved in excavations of Māori occupation sites (some of which contained jawbones of tuatara), reported several sightings of live tuatara, including one that he made himself in about 1908. McLeod tried to catch the animal, but it disappeared into a large heap of quarried stone.[12] Intriguingly, today's Miramar Peninsula was an island in the days of early Māori settlement. Later uplift during earthquakes, combined with sediment deposition, led to a land connection between the island and nearby mainland. Perhaps this delayed connection was sufficient to stem the arrival of introduced mammals to Miramar, allowing a relict population of tuatara to survive until European arrival.

On the other hand, reports of tuatara on the mainland have typically been isolated specimens of adults: no breeding population of tuatara, with juveniles, has ever been confirmed on any part of the North or South island since European arrival. Two early explorers who were in a good position to see wild populations, if they still existed, were the missionary-naturalists William Colenso and Richard Taylor. Both arrived in the Bay of Islands during the 1830s, and later travelled widely through the North Island. Both knew what tuatara were (Colenso had kept one from Karewa Island captive during the early 1840s) and both later published on the biology of tuatara, mentioning Māori traditions of their past presence on the North Island. However, there is no indication that either man saw tuatara living there wild.[13] Other early European explorers with interests in natural history, including Ernst Dieffenbach in the North Island, Edward Shortland in the South Island[14] and Andreas Reischek in both islands, failed to confirm the existence of mainland tuatara. Given Reischek's reputation as a compulsive collector, if not plunderer, it seems certain that he would have publicised the existence of a mainland specimen had he found one.

In summary, although there are many reports of tuatara seen on the mainland in the last two centuries, none provides evidence of natural breeding populations and only a few provide confidence about species identity and plausible natural habitat. Sadly, even among the latter, such as some from the Wellington region during the 1800s and early 1900s, none of the early specimens still exists as preserved material. Had preserved specimens been available today, genetic comparisons with older mainland fossils from natural deposits and with natural populations on offshore islands would be feasible. Lacking such evidence, we cannot rule out the possibility that such animals escaped following collection from offshore islands.

Box 5.1 Fig. 1 While travelling in the North Island during 1839–1841, Dieffenbach endeavoured to find tuatara, searching in places where the reptile had been reported. Although he climbed Mt Taranaki, where Māori reported traditions of man-eating reptiles, the only tuatara he obtained came from Karewa Island.

Source: *Taranaki or Mt Egmont and war canoe (early morning).* Plate 2 by George French Angas (J. W. Giles lithog. 1846). Alexander Turnbull Library, Wellington, New Zealand, ref. PUBL-0014-02.

Box 5.1 Fig. 2 During the 1800s, tuatara were reported from several mainland locations around Wellington (Port Nicholson), including Mt Victoria, Evans Bay and the Miramar Peninsula (centre left), the Hutt Valley (centre right, in the distance), and Makara (towards coastline, over the hills to the left). Although the habitats seem plausible, we cannot eliminate the possibility that such animals had been collected from islands with natural populations nearby, such as Matiu/Somes Island (centre rear).

Source: *Birdseye view of Port Nicholson in New Zealand*, lithograph hand-coloured by Thomas Allom, from a drawing by Charles Heaphy in 1839, published by Trelawney Saunders, London, c. 1843. Alexander Turnbull Library, Wellington, New Zealand, ref. C-029-006-b.

Colenso was a difficult, self-righteous character; despite his vocation, he fathered a child with a Māori servant girl, which led to his dismissal from the mission, the final breakdown of his marriage and a temporary withdrawal from society. Eventually, he entered local politics and spent his later years adding to his scientific publications.[24] Many of these were read before the Hawke's Bay Philosophical Society and then published in the fledgling scientific journal of New Zealand, the *Transactions and Proceedings of the New Zealand Institute*.

Colenso's observations on tuatara extend from his early years in the Bay of Islands. In 1844 he published, in a British journal, one of the first descriptions of a live tuatara (in an age when double publication was seemingly permissible, essentially the same article was published in 1846 in a Tasmanian journal).[25] He recorded that his captive specimen was one of three taken from

Karewa Island and that he had managed to keep it alive for nearly 3 months before preservation (he did not record the fate of the other two specimens, but see below for a later dispute that may have involved one of them).[26] He was unable to classify his tuatara as to family, suggesting that its thin, extensible and undivided tongue had features overlapping with both lacertid and iguanid lizards. Colenso made other contributions, publishing collections of Māori traditions involving tuatara; these are significant from a cultural perspective and also contain biological information about past distribution and harvesting on the mainland.[27] In his final years Colenso was shown bones of tuatara from a Napier quarry, and described these as being from a possible new species, *Sphenodon diversum*.[28] This evidence confirmed the past existence of tuatara on the North Island, although later work has not provided support for a distinction from *S. punctatus*.

Colenso's contemporary, **Reverend Richard Taylor**, was also English-born and educated (Figure 5.6). Taylor's contribution to the natural history of tuatara is less significant than Colenso's, and probably derivative of both Colenso and Dieffenbach (of whom more below). Taylor arrived in New Zealand in 1839, and after a short tour of the East Coast ran a mission school in the Bay of Islands. In 1843 he became a missionary in the Wanganui district, where he largely remained until his death in 1873. Like Colenso, he travelled great distances over rugged terrain to fulfil his parish duties, maintaining interests in natural history as he went.

Taylor wrote two books with information on tuatara, *A leaf from the natural history of New Zealand* (1848), and *Te Ika a Maui, or, New Zealand and its inhabitants, illustrating the origin, manners, customs, mythology, religion, rites, songs, proverbs, fables, and language of the natives* (1855, 2nd edition 1870). In *A leaf*, he wrote:

> Several varieties of the lizard were in existence until lately ... The largest kind I have seen is the *ruatara*, which is about eighteen inches [45 centimetres] long; but, if native reports are to be credited, very large lizards have existed, which were as terrible to them as ancient legends represent the dragons to have been, which are said formerly to have inhabited every part of Christendom; and in some parts the natives affirm larger lizards than any we have seen are still in existence. The *ruatara*, though viewed with extreme dread by the natives, appears to be perfectly harmless. It has a large benevolent looking eye, rounded teeth, and a serrated white comb along its back. It burrows, and is extremely slow in all its movements.[29]

This and other information on natural history was deleted from a later 'new and enlarged edition' of *A leaf*, which instead expanded the Māori–English dictionary included in the first edition.[30] Both editions of the dictionary include 'ruatara', 'tuatara' and other Māori names for the guana or 'great fringed lizard'.

Much the same material on tuatara appeared in expanded form in Taylor's first edition of *Te Ika a Maui*. The description included an incorrect scientific name, '*tiliqua Zealandica*', for tuatara (*Tiliqua zelandica* had been described, in a British publication in 1843, as a small New Zealand skink). In the second edition of *Te Ika a Maui*, this name was updated to '*tiliqua Zealandica vel Hatteria punctata*', which shows some awareness of overseas developments (tuatara had been named as *Hatteria punctata* in 1842), but restated the incorrect name (the latter was finally dropped from Taylor's *Maori and English Dictionary* of 1872). In both editions of *Te Ika a Maui*, Taylor wrote of tuatara that 'the teeth are rounded, and the tongue triangular; its toes are slender; it

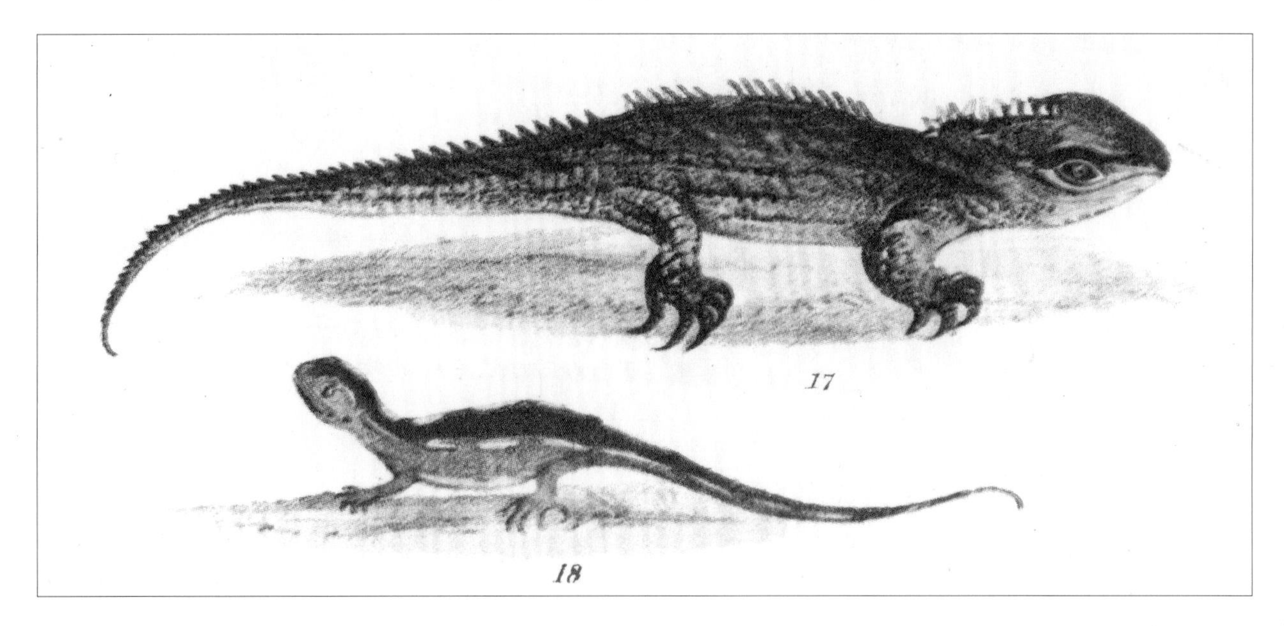

Fig. 5.7 Watercolours of a tuatara (top) and a green gecko (*Naultinus* sp., bottom), based on sketches by the Reverend Richard Taylor. This illustration of a tuatara was the first to be published in New Zealand, and the first in colour worldwide.

Source: Taylor, R. 1855. *Te Ika a Maui, or, New Zealand and its inhabitants, illustrating the origin, manners, customs, mythology, religion, rites, songs, proverbs, fables, and language of the natives*. Wertheim & MacIntosh, London.

lays on its back when basking in the sun, and burrows.' The comment about lying on the back while basking is at best confusing, and suggests that Taylor was not overly familiar with live tuatara. Perhaps Taylor's most significant contribution is the watercolour in his first edition of *Te Ika a Maui* (Figure 5.7; it was dropped from the second edition). Although naïve in style, it was the first published illustration of tuatara by a New Zealand author, and the first in colour worldwide.

The arrival of **Ernst (Ernest) Dieffenbach** in 1839 marks the entry of a professional scientist into New Zealand, as well as the first of a series of German and Austrian contributions to the biology of tuatara in the 19th century (Figure 5.8). Medically trained and with broad scientific interests, Dieffenbach was appointed as naturalist to the New Zealand Company, a group of land investors promoting settlement by Europeans looking for a life less restricted by the class distinctions that prevailed in England. Dieffenbach, accompanied by the artist Charles Heaphy and two company representatives, sailed to New Zealand in 1839 to help establish New Zealand's potential for European settlement.

Dieffenbach was intellectually able, sympathetic to Māori, willing to learn their language, and physically robust. These features combined to produce a new standard of scientific exploration in New Zealand. He explored briefly in

Fig. 5.8 Ernst Dieffenbach (1811–1855), the naturalist for the New Zealand Company, explored widely in the North Island between 1839 and 1841. Eventually he obtained a tuatara, from Karewa Island, which was taken to London and described in 1842 as *Hatteria punctata* by J. E. Gray at the British Museum. It remains the type specimen for *Sphenodon punctatus*. In this engraving from Dieffenbach's book *Travels in New Zealand; with contributions to the geography, geology, botany, and natural history of that country*, Dieffenbach (second from left, wearing a broad-brimmed hat) is shown meeting Te Waru.

Source: *Te Waro*, by Joseph Jenner Merrett (L. Haghe, lith. Day & Haghe, London, J. Murray, 1843). Alexander Turnbull Library, Wellington, New Zealand, ref. A–259–010.

Queen Charlotte Sound and then extensively in the North Island, including the Wellington–Hutt Valley region, in Taranaki (completing the first European ascent of Mt Taranaki), Rotorua, Lake Taupo and Tauranga. During a period in Northland in the summer of 1840–1841, he travelled with Richard Taylor and lived as a neighbour of Colenso's in the Bay of Islands.[31] As he went, Dieffenbach made detailed observations on geology, climate, botany, zoology and the Māori people in reports for his employers. After about 2 years of working for the New Zealand Company, including periods in the Chatham Islands and Australia, he tried to persuade the fledgling government in New Zealand to retain his services (expenses only!) for similar work in the South Island. His offer was turned down, and he returned to England in October 1841. The loss of such an able and impartial scientist and ethnologist after only a brief period spent in New Zealand has been widely regretted;[32] this loss extends to the additional research he might have contributed on tuatara.

Dieffenbach published a detailed account of his experiences in New Zealand in his book *Travels in New Zealand; with contributions to the geography, geology, botany, and natural history of that country*.[33] His book contains important references to the probable extinction of tuatara on the North Island. When proposing to climb Mt Taranaki, the natives 'tried much to dissuade me from the attempt, by saying that the mountain was "tapu", that there were ngarara (crocodiles) on it, which would undoubtedly eat me'.[34] However, there were no tuatara to be seen there, or anywhere in the North Island, despite his 'looking for it at the places where it was said to be found, and offering great rewards for a specimen'.[35] Finally, Dieffenbach attempted to reach Karewa Island in the Bay of Plenty by boat, but was obliged to put back on account of a heavy swell. 'It is a mere rock', he noted, 'but possesses some interest as being the only spot, that I know of, in New Zealand where the new species of reptiles [in footnote 'Hatteria punctata, Gray'] exist which were brought over by me to England.'[36]

The specimen that Dieffenbach eventually obtained, and took to England, was one of the first to reach that country. It was formally described by the scientist John Edward Gray in 1842, and listed in Gray's *Fauna of New Zealand*, which was included as Chapter X of Dieffenbach's *Travels* in 1843 (see the following section). According to Dieffenbach, the specimen had been given by the Reverend W. Stack, in Tauranga, to Dr Johnson, the colonial surgeon. However, this was later disputed by Colenso, who claimed that he had given the specimen to Dieffenbach. Colenso also reported having informed Dieffenbach about the acrodont teeth of tuatara: 'I was the first to point out this curious novelty; and this I did first to Dr. Dieffenbach (in 1841) from my living specimen, which I had and kept alive for several months'.[37]

Dieffenbach was perceptive in noting that the presence of tuatara on Karewa Island, as well as flightless birds on Hauturu/Little Barrier Island, implied that such islands were once connected to the North Island.[38] He was also one of the first Europeans to consider the potentially harmful impacts of European mammals on the native fauna. He attributed the disappearance of tuatara from the North Island to both human (Māori) predation and the introduction of pigs, although pigs perhaps arrived too late to have a major impact.

A North Island missionary who contributed in a small and accidental way to increased knowledge of the distribution of tuatara was **William Richard Wade**. During a sea voyage in 1838, his party was briefly shipwrecked on 'the southernmost and largest' of the Aldermen Islands (on Ruamahuaiti). Wade commented on the 'lizard' fauna, noting that 'we caught two of the guana kind, one of them measuring about fourteen inches [35 centimetres]' (he did not record whether the hungry party ate them). Wade added that 'by native account, many of the insular rocks abound with guanas. The island Karewa, off Tauranga, is said to swarm with them.'[39]

A valuable but little known report on the natural history of tuatara from the Bay of Plenty came soon afterwards from a group identified simply as '**A Party of Officers of the 58th Regiment**'.[40] The group landed on one of the Rurima Rocks islands in 1851 and collected nearly 40 tuatara, a few of which reached Auckland. The officers' account, published in the *New Zealander* newspaper, described physical appearance in detail, reported on Māori beliefs including past distribution, and made observations on hearing, respiration, body temperature, diet, cannibalism and other aspects of behaviour.

European exploration was much slower to begin in the South Island than in the North, reflecting the smaller population of Māori and reduced attention from missionaries. Once again, it was a churchman, the **Reverend James West Stack**, who first published Māori traditions attesting to the past presence of tuatara. According to Stack's Māori informants, a large form of ngārara (2–3 feet (60–90 centimetres) long, with a crest from the nape of the neck to the tail) was still being eaten in the eastern Canterbury region of the South Island within living memory.[41] However, there is no indication that Stack, who lived and travelled extensively around Kaiapoi, Christchurch and Banks Peninsula, saw wild tuatara himself.

The British taxonomists: 1831–1875

The missionaries and other Europeans who explored New Zealand during the early 1800s played an important role in bringing tuatara to scientific attention. Untrained as they were in taxonomic skills, however, it was inevitable that specimens would need to be sent to England for formal description and analysis. Remarkably, it was a long and tortuous 38 years from first description before the genus was recognised as both *Sphenodon* and a New Zealand endemic, and more than a further half-century before this scientific name became universally used in the scientific literature.

John Edward Gray (Figure 5.9) was the first to describe tuatara – and unwittingly he did so twice! From 1824 Gray had been responsible for cataloguing the reptile collection at the British Museum, and in 1840 became Keeper of the Zoological Collections there.[42] His first publication on the taxonomy of tuatara, *Note on a peculiar structure in the head of an agama*, was published in 1831 in *Zoological Miscellany*, Gray's privately published journal for museum acquisitions (Figure 5.10). The description was based solely on a skull, held in the collection of the Royal College of Surgeons in London, and no information on the skull's source (not even country) was given. Despite these shortcomings, Gray's description of the skull was significant: he recognised that the skull structure prevented the lower jaw from moving side to side, and that there was a double tooth row towards the rear of the upper jaw, forming a groove into which the lower jaw fitted. Although noting affinities with the lizard genera *Agama* and *Uromastyx* (both today placed in the Agamidae, a family primarily of Africa, Asia and Australia), Gray concluded that 'This Skull will doubtless form the type of a new genus, which I propose to call *Sphænodon*.'[43]

Fig. 5.9 John Edward Gray (1800–1875) was employed at the British Museum for half a century (from 1824 until 1874), becoming Keeper of the Zoological Collections from 1840. Gray's zeal for obtaining new specimens helped make the museum's zoological collections the most extensive in the world. Gray provided the first two descriptions of tuatara, as *Sphænodon* (now *Sphenodon*) in 1831, and as *Hatteria punctata* in 1842. In 1869 he reported that these two descriptions represented the same species, and thus the generic name *Sphenodon* had priority.

Source: John Edward Gray. ©The Natural History Museum, London, ref. 1304.

Gray's next specimen of tuatara arrived at the British Museum in 1842, courtesy of Dieffenbach and preserved in spirits. Not wishing to dissect the only example of what he thought was a new species, Gray described the reptile solely on the basis of external features, including head shape, scale pattern and the spiny crest (Figure 5.11). Without examination of

Note on a peculiar structure in the Head of an Agama. By J. E. Gray.

In a skull of an animal allied to *Agama*, or *Uromastyx*, in the College of Surgeons, I have observed that the Ramus (*Os. Complementaire*, Cuv. Os. Fos. fig. c.) of the lower jaw rubs against the lateral processes of the

14 *Zoological Miscellany.*

pterygoid bones, so as to prevent the lower jaw from moving from side to side, and that in the species under consideration the hinder teeth of the upper jaw has a series of teeth about half the length of the outer series placed on a ridge just on the inner edge of the outer teeth, leaving a groove between the two series for the lower jaw to fit into. This Skull will doubtless form the type of a new genus, which I propose to call *Sphænodon.*

Fig. 5.10 The first description of tuatara, as *Sphænodon* (now *Sphenodon*), based on a skull of unknown provenance.

Source: Gray, J. E. 1831. Note on a peculiar structure in the head of an agama. *Zoological Miscellany* 1: 13–14. The Natural History Museum, London, ref. 547422 and 54743, images combined with background adjustments.

72 *Zoological Miscellany.*

Descriptions of two hitherto unrecorded species of Reptiles from New Zealand; presented to the British Museum by Dr. Dieffenbach. By J. E. Gray.

Hatteria. Fam. *Agamidæ.* Head quadrangular, covered with small scales; throat with a cross fold; nape and back with a crest of compressed spines: body covered with small scales; belly and under side of the tail with large, squarish, keelless, flat scales, placed in cross series: tail compressed, triangular, covered with small scales, and with a ridge of large compressed spines: legs strong; toes 5.5, short, strong, cylindrical, slightly webbed at the base, covered above and below with small scales; claws short, blunt: femoral pores none: preanal scales small, a few of them are pierced in the centre.

Hatteria punctata, n. s. Olive; sides and limbs with minute white specks, beneath yellowish: the spines of the nuchal and dorsal crests yellow, of the caudal brown; the scales of the back, head, tail and limbs small, granular, nearly uniform; the irregular folds in the skin are fringed at the top with a series of rather large scales; an oblique ridge of larger scales on each side of the base of the tail, and a few shorter longitudinal ridges of rather smaller ones on each side of the upper part of the tail.

Inhabits New Zealand.

Dr. Dieffenbach observes that the species "lives in holes, especially on the slopes of the sand hills of the shore. The older missionaries say it was formerly very common, and the natives lived upon it, but for the last fifty years it has been scarcely ever seen. This specimen was found on a small rocky island, two miles from the coast, in the Bay of Plenty, and was given to Dr. Dieffenbach alive, but shortly died, as it would not eat anything that was offered to it. It is extremely sluggish in captivity, and could be handled without any attempt at resistance or biting." The natives called it "*Tuatera.*"

Fig. 5.11 The second description of tuatara, as *Hatteria punctata*, based on a specimen preserved in spirits and supplied by Dieffenbach.

Source: Gray, J. E. 1842. Descriptions of two hitherto unrecorded species of reptiles from New Zealand; presented to the British Museum by Dr. Dieffenbach. *Zoological Miscellany* 4: 72 (in part). The Natural History Museum, London, ref. 54741, with background adjustments.

the skeleton or dentition, the specimen's affinity with the skull of *Sphenodon* went unnoticed. Gray published his description in *Zoological Miscellany* in 1842, naming the tuatara as *Hatteria punctata* (Family Agamidae), a 'hitherto unrecorded species … from New Zealand'.[44] Brief observations from Dieffenbach were included, including the specimen's source from a small rocky island in the Bay of Plenty. The specimen, known today as the type specimen for *Sphenodon punctatus*, remains in the collection of the Natural History Museum, formerly the British Museum (Natural History).[45]

The following year, Gray expanded slightly on this description in his *Fauna of New Zealand* (Chapter X of Dieffenbach's *Travels*). The additions included a reference to the species as the 'Gigantic lizard' of Cook's third voyage and the 'guana' of Polack's *New Zealand*.[46] Dieffenbach's own notes were slightly expanded, naming Karewa Island as the source.

Gray's chapter in Dieffenbach's *Travels* also mentioned the existence of a 'young specimen' of the same species in the Museum of Haslar Hospital, Gosport. The Gosport specimen was brought back to England in September 1843 by the British Government's *Erebus* & *Terror* expedition. The primary objective of this voyage had been to explore the Antarctic region, but stops were also made in Australasia, including a period in the Bay of Islands from August to November 1841. The naturalists on board met there with local missionaries including Williams, Taylor and Colenso, and it was probably with the missionaries' assistance that the tuatara was obtained. This specimen is now also at the Natural History Museum, but no information on its collection is recorded.[47]

It was Gray's task to write up the reptile section of the zoological findings from the *Erebus* & *Terror* expedition.[48] His first instalment to the reptile section was issued in 1845, but the section covering Agamidae had not been reached and no text description of tuatara was provided.[49] The issue did, nonetheless, include some exquisite lithographic plates, most if not all by George Henry Ford, and one of a tuatara (as '*Hatteria punctata*') was among these (Figure 5.12). This is the first-known published illustration of a tuatara, and was probably based on Dieffenbach's specimen. The plate also included a skull, drawn from the side and from below; very likely, this skull came from a specimen from Matiu/Somes Island sent to the British Museum in 1842,[50] which Gray listed in his *Catalogue of the specimens of lizards in the collection of the British Museum* as having been prepared as a stuffed skin and skeleton.[51] The skin and skeleton are still in the museum's collection.[52] In the illustration, the double row of teeth in the upper jaw is unmistakeable, yet Gray still did not note the connection to his description of 1831.

Meanwhile, **Richard Owen**, Hunterian Professor and Conservator at the Hunterian Museum of the Royal College of Surgeons in London (Figure 5.13), was also busy with a tuatara skull – the very same skull that Gray had described in 1831. Owen was on the verge of becoming the leading British anatomist of his generation; he had introduced the term 'Dinosauria' in 1842 to great acclaim, and had established his fame in the context of colonial natural history for descriptions of monotremes, marsupials and moa during the 1830s and early 1840s. With respect to tuatara, however, Owen seems to have missed an opportunity. To be fair, the primary subjects of his 25-page

Fig. 5.12 The first-known published illustration of a tuatara, a lithograph by 'J. Ford' (probably in error for G. H. Ford). It first appeared in 1845 in J. E. Gray's initial instalment towards *The zoology of the voyage of H.M.S.* Erebus & Terror. It was also included in Gray's *The lizards of Australia and New Zealand in the collection of the British Museum*, published in 1867. The main illustration is probably based on the preserved specimen that Dieffenbach delivered to London in 1842. The eyes are rather small and sunken (an artifact of preservation), but the illustration is otherwise surprisingly life-like. Despite the illustration of a skull in which the double row of teeth in the upper jaw is unmistakeable, Gray did not make the connection to his description in 1831 of this feature in *Sphenodon* until 1869. In 1867 Günther pointed out a small error in the illustration: the artist had 'introduced an erect process of the lower jaw, just below the tympanic condyle; this, however, is merely the remainder of a dried ligament.' (Günther 1867, p. 595).

Source: ©The Natural History Museum, London, ref. 54552.

publication of 1845 were large extinct reptiles from Africa and England. In passing, Owen compared the skull structure of these fossils with that of a 'lacertilian' casually referred to as '*Rhynchocephalus*' (literally, the somewhat unflattering 'snout-head', in reference to the downward-projecting premaxillary bones armed with incisor-like teeth), without describing a species.[53]

Despite his cursory identification of what was in fact the skull of a tuatara, Owen was a brilliant anatomist. He correctly noted that the skull of his '*Rhynchocephalus*' had a two-arched (diapsid) structure and that the

Fig. 5.13 Richard Owen (1804–1892), with moa femur in hand and wearing the robes of the Hunterian Professor of the Royal College of Surgeons. Owen was already famous for his descriptions of moa (large, flightless and extinct New Zealand birds) when he referred to tuatara as *Rhynchocephalus* in 1845, overlooking Gray's earlier descriptions.

Source: Sir Richard Owen, c. 1846. ©The Natural History Museum, London, ref. 1350.

teeth were acrodont. He also provided illustrations from the side, above and below – the latter clearly revealing the double tooth row, although Owen made no comment on this. Rather, he emphasised similarities in the structure of the premaxillary bones with *Rhynchosaurus*, an extinct Triassic reptile from England (rhynchosaurs, however, are no longer considered closely related to tuatara).[54]

What is surprising about Owen's work is that he did not recognise the skull of his *Rhynchocephalus* as Gray's *Hatteria punctata*, let alone *Sphenodon*. Whether this was due to ignorance (puzzling, given that the two London researchers were working only about a kilometre from each other, and that Gray had already described the very same specimen) or arrogance (the highly competitive Owen was not above deliberately overlooking, or incorporating, the work of others on other occasions) is unknown. Undoubtedly, the skull of a small, extant 'lizard' was less likely to attract public and peer recognition than the bones of extinct giant reptiles. There is a possibility that Owen may have also received, but overlooked, a complete specimen of tuatara sent to him from New Zealand. A letter sent by William Wakefield of the New Zealand Company to Owen in 1845 refers to having collected for him 'a curious species of lizard, about 18 inches [45 centimetres] long and having a row of white spikes along its back, caught on Somes Island, [now known as Matiu/Somes Island, in Wellington Harbour] Port Nicholson'.[55] If Owen received the specimen, there is no obvious mention in his publications.

One of Owen's duties at the Royal College of Surgeons was to catalogue the museum's collections. In 1853 he issued a catalogue covering the fish, reptiles, birds and marsupials in which two items from the skeleton of tuatara were listed. One was a skull (presumably the same specimen as previously described) and the other comprised the five anterior-most vertebrae. Both were listed under 'Family Geckotiæ. Genus *Rhynchocephalus*' and once again there was no reference to Gray's previous work. This time, the unique double tooth row of the skull was noted: 'The ... palatal plate ... is of unusual breadth as compared with the Lizards generally, and presents the unusual peculiarity of a dentigerous ridge parallel with the posterior half of the alveolar border. It is situated close to the inner side of this border, leaving only space sufficient for the reception of the teeth of the under jaw.'[56] Owen also noted, for the first time, the deep cavities on both anterior and posterior surfaces of the centrum of the vertebrae (the amphicoelous condition).

The year 1867 was a momentous one for tuatara classification. Back at the British Museum Gray was frustrated with progress on publication of the zoological findings of the *Voyage of H.M.S.* Erebus *&* Terror. Finally, he decided to publish his contribution in a separate document, *The Lizards of Australia and New Zealand in the collection of the British Museum*.[57] This report included most of the original lithographic plates from the 1845 instalment of the *Voyage*, including Ford's illustration of a tuatara. For the first time, synonymy of Owen's *Rhynchocephalus* with Gray's *Hatteria punctata* was noted, although there was still no reference to *Sphenodon* (Figure 5.14).

112. **Hatteria punctata** (T. 20.), *Gray, l. c.* 249; *Zool. Misc.* 72; Gigantic Lizard, *Cook*; Rhynchocephalus, *Owen, Trans. Geol. Soc.* vii. 80, t. 6. f. 5, 6, 7, 1845 (not described). *Hab.* New Zealand.

RHYNCHOCEPHALIA.

3. HATTERIA PUNCTATA. Plate 20.

Hatteria punctata, Gray, Zool. Misc. p. 72; Günth, Phil. Trans. 1867, pp. 595—629; pls. 26—28.
New Zealand (North Island).

Fig. 5.14 Further listings of tuatara in works by Gray and Günther. Top: Gray's listing of tuatara (under Family Agamidae) from *The lizards of Australia and New Zealand in the collection of the British Museum* (1867). This recognised that Owen's *Rhynchocephalus* was synonymous with Gray's *Hatteria punctata*. Bottom: Günther's listing of tuatara under the recently created order Rhynchocephalia, from *The zoology of the voyage of H.M.S.* Erebus & Terror (1875). Curiously, tuatara were still listed as *Hatteria*, even though Gray had argued in 1869 that the name *Sphenodon* had priority.

Source: Gray, J. E. and Günther, A. 1845–1875. *Lizards of Australia and New Zealand.* Facsimile reprint in herpetology [1995], Society for the Study and Amphibians and Reptiles, Ithaca.

Much more significantly, 1867 was the year that **Albert Günther** (Figure 5.15) published his lasting and essentially sole contribution to tuatara biology, a monograph of 34 pages and 34 figures that established tuatara as a reptile distinct from lizards, snakes, crocodiles and turtles, and deserving of its own order. Günther was German-born and educated, and in 1857 had been appointed to assist Gray at the British Museum. Eventually, in 1875, he would succeed him as Keeper of the Zoological Collections. By 1867 Günther had specimens of tuatara additional to those of Gray and Owen at his disposal, and was thus able to describe features of the entire skeleton, including the pattern of tooth formation and wear through juvenile and adult life, as well as the internal organs of two males.

Günther confirmed some of the unusual features already noted by Gray and Owen, including the two-arched (diapsid) skull, acrodont teeth, palatine tooth row and biconcave (amphicoelous) vertebrae. He also described many additional features of significance, including the fixed nature of the quadrate (i.e. firmly united to the skull and bones of the palate). As Günther noted, this was compensated for by increased mobility of the lower jaw (the lower jaw could move in a forwards direction while closed to achieve a cutting and sawing effect – a type of movement now known as proal, a subcategory of propalinal). Günther acknowledged that there were many features aligning the tuatara with lizards (such as the structure of the sternum, pelvis and limbs, some aspects of the vertebrae, the single-headed ribs, the ability to drop and regrow the tail, the structure of the heart and the transverse 'anal cleft' or vent). He argued, however, that tuatara showed substantial differences (such as an unusual skull and dentition, uncinate processes on the ribs, 'abdominal ribs' – now known as gastralia, lack of a tympanic cavity, and lack of an intromittent organ in males), meaning that the species could not possibly be an agamid lizard. In all these respects, Günther's work remains accurate today. Günther was also the first to report that tuatara consumed birds (the feathered remains of a young bird, as well as a fragment of a beetle, were found in the digestive

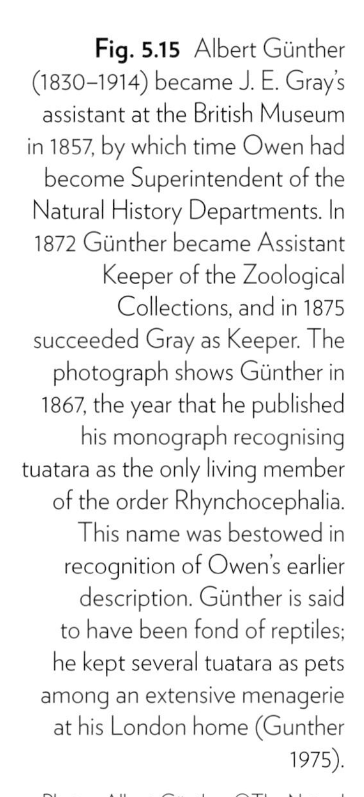

Fig. 5.15 Albert Günther (1830–1914) became J. E. Gray's assistant at the British Museum in 1857, by which time Owen had become Superintendent of the Natural History Departments. In 1872 Günther became Assistant Keeper of the Zoological Collections, and in 1875 succeeded Gray as Keeper. The photograph shows Günther in 1867, the year that he published his monograph recognising tuatara as the only living member of the order Rhynchocephalia. This name was bestowed in recognition of Owen's earlier description. Günther is said to have been fond of reptiles; he kept several tuatara as pets among an extensive menagerie at his London home (Gunther 1975).

Photo: Albert Günther, ©The Natural History Museum, London, ref. 41834.

tract of one specimen), although his conclusion from 'the feebleness of its claws' that tuatara could not burrow for themselves was incorrect.

Respect for, or deference towards, Owen's previous work is evident throughout Günther's monograph. In 1856 Owen had been appointed to a specially created position (Superintendent of the Natural History Departments) at the British Museum, and, along with Gray, had become one of Günther's mentors. Effectively now Gray's and Günther's superior, and competitive in character, Owen was not someone who should be offended lightly. Gray was also nearing retirement and, if Günther was to succeed him, Owen's continued support was essential. Günther followed Owen in comparing the premaxillaries of tuatara with the fossil genus *Rhynchosaurus,* but noted that in absence of a complete skull for the latter, 'a real affinity … can only be conjectured',[58] a caution that later proved prescient. Günther completed his monograph by placing tuatara in a new group, the order Rhynchocephalia, noting that 'In this way the name *Rhynchocephalus* may be preserved, which, otherwise, must give way to the prior *Hatteria*.' Thus, Owen's contribution received continued notice.[59]

It was Günther, too, who discovered the final piece of the jigsaw: that Gray's *Hatteria* and *Sphenodon* were one and the same. The announcement was left to Gray, in an article succinctly entitled 'Sphenodon, Hatteria, *and* Rhynchocephalus'.[60] Gray was by this time nearing retirement; in ill health, he was frustrated by years of disagreements with Owen over management of the British Museum's collection. He did not hold back from implied criticism of Owen, on counts of both claiming the credit for donating the specimen that he, Gray, had described and for subsequently having lost part of it:

> The skull so named [as *Rhynchocephalus*, by Owen] is evidently the same as that I described in ... 1831, as *Sphenodon*, though the specimen is said in the Catalogue [published by Owen in 1853] to have been presented by Prof. Owen, whose name certainly was not attached to the specimen when I described it. The specimen is still in the collection, but without the lower jaw, which was with it in 1831.[61]

Sadly, this skull, the subject of so much close attention, is no longer in the collection of the Royal College of Surgeons; it was probably destroyed when the museum was bombed during 1941.[62]

Gray had never explained the origin of the name *Hatteria*: possibly it came from the Old English hœt, meaning hood or cowl, in reference to the fold (of skin) at the throat that Gray had mentioned in his description of 1842.[63] However, Gray has been described as 'an inveterate coiner of scientific names, many of which have no meaning in any language',[64] so we may never know the true origin. Regardless, discovery that the name *Sphenodon* had priority over *Hatteria* certainly met with approval, according to a report the following year in the new British journal *Nature*. P. Sclater, the Secretary of the Zoological Society of London, in commenting on the purchase of a live tuatara by the Zoological Gardens earlier that year, wrote that the species had been referred to as:

> *Hatteria* (!) *punctata* ... until lately, when it was most fortunately discovered that the generic term *Sphenodon* had been previously applied to a specimen of its skull in the museum of the College of Surgeons. It has thus become possible not to be obliged to employ so vile and barbarous a term as *Hatteria* for the name of this important animal.[65]

Günther made a subsequent reference to tuatara 8 years later in 1875, when he completed the section on reptiles in the long-awaited *Zoology of the Voyage of H.M.S.* Erebus *&* Terror. So many changes and discoveries about Australian and New Zealand reptiles had been made following Gray's first instalment in 1845 that Günther found it necessary to re-present and update the entire list of species, rather than just those that had not been covered by Gray. His listing for tuatara updated the ordinal classification to Rhynchocephalia, but curiously continued with the now-outdated *Hatteria* rather than *Sphenodon* as the name for the genus (Figure 5.14). Following Günther's work, the name *Hatteria* received continued use within the numerous publications on the anatomy of tuatara emerging from continental Europe into the first half of the 20th century. Uptake of the name *Sphenodon* was much more rapid within New Zealand, where research on tuatara was now fired by its recognition as the sole living representative of a most curious order of reptiles.

Naturalists and collectors: 1870–1895

Much had changed in New Zealand by the 1870s (Figure 5.16). Annexed as a British Colony in 1840, New Zealand subsequently experienced an influx of British settlers, leading to new towns and industries. Wealth from the discovery of gold and from sheep farming helped to fund the first scientific institutions. The first national museum, the Colonial Museum, began in the new capital, Wellington, in 1865. By 1877 five towns had museums and two had universities or university colleges. Philosophical societies and learned institutes appeared in many towns. From 1867 these were amalgamated into the New Zealand Institute, which began publishing the *Transactions and Proceedings of the New Zealand Institute* (hereafter *TNZI*) from 1868 (the New Zealand Institute became the Royal Society of New Zealand in 1933).

Günther's publication announcing the significance of tuatara stimulated a flurry of activity among members of the Wellington Philosophical Society. First off the mark was a medical man, **F. J. Knox**, who somewhat querulously challenged the minutae of Günther's anatomical descriptions based on his own dissections dating from 1842, but provided useful information about the past presence of tuatara on Matiu/Somes Island.[66]

Fig. 5.16 By the 1870s Wellington, the new capital of New Zealand, had a developing scientific community, several of whose members published information about tuatara.

Photo: Wellington (1870s), Hocken Collections Uare Taoka o Hākena, University of Otago, c/n E6213/31.

BOX 5.2

Tuatara as pets and garden curiosities

Many older New Zealanders have grown up with the sense that tuatara were animals restricted to remote and unvisitable islands, or statue-like exhibits in zoo enclosures. It comes as a surprise to learn just how frequently, in colonial times or shortly afterwards, that tuatara were kept as pets, or released to the outdoors as desirable garden inhabitants.

Soon after the evolutionary status of tuatara as a rhynchocephalian reptile was announced, it became rather fashionable among European New Zealanders to have a pet tuatara.[1] Walter Buller, writing in Wellington in 1877, tells us that 'specimens preserved in cages are scattered among lovers of the curious in this city'.[2] In 1882 Captain Fairchild, master of the government steamer servicing the lighthouses, was reported to have cited 'one or two instances of ladies in a Southern city who have them in their home in a domesticated state ... so much so that it is quite the correct thing for them to nestle in their mistress' lap, or, running up her arms, perch on her shoulders, and press their heads against the lady's damask cheek'.[3] Other accounts, from the 1890s, tell of a pet tuatara that had free run of the Invercargill Athenaeum. It was befriended by the cleaning lady, who made a point of dusting it off in the morning along with the rest of the library's 'furniture'.[4] At the time of its death in 1899, it was regarded as 'a most intelligent creature', although somewhat 'bothered' by 'recent alterations in the building'.[5]

T. E. Donne, Superintendent of the Department of Tourist and Health Resorts from 1901, gave an affectionate account of a pair of tuatara collected from Stephens Island and kept in his Wellington office as pets. During the working day they were permitted to emerge from their 'sleeping box'. The female gradually became accustomed to Donne's presence to the point where she would climb up his legs to his shoulder and rest there on a daily basis, diving into his V-necked vest if a stranger entered the room. 'I liked her and apparently she liked me', wrote Donne.[6] In summer he transferred the pair of tuatara to a hutch in his garden. The female laid eggs, which hatched, but the hatchlings were eaten by his pet dog. How long the adults survived is not known, but the mounted remains of two tuatara were eventually housed, in a glass case, in Donne's later home in London.[7]

Many writers, like Donne, felt that they were recognised by their pet tuatara, and that strangers were not so well tolerated. Reports from employees of the lighthouse service certainly give this impression. Mr Campbell, the Chief Engineer of the government steamer Hinemoa in 1890, kept a pet tuatara on board: 'It was allowed to roam the ship, but always came to him when he whistled for it. It slept in the engine room and was so tame it fed out of the hand.'[8] A lighthouse keeper on Cuvier Island in 1919, D. J. Grindlay, befriended a wild tuatara from the population that lived in his garden. 'Jimmie, as I call him ... can be seen every evening at sunset going into the creek for his well loved and well earned bath. "How he does love it!" Strange as it seems, Jimmie takes no more notice of my presence than if I were one of his fellow creatures, but let a stranger come near and he is off, not waiting to wash the dirt off his back. I have stood talking to this reptile while he has been bathing for fully half an hour.'[9] Such observations suggest that tuatara learn to discriminate between the sounds of different human voices or whistles[10] (see Chapter 7 for more on hearing in tuatara).

Another who kept a pet tuatara was one Percy Isaac. According to the Bay of Plenty Times in 1909, Isaac had 'an embarrassing habit of producing his peculiar pet from various portions of his clothing'.[11] On one occasion he absent-mindedly pulled out his tuatara (somewhat unkindly described as 'that living heap of hideousness') and placed it on a shop counter, whereupon the draperesses present went into 'violent fit(s) of hysterics'. Close friendship was not without other liabilities – for both owners and tuatara. One young girl who was too familiar with her father's pet was eventually bitten on the nose; sad to say, the tuatara, which could not be induced to release its hold, was beheaded.[12]

Tuatara were sometimes turned outdoors – although there, if not enclosed in a hutch, they had to fend for themselves. Some brought in about 1882 by Captain Fairchild to Auckland were to be given to a nurseryman 'in whose grounds they may find congenial employment in searching for slugs'.[13] In 1884 E. C. Gold-smith, the District Surveyor in Tauranga, collected 20 from Karewa Island, turning many of them out in his garden. There they 'thrived very well, and were quite at home ... The cats and dogs did not take any notice of them at all ... They are of great use in getting a place rid of rats'.[14]

Gold-smith's suggestion that introduced mammals caused no harm is remarkable; other released tuatara (such as Donne's hatchlings) were not so fortunate. Even adults were not immune. In the early 1900s a woman found a large specimen (probably an escapee from a colony maintained by Colin Norris) in her garden in Tauranga, but, 'after giving fright to various people on account of its large size, it was ultimately killed by a dog.'[15]

Another mainland colony managed to survive for about 60 years or more (possibly with repeated supplementation) at French Pass. Mrs A. E. Moss, the schoolteacher at the small settlement of Elmslie Bay farmed by Wallace Webber and his wife Elizabeth, heard of the reptiles in the early 1890s from workers

Box 5.2 Fig. 1 Tuatara were exhibited during the early 1900s at Troutdale Farm, a popular picnic spot maintained by Andrew Johnson in Christchurch, but whether they were housed indoors or outdoors, whether they bred and how long they survived is all unknown.

Photo: Johnson's Fishponds, the Aquarium at 105 Clarendon Terrace, Opawa, c. 1900. Christchurch City Libraries, ref. CCL PhotoCD 3 IMG 0038.

Box 5.2 Fig. 2 A tuatara and companion at Johnson's gardens in Opawa, Christchurch. The caption from this postcard of the early 1900s reveals a then-popular but incorrect belief about the evolutionary significance of tuatara.

Photo: Muir & Moodie postcard No. 49R, supplied by James Taylor.

constructing the new lighthouse on nearby Stephens Island.[16] Three tuatara were duly brought from Stephens at her request 'so that she might study their habits.'[17] The tuatara were initially kept in a box and cared for by George Webber, son of Wallace and Elizabeth. Soon afterwards, George learnt from the German scientist Hugo Schauinsland, whom he was assisting on Stephens Island, how to distinguish the sexes. Realising that the three at French Pass were all male, George collected two more from the Trio Islands. One laid eggs in the vegetable garden, and although some hatchlings resulted, at least one was killed by a cat. Two tuatara were said to be present at French Pass in 1936 (living in the garden of the old homestead),[18] and four (one at the lighthouse and three at the boarding house) in the 1950s.[19] Three were seen in 1980,[20] but no later confirmed reports are known.[21]

In an early attempt to see whether tuatara from Cook Strait could acclimatise further south, tuatara were on display from as early as 1903 at Troutdale Farm, a popular picnic ground, aquarium and gardens maintained by Andrew Johnson in Christchurch.[22] The tuatara were probably collected from Stephens Island (some shown in 1906 at the New Zealand International Exhibition in Christchurch's Hagley Park may have been added to the collection).[23] Johnson's original request to hold tuatara had been for the purposes of 'turning tuatara out ... with a view to their acclimatisation in the mainland'.[24] Sadly, we know next to nothing about the long-term success of this or most other attempts. A few tuatara from such colonies probably escaped, contributing to sporadic reports of wild tuatara surviving on the main islands; most, however, would have succumbed eventually to introduced mammals or old age.

Another medical man, **A. K. [Alfred] Newman**, contributed a much more detailed and convincing report about tuatara in 1878, with new information about past distribution on the mainland and the habits of captive specimens (including slow growth and the ability to live for months without eating). Newman also clarified and corrected some issues of anatomy raised by Knox, provided detailed information on musculature, and plausibly suggested that the gastralia ('abdominal ribs'), rather than assisting with a legless, snake-like locomotion when required, as proposed by Günther, probably served to support and protect the abdominal cavity when the latter was distended by eggs.[67] Importantly, Newman provided references to nesting behaviour in the wild and in captivity, and the first descriptions of the reproductive anatomy of adult females, including that of a female with oviducal eggs. Unfortunately for biology, Newman later became a politician and wrote no more on tuatara.

Another and ultimately more prominent member of the Wellington scientific fraternity was **Walter Buller** (eventually Sir Walter; Figure 5.17). Buller was born at a mission station in Hokianga, Northland, in 1838, and had an administrative and legal career working first as a native (Māori) interpreter, then as a magistrate and later as a barrister.[68] Alongside these activities, he maintained a vigorous interest in natural history, publishing seven papers between 1871 and 1896 that referred to tuatara. Buller was ambitious and self-promoting, and although best known for dominating New Zealand ornithology for several decades, the prestige of working on a reptile that had just been brought to international attention by Günther in 1867 cannot have escaped him. In fact, Günther and Gray were among Buller's initial sponsors to become a Fellow of the Royal Society of London, and his nomination was, with Günther's help, eventually successful in 1879.[69]

Fig. 5.17 Walter Lawry Buller (1838–1906) was a New Zealand-born lawyer and naturalist, who played a leading role in the scientific community that developed in Wellington from the 1860s to the 1880s. Although primarily interested in ornithology, Buller helped stimulate an increased local interest in tuatara.

Photo: Sir Walter Buller, photographed in 1905 by Walter A. Cox. Alexander Turnbull Library, Wellington, New Zealand, ref. A-087-012.

Buller's contributions included a description of the Brothers Island tuatara as a new species, *Sphenodon guntheri,* which Buller named in recognition of Günther.[70] Günther was privately not convinced, writing in 1885 to James Hector, New Zealand's senior government scientist and Director of the Colonial Museum, that 'Buller writes me in every letter that he gave you a *Hatteria guntheri* for me. I do not believe in the species, but I should like to see it'.[71] (Although *S. guntheri* was reinstated as a distinct species in 1990, Günther's assessment was vindicated in 2010 when the Brothers Island population was once again subsumed within *S. punctatus*.[72]) Buller also provided observations of tuatara in captivity,[73] described tuatara from East Island as a new variety[74] (a distinction that remains unconfirmed today), and collated new information about previous and then-current distributions.[75] Buller was an energetic and efficient networker who helped

BOX 5.3

The making of a myth: the 300-year-old tuatara in the kūmara pit

Humans have a fascination with animals that live longer than we do, and reptiles – such as giant tortoises – provide some notable examples. Tuatara have entered the popular literature as another example, based on reports that Māori tradition recounted a tuatara living for 300 years in a kūmara pit[1] on Motiti Island. A careful examination of earliest written sources reveals that a fascination with extreme longevity may have run away with the facts!

The first published reference to a Māori tradition of a tuatara in a kūmara pit stems from a letter from Gilbert Mair to his brother-in-law, Walter Buller – though without any claim of a 300-year time frame. According to Buller, Mair wrote in 1876:

> Regarding the probable age to which these funny creatures live, I should have mentioned that I have seen an unusually large one which has been kept in an old kumara pit on Flat Island (Motiti) for over three generations. Could the old fellow only speak, what an interesting account he might give us of the fight on 'bare Motiti,' when the famous *tohunga*, Te Haramiti, and 170 of his warriors (the Ngatikuri) were killed and eaten by the Ngaiterangi...[2]

Mair's unpublished correspondence from the same year to Thomas Cheeseman, Curator of what became the Auckland Museum, provides further detail about what was presumably the same tuatara.

> I believe [tuatara] live to great age. For the last 11 years [For at least 11 years?][3] I have been told of one, living in an old kumara pit on flat island (Motiti) and told it had been there as long as the Maori residents of the Island could remember. Last week I induced the natives to bring it over in a box; but nothing will induce them to sell it. It was an unusually large one (27 inches in length, and 4 in girth).[4] In 1826 Te Haramiti's pa was taken by the Ngaiterangi but the victorious party were frightened by the big 'Tuatara'. The natives inform me that it is called after one of their ancestors, and has lived on Motiti for three generations.[5]

Box 5.3 Fig. 1 Captain Gilbert Mair (1843–1923) collected early information on the biology of tuatara in the Bay of Plenty, assisting his brother-in-law Walter Buller with the latter's publications. Gilbert Mair was told by Māori of a tuatara living in a kūmara pit on Motiti Island, but claimed only that Māori reported a lifespan for it of three generations, not 300 years.

Photo: Gilbert Mair standing by the grave of Captains Travers and White, photographed in 1921 by James Cowan. Alexander Turnbull Library, Wellington, New Zealand, ref. PAColl-4306, J. Coulthard Collection.

It seems likely that Mair obtained the specimen soon afterwards, based on an account given much later (1913) to the Minister for Internal Affairs:

> In 1867, I procured one from Motiti Island, which had lived in a deserted kumara pit, since the year 1832, when Te Haramiti's war party attacked the native dwellers there. This specimen still lives, I am told in Kew Gardens whither I sent it to Sir Walter Buller, after keeping it many years in a small box at Tauranga.[6]

Although there are some discrepancies in dates between these accounts (unsurprising given the long passage of time), the broad claims relevant to longevity are consistent. First, a large tuatara seen by Mair, probably in 1876, was said by local Māori to have been living on Motiti Island for 'three generations'; second, the tuatara was said to have been already 'large' at the time of Te Haramiti's attempted invasion (which occurred in 1832); and third, a tuatara procured by Mair from Motiti Island (perhaps the same individual throughout, although this cannot be certain) was sent to Walter Buller, and may have been still alive in 1913. Given these dates, it is conceivable that an individual tuatara from Motiti Island lived as an adult from at least 1832 to 1913 – a period of 81 years.

At first, Mair's reference to 'three generations' was faithfully reported, or at least not too wildly distorted. Drummond and Hutton, for instance, wrote in a popular textbook in 1902 that a tuatara 'kept by some Maoris in a kumara pit was known to be at least sixty years old'.[7] Somewhere between then and 1935, however, the myth of '300 years' began[8] – perhaps because of a mismatch between 'generations' and 'centuries'. In 1935 Falla wrote, without citing a source: 'the Maoris have a record of one living in a shell pit at Motiti Island for nearly three hundred years'.[9] Falla's interpretation has been repeated by many later writers,[10] my co-authors and I included,[11] though others have rightly pointed out that claims of such a lifespan are unsubstantiated.[12] Increasingly, it seems they are also extremely doubtful.

Today, the best available evidence points to a lifespan of about 100 years[13] being possible for tuatara – similar to that of humans, and consistent with Mair's account, once we allow for a period of about 25 years to reach maximum adult size. The oldest known individuals in the wild (on Stephens Island) are now an estimated 80–90 years of age,[14] based on their size at the time when first marked, and the period of time elapsing between recaptures. Although in 2007 it was suggested, based on interviews with contemporary Māori elders from regions outside the Bay of Plenty, that longevity of up to 300 years represents traditional knowledge about tuatara and a novel hypothesis about lifespan,[15] it is also possible that an early misinterpretation, probably stemming from Falla, has entered the imagination of many New Zealanders to become a widely held local tradition.

Box 5.3 Fig. 2 Major William Gilbert Mair (1832–1912), an older brother of Gilbert Mair, collected and published information on the biology of tuatara on islands in the Bay of Plenty.

Photo: William Mair, soldier with the Colonial Defence Force and New Zealand Militia, photographed c. 1862 by Hartley Webster. Alexander Turnbull Library, Wellington, New Zealand, ref. F-92933-1/2, PA2-1870.

bring an increased local awareness of overseas developments: just 4 months after publication of Sclater's article in 1870 publicising the fact that tuatara were now recognised in England as *Sphenodon*, not *Hatteria*, Buller brought this to the attention of the Wellington Philosophical Society.[76] On the less positive side, he initiated a surprising myth that the 'male is considerably smaller than the female';[77] this and similar statements contributed to confusion about sexing of tuatara that lasted until the 1950s (see Chapter 8, Box 8.1).

Much has been written about Buller's ambivalent and even contradictory attitudes toward conservation, especially the conservation of New Zealand birds.[78] With respect to tuatara, Buller praised the New Zealand government for passing legislation to protect tuatara in 1895,[79] and helped to promote Hauturu/Little Barrier Island (on which tuatara were present) as a sanctuary for flora and fauna. However, like many of his generation, he believed that New Zealand's indigenous animals, plants and people would eventually be overwhelmed by European immigrants (biota and people), which were perceived to be more vigorous.[80] He was publicly critical of overseas collectors of tuatara but was himself a collector and supplier, albeit on a smaller scale. In 1894 he was criticised for proposing to send a newly hatched tuatara from Stephens Island 'Home' (to Britain) when a colleague felt that the specimen should be provided to the Colonial Museum (Buller acceded to the request).[81] By 1893 Buller had released three tuatara on an island in Lake Papaitonga (with the intention of shielding kiwi on his private estate from Māori interest),[82] but he left for England in 1899 and never returned, leaving the fate of the tuatara unrecorded.

Buller was undoubtedly assisted in his work on tuatara by his brothers-in-law **Captain Gilbert Mair** and **Major William G. Mair**. Both were based in the Bay of Plenty. Gilbert Mair, well known as a soldier with strong Māori connections, supplied Buller with tuatara from Karewa Island, provided him with field observations such as the presence and habits of tuatara on Moutoki Island (in the Rurima Rocks group),[83] and later escorted him on a collecting trip to Moutoki.[84] Famously, Gilbert Mair was quoted by Buller about an unusually large tuatara he had seen, 'which has been kept in an old kumara pit on Flat Island (Motiti) for over three generations'.[85] (This report seems subsequently to have been misinterpreted, leading to questionable assertions that Māori tradition recorded life spans for tuatara of up to 300 years, as described in Box 5.3.) In later years Gilbert Mair contributed to national discussions about how best to conserve tuatara, providing information to government scientists about the islands on which tuatara were to be found.[86] His brother, William Mair, published information in his own right about tuatara from Moutoki Island, and provided specimens for display at the Wellington Philosophical Society and the Auckland Museum.[87]

A very different and complementary contributor to tuatara biology during the 1880s from Sir Walter Buller was the Austrian **Andreas Reischek** (Figure 5.18). Reischek spent the period from 1877 to 1889 in New Zealand, working periodically as a highly regarded taxidermist at the Canterbury and Auckland

Fig. 5.18 Andreas Reischek (1845–1902), an Austrian-born naturalist-collector and taxidermist, spent 12 years in New Zealand. He visited at least four offshore islands or island groups that supported tuatara, published early observations on behaviour, and collected many specimens to sell. He was often assisted by his dog, Caesar (possibly the dog shown in this photo).

Photo: Andreas Reischek, photographed c. 1870s by Adele (Vienna, Austria). Alexander Turnbull Library, Wellington, New Zealand, ref. F-5238-1/2, H. von Haast Collection.

museums and as a supplier of specimens to private collections.[88] His major impact on New Zealand's natural history, however, was as a collector. Unlike Buller, to whom he supplied birds and information, he was a man for the field. Accompanied by his dog, Caesar, he explored widely – often under atrocious conditions: in Fiordland, from Christchurch to the West Coast and back, the King Country, Mt Ruapehu, the Bay of Plenty, Kaipara Harbour and Northland. Most significantly for tuatara, he spent time on several northern offshore islands.

In the first of his two articles for *TNZI* that contain information on tuatara, Reischek described the habits of tuatara on the Marotere (Chickens) Islands.[89] He described the way in which tuatara and burrow-nesting seabirds shared their burrows, a supposed case of commensalism that was commented on with great interest in the British journal *Nature* soon afterwards.[90] However, Reischek's view that the seabirds were living 'apparently on the best of terms with [tuatara]'[91] was optimistic in light of his subsequent observations of predation by tuatara on seabirds on Karewa Island. Other notes provided in this paper touched on aspects of nocturnal behaviour, egg production, diet, a bite to his finger ('the wound healed very slowly and was rather painful')[92] and his failure to find tuatara on Hen Island (now Taranga Island) or on Hauturu.

In a later article, Reischek reported success, with the help of his dog, in finding tuatara, or the 'fringe-back lizard' as he called it, on second visits to Taranga and Hauturu.[93] In displaying his specimens before a meeting of the Auckland Institute, he suggested that tuatara on Taranga varied from those on the Chickens and might be the same as Buller's '*Ginteri*' [*S. guntheri*], whereas those on Hauturu were an unusual brick-red colour with brown bars which he considered 'a very well marked local form'.[94] Tuatara with similar colouration have since been seen on other northern islands (see Chapter 1, Figure 1.3), and neither of Reischek's suggestions about the distinctiveness of tuatara from Taranga and Hauturu have stood up to genetic scrutiny.[95] Possibly Reischek was trying to inflate the sale value of his specimens by exaggerating their distinctiveness.

Like Buller, Reischek could see the destructive impacts of human society on nature. He commented at one point that 'civilised man can be the worst vermin of the whole earth',[96] yet such reflections did not prevent him from collecting and preserving the rarest examples of animal life. When he eventually returned home to Vienna (and his long-suffering wife) in 1889, he was in possession of what is said to have been 'the largest single collection of ethnological and natural history specimens ever taken to Europe from New Zealand'.[97] It was

Reischek's intention that his collection would make his fortune, but in the event it sold for relatively little and he died in poverty. The number of tuatara collected by Reischek is uncertain but probably totalled some hundreds; almost certainly, his activities were the major stimulus for local biologists who subsequently advocated for legal protection of tuatara (see Chapter 10).

Buller, Newman and Reischek were the last of a generation of amateur naturalists to write about tuatara. Collectively, they expanded the information available on earlier and then-current distributions, published notes on the appearance and behaviour of wild and captive specimens, and preserved samples in spirits, some of which survive today. As part of a fledgling scientific community in New Zealand in which new, Darwinian ideas about natural selection were readily accepted, Buller and Newman also used tuatara as an example to begin exploring issues of sexual selection and variation among island populations.[98] Beyond this, they barely ventured.

Professional scientists and the lure of embryonic development: 1885–1900

It was a new generation of professionally trained scientists from overseas, both those arriving to staff the new university colleges and some visiting as individuals from Germany, who made the final contributions to research on tuatara within New Zealand during the 19th century. Embryology was the new focus, both for its own sake and for its contribution to understanding the evolution of reptiles and other vertebrates. Many biologists came armed with knowledge of histological techniques, in which preserved pieces of tissue could be sectioned, stained and examined with microscopes to reveal details at the cellular level. But to get embryos for examination, eggs at different stages of development had to be found.

The first attempt was made by **A. P. W. [Algernon] Thomas**, Professor of Geology and Botany at Auckland University College, with his colleague **T. Jeffery Parker**, Professor of Biology at the University of Otago in Dunedin and curator of the Otago Museum. Their approach was to visit Karewa Island in February 1885 (accompanied by Reischek) at what they believed was the time of nesting.[99] Efforts to find eggs of tuatara inside residential burrows were fruitless, so adult tuatara were collected for subsequent maintenance at Thomas's home in Auckland. This represents the first-known scientific attempt at captive breeding of tuatara, but although eggs and hatchlings were eventually produced, few details are available.[100] Thomas, however, did publish accurate information on how to identify the sex of adult tuatara, in both New Zealand and British journals.[101] This work corrected Buller's earlier assertion that females were the larger sex, but unfortunately, it was overlooked for decades.

Parker also collected at least 18 tuatara with the intention of obtaining eggs,[102] but subsequently turned his attention to the embryonic development of kiwi (*Apteryx*). Stuffed male and female tuatara were exhibited by Parker as part of a diorama at the New Zealand and South Seas Exhibition held during 1889 and 1890 in Dunedin;[103] these and other similar early exhibits (Figure 5.19) shared new knowledge about the biology of tuatara in their natural habitats with the New Zealand public.[104] Prior to his untimely death in 1897 at the age of 47, Parker and his Dunedin colleague G. M. Thomson also advocated strongly for the legal protection of tuatara (see Chapter 10), and at least some tuatara from Parker's collection were supplied as preserved specimens to overseas researchers.

Two of Parker's preserved females reached the Anatomical Institute in Freiburg, Germany, where they became the subjects of a study of female reproductive organs by a Japanese researcher, **Gakutaro Osawa**.[105] The resulting publication was just one among an extraordinary volume of material (totalling over 400 pages plus plates) published in German by Osawa on the anatomy of tuatara between 1896 and 1899. Additional topics included the integument, abdominal organs, male reproductive anatomy, sense organs and structure of the eye.[106]

Back in New Zealand, a new approach to obtaining embryos was needed – why not arrange for a local person to help by collecting eggs on an island on which the time of nesting and location of nests were known? The first to

Fig. 5.19 Visits to offshore islands by Professors A. P. W. Thomas, T. J. Parker and others in the 1880s led to new information about the natural history of tuatara, including relationships with seabirds. At least six male tuatara are present in this diorama (location unknown) from 1889 or 1890, and a further six in a partner image (BB 4635) from the Burton Brothers collection.

Source: Alexander Turnbull Library, Wellington, New Zealand, Pearson Album, BB 4633, ref. PA1-f-048-30-1.

explore this approach was **Hugo Schauinsland**, Professor and Director of the Ūbersee-Museum in Bremen, Germany (Figure 5.20). Schauinsland visited New Zealand as part of an expedition to the South Pacific, funded by the German government, to collect biological and ethnographic material.

Schauinsland spent 3 weeks from late December 1896 until early January 1897 based at French Pass, visiting both Stephens Island and the Trios Islands (Kuru Pongi; specifically, Middle Trio) by open boat ('an undertaking which … could almost have cost me my life', he later wrote).[107] He was ably assisted by his wife (an expert at skinning) and by a young local man, George Webber, who much later published his recollections of the period.[108] Tuatara (although not their eggs) had become legally protected in 1895, but Schauinsland's collecting had the support of the New Zealand government. This included the assistance of the Marine Department's keepers who staffed the new lighthouse on Stephens Island, operational since 1894. Following Schauinsland's departure for Germany, Webber sent him more eggs, collected from Stephens during the nesting season the following November.

Fig. 5.20 The German scientist Professor Dr. Hugo Schauinsland (1857–1937) visited Stephens Island and Middle Trio Island in Cook Strait during 1896 and 1897. Schauinsland subsequently published nine papers with information on embryonic development in tuatara.

Photo: Portrait of Professor Hugo Schauinsland, artist unknown, pre-1914. Photograph by Gabriele Warnke, supplied by Peter-René Becker, © Ūbersee-Museum, Bremen.

Schauinsland was extremely productive with the material that he collected, publishing nine papers with information on tuatara. Unfortunately, all were in German, meaning that his findings on tuatara remained poorly known within New Zealand for many years (as was also the case with information about his bird and lizard collections).[109] Schauinsland's first two papers on tuatara, produced within 3 years of his visit,[110] include accurate and detailed information on external differences between the sexes, nesting behaviour, clutch size, egg size and some aspects of embryonic development, including development of the parietal eye. Schauinsland also accurately inferred that the incubation period was extremely long (about 1 year), that not all females nested in the same year, and that longevity was probably great. A third publication combined and repeated the information in the first two, adding further detail and illustrations.[111] Similarities to development in turtles and lizards were noted. His later papers[112] include aspects of development of the skeleton, nerves, ears and teeth,[113] extra-embryonic membranes[114] and vertebral column, ribs and sternum.[115] Thirteen tuatara (12 mounted specimens and one in alcohol) plus numerous tissue sections on slides from Schauinsland's work remain today in the Ūbersee Museum in Bremen (Figure 5.21).[116]

About two years after Schauinsland's visit, another German researcher, **Georg Thilenius** (Figure 5.22), visited both Stephens and Karewa Islands with a similar goal of collecting eggs for studies of embryonic development. Thilenius was trained in natural history and medicine, and in 1897 was commissioned by the Prussian Academy of Sciences to make a 2-year expedition to New Zealand and the South Sea islands to collect ethnographic and zoological material.[117] Afterwards, in 1904, Thilenius became Director of the Museum of Ethnology in Hamburg, Germany, and in 1920, Rector of

Hamburg University. This subsequent focus on anthropology helps explain why Thilenius apparently published only a single publication on tuatara.[118]

Thilenius visited Stephens Island from 31 October to 30 November 1898 with an assistant.[119] This was apparently the more productive of the two islands, both because the visit coincided with the time of nesting and because the lighthouse keepers' children were able to provide directions to nesting areas. Thilenius dissected females and collected eggs, and the following year a keeper reported supplying a further 50 eggs to him in accordance with government instructions.[120] From his observations and collections on Stephens, Thilenius was able to provide accurate information on the nesting season, movement of females to nesting areas, nesting behaviour, clutch size, swelling of eggs during development, egg losses during incubation, embryonic development and behaviour of hatchlings. He also correctly observed that, as incubation can last 12–14 months, nesting by females in a given year can expose eggs laid the previous year. Although Thilenius did not comment on differences in appearance between the sexes, his observation that males were more conspicuous by day than females suggests that he could accurately tell the difference.

As far as Karewa Island was concerned, Thilenius's activities are more controversial. Thilenius himself wrote only that he visited the island during December to January (year unstated) and was unable to find nests there.[121] However, contemporary newspaper accounts and unpublished government

Fig. 5.21 Schauinsland's diorama of tuatara and seabirds, based on his studies on Cook Strait islands in the summer of 1896–1897, is still on display in Bremen.

Photograph by Gabriele Warnke, supplied by Peter-René Becker, © Übersee-Museum, Bremen.

files reveal concern that Thilenius had exceeded his permit for collection, apparently holding some dozens of tuatara in Tauranga for observation.[122] The captive tuatara were maintained in collaboration with a local boatman, Colin Norris, who may have eventually sold the tuatara. Other possible fates of the tuatara from the Norris–Thilenius colony are discussed in Chapter 10.

The final and most comprehensive contributor to discoveries about embryonic development in tuatara during the late 1800s and early 1900s was **Arthur Dendy** (Figure 5.23). Dendy was born and educated in England. From 1886 to 1887 he assisted in the Zoological Department of the British Museum (Natural History) with research on sponges; he would have met Günther during this time. After 7 years at the University of Melbourne, Australia, Dendy came to Canterbury College, Christchurch, in 1894, as a lecturer (soon afterwards professor) in biology. Dendy departed for England in 1901, and from 1903 to 1905 was a professor of zoology in South Africa. In 1905 he became Professor of Zoology at King's College, University of London, where he remained until his death in 1925.[123]

Fig. 5.22 Georg Thilenius (1868–1937), later a famous German ethnologist, published early observations on the nesting of tuatara based on a visit to Stephens Island in 1898.

Photo: Georg Christian Thilenius, by Rudolf Dührkoop, 1905.

Dendy began his work on tuatara in Christchurch at the suggestion of his colleague in England, Professor G. B. Howes. Howes later assisted with the publication of Dendy's articles in British journals, and himself received tuatara specimens from Dendy in which he studied skeletal development. Dendy's initial interest was to compare development of the parietal eye in tuatara with similar observations on Australian skinks.

Having obtained permission from the government to collect, Dendy planned to visit Stephens Island himself. However, in writing to the Principal Lighthouse Keeper in July 1896, he found an enthusiastic collaborator whose contributions made a visit to Stephens Island unnecessary. The keeper, Patrick Henaghan, would find the eggs, pack them in sand in tin cans, and send them off intermittently to Dendy by boat. Later they refined the process to obtain embryos at a greater range of stages: when Henaghan found a nest, he would keep the eggs on the island to continue their development, and then send them on to Dendy at intervals in small quantities, 'a plan which we find to work admirably', wrote Dendy.[124] Dendy was extremely grateful for Henaghan's contribution, stating that 'I cannot sufficiently express my gratitude to him for the magnificent supply of material which he obtained for me, and also for the valuable information which he gave me from time to time in his interesting letters regarding the habits of the tuatara.'[125]

Dendy published eight main papers on aspects of development in tuatara, the first five of which appeared within 4 years of his receiving eggs. His first two papers[126] were brief articles summarising some of the discoveries from his later, more extensive monograph on development.[127] In the latter he quoted from Henaghan's letters with information on the time and places

Fig. 5.23 Arthur Dendy (1865–1925) was appointed as a lecturer in biology at Canterbury College, Christchurch (now the University of Canterbury) in 1893. He was promoted to professor in 1894, resigning in 1903. While at Canterbury College, and provided with tuatara eggs by Patrick Henaghan, Principal Lighthouse Keeper on Stephens Island, Dendy began an extensive programme of research into embryonic development. According to an obituary, Dendy would have considered his work on the morphology and classification of sponges to be his most important contribution to knowledge. Nevertheless, the obituary continued, Dendy had 'laid all zoologists under the deepest of gratitude to him for his fine work in collecting, preserving and examining the developmental stages of that primitive reptile the tuatara.' (Quoted from E. W. M. [E. W. MacBride] 1926. Arthur Dendy 1865–1925. *Proceedings of the Royal Society B* 99, p. xxxv.)

Photo: 'Arthur Dendy, Professor of Biology, 1893–1903', from Hight, J. and Candy, A. M. F. 1927. *A short history of the Canterbury College (University of New Zealand)*. Whitcombe and Tombs, Auckland. Hocken Collections Uare Taoka o Hākena, University of Otago, S08-506.

of nesting and the structure of nest cavities. Dendy concluded that development extended over about 13 months from nesting in November of one year until hatching in about December of the next, though most development had occurred before winter. He provided a staging scheme for tuatara embryos (still the only scheme available today) and described in detail the formation of the germ layers, the extra-embryonic membranes and the principal organ systems, with copious illustrations. In a related publication, he described the formation of the parietal eye and other components of the pineal complex in the same embryos.[128] His fourth paper of 1899 was a brief summary of life history for a local audience.[129]

Dendy reported difficulties in writing up his work in Christchurch without access to a 'proper scientific library'[130] and, in 1901, he returned to England. Throughout the following decade in South Africa and London, he continued to section more material and, in 1907, received six male tuatara shipped from New Zealand. These were used to provide information on the pineal complex of adults in comparison with embryos, his original sections having fortunately been salvaged from a shipwreck en route from New Zealand to South Africa. Dendy first investigated the blood supply within the brain[131] before writing his final, exhaustive memoir on the pineal complex of tuatara.[132] A 10-page 'abstract' of the main findings of the latter paper was also published.[133] Dendy's collection of preserved embryos remains at King's College, University of London, today, although the catalogue has been lost.[134] Four of his specimens of eggs or embryos are also in the collection of the Natural History Museum in London.

Apart from the growing interest in the descriptive embryology of tuatara (Schauinsland, Thilenius and Dendy are still the major source references today), the research period from 1885 to 1900 is notable for several reasons. One is the transition from Karewa to Stephens Island as the primary source of research material, reflecting the relative ease of studying tuatara and their eggs there. A second is, by today's standards, negligible awareness among researchers with similar interests. Schauinsland and Dendy did not cite each other's works until towards the end of their own series of publications, and Dendy, at least, never cited Thilenius's paper.[135] Despite tuatara having become legally protected in 1895 and their eggs from 1898, the fledgling colonial government – the central authority responsible for issuing the permits – apparently had insufficient expertise or resources to guide the research effort, a situation for which it received some public criticism at the time.[136]

Research from 1900 until the present: some highlights

Soon after Günther's announcement of 1867, new monographs on the anatomical features of tuatara began to pour out of European and eventually North American institutions. The flood continued into the early decades of the 20th century: during 1900–1939, over 1000 printed pages on original aspects of the anatomy of tuatara were published. Material sent or taken by Dendy to England resulted in enormous productivity, alone accounting for over 640 pages on the skeleton,[137] tail,[138] teeth,[139] Jacobson's organ,[140] retina,[141] auditory apparatus,[142] pituitary body and brain,[143] hepatic venous and excretory system,[144] blood vascular system,[145] primordial germ cells,[146] embryonic membranes[147] and spermatogenesis.[148]

Many of these findings were compiled in 1931 into a substantial monograph (in German) on 'Rhynchocephalia' by **Otto von Wettstein**, Curator of the Herpetological Collection of the Natural History Museum in Vienna.[149] A few inaccuracies in von Wettstein's account suggest that he was not personally familiar with either live tuatara or New Zealand's geography.[150] Nonetheless, it was an impressive collation of the existing literature, though sadly not widely known within New Zealand. Later, von Wettstein described the tuatara of Hauturu/Little Barrier Island as a new subspecies, *Sphenodon punctatus reischeki*,[151] although genetic evidence argues against this proposal.[152]

Within New Zealand, field research on tuatara came to a virtual standstill for the first two decades of the 1900s, in part because new laws restricted collection but also because economic and social factors favoured more applied research topics such as agriculture.[153] An expedition to Stephens Island in 1922, by staff of the Dominion Museum and the State University of Iowa, resulted in four living and one preserved tuatara contributing to studies in Iowa on brain morphology, the digestive and muscular systems, and behaviour in captivity.[154] A further specimen was collected in 1927 by herpetologists from the University of Michigan.[155] Anatomical research (with an emphasis on brain structure) was pursued in New Zealand during the 1920s and 1930s under the leadership of Professor W. P. Gowland in the Department of Anatomy, University of Otago;[156] Gowland's histological collection is now held by the Otago Museum. Popular articles with new observations on the biology and distribution of tuatara were provided from the 1930s by Robert Falla, an ornithologist with the Auckland Museum, and later Canterbury and Dominion Museums.[157]

It was not until 1949 when Karl Schmidt, a distinguished American herpetologist, visited Stephens Island and the Trios Islands accompanied by **William (Bill) Dawbin** (Figure 5.24) that the potential for long-term field research was fully recognised.[158] Dawbin was a lecturer in zoology at Victoria University of Wellington and primarily a marine mammalogist, but Schmidt's visit stimulated him to initiate a research programme on the field ecology of tuatara.[159] Between 1949 and 1955, Dawbin and his teams of volunteers visited Stephens Island on 18 occasions, toe-clipping about 750 tuatara for studies of growth, longevity and general biology. In 1956 Dawbin shifted to the University of Sydney, Australia, where he maintained a captive colony of tuatara for further studies with or by his students on topics including nitrogen excretion,[160] water balance[161] and anatomy of the heart.[162] Dawbin, assisted on at least some occasions by his wife,[163] visited Stephens Island nine more times between 1956 and 1980 to obtain long-term recaptures for his growth studies and to assist accompanying researchers.[164] His own studies on tuatara resulted in several popular articles,[165] culminating in a summary of life history, growth and longevity[166] and a significant review of the literature on anatomy, ecology, behaviour and other topics.[167]

Since the late 1800s, few specimens of tuatara have been available to researchers outside New Zealand. Utilising some obtained during the 1960s, **Manfred Gabe** and **Hubert Saint Girons** (Figure 5.25) made elegant studies of the histological structure of numerous organs;[168] Saint Girons also collaborated on field studies of thermoregulation in wild populations.[169] **Carl Gans** at the University of Michigan and his colleagues contributed studies of the ear and hearing,[170] visual function,[171] respiratory pattern,[172] mastication[173] and copulatory behaviour.[174] Following a field trip to Stephens Island, Gans

Fig. 5.24 William (Bill) Dawbin (1921–1998) led field studies on Stephens Island from 1949 to 1980, resulting in the first scientific estimates of growth and longevity in tuatara. Dawbin's 27 visits to Stephens Island resulted mainly in popular articles, but helped stimulate other research by colleagues and showed that long-term field studies were possible. Shown here measuring jaw dimensions of a tuatara during the early 1950s, Dawbin was renowned for placing strenuous demands on his volunteers, one of whom is said to have fallen asleep with her arm down a burrow.

Photo: collection of Colleen King.

contributed a seminal article on the inappropriateness of viewing tuatara as a 'maladapted relict'.[175]

Since Dawbin's pioneering studies, field research in New Zealand on the ecology, reproduction, physiology and genetics of tuatara has expanded greatly, with an increasingly conservation-oriented focus during the late 1900s and early 2000s. Within the New Zealand Wildlife Branch (later Wildlife Service) of the Department of Internal Affairs, **Ian Crook** initiated studies from 1970 into interactions of tuatara with rats and petrels[176] and provided a brief review of general biology (Figure 5.26).[177] **Donald Newman** continued from the late 1970s with studies including captive maintenance and breeding,[178] distribution and population status,[179] burrowing and relationships with petrels,[180] effects of rats and mice,[181] and reproductive frequency in females.[182] A conference on 'New Zealand Herpetology' organised by Newman and others in 1980[183] led to an edited volume of proceedings,[184] including 10 articles on tuatara; Newman also wrote several semi-popular accounts.[185]

Turning to other popular works, a book on the New Zealand herpetofauna by **Richard Sharell** in 1966 (revised in 1975) included original observations on tuatara.[186] The biology of tuatara was covered in a review of the New Zealand herpetofauna by **Joan Robb** in 1973;[187] this was followed by a small but valuable semi-technical book on tuatara in 1977[188] and a widely read popular account covering tuatara and the New Zealand herpetofauna generally in 1980 (revised in 1986).[189] Other, briefer popular accounts have been written in recent years from the perspectives of tuatara curator **Lindsay Hazley** at the Southland Museum and Art Gallery,[190] various scientists[191] and others.[192]

Formal research on tuatara has continued to blossom over the last few decades, and publications have become too numerous to cite individually.

Fig. 5.25 Tuatara researchers and associates attending the First World Congress of Herpetology, held in 1989 at the University of Canterbury in Kent. From left to right: Marie-Charlotte Saint Girons, Don Bradshaw, Charles Daugherty, Donald Newman, Lorenzo Alibardi, Louis Guillette, James Gillingham, Mary McIntyre, Chris Carmichael, Kathy Packard, Gary Packard, Nicola Grimmond, Alison Cree, Michael Thompson, Hubert Saint Girons.

Photo: collection of Alison Cree.

Fig. 5.26 Ian Crook, shown here investigating burrows on Stephens Island in 1972, pioneered studies on tuatara by the Wildlife Branch (later the Wildlife Service) of the Department of Internal Affairs.

Photo: Pip Aplin.

Victoria University of Wellington has maintained a vigorous programme on the biology of tuatara since 1985, led until the early 2000s by **Charles Daugherty** (a professor of ecology) and more recently by senior lecturer **Nicola Nelson**, with significant technical support from **Susan Keall**. Scientists and students from the University of Otago, University of Auckland and other New Zealand universities, and from many overseas institutions have also contributed to recent research, both in field studies and on material sent overseas. These endeavours were aided in 1987 by the formation of a local professional society for herpetological research, the Society for Research on Amphibians and Reptiles in New Zealand (SRARNZ). As is typical of other areas of biological study in New Zealand, a substantial entry of women[193] and to a smaller extent of Māori[194] into published research on and management of tuatara is evident.

Many 20th-century researchers are alive and their careers still active, and it is not the intention to give full personal accounts here. Rather, the enormous expansion of research on the biology of live tuatara will be evident from the numerous citations in Part 2 of this book. Part 3, on conservation, will continue from the current chapter to explore historical developments within New Zealand from the late 1800s. These developments have culminated recently in an active, government-coordinated recovery programme, centred around recovery plans published by the Department of Conservation (successor to the Wildlife Service) in collaboration with university scientists and others, in 1993[195] and in 2001.[196]

Endnotes to Chapter 5

See the bibliography for full references to articles on tuatara, which are cited here at first mention by surname and year only. References on other topics are cited here in full at first mention, and then using 'op. cit.'. Except where required for clarity, the convention of parentheses around years has been dispensed with for economy of space.

1 Colenso, W. 1844. *Excursion in the northern island of New Zealand: in the summer of 1841–42.* Launceston, V. D. Land, p. 93.
2 Salmond, A. 1991. *Two worlds: first meetings between Maori and Europeans 1642–1772.* Viking, Auckland. See p. 75.
3 Islands that would have had tuatara at the time, including Stephens Island, the Trio Islands and the Chetwode Islands, were drawn by Isaac Gilsemans, cartographer on Tasman's 1642 voyage. See p. 97 of Anderson, G. 2001. *The merchant of the Zeehaen: Isaac Gilsemans and the voyages of Abel Tasman.* Te Papa Press, Wellington. The presence of tuatara on these islands would remain unknown to Europeans for at least 200 years.
4 Mackay, D. 2010. Cook, James 1728–1779. In: *Dictionary of New Zealand Biography.* http://www.teara.govt.nz/en/biographies/1c25/1 (accessed 16 July 2012).
5 Beaglehole, J. C. 1955. *The journals of Captain James Cook on his voyages of discovery. Vol. I. The voyage of the* Endeavour *1768–1771.* Published for the Hakluyt Society at the University Press, Cambridge; David, A. C. F., Joppien, R. and Smith, B. 1988. *The charts and coastal views of Captain Cook's voyages.* Hakluyt Society in association with the Australian Academy of the Humanities, London.
6 William Anderson, the surgeon-naturalist on the final voyage in 1776, quoted by Beaglehole, J. C. 1967. *The journals of Captain James Cook on his voyages of discovery. Vol III. The voyage of the* Resolution *and* Discovery *1776–1780, Part 2.* Published for the Hakluyt Society at the University Press, Cambridge (see p. 808 of Appendix 1 therein).
7 Beaglehole 1967, op. cit. See pp. 74–75.
8 Shortland, E. 1856. *Traditions and superstitions of the New Zealanders: with illustrations of their manners and customs.* Longman, Brown, Green, Longmans and Roberts, London; Colenso 1879.
9 Andrews, J. R. H. 1986. *The southern ark: zoological discovery in New Zealand 1769–1900.* Century Hutchinson New Zealand, Auckland. See p. 20. See also Salmond 1991, op. cit., p. 355.
10 Millener 1981.
11 Salmond 1991, op. cit.
12 Salmond 1991, op. cit. See p. 407.
13 Parsonson, G. S. 2010. Marsden, Samuel 1765–1838. In: *Dictionary of New Zealand Biography.* http://www.teara.govt.nz/en/biographies/1m16/1 (accessed 16 July 2012).
14 Nicholas 1817.
15 Nicholas 1817. See pp. 124–125.
16 Kendall, T. 1820. *A grammar and vocabulary of the language of New Zealand.* Church Missionary Society, London. Facsimile published 1980 by University Microfilms International, Michigan, p. 218.
17 Andrews 1986, op. cit. See p. 85.
18 Wright, O. 1950. *New Zealand 1826–1827, from the French of Dumont d'Urville: an English translation of the* Voyage de l'Astrolabe *in New Zealand waters.* Wingfield Press, Wellington. See p. 195.
19 Chisholm, J. 2010. Polack, Joel Samuel 1807–1882. In: *Dictionary of New Zealand Biography.* http://www.teara.govt.nz/en/biographies/1p18/1 (accessed 16 July 2012).
20 Polack 1838, Vol. I, p. 317.
21 Polack 1838, Vol. I, p. 221.
22 Polack 1840, Vol. II, p. 284.
23 Chisholm 2010, op. cit.
24 Bagnell, A. G. and Peterson, G. C. 1948. *William Colenso, printer, missionary, botanist, explorer, politician: his life and journeys.* A. H. and A.W. Reed, Wellington; See also Mackay 2010, op. cit.
25 The original manuscript (primarily about moa) was completed in 1842 and submitted to the *Tasmanian Journal of Natural Science*, but problems in Tasmania meant that publication was delayed until 1846. By this time Richard Owen had assisted Colenso to have the paper published in *Annals and Magazine of Natural History* in 1844. See Andrews 1986, op. cit., pp. 127–129.
26 Colenso 1844 (see bibliography).
27 Colenso 1879.
28 Colenso 1886.
29 Taylor 1848. See p. viii.
30 Taylor, R. 1870. *Maori and English dictionary* [revision of *A leaf from the natural history of New Zealand*, 1848]. George T. Chapman, Auckland.
31 Bagnell and Peterson 1948, op. cit. See p. 98.
32 McLean, D. 2011. Dieffenbach, Johann Karl Ernst 1811–1855. In: *Dictionary of New Zealand Biography.* http://www.teara.govt.nz/en/biographies/1d13/1 (accessed 16 July 2012).
33 Dieffenbach 1843, in two volumes.
34 Dieffenbach 1843, Vol. I. See p. 140 (elsewhere his book includes 'narara' and 'tuatera' as native names for the 'gigantic lizard' or 'guana': see Vol. II, pp. 204–205).
35 Dieffenbach 1843, Vol. II. See p. 205.
36 Dieffenbach 1843, Vol. I. See p. 405.
37 Colenso 1886, p. 119.
38 Dieffenbach 1843, Vol. I. See pp. 233–234.
39 Wade 1842. See p. 178.
40 A Party of Officers of the 58th Regiment 1852. As noted in the bibliographic entry, the article was reprinted in 1982 in a more accessible source. The 58th Regiment was part of the British Army. Its surgeon, Arthur Thomson, compiled statistical observations on disease among Māori and on the local climate. Although his later book on New Zealand contains no detailed reference to tuatara, he was presumably involved with the article published in 1852.
41 Stack 1875.
42 Shea, G. M. 1995. John Edward Gray, Albert Günther, and the lizards of Australia and New Zealand. In: Gray, J. E. and Günther, A. (eds). *The lizards of Australia and New Zealand.* Society for the Study of Amphibians and Reptiles, Ithaca, New York, pp. iii–viii.
43 Gray 1831.
44 Gray 1842, p. 72.
45 Specimen BMNH 1946.1.22.3; Colin McCarthy, Department of Zoology, Natural History Museum, pers. comm. 14 August 2003.
46 Gray 1843.
47 Specimen BMNH 1855.10.16.37; Colin McCarthy pers. comm. 14 August 2003.

48 Shea 1995, op. cit.

49 Gray 1845a.

50 Knox 1870.

51 Gray 1845b.

52 BMNH 1844.20.29.11.

53 Owen 1845.

54 Benton, M. J. 2005. *Vertebrate palaeontology*. 3rd edn. Blackwell, Oxford. See pp. 146–147.

55 Temple, P. 2002. *A sort of conscience: the Wakefields*. Auckland University Press, Auckland. See p. 400. Whether this is the same as either the specimen from Somes Island recorded as being collected in 1842 (Knox 1870) or the skin and skeleton from that island that found their way to the British Museum soon afterwards (Gray 1845b) is unclear.

56 Owen 1853. See p. 143.

57 Gray 1867.

58 Günther 1867, p. 603.

59 Some modern palaeontologists have argued that, because of the confusion with rhynchosaurs, the term 'Rhynchocephalia' should be dropped entirely in favour of 'Sphenodontia'. More recently others have argued that a redefined 'Rhynchocephalia' is acceptable, and preferable historically. The latter position is followed in this book. See Chapter 1, Box 1.1 for further discussion.

60 Gray 1869.

61 Gray 1869, p.167.

62 Simon Chaplin, Senior Curator of the Museums of the Royal College of Surgeons of England, London, pers. comm. 15 August 2003.

63 My thanks to Kenneth Cree for this suggestion. Another and perhaps less likely suggestion, from the nature writer James Drummond, was that the name *Hatteria* was 'the Latinised form of a supposed Maori name of the reptile' (Drummond 1917).

64 Adler, K. 1989. Herpetologists of the past. In: Adler, K. (ed.). *Contributions to the history of herpetology*. Society for the Study of Amphibians and Reptiles, Oxford, Ohio. See p. 34.

65 Sclater 1870, p. 148. Sclater was, however, unconvinced that the distinctiveness of tuatara merited recognition in an order separate from lizards; see Sclater 1871. Osawa (1898d) was also unconvinced.

66 Knox 1870.

67 Newman 1878.

68 Galbreath, R. 2010. Buller, Walter Lawry 1838–1906. In: Dictionary of New Zealand Biography. http://www.teara.govt.nz/en/biographies/1b46/1 (accessed 16 July 2012).

69 Galbreath, R. 1989. *Walter Buller: the reluctant conservationist*. GP Books, Wellington.

70 Buller 1877.

71 A. Günther to J. Hector, 22 February 1885. NMNZ MU 95 Box 2, Item 115 1885/200. Hector, however, was publicly supportive of Buller, noting in 1879 that 'With regard to the tuatara, he agreed with Dr. Buller that the two species – *Sphenodon punctatum* and *S. guntheri* – would hold good.' *Transactions and Proceedings of the New Zealand Institute* 11: 521.

72 Hay et al. 2010.

73 Buller 1871, 1879.

74 Buller 1878, 1879.

75 Buller 1895.

76 Buller 1871.

77 Buller 1871; reiterated in Buller 1877.

78 Galbreath 1989, op. cit.

79 Buller 1896.

80 Galbreath 1989, op. cit.; 2010, op. cit.

81 Anon. 1895.

82 Buller 1893.

83 Buller 1871, 1877.

84 Mair, G. *Diary and notes Nov. 1885–Mar. 1886* [Diary 36]. Alexander Turnbull Library, Wellington, New Zealand.

85 Buller 1878.

86 Gilbert Mair to H. D. Bell, 8 June 1913. ANZ IA 1 46/18/4 pt 1.

87 Anon. 1872; Mair 1873; Buller 1878.

88 King, M. 1981. The collector: a biography of Andreas Reischek. Hodder and Stoughton, Auckland; Prebble, R. G. 2010. Reischek, Andreas 1845–1902. In: Dictionary of New Zealand Biography. http://www.teara.govt.nz/en/biographies/2r14/1 (accessed 16 July 2012).

89 Reischek 1882.

90 Anon. 1882a.

91 Reischek 1882, p. 275.

92 Reischek 1882, p. 276.

93 Reischek 1886.

94 Reischek 1886, p. 110.

95 Hay et al. 2003, 2010; MacAvoy et al. 2007.

96 Reischek 1971, pp. 86–87.

97 Prebble 2010, op. cit.

98 Newman 1878; Buller 1895.

99 Thomas 1890, 1891.

100 Anon. 1955a, b.

101 Thomas 1890, 1891.

102 T. J. Parker to J. Hector, 17 June 1885: 'I have about 18 still alive Thomas probably about twenty.' NMNZ Series No. MU000147 Location R–1M03–130E Box 6, Item 118 (other number: 85/2010).

103 Parker 1891 (see also Chapter 10, Box 10.1).

104 Another early diorama of tuatara and seabirds, based on collections by C. F. Adams from Karewa Island, was on display at the Auckland Museum by 1886 (Gill, B. J. 1999. History of the land vertebrates collection at Auckland Museum, New Zealand, 1852–1996. *Records of the Auckland Museum* 36: 59–93). However, the photo of the diorama in Gill's article seems to differ from that in Figure 5.19.

105 Osawa 1898b.

106 Osawa 1896, 1897, 1898a, b, c, d, 1899.

107 Schauinsland 1898b, p. 69, in translation.

108 Webber 1953. Webber's account (written many years later) appears to contain some inaccurate dates.

109 Dawson, G. E. and Dawson, E. W. 1958. Birds of the Cook Strait Islands, collected by Professor Hugo Schauinsland in 1896 and 1897. *Notornis* 8: 39–49.

110 Schauinsland 1898a, b.

111 Schauinsland 1899.

112 As described by Moffat 1985.

113 Schauinsland 1900, 1903a.

114 Schauinsland 1903b, c, 1906a.

115 Schauinsland 1906b.

116 Peter-René Becker, Übersee Museum, pers. comm. 2 September and 25 October 2003.

117 Beasley, H. G. 1938. Georg Thilenius: born October, 1868–28th December, 1937. *Man* 38: 75.

118 Thilenius 1899.

119 Thilenius 1899; Schmidt 1952.

120 Schmidt 1952.

121 Thilenius 1899.

122 Untitled summary of events regarding collections made during 1896–1898 by Thilenius, Schauinsland and Dendy. NMNZ MU 000001 Box 23, Item 14, File No. 14/6/9. See also articles during 1898 in the *Bay of Plenty Times and Thames Valley Warden*, 18 May, p. 2; 29 May, p. 2; 16 September, p. 2.
123 E. W. M. [E. W. MacBride] 1926. Arthur Dendy 1865–1925. *Proceedings of the Royal Society B* 99: xxxiii–xxxv.
124 Dendy 1899b, p. 3.
125 Dendy 1899b.
126 Dendy 1898, 1899a.
127 Dendy 1899b.
128 Dendy 1899c.
129 Dendy 1899d.
130 Dendy 1899c.
131 Dendy 1909.
132 Dendy 1911.
133 Dendy 1910.
134 Jill Sales, Honorary Curator of the Kings College Natural History Collection, pers. comm. 5 August 2003.
135 Though Thilenius apparently visited Dendy in Christchurch in 1898, after being informed of his work (see *Bay of Plenty Times and Thames Valley Warden*, 3 October 1898, p. 2).
136 Anon. 1899b; Benham 1899.
137 Howes and Swinnerton 1901.
138 Woodland 1921.
139 Harrison 1901a, b.
140 Broom 1906.
141 Bage 1912.
142 Wyeth 1920, 1924a.
143 Hines 1923; Wyeth and Row 1923; Wyeth 1924b, 1925.
144 Tribe and Fisk 1940.
145 O'Donoghue 1920.
146 Tribe and Brambell 1932.
147 Fisk and Tribe 1949.
148 Hogben 1921.
149 von Wettstein 1931.
150 For example, the Waimakariri River is mistakenly placed in the North Island, and the Brothers and Trio Islands appear to be amalgamated into one island group, the 'Three Brothers'.
151 von Wettstein 1943.
152 Hay et al. 2003, 2010.
153 Andrews 1986, op. cit.
154 Krull 1923; Byerly 1925; Keck 1925; Nutting 1926; Christensen 1927.
155 Dr Frank N. Blanchard and his wife Dr Frieda Cobb Blanchard. Frieda Blanchard wrote an entertaining account of their adventure in *National Geographic* (Blanchard 1935).
156 Cairney 1926a, b; Durward 1930; Hindenach 1931; Walls 1935.
157 Falla 1935, 1937, 1960.
158 Schmidt 1952; Bogert 1953a, b.
159 Dawbin and Schmidt also facilitated the shipment of four tuatara to overseas zoos for observation; see Oliver 1953.
160 Dawbin and Hill 1969; Hill and Dawbin 1969.
161 Hill 1982.
162 Simons 1965.
163 'My wife has worked on several expeditions to Stephens Island and it would help me greatly and add much to the results if she came to co-ordinate catering and also the field work of a group of helpers'; W. H. Dawbin to Secretary for Internal Affairs, 18 November 1964, ANZ IA 1 46/14/26.
164 See Newman 1982a for some resulting publications.
165 Dawbin 1949, 1953, 1962a, 1974.
166 Dawbin 1982a.
167 Dawbin 1982b.
168 Gabe and Saint Girons 1964a, b, c, 1965a, b, 1967, 1969, 1972, 1973, 1976.
169 Saint Girons 1980, 1985; Saint Girons et al. 1980. See also Newman, D. G. 2000. Hubert Saint Girons PhD Hon FRSNZ 1926–2000. *Yearbook of the Academy Council 2000*, Royal Society of New Zealand, Wellington.
170 Gans and Wever 1976.
171 Ireland and Gans 1977.
172 Gans and Clark 1978.
173 Gorniak et al. 1982.
174 Gans et al. 1984.
175 Gans 1983.
176 Crook 1970, 1973a, b, 1974, 1976.
177 Crook 1975.
178 Newman et al. 1979; Newman 1982d.
179 Newman 1982b; Newman and McFadden 1990a.
180 Newman 1978, 1982c, 1987a.
181 Newman 1983, 1986, 1988; Newman and McFadden 1990b.
182 Newman and Watson 1985; Newman et al. 1994.
183 Newman 1980.
184 Newman 1982a.
185 Newman 1987b, 1998.
186 Sharell 1966, revised 1975.
187 Robb 1973.
188 Robb 1977.
189 Robb 1980, revised 1986.
190 Hazley 1982.
191 Cree and Daugherty 1990a; Daugherty et al. 1990a; Newman 1998; Cree 2002; Towns 2004.
192 For example, Lutz 2006, an American tourist in New Zealand.
193 For example, Nicola Nelson and Susan Keall (mentioned in text), postdoctoral researchers Jennifer Hay and Hilary Miller, numerous postgraduate students, and the author of this book.
194 For example, Ramstad et al. 2007a, b.
195 Cree and Butler 1993.
196 Gaze 2001.

Endnotes to boxed text

For explanation of citation format in box endnotes, see endnotes to the main text.

Box 5.1

1 A bone from Earthquakes in the South Island is dated at 500 years before present (BP, where present is by convention 1950), with a standard error of ± 100 years, in supplementary material of the following: Wilmshurst, J. M., Anderson, A. J., Higham, T. F. G. and Worthy, T. H. 2008. Dating the late prehistoric dispersal of Polynesians to New Zealand using the commensal Pacific rat. *Proceedings of the National Academy of Sciences* 105: 7676–7680. Radiocarbon dates of 649 years BP for a bone from Northland and 659 years BP for a bone from Southland were reported in supplementary material of Hay et al. 2008.
2 B. Sladden to Undersecretary, Department of Internal Affairs, 7 November 1935, ANZ IA 1 52/30 pt 1 Sanctuaries – Karewa Island.
3 Anon. 1979. Another tuatara found on a street in Auckland in 1893 was thought to be from a group that escaped from Government House (Anon. 1893).
4 Anon. 1915a.
5 Bruce Thomas, Ka Mate Traps Ltd, Nelson, pers. comm. 27 February 2008.
6 Reported locations (usually of lone specimens) include Plimmerton Beach in 1922 (Anon. 1922) as well as Invercargill in 1951, Tutukaka in 1956, the Mangaiti District in 1957 and Konini near Pahiatua in 1959 (the latter four examples in ANZ IA 1 46/18/4 pt 2). A large adult male was found at Tutukaka in 1995 (*Reuters News*, 26 May 1995). Earlier reports, of uncertain validity, include Hare Hongi's account of tuatara in Northland forest in the late 1800s (Hongi 1922), a young man's 'find' of a tuatara living in the cracks of a boulder near Akaroa (Anon. 1908b) and a report, attributed to two prospectors, of tuatara seen in 'the country at the back of the Thames' prior to about 1922 (Goodlet c. 1922, p. 16).
7 Tony Whitaker (Whitaker Consultants Ltd, Motueka, pers. comm. 24 February 2008) visited Mokopuna Island (off Matiu/Somes Island, in Wellington Harbour) in the 1960s when 'tuatara' were reported: the specimens were described as about 18 inches (45 centimetres) long and having a crest, but proved to be common geckos (a species within the *Hoplodactylus maculatus* complex) – a much smaller and crestless reptile. Similarly, I was once persuaded to visit a Wellington headland where a caller insisted that juvenile tuatara were to be found; these specimens also proved to be common geckos. Other New Zealand herpetologists have had similar experiences.
8 Herbert W. William [to Hector], 27 July 1900, NMNZ MU 147 Box 8, Item 104.
9 Hector's draft reply of 30 July 1900, NMNZ MU 147 Box 8, Item 104.
10 Buller 1895.
11 Various newspaper clippings in *Fildes #635*, Beaglehole Collection, Victoria University of Wellington Library; McLeod 1922.
12 McLeod 1922.
13 For example, in reporting the discovery of bones from tuatara in a quarry near Napier, Colenso merely noted that 'The old Maoris always said that the tuatara (*Sphenodon* sps. [*sic*]) formerly inhabited the headland off the New Zealand coast (as well as the islets lying off it), which the finding of this specimen proves' (Colenso 1886, p. 122.).
14 For example, during 1843 and 1844, Shortland travelled north on foot from Otago to Banks Peninsula (a region where tuatara were once present), but his observations on natural history contain no reference to tuatara. Shortland, E. 1851. *The southern districts of New Zealand: a journal, with passing notices of the customs of the aborigines*. Longman, Brown, Green and Longmans, London.

Box 5.2

1 For references to tuatara and other reptiles as pets among Māori, see Beattie, H. 1920a. Nature-lore of the southern Maori. *Transactions and Proceedings of the New Zealand Institute* 52: 53-77; Stack 1875.
2 Buller 1877, p. 322.
3 Anon. 1882b, p. 5. But not everyone was as enthusiastic. A colony of tuatara in a coal cellar took a visiting milkman by surprise in Ponsonby, Auckland in 1887 (apparently the tuatara had been the pets of a previous occupant, a naturalist). In a subsequent newspaper report, the tuatara were considered as 'ugly enough in all conscience to warrant people who are unacquainted with their habits not wanting them as neighbours' (Anon. 1887, pp. 9–10).
4 Anon. 1896.
5 Anon. 1899a, p. 34.
6 Donne 1942, p. 102.
7 Donne 1942. Donne was 82 when this article was published, having left New Zealand about 30 years earlier. Some of the dates and sequences of events regarding his visit to Stephens Island seem inaccurate.
8 Sewell 1931.
9 D. J. Grindlay to Under Secretary for Internal Affairs, 17 March 1919, ANZ IA 1 46/18/8 pt 1.
10 Buller 1878.
11 Anon. 1909, p. 2.
12 Anon. 1882b.
13 Anon. 1882b, p. 5.
14 Gold-Smith 1885, p. 424.
15 W. J. Phillipps to the Director, Dominion Museum, 9 April 1929. ANZ IA 1 52/30 pt 1 Sanctuaries – Karewa Island.
16 Webber 1953.
17 Anon. 1936. A century at Elsmlie Bay. *Weekly News*, 1 January, pp. 20–21. Copy in: Teviotdale, D. c. 1930–1948. Scrapbooks containing newspaper clippings, photographs and manuscript notes on the Otago Museum and New Zealand history, particularly Maori history and archaeology. *Teviotdale Clippings, Vol. 5*, Hocken Library, Dunedin.
18 Anon. 1936, op. cit.
19 *Extracts from Report on bird life around French Pass, received from A. Wright, Relieving Keeper, Cape Brett, 2/25 May*. [undated, but filed with items from the 1950s.] ANZ IA 1 46/18/4 pt 2. Schmidt 1952 reported five tuatara.
20 One was seen between the school and schoolhouse, and two on the road above the reserve (Royden Smith, pers. comm. 7 April 2009). Smith was then the French Pass schoolteacher.
21 Some Māori elders of this region have suggested that tuatara might still be present at French Pass, but recent surveys provide no evidence (Ramstad et al. 2007a).
22 Hutton and Drummond 1904.
23 Sixteen tuatara collected from Stephens Island for the Tourist Department in 1906 (Schmidt 1952) may have been destined for

the International Exhibition in Christchurch that year. Tuatara from the exhibition were subsequently passed on to the Canterbury Acclimatisation Society; Andrew Johnson was a former curator of the society. See p. 49 (photo of tuatara) and p. 123 of Lamb, R. C. 1964. *Birds, beast & fishes: the first hundred years of the North Canterbury Acclimatisation Society*. North Canterbury Acclimatisation Society, Christchurch.

24 Colonial Secretary memo re A. M. Johnson per C. A. Baker, Opawa. ANZ IA I 46/18/4 pt 1. For more on the historical context of collection for Johnson's colony, see Chapter 10.

Box 5.3

1 A storage pit for sweet potatoes.

2 Buller 1877.

3 Handwriting difficult to decipher.

4 The length as reported by Mair (27 inches, or about 690 mm) was indeed remarkable, but the girth (4 inches, or about 100 mm) seems disproportionately small.

5 G. Mair to T. Cheeseman, 13 March 1876, Auckland Museum Archives MA 95/38/7 AV2.1 Correspondence 1868–99–Ma–Mv. The date of 1826 for the attempted invasion of Motiti Island by Te Haramiti (of Ngāpuhi), thwarted by the arrival of Ngāiterangi warriors, seems in error – published sources refer to 1832, as in Mair's later letter of 1913.

6 G. Mair to H. D. Bell, Minister for Internal Affairs, 8 June 1913, ANZ IA 1 46/18/4 pt 1.

7 Drummond and Hutton 1902, p. 57.

8 In one extreme claim, a Mr G. G. Norris, (presumably Greg Norris, younger brother of Colin Norris, a popular authority on tuatara in Tauranga) was reported in 1908 as estimating the age of a pair in his colony as from 700 to 1000 years (Anon. 1908a).

9 Falla 1935, p. 4.

10 For example, Dawbin 1949; Sharell 1966.

11 Daugherty et al. 1990a.

12 Robb 1977.

13 As suggested by Schauinsland 1898b and Dawbin 1974, 1982a, for example.

14 Nicola Nelson, Victoria University of Wellington, pers. comm. 26 January 2011.

15 Ramstad et al. 2007a.

PART TWO

BIOLOGY OF TUATARA TODAY

CHAPTER SIX

Island populations and histories of study

Biologists have often focused on Stephens Island (Takapourewa) for information on the ecology of tuatara. But a true appreciation of today's natural habitats involves populations on nearly three dozen islands, some with a longer history of written study than on Stephens Island.

Introduction

This chapter describes the island habitats of tuatara, including geographic distribution, the status of natural populations and the history of study. Recent translocations of tuatara to new islands and to a mainland site are also mentioned under the relevant source island (and discussed further as conservation approaches in Chapter 10). Boxes in this chapter explore geographic variation in genetics and in body size, the experience of studying tuatara on a seabird-dominated island, and the history of collections of tuatara from small islands in the Bay of Plenty. For more information on the habits, diets and behaviour of tuatara, see Chapters 7, 8 and 9. For updates since 2012, see Chapter 11.

Distribution and general characteristics of natural populations on islands

As sea levels rose about 10,000 years ago at the end of the Pleistocene Epoch, tuatara became isolated on the North and South Islands and on dozens of smaller islands around the mainland coast. Later, following human arrival, tuatara became extinct on the mainland, as well as on about half of all offshore islands for which there is some evidence for past presence (Chapter 3). Today, tuatara remain naturally present on about 32 offshore islands (Table 6.1), an area of land representing less than about 0.5% of the former pre-human distribution.[1]

The latitudinal range of surviving natural populations extends from the Poor Knights Islands at 35°28'S, to North Brother Island at 41°07'S (Figure 6.1).

TABLE 6.1

Existing wild populations of tuatara as of 2011

Within each category (natural and translocated), populations are listed from north to south.

NATURAL POPULATIONS ON ISLANDS

Location	Area (ha)	Approximate size of tuatara population[1]	Introduced mammals present/eliminated[2]	Comments on population trends, including genetic diversity[3]
Poor Knights Islands				
Tawhiti Rahi Island	163	Hundreds or low thousands	–	Relatively low microsatellite heterozygosity for size of island
Motu Kapiti Island	1.3	Few tens	–	
Stack B (inf.)	0.8	Few tens	–	
Aorangi Island	110	Hundreds or low thousands	Pigs removed 1936	Relatively low microsatellite heterozygosity for size of island
Aorangaia Island	6.3	Few tens	–	
Hen and Chickens Islands				
Mauitaha Island	26	Fewer than ten	Pacific rats present	Doomed to extinction
Lady Alice Island	155	High hundreds	Cattle removed by about 1924; Pacific rats eliminated 1994	Increasing after recent recruitment failure.
Coppermine Island	80	Low hundreds	Pacific rats eliminated 1997	Increasing after recent recruitment failure
Whatupuke Island	102	Low hundreds	Pacific rats eliminated 1993	Increasing after recent recruitment failure
Taranga Island	500	Few tens	Eradication of Pacific rats began in 2011[4]	Recent recruitment failure
Hauturu/Little Barrier Island	3083	c. 134 in wild, most being captive-reared offspring released 2006–2010 (7 adults and 7 off-spring remain captive on island, 31 eggs incubating off-island)	Pigs and dogs present during 1800s; cats eliminated 1980, Pacific rats eliminated 2004	Expected to increase after recent recruitment failure, though genetic diversity reduced in captive-reared offspring (many are full or half-sibs)
Cuvier Island (Repanga Island)	170	Few tens, including 39 captive-reared offspring released 2001–2010 (4 adults and 7 juveniles remain captive off-island)	Sheep and cattle removed 1960s, goats eliminated 1961, cats eliminated 1964, Pacific rats eliminated 1993	Recovery uncertain, but expected to increase following recent recruitment failure
Mercury Islands (Iles d'Haussez)				
Red Mercury Island (Whakau)	225	Few tens (including 24 captive-reared offspring released 1996–2001)	Pacific rats eliminated 1992	Recovery uncertain, but expected to increase following recent recruitment failure
Kawhitu or Stanley Island	100	Few tens, including 17 captive-reared offspring released 2003–2010 (7 adults, 10 juveniles and 4 incubating eggs remain captive off-island)	Pacific rats and European rabbits eliminated 1991	Recovery uncertain following recent recruitment failure; may benefit from supplementation
Atiu or Middle Island	13	Low thousands	–	
Green Island	2.3	Hundreds	–	

NATURAL POPULATIONS ON ISLANDS (cont.)

Location	Area (ha)	Approximate size of tuatara population[1]	Introduced mammals present/eliminated[2]	Comments on population trends, including genetic diversity[3]
The Aldermen Islands				
North Stack (inf.)	0.7	Tens	–	
Ruamahuanui Island	33	Hundreds or low thousands	–	
Middle Chain Stack (inf.)	0.4	Tens	–	
Hongiora	16	Low hundreds	–	
Nga Horo Island	3.4	Tens to low hundreds	–	
Half Island	1.5	Tens to low hundreds	–	
Ruamahuaiti Island	25	Low thousands	–	
Karewa Island	5.0	Hundreds	–	
Motunau Island (Plate Island)				
Northern part	0.8	Tens	–	
Southern part	2.8	Hundreds	–	
Rurima Islands				
Moutoki Island	0.8	Low hundreds	–	Low variation in microsatellites and mtDNA
Stephens Island (Takapourewa)	150	At least 30,000	Cats eliminated by about 1922, cattle removed by 1988, sheep removed by 2004, pig and dogs once briefly present	Exceptionally high density and total numbers; high genetic diversity
Trio Islands (Kuru Pongi)				
North Trio Island (inf.)	1	Few tens	–	
Middle Trio Island (inf.)	20	Low thousands	– [weka removed about 1964]	
South Trio Island (inf.)	2	Few tens	–	
The Brothers				
North Brother Island (inf.)	4	Low to mid-hundreds	–	Very low genetic diversity

TRANSLOCATED POPULATIONS

Location	Area (ha)	Approximate size of tuatara population[1]	Introduced mammals present/eliminated[2]	Comments on population trends, including genetic diversity[3]
Tiritiri Matangi Island	197	60 adults released 2003, sourced from Atiu Island	European rabbits and cats once present, Pacific rats eradicated 1993	Survival and recruitment noted (see Chapter 10)
Mayor Island (Tuhua)	1277	30 adults released 2007, sourced from Karewa Island	Pigs eradicated 1999, cats eradicated 2000, Pacific rats and Norway rats eradicated by 2003	Survival noted (see Chapter 10)
Moutohora Island	143	32 adults released 1996; 30 large juveniles or small adults released 2008, sourced from Moutoki Island	Cattle and sheep once present, goats eradicated 1977, Norway rats eradicated 1986, rabbits eradicated 1987	Survival and recruitment noted (see Chapter 10)

Table 6.1 (cont.)

TRANSLOCATED POPULATIONS (cont.)

Location	Area (ha)	Approximate size of tuatara population[1]	Introduced mammals present/eliminated[2]	Comments on population trends, including genetic diversity[3]
Wakaterepapanui Island	61	89 adults and juveniles released 2003; 343 captive-reared juveniles released 2004; all sourced from Stephens Island	Sheep once present, Pacific rats and Norway rats eradicated 1999	Survival noted (see Chapter 10)
Titi Island	32	18 adults and 50 captive-reared juveniles released 1995, sourced from North Brother Island	Norway rats eradicated 1975	Survival and recruitment noted (see Chapter 10)
Long Island	142	53 captive-reared juveniles released 2007, sourced from North Brother Island	Pacific rats eradicated 1997	Unknown (see Chapter 10)
Matiu/Somes Island	25	20 adults and 35 captive-reared juveniles released 1998, sourced from North Brother Island	Ship rats eradicated 1990, agricultural and quarantined mammals once present	Survival and recruitment noted (see Chapter 10)
Zealandia™ (formerly Karori Wildlife Sanctuary), Wellington	225	70 adults released 2005 and 130 released 2007, sourced from Stephens Island	Many mammals eradicated (but not mice)	Survival and recruitment noted (see Chapter 10)

Notes

1. Updated from Gaze 2001. For additional sources, see main text.
2. For scientific names, see main text. Most dates of eradications from relevant chapters in King, C. M., (ed.) 2005. *The handbook of New Zealand mammals*. Oxford University Press, Auckland.
3. Information on genetic diversity from MacAvoy et al. 2007 and Hay et al. 2010.
4. Towns, D. R., West, C. J. and Broome, K. G. 2012. Purposes, outcomes and challenges of eradicating invasive mammals from New Zealand islands: an historical perspective. *Wildlife Research* doi.org/10.1071/WR12064. There has been no sign since of rats on Taranga; however, definitive confirmation of rat-free status awaits surveys in 2013 (Keith Hawkins, Department of Conservation, pers. comm. 18 September 2012).

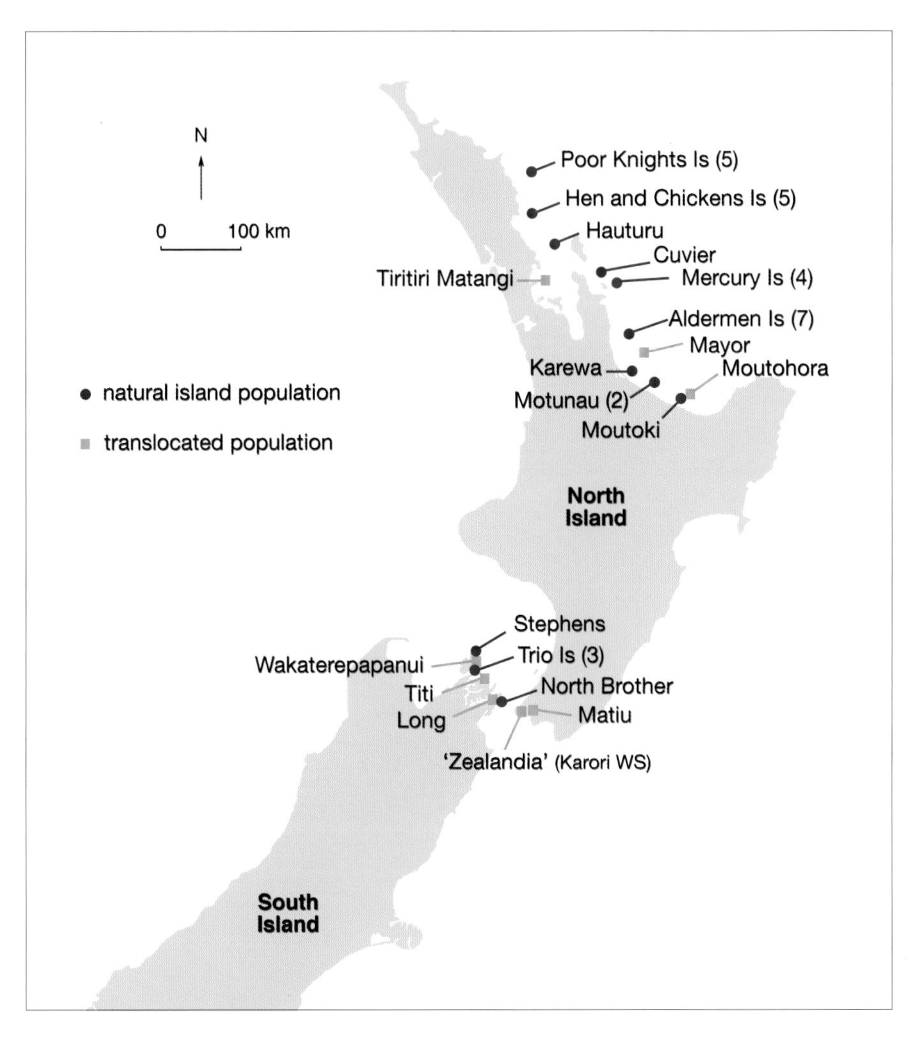

Fig. 6.1 Distribution of free-roaming tuatara to 2011. Natural populations exist on 32 islands (dark blue circles; the number of islands with tuatara in each major group is indicated). In addition, between 1995 and 2011 tuatara were translocated to seven islands and one large mainland sanctuary (light blue squares).

In size, Hauturu/Little Barrier Island is by far the largest (at 3083 hectares) and highest (to 722 metres above sea level). The remaining 31 islands range downwards from Taranga Island, at 500 hectares and 460 metres above sea level, to several islets or stacks that are less than 1 hectare in area and just a few tens of metres above sea level. Some islands are separated by only a few dozen metres from adjacent islands with tuatara, whereas others are more isolated.

Tuatara survive naturally today on islands about 5–23 kilometres from the mainland, falling within two geographic groups: 27 off the northeast coast of the North Island and five in Cook Strait. This geographic separation is paralleled by genetic differences. Populations in the northern group are known as northern tuatara (formally the subspecies *Sphenodon punctatus punctatus*), whereas those in the southern group are known as Cook Strait tuatara (a group within *S. punctatus* not formally named as a subspecies, although effectively treated as such). The population on North Brother Island, for a time managed as a second species of tuatara, is now considered to fall within the Cook Strait group of *S. punctatus*.[2]

BOX 6.1

Genetic variation within and among island populations of tuatara

The application of molecular biological tools to ecology has led to new insights into relationships among tuatara on different islands. Studies during the 1990s initially focused on allozyme electrophoresis, a method that compares the mobility of variant forms of enzymes from blood or other tissues through a gel carrying an electric current. This approach is inherently conservative, since not all variations in DNA result in variations in proteins, and not all protein variants are distinguishable from each other. Nonetheless, notable differences among 24 island populations were found. Results from a study led by Charles Daugherty,[1] supported by later work by Mark Brown,[2] revealed genetic separation between northern populations and those in western Cook Strait. Unexpectedly, the earlier study also revealed differences that were considered large enough to warrant reinstatement of the North Brother population (in eastern Cook Strait) as a separate species from all other populations.[3]

Box 6.1 Fig. 1 Analyses of blood samples (which can safely be collected from vessels in the tail, as demonstrated here by Claudine Tyrrell) have yielded information on genetic structure within and among populations of tuatara. Recent work shows that DNA can also be obtained from swabs of the cloaca or mouth, making collection of samples for some genetic analyses easier and less invasive (see Miller 2006).

Photo: collection of Claudine Tyrrell.

Newer techniques have allowed variation in DNA to be analysed directly, across an often wider range of islands. The type of DNA includes that found in small cellular organelles known as mitochondria (hence, mitochondrial DNA or mtDNA), as well as sections of chromosomal DNA from the nucleus. The overall patterns for tuatara support the separation between northern populations of tuatara and those in Cook Strait,[4] but not the recognition of the North Brother population as a separate species. In terms of mtDNA in particular, the North Brother population is grouped within other Cook Strait populations. This population is now considered as falling within the species *S. punctatus*, albeit with its own distinctive genetic history.[5]

Among the northern populations, analyses of minisatellites and microsatellites (segments of nuclear DNA made up of repeating sequences of a small number of base pairs) confirm that tuatara within island groups tend to be genetically more similar to each other than to those in other island groups. In fact, it is now possible to identify many individuals as coming from specific island groups, or even specific islands,[6] an outcome that may prove useful for inferring the source of tuatara that turn up in illegal collections or on the mainland. Within a population, variation within the fast-evolving minisatellite and microsatellite regions also suggests that tuatara living closer to each other may be more closely related, and that some large males may dominate mating.[7]

Genetic approaches also lead to inferences about the past history of populations. In general, populations with large numbers of individuals, which tend to be those on large islands, maintain greater genetic diversity than do small populations, which tend to be on small islands. However, on medium-sized or large islands such as Kawhitu, Cuvier (Repanga), Red Mercury (Whakau) and Hauturu/Little Barrier Islands, where the past presence of rats has led populations of tuatara to crash to a few tens of adults at most, diversity in microsatellites can still be moderately high.[8] On the other hand, genetic diversity is very low on North Brother Island, despite the adult population currently numbering about 300–400.[9] This

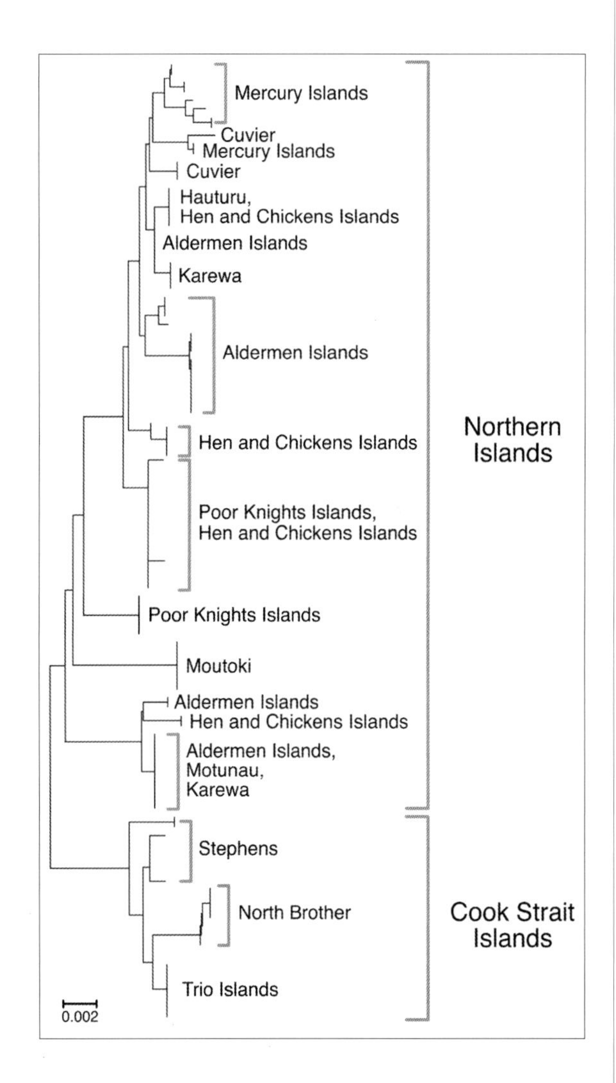

Box 6.1 Fig. 2 Minimum evolution tree, based on analysis of the mitochondrial control region and flanking tRNA sequence for 106 tuatara representing 26 populations. This tree supports a major distinction between all northern and Cook Strait islands, but not between North Brother Island and remaining populations. Scale bar = Kimura-2 distances; see original source for island names within island groups and for confidence intervals for branch nodes.

Modified, with simplification, from Hay et al. 2010.

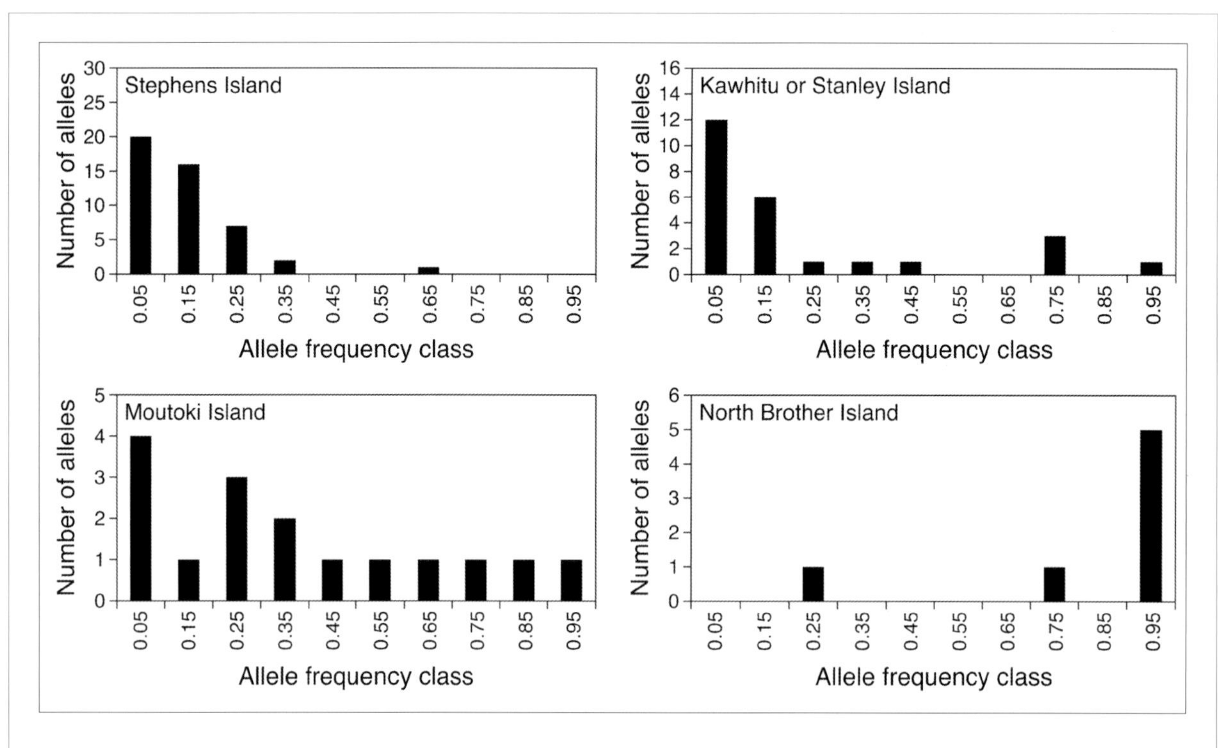

Box 6.1 Fig. 3 A way of illustrating levels of genetic variation within a population is to plot frequency distributions for alleles (variant forms of a gene). These examples, from a study of alleles at six microsatellite loci (specific locations on a chromosome) by Elizabeth MacAvoy and colleagues, show marked differences among populations of tuatara. Populations with a 'normal history' such as those on Stephens Island (Takapourewa) have a left-skewed or L-shaped distribution (many alleles are present, with most occurring in relatively few individuals). A similar profile is seen on Kawhitu or Stanley Island, despite the currently small size of this population. The naturally small population on Moutoki Island is less strongly left-skewed. In contrast to these three islands, the distribution for North Brother Island is strongly right-skewed, indicating a marked reduction in genetic diversity (few alleles are present, with most being 'fixed', or invariant among individuals).

Modified from MacAvoy et al. 2007.

low variation implies a past population 'bottleneck' (an extreme reduction in the number of reproducing individuals), but whether this resulted from a low number of individuals founding the original population,[10] from genetic drift subsequent to founding, from over-collection in the 1800s[11] or from some combination of these[12] is unclear. The low levels of variation in the North Brother population extend to genes of the major histocompatibility complex (MHC), genes that enable animals to combat pathogens, and hence resist disease.[13] Further study may reveal whether tuatara on North Brother Island have functional deficits arising from inbreeding. Regardless, the wisdom of using this population as a sole source for translocations to establish tuatara on other islands must now be questioned.[14]

The figure of 32 for the number of islands with natural populations of tuatara is imprecise, because other small populations may exist. New Zealand has more than 700 offshore islands, many of which are visited infrequently, and not necessarily under conditions when tuatara would be seen. Some islands are privately owned and the presence of tuatara may go unreported. Although it is unlikely that tuatara exist on all, several islands have unconfirmed reports from the past century suggesting the recent presence of tuatara. These include three islands in the Northland region,[3] three in the Auckland region,[4] several in the Bay of Plenty[5] and several in Marlborough.[6] The most recent confirmed discoveries of tuatara in new locations occurred in 1986, when a relict population was discovered on Kawhitu or Stanley Island in the Mercury Islands group,[7] and in 1994, when a single individual was found on Mauitaha Island in the Hen and Chickens Islands.[8]

Habitats and climates on naturally inhabited islands

Islands with natural populations of tuatara are frequently cliff-bound. Climates are temperate and largely if not completely frost-free (at corresponding mainland latitudes, mean annual temperatures range within bands of 14.1–16.0°C in the north to 12.1–14.0°C in Cook Strait).[9] Salt spray affects the coastline of large islands and the entire habitat of small islands. Permanent standing or running fresh water is rare, and often absent from small islands.

Most of the larger islands with tuatara are forest-clad today, though the vegetation usually shows signs of past disturbance. On the larger northern islands, tuatara often live in forest dominated by pōhutukawa (*Metrosideros excelsa*), māhoe (*Melicytus ramiflorus*), karo (*Pittosporum crassifolium*) and kānuka (*Kunzea ericoides*),[10] whereas on the larger Cook Strait Islands, the dominant species in forest inhabited by tuatara include kohekohe (*Dysoxylum spectabile*), māhoe, ngaio (*Myoporum laetum*), taupata (*Coprosma repens*) and akiraho (*Olearia paniculata*).[11] The smallest and least modified islands often have little or no forest canopy, the cover instead consisting of low, wind-shorn plants such as taupata.[12] Tuatara have been found among flax bushes (harakeke, *Phormium tenax*),[13] and have survived in the highly modified habitat of sheep pasture on Stephens Island (Takapourewa), although at much lower density than in forest.[14] Large tuatara have often taken up residence under dwellings on Stephens and other islands with lighthouses.[15]

Islands with natural populations of tuatara remain important sanctuaries for many endemic animals and plants that have become uncommon or extinct on the mainland. The visible fauna of islands with tuatara is dominated by seabirds, reptiles and invertebrates. Although the composition varies, rat-free islands with tuatara typically host several species of burrow-nesting procellariiform seabirds (four to eight species on large islands), and lizards (seven to ten species on some islands). The lizards (geckos and skinks) include both terrestrial and arboreal, diurnal and nocturnal species, but all are much smaller than adult tuatara.

BOX 6.2

Studying tuatara on 'Planet Seabird'

Eminent biologist Jared Diamond once commented that, from an international perspective, studying the biota of New Zealand is like studying life on another planet.[1] For a tuatara biologist living on the New Zealand mainland, a visit to an offshore island inhabited by burrow-nesting seabirds, and free of introduced mammals, provides that sense of 'other world' experience. It is here, on islands populated by colonial-nesting seabirds including petrels, prions and shearwaters, that tuatara are found in their greatest abundance.

On our field trips to study tuatara, we typically arrive by day. As we enter the forest, we see only indirect indicators of the birds' presence – a few feathers or carcasses, the odd patch of pink, regurgitated stomach contents or white guano on the ground. A more careful look reveals abundant burrow entrances in the freshly turned earth – more than one every square metre in places – as well as a surprisingly open forest floor, with a scattering of broken foliage and few seedlings. But over and above these visual signs, it is the pungent aroma that insists this is an ecosystem like no other: resembling a combination of vomit and freshly turned earth, this is the smell of fertility, seabird-style.

With the approach of dusk during the breeding season, the seabirds make their physical presence known. As the night closes in and the mist gathers, thousands of wheeling and wailing birds circle overhead. Eventually, the rain of bodies begins, the birds

crashing though the canopy in a shower of fresh foliage. Petrels, so elegant on wing and sea, are clumsy on land and easily distracted by light – naturalist and collector Andreas Reischek recorded, on a field trip in 1885, having to keep them from walking into the frying pan over his party's camp fire.[2] As we set out with headlamps to search for tuatara we, too, must take care, to avoid stepping on courting birds or crashing though the soft earth, which would destroy the subterranean apartments below. Our headlamps must also be kept in constant motion – should we forget, a face full of feathers is the likely reminder.

A night spent on an island with colonial-nesting seabirds is, above all, an aural experience. Typically, there are several species breeding on an island with tuatara, each with its own distinctive and incessantly repeated calls. Māori names give some idea of the vocalisations, for example kuaka (common diving petrel, *Pelecanoides urinatrix*), pakahā (fluttering shearwater, *Puffinus gavia*) and ōi (grey-faced petrel, *Pterodroma macroptera gouldi*). But the full variety produced by the asthmatic inhalations and exhalations, rising and falling pitches and varied paces is overwhelming, the calls often becoming more strident, even hysterical, when the birds are disturbed by their neighbours or by tuatara. Schmidt wrote in 1953 that the 'crowing and cackling, mewing and caterwauling, cooing and clucking, squeaking and crying, yelling and wailing'[3] of flesh-footed shearwaters (*Puffinus carneipes*) on Karewa Island was among 'the most astonishing nocturnal phenomena in all my naturalist's experience of nocturnal voices'.[4] In 1967 Thoresen recorded that the calls of the same species made in chorus on Green Island 'gave the impression that a hundred children were loose in a doll shop – all squeezing the "ma-ma" dolls simultaneously'.[5]

Amid the mêlée, it can be hard to stay focused on our objective. With headlamps moving and ears attuned, we continue our search for tuatara. What's that sound: a gecko plopping to the ground, a giant centipede rattling along, or a tuatara making a stop-start scuffle? With a quick lunge, our first capture is made. After observations are complete, each tuatara

Box 6.2 Fig. 1 (opposite) The forest floor on Green Island – habitat of tuatara and seabirds – by day (October 1996).

Photo: Alison Cree.

Box 6.2 Fig. 2 At night in spring, several species of seabirds, including diving petrels with their distinctive blue feet, share the forest floor with tuatara on Green Island (October 1996).

Photo: Alison Cree.

Box 6.2 Fig. 3 Claudine Tyrrell with an adult tuatara of reddish hue, on Lady Alice Island at night in December 1997.

Photo: Alison Cree.

is released, with a pen mark to prevent its recapture on future searches. Tuatara roam most widely from their burrows at night – up to 63 metres in the case of an adult male[6] – and while normally seen on the ground, they also sometimes climb on gently sloping tree trunks and branches, probably to hunt for wētā and other crevice-dwelling insects.

In the early hours of the morning, when air temperature falls to its lowest and the activity of tuatara wanes, we too head for home – hoping that our hut or tent is not positioned above the chamber of a pair of cackling shearwaters or, worse, a braying penguin. Eventually, with the approach of dawn, the terrestrial activity of the seabirds diminishes. As the birds make their way to their take-off points – gently sloping trees or cliff tops – and depart for the sea, the forest falls quiet once more.

Then it is the turn of the day birds to begin their morning chorus. As the sun rises and warms the land, tuatara re-emerge to bask. But a sun-warmed tuatara is fast, acutely sensitive to disturbance (especially sound) and never far from its burrow, so an attempt at capture is more often than not met with the flick of a disappearing tail. We must be patient, and wait for nightfall and the return of the seabirds for the greatest chance of a capture – and for our best opportunity to appreciate what much of the New Zealand mainland would have been like, before the arrival of land mammals that prey on the fauna of the night.

The forest floor and vegetation support a diverse mix of invertebrates, including some large-bodied orthopteran insects known as wētā (of the tree, giant, ground, cave and/or tusked varieties), large terrestrial beetles and, on northern islands, a centipede (weri, *Cormocephalus rubriceps*) up to a quarter of a metre long.[16] Invertebrates, small seabirds and occasional lizards constitute the main food items of tuatara.[17]

All tuatara islands show some evidence of modification by humans. Many, and probably all, were known to and visited by Māori in pre-European times. Some were perhaps visited only occasionally, mainly to harvest tītī (the chicks, also known as muttonbirds, of the burrow-nesting grey-faced petrel *Pterodroma macroptera* or the sooty shearwater *Puffinus griseus*) and kaimoana (seafood). Others, especially the larger islands, were inhabited on a permanent basis. Archaeological remains can include stone retaining walls and terraces, pits for crop storage, middens and wāhi tapu (burial sites). Some islands were still inhabited by Māori in early European times, and their significance remains high among Māori today. With the arrival of Europeans, new and sometimes more intensive styles of resource use began. A handful of strategically sited islands became locations for lighthouses, some (especially the larger islands) were farmed, and logging and mining occurred on a few.

Consequently, many islands with tuatara, especially the larger ones, have experienced substantial modification. Forest burning provided sites for human settlement and eased access for mutton-birding. Pacific rats (*Rattus exulans,* known locally as kiore) became established, probably in pre-European times in most cases, on nine of today's islands with natural populations of tuatara; as of 2012 they remain on one of these islands. Various well-known European mammals (pigs, cats, dogs, cattle, sheep, goats and/or rabbits)[18] have been present at times on some tuatara islands, but all were eliminated by the late 1900s. No tuatara population survives where Norway rats (*Rattus norvegicus*) or ship rats (*Rattus rattus*) have been introduced. Other introduced mammals that harm wildlife on the New Zealand mainland (three species of mustelids and an Australian possum)[19] have fortunately never established on islands known to have tuatara.

Even the least-modified islands with tuatara are not perfect microcosms of mainland conditions from pre-human times. Some forest types in which mainland tuatara once occurred (kauri, *Agathis australis,* and southern beech, *Nothofagus* spp.) are rare or absent on offshore islands with tuatara. Small native vertebrates that probably were once eaten by tuatara on the mainland, including frogs (*Leiopelma* spp.) and short-tailed bats (*Mystacina* spp.), are absent from almost all islands with tuatara. European birds, including common starlings (*Sturnus vulgaris*) and Eurasian blackbirds (*Turdus merula*), which are predators of lizards[20] and probably of small tuatara, have established on many offshore islands. The swamp harrier (*Circus approximans*), a predator of larger tuatara and a frequent visitor to offshore islands, may have established in New Zealand from Australian stock only in human times after forest clearance provided open habitat.[21] Introduced weeds and invertebrates (probably often transported by birds or humans) have established on some offshore islands, and in recent years cannabis (*Cannabis sativa*) has been illegally cultivated on some northern islands with tuatara.[22] Biosecurity precautions are in place to reduce the risk of introducing further pests and weeds of concern that are present on the nearby mainland.[23]

Population status

The number of tuatara present in natural populations is difficult to say with certainty. Most islands are surveyed infrequently. Uneven terrain, impenetrable scrub and vulnerable seabird burrows limit coverage during surveys on islands, especially on large ones. Weather conditions have a strong effect on the emergence of tuatara, and observers vary in their skills at seeing and catching tuatara (and, it must be said, in staying awake to search for them). Therefore, the estimates of population size given in Table 6.1, while valuable, are often 'best guesses', based on literature reports and the field observations of experienced herpetologists. Major sources of information over recent decades have included surveys during the late 1960s to the 1980s led by Ian Crook and Donald Newman of the Wildlife Service (Figure 6.2),[24] and a survey during 1989 to 1991 of geographic variation in genetics and reproduction, initiated by Charles Daugherty of Victoria University of Wellington (Figure 6.3).[25] Recent surveys have been led by the Department of Conservation, or by university staff members and students.[26]

Fig. 6.2 Ecological studies on tuatara of northern and Cook Strait islands were led in the 1970s and 1980s by scientists from the Wildlife Service. Here, Donald Newman holds a tuatara on Lady Alice Island in 1982.

Photo: Rufus Wells.

To summarise, eight natural populations of tuatara are on medium-sized or large (13–163-hectare), rat-free islands where populations are likely to number in their hundreds, thousands or in one case, tens of thousands. A further 14 populations are on small (≤ 6.3 hectares), rat-free islands or stacks where tuatara may be present in only a few tens or hundreds. Finally, nine populations are on medium-sized or large (26–3083-hectare) islands where rats have been present until recently and where tuatara are usually present as relict populations (tens or low hundreds in most cases, as a result of impaired recruitment). The status of populations on islands inhabited by Pacific rats has been a major concern of recent decades.[27] However, rats have now been removed from all but one island with tuatara.[28]

The island habitats of tuatara are managed as nature reserves, wildlife sanctuaries or wildlife refuges.[29] Almost without exception, written permission is required from the Department of Conservation before landing. A few islands remain Māori-owned,

with permission from the owners also required to visit. Some other islands are the subject of Māori land claims under the Treaty of Waitangi. In most if not all cases, formal or informal management agreements exist between the Department of Conservation and local iwi (tribes), and permission from both parties is effectively required to visit and conduct research.

Each island with tuatara is distinctive in terms of its characteristics and history. Below is a summary of the features of the natural populations, including some highlights from past research and suggestions for future study. Islands are discussed from north to south.

Fig. 6.3 PhD student Mark Brown (left) and field assistant Gerry Brackenbury (right) measure a tuatara captured on Coppermine Island in 1989. This work formed part of a geographical survey of genetics and reproduction led by researchers from Victoria University of Wellington.

Photo: Alison Cree.

The Poor Knights Islands

Named by Captain Cook, the Poor Knights Islands (35°28'S, 174°44'E) lie about 22 kilometres northeast of Tutukaka on the Northland coast (Figure 6.4). The group's volcanic origins have resulted in impressive lava flows as well as blowholes, sea arches and sea caves.[30] Subtropical and temperate waters meet nearby, favouring a rich sea life for which the surrounding marine reserve is renowned.

Tradition records that the Poor Knights were inhabited by Māori for many generations, and evidence including stone walls and terraces remains on the two main islands.[31] Pigs were present from the late 1700s on the large southern island of Aorangi, and may have been the root cause of a massacre of the

Fig. 6.4 Lying at about 35°S, the northernmost islands with tuatara are those in the Poor Knights Islands. This photo, taken from the northern island of Tawhiti Rahi in 1989, shows the author looking south toward Aorangi, the other main island with tuatara in the group.

Photo: Mary McIntyre, collection of Alison Cree.

human population in about 1822.[32] A tapu (restriction) was then placed on the islands and the surviving people departed. The pigs became feral, destroying seedlings and seabirds until removed in 1936.[33] The forest has since been regenerating[34] and shearwater populations have expanded rapidly.[35] Despite their long history of occupation, the islands have remained rat-free.

Tuatara are known from five islands in the Poor Knights. Two (Tawhiti Rahi and Aorangi; Figure 6.5) are large (163 and 110 hectares, respectively) and three (Aorangaia, Motu Kapiti and Stack B) are very small (6.3, 1.3 and 0.8 hectares, respectively). A sixth small island (Archway) appears to offer suitable habitat; although 'sign' of tuatara (probably moult skins and/or inferred faeces) has been reported, subsequent brief surveys have not confirmed the presence of tuatara there. On one or both of the two largest islands, tuatara have been described as 'common', 'plentiful' or 'abundant';[36] however, during my only visit in warm, dry weather in 1989, tuatara were not as visible as on other large, rat-free islands.

Both sexes of tuatara on the two main islands of the Poor Knights are known to be exceptionally large bodied. However, apart from reports related to egg production in females,[37] and on genetic variation,[38] there have been few studies on the group's populations. Ecological topics that would be of interest include relationships between tuatara and the large and aggressive Buller's shearwaters (*Puffinus bulleri*), which breed here in dense colonies but nowhere else in the world, and with the terrestrial centipedes that here

reach a particularly large length (up to 25 centimetres[39]). Information on the responses of tuatara on these most northerly of islands to hot and dry weather would be valuable in helping predict the future effects of global climate change.

Fig. 6.5 Tuatara on Aorangi Island in the Poor Knights live in forest amid spectacular surroundings of volcanic origin. Here, researchers rest after a night spent searching for tuatara.

Photo (1989): Alison Cree.

The Hen and Chickens Islands

Taranga Island (previously known as Hen Island) and Marotere (the Chickens) Islands lie about 50 kilometres south of the Poor Knights, at about 35°55'S, 174°44'E. Taranga, the largest island at 500 hectares, lies about 8 kilometres south of the Chickens, and about 14 kilometres southeast of the mainland at Bream Head. The Māori name Taranga refers to the mother of the demi-god Maui; the Chickens Islands are viewed as her sons, 'cast adrift on her loincloth because of their disruptive behaviour'.[40] The three largest of the Chickens Islands (Lady Alice, Whatupuke and Coppermine Islands) are also known locally as Mauimua, Mauiroto and Mauipae, respectively.[41]

The Hen and Chickens Islands, especially Taranga Island, show extensive evidence of past Māori occupation.[42] Pacific rats became established on all five islands with tuatara, perhaps as recently as the early 1800s on Lady Alice Island at least.[43] Of the islands with tuatara, Lady Alice, Whatupuke and Coppermine had rats eradicated between 1993 and 1997, and eradication has been underway on Taranga since 2011.[44] However, rats remain on the smallest island (Mauitaha).

BOX 6.3

Tuatara large and small

For decades researchers have been drawn to Stephens Island (Takapourewa) as an outdoor laboratory for studying tuatara. What we are now learning from wider sampling is that tuatara from Stephens Island are not fully representative of all islands, genetically, ecologically or morphologically.

Take body size, for instance. Tuatara vary substantially among islands in mean snout–vent length, with those on Stephens Island being in the mid-range for both males and females. Of populations on other rat-free islands, tuatara on Tawhiti Rahi Island in the Poor Knights Islands are the longest, whereas those on the two parts of Motunau Island are the shortest. In fact, the largest tuatara on Motunau only slightly exceed the size at maturity on Tawhiti Rahi. The difference in snout–vent length (and mass) is so extreme that small adult tuatara from Motunau could well be 'on the menu' if they were brought into contact with large males from Tawhiti Rahi.

To what may we attribute such variation in size among islands? One pattern often seen among endothermic vertebrates (mammals and birds) is for body size to increase with latitude, a pattern known as Bergmann's Rule. This pattern is interpreted as an evolutionary response to colder climates: larger bodies are better at conserving body heat. Ectotherms do not always show the same pattern, however, with one recent analysis for lizards and snakes showing just the opposite: the majority of species declined in body size with latitude.[1] Tuatara on islands also show weak evidence of a decline in size with latitude, although the pattern is statistically significant only for females.[2]

More convincingly, however, tuatara on islands increase in mean snout–vent length with island size. In other words, large tuatara tend to be found on large islands, and small tuatara on small islands.

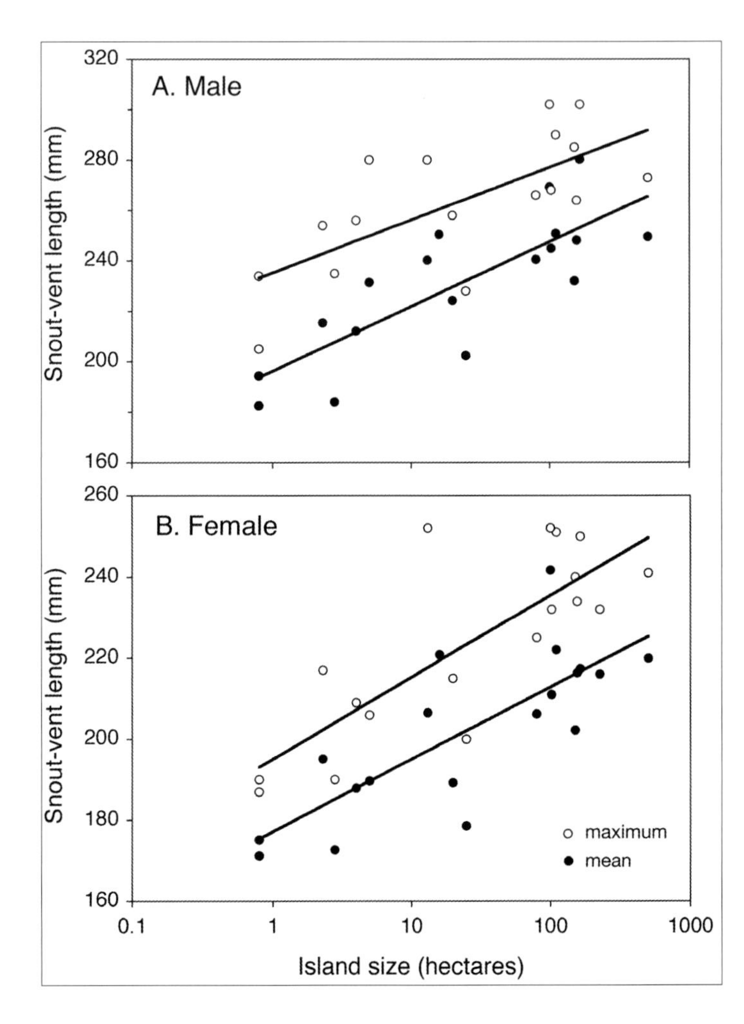

Box 6.3 Fig. 1 Mean snout–vent length and maximum snout–vent length increase significantly with island size in both in adult male (A) and adult female (B) tuatara.[10]

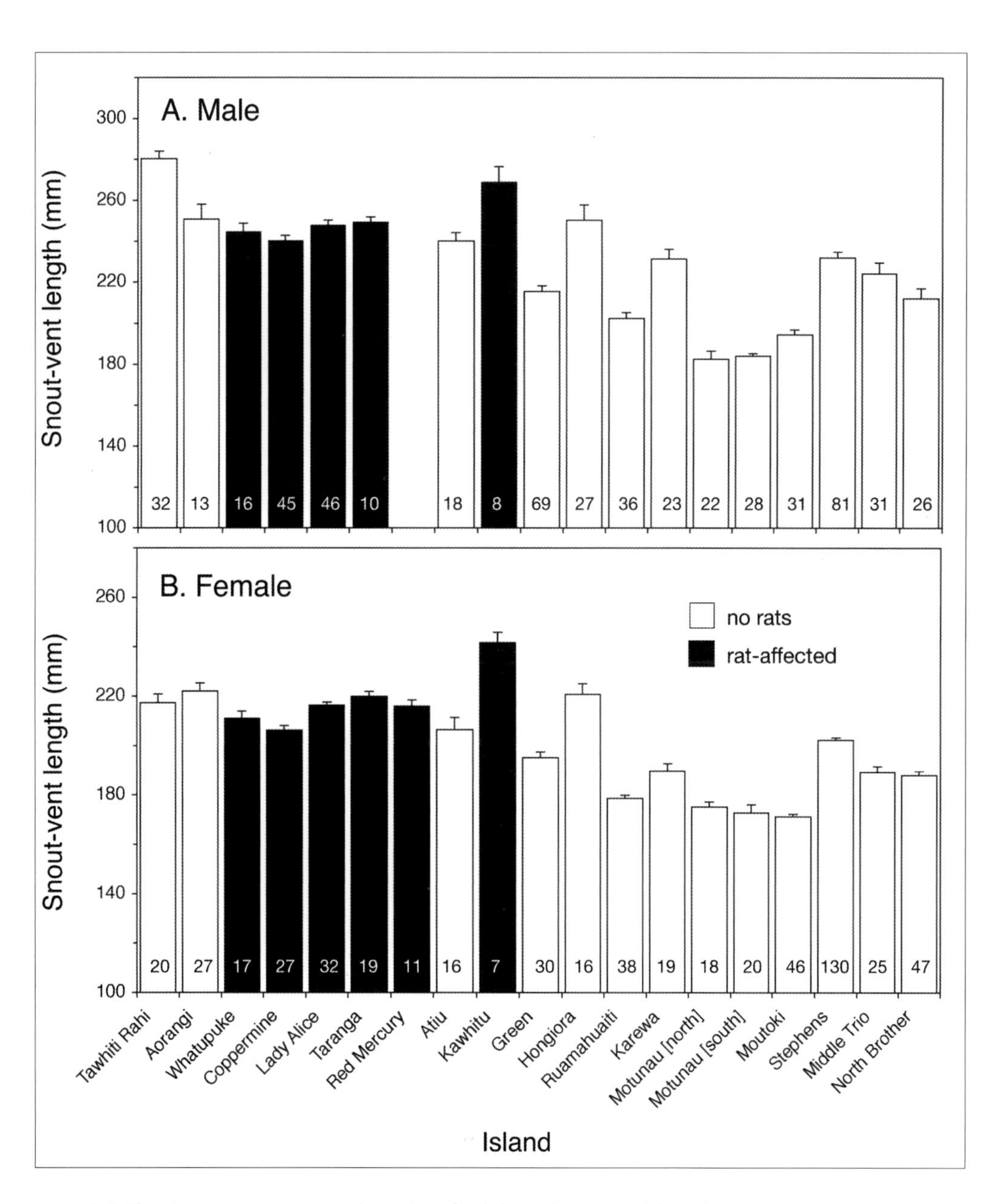

Box 6.3 Fig. 2 Mean snout–vent lengths of adult male (A) and female (B) tuatara vary significantly among islands. Islands are arranged on the x-axis from north to south. Sampled individuals were those for which sex and maturity were inferred from signs of external dimorphism. Sample sizes are given at the base of the bars, and vertical bars indicate the standard error. Only samples with seven or more individuals measured between 1986 and 2005 are included (in the case of Kawhitu and Karewa Islands, records from separate trips, excluding recaptures, were combined).

Box 6.3 Fig. 3 Tuatara on Tawhiti Rahi Island are the largest of any on rat-free islands. Adult males, which reach up to 302 millimetres in snout–vent length and 1110 grams in mass, more than fill the hand of a human male captor. The photo shows a large adult male held by field assistant John Newton in 1989.

Photo: Alison Cree.

Box 6.3 Fig. 4 Tuatara on Motunau Island, which reach up to 235 millimetres in snout–vent length and 435 grams in mass, are the smallest of those on any islands. The photo shows an adult male held by John Heaphy of the Department of Conservation in 2005.

Photo: John Heaphy.

This pattern is highly significant for both males and females, whether it is plotted as mean snout-vent length[3] or as maximum snout-vent length.[4] What might explain such a pattern? Large islands clearly offer a greater diversity of habitats, including forest, which is likely to support a greater diversity of prey species. Thus, at least part of the explanation is likely to be that food supplies are more abundant and more stable on large islands.[5]

Are the differences in body size among islands in part genetically determined, or solely the result of an individual's lifetime experience? This question is difficult to answer, but recent information from translocations suggests that the differences are not entirely genetic. When tuatara from North Brother Island (4 hectares) were shifted to a larger island (Titi Island, 32 hectares) as part of a restoration project, the density of tuatara was initially very low. Under these conditions, some of the translocated adults grew unexpectedly by up to 39 millimetres in snout-vent length, despite not having grown for the previous 7 years on North Brother Island.[6] This response could result from the translocated individuals experiencing lower competition for food and space than on their source island. (A translocation in the reverse direction – putting large-bodied tuatara on a very small island – would be ethically questionable, given that tuatara are unlikely to shrink to match a small island's more limited resources.)

As with any analysis, a comparison of sizes of tuatara among islands depends on the quality of the data available. Comparisons of maximum size are vulnerable to possible misidentifications of sex, especially for inferred 'females' in which differences of opinion between researchers on different field trips occasionally emerge. Comparisons of mean snout-vent length require reasonably large sample sizes to be of value. They also involve the assumption that past harvesting has not distorted the sizes of animals available. The latter assumption is likely to be valid for recent studies on rat-free islands, but the same cannot be said of islands with rats. The recruitment failure that occurs in the presence of Pacific rats (*Rattus exulans*) means that samples of adult tuatara eventually become skewed towards just a few, very large animals. This scenario probably explains the very large snout-vent length seen in the figure for tuatara on Kawhitu or Stanley Island. Small adults were clearly absent at the time of sampling in about 1990, by which time the population was on the verge of collapse.[7]

Other factors that could plausibly influence mean snout-vent lengths are clumping of tuatara of particular sizes in different habitats (combined with uneven search effort across habitats), or a dramatic change in habitat over time. For example, one might speculate that snout-vent length on highly modified islands such as Stephens Island will increase as forest cover increases[8] (such an increase is even more likely for body condition index, or mass relative to snout-vent length, for which there is evidence of a decline over recent decades following the loss of forest cover[9]). Another question is whether the differences between tuatara from large, forested islands and those from small islands with no forest extend beyond snout-vent length. In other words, could there be differences in the proportions of body parts that amount to different morphotypes of tuatara? Only further study will tell.

Taranga Island is a dissected volcanic cone rising to 460 metres above sea level. Much of the terrain is rugged and exposed rocky pinnacles make the island a distinctive landmark. Taranga is clad in regenerating hardwood and coastal forest, including kānuka and pōhutukawa.[45] An automatic light beacon is present, but European influence appears to be small.

Written study of tuatara on Taranga Island began in the 1880s. The collector Andreas Reischek considered the tuatara there to vary in colour from those on the Chickens, suggesting that they might be specimens of the species then recognised by Walter Buller as *S. guntheri*.[46] This level of distinction for the Taranga population has not been supported by further observations of colouration[47] or studies of genetic variation.[48] On the other hand, all reports from Reischek's onwards consistently indicate a relict status for the population. No tuatara were seen by Wildlife Service staff in 1949,[49] and no more than five during a survey by the Auckland University Field Club in about 1954.[50] Only one or two were seen on each of four, month-long visits by Wildlife Service staff between 1963 and 1968.[51] More frequent surveys since the 1980s have yielded sightings almost exclusively of adults.[52] The population (estimated to comprise about 50 adults) is close to extinction, but is expected to recover following the removal of rats.[53] When recruitment resumes, this large island with varied topology will once again become valuable habitat for tuatara. Taranga is of special importance for northern tuatara, given the vulnerability

Fig 6.6 Regenerating coastal forest inhabited by tuatara on Coppermine Island, in the Hen and Chickens Islands.

Photo (1989): Alison Cree.

of smaller islands to the rising sea levels and possible drought anticipated from global climate change.

The three largest of the Chickens Islands experienced more substantial modification than Taranga Island in European times. On Lady Alice Island (155 hectares), burning was extensive during the 1800s and partial in the early 1900s;[54] cattle were present between about 1890 and 1924, and flax was harvested during this time.[55] Short-lived attempts at mining occurred on Coppermine Island (80 hectares) in 1849 and 1896, and an automatic light beacon was established in 1913. Whatupuke Island (102 hectares) has probably suffered the least from past burning.[56] All three islands are now clad in coastal forest with hardwood, often regenerating through kānuka and pōhutukawa (Figures 6.6 and 6.7).[57]

Despite greater changes to the vegetation, tuatara are at far higher densities on the three largest of the Chickens Islands than on Taranga Island. Tuatara on Lady Alice Island were collected by Reischek in the 1880s,[58] and were present in sufficient abundance on Coppermine Island in 1898 for a party soon to collect 'half a sugar bag full, twenty-six in all'.[59] Captain Fairchild, master of the government steamers between 1871 and 1898, is reported to have dug pitfalls for tuatara, collecting the trapped animals on subsequent trips.[60] Fairchild's successor, Captain Bollons, used the same technique to collect tuatara whilst servicing the light beacon on Coppermine.[61] Tuatara

Fig. 6.7 West Bay of Lady Alice Island has shown substantial increases in abundance of juvenile tuatara since Pacific rats (*Rattus exulans*) were removed, suggesting that the warm slopes and sand dunes here are important sites for nesting.

Photo (1997): Alison Cree.

Fig. 6.8 Tuatara were discovered on Mauitaha Island, in the Hen and Chickens Islands, in 1994 (reported in Tennyson and Pierce 1995). This photograph shows the only confirmed individual, an adult male.

Photo: Alan Tennyson.

appeared to be 'common' on Coppermine in 1965, especially in kānuka on the eastern end of the island[62], and on Lady Alice, in 1953, ornithologists were 'continually finding tuataras [*sic*], the numbers of which were remarkable'.[63]

Nonetheless, juvenile tuatara were absent in the first formal survey on the three Chickens Islands in the late 1960s or early 1970s,[64] and almost completely so in a later survey in 1979.[65] Further surveys in 1989[66] and on Coppermine Island in 1996–1998[67] showed the same pattern. However, since the late 1990s, the Chickens have become model islands for demonstrating the increase in juvenile recruitment and/or body condition index of adults (mass relative to snout–vent length) that follows the elimination of Pacific rats.[68] In surveys about 6–8 years after rat removal, over 40% of the tuatara captured on Lady Alice and Whatupuke were juveniles. Adult males and females on Coppermine had relatively low body condition before the removal of rats, but a substantial improvement in body condition has since been observed and is expected to lead to improved recruitment.[69] A reminder that restoration islands remain vulnerable to incursions by rodents came in January 2009, when sign of rats was detected on Lady Alice and Whatupuke. After 4 months of effort costing in excess of $100,000, a single male ship rat was killed on each island.[70]

Mauitaha Island (26 hectares), the most recently discovered to have tuatara, cannot be said to support a viable population. No tuatara were found on this rat-inhabited island during a survey by the Wildlife Service in the early 1970s.[71] To his surprise, an ornithologist searching burrows for petrels in 1994 found a tuatara in his grasp (Figure 6.8).[72] The specimen was an adult male; a later sighting of a tuatara, in 1995, may have been the same individual.[73] Sadly, no thorough surveys have been conducted and the continued presence of Pacific rats makes extinction of tuatara on Mauitaha inevitable.

Research on the Hen and Chickens Islands has focused on the more abundant and accessible populations of tuatara on the three largest of the Chickens, especially Lady Alice Island. Apart from studies noted above that have documented recruitment patterns, themes from recent research have included studies of female reproduction in the presence of Pacific rats,[74] nutritional physiology and overlap in diets between tuatara and Pacific rats,[75] relationships with seabirds,[76] growth and ageing using skeletochronology,[77] kinship and genetic divergence using minisatellite DNA profiling[78] and techniques for estimating population density.[79] Current research is exploring the stress hormone responses of the Taranga population to the removal of Pacific rats.[80]

Hauturu/Little Barrier Island

Hauturu ('resting place of the wind'; 36°12'S, 175°06'E) lies at the entrance to the Hauraki Gulf, about 22 kilometres from the mainland and about 40 kilometres southeast of the Hen and Chickens Islands. Also known from Cook's time onwards as Little Barrier Island, Hauturu is the dissected cone of an extinct volcano, roughly circular in outline with streams radiating out from the central peaks (Figure 6.9). It is by far the largest (3083 hectares) and highest (722 metres above sea level) of all the tuatara islands. It also has a greater range of forest types, including kauri, than on other islands with tuatara (Figure 6.10).[81]

Hauturu was occupied by Māori in prehistoric and early European times. The human population was sizeable, with occupation possibly extending over several centuries.[82] Forest clearing or burning occurred over the lower, more

Fig. 6.9 At 3083 hectares, Hauturu/Little Barrier Island is the largest and most rugged of islands with tuatara. The relict population has been enhanced following captive breeding on the island and the eradication of Pacific rats. In 2006 the party shown here gathered with others to help celebrate the release of captive-bred juveniles to the wild.

Photo: Alison Cree.

Fig. 6.10 The habitat of tuatara on Hauturu includes a distinctive podocarp–broadleaf forest with emergent kauri. Lush shrub and ground layers are present in sites where burrowing seabirds are absent. The blue tape indicates where a tuatara was found during surveys in 1991 and 1992.

Photo: Tony Whitaker.

gentle slopes and Pacific rats became established. During the late 1800s parts of the island were logged and farmed (for cattle and sheep), and dug for the hardened resin (gum) of the kauri tree.[83] Pigs and dogs became established. Nonetheless, the island retained significant wildlife values and was purchased (compulsorily, from its Māori owners) by the government as a reserve in 1894.[84]

Hauturu has since been managed primarily for rare birds. Forest regeneration has occurred rapidly on the lower slopes, and thick virgin forest remains on the steeper, inland slopes.[85] Feral cats, which had been introduced by the 1870s, were eliminated by 1980 following an intensive campaign.[86] The island retains rare bats (including *Mystacina tuberculata*, the lesser short-tailed bat, and *Chalinolobus tuberculatus,* the long-tailed bat), invertebrates (including *Deinacrida heteracantha*, the largest species of giant wētā) and 12 species of lizards (including the endangered chevron skink, *Oligosoma homalonotum*).[87] Two Department of Conservation rangers reside on the island.

The opportunity to save the tuatara of Hauturu was almost lost. Reischek, the first to write about tuatara here, was unsuccessful in finding any in 1880, a result he attributed to the presence of feral pigs.[88] On later visits he found (and collected)[89] a few tuatara on the eastern part of the island, but concluded that they were 'very rare'.[90] Reports by the resident caretaker, Robert Nelson, from 1913 when he was appointed a Special Protector of the tuatara until 1932, indicate only sporadic sightings, sometimes only one in a year.[91]

Anecdotal reports from then until the 1970s indicate occasional sightings, although many may have been of one particular animal at a well-known spot at West Landing.[92] From 1977 until 1990, there were no further sightings, and it seemed the population might be functionally, if not completely, extinct.[93]

Momentum grew for thorough searches, involving trained herpetologists. In part, this was driven by Reischek's suggestion that the tuatara of Hauturu differed from others 'in colour, form, scales, and touch of the skin'.[94] A later review of tuatara by von Wettstein (of the Natural History Museum of Vienna) had suggested that the population should be considered as a subspecies,[95] and in 1943 it had been named as such (*Sphenodon punctatus reischeki*).[96]

Two surveys were eventually carried out, and their success exceeded expectations: four tuatara were found in February 1991, and a further four in October–November 1992 (Figure 6.11).[97] All were large adults in fine health, and the sexes were equally represented. Eight tuatara may sound a slim result for such a large island, but the effort was considerable. Each of the survey teams involved 10 or more people searching by day and night for 7–14 days in rugged terrain. On average, 143 person-hours of searching were required to make each night-time capture of a tuatara.[98] On the other hand, the hard-won captives revealed no evidence from their appearance, blood proteins or DNA to justify sub-specific status,[99] and the population is today considered as one of northern tuatara.

Fig. 6.11 After many hours of effort in 1991, a search party captured several adult tuatara from the relict population on Hauturu for captive breeding. Shown from left to right are Derek Brown (Department of Conservation), and Susan Keall and Charles Daugherty (Victoria University of Wellington).

Photo: Jennifer Hay, collection of Susan Keall.

Given the geographic isolation of some of the surviving adults, the absence of juveniles and the continued presence of Pacific rats, a decision was made to bring all eight tuatara into an outdoor, rat-proof enclosure for captive breeding on Hauturu.[100] Clutches have been produced by all of the females. Eggs have been incubated off-site at Victoria University of Wellington, with offspring representing both sexes[101] returned to Hauturu for captive rearing.[102] Following eventual agreement between Ngātiwai and the Department of Conservation, rats were eliminated from the island in 2004. Between 2006 and 2010, 131 captive-reared offspring were large enough to be released into the wild.[103] Genetic analyses revealed that 78% of the first 121 offspring produced had been sired by one male;[104] this male (which also became the heaviest in captivity) has since been removed from the captive colony and released to the wild. As of December 2010, seven of the original adults, seven offspring and about 31 incubating eggs remained in captivity, and two other adults had recently been seen in the wild.[105] As a means of introducing more genetic diversity to the population, the possibility of supplementation with animals from other populations has been raised.[106]

Cuvier Island (Repanga Island)

Named after the great comparative anatomist of early 19th-century France, Cuvier Island (36°26'S, 175°46'E) lies about 62 kilometres southeast of Hauturu and about 23 kilometres east of the Coromandel Peninsula. Despite being relatively large (170 hectares), and having the facilities of a staffed lighthouse station from 1889 until 1982, this remote island has one of the least-studied populations of tuatara.

Known also as Repanga Island, after a figure from Māori mythology, Cuvier Island was occupied by Māori over several centuries, perhaps intermittently, until the 1800s.[107] The forest was burnt and Pacific rats became established. The island was reportedly sold by its Māori owners in the 1880s. Most of the land passed into private hands, but about 26 hectares at the eastern end were acquired by the Crown for a lighthouse station. Cattle and sheep soon roamed widely, and both goats and cats became established as feral populations. During World War II a road was bulldozed through the forest to the highest point and a radar station (now derelict) was built.[108]

The remainder of the island was purchased by the Crown in about 1956, allowing the Wildlife Service to rehabilitate the island as a nature reserve.[109] Goats were eradicated by 1961 (about 500 were shot over 2 years[110]) and feral cats by 1964. A new fence between the lighthouse station and the reserve was built in 1963, and cattle and sheep were removed from the latter. Forest regeneration occurred rapidly.[111] The lighthouse was automated in 1982 and farm animals were removed. Apart from occasional visits by conservation workers, the island has since been uninhabited by humans. Pacific rats were eradicated in 1993.

The history of the relict population on Cuvier Island is similar to that of the population on Hauturu. The lighthouse keepers were appointed as Special Protectors of the tuatara in 1913. Their initial reports indicate sporadic sightings of tuatara (including eggs and hatchlings), coupled with concern about the numbers of rats and feral cats, the latter of which they began to shoot.[112] In 1944 a keeper reported that tuatara appeared 'less numerous than seven

years ago and were anything but numerous then.'[113] Tuatara biologist William (Bill) Dawbin searched for tuatara in 'ideal conditions' in 1957 but found only one, a large animal.[114] Occasional sightings of adults continued during the 1960s and 1970s, including a few seen living under the lighthouse keepers' houses, but juveniles were lacking.[115] Although seven tuatara were known to occur on the island in 1979, surveys in late 1979 and 1980 located only two, both adults.[116]

A recovery programme was mounted in 1989, when three adult males and a female were removed for captive breeding at mainland zoos (Figure 6.12).[117] Another male and a female were caught in 1991 and added to the captive group.[118] Adults were thus brought into close contact, and away from possible risk, while rats were poisoned on the island. By 2010 39 captive-bred juveniles had been released on to Cuvier Island;[119] four adults and seven juveniles survived in captivity, and at least seven other adults had recently been sighted on the island.[120]

Fig. 6.12 Six of the last remaining tuatara on Cuvier Island (Repanga Island) were removed between 1989 and 1991 for captive breeding at Auckland Zoo. The adults (including those held here by Phil Thomson of the Department of Conservation and Susan Keall) have produced over 30 offspring. The juveniles were repatriated to their source island once rats were eradicated.

Photo: Alison Cree.

The Mercury Islands (Iles d'Haussez)

The Mercury Islands (36°36'S, 175°51'E) lie about 20 kilometres south of Cuvier Island and about 8 kilometres east of the Coromandel Peninsula (Figure 6.13). Best known from Cook's time onwards as the Mercury Islands, the group's alternative name, Iles d'Haussez, was bestowed by the explorer

Fig. 6.13 Aerial view of the Mercury Islands, looking east from Red Mercury Island (Whakau) in the foreground. In order of proximity to Red Mercury Island are Double (Moturehu), Kawhitu or Stanley, Atiu, Green and Korapuki Islands. Great Mercury Island (Ahuahu; privately owned) is in the background right of centre, with the coast of the North Island in the distant background. Tuatara are present on Red Mercury, Kawhitu, Atiu and Green islands, and bones have been found on Korapuki Island. Tuatara probably occurred on all islands in the group before the arrival of introduced mammals.

Photo: GNS Science/Lloyd Homer (GNS 9360).

Dumont D'Urville to recognise a French minister of marine. Tuatara survive on four of the seven main islands.

The two smallest of the Mercury Islands with tuatara, Atiu or Middle Island (13 hectares) and Green Island (2 hectares; Figure 6.14), have remained free of all introduced mammals. Although there is evidence of visits by Māori,[121] Atiu and Green are among the least modified of all islands with tuatara and retain a rich fauna of seabirds, reptiles and invertebrates.[122] Atiu, in particular, is home to 10 species of lizards and has the only natural population of a large and distinctive tusked wētā (*Motuweta isolata*).

Tuatara are abundant on Atiu and Green Islands, and since formal reports began in the 1960s, juveniles have frequently been seen.[123] On Atiu Island, diets have been inferred from faecal samples,[124] and the island was the source for a transfer of 60 adult tuatara in 2003 to Tiritiri Matangi Island in the Hauraki Gulf.[125] Tuatara on Green Island have been part of recent studies comparing nutrition, reproduction and physical condition between rat-free and rat-inhabited islands.[126] Otherwise, these populations have been little studied. Owing to the near-pristine state of the islands and the fragile, heavily burrowed forest floor, access is tightly restricted.

The modern history of tuatara on the remaining two islands in the group resembles that of Cuvier Island in many respects. Kawhitu or Stanley Island, and Red Mercury Island (Whakau), are moderately large islands (100 and 225 hectares, respectively) with evidence of past Māori occupation.[127] Pacific rats were established before or during early European times. Forest burning continued into the 1900s,[128] with extensive burning on Red Mercury recorded in 1934.[129] Today, the islands are covered in regenerating forest dominated by pōhutukawa, matipo (also known as matipou or māpou, *Myrsine australis*) and/or māhoe.[130] Unlike Cuvier Island, however, neither Kawhitu nor Red Mercury was a lighthouse station and only one species of European mammal was introduced: European rabbits were present on Kawhitu by about the early 1900s.[131]

On Red Mercury Island occasional sightings of tuatara were reported from the 1950s to the 1970s.[132] On Kawhitu Island the population was so depleted that tuatara were not discovered until 1986,[133] despite occasional visits by herpetologists and ornithologists during at least the previous decade.[134] Extensive surveys during 1989 and 1990 confirmed that a few large adults were present on each island. In two surveys on Red Mercury (March 1989, October 1990), 12 individuals (1 male and 11 females) were captured.[135] In five surveys on Kawhitu between February 1989 and November 1990, 21 tuatara (10 males and 11 females) were captured.[136] All were large adults of at least 200 millimetres in snout–vent length, a strong indication of recruitment failure.

The Kawhitu Island and Red Mercury Island populations were then the subjects of rescue programmes similar to that on Cuvier Island.[137] Of the tuatara captured in 1990, seven males and eight females from Kawhitu, plus one male and nine females from Red Mercury, were removed to temporary captivity at North Island zoos.[138] Meanwhile, rats and rabbits were eliminated from Kawhitu in 1991, and rats from Red Mercury in 1992.[139] The Red

Fig. 6.14 Like most islands with tuatara, Green Island in the Mercury Islands is managed as a nature reserve, limiting access to those with permits from the Department of Conservation. Tuatara live in the forest but sometimes forage for crabs along the seashore at night.

Photo (1996): Alison Cree.

Mercury population has been the more productive in captivity: including hatchlings from eggs induced at capture (three of nine females oviposited in response to an oxytocin injection),[140] 24 juveniles have been raised.[141] These and the original adults were released on Red Mercury Island between 1996 and 2001, ending the captive propagation. Some adults have dispersed up to 1 kilometre from the release site, leading to concern that they may have become too physically separated to mate,[142] but whether homing was involved is unclear. On Kawhitu a total of 17 captive-bred juveniles were released between 2003 and 2010, and seven surviving adults, 10 juveniles and four incubating eggs remained in captivity.[143] There are several indications that the Kawhitu adults were in relatively poor condition compared with those from Red Mercury, perhaps because of advanced age. These include physical appearance (several males seemed to be very long, lean and scarred at capture), relatively low plasma testosterone concentrations in adult males prior to capture,[144] the lack of eggs from 12 females treated with oxytocin at capture,[145] poor hatching success in eggs laid in captivity,[146] plus an overall mortality of 54% for adults in captivity to 2010.[147] Plans are in development to supplement the released juveniles on Kawhitu with tuatara from a healthy neighbouring population on the Mercury Islands,[148] and to reintroduce tuatara to Korapuki Island,[149] an island within the same group on which bones of tuatara have been found.[150]

Fig. 6.15 (opposite) The French explorer Dumont D'Urville, who did not land, is said to have described the Aldermen Islands as 'a confused mass of dry and barren rocks, of which two or three might deserve the name of islets' (Sladden, B. and Falla, R. A. 1927. Alderman Islands. A general description, with notes on the flora and fauna. *New Zealand Journal of Science and Technology* 9: 193–205; see p. 196). Of the islands visible here, Ruamahuaiti (the largest visible island, in the left foreground), Hongiora (once known as Flat Island, in the distance behind Ruamahuaiti), and Half and Nga Horo Islands, in between, all have tuatara. Although Middle Island (the chain-like island of jagged profile, to the right rear) has no tuatara, a translocation (presumed reintroduction) is likely now that Pacific rats have been removed.

Photo: GNS Science/Lloyd Homer (GNS 9183).

The Aldermen Islands

Lying about 18 kilometres east of Pauanui on the Coromandel Peninsula and about 38 kilometres south of the Mercury Islands, the Aldermen Islands (36°58'S, 176°05'E) consist of six small islands and a handful of smaller stacks (Figure 6.15). Given their modest size and scarcity of fresh water, it is unlikely that these islands have ever had permanent settlement,[151] but evidence for intermittent Māori occupation exists.[152] The islands have been burnt, probably for mutton-birding;[153] however, regeneration of the forest was apparent by the middle decades of the 20th century.[154] The islands were donated to the Crown by local Māori in 1969.[155]

In 1838 a missionary, the Reverend Wade, was part of a group shipwrecked overnight on the second-largest island, Ruamahuaiti Island (25 hectares). Wade's written account brought the presence of tuatara on the Aldermen Islands to European attention.[156] Tuatara also occur on three other islands in the group (Ruamahuanui, Hongiora and Nga Horo[157] Islands, at 33, 16 and 3 hectares, respectively), as well as on three islets or stacks of 1.5 hectares or less (Half Island, and two known informally as North Stack and Middle Chain Stack). All are rat-free. The remaining larger island of the group, Middle Island (once known as Middle Chain Island), was inhabited until recently by Pacific rats; tuatara probably once existed there,[158] and plans are in development to reintroduce them.[159]

There has been little research on the tuatara of the Aldermen Islands. In the 1920s Sladden and Falla considered tuatara to be common in all the populations. On Ruamahuaiti Island, tuatara are abundant in coastal scrub (*Coprosma*/ngaio) fringing the beach margin, and also extend uphill into forest dominated by *Pittosporum* and *Melicytus*[160] with estimated densities of 220–335 per hectare.[161] Observers have repeatedly observed that tuatara on Ruamahuaiti are of a smaller size than those on Hongiora.[162] This was formally shown by Newman and McFadden in 1990,[163] who did not agree with an earlier inference[164] that the Hongiora population was declining in abundance.

Islands in the Bay of Plenty

Nestled close to shore in the Bay of Plenty are three small, mammal-free islands with notable populations of tuatara. These are Karewa Island (37°32'S, 176°07'E), Motunau Island (Plate Island) (37°40'S, 176°34'E) and Moutoki Island (37°50'S, 176°53'E). All three lie in an area with a long history of human settlement. They were probably visited by Māori in prehistoric times, and feature in Māori traditions regarding relationships between humans and tuatara, although their small size makes prolonged visits unlikely.

BOX 6.4

Were tuatara from the Bay of Plenty over-collected?

Proximity to settlements on the North Island coast made some of the Bay of Plenty islands attractive sources of tuatara for early collectors. Many collectors came from overseas, especially following Albert Günther's monograph of 1867 recognising tuatara as a rhynchocephalian reptile. New Zealanders also harvested tuatara, for study, companionship and sale.

Of the three small Bay of Plenty islands with tuatara, Karewa and Moutoki islands were the most popular and accessible early sources. Recorded harvests from Karewa Island include the following.

- In the early 1840s what became the type specimen for *Sphenodon punctatus* was collected for the New Zealand Company's naturalist Ernst Dieffenbach,[1] and one or two other specimens were studied in captivity by the missionary-printer William Colenso.[2]
- In the 1870s at least nine tuatara were received by the lawyer-naturalist Walter Buller,[3] who also referred to another two being collected for the Zoological Gardens in London.
- Three tuatara were collected in 1881 by the travelling natural-history salesman Henry Augustus Ward of New York.[4]
- In 1882 Captain Fairchild, Master of a government steamer, was reported to have collected 19 tuatara, some of which reached the Auckland Museum.[5]
- Seven tuatara were brought to Tauranga in 1883 by Captain Mair (Buller's brother-in-law).[6]
- Twenty individuals were collected in 1884 by E. C. Gold-Smith, the district surveyor for Tauranga, who released them under fir trees at his home.[7]
- During 1885 and 1886, at least 41 tuatara were collected by Professor A. P. W. Thomas for breeding studies in Auckland (possibly some of the first 29 were returned on Thomas's second visit, after he realised that they all were male).[8]
- In 1885 Professor T. J. Parker of Dunedin collected 18 or more specimens,[9] and Andreas Reischek, who accompanied Thomas and Parker,[10] collected probably dozens more for export to Europe.[11]
- Four or five tuatara collected in 1886 by C. F. Adams were later displayed at the Auckland Museum.[12]
- In 1897 the German naturalist Georg Thilenius had permission to export six tuatara from Karewa, and also collected about 60 more that were held in Tauranga for observation.[13]
- In 1905 Colin Norris, a boatman from Tauranga, offered tuatara for sale, some of which came from a group of 125 said to have been collected prior to legal protection and others that came from Thilenius's collection;[14] a contact of Norris's later suggested that Norris's tuatara were sourced from the Aldermen Islands and Rurima Rocks.[15]

Uncertainty about the total numbers removed from Karewa Island will always remain. Other islands may have been involved in some recorded collections, some collected animals may have been returned to Karewa,[16] and many harvests probably went unrecorded. Taking the uncertainties into account, it seems likely that at least 200 tuatara, and perhaps 300 or more, were removed from Karewa between about 1840 and 1900.

Further down the coast tuatara were so abundant and accessible on tiny Moutoki Island, in the group known as the Rurima Rocks, that the site was sometimes described as 'Lizard Island'.[17] In 1851 nearly 40 tuatara were collected from an unnamed island in the Rurima group by officers of the 58th Regiment.[18] Other harvesting included: at least 43 collected from the Rurima Rocks in the 1870s and 1880s by Captain Gilbert Mair,[19] and at least one collected from Moutoki by Major William Mair;[20] between about 17 and 28 collected in 1881 by Ward;[21] 23 or 24 collected by the government for translocation to Mokopuna Island in 1920; six collected for the Wellington Zoological Gardens and six for the British Empire Exhibition in 1923;[22] and four collected for study by Professor W. P. Gowland at the University of Otago in 1926.[23] Thus, between 1851 and 1926, it is likely that at least 140 tuatara were collected from Moutoki, and there was probably other unrecorded harvesting.

These minimum estimates for collections from Karewa and Moutoki islands would represent sizeable proportions of the populations on these small islands, and could be expected to have had some impact. However, the first formal surveys of populations were not made until the late 1960s or early 1970s, when researchers from the Wildlife Service measured the body sizes of tuatara. The size-class distributions (combined for the three Bay of Plenty islands, including Motunau Island) suggested a predominance of relatively small animals compared with other rat-free islands,[24] leading Ian Crook to conclude in 1975 that there had been a 'major shift in age distribution' attributable to early collection.[25]

The next major survey of the three islands was in 1989, led by researchers from Victoria University of Wellington and staff of the Department of Conservation. Although the size-class distributions they obtained might at first glance indicate an increase in the abundance of large animals, in support of Crook's suggestion, the difference is not statistically significant.[26] Furthermore, we now know from this and more recent surveys that tuatara on Karewa Island are naturally somewhat larger than on Moutoki and Motunau islands (a pattern consistent with larger island size; see Box 6.3). Thus, the small upward shift seen in the sample from 1989 can be

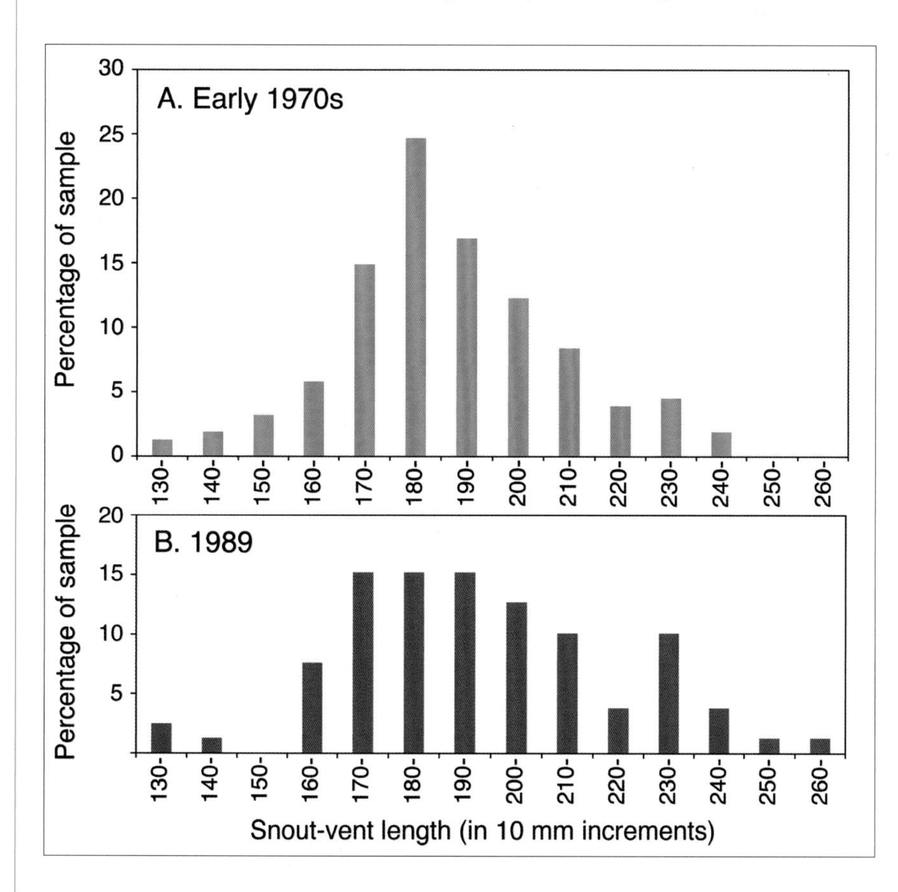

Box 6.4 Fig. 1 Percentages of tuatara in different size classes in two surveys of three Bay of Plenty islands (combined samples for Karewa, Motunau and Moutoki islands). A: Data from Wildlife Service survey in the late 1960s or early 1970s (154 tuatara; Crook 1973). B: Data from survey by Victoria University of Wellington in 1989 (79 tuatara; data courtesy of C. H. Daugherty).

explained simply by the larger proportion of animals sampled from Karewa.[27] We also know that the shape of population-size class distributions on the same island can change somewhat between collections. For example, in 1989 the modal size in the sample from Moutoki was 210–219 millimetres in snout–vent length (accounting for 23% of a sample of 30 animals),[28] whereas in 1994 the mode was lower, between 170 and 179 millimetres (accounting for 29% of a sample of 89 animals).[29] This difference is too great to be explained by variation in measurement technique. It probably arose through changes in the proportion of small animals captured, including a higher proportion of adult females in the later sample.[30] Such differences may arise through the effects of weather or season on emergence (for example, in the driest, coldest conditions, the animals most likely to be seen are large males).[31]

In summary, although the known collections do suggest that tuatara on Karewa and Moutoki islands were heavily harvested during the 1800s and early 1900s, the population size–class structures had probably already recovered by the time of Crook's survey.[32] Certainly, by the 1990s or 2000s it was apparent that the abundance of tuatara on both islands was sufficient to harvest animals once again. The knowledge that these small populations have survived earlier harvesting helped provide confidence that, at least in terms of population structure, occasional cropping for translocations would not necessarily be harmful for the source populations. Indeed, monitoring in 2004, eight years after the first translocation of 32 adults to Moutohora Island, suggested that the juvenile tuatara remaining on Moutoki Island had benefited through reduced competition and predation.[33] Nonetheless, these small populations have since been shown to have relatively low diversity in some genetic markers,[34] suggesting that caution is required before further harvesting.

Karewa Island is a rugged, 5-hectare islet lying about 6 kilometres from the coast near Tauranga (Figure 6.16). The Māori name means 'to uplift'; when seen from a distance, the island appears suspended above the sea.[165] According to tradition, a chief's daughter named Taurikura transformed herself into a tuatara in humiliation after failing to respect her elders, and became the ancestor of tuatara on Karewa.[166] The island has little real forest. Instead, the vegetation is dominated by low thickets of taupata,[167] the condition of which has deteriorated over recent years, perhaps because of drought.[168] The deep soils are heavily burrowed by flesh-footed shearwaters (toanui or tuanui, *Puffinus carneipes*), making them easily damaged during surveys.

Karewa Island's long written history as a collection site for tuatara dates from the early 1840s. The type specimen for the living species of tuatara (*Sphenodon punctatus*) was procured from here by Dieffenbach, and described by J. E. Gray of the British Museum in 1843.[169] Karewa was also the site of some of the earliest published fieldwork on tuatara, by Professor A. P. W. Thomas of Auckland University College,[170] who visited in 1885 accompanied by Professor T. Jeffery Parker of the University of Otago and the collector Andreas Reischek.[171] In the course of attempting captive breeding in Auckland with animals he collected, Thomas made valuable observations that corrected earlier misconceptions about patterns of sexual dimorphism (Chapter 8, Box 8.1).[172] Observations by Thomas and Parker on the island also corrected an earlier, idealistic view of Reischek's[173] that tuatara and seabirds lived together without interference.

Following legal protection of tuatara in 1895, and recognition of Karewa Island as a wildlife sanctuary in 1917, collection pressure diminished. Bernard Sladden, an honorary ranger during the 1920s, considered tuatara as 'plentiful,'[174] and Karl Schmidt of the Chicago Natural History Museum, who visited in 1949, considered there to be some hundreds, at least.[175] Surveys during 1989[176] and in 2005[177] also suggested a population size of several hundreds. Tuatara remained sufficiently abundant on Karewa in 2007 for a harvest by the Department of Conservation of 30 adults, transferred to Mayor Island (Tuhua) to establish a population there.[178] Local Māori have expressed interest in resuming a muttonbird harvest from Karewa, but this has not been supported by the Department of Conservation.[179]

Further to the southeast, Motunau Island[180] is more distant from the coast (12 kilometres) than either Karewa or Moutoki Islands. Despite having a single name in Māori and in English, Motunau comprises two main islands (of 0.8 and 2.8 hectares, both with tuatara) surrounded by several smaller islets (Figure 6.17).[181] The populations of tuatara on the two main islands are effectively separate, being divided by a sea gap of at least 2 metres, although their time of separation must be more recent than for other islands. Motunau has a cover of low coastal forest dominated by karo, the vegetation suggesting that there has been past burning associated with mutton-birding (Figure 6.18).[182] The Māori owners (Ngāti Whakahemo, as represented by the Pukehina Tribal Committee) continue to retain rights to and administer an occasional muttonbird harvest.[183]

Fig. 6.16 (opposite) Karewa Island was the site of many early collections of tuatara. This view, looking west from the summit, shows the low forest that covers much of the island.

Photo (2007): John Heaphy.

Fig. 6.17 Strictly speaking, Motunau Island is a composite of two islands separated by a narrow channel of water. This view shows the smaller, northern half in front of the larger, more obviously forested, southern half (the channel between the two halves is not visible).

Photo (2005): John Heaphy.

The tuatara of Motunau Island have received less attention than those on neighbouring islands, with most recent work being limited to unpublished surveys.[184] Based on a mark-recapture survey in March 1988, the combined population of the two islands was estimated as 500 animals.[185] However, erosion of the islands continues, especially on the smaller northern island, which is now thought to support only a few tens of tuatara on about 0.2 hectares of habitat.[186] These concerns have led to discussion within the Tuatara Recovery Group about management options, including a possible salvage translocation.[187] Discussion with the Māori owners is ongoing.

The tiny Moutoki Island (0.8 hectares) lies about 8 kilometres from shore near the town of Whakatane. Clad in low coastal forest incorporating pōhutukawa, taupata and coastal māhoe (*Melicytus novae-zelandiae*),[188] Moutoki is part of the Māori-owned Rurima Rocks, a wildlife refuge administered by trustees from Ngāti Awa. According to local tradition, two children from the mainland settlement of Opotiki were in the habit of taking their pet tuatara with them when fishing from their canoe. A storm arose and the boys were drowned, but their tuatara swam on to establish their kind on the Rurima Rocks.[189]

Like Karewa Island, either Moutoki Island, or the Rurima Rocks generally, have experienced many visits associated with tuatara. In 1852 a detailed account was published by 'A Party of Officers of the 58th Regiment', who stated that tuatara were present on two islands in the group (Moutoki and Rurima).[190] They collected tuatara from one of the islands (not named);

Moutoki seems the most likely in that it is the only one now known to have tuatara, but at least one later author has taken the location to be Rurima Island.[191] Pacific rats were eradicated from Rurima Island by 1984,[192] although their time of arrival is not known. Possibly, tuatara disappeared from Rurima Island only within recent human history.

Despite its small size, the Moutoki Island population (recently estimated as about 120–300 tuatara)[193] has been the source for several translocations. Concern about the impacts of human visitors on Moutoki (then without sanctuary status) led government officials to make an early translocation from the island in 1920.[194] About 23 or 24 tuatara were transferred to Mokopuna Island in Wellington Harbour.[195] However, Mokopuna was frequently visited by people, monitoring was limited, an absence of mammals went unconfirmed and the tuatara disappeared within a few years (Chapter 10, Box 10.6).[196] With superior planning and a more promising outcome, Moutoki was used as the source for a translocation of 32 tuatara to nearby Moutuhora Island (formerly known as Whale Island) in 1996, and a further 30 in 2008.[197] Recovery of the population on Moutoki after the first translocation was monitored by Jonathon Ruffell, an MSc student from the University of Auckland. Encouragingly, the results suggested that the removal of competing adults allowed recruitment and body condition to increase among the tuatara remaining on Moutoki.[198]

Fig. 6.18 The tangled forest habitat of tuatara on the larger, southern part of Motunau Island. The white plastic structure on the trunk to the left is a bait station for detecting unwanted rodents that could arrive on the island.

Photo (2005): John Heaphy.

Stephens Island (Takapourewa)

Often described as 'the jewel in the crown' of New Zealand's conservation estate, Stephens Island (40°40'S, 174°00'E) is unquestionably a global gem for its exceptional abundance of tuatara. This rugged and cliff-bound island of 150 hectares occupies a strategic position in western Cook Strait, about 3 kilometres northeast of Rangitoto ke te Tonga (D'Urville Island) (Figures 6.19, 6.20). It was named after Phillip Stephens, secretary of the British Admiralty, by Cook in 1770, for whom it provided a marker of circumnavigation of the South Island.[199] It was apparently not permanently inhabited by Māori, and indeed old Māori names for the island, Ti-tapua or Ti-tapu, suggest that visits were once discouraged by the presence of tuatara and other reptiles.[200] Stephens Island was included in land gifted as a token of peace during intertribal warfare in the 1820s, and the island retains significance today to Ngāti Koata and Ngāti Kuia as a boundary marker for the claim. Visits were made to the island to harvest muttonbirds and other food resources, to train tohunga (priestly experts) and as a place of wānanga (learning).[201]

The modern Māori name for the island, Takapourewa ('floating matipo trees') refers to the thick mantle of vegetation that once cloaked the island to the cliff-tops. The island was still forest-clad when Dieffenbach sailed past in 1839.[202] But, with the acquisition of the island by the Crown in 1891,[203] the vegetation changed rapidly. Clearance of land to provide for the lighthouse station (completed in 1894), dwellings and farm stock made the remaining

Fig. 6.19 During the 1980s the pastured northern slopes of Stephens Island (Takapourewa) provided valuable habitat for studies of tuatara, notably those on nesting. This photograph illustrates the view that greeted researchers on arrival by helicopter in November 1988, with well-known landmarks labelled. The No. 3 lighthouse keeper's house (since removed) and, more recently, the No. 2 house have provided a field base for research. Since automation of the lighthouse, a Department of Conservation presence on the island has assisted the conversion of some pasture to forest, especially between the Palace and the Ruston winch-house, and between Keeper's Bush and Ruston Bush.

Photo: Alison Cree.

forest vulnerable to gales, and by 1922 most of the forest cover had been lost.[204] Cats, feral by 1895, did considerable damage to native birds and to tuatara, but their elimination was complete by about 1922.[205]

By the 1950s, the southern part of the island was fenced to eliminate stock and left to regenerate. The lighthouse was automated in 1989 and Crown management of the island passed to the Department of Conservation. A field officer lives on the island as a permanent caretaker. A co-management agreement was established with Ngāti Koata in 1994, under which the island (apart from a small area retained for lighthouse purposes) is administered as the Takapourewa Nature Reserve (Figure 6.21).[206] The deed of settlement requires consultation with Ngāti Koata over conservation management strategies, other plans and programmes, and access permits, although the exact meaning of 'consultation' and whether and how Ngāti Kuia should be involved have been the subject of ongoing debate.[207] Today, cattle and sheep have been removed, problem weeds are being eradicated and a revegetation programme is underway. The proposed management goal is to restore the island towards its pre-human condition, with appropriate maintenance of historic and cultural features.[208]

Fig. 6.20 Getting ashore to study tuatara on cliff-bound Stephens Island is not easy. Here, Nicola Nelson from Victoria University of Wellington prepares to make the leap.

Photo: Brett Robertson/Victoria University of Wellington.

Complete ecological restoration is, of course, a heroic task. Some treasures, such as the famously extinct Lyall's wren (*Traversia lyalli*) and the South Island piopio (*Turnagra capensis*), can never be replaced, and rare forest species such as South Island saddlebacks (tīeke, *Philesturnus carunculatus*) have not so far been returned. Yet, the island today retains seven species of endemic lizards including the nationally vulnerable striped gecko (*Toropuku* (formerly *Hoplodactylus*) *stephensi*),[209] the critically endangered Hamilton's frog (*Leiopelma hamiltoni*), and several rare or threatened insects and plants including the Cook Strait giant wētā (*Deinacrida rugosa*), Cook's scurvy grass (nau, *Lepidium oleraceum*), the large-leaved milk tree (tūrepo, *Streblus banksii*) and the Cook Strait kōwhai (*Sophora molloyi*).

Above all, the island's international reputation comes from the thriving population of tuatara. This is by far and away the most abundant of all

Fig. 6.21 In 1994 a historic agreement was signed between Ngāti Koata and the Department of Conservation (DOC), signalling a new era of co-management of the Takapourewa Nature Reserve on Stephens Island. The ongoing relationship acknowledges that tuatara are a taonga (treasure) to Ngāti Koata and important to the tribe's 'identity, culture, and spiritual well-being'. Representatives of both parties gathered on the island in 1996 to recognise the event. Included are the then-Director-General of DOC Bill Mansfield (fourth from left), and the late Jim Elkington (Ngāti Koata signatory, third from right).

Photo: collection of Peter Lawless.

populations, with an estimated 30,000–50,000 individuals.[210] Tuatara occur in all major habitats on the island,[211] in the remnant forest, in the regenerating shrublands and vinelands, in the sheep pasture, under the houses and – indeed – under your feet when you walk out the field station door. Densities can exceed 2000 per hectare in the forest,[212] but even highly modified habitat such as sheep pasture still supports modest densities and has become a popular site for nesting. Although my colleagues Charles Daugherty, Michael Thompson and I once suggested that development of sheep pasture might have led to greater nesting success for tuatara and hence an increase in total numbers,[213] this is speculative. An anecdotal observation from the 1890s attests to tuatara even then being 'so very common that if one went out at night with a lantern, it was almost impossible to avoid treading on them'.[214]

The Stephens Island population is the most intensively studied of all tuatara populations, reflecting not only the abundance of animals, but also access to research facilities and to less fragile habitat (sheep pasture) than on many islands. Lighthouse operations meant that transport, accommodation, communications, food (endless mutton), rainwater, local knowledge and eventually electricity were available to researchers, and enthusiastic accounts by visitors spread this reputation internationally.[215] Significant early research on the island included investigations in the 1890s of sexual dimorphism, nesting ecology and embryonic development by the German biologists Hugo Schauinsland

and Georg Thilenius, with eggs also supplied to Englishman Arthur Dendy for study in Christchurch. Research during the 1950s to 1980s by Dawbin and his associates focused on growth, life history and thermoregulation.[216] During the 1970s and 1980s, studies on density, burrow use, growth, thermoregulation, reproduction and relationships with petrels were made by scientists from the Wildlife Service (later the Department of Conservation) and their colleagues.[217] More recently, researchers from Victoria University of Wellington and elsewhere have studied topics including diets,[218] nutrition,[219] metabolism,[220] reproductive biology (including behaviour, hormone cycles and nesting ecology),[221] stress endocrinology,[222] spatial ecology,[223] genetic variation[224] and parasitology (Figure 6.22).[225] Researchers helped monitor the outcome for 200 tuatara translocated in 2005–2007 from Stephens Island to Zealandia™ (formerly Karori Wildlife Sanctuary) in the North Island (Wellington).[226] Numerous studies on captive tuatara or preserved tissues have also involved animals from Stephens. Thus, research on tuatara on or from this island forms the bulk of our knowledge on tuatara in general.

Fig. 6.22 Since the 1940s, researchers from Victoria University of Wellington (VUW) and their collaborators have made a major contribution to studies on tuatara of Stephens Island. Shown here from left to right are Charles Daugherty (VUW), Nicola Nelson (VUW), Toniah Hsieh (Arizona State University, kneeling), Bryan Simpkins (Ngāti Koata), Anaru Paul (Ngāti Koata, at rear), Mike Thompson (formerly VUW, now University of Sydney, kneeling), Dale DeNardo (Arizona State University) and Susan Keall (VUW).

Photo (1998): Brett Robertson/Victoria University of Wellington.

The Trio Islands (Kuru Pongi)

The Trio Islands (Kuru Pongi; 50°40'S, 174°00'E) lie in the western Marlborough Sounds, about 6 kilometres east of D'Urville Island and 18

Fig. 6.23 Coastal forest on the west-facing slopes of Middle Trio Island supports a little-studied population of tuatara.

Photo (1988): Alison Cree.

kilometres south of Stephens Island. The main island, known informally as Middle Trio Island, is a forest-clad island of 20 hectares (Figure 6.23), but the two remaining islands with tuatara, known informally as North Trio and South Trio Islands, are steep rocky islets of 1–2 hectares with small patches of scrub and shrubs near the tops (Figure 6.24).

The Trio Islands appear not to have been permanently inhabited by Māori, but have a long history of periodic visits for mutton-birding, perhaps spanning several centuries.[227] These and other visits probably reflect that the group is more accessible than Stephens Island.[228] The main island has been burned to assist mutton-birding, with a large burn in 1917 or 1918 causing widespread consternation.[229] The mutton-birding harvest subsequently

Fig. 6.24 North Trio Island (area: 1 hectare), like other very small islands with tuatara, has abundant seabirds and a cover of low vegetation, but little real forest. It remains uncertain whether tuatara have survived on such islets indefinitely since sea levels rose at the end of the last glaciation, or have occasionally dispersed there, perhaps after being washed to sea from larger islands nearby.

Photo (1988): Alison Cree.

diminished, apparently subject in recent decades to a self-imposed rāhui or harvesting ban,[230] and forest cover has regenerated.[231] The islands have no history of introduced mammals. Weka (*Gallirallus australis*), believed to have been introduced to the island about 1950, were eradicated by about 1964.[232]

The Trio Islands have received little attention from tuatara researchers, most visitors subsuming information on tuatara there within general accounts that also include Stephens Island.[233] Collections of tuatara were made during the summer of 1896/97 from Stephens and the Trios by Schauinsland and local assistant George Webber,[234] with some being released on the mainland at French Pass (see Chapter 5, Box 5.2).[235] On Middle Trio Island, tuatara were considered to be 'common' in the early 1970s,[236] with density in forest near the top considered by Dawbin as similar to that on Stephens.[237] This was also my impression during a brief visit in November 1988.

North Brother Island

North Brother Island (an informal name; also known as Little Brother; 41°07'S, 174°27'E) is the southernmost of all islands with natural populations of tuatara. The larger and more inaccessible South Brother Island (Big Brother) is not known to have tuatara, despite the absence of rodents.[238] The Brothers are known to local Māori as Ngawhatu Kai-ponu ('the eyeballs that stand witness'), in reference to a tradition of a giant octopus killed nearby then placed in Cook Strait as a sign of dangerous waters.[239]

North Brother Island is essentially a conical rock of about 4 hectares, cliff-bound on almost all but the northern face (Figures 6.25, 6.26). Tuatara are most abundant in the top 2.2 hectares, where the vegetation consists primarily of plants less than 1 metre high, including taupata, koromiko (*Hebe elliptica*) and native iceplant (horokaka, *Disphyma australe*).[240] Māori activity on the island is not recorded. The presence of tuatara, and opportunities for collection, first became publicised during construction of a lighthouse, completed in 1877. Addressing a meeting of the Wellington Philosophical Society in July 1876, Walter Buller reported that 'over 20 specimens of all sizes [of tuatara] ... have been received at the Colonial Museum ... Another batch has been forwarded to England ... and other specimens preserved in cages are scattered among lovers of the curious in this city'.[241]

After examining several specimens, Buller concluded that they represented a distinct species from tuatara on islands in the Bay of Plenty. He emphasised differences in colouration, describing the Brothers tuatara as 'irregularly spotted or marked all over the body with pale yellow' (Figure 6.27). This variegated appearance was said to be especially noticeable in a half-grown animal (still preserved, although its colours have faded, at Canterbury Museum[242]). Buller named the population from North Brother Island as a new species (*Sphenodon guntheri*, honouring Albert Günther of the British Museum).[243] He later reported that captive specimens seemed less lethargic and timid, and more aggressive with each other, than those from Karewa Island.[244] A report

Fig. 6.25 Tuatara on North Brother Island survive in low vegetation on the north-facing slope, between the paths and buildings created by lighthouse operations.

Photo (1998): Alison Cree

in 1877 described aspects of internal anatomy, and commented that collection during lighthouse construction had resulted in the population becoming 'now almost, if not quite, exterminated by curiosity-hunters'.[245]

Interest in study of the population, and its status as a possible second species, then waned. Legislation to protect tuatara in 1895 and their eggs in 1898 referred only to '*Sphenodon punctatum*'.[246] In 1931 the status of the Brothers Island population was listed in von Wettstein's compilation as a sub-species, *Sphenodon punctatus guntheri*.[247] However, that distinction was later considered in New Zealand to be unjustified,[248] and official conservation documents of the 1980s listed only the tuatara *S. punctatus*.[249]

Results in 1990, from a survey of genetic variation among 24 populations of tuatara, initially provoked a reassessment of that view.[250] Blood samples analysed for 25 allozyme loci suggested that the population on North Brother Island was sufficiently different to represent a separate species, albeit a morphologically cryptic one. However, further analysis, of DNA, has not revealed the same pattern of differentiation among islands. Instead, the consensus is that the Brothers population falls within others from Cook Strait, and should be treated as the same species.[251] Genetic variation within the population is very low, implying that the population has experienced a past bottleneck and may be inbred.[252] These findings necessitate a reassessment of conservation actions for this population (see below).

Meanwhile, the apparent status of the population as a unique species of tuatara in 1990 roused much interest in its ecology. An analysis of 12 surveys between 1988 and 1997 led Niçola Nelson and colleagues to conclude that the adult population numbered about 349.[253] These surveys also yielded a significant male bias in captures, a pattern that has since become more extreme: in 2008 more than three times as many males as females were captured.[254] Although some bias in sex ratio might be explained by differences in emergence behaviour (for example, with weather[255]), the bias is large and may not fully be explained by such influences. Whether a sex bias exists at hatching is not known. Clearly, the limited habitat (including a lack of forest) makes nest temperatures on North Brother Island vulnerable to global warming,[256] which could theoretically lead to such a strong bias in hatchling sex ratio that the population would eventually collapse.[257] An apparent long-term decline in body condition (of females in particular) has been reported,[258] and the proportion of females that produce eggs in a given year seems low compared with that for the nearby population on Stephens Island.[259] Recaptures of individual tuatara, some of which were first marked by Richard Barwick in the late 1950s, confirm that females reach the minimum size of maturity in about 23 years,[260] with longevity of at least 60–70 years being possible.[261]

Given the vulnerability of the population on North Brother Island, and its then-apparent genetic distinctiveness, a captive egg-incubation programme

Fig. 6.26 Until the early 1990s, when the automation of lighthouses was complete, access to several islands with tuatara often relied on the expertise of the crew of the lighthouse tender. Here, biologists prepare to land on North Brother Island in 1988 using a transportation device known as a Billy Pugh. A similar, pulse-raising experience was described by Gerald Durrell, who visited North Brother Island to film tuatara (Durrell, G. 1966. *Two in the bush*. Fontana/Collins, Glasgow).

Photo: Alison Cree.

was initiated at Victoria University of Wellington in 1989.[262] Captive-reared juveniles, plus in some cases wild adults, have been released on three islands in the Cook Strait–Wellington region (Titi, Matiu/Somes and Long Islands),[263] and permanent captive colonies have been established in New Zealand and in the US.[264] Since the discovery of very low genetic variation in the source population, the possibility that one or more of the newly established populations on islands might benefit from 'genetic rescue' (the addition of animals from genetically distinct populations to boost measures of performance) has been raised.[265] The possibility of harvesting 'excess' males from North Brother Island for release elsewhere (particularly if predictions of high levels of harassment toward females are confirmed) has also been put forward.[266]

Fig. 6.27 A posed photo of an adult female tuatara on North Brother Island. Differences in colouration from tuatara on other islands were once used to justify recognition of this population as a separate species, but this status is no longer supported.

Photo: Alison Cree.

Endnotes to Chapter 6

See the bibliography for full references to articles on tuatara, which are cited here at first mention by surname and year only. References on other topics are cited here in full at first mention, and then using 'op. cit.'. Except where required for clarity, the convention of parentheses around years has been dispensed with for economy of space.

1 Cree and Butler 1993.

2 Hay et al. 2010.

3 For Nukutaunga Island (in the Cavalli Islands), the past presence of tuatara was reported to the Wildlife Service in 1953 (ANZ IA 1 46/18/4 pt 2). For Motukokako Island (Piercy Island), muttonbirders reported pulling what may have been a tuatara from a burrow (Cameron, E. K. and Taylor, G. A. 1991. Flora and fauna of Motukokako (Piercy Island), Cape Brett, northern New Zealand. *Tane* 33: 121–141). (Note also that in 1922, keepers at the adjacent Cape Brett Lighthouse Reserve reported seeing tuatara; B. W. Millier to Undersecretary of Internal Affairs, 8 November 1922. ANZ M1 25/611 pt 1.) For Archway Island (in the Poor Knights Islands), 'sign' of tuatara was reported (Newman 1982b).

4 For 'Green or Lizard Islet' (in the Mokohinau Islands), 'sign' of tuatara was reported to the Wildlife Service in or prior to 1952 (report dated 18/2/52, ANZ IA 1 46/18/4 pt 2), but no tuatara were seen in subsequent surveys (McCallum 1980). For Aiguilles Island (close to Great Barrier Island), the past presence of tuatara was reported to the Wildlife Service in 1953 (Extract of report by Mr L. C. Bell 12 October 1953. ANZ IA 1 46/18/4 pt 2). Concerning Ohinauiti Island (near the Mercury Islands), a report in 1963 noted that 'tuatara were said to be found on Little Ohena' (Skegg 1963).

5 Tuatara were reported on Motuoneone (near Mayor Island (Tuhua)) by Gold-Smith (1885), and on 'a rock close to Mayor Island' in 1898 (*Bay of Plenty Times and Thames Valley Warden*, 18 May 1898, p. 2). Specimen(s) were seen 'from ... small island off Mayor Island' by Cheeseman in the early 1900s (T. F. Cheeseman to A. Hamilton, 22 March 1913. NMNZ MU 000207 Box 4, Item 1). However, when Motuoneone was scaled in 1989 and again more recently, no tuatara were seen by day or night (Whitaker, A. H. 1989. *Victoria University Tuatara Research Project: Survey of the Mercury Islands and Motuoneone Island – March 1989*. Unpublished report to the Department of Conservation; John Heaphy, Department of Conservation, pers. comm. 6 August 2003).

6 For Glasgow Island, reports of the presence of tuatara were received by the Wildlife Service in 1957 (B. D. Bell to Conservator, 24 October 1957. ANZ IA 1 46/18/4 pt 2). For islands in the Cook Strait–Marlborough region with a Māori tradition of tuatara, see Ramstad et al. 2007a.

7 T. Lovegrove, pers. comm. in Cree et al. 1995a.

8 Tennyson and Pierce 1995.

9 Mean annual temperatures 1971–2000 downloaded 6 November 2010 from http://www.niwa.co.nz/education-and-training/schools/resources/climate/overview.

10 For example, Newman and McFadden 1990a; Towns et al. 2007.

11 For example, Dawson, J. W. 1954. Trio and Stephens Islands: home of the tuatara. *Wellington Botanical Society Bulletin* No. 27: 2–8; Newman 1982c.

12 Sladden 1924; Schmidt 1952.

13 Sladden and Falla 1928.

14 For pasture: Newman 1986; Carmichael et al. 1989; Markwell 1997; Moore 2008. For forest: Newman 1987a; Carmichael et al. 1989; Markwell 1997; Moore et al. 2009b.

15 Pers. obs.

16 Warne, K. 2006. Poor Knights – rich seas. *New Zealand Geographic* No. 78 (March–April): 52–75.

17 Walls 1981; see Chapter 7 for more information on diets.

18 Scientific names as follows: pig (*Sus scrofa*), cat (*Felis catus*), dog (*Canis familiaris*), cattle (*Bos taurus*), sheep (*Ovis aries*), goat (*Capra hircus*) and rabbit (*Oryctolagus cuniculus*).

19 The ferret (*Mustela furo*), stoat (*M. erminea*), weasel (*M. nivalis*) and brushtail possum (*Trichosurus vulpecula*).

20 Bell, B. D. 1996. Blackbird (*Turdus merula*) predation of the endemic copper skink (*Cyclodina aenea*). *Notornis* 43: 213–217; Thompson, M. B. 2000. *Oligosoma* spp. (New Zealand skinks): predation. *Herpetological Review* 31: 175.

21 Holdaway, R. N., Worthy, T. H. and Tennyson, A. J. D. 2001. A working list of breeding bird species of the New Zealand region at first human contact. *New Zealand Journal of Zoology* 28: 119–187.

22 Pers. obs.

23 Towns, D., Parrish, R. and Resource Management Unit, Ngatiwai Trust Board. 2003. *Restoration of the principal Marotere Islands*. Department of Conservation, Wellington.

24 For example, Crook 1970, 1973a; Newman 1982b, e; Newman and McFadden 1990a.

25 For example, Cree and Thompson 1988; Daugherty et al. 1990b; Brown et al. 1994, 1995; Cree et al. 1995a; Hay et al. 2003, 2010.

26 See references under individual islands.

27 For example, Crook 1973a; Whitaker 1978; Cree et al. 1995a.

28 Towns et al. 2001, 2007. The island where rats remain is Mauitaha Island. On nearby Taranga Island, which is treated here as rat-free, eradication of rats began in 2011 (Towns, D. R., West, C. J. and Broome, K. G. 2012. Purposes, outcomes and challenges of eradicating invasive mammals from New Zealand islands: an historical perspective. *Wildlife Research* doi.org/10.1071/WR12064). No sign of rats has been seen since on Taranga, although definitive confirmation of rat-free status will await surveys in 2013 (Keith Hawkins, Department of Conservation, pers. comm. 18 September 2012).

29 Gaze 2001.

30 Warne 2006, op. cit.

31 McCallum, J. 1983. A review of field club research on the northern offshore islands. *Tane* 29: 223–245.

32 Warne 2006, op. cit.

33 Veitch, C. R. and Bell, B. D. 1990. Eradication of introduced animals from the islands of New Zealand. In: Towns, D. R., Daugherty, C. H. and Atkinson, I. A. E. (eds). *Ecological restoration of New Zealand islands*. Conservation Sciences Publication No. 2. Department of Conservation, Wellington, pp. 137–146.

34 Oliver, W. R. B. 1925. Vegetation of Poor Knights Islands. *New Zealand Journal of Science and Technology* 7: 376–384; McCallum, J. 1981. Birds of Tawhiti Rahi Island, Poor Knights Group, Northland, New Zealand. *Tane* 27: 59–66.

35 Harper 1983.

36 Chambers, B. S. 1956. Birds of the Poor Knights Island. *Tane* 7: 66–67; Kinsky, F. C. and Sibson, R. B. 1959. Notes on the birds of the Poor Knights Islands. *Notornis* 8: 132–142; Whitaker, A. H. 1968. The lizards of the Poor Knights Islands, New Zealand. *New Zealand Journal of Science* 11: 623–651; Whitaker 1978. McCallum (1981) saw few tuatara on Tawhiti Rahi in September 1980, although temperatures were low.

37 Brown et al. 1994; Tyrrell et al. 2000.

38 Hay et al. 2003, 2010; MacAvoy et al. 2007.

39 Warne 2006, op. cit.

40 Towns et al. 2003, op. cit, p. 6.

41 Towns et al. 2003, op. cit.

42 Hayward, B. W., Moore, P. R. and Newman, M. 1978. Archaeological site survey of Hen Island (Taranga). *Tane* 24: 137–157; McCallum 1983, op. cit.

43 Brook, F. J. 1999. Changes in the landsnail fauna of Lady Alice Island, northeastern New Zealand. *Journal of the Royal Society of New Zealand* 29: 135–157.

44 Towns et al. 2012, op. cit.

45 Towns et al. 2007.

46 Reischek 1886.

47 Hard 1954.

48 Daugherty et al. 1990b; Hay et al. 2003, 2010.

49 Extract from "Report on visit to Hen Island 17-5-49/6-6-49" by L.C. Bell, ANZ IA 1 46/18/4 pt 1.

50 Hard 1954.

51 D. V. Merton, pers. comm. to Crook 1973a.

52 Cree and Butler 1993; Parrish, R. 1994. *Report on visit to Taranga (Hen) Island.* Unpublished report to the Northland Conservancy of the Department of Conservation, ref. R07 003; Cree et al. 1995a; Parrish, R. Stringer, I. and McManus, S. 2000. *Hen Island visit report 12–18 April 2000.* Unpublished report to the Northland Conservancy of the Department of Conservation; Tyrrell 2000; Tyrrell et al. 2000; Parrish, R. 2005. *Taranga Island tuatara monitoring report 31 March – 4 April 2005.* Unpublished report to the Northland Conservancy of the Department of Conservation.

53 Towns et al. 2007.

54 Skegg, P. D. G. 1965. Historical notes on the offshore islands recently visited by A. U. Field Club. *Tane* 11: 93–99; Cameron, E. K. 1984. Vascular plants of the three largest Chickens (Marotere) Islands: Lady Alice, Whatupuke, Coppermine: north-east New Zealand. *Tane* 30: 53–75.

55 McCallum, J., Bellingham, P. J., Hay, J. R. and Hitchmough, R. A. 1984. The birds of the Chickens Islands, northern New Zealand. *Tane* 30: 105–124.

56 Cameron 1984, op. cit.

57 Towns et al. 2007.

58 Reischek 1882; tuatara from 'Morotiri' Island are recorded in the collection of the Naturhistorisches Museum Wien (Museum of Natural History, Vienna), but the number is uncertain (catalogue information supplied by Franz Tiedemann, curator, and Christa Krey, Victoria University of Wellington).

59 Skegg 1965, op. cit., p. 94.

60 Cowan 1908.

61 Skegg 1965, op. cit.

62 Atkinson 1968.

63 Chambers, B. S., Chambers, S. and Sibson, R. B. 1955. Notes on the Hen and Chickens Islands. *Notornis* 6: 152–157; see p. 153.

64 Crook 1973a.

65 Newman, D. G. 1980. *An ecological reconnaissance of islands in the Hen and Chickens Group, 20 November–3 December 1979.* Unpublished report to the Wildlife Service, Department of Internal Affairs, Wellington.

66 Cree et al. 1995a.

67 Tyrrell 2000; Tyrrell et al. 2000.

68 Parrish, R. and Ussher, G. T. 2002. *Reptile conservation in the Marotere (Chicken) Islands, Northland, NZ.* Unpublished report to the Northland Conservancy of the Department of Conservation; Ombler 2004; Parrish, R. 2005. *Coppermine Island tuatara monitoring report 2005.* Unpublished report to the Northland Conservancy of the Department of Conservation; Towns et al. 2007.

69 Towns et al. 2007.

70 Anon. 2009. *Minutes of the Tuatara Recovery Group Meeting 23–24 November 2009, DOC Nelson/Marlborough Conservancy Office, Nelson.* Unpublished report, Department of Conservation. DOCDM–525029.

71 Crook 1973a.

72 208 mm SVL; Tennyson and Pierce 1995.

73 By 2003, only one or two adults were known to remain (Towns et al. 1993, op. cit.).

74 Newman and Watson 1985; Brown et al. 1994; Newman et al. 1994; Cree et al. 1995a; Tyrrell et al. 2000.

75 Newman and McFadden 1990a, b; Ussher 1995, 1999a, d; Blair 1998; Blair et al. 1999, 2000a.

76 Pierce 2002.

77 Castanet et al. 1988.

78 Finch and Lambert 1996.

79 Cassey and Ussher 1999.

80 By Lindsay Mickelson (Nicola Nelson, Victoria University of Wellington, pers. comm. 22 November 2010).

81 Whitaker, A. H. and Daugherty, C. H. 1991. *Research on the tuatara (*Sphenodon punctatus*) of Little Barrier Island, 5–12 February 1991.* Unpublished report to the Department of Conservation, Wellington.

82 Hayward, B. W. 1982. Prehistoric archaeological sites on Little Barrier Island, New Zealand. *Tane* 28: 67–78.

83 McCallum 1983, op. cit.

84 Hamilton, W. M. 1961. *Little Barrier Island (Hauturu).* 2nd edn. New Zealand Department of Scientific and Industrial Research Bulletin No. 137. Wellington.

85 Hayward 1982, op. cit.

86 Veitch and Bell 1990, op. cit.

87 Whitaker and Daugherty 1991, op. cit.

88 Reischek 1882.

89 The collection of the Natural History Museum of Vienna includes six specimens from Taranga Island collected by Reischek (catalogue information supplied by Franz Tiedemann, curator at the Natural History Museum of Vienna, and Christa Krey, Victoria University of Wellington).

90 Reischek 1886.

91 Reports from R. Nelson 1913–1932 in ANZ IA Series 1 46/14/17.

92 Whitaker and Daugherty 1991, op. cit.; Whitaker, A. H. 1993. *Research on the tuatara (*Sphenodon punctatus*) of Little Barrier Island, 6–20 October 1992.* Unpublished report to the Department of Conservation, Auckland.

93 McCallum and Harker 1982; Daugherty et al. 1990b.

94 Reischek 1886, p. 109.

95 von Wettstein 1931.

96 von Wettstein 1943.

97 Whitaker and Daugherty 1991, op. cit.; Whitaker 1993, op. cit.
98 Whitaker 1993, op. cit.
99 Whitaker and Daugherty 1991, op. cit.; Hay et al. 2003, 2010.
100 Whitaker and Daugherty 1991, op. cit.; Whitaker 1993, op. cit.; Cree et al. 1994; Towns et al. 2001.
101 Nelson et al. 2004a; Mitchell et al. 2006.
102 Keall and Daugherty 1997; Moore et al. 2008b; Keall et al. 2010.
103 Shane McInnes, Department of Conservation, pers. comm., 20 December 2010.
104 Moore et al. 2008b.
105 Moore et al. 2008b; Shane McInnes, pers. comm., 20 December 2010 and 21 February 2011.
106 Moore et al. 2008b.
107 Beever, R. E., Parris, B. S. and Beever, J. S. 1969. Studies on the vegetation of Cuvier Island. I. The plant communities and a vascular plant species list. *Tane* 15: 53–68; Merton, D. V. 1970. The rehabilitation of Cuvier Island. *Wildlife: A Review* 2: 5–8.
108 Beever et al. 1969, op. cit.
109 Merton 1970, op. cit.
110 Beever et al. 1969, op. cit.
111 Blackburn, A. 1967. A brief survey of Cuvier Island. *Notornis* 14: 3–8; Beever et al. 1969, op. cit.
112 A. Hamilton to Minister of Internal Affairs 15 May 1913, ANZ IA 1 46/18/4 pt 1; Thomson 1915; R. S. Sutherland to Secretary of Marine, 9 April 1923, ANZ M1 25/611 pt 2; G. H. Brown to Secretary of Marine, 15 February 1924, ANZ M1 25/611 pt 2; H. B. Jamieson to 'Sir', 30 January 1937, ANZ M1 25/611 pt 3.
113 E. Bowley to unrecorded recipient, 20 September 1944. ANZ IA 1 46/18/4 pt 1.
114 W. H. Dawbin to Secretary of Internal Affairs, 1 March 1957, ANZ IA 46/18/4 pt 2.
115 Merton 1970, op. cit.; Crook 1973a; Reed, S. M. 1976. Report on Cuvier Island, January 1976. *Notornis* 23: 259–262.
116 McCallum and Harker 1981.
117 Cree et al. 1994; Keall et al. 2010.
118 Chris Smuts-Kennedy, Department of Conservation, pers. comm., 25 January 2000. See also Towns et al. 2001.
119 Sources include minutes of Tuatara Recovery Group meetings; Barbara Blanchard, Wellington Zoo, pers. comm. 25 July 2006 and 28 October 2009; Leigh Marshall, Department of Conservation, pers. comm. 29 August 2006; Susan Keall, Victoria University of Wellington, pers. comm. 23 October 2009; Rob Chappell, Department of Conservation, pers. comm. 22 December 2010.
120 Rob Chappell, pers. comm. 22 December 2010.
121 Cameron, E. K. 1990. Flora and vegetation of Middle Island, Mercury Islands Group, eastern Coromandel, northern New Zealand. *Journal of the Royal Society of New Zealand* 20: 273–285.
122 Towns et al. 1990.
123 Skegg 1963; Thoresen 1967; Crook 1973a; Cree et al. 1995a.
124 Southey 1985.
125 Ruffell 2005.
126 Blair et al. 2000a; Tyrrell 2000; Tyrrell et al. 2000.
127 Edgar, A. T. 1962. A visit to the Mercury Islands. *Notornis* 10: 1–15; Moore, P. R. 1972. Maori artifacts from Red Mercury Island. *Tane* 18: 119–121; Taylor, G. A. and Lovegrove, T. G. 1997. Flora and vegetation of Stanley (Atiu) Island, Mercury Islands. *Tane* 36: 85–111.
128 Towns et al. 1990.
129 Millener, P. R. 1972. Auckland University Field Club Scientific Camp: Red Mercury (Whakau) Island, August, 1971. *Tane* 18: 5–7; McCallum 1983, op. cit.
130 McCallum 1983, op. cit.; Taylor and Lovegrove 1997, op. cit.
131 Towns et al. 1990, op. cit.
132 Bell, L. 1951. *Report on visit to offshore islands of the Coromandel Peninsula 20.2.51 to 3.3.51*. ANZ IA 1 46/18/4 pt 2; Edgar 1962, op. cit.; Crook 1973a.
133 Tim Lovegrove, pers. comm. in Cree et al. 1995a.
134 Crook 1973a; Taylor and Lovegrove 1997, op. cit.
135 Whitaker, A. H. 1989. *Survey of the Mercury Islands and Motuoneone Island – March 1989*. Unpublished report, Victoria University of Wellington, Wellington; Whitaker, A. H. 1991a. *Report on a visit to Red Mercury Island, Mercury Island Group, Coromandel, 10–18 October 1990*. Unpublished report to the Department of Conservation, Wellington; Cree et al. 1995a and unpubl. data.
136 Whitaker, A. H. 1991b. *Report on a visit to Stanley Island, Mercury Island Group, Coromandel, 20–27 November 1990*. Unpublished report to the Department of Conservation, Wellington; Keall et al. 2010.
137 Towns et al. 1990, 2001; Cree et al. 1994; Keall et al. 2010.
138 Cree and Butler 1993.
139 Cree and Butler 1993; Towns et al. 2001.
140 Whitaker 1991a, op. cit.; Cree and Butler 1993.
141 From 35 eggs hatched (Susan Keall, pers. comm. 8 March 2011).
142 Two additional males were found on Red Mercury Island in 1997, but they were at least 400 m away from the nearest female. Ussher, G. 1997. *Translocation success of tuatara (*Sphenodon punctatus*) to Red Mercury (Whakau) and Moutohora (Whale) Island.* Unpublished report to the New Zealand Department of Conservation.
143 Sources for Red Mercury and Kawhitu performance in captivity and translocations (to December 2010) include minutes of Tuatara Recovery Group meetings; Barbara Blanchard, Wellington Zoo, pers. comm. 25 July 2006, 28 October 2009, 11 January 2011 and 11 March 2011; Leigh Marshall, Department of Conservation, pers. comm. 29 August 2006; Susan Keall, Victoria University of Wellington, pers. comm. 23 October 2009 and 8 March 2011; Rob Chappell, Department of Conservation, pers. comm. 10 January 2011.
144 Cree et al. 1995a.
145 No eggs were obtained from any of four females from Kawhitu treated with oxytocin in October 1989, one female treated in October 1990, or seven females treated in November 1990 (Whitaker 1991b, op. cit.; pers. obs.). When examined laparoscopically at about 3 years after capture, six of seven females from Kawhitu had ovaries dominated by regressed or atretic follicles; in only one female were the majority of follicles in apparently healthy, vitellogenic condition (pers. obs.). Some eggs were eventually obtained from most of these females, but zoo records reveal a relatively poor response to oxytocin treatment, with one female believed to have carried a partial clutch of shelled eggs over 1–2 years (Tyrrell 2000).
146 Keall and Daugherty 1997; Keall et al. 2010; see also Tyrrell 2000, p. 48.
147 Only seven of 15 adults remained alive in 2010 (Barbara Blanchard, pers. comm. 20 January 2011).
148 Anon. 2009, op. cit.
149 Rob Chappell, pers. comm. 22 December 2010.
150 Hicks, G. R. F., McColl, H. P., Meads, M. J., Hardy, G.

S. and Roser, R. J. 1975. An ecological reconnaissance of Korapuki Island, Mercury Islands. *Notornis* 22: 195–220; Mike Meads, pers. comm. in Cree et al. 1995a; David Towns, Department of Conservation, pers. comm. 3 September 2003.

151 Sladden, B. and Falla, R. A. 1927. Alderman Islands. A general description, with notes on the flora and fauna. *New Zealand Journal of Science and Technology* 9: 193–205.

152 Moore, P. R. 1973. Evidence of former Maori occupation of the Aldermen Islands. *Tane* 19: 21–29.

153 Sladden and Falla 1927, op. cit.

154 Court, D. J., Hardacre, A. K. and Lynch, P. A. 1973. The vegetation of the Aldermen Islands: a reappraisal. *Tane* 19: 41–60.

155 McCallum 1983, op. cit.

156 Wade 1842.

157 Nga Horo Island was until recently also known as Hernia Island.

158 Crook 1973a.

159 Rob Chappell, pers. comm. 22 December 2010.

160 Towns and Hayward 1973.

161 G. Ussher, pers. comm. to C. Tyrrell in Tyrrell et al. 2000 (see p. 34).

162 Sladden and Falla 1928, op. cit.; W. H. Dawbin to the Controller, Wildlife Section, Internal Affairs, 15 March 1965. ANZ IA 1 46/14/26.

163 Newman and McFadden 1990a.

164 Crook 1973a.

165 Sladden 1924.

166 The legend of Pyes Pa: http://www.pyespa.school.nz/history.htm (accessed 18 July 2012).

167 Sladden 1924; Schmidt 1952.

168 Heaphy, J. 2005. *Karewa tuatara survey – March 2005*. Unpublished report to the Tuatara Recovery Group, Department of Conservation TAUAO–2374.

169 Dieffenbach 1843, Vol. 2, p. 405; Gray 1842, 1843.

170 Thomas 1890, 1891. See also Chapter 10.

171 Reischek 1886.

172 Thomas 1890, 1891.

173 Reischek 1882; see also Anon. 1882a.

174 Sladden 1924.

175 Schmidt 1952. Additionally, William Dawbin, in correspondence with the Wildlife Service in 1965, referred to a letter received (ref 46/29/33) containing the results of a tuatara census on Karewa, presumably by Wildlife Service staff, but details are not known. W. H. Dawbin to Wildlife Service 1 July 1965. ANZ IA 46/14/26.

176 Unpublished survey results of A. H. Whitaker/VUW.

177 Heaphy 2005, op. cit.

178 J. Heaphy, Department of Conservation, pers. comm. 4 December 2008.

179 Heaphy 2005, op. cit.

180 Not to be confused with Motunau Island off the coast of North Canterbury, on which tuatara bones have been found (see Chapter 3).

181 Taylor 1991.

182 Taylor 1991.

183 Jansen, P. 1988. *Plate Island survey 17–20 March 1988*. Unpublished report to the Department of Conservation, Bay of Plenty; J. Heaphy, Department of Conservation, pers. comm. 14 March 2011.

184 Jansen 1988, op. cit.; unpublished survey results of A. H. Whitaker/VUW.

185 Jansen 1988, op. cit.; Taylor 1991.

186 Anon. 2009, op. cit.

187 For example, to Mayor Island (to supplement those already translocated there from Karewa Island), provided that monitoring of those already translocated to Mayor is positive (Anon. 2009, op. cit.).

188 Whitaker, A. H. 1989. *Victoria University tuatara research project: report on a visit to Moutoki Island, Rurima Rocks Group, Bay of Plenty, 19–21 January 1989*. Unpublished report to the Ngati Awa Trust Board and Tuwharetoa Maori Trust.

189 Kingsley-Smith 1966.

190 A Party of Officers of the 58th Regiment 1852. Dr Arthur Saunders Thomson, military surgeon to the regiment, is known to have published on medical and scientific matters while in New Zealand and probably helped prepare the publication. Buller (1877) referred to Dr Thomson in connection with this publication, although Buller incorrectly named the island as Karewa.

191 Parham 1982. The description, including ease of landing by canoes, seems suggestive of Rurima Island (J. Heaphy, pers. comm. 28 September 2005).

192 Atkinson, I. A. E. and Towns, D. R. 2005. Kiore. In: King, C. M. (ed.). *The handbook of New Zealand mammals*. Oxford University Press, Auckland, pp. 159–174.

193 Whitaker 1989, op. cit.; Garrick, A. 1996. *Preliminary report on the status of tuatara on Moutoki, Rurima Rocks, Bay of Plenty*. Unpublished report to the Department of Conservation, Rotorua (file RWL: 030); Ruffell, J. [undated] *Preliminary report on the status of the Moutoki Island tuatara population following harvesting*. Unpublished report to the Department of Conservation; Ussher 1997, op. cit. See also Ruffell 2005.

194 W. Hill to General Manager of the Government Tourist Bureau, Department of Tourist and Health Resorts, 13 February 1920. ANZ IA 46/18/61.

195 Mokopuna Island was also known as Leper Island during the 1900s; see Thomson 1920.

196 W. Cleverly, Principal Keeper on Somes, to Secretary of Marine, 22 January 1924. ANZ M 1 25/611 pt 2. See also Box 10.2.

197 Owen 1998; Ussher 1997, op. cit.; Ussher, G. 2002. *Reintroduction of tuatara (*Sphenodon p. punctatus*) to Moutohora (Whale Island): population status after 5 years*. Unpublished report to the Bay of Plenty Department of Conservation, Rotorua, New Zealand. 47 p.; Heaphy, J. 2009. *2009 ECBOP Conservancy report to Tuatara Recovery Group*. Unpublished report, Department of Conservation DOCDM-508596.

198 Ruffell 2005.

199 Baldwin, O. 1979. *Story of New Zealand's French Pass and d'Urville Island. Book 1: History*. Fields Publishing House, Plimmerton.

200 Baldwin 1979, op. cit.; Beattie 1994.

201 New Zealand Waitangi Tribunal. 2008. *Te Tau Ihu o te Waka a Māui: report on northern South Island claims. Vol. 1*. Waitangi Tribunal Report. Legislation Direct, Wellington; New Zealand Waitangi Tribunal. 2011. *Ko Aotearoa tēnei: a report into claims concerning New Zealand law and policy affecting Māori culture and identity*. Wai 262 Waitangi Tribunal Report. Legislation Direct, Wellington (see pp. 302–303).

202 Dieffenbach 1843, Vol. 1, p. 22. See also Baldwin 1979, op. cit.

203 The earliest specimen of a tuatara in the Museum of New Zealand Te Papa Tongarewa was collected from Stephens Island in 1886, even before lighthouse construction began (RE.000881; collected by W. H. Strand of Lower Hutt); Colin Miskelly, Museum of New Zealand, pers. comm. 15 February 2011 (see also Miskelly 2011).

204 For a detailed history of the loss of vegetation, arrival of cats and demise of native birds, see Medway, D. G. 2004. The land bird fauna of Stephens Island, New Zealand in the early 1890s, and the cause of its demise. *Notornis* 51: 201–211.
205 Schmidt 1952; Galbreath, R. and Brown, D. 2004. The tale of the lighthouse-keeper's cat: discovery and extinction of the Stephens Island wren (*Traversia lyalli*). *Notornis* 51: 193–200.
206 *Deed of settlement between Her Majesty the Queen and Ngati Koata no Rangitoto ki te Tonga Trust and James Hemi Elkington, dated 29th day of November 1994.*
207 New Zealand Waitangi Tribunal 2008, op. cit.
208 Millar, I. and Gaze, P. 1997. *Island management – a strategy for island management in Nelson/Marlborough Conservancy.* Occasional Publication No. 31, Department of Conservation, Nelson.
209 Nielsen, S.V., Bauer, A. M., Jackman, T. R., Hitchmough, R. A. and Daugherty, C. H. 2011. New Zealand geckos (Diplodactylidae): cryptic diversity in a post-Gondwanan lineage with trans-Tasman affinities. *Molecular Phylogenetics and Evolution* 59: 1–22.
210 Newman 1982b.
211 East et al. 1995.
212 Moore et al. 2009b.
213 Cree and Daugherty 1990a; Thompson and Daugherty 1992.
214 The Collector 1913, p. 4.
215 For example, Blanchard 1935; Donne 1942; Schmidt 1952; Sharell 1966; Goellner 1985.
216 Bogert 1953a; Dawbin 1949, 1962a, 1982a, b; Barwick 1982; Heatwole 1982.
217 Crook 1974, 1975; Newman 1978, 1982a, e, 1987a; Saint Girons et al. 1980; Newman and Watson 1985; Saint Girons and Newman 1987; Castanet et al. 1988; Newman et al. 1994.
218 Fraser 1993 (see also Walls 1978, 1981, 1983); Cree et al. 1999; Refsnider et al. 2008.
219 Cartland-Shaw et al. 1998.
220 Thompson and Daugherty 1998.
221 For example, Bradshaw et al. 1988; Guillette et al. 1990; Thompson et al. 1991, 1996; Brown et al. 1991b; Cree et al. 1991a, 1992; Gillingham et al. 1995; Nelson et al. 2004c, d; Refsnider et al. 2009. For semi-popular accounts by Victoria University of Wellington researchers in the 1980s, see Cree and Thompson 1988; Cree and Daugherty 1990a; and Thompson and Daugherty 1992.
222 Tyrrell and Cree 1998; Cree and Tyrrell 2001.
223 East et al. 1995.
224 Miller et al. 2007a; H. C. Miller et al. 2009; Moore et al. 2008a, 2009a, b.
225 Godfey et al. 2008, 2010a, b.
226 McKenzie 2007.
227 Campbell, D. J. 1967. *The Trio Islands, Marlborough Sounds: an ecological study of a bird modified island.* Unpublished MSc thesis, Victoria University of Wellington, Wellington.
228 Baldwin, O. 1983. *Story of New Zealand's French Pass and d'Urville Island. Book 3.* Fields Publishing House, Plimmerton.
229 Campbell 1967, op. cit.; Baldwin 1983, op. cit.
230 Millar and Gaze 1997, op. cit.
231 Campbell 1967, op. cit.
232 Campbell 1967, op. cit.; Veitch, C. R. and Bell, B. D. 1990. Eradication of introduced animals from the islands of New Zealand. In: Towns, D. R., Daugherty, C. H. and Atkinson, I. A. E. (eds). *Ecological restoration of New Zealand islands.* Conservation Sciences Publication No. 2. Department of Conservation, Wellington, pp. 137–146.
233 Dawson 1954, op. cit.
234 Schauinsland 1898b; Webber 1953.
235 Webber 1953.
236 Crook 1973a (see also Crook 1970).
237 See Discussion p. 294 in Newman 1982a.
238 Newman 1878; Crook 1970; P. Gaze, Department of Conservation, pers. comm. 21 May 2010.
239 http://www.theprow.org.nz/The-Brothers-Islands-and-Lighthouse/ (accessed 27 November 2010).
240 Cree et al. 1991b; Thompson et al. 1992; see also Gaston and Scofield 1995.
241 Buller 1877, p. 322.
242 Freeman and Freeman 1995.
243 Günther, however, was unconvinced that the population deserved recognition as a separate species; see Chapter 5.
244 Buller 1878.
245 Newman 1878, p. 224.
246 Published in the *New Zealand Gazette* 1895, 1898.
247 von Wettstein 1931.
248 Dawbin 1962a; Robb 1977.
249 Daugherty et al. 1990b for brief review.
250 Daugherty et al. 1990b.
251 Hay et al. 2010.
252 Aitken et al. 2001; MacAvoy et al. 2007; Miller et al. 2008a; Hay et al. 2010.
253 With a 95% profile likelihood interval of 294–427; Nelson et al. 2002a.
254 Wilson 2010.
255 Dawbin 1962a.
256 Mitchell et al. 2008.
257 Mitchell et al. 2010.
258 Hoare et al. 2006; Wilson 2010.
259 Cree et al. 1991b; Mitchell et al. 2010. Mean clutch size is also smaller on North Brother Island, but this is to be expected given the smaller body size of adults than on Stephens Island (see Chapter 8 for the effects of snout–vent length on clutch size).
260 Mitchell et al. 2010.
261 Thompson et al. 1992; Nelson et al. 2002a.
262 Cree et al. 1991b, f; Cree and Butler 1993.
263 Nelson 1998; Merrifield 2001a, b; Nelson et al. 2002b; K. A. Miller et al. 2010.
264 Cree et al. 1994.
265 K. A. Miller et al. 2009; K. A. Miller et al. 2010.
266 Wilson 2010.

Endnotes to boxed text

For explanation of citation format in box endnotes, see endnotes to the main text.

Box 6.1

1 Daugherty et al. 1990b.
2 Brown et al. 1995.
3 Daugherty et al. 1990b.
4 Hay et al. 2003, 2010; MacAvoy et al. 2007. See also H. Miller et al. 2010.
5 Hay et al. 2010.
6 Aitken et al. 2001; MacAvoy et al. 2007; Hay et al. 2010.
7 Finch and Lambert 1996; Moore et al. 2008b.
8 MacAvoy et al. 2007; Hay et al. 2010.
9 Nelson et al. 2002a; Wilson 2010.
10 MacAvoy et al. 2007.
11 Hay et al. 2003.
12 Hay et al. 2010.
13 H. C. Miller et al. 2007a, 2008a, 2010.
14 K. A. Miller et al. 2009, 2010.

Box 6.2

1 Diamond, J. M. 1990. New Zealand as an archipelago: an international perspective. In: Towns, D. R., Daugherty, C. H. and Atkinson, I. A. E. (eds). *Ecological restoration of New Zealand islands*. Conservation Sciences Publication No. 2. Department of Conservation, Wellington, pp. 3–8.
2 Reischek 1971, p. 100.
3 Schmidt 1953, pp. 157–158.
4 Schmidt 1953, p. 157.
5 Thoresen 1967, p. 185.
6 Newman 1982c.

Box 6.3

1 Ashton, K. G. and Feldman, C. R. 2003. Bergmann's Rule in non-avian reptiles: turtles follow it, lizards and snakes reverse it. *Evolution* 57: 1151–1163.
2 $F_{1,16} = 3.868$, $r^2 = 0.195$, $p = 0.067$ for males; $F_{1,17} = 4.999$, $r^2 = 0.227$, $p = 0.039$ for females.
3 $F_{1,16} = 28.366$, $r^2 = 0.639$, $p < 0.001$ for males; $F_{1,17} = 26.976$, $r^2 = 0.613$, $p < 0.001$ for females.
4 $F_{1,15} = 14.254$, $r^2 = 0.487$, $p = 0.002$ for males; $F_{1,16} = 28.326$, $r^2 = 0.639$, $p < 0.001$ for females.
5 Interestingly, the same pattern seems to hold for Duvaucel's geckos (*Hoplodactylus duvaucelii*) from some of the same islands as for tuatara (Cree 1994), although more detailed sampling is required.
6 Nelson et al. 2002b.
7 Cree et al. 1995a.
8 Schauinsland, who visited in the summer of 1896/97, reported that males on Stephens Island 'reach a length of ¾ meter' (as translated by Schmidt 1952), but no males exceeding 611 mm in total length have been seen in recent decades according to the database of Christmas et al. 1996 (see also Moore et al. 2007).
9 Moore et al. 2007.
10 Sources for the box figures include VUW survey data from 1989–1991 (Tawhiti Rahi; Karewa; Moutoki; Middle Trio; unpublished data courtesy of C. H. Daugherty); Tyrrell et al. (2000) for Aorangi, Coppermine, Lady Alice, Green and Ruamahuaiti; Cree et al. (1995a) for Taranga, Red Mercury (Whakau), Atiu and Kawhitu; for Taranga: Towns et al. (2007); for Hongiora: Newman and McFadden (1990a); for Karewa: Heaphy, J. 2005. *Karewa tuatara survey* – March 2005. Unpublished report to the Tuatara Recovery Group, Department of Conservation TAUAO–2374; for Motunau: Jansen, P. 1988. *Plate Island survey 17–20 March 1988*. Unpublished report to the Department of Conservation, Bay of Plenty; for Moutoki: Garrick, A. 1996. *Preliminary report on the status of the tuatara on Moutoki, Rurima Rocks, Bay of Plenty*. Unpublished report, Department of Conservation, Rotorua (file RWL: 030); for Stephens Island: Cree (pers. obs.) and Cartland-Shaw et al. (1998); for North Brother Island: Cree et al. (1991b).

Box 6.4

1 Dieffenbach 1843, Vol. 2, p. 405; Gray 1843.
2 Colenso 1844.
3 Buller 1871, 1877.
4 Ward 1882; Jordan, T. 1989. Visit of an American naturalist to the Bay of Plenty, 1881. *Historical Review (Whakatane and District Historical Society)* 37: 23–32.
5 Anon. 1882b; Auckland Museum "Blue Book" (Brian Gill, Auckland Museum, pers. comm. 9 August 2000).
6 *Bay of Plenty Times and Thames Valley Warden*, 15 January 1883, p 2.
7 Gold-Smith 1885.
8 Thomas 1890, 1891.
9 Four months after the field trip, Parker reported having about 18 still alive. T. J. Parker to J. Hector, NMNZ Series No. MU000147 Location R-1M03-130E Box 6 Item 188 (other number 85/2010).
10 Reischek 1886.
11 For example, the collection of the Naturhistorisches Museum Wien (Museum of Natural History, Vienna) (NMW) includes seven tuatara collected from Karewa Island by Reischek, two collected from Karewa by persons unknown, 10 from 'Karewa and Morotiti' islands collected by Reischek, plus a further 48 from New Zealand (islands unrecorded), of which 14 were recorded as collected by Reischek (catalogue information supplied by Franz Tiedemann, curator NMW, and Christa Krey, Victoria University of Wellington). For an indication that the Thomas/Parker/Reischek party collected 90 or more from Karewa Island (which would leave perhaps 43 for Reischek, after subtracting those reported by Thomas and Parker), see O'Sullivan, M. J. 1953. *Algernon Phillips Withiel Thomas, 1856–1937: educationalist, scientist, horticulturist*. Auckland. See also Chapter 10 for indications that Parker was concerned in 1885 about the scale of collecting for export, probably by Reischek in particular.
12 Gill, B. J. 1999. History of the land vertebrates collection at Auckland Museum, New Zealand, 1852–1996. *Records of the Auckland Museum* 36: 59–93.
13 Anon. [undated]. [Summary of events regarding collections made during 1896–1898 by Thilenius, Schauinsland and Dendy.] NMNZ MU000001 Box 23, Item 14, File No. 14/6/9. See also articles during 1898 in the *Bay of Plenty Times and Thames Valley Warden*: 18 May, p. 2; 29 May, p. 2; 16 September, p. 2.
14 ANZ IA 1 46/18/61. Goodlet (c. 1922) referred to 150.
15 As 'Ruarima' (Goodlet c. 1922).

16 Thilenius undertook to do this when questioned by police about his collection (see Anon. Undated, op. cit.), but Norris's later comment makes it unlikely that all specimens were returned.
17 Parham 1982.
18 A Party of Officers of the 58th Regiment 1852. Sadly, only three of the tuatara reached Auckland alive; the others died, escaped or were cannibalised.
19 In 1875 Gilbert Mair sent 13 from the Rurima Rocks to Thomas Cheeseman at the Auckland Museum (G. Mair to T. Cheeseman, 12 December 1875, Auckland Museum Archives MA 95/38/7 AV2.1 Correspondence 1868–99–Ma–Mv). In 1876 Buller reported at least five collected by Gilbert Mair from the Rurima Rocks (Buller 1877). In 1886 25 were collected by Gilbert Mair from Moutoki for Buller (see entries for 19–22 January in Mair, G. 1886. *Diary and Notes Nov 1885–Mar 1886* [Diary 36]. Alexander Turnbull Library, Wellington.)
20 Anon. 1872; Mair 1873; Buller 1877.
21 Ward 1882; Jordan 1989, op. cit.
22 W. Hill to General Manager of the Department of Tourist and Health Resorts, 4 April 1923. ANZ IA 1 46/18/61.
23 W. Hill to General Manager of the Department of Tourist and Health Resorts, 17 July 1926. ANZ IA 1 46/18/61.
24 Crook 1973a.
25 Crook 1975, p. 342.
26 Kolmogorov-Smirnov $Z = 0.858$; $p = 0.454$.
27 In Crook's sample, 18% of tuatara from the three islands were from Karewa, as compared with 39% in the 1989 survey.
28 Whitaker, A. H. 1989. *Report on a visit to Moutoki island, Rurima Rocks group, Bay of Plenty, 19–21 January 1989*. Unpublished report prepared for the Ngati Awa Trust Board and Tuwharetoa Maori Trust.
29 Garrick, A. 1996. *Preliminary report on the status of tuatara on Moutoki, Rurima Rocks, Bay of Plenty*. Unpublished report, Department of Conservation, Rotorua (file RWL: 030).
30 The ratio of adult males to females (≥ 160 mm snout-vent length) on Moutoki in the survey of January 1989 (21:9) differed significantly from that in December 1994 (31:47; chi-square = 7.945, df = 1, $p = 0.005$).
31 pers. obs. See also Moore et al. 2010.
32 This interpretation also seems consistent with growth rates on other islands (Dawbin 1982a; Castanet et al. 1988), which suggest that 70 years would be more than sufficient for large and relatively old animals to reappear in the populations.
33 Ruffell 2005.
34 MacAvoy et al. 2007; Hay et al. 2010; H. C. Miller et al. 2010. However, the markers indicating low genetic diversity are not always the same for the two islands.

CHAPTER SEVEN

Ecology, feeding and behaviour

Tuatara are notorious for doing nothing. But, as field biologist Bill Dawbin noted of wild tuatara in 1949, 'Anyone who catches a glimpse of the rapid movement will immediately lose any preconceived ideas about their sluggishness'.[1]

Introduction

Early studies of tuatara abound with references to 'sluggish' behaviour.[2] Given that most early work involved captive adults held under less than optimal conditions, such views are not surprising. Overfed tuatara, held alone in small, dry cages without basking opportunity, are almost guaranteed to seem inactive, especially if observed only by day.

The experience of those of us who, like Bill Dawbin, have studied wild tuatara in their natural habitats is very different. There is no denying that these animals are economical in their movements. But if tuatara are disturbed while basking, they dive quickly for their burrows, and on warm, misty nights, when tuatara roam and hunt for food, most run at least briefly when chased. Attempts at capture are not always successful, and careless handling is likely to be rewarded with vigorous thrashing, a raking clawing and a swiftly delivered, excruciatingly vice-like bite.[3]

This chapter covers the topics of burrowing behaviour, daily and seasonal patterns of activity, density and biomass, sound production and hearing, and locomotion and speed. A large section exploring diets and feeding behaviour is followed by discussions of predators, parasites and abnormalities. Topics explored in boxes include relationships with seabirds, anatomy and function of the digestive system, and effects of introduced rats. Information is presented first for tuatara from Stephens Island (Takapourewa), including for those in captivity where relevant, then for other Cook Strait populations, and finally for northern populations. Ideas for new research are presented at the chapter's end. Reproductive and thermoregulatory behaviour are mentioned here only in passing, being covered in more detail in Chapters 8 and 9, respectively.

BOX 7.1

Tuatara and seabirds: mutual benefit or reptilian advantage?

Ever since 1867, when downy feathers in the intestines of a preserved tuatara were reported by Albert Günther, the relationships between tuatara and seabirds have aroused curiosity. One of the first natural historians of tuatara, Andreas Reischek, ignored Günther's observation and emphasised a mutually beneficial relationship. Seabirds, wrote Reischek, lived sociably with tuatara and were 'apparently on the best of terms' with them.[1] In fact, after being bitten by a tuatara from a burrow shared by a seabird, Reischek concluded that tuatara were defending the seabirds' nests.[2]

This notion of harmony was initially repeated with enthusiasm both locally and internationally,[3] but some observers were more sceptical. Professor A. P. W. Thomas, who visited Karewa Island with Reischek in 1885, noted tuatara with seabird chicks in their mouths on four occasions. He concluded that 'it is quite a mistake to suppose that any friendly relation whatever exists between the tuataras [*sic*] and the birds ... the idea of friendliness is quite out of place'.[4]

Today, far from being considered nest guardians, tuatara are known or suspected to prey on the eggs and/or chicks of at least eight species of seabirds. From smallest to largest,[5] these are the tiny, white-faced storm petrel (takahikare, *Pelagodroma marina*, 40–70 grams),[6] common diving petrel (kuaka, *Pelecanoides urinatrix*, 110–150 grams),[7] fairy prion (tītī wainui, *Pachyptila turtur*, 90–175 grams),[8] Pycroft's petrel (*Pterodroma pycrofti*, 128–198 grams),[9] little shearwater (*Puffinus assimilis haurakiensis*, 191–242 grams),[10] fluttering shearwater (pakahā, *Puffinus gavia*, 230–415 grams),[11] Buller's shearwater (*Puffinus bulleri*, 412 grams)[12] and flesh-footed shearwater (tuanui, *Puffinus carneipes*, 580–750 grams).[13] Tuatara have also been reported to share the burrows of, although not necessarily

Box 7.1 Fig. 1 A tuatara on Stephens Island (Takapourewa) mouthing fragments of a failed prion egg (the egg was attacked where it lay on the forest floor). Tuatara probably also consume viable eggs from nest chambers.

Photo: Rod Morris.

prey on, the grey-faced petrel (adults ōi, chicks tītī, *Pterodroma macroptera gouldi*, 460–750 grams)[14] and sooty shearwater (adults hākoakoa, chicks tītī, *Puffinus griseus*, 650–950 grams),[15] and have been seen entering the nest cavities of the little penguin or kororā (*Eudyptula minor*, about 1 kilogram).[16]

Not surprisingly, adult seabirds show little tolerance for tuatara that attempt to enter their burrows, defending their nests with vigorous pecking and squawking. Buller's shearwater is especially aggressive, and tuatara that share the Poor Knights Islands with this species sometimes come off second best. One ornithologist described a very large male tuatara (700 millimetres in length) rapidly reversing out of the burrow of a Buller's shearwater 'with the screaming occupant's beak embedded in its snout'.[17] Others reported seeing a tuatara and a Buller's shearwater 'matched in fierce combat. The tuatara was severely torn about the neck. On another occasion we found a large tuatara lying dead on a slope riddled with burrows.'[18] To some extent, tuatara can avoid such interactions by living in separate side passages or chambers within complex burrows, as Reischek had noted on the Marotere (Chickens) Islands.[19] However, as noted elsewhere in this chapter, aggression from seabirds may contribute to a relatively high incidence of eye damage in tuatara on some northern islands.

Despite the risk, there are obvious benefits to tuatara from living with burrow-nesting seabirds. Bird-made burrows are convenient residences, and eggs and chicks provide a valuable source of protein, especially in summer when metabolic demands of tuatara are probably highest. Seabird tissues are rich in so-called 'marine', long-chain polyunsaturated

Box 7.1 Fig. 2 Tuatara reach their maximum body mass (just over 1 kilogram) on the Poor Knights Islands, but are still vulnerable to attack from aggressive Buller's shearwaters (*Puffinus bulleri*), which reach only about half this mass.

Photo: Rod Morris.

Box 7.1 Fig. 3 Burrow-nesting seabirds, such as these fairy prions on Stephens Island, are sometimes known as 'ecosystem engineers' for their role in creating new habitats – specifically burrows – that tuatara and other animals can use. Their ground-level activities also maintain the forest floor in an open state, making hunting and social communication easier for tuatara.

Photo: Alison Cree.

fatty acids; these might bring specific health benefits for tuatara (see Box 7.3), although further study is required. Ingested chicks might also be a source of moisture in summer when rainfall is infrequent. [20]

Other benefits to tuatara are more indirect and stem from the important roles that seabirds play in structuring their island ecosystems.[21] Seabirds bring nutrients to the soil in forms such as excrement (guano), regurgitated food, carcasses and failed eggs. Burrow-nesting seabirds also transfer foliage to the ground as they crash-land through the forest canopy. They then incorporate this and other material into the soil by burrowing, simultaneously fertilising, aerating and homogenising the soil to a depth of about half a metre.[22] These actions help explain the very high densities of invertebrates and lizards (potential prey for tuatara) that occur on islands with burrow-nesting seabirds.[23]

There are additional benefits to tuatara from seabirds. By shuffling about, seabirds keep the forest floor open; tuatara gain by being able to hunt more effectively for food and being able to communicate visually during the mating season.[24] A recent novel suggestion is that tuatara might even receive a thermal benefit from sharing burrows with seabirds: remote sensing revealed that some tuatara in burrows on Lady Alice Island had unexpectedly high body temperatures, best explained by proximity to the warm bodies of seabirds.[25]

Do tuatara need seabirds to survive? The answer so far seems to be a qualified 'no'.[26] Tuatara in captivity have survived and bred without access to seabirds, and limited information suggests to palaeoecologists that pre-human densities of tuatara and seabirds on mainland New Zealand were not closely related.[27] Tuatara on Stephens Island (Takapourewa) seem to avoid the areas of highest seabird density, instead being most common either where seabird densities are moderate, or on the periphery of high-density colonies.[28] Nevertheless, when one compares among islands, populations of tuatara show strongest evidence of recruitment and highest densities where small seabirds are also present.[29] This pattern may be partly confounded by a negative effect of rats on both small seabirds and tuatara. However, the evidence indicates that seabirds do bring many benefits, both direct and indirect, for tuatara – without receiving any obvious advantage in return.

Fig. 7.1 A tuatara pauses before entering its burrow on Stephens Island (Takapourewa), perhaps aware of the camera inside.

Photo: Michael Schneider, reproduced from Daugherty, C., Cree, A. and Schneider, M. 1990. Tuatara: a survivor from the dinosaur age. *New Zealand Geographic* 6: 66–86.

Burrows, burrowing and other retreat sites

Tuatara dig strongly and can make their own burrows. Often, however, they use burrows made by nesting seabirds (Figures 7.1–7.3). Shared use of burrows with the fairy prion (tītī wainui, *Pachyptila turtur*) has been studied intensively in forest on Stephens Island with the aid of inspection hatches, vertical shafts, mirrors and video cameras.[4] In two patches of forest studied in the 1980s by Donald Newman,[5] density of burrow entrances (whether made by tuatara or seabirds) ranged from 0.6 to 2.3 per square metre (m^2). Some burrows had more than one entrance, yielding an overall density of burrows (as opposed to entrances) ranging from 0.45 to 1.0 per m^2. Similar estimates for burrow density in forest (mean of 0.9 per m^2 for five patches,[6] with values up to 1.61 per m^2 in some patches[7]) were later obtained using a video camera 'burrowscope'.

Tuatara have essentially sole use of burrows during autumn and winter when fairy prions are at sea, but they overlap with prions in the use of burrows during the prion nesting season in spring and summer. Tuatara that enter burrows used by nesting prions are pecked at vigorously. Burrows are occupied by a maximum of one tuatara at a time, and are defended from other tuatara that intrude, but the same burrow is sometimes used by several tuatara at different times. Most adult tuatara appear to use several burrows, but typically these are within about 1 metre of each other. Juveniles or 'small animals' show less site fidelity than adults.[8]

Until recently on Stephens Island, adult tuatara also made use of logs, timber, corrugated roofing iron, other debris and even the foundations of buildings as retreat sites or as covers for hidden burrows (most human debris has now been removed from natural habitats). Tree hollows up to about 2 metres above ground are also sometimes used. Tuatara probably cannot climb

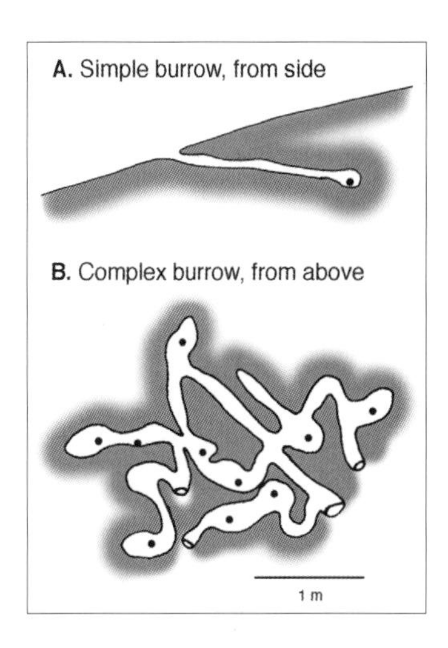

Fig. 7.2 Tuatara often use burrows made by seabirds such as fairy prions (*Pachyptila turtur*). On Stephens Island, fairy prion burrows may be simple (with a single nesting chamber and one or two entrances), as shown from the side in A, or complex (with many entrances leading to several chambers), as shown from above in B. Burrows are typically about 0.2–2 metres in length from the entrance, with a twist or bend excluding light from the prion's nesting chamber. The scale applies to both diagrams (solid circle = prion nest).

Modified from Walls 1978 and Newman 1982c.

Fig. 7.3 Forest floor on Stephens Island – habitat of tuatara and seabirds. Burrows may reach a density of 1.0 per m^2, and burrow entrance density more than double this. Seabird excrement, a seabird carcass and numerous fallen leaves lie on the forest floor.

Photo: Alison Cree.

Fig. 7.4 (above left) A tuatara, emerged by day about 2 metres up a tree trunk on Stephens Island, is admired by a past MSc student, Linda Cartland-Shaw.

Photo: Alison Cree.

Fig. 7.5 (above right) Although primarily terrestrial, tuatara roaming at night sometimes climb short distances – such as this adult straddling the stems of kawakawa (*Macropiper excelsum*) on Stephens Island.

Photo: Rod Morris.

far up vertical tree trunks, but they negotiate wide, sloping branches with the aid of their strong claws (Figures 7.4–7.6). Burrows in pasture are often constructed at the margins of large rocks or boulders (Figure 7.7). Hatchling tuatara have frequently been found under cover (rocks, timber or roofing iron) in pasture.[9] In captivity, hatchlings of Stephens Island origin begin digging burrows between about 2 weeks and 5 months of age.[10]

Use of burrows has not been studied as intensively on other islands. On North Brother Island, burrows occupied by tuatara were mostly single-chambered, not occupied by seabirds, and ranged in length from 0.25 to 0.5 metres.[11] On the Hen and Chickens Islands, tuatara shared use of burrows with three species of seabirds,[12] but the tuatara and birds were probably often in separate chambers of complex burrows.[13] Burrows used by tuatara measured about 0.6–1.2 metres in length, and, as on Stephens Island, there was never more than one tuatara in each.[14] On rocky islands, adult tuatara often inhabit crevices beneath rocks and boulders.[15]

Daily activity patterns

Adult tuatara are primarily nocturnal, roaming most widely from their burrows at night.[16] In suitable habitats and weather, however, they also emerge

Fig. 7.6 Small adults, such as this female basking nearly 2 metres above ground on Stephens Island, sometimes make use of hollowed trees as retreat sites.

Photo: Alison Cree.

Fig. 7.7 A large and possibly old adult male basks in the entrance of a burrow beneath a boulder in pasture on Stephens Island. Tuatara in pasture are more vulnerable to predation by harriers (*Circus approximans*) and less likely to emerge fully to bask than in forest. They are also present at lower density, and more likely to be male (Moore 2008).

Photo: Alison Cree.

Fig. 7.8 A female tuatara basks in dappled sunlight in forest on Stephens Island. The red patches on the shoulders and pelvic region are chigger mites (*Neotrombicula* spp.).

Photo: Michael Schneider, reproduced from Daugherty, C., Cree, A. and Schneider, M. 1990. Tuatara: a survivor from the dinosaur age. *New Zealand Geographic* 6: 66–86.

during the day (Figure 7.8), leading to a suggestion that they be described as diurno-nocturnal.[17]

Daily activity patterns were compared between tuatara in forested and in pasture habitats on Stephens Island by James Gillingham and Tracy Miller.[18] This study focused on adult males during the late summer–autumn mating season (January to March). In the forest, tuatara emerged in the morning but spent much of the period between 6 a.m. until noon sitting motionless on one spot (Figure 7.9A). Feeding, aggressive interactions and burrow excavations were infrequent during this time, and courtship was not seen. However, between noon and 6 p.m., these activities collectively increased in frequency, and courtship also occurred. Greatest activity was seen between 6 p.m. and midnight.

In the same study, and in striking contrast, daytime activity among tuatara living in the pasture was negligible (Figure 7.9B).[19] Tuatara in pasture remained in their burrows by day, occasionally venturing as far as the entrance. Reduced daytime emergence in pasture is widely attributed to the presence of predatory birds such as the Australasian harrier (kāhu, *Circus approximans*), a species not known to enter forest. Tuatara living in pasture emerged soon after sunset, with social interactions occurring from then until about 1 a.m. Little activity occurred between 1 a.m. and dawn, when animals returned to their burrows. In both forest and pasture, aggressive and courtship behaviours accounted for much of the activity time of male tuatara in late summer.

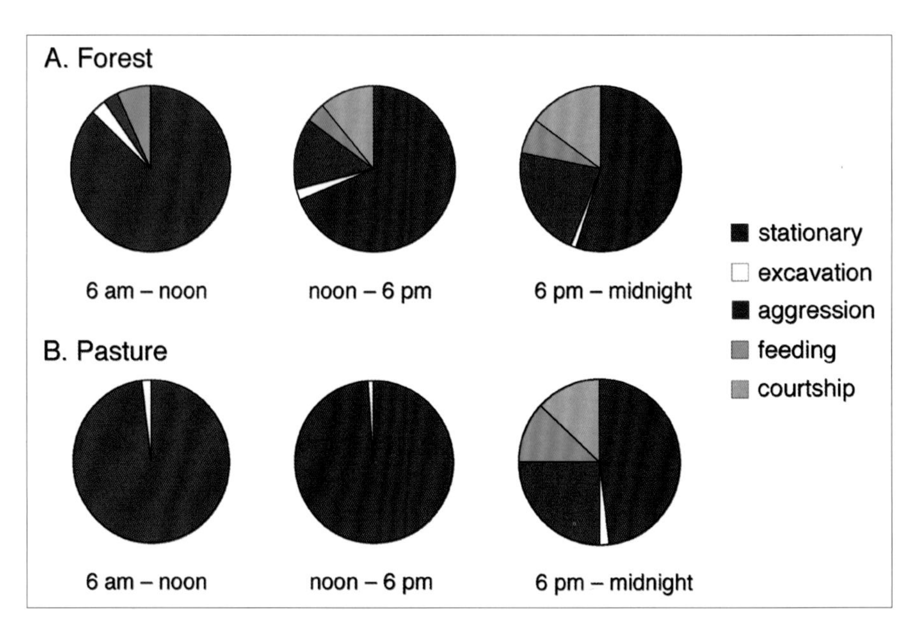

Fig. 7.9 Frequency of daily activities among adult male tuatara during late summer on Stephens Island. Tuatara in forested habitat (A) are more active by day than those in pasture (B).

Modified from Gillingham, J. C. and Miller, T. J. 1991. Reproductive ethology of the tuatara *Sphenodon punctatus*: applications in captive breeeding. *International Zoo Yearbook* 30: 157–164. John Wiley & Sons.

Hatchlings on Stephens Island apparently disperse from nesting areas during the daytime but become increasingly nocturnal within a few months.[20] Juveniles in captivity are also mostly diurnal during their first few months.[21] By day, they are skittish, scampering quickly to catch food or to reach cover following disturbance,[22] but they become more nocturnal[23] (and also more aggressive[24]) with age. Daily and seasonal patterns of activity for two captive tuatara of Stephens Island origin (small females of subadult or small adult size) were generally similar to those of wild tuatara.[25]

Activity patterns have not been studied in detail in adult females (except at nesting: see Chapter 8), nor in tuatara generally on other islands.

Nocturnal and seasonal emergence

On Stephens Island, nocturnal emergence of tuatara peaks in mid to late summer, when air temperatures are warmest and invertebrate prey items are most abundant.[26] Across the year, emergence on any given night is greatly affected by air temperature, being highest on the warmest nights (about 20°C) and negligible on the coldest (about 7°C; Figure 7.10).[27] Nocturnal emergence is enhanced by moisture but inhibited by wind. Overall, tuatara are most often seen (and are most lively and aggressive) on warm, calm, misty evenings in the rain, and are least often seen on cold, dry, clear and windy nights.[28] Activity declines as the night progresses,[29] a pattern attributed to the loss of body heat gained during the day.[30]

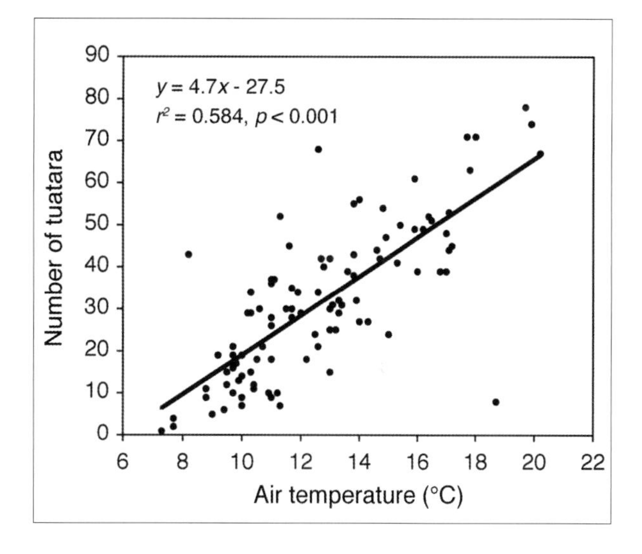

Fig. 7.10 Warm nights encourage emergence of tuatara at night. The number emerged is shown in relation to air temperature during transect counts (c. 9.30 pm–10.30 pm New Zealand Standard Time) across a year on Stephens Island. The regression is significant ($p < 0.001$).

Activity of tuatara on Stephens Island is greatly reduced during winter but, in suitable weather, emerged animals can still be seen, captured and studied.[31] Similarly, northern tuatara on Green Island remained active during winter.[32] Tuatara experience short periods of torpor (dormancy) in cold weather but, like other reptiles, do not hibernate in the physiological sense that mammals do. Body temperatures and activity are typically reduced in winter, not because of a radical alteration of a thermal set-point but mainly because solar radiation is limited.

Density, biomass and home range

Estimating the abundance of wild tuatara is not easy. From a practical point of view, island populations are ideally surveyed at night, in warm and humid weather to encourage emergence of tuatara, while avoiding a full moon (Figure 7.11). Researchers also need suitable sea conditions to get ashore, and must be adept at negotiating difficult terrain. On some rat-free islands, the ground is so heavily burrowed by seabirds that merely walking about results

Fig. 7.11 The author about to capture a tuatara on Stephens Island at night.

Photo: Michael Schneider.

in numerous burrow collapses.[33] To avoid damage, boardwalks have sometimes been constructed. In addition, although surveys under ideal conditions reveal the presence of large juvenile and adult tuatara, they underestimate the abundance of hatchlings and small juveniles, which are more secretive and diurnal. Finally, once the numbers are in hand, one must decide on the most appropriate mathematical model for analysing the results.[34]

BOX 7.2

Tuatara and Pacific rats: the subtle slide to extinction

The Pacific rat's place in the fauna of New Zealand, and its impacts on tuatara and other organisms, have long been sources of debate. Pacific rats (*Rattus exulans*) originated in southeast Asia and eventually became widespread throughout the Pacific as a commensal with humans. Once also known as Polynesian rats, Pacific rats were introduced to New Zealand about AD 1280 with early Polynesian settlers,[1] and became known locally by the Māori name 'kiore'. Traditions developed within Māori culture about tuatara and about kiore, but, as far as is known from 19th-century records, were silent about interactions between kiore and tuatara.

For about a century the fledgling scientific community in New Zealand seemed to consider Pacific rats as merely a benign herbivore. During the 1950s and 1960s, it remained common for scientists and government officials to refer to the species erroneously as the 'native rat' or even as 'the harmless native species'.[2] But by the 1970s it was apparent, in both Hawaii and New Zealand, that Pacific rats were predators of ground-nesting seabirds, including their eggs up to 55 millimetres in length.[3] Yet, the very idea that any predator could cause extinction of a prey species was discounted within New Zealand by several senior and influential biologists. This mindset, discussed by Trevor Worthy and Richard Holdaway in 2002 in the context of extinctions of New Zealand birds, may have been influenced by northern hemisphere studies in which predator and prey populations (such as the famous lynx and snowshoe hare system in North America) were modelled to oscillate but without the prey species ever becoming extinct.[4] However, as Worthy and Holdaway point out, this analogy fails for New Zealand animals of Gondwanan origin. Tuatara, for instance, evolved in an environment where, for most of the last 80 million years, the only significant predators have been birds that hunted mainly using vision. This is unlikely to have prepared tuatara for co-existence with a fast-breeding rat that hunts at night using both smell and sight.

First recognition that Pacific rats might have harmful effects on tuatara has been widely attributed to Ian Crook's research from the 1970s,[5] but unpublished documents reveal that the relationship had been recognised around 50 years previously. In 1914 a lighthouse keeper on Cuvier Island (Repanga Island) reported to the Department of Internal Affairs (the government department then responsible for wildlife management): 'I think the rats and the cats destroy a good many [tuatara] as there seems to be plenty of both on the Island ...'[6] Another keeper suggested in 1924 that a concerted effort to poison and trap rats had brought positive results for tuatara: 'The increase in the number of lizards [*sic*] is due entirely, I believe, to the decrease in the number of rats, for it is only some months after many rats have been destroyed, that Tuatara eggs or young have been seen at all.'[7]

Even more tellingly, in 1951 Logan Bell, a senior field officer with the Wildlife Division of Internal Affairs, reported on the 'Tuas' he had seen during field trips to islands off the Coromandel Peninsula:

> Tuataras are present in fair numbers on several of these islands, but an interesting point which struck me was that if tuataras were plentiful on an island there would be no rats and if rats were plentiful there would be no tuataras ... on Little Barrier and Hen Island where rats are plentiful Tuas are very scarce indeed. It would appear that these few are the old surviving adults that grew up before the rat invasion.[8]

These intimations that Pacific rats inhibited recruitment of tuatara seem initially to have fallen on deaf ears, at least within wildlife administration.

However, among biologists who had been in quick succession to offshore islands with and without rats, suspicions were rising.[9] By the early 1970s, further surveys had been carried out by Crook and others (Crook having been recently appointed as a scientist to the Wildlife Service), and these provided quantitative evidence that tuatara populations were not self-sustaining where Pacific rats were present.[10]

By the late 1990s the following evidence, entirely circumstantial but collectively compelling, had been accumulated by many researchers to indicate a negative impact on tuatara from Pacific rats.

- Tuatara became extinct on mainland New Zealand (North and South Islands) in the presence of Pacific rats, the decline beginning prior to widespread habitat change or the introduction of most other mammalian species.[11]
- Tuatara became extinct during European times on at least four offshore islands in the known presence of Pacific rats (East or Whangaokeno, plus three islands in the Mokohinau Islands).[12]
- Tuatara became extinct on at least four more islands, during pre-European or early European times, in the probable presence of Pacific rats (Korapuki, Motuopao, Tiritiri Matangi, and Rangitoto ke te Tonga (D'Urville Island)).
- Tuatara are absent from (and suspected to have become extinct on) at least three islands in island groups where tuatara are otherwise present and where the habitat seems suitable apart from the presence of Pacific rats (Double, Middle (Aldermen Islands), Rurima).[13]
- On offshore islands where tuatara were present with Pacific rats, tuatara were usually sighted infrequently or rarely, indicating low density.[14]
- On offshore islands with Pacific rats, tuatara that were sighted were almost exclusively adults, implying reduced recruitment.[15]
- In the worst affected populations, the few remaining tuatara appeared aged, with possibly declining reproductive activity.[16]

Thus, by the late 1990s a gradation had emerged. At one, irretrievable end of the spectrum were populations of tuatara, such as those on the Mokohinau Islands, which had become extinct in the presence of Pacific rats. In the intermediate but salvageable range were endangered populations, such as those on Hauturu/Little Barrier, Red Mercury (Whakau) and Kawhitu or Stanley Islands, where fewer than about 20 tuatara (all large adults) were known, as well as less-affected populations such as those on Coppermine and Whatupuke islands, where at least some dozens of small and large adults survived. At the least-affected end of the spectrum was Lady Alice Island, where juvenile tuatara were still present although the overall density of tuatara was low.[17] This variation among islands probably reflects factors such as the time that Pacific rats were first introduced (unknown for most islands, but perhaps as recent as the early 1800s for Lady Alice[18]) and island size (populations on smaller islands seem the most vulnerable, perhaps because of a more limited range of habitats, including cliff-side refuges), as well as the past presence of other introduced mammals and the past degree of habitat disturbance.[19]

In circumstantial situations such as this, the associations are not always perfect: just as some people who have never smoked get lung cancer, tuatara are absent from several islands that appear suitable and that have never had any introduced mammals.[20] Furthermore, associations, even strong ones, do not themselves prove cause and effect. Nonetheless, by the 1990s it was widely (though not universally[21]) accepted that Pacific rats have insidious effects on ecosystem structure sufficient to cause extinctions of tuatara and should be removed.

Pacific rats in New Zealand do not just consume plant material and affect forest composition. They also eat a broad variety of animal foods including invertebrates, lizards, and seabird eggs and chicks,[22] and thus have the potential to compete with tuatara for food (see main text). Pacific rats almost certainly also prey on small tuatara and eggs. (Eggs of tuatara may be especially vulnerable, being smaller and less rigidly calcified than the seabird eggs known to be eaten by Pacific rats; they are also only shallowly buried and sometimes remain exposed in nests for a

Box 7.2 Fig. 1 Pacific rats feeding on the carcass of a fish on Hauturu/Little Barrier Island. The impacts of this small rat were much debated during the 20th century, but new evidence unequivocally confirms that these rats cause recruitment failure in tuatara.

Photo: Rod Morris.

day or more before being covered.) Although direct predation on tuatara by Pacific rats has never been observed, a dead hatchling on Lady Alice Island showed wounds consistent with an attack.[23] Given that Pacific rats eat lizards, and that even mice can kill and eat small lizards,[24] wound adult tuatara in captivity[25] and apparently remove tuatara eggs,[26] there is little doubt that direct predation by Pacific rats on small tuatara must occur. The tables seem to be turned only infrequently: tuatara were reported to hunt rats on Cuvier Island by a lighthouse keeper, and on a few occasions, tuatara scats have been found containing fur of Pacific rats (see main text).

One thing that Pacific rats do not do is reduce the number of eggs produced by nesting tuatara, at least in the short-to-medium term. A suggestion that Pacific rats might have such an effect arose during the 1980s, when tuatara on Lady Alice Island (inhabited by Pacific rats at the time) were observed to have a smaller clutch size for a given snout–vent length than tuatara on Stephens Island; it was plausibly suggested that this might result from competition with rats for food.[27] However, clutch size had not increased on Lady Alice by 3–4 years after removal of rats, and small egg clutches are now known to be characteristic of northern populations of tuatara whether or not Pacific rats are present;[28] the reason for this difference remains unclear.

Ideally, the circumstantial evidence for harmful impacts of Pacific rats on tuatara would be supported by ecologically realistic experiments to prove a 'cause and effect' relationship. Pacific rats could,

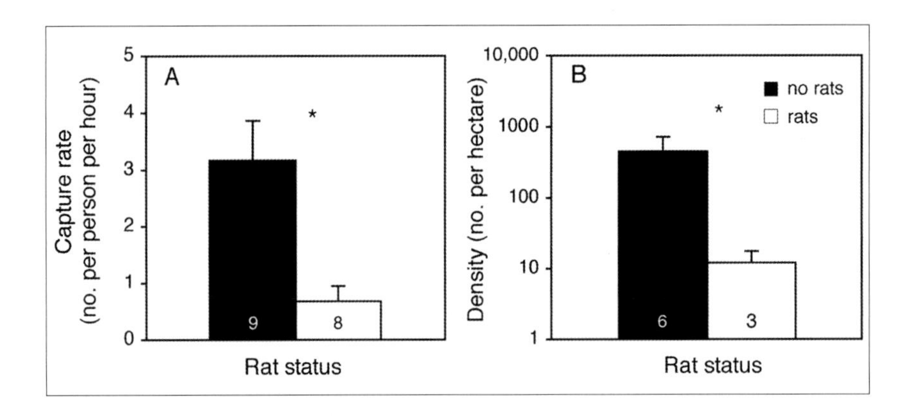

Box 7.2 Fig. 2 A: Mean capture rates (number per person per hour) for tuatara since 1980 have remained significantly lower on islands where Pacific rats were present than on rodent-free islands, supporting earlier work by Crook (1973). B: Although formal estimates of density (number per hectare) are available for fewer islands, these also support a trend of reduced density of tuatara where rats were present. Data for A and B are for mainly forested or scrub habitats (sources to 2007 listed in Appendix 4). Where only encounter rate was reported, capture rate has been assumed to be half. Where a range for density was reported, the mid-point has been used. Note log scale for density in B. Sample sizes are given at base of bars. Vertical lines above the mean indicate the standard error, and asterisks indicate a statistically significant difference ($p < 0.05$).

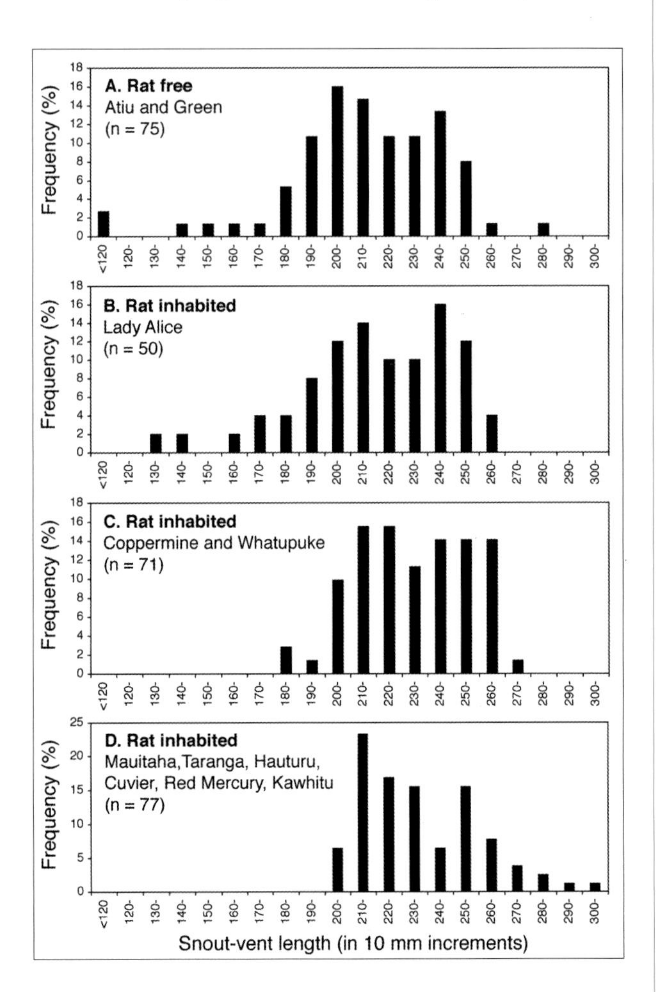

Box 7.2 Fig. 3 Population size-class structures show evidence of reduced recruitment of northern tuatara in the presence of Pacific rats. A: rodent-free; B, C, D: rat-inhabited (all surveys during 1980s–1990s). Populations in B, C and D are grouped to show increasing trends in recruitment failure. Juvenile tuatara (less than about 170–180 millimetres snout-vent length, SVL) are absent in samples from most rat-inhabited islands (C and D). Populations at the greatest risk of extinction (D) have size-class distributions strongly skewed to the right, lacking both small adult tuatara (170–200 millimetres SVL) as well as juveniles. All tuatara islands are now free of rats, with the exception of Taranga (on which eradication is underway) and Mauitaha Islands in D.

Sources: Whitaker, A. H. and Daugherty, C. H. 1991. *Research on the tuatara (*Sphenodon punctatus*) of Little Barrier Island, 5–12 February 1991*. Department of Conservation, Wellington; Whitaker, A. H. 1993. *Research on the tuatara (*Sphenodon punctatus*) of Little Barrier Island, 6–20 October 1992*. Department of Conservation, Auckland; Cree et al. 1995a and unpubl. data; Tennyson and Pierce 1995; C. Smuts-Kennedy, Department of Conservation, pers. comm. 25 January 2000.

in principle, be added experimentally to tuatara islands that have until now remained rat-free – but there are practical and ethical difficulties with such a proposal.[29] Fortunately, the alternative approach has been made possible: techniques now exist for eliminating rats from islands of up to several thousand hectares in size.[30]

As a result, attention has turned to testing the effects of eliminating rats on the recovery of tuatara and other rare species – with some spectacular results. There have been increases in the capture rates for vulnerable invertebrates, skinks and geckos, and in the nesting success of small, burrow-nesting seabirds.[31] In surveys by David Towns and colleagues, the abundance of juvenile tuatara (as a proportion of all tuatara captured) has also risen, sometimes dramatically.[32] On the three Chickens Islands, for example, while rats were present, juveniles never accounted for more than 8% (and typically 0%) of each sample, whereas by 6–10 years after rat removal, juveniles accounted for about 12–43% of each sample (a 3.5- to 17-fold increase). In comparison, on nearby Taranga Island (considered a control in that Pacific rats remained present throughout), no juvenile tuatara were found over a period of at least 21 years. The few remaining adults on Taranga Island have also shown no increase in body condition index (mass adjusted for snout–vent length) over this time, whereas adults on the Chickens have often increased in body condition index since rats were removed. The magnitude of this effect varies with sex, season and island, but in the most dramatic case, the mean index of body condition increased, in females in autumn on Coppermine Island, by 45% over the value when rats were present.[33]

The evidence is thus now unequivocal: removal of rats benefits tuatara. The result – more juvenile tuatara, and adults in better condition – means that populations on the Chickens Islands now face a more secure future. These results have helped an agreement to be reached between Ngātiwai and the Department of Conservation to eliminate Pacific rats from Taranga Island; eradication began in 2011[34] and is expected to be declared successful in 2013.[35] The nearby but smaller island of Mauitaha, on which only one tuatara has been sighted over the past 17 years,[36] is to retain rats. Thus, the extinction of any relict population of tuatara there is inevitable.

Box 7.2 Fig. 4 Richard Parrish of the Department of Conservation, Northland, holding one of the many juvenile tuatara that appeared on Lady Alice Island following removal of Pacific rats. Six years after rat eradication, juveniles accounted for over 40% of a sample of 51 tuatara.

Photo: Martin Hill.

Given the difficulties (at least in the days before global-positioning devices) of repeatedly surveying tuatara over a known search area, for many islands the only estimates of tuatara abundance come from measurement of 'catch per unit effort' – the number of tuatara seen or caught per person searching per hour. Since 1980 these simple estimates of relative abundance have been reported for 18 islands, including several habitat types on Stephens Island (Appendix 4).[35] The lowest capture rate reported was 0.01 tuatara per person per hour on Hauturu/Little Barrier Island while Pacific rats (kiore, *Rattus exulans*) were still present, a value equivalent to one person searching for 100 hours to find one tuatara.[36] The highest reported capture rates ranged from 9.8 to 25.6 tuatara per person per hour across different nights on rat-free Stephens Island.[37] Sometimes effort has been reported as encounter rate, which overestimates capture rate but probably by not more than double.[38] Given that such estimates are vulnerable to variation in search effort, their value has sometimes been questioned.[39] Nonetheless, a significant correlation exists between capture rates and formal estimates of density (number of tuatara per hectare) now available for several islands from mark-recapture or line-transect techniques (Figure 7.12).[40] This relationship suggests that, in the absence of other information, capture rates recorded by experienced searchers in appropriate weather should not be overlooked as an initial indication of relative abundance.

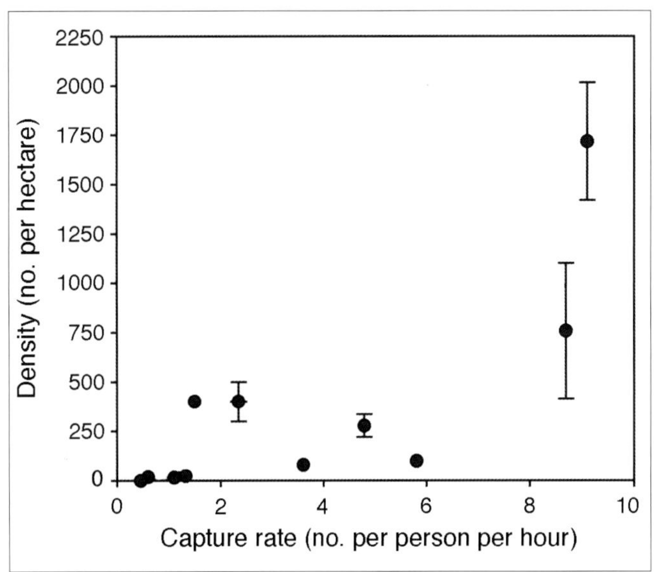

Fig. 7.12 Formal estimates of the density of tuatara (number per hectare) increase significantly with estimates of abundance based on capture rate (number per person per hour). Estimates are for seven islands (including three habitat types on Stephens Island and two on Atiu Island; midpoint and range are shown where more than one estimate within the same habitat type is available). Three values reported as encounter rates have been halved here as conservative estimates of capture rate. The correlation is significant ($p < 0.01$). For sources (studies between 1980 and 2007), see Appendix 4.

Formal estimates of density are available for nine island populations, including several habitat types on Stephens Island (Appendix 4). Three points stand out from these estimates. First, the density of tuatara in forest on Stephens Island is enormous (1420–2732 tuatara per hectare),[41] at least three-fold and sometimes more than ten-fold higher than in forest or scrub on other rodent-free islands. Second, density estimates in sheep pasture (415–1100 tuatara per hectare) on Stephens Island,[42] while that habitat was still present and grazed, were relatively low compared with that in forest. Finally, the density of tuatara on northern islands in the presence of Pacific rats has been worryingly low (1.1–26 tuatara per hectare).[43]

Just why are densities of tuatara in forest on Stephens Island so high compared with other forested islands? Given the cool, windy conditions that prevail in Cook Strait, at the southern end of the current distribution of tuatara, one might expect densities to be low, but this is not the case. Seabirds, invertebrates and lizards are all abundant on Stephens, and the high density of tuatara probably to some extent reflects this ample supply of food. Past human-induced changes to the island's ecosystem may also have played a role. Much forest was lost and cats were present during the late 1800s and early 1900s, resulting in local extinction of several species of forest birds that would have competed with tuatara for insects[44] and/or preyed on at least small tuatara (see later section on predation). Simultaneously, open sites for nesting by tuatara became widely available, favouring recruitment. Although the reasons for variations in density among rat-free islands remain speculative, they probably include the extent to which nutrients are brought ashore by seabirds, as well as the effects of past habitat modification and the past presence of introduced mammals.

Densities of tuatara in forest on Stephens Island are high, not just in comparison with other tuatara populations but also with other reptiles of similar body size.[45] Conversely, territory size, or home range, is unusually small.[46] High density, in combination with large body mass (weight), makes for an exceptionally high biomass of tuatara tissue. Mean body masses of tuatara reported by Donald Newman in 1982 for two patches of forest were 343–423 grams, yielding estimates for biomass of 231–684 kilograms per hectare.[47] These values (equivalent to about 23–68 tonnes of tuatara per km^2)[48] are larger than known for any other species of lepidosaurian reptile. Indeed, they are more than three times higher than the maximum value (about 70 kilograms per hectare) for total herpetofaunal biomass (the combined biomass of all species of lepidosaurian reptiles and amphibians present) reported for about 75 sites around the world in a recent compilation.[49] As Graham Hardy commented in 1982, when the biomass of lizards in forest on Stephens Island is eventually calculated and added to that available for tuatara, the total herpetofaunal biomass there is likely to prove 'very spectacular indeed'.[50]

Like other reptiles, tuatara are able to produce biomass with high efficiency compared with terrestrial endotherms (mammals and birds, the so-called 'warm-blooded' animals) because so little ingested energy is required for metabolism.[51] An additional consequence of ectothermy (loosely and inaccurately known as 'cold-bloodedness') is that reptiles are able to tolerate wide swings in food availability, so-called 'feast or famine' situations.[52] In a classic example of the latter, Captain Gilbert Mair obtained an adult tuatara from an island in the Bay of Plenty in 1872. It was placed, in a box with air slots, under an old storehouse in Whakatane with instructions that it be placed on the next boat for Tauranga. The box was inadvertently overlooked for more than 3 months until Mair enquired what had become of it – whereupon the tuatara was found alive inside and seemingly none the worse for wear.[53] Living in low-temperature environments, tuatara have low mass-specific metabolic rates even by reptilian standards (see Chapter 9), although an early popular belief that they could survive on air[54] is overstating the case! Exceptionally low activity temperatures and hence metabolic rates probably help tuatara on Stephens Island maintain higher biomasses than reported for other lepidosaurian reptiles (Figure 7.13).

Fig. 7.13 Compared with mammals and birds, reptiles are considered as 'low-energy' animals because of their ability to survive and reproduce on relatively small amounts of food. Tuatara are exceptional in this regard, especially if body temperatures and activity levels remain low. This Stephens Island male has patches of red chigger mites on its neck.

Photo: Michael Schneider, reproduced from Daugherty, C., Cree, A. and Schneider, M. 1990. Tuatara: a survivor from the dinosaur age. *New Zealand Geographic* 6: 66–86.

North Brother Island has an estimated biomass of 23 kilograms per hectare for tuatara, and 29.75 kilograms per hectare for terrestrial and arboreal lizards.[55] The lower biomass of tuatara there than on Stephens Island reflects both lower density (Appendix 4) and smaller body size (Chapter 6, Box 6.3). These differences probably result from a combination of small island size, lack of forest and more limited diet (for example, tree wētā seem to be absent from North Brother Island). An index of body condition (mass adjusted for variation in snout–vent length) of tuatara has also declined in recent decades on North Brother Island.[56] Initially, this was linked with low genetic diversity in the population on North Brother,[57] but a significant decline in body condition over the last few decades has also been reported on Stephens Island,[58] a population with substantially greater genetic diversity.[59] The decline on Stephens Island is attributed to past habitat modification, including loss of about 80% of the forest.[60] The revegetation now under way may lead to a reversal of this trend.

As noted above, during the late 1900s when sheep still grazed the northern end of Stephens Island, density of tuatara was lower there than in forest. This difference has several explanations. First, the quantity and variety of potential food for tuatara (especially invertebrates) was lower in pasture than forest,[61] resulting in differences in diets (see later section).[62] Mean home ranges of adult males were also significantly larger in the pasture (86.7 m^2) than in the forest (15.7 m^2),[63] suggesting that tuatara in the pasture had to roam more widely to obtain the food that was present. Tuatara in pasture were more vulnerable to aerial predators and emerged less frequently to bask, which is likely to have slowed the rate at which food was digested and assimilated into body tissues. Tuatara in pasture may also have experienced the impacts of drought more severely. It follows from all of these differences[64] that tuatara in pasture may have shown a steeper decline in body condition index over recent decades than tuatara resident in forest, although this has yet to be examined.

Overall, despite pasture being an attractive habitat for nesting on Stephens Island, it seems to have been a relatively poor-quality habitat for residence by adult tuatara. As mentioned above, former pasture on the island is now being revegetated (the last sheep were removed in 2004), and availability of forest for adult tuatara is increasing. Changes in the overall size of the population are hard to predict, as it seems likely that recruitment will decline owing to the loss of nesting habitat. On the other hand, those tuatara that are present will probably be heavier, relative to their snout–vent length, than at present.

Vocalisations and responsiveness to sound

Tuatara lack an external ear opening and a functional tympanic membrane. Their middle ear cavity is also not the open, air-filled space seen in most reptiles, instead being filled with connective and adipose tissue.[65] Although these features were once considered ancestral for reptiles, comparisons with basal rhynchocephalians indicate that they are derived features in tuatara,[66] perhaps having evolved alongside a burrowing habit. Although there is no sign on the external surface of the body, the inner ear is well developed.[67]

Tuatara appear from anecdotal accounts to be very responsive to sound. An early (1903) newspaper article reported that tuatara at the 'Opawa fisheries' in Christchurch 'seem to be susceptible to music. They will come out of their holes in the rocks to hear a song, when nothing else will induce them to appear. They prefer a good rousing chorus rather than a solo'.[68] Another early account referred to a captive tuatara being 'greatly attracted by musical sounds, piano-playing invariably brought it from its sleeping-box'.[69] An adult male held in a vivarium in a Polish research institution was noted for emerging during evening seminars; it would turn its head in the direction of whoever was speaking.[70] In 1928 Sladden and Falla reported that tuatara on the Aldermen Islands 'appear to be unable to resist the attraction of any unusual sound, and five or six at a time have emerged from their burrows near our party while we rested for a meal'.[71]

Although many of us have had similar anecdotal impressions, scientific study of hearing in tuatara is essentially limited to the work of Carl Gans and Ernest Glen Wever in 1976. These authors measured cochlear potential in a specimen that was probably from Stephens Island.[72] The sensitivity of the ear, between 100–900 hertz, was considered surprisingly good in low tones for an animal lacking a tympanic membrane, with a peak response around 200–400 hertz (Figure 7.14A).[73] This sensitivity corresponds well with the dominant frequency of the harsh croaks often emitted by tuatara on handling (maximal sound pressure around 300 hertz; Figure 7.14B).

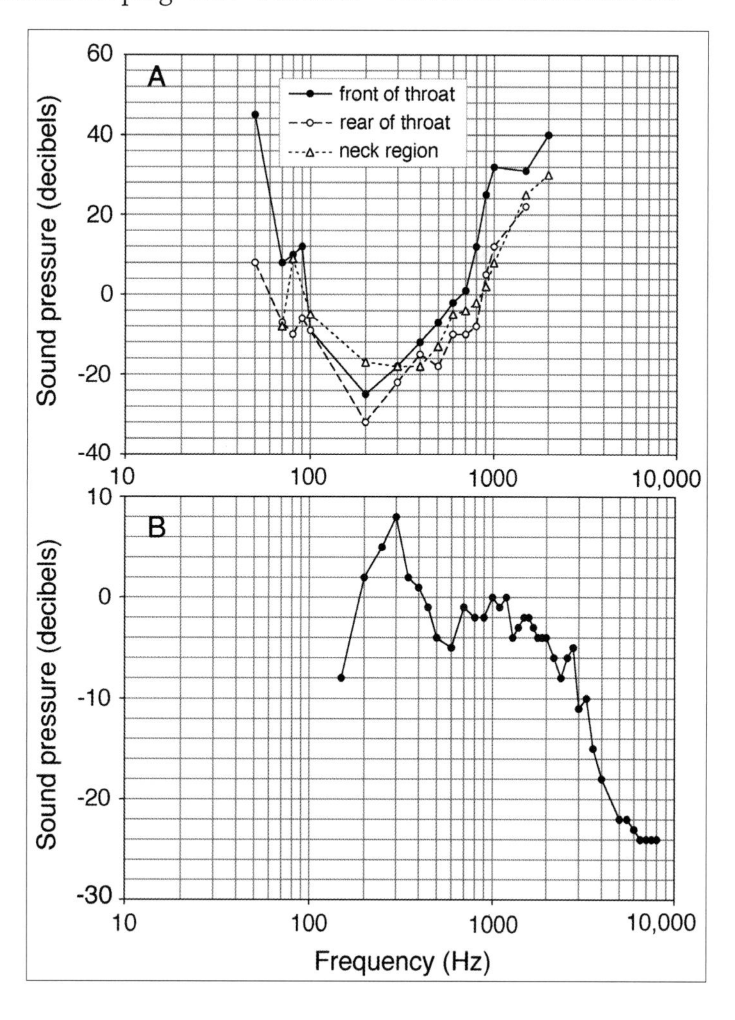

Fig. 7.14 A: Aerial sensitivity curves of the ear of a tuatara (sex not stated; probably an individual from Stephens Island), measured as the lowest sound pressure required for a standard response of the cochlea. Sensitivity of hearing is greatest at about 200–400 hertz (Hz), regardless of where the aerial sound stimulus is applied (see key). A vibratory stimulus (mechanical vibrator in contact with skin over the quadrate bone) gave similar results. B: Distribution of sound frequencies in a call made by a tuatara when handled. The area of maximum sound pressure in the call corresponds with the area of greatest hearing sensitivity in A.

Modified from Gans and Wever 1976.

These croaks have been likened to the croaks made by young alligators,[74] the grunts of frightened pigs,[75] the croaks of frogs[76] and a 'roar like the M.G.M. lion'.[77] Results from an earlier study[78] had suggested that tuatara would be largely insensitive to the low-frequency components of their own croaks, but this misinterpretation appears to have arisen through technical limitations of the research.[79] In addition to the croaks emitted on handling and during fighting, adult tuatara on Stephens Island produce softer vocalisations,[80] including one heard during courtship.[81] No specific studies have been made of sound production or detection in different sexes, in juveniles or in other populations.

Locomotion and speed

Tuatara typically walk with a ground-hugging crawl, the body and tail being thrown into pronounced sideways bends (Figure 7.15). The gait is of a diagonal couplet type, in which the hind foot leaves and contacts the ground before the forefoot on the opposite side of the body. This sequence of limb movement is unusual among vertebrates, though not unique.[82] The body is raised and suddenly lowered throughout the stride (in technical terms, there is a high vertical displacement of the centre of mass), making locomotion somewhat jerky.[83] Tuatara can run rapidly for a short distance but tend to do so in a stop-start manner[84] and, at least in captivity, without a change in gait.[85] Presumably, like other reptiles,[86] they tire quickly as tissues become anaerobic.

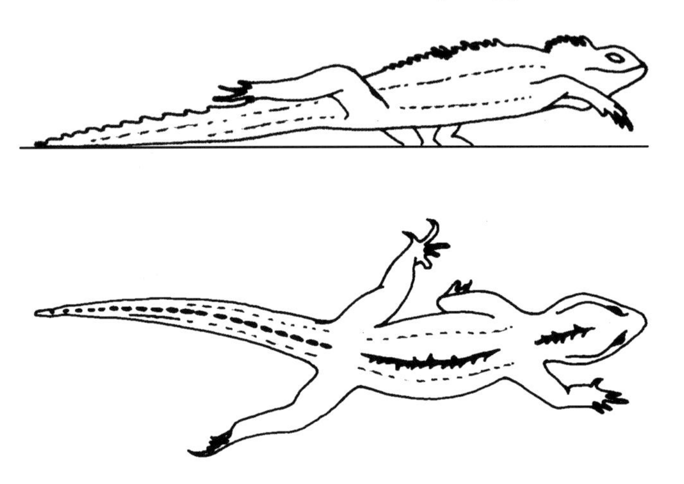

Fig. 7.15 Tuatara have an unusual gait in that the hind foot leaves and contacts the ground ahead of the forefoot on the other side. Note how the belly leaves the ground whereas the tail drags on the ground. Raising and sudden lowering of the body produce a jerky progression.

Locomotion in an adult male tuatara, redrawn by Claudine Tyrrell from Pepperell 1982.

Running speeds of tuatara have been measured only in captivity, for animals of Stephens Island origin. Although not always stated, it seems likely that most if not all measurements were obtained by day. The maximum speed reported in a study in 1982 was 0.9 metres per second (m/s); the length of raceway, size of animal, degree of prompting and time of day were not mentioned, but juveniles were said to run at greater speed (both relative and absolute) than adults.[87] A more recent study recorded a (surprisingly low) maximum speed of < 0.3 m/s in three adult-sized (> 700 grams) tuatara on a 3-metre track.[88] Sprint speed (the fastest 0.25 m over a 1.5-metre racetrack) was recently examined in captive-raised juveniles, during daylight, by Nicola Nelson and colleagues.[89] In Nelson's study, sprint speed increased with age after hatching (1 month versus 10 months) and also varied with egg-incubation condition. By 10 months of age, tuatara arising from eggs that were naturally incubated were significantly faster on average (0.84 m/s) than those arising from artificially incubated eggs (0.67 m/s).[90] Juveniles from naturally incubated eggs were also more likely to have lost their tails, which perhaps influenced their speed. In these small juveniles, speeds were within the range for lizards of similar body size.[91]

Nothing is known of nocturnal running speeds or endurance in wild tuatara. Given a tendency for captive adults to be overweight with limited opportunity for exercise, faster speeds in the wild than in captivity may be anticipated. Nonetheless, even wild tuatara have a clear inability to sustain rapid speeds for prolonged periods. Thus, it is truly remarkable that a new supercar, intended by its makers to be able to exceed 430 kilometres per hour and thus become the world's fastest production car, was named the Tuatara in 2011.[92]

Diets

Food and feeding behaviour of tuatara have received enormous attention, reflecting the willingness of tuatara to move when food is placed in front of them (Figure 7.16), the significance of jaw structure to the evolution of rhynchocephalians, and the importance of diets (such as in the presence of introduced rats, or in captivity) to conservation.

Tuatara on Stephens Island are primarily insectivorous, although a wide variety of invertebrates as well as a few small vertebrates are consumed (Table 7.1). In the classic study by Geoff Walls on Stephens Island in 1981, still the most detailed analysis of tuatara diets on any island, the contents

Fig. 7.16 Tuatara readily feed on warm days if food (such as this unlucky tree wētā, *Hemideina crassidens*) is tossed in front of their burrow. Moller (1985) estimated that tree wētā, which reach nearly 15 grams in body mass, have a population biomass of 42 kilograms per hectare in forest on Stephens Island, and that adult females coming to ground to lay eggs may be the most vulnerable to predation by tuatara. At night, tuatara also hunt wētā living in holes in trees.

Photo: Alison Cree.

Prey items are recorded by family (F), order (O), subclass (SC) or class (C), and all are consumed as adults unless otherwise stated.

TABLE 7.1

Animals recorded in the diet of tuatara on Stephens Island

Item	Taxonomic group
INSECTS	
Beetles (O. Coleoptera)	
Darkling beetles (*Mimopeus* spp. as adults and larvae; *Artystona* spp.)	F. Tenebrionidae
Chafer beetles (*Odontria* spp. as adults and larvae)	F. Scarabaeidae
Ground beetles (adults and larvae; at least 3 species)	F. Carabidae
Click beetles (*Elatichrosis* sp.; *Amychus granulatus*)	F. Elateridae
Weevils (at least 3 species)	F. Curculionidae
Ladybirds (unidentified species)	F. Coccinellidae
Leaf beetles (unidentified species)	F. Chrysomelidae
Wētā (O. Orthoptera)	
Tree wētā (*Hemideina crassidens*)	F. Anostostomatidae
Giant wētā (*Deinacrida rugosa*)	F. Anostostomatidae
Soil wētā (*Hemiandrus anomalis*)	F. Anostostomatidae
Cave wētā (*Pachyrhamma fascifer*)	F. Rhaphidophoridae
Other insects	
Moths (adults, larvae, pupae; several species)	O. Lepidoptera
Blowflies (adults and larvae; at least 2 species)	O. Diptera
Craneflies (adults, pupae; several species)	O. Diptera
Other unidentified dipterans	O. Diptera
Shield bug (*Cermatulus nasalis*)	O. Hemiptera
Other unidentified bugs (at least 3 species)	O. Hemiptera
Cockroach (*Celatoblatta pallidicauda*)	O. Blattodea
Earwig (*Anisolabis littorea*)	O. Dermaptera
Ants (at least 2 species)	F. Formicidae
Stick insect (*Micrarchus* sp.)	O. Phasmatodea
Caddisflies (adults and larvae; several species)	O. Trichoptera
ARACHNIDS	
Spiders (at least 3 species)	O. Araneae
Harvestmen (at least 3 species)	O. Opiliones
Tuatara tick (*Amblyomma sphenodonti*)	O. Acari
Pseudoscorpions (unidentified species)	O. Pseudoscorpiones
MYRIAPODS	
Millipedes (at least 2 species)	SC. Diplopoda
Centipedes (unidentified species)	SC. Chilopoda
EARTHWORMS	
At least 2 species	C. Oligochaeta
OTHER INVERTEBRATES	
Woodlice (unidentified species)	O. Isopoda
Sandhoppers (unidentified species)	O. Amphipoda
Slugs (unidentified species)	O. Stylommatophora
Snails (at least 2 species)	O. Stylommatophora
AMPHIBIANS	
Hamilton's frog (*Leiopelma hamiltoni*)	F. Leiopelmatidae
REPTILES	
Skinks (*Oligosoma lineoocellatum* and at least 1 other species)	F. Scincidae
Common gecko (*Woodworthia* sp., formerly *Hoplodactylus maculatus*)	F. Diplodactylidae
Tuatara (*Sphenodon punctatus*) (juvenile)	O. Rhynchocephalia
BIRDS	
Fairy prion (*Pachyptila turtur* as eggs and chicks)	F. Procellariidae
Fluttering shearwater (*Puffinus gavia* as chicks)	F. Procellariidae
Sparrow (*Passer domesticus*)	F. Ploceidae

Sources: Walls 1981; Newman 1977; Moller 1985; Carmichael et al. 1989; Fraser 1993; Moore and Godfrey 2006. Invertebrate taxonomy updated from Walls 1981 with assistance from Anthony Harris, Otago Museum, and Barbara Barratt, AgResearch Ltd. Plant material and soil are found in scats but were not listed here.

of scats (faecal pellets) were examined (Figure 7.17).[93] Scats were collected across all four seasons of the year and from habitats all over the island (forest, scrub, edge and pasture). Although not stated, it is likely that the 392 pellets were dominated by samples from adult tuatara, since adults produce larger and more visible scats.[94] The most frequently occurring prey items were darkling beetles in the genus *Mimopeus* (present in at least 52% of scats; Figure 7.18), weevils (in at least 52%), spiders (45%), earthworms (37%), ground beetles (34%) and adult moths (27%). Tree wētā (*Hemideina crassidens*) were present in 15% of scats. Vertebrate prey in the form of seabird chicks (in 10% of scats), seabird eggs (7%), reptiles (in at least 3%) and passerine birds (0.3%) appeared only infrequently.[95]

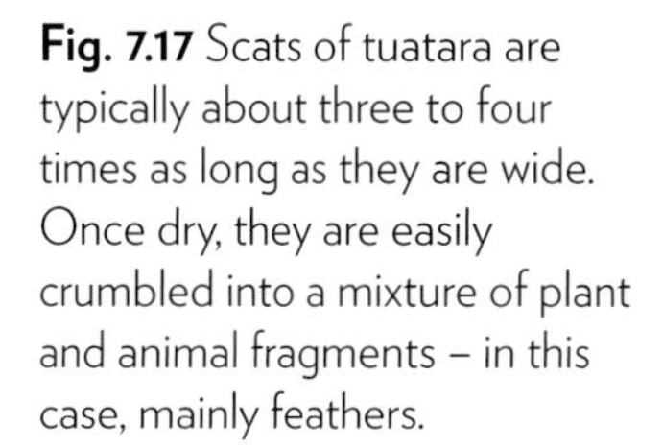

Fig. 7.17 Scats of tuatara are typically about three to four times as long as they are wide. Once dry, they are easily crumbled into a mixture of plant and animal fragments – in this case, mainly feathers.

Photo: Alison Cree.

Two subsequent studies of diets on Stephens Island have yielded information from stomach flushing (Figure 7.19). This method reveals a higher proportion by volume of soft-bodied invertebrates than does analysis of scats.[96] In one report, a significant difference in diets of tuatara between forest and pasture habitats was noted.[97] Tuatara in the forest consumed mainly tree wētā and the darkling beetle *M. opaculus*, whereas those from pasture consumed mainly isopods, craneflies and small beetles. In both habitats, tuatara consumed large numbers of adult moths and arachnids.

Fig. 7.18 The darkling beetle *Mimopeus opaculus* (about 20 mm long) is a major food item of tuatara on Stephens Island.

Photo: Alison Cree.

Fig. 7.19 Stomach flushing is a standard technique for studying the diets of large and medium-sized reptiles. This Stephens Island tuatara has regurgitated a tree wētā previously fed to it, confirming the success of the procedure. Its eye is obscured by a nictitating membrane.

Photo: Alison Cree.

In another study employing stomach flushing and scat analysis, diets of tuatara of different life-history stages from the forest or forest margins of Stephens Island were compared.[98] Adult males, adult females and large juveniles (the latter ranging in snout–vent length from 100 to 170 millimetres) had similar diets to each other, and once again invertebrates, especially beetles, were the most common prey. However, an earlier suggestion that carabid beetles were rejected by tuatara as distasteful[99] was not confirmed: these insects were found in scats, and one was seen being eaten. Diets of small juvenile tuatara (50–100 millimetres snout–vent length), a size range that includes hatchlings, were also examined. Faecal samples from small juveniles living in pasture contained only small invertebrates, including beetles, harvestmen and snails.[100]

The eggs and chicks of seabirds, especially the fairy prion, are the most frequently occurring vertebrate item in the diet of tuatara on Stephens Island. They are also the only strongly seasonal component, being eaten in spring and summer during the prions' nesting season (Figure 7.20). In summer, evidence of fairy prion remains (eggshell, bone, beak, claw, feather or down) was reported for at least 33% of tuatara scats by Walls in 1981, in 67% of scats by Moller in 1985, and in 45% of scats from adult males (lesser proportions for females and large juveniles) by Fraser in 1993. Walls estimated that tuatara were directly responsible, through predation and interference, for the loss of more than 25% of the eggs and chicks of fairy prions during the nesting season of 1974/75.[101] Similar or slightly lower losses to tuatara were estimated during the nesting seasons of 1992/93[102] and 1994/95.[103]

BOX 7.3

Seabirds as a source of omega-3 fatty acids

Seabirds are a rich source of nutrients known as long-chain polyunsaturated fatty acids. Tissues of fairy prions *(Pachyptila turtur)*, for example, contain notable amounts of eicosapentaenoic acid (EPA, C20:5 n-3) and docosahexaenoic acid (DHA, C22:6 n-3).[1] These are called omega-3 or n-3 fatty acids because the position of the first double bond in the carbon chain is after the third carbon. (In EPA, the 'C20' indicates that there are 20 carbon atoms in the chain, and the '5' indicates that there are five double bonds.)

Seabirds obtain these so-called 'marine' fatty acids by eating krill, which are small sea-dwelling crustaceans. In the same way, tuatara can obtain these fatty acids by eating seabirds. Linda Cartland-Shaw, Murray Skeaff, Nicola Grimmond and I inferred this in part from the way that levels of EPA and DHA, expressed as a percentage of total fatty acids, vary in the plasma of wild tuatara. On Stephens Island, plasma levels of EPA and DHA are especially high in adult male tuatara,[2] which also appear, from their behaviour and from analyses of scats,[3] to be the most frequent predators of seabirds. Plasma levels of EPA and DHA also peak in summer, the time of year when seabird eggs and chicks are most available.[4]

Tuatara held in New Zealand zoos are fed diets as varied as possible, but typically have no access to seabirds (they are also fed insects different from those consumed in the wild). Not surprisingly, captive tuatara also have lower percentages of EPA and DHA in their plasma lipids than wild tuatara on Stephens Island, and a lower percentage of long-chain n-3 polyunsaturated fatty acids in their plasma overall.[5] Captive tuatara also differ from their wild counterparts on Stephens Island in having higher plasma concentrations of cholesterol and triacylglycerols,[6] again probably influenced by differences in diets.

To see whether the same patterns held up for

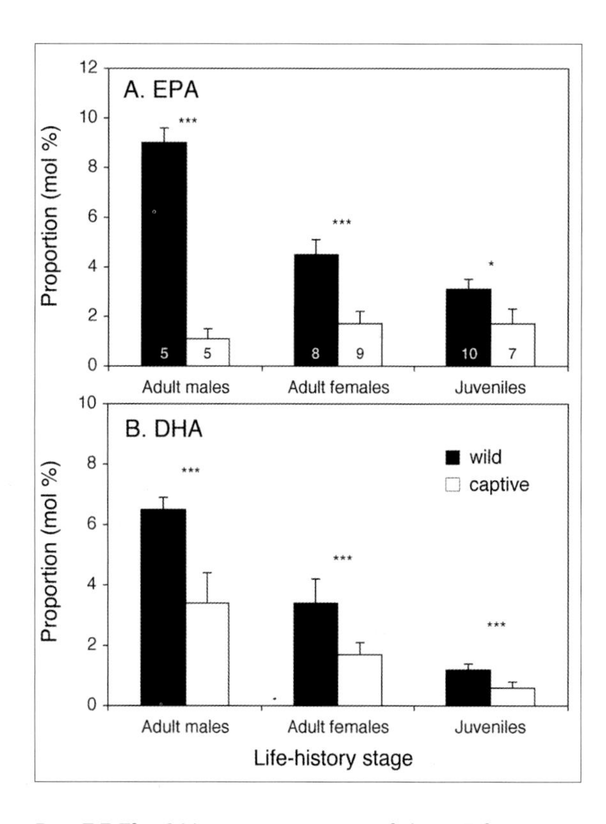

Box 7.3 Fig. 1 Mean percentages of the n-3 fatty acids eicosapentaenoic acid or EPA (A) and docosahexaenoic acid or DHA (B) in plasma phospholipids of tuatara in February (summer). In all comparisons, values (calculated as a percentage of total fatty acids) were significantly higher in wild tuatara on Stephens Island than in captive tuatara of Stephens Island origin held in zoos, museums and wildlife parks. Among wild tuatara, values were also higher in adult males than in females or juveniles. Sample sizes are given at base of bars in A. Vertical lines above the mean indicate the standard error, and asterisks indicate a statistically significant difference (* = $p < 0.05$; *** = $p < 0.001$).

Drawn from data in Cartland-Shaw et al. 1998.

fatty acids in tuatara on northern islands, Tracy Blair, Murray Skeaff and I compared percentages of EPA and DHA in plasma between two populations that differed in the abundance of seabirds (and, inevitably, in the presence of Pacific rats, *Rattus exulans*).[7] During late summer, plasma levels of EPA and DHA were higher in tuatara on rat-free Green Island (where seabirds are abundant) than on rat-inhabited Coppermine Island (where seabirds were infrequent). This is consistent with the hypothesis that seabirds are a source of EPA and DHA in the plasma. Plasma levels of DHA were also higher in adult male tuatara than in females, again fitting predictions about relative levels of seabird consumption.[8] In addition, tuatara on Green Island were feeding on intertidal crabs, and this may have contributed to differences in fatty acid composition.

Another way of obtaining information about marine food sources in the diet is to examine the ratios of stable carbon isotopes (the relative abundance of ^{13}C and ^{12}C, two non-radioactive forms of carbon) in animal tissues. Working with Graeme Lyons, we confirmed that muscle of fairy prions is enriched in ^{13}C compared with the wild insects eaten by tuatara on Stephens Island, providing a marker that can be detected in a tuatara's tissues. Blood cells of adult males on Stephens Island are enriched in ^{13}C compared with those of adult females and juveniles, again suggesting that adult males consume more seabirds. On average, about 45% of carbon in tissue of wild adult male tuatara in February appears to be derived from fairy prions, although there is much variation among animals.[9]

Thus, the composition of plasma lipids and fatty acids in tuatara varies not only with life-history stage and season, but also among wild populations and between wild tuatara and those in captivity. These differences probably reflect differences in the proportion of seabirds, as well as other foods, in the diet. What are the consequences of this variation in nutrition?

In other animals that have been studied more closely, long-chain polyunsaturated fatty acids such as EPA and DHA have important consequences for health. In mammals, they bring cardiovascular benefits by helping to reduce plasma concentrations of cholesterol and/or triacylglycerol, and in embryos, they are especially important for neural development; in fish, they can help improve growth.[10] We investigated the possibility of benefits in captive

Box 7.3 Fig. 2 Among tuatara, it is adult males that are most often seen preying on seabird chicks – such as this male that has dragged a chick from its nest chamber by day on Stephens Island. Seabirds provide tuatara with a rich source of EPA and DHA, which may provide health benefits although further study is needed. The red marks around the tuatara's eyes and neck are chigger mites.

Photo: Clare Allen.

juvenile tuatara (of Stephens Island stock) originating from New Zealand zoos. The young tuatara were fed a diet of insects and insect larvae, supplemented with either fish oil (containing EPA and DHA) or olive oil (as a control). After 2 years, the composition of EPA and DHA had risen markedly in the plasma of the tuatara receiving the fish-oil supplement, but no differences were detected in growth, resting metabolic rate or food conversion efficiency between the two groups of tuatara. Plasma concentrations of cholesterol and triacylglycerol fell in both groups, probably because total food intake was restricted.[11]

Thus, the jury is out on the importance of long-chain n-3 fatty acids in the diet of tuatara. Unfortunately, the delayed maturity and long life span of tuatara mean that decades of study would be required to fully understand the significance of these fatty acids and other dietary components for growth, reproduction (including quality of sperm and eggs) and survival.[12] In the meantime, it is possible to provide tuatara in captivity with a source of marine fatty acids by offering them fish-based cat food as part of a varied diet – in our laboratory colony, this is readily eaten.

Tuatara enter the nest chambers of fairy prions on Stephens Island and consume eggs. They have been observed lapping yolk from eggs that were perhaps accidentally broken,[104] but the complete disappearance of eggs from burrows and the presence of eggshell in scats argues for deliberate consumption of entire eggs.[105] Chicks may be captured either in the burrow or above ground in late summer during fledging. Although headless prion chicks are common on the forest floor on Stephens Island during late summer, there is no convincing evidence from stomach or scat contents of tuatara that heads are preferentially eaten.[106] Juvenile tuatara are probably unable to kill fledgling prions themselves, but they (and adult tuatara) have been seen feeding on

Fig. 7.20 A male tuatara feeding on a fledgling fairy prion on Stephens Island. Chicks and fledglings are caught either in the burrow or on the forest floor. The bird is seized (often by the head, but sometimes by a wing, foot or other body part) and held in the jaws until it ceases struggling and dies.

Photo: Alison Cree.

Fig. 7.21 Juvenile tuatara up to at least 120 mm in snout–vent length on Stephens Island are vulnerable to cannibalism by adult males – the probable cause of this individual's demise.

Photo: Alison Cree.

carcasses of chicks that had been dead at least 24 hours; such scavenging could account for the presence of seabird remains in the scats or stomach contents of juvenile tuatara as small as 124 millimetres snout–vent length.[107] Adult prions are evidently too pugnacious to be often killed by tuatara.[108]

Other vertebrates eaten by tuatara on Stephens Island include skinks, geckos and tuatara.[109] Although geckos and diurnal skinks seem to be infrequently caught in the wild, they fall easy prey to tuatara when incapacitated or confined. This vulnerability was turned to gruesome advantage by collectors of tuatara in the 1800s: in one instance, 26 geckos were placed in the same container as several (possibly seven) tuatara from Karewa Island to provide them with food; only two geckos had survived when the container arrived in Wellington.[110] A threatened amphibian, Hamilton's frog (*Leiopelma hamiltoni*), fell prey to wild tuatara in the past,[111] but tuatara have now been excluded by fencing from the small area of frog habitat (the 'frog bank') on the top of Stephens Island.

Like many other species of reptiles,[112] tuatara on Stephens Island are also cannibalistic (Figure 7.21). Juvenile tuatara are cryptic in colour and differ in habitat from adults, which led Walls to suggest that cannibalism is infrequent.[113] However, recent observations suggest that cannibalism in tuatara may be more common than once thought. In a study by Mary McIntyre, movements of small juveniles were tracked using cotton spools. The thread from one small juvenile led under a rock, straight into the mouth of a large adult male tuatara.[114] In another instance, the remains of a juvenile tuatara less than 100 millimetres long were found in a scat.[115] Other cases involved larger victims. James Fraser, a student of mine, once saw a large male tuatara dragging a juvenile tuatara, partly eaten, into its burrow (from the recovered hindquarters, we estimated snout–vent length as about 120 millimetres).[116] On another occasion, I disturbed a large male with a juvenile of about 120 millimetres snout–vent length in its mouth; only a few days earlier I found a dead juvenile of similar size, uneaten but with crush wounds indicating an attack by tuatara (Figure 7.21). Overall, adult males appear the most frequent perpetrators, with most instances occurring in late spring and summer (November–February) when activity of tuatara is generally high and predation on seabirds also occurs. Adult male tuatara have elevated plasma concentrations of testosterone at this time,[117] and perhaps this hormone stimulates heightened aggression, not just towards other adult males at territorial boundaries[118] but also towards any animals that are not adult female tuatara.

Soil and plant material (including leaves, flowers, seeds, twigs and grasses) is almost universally present in scats of tuatara on Stephens Island.[119] Collectively, this material accounted on average for up to 63% by volume of stomach contents in some samples.[120] Although soil and plant material is probably ingested accidentally during capture of ground-dwelling prey,[121] deliberate ingestion has not been discounted.[122] Intriguingly, a lighthouse keeper on Stephens Island reported in about 1959 that tuatara were 'very partial to his tomatoes, which they strip from his plants.'[123] Although this report initially surprises, it seems consistent with a recent inference that tuatara deliberately consume the aromatic orange fruit spikes that fall to the ground from kawakawa plants (*Macropiper excelsum*).[124]

Diets of tuatara on other islands have not been as closely examined. Tree wētā are scarce or absent from North Brother Island and probably several other small islands without tall forest,[125] and hence are assumed to be absent from the diets of tuatara there. As on Stephens Island, tuatara on North Brother Island probably consume the contents of fairy prion eggs left unattended.[126] Cannibalism has been reported for tuatara from North Brother Island in captivity: an adult killed and partly consumed a 'half-grown' specimen.[127]

Diets of wild tuatara on northern islands include a high proportion of invertebrates, especially beetles.[128] On rat-free Atiu or Middle Island (in the Mercury Islands), 90.6% of a sample of 117 scats contained the remains of tenebrionid beetles.[129] On Lady Alice Island (in the Hen and Chickens Islands), during the time that Pacific rats were present, the stomach contents of adult tuatara (sex not reported) were examined from forested habitats.[130] Samples collected from spring, summer and autumn yielded only invertebrate prey; beetles, insect larvae, arachnids and wētā were the most frequently occurring prey items. Although it was suggested that there was little difference in the diets of tuatara between Lady Alice Island and rodent-free Stephens Island,[131] this seems somewhat at odds with the absence of tenebrionid beetles, seabirds and reptiles in these samples.

Small to medium-sized seabirds are consumed by tuatara on many and probably all northern islands where they are abundant, although the species differ from those on Stephens Island. Predation on birds (chicks or adults) was evident in 19.7% of scats from tuatara on Atiu, and the remains of eggs were also evident in 4.3% (times of year not stated).[132] On Lady Alice and Coppermine islands, after the removal of Pacific rats, predation on seabird chicks (Pycroft's petrel, *Pterodroma pycrofti*, and/or little shearwater, *Puffinus assimilis haurakiensis*) by tuatara was suspected, but not at a level that significantly affected the breeding success of the seabirds.[133] An introduced passerine (a chaffinch, *Fringilla coelebs*) was captured by a tuatara on the Aldermen Islands.[134]

Remains of reptiles, including skinks, geckos and one small tuatara, were found in 14.5% of scats on Atiu.[135] Based on these results, plus observations of attempted predation, bite scars and missing feet, Ian Southey considered in 1985 that three species of nocturnal skinks were preyed on heavily by tuatara (these vulnerable species of *Oligosoma*, then described as *Cyclodina* spp., are not present on Stephens Island). Partly stunned skinks (probably *Oligosoma smithi*) were readily eaten when thrown to tuatara on Green Island.[136] A lighthouse keeper reported in 1919 that tuatara hunted rats on Cuvier Island (Repanga Island).[137] Fur of the Pacific rat was subsequently found in a tuatara scat on this island[138] and on Whatupuke Island.[139]

On small northern islands, tuatara have greater opportunity to prey on marine foods than on Stephens Island, which is largely cliff-bound. For example, tuatara forage nocturnally on the beach below high-tide level on Green Island[140] and Moutoki Island,[141] and crabs were present in scats on Green[142] and Atiu.[143] On Moutoki, tuatara were seen feeding in forest on dead fish, presumably carrion from a colony of black-backed gulls (karoro, *Larus dominicanus*).[144] Tuatara on the Chickens Islands feed on remnants of fish and crustaceans brought into the burrows by petrels.[145] Although tuatara feed with relish on small fish in captivity,[146] an early suggestion that they devour fish that they catch themselves in the wild[147] seems doubtful. As on Stephens Island, scats from northern tuatara include plant material: 'fruits' were present in 21.4% of scats on Atiu, for instance.[148]

Prey capture and ingestion

> 'Although the reptile appears to be cumbersome, its movements are of a lightning-like quickness, so needful to it in obtaining food.'[149]

In feeding seasonally on seabirds, tuatara show dietary similarities with some large snakes, such as the gull-feeding tiger snakes (*Notechis scutatus*) of Carnac Island, Australia.[150] However, the manner of prey capture and ingestion in tuatara is very different. Tuatara do not tongue-flick, envenomate their prey, or ingest and swallow prey items whole through highly kinetic skulls. Nor does prey detection rely heavily on chemoreception via the olfactory system or vomeronasal (Jacobson's) organ, a sensory structure in the roof of the mouth. In fact, tuatara have a curious and distinctive manner of food-handling seen in no other living vertebrate.[151]

Based on observations of tuatara on or from Stephens Island, and supplemented by reports for tuatara from North Brother and northern islands, a typical feeding event on small moving prey proceeds as follows.[152] Once prey has been detected (as discussed below) the head is turned, followed if necessary by an approach at a fast walk. The tuatara's head is cocked (tilted to one side and downwards toward the prey) and the prey may be watched intently for some seconds. In juveniles in our laboratory colony, a hind leg will sometimes be repeatedly twitched (the foot being slightly raised and lowered) during this time. Eventually, the tuatara lunges rapidly at the prey, arching the neck so as to approach the prey from above. The fleshy, un-forked tongue, which has numerous taste buds,[153] is partly extended to assist with the capture of small prey items (Figure 7.22).[154] Larger items such as tree wētā (or mice in captivity) are impaled with a swift grip of the anterior incisor-like teeth.[155] The prey item is manipulated in the mouth with the aid of the tongue or by vigorously shaking the head from side to side.[156]

Tuatara appear somewhat clumsy and inaccurate in their initial lunge: many prey items escape, and soil and vegetation are often grabbed in the attempt. Once an insect is caught, however, its fate is more or less certain. Strong jaw muscles confer a vice-like grip, and the hapless prey is slowly macerated in a 'crush and shear' sequence unique among living vertebrates. First, the tooth row of the lower jaw closes on the prey, often with an audible crunch as the exoskeleton is perforated. The lower jaw then moves forward in the groove between the two tooth rows in the upper jaw, producing a shearing force that can rip an insect exoskeleton apart or decapitate small vertebrates.[157] The forward movement of the lower jaw while the mouth is closed forms the characteristic propalinal (more specifically, proal) jaw action of advanced sphenodontians.[158] The lower jaw then drops and is pulled back up again to begin the next stroke.

Fig. 7.22 A tuatara on Stephens Island with its tongue extended. Though bulky, and attached to the floor of the mouth for much of its length, the tongue is free on its lateral edges and tip. The velvety surface has long, filamentous papillae, giving the tongue 'the appearance of a deep-pile carpet', in the words of Gorniak et al. (1982: 346). The presence of an un-forked, extensible tongue used in prey capture is considered an ancestral characteristic among lepidosaurian reptiles.

Photo: Michael Schneider.

During several minutes of intermittent mastication, in the course of which the prey may be moved to the opposite side of the mouth, bits of prey such as the legs of wētā (or the head or limbs of seabirds) may fall off outside the mouth. Bouts of propalinal shearing are also interspersed with periods of conspicuous resting and breathing. In an adult tuatara feeding on a medium-sized cockroach in captivity, about 35–45 jaw strokes in total were interspersed with about four periods of breathing, each period involving 5–13 breaths.[159] Eventually, the mushy, saliva-covered remainder of the crushed prey item is swallowed. One wild adult male spent about 2 hours ingesting a house sparrow.[160]

As for the cues that tuatara use to detect prey, movement is undoubtedly an important stimulus.[161] A stationary prey item is often overlooked, even if literally underfoot such as a Hamilton's frog, in one instance.[162] On the other hand, tuatara can easily be tempted to bite at inanimate objects, such as sticks or pieces of cardboard moved about by humans.[163] Feeding attempts are possible at very low light intensity (about one-fiftieth that of full moonlight) but cease in total darkness.[164] Nonetheless, tuatara are able to consume seabird eggs, scavenge on chick carcasses, retrieve pieces of food that have dropped from the mouth, and feed on dead insects, pieces of meat and fish-based cat food in captivity,[165] suggesting some chemosensory detection.[166] Recent tests have provided support. In our laboratory, juvenile tuatara spent more time near cotton swabs containing the odour of mealworms (insect prey) than swabs presented simultaneously with the odour of celery (pungency control) or water. Swabs with the odour of mealworms were also bitten frequently (sometimes for over 20 minutes), whereas swabs with celery odour or water were never bitten.[167] Initial detection of some prey by sound may also occur.[168]

During the day, wild adults take advantage of prey that passes in front of the burrow entrance or near to where they are basking.[169] On the basis of such passive behaviour, tuatara have sometimes been categorised as 'sit-and-wait' or 'ambush' predators.[170] However, adults can travel up to 63 metres from their burrows[171] and often appear at night to be hunting or actively foraging. On Stephens Island, tuatara enter seabird burrows to consume eggs and chicks,[172] and climb trees to inspect hollows occupied by tree wētā.[173] On Cuvier Island in 1919, a lighthouse keeper provided a dramatic description of tuatara hunting rats:

> I like to watch this animal chasing a rat; the latter which are plentiful & live in the long grass. Learned people who have studied the Tuatara Lizard's [*sic*] say they are very slow in their movements. But I cannot agree on that point for I have stood and watched one chasing a rat over my garden and [it] has caught it within 12 yards. How they do rattle when they are moving in this way ...[174]

These observations reveal that the foraging strategy of adult tuatara (and perhaps adult males in particular) combines elements of both ambush capture and active foraging.[175]

Given the profound differences in skull structure that have evolved between rhynchocephalians and lizards, interest has grown in comparing feeding efficiency between tuatara and contemporary lizards. In our laboratory, Anne Besson observed that juvenile tuatara were at least as quick to capture mealworm prey at 20°C and 12°C as were two species of geckos and a skink. Prey-handling time (the time between first seizure of the prey to jaw closure following swallowing) was also similar between tuatara, the geckos and the skink (and declined in all species between 20°C and 5°C).[176] However, in a study elsewhere, the duration of the transport stage prior to swallowing was longer for tuatara eating mealworms than for similar-sized agamid lizards.[177] Bite force also appears lower in tuatara than in agamids, a difference attributed in part to limitations in jaw musculature,[178] which could be a constraint of having large eyes for nocturnal activity.[179] However, it is unclear whether maximum bite force has yet been measured in tuatara; more comparisons are needed of feeding efficiency between tuatara and lizards at the same temperature with varied items of prey.

Food intake and gut passage time

Wild tuatara feed throughout the year, although the rate of food consumption is probably low, especially during winter. On both Stephens and Lady Alice Islands, the proportion of tuatara that yielded food items when stomach-flushed varied among seasons or habitats, but overall appeared to be about 60%.[180] Thus, about 40% had apparently empty stomachs. This figure is within the range for nocturnal lizards, which 'run on empty' more often than diurnal lizards.[181]

On Stephens Island the proportion of tuatara with food in their stomachs varied from a high of 75% in summer to a low of 46% in winter, which matches trends in insect availability and tuatara activity.[182] In all four seasons of the year, the mean volume of food retrieved by stomach flushing was small (usually $< 1\ cm^3$). However, if tuatara that have recently eaten retreat to their burrows, then estimates of feeding frequency and food volume from tuatara emerged on the forest floor at night could be unrepresentative.[183] The possibility that stomach flushing might fail to dislodge very large prey items also needs investigation.

Observations from stomach flushing and the frequency of scat production in wild tuatara provide estimates of the minimum time for food to pass through sections of the digestive tract.[184] To determine time taken for ingested food to disappear from the stomach, adult males were held in an unheated room at the field station on Stephens Island, fed a darkling beetle and then stomach-flushed at set time periods after ingestion. In November (spring), beetle remains were obtained in stomach flushings up to 48 hours after ingestion, whereas in January (summer), beetle remains were obtained at 20–24 hours but not after longer time periods. Among other adult tuatara held with access to water but not fed, scats were produced up to 7 days after

BOX 7.4

Structure and function of the digestive system

The anatomy and function of the digestive system of tuatara are discussed briefly here because of their relationship with feeding behaviour and diets. The digestive system includes the alimentary tract, which runs from the mouth to the vent, plus associated glandular structures including the liver and pancreas. The digestive anatomy of tuatara has been the subject of numerous studies dating mainly from 1867 until the early 1900s.[1] More recent work includes histology, histochemistry and biochemistry of digestive tissues,[2] and reviews of earlier studies.[3] Sources of specimens include Stephens Island (Takapourewa), North Brother Island, Karewa Island and perhaps others.

The teeth and jaws of tuatara are distinctive and have been described in Chapter 1. In contrast, the soft tissues of the digestive system are broadly similar to those of other reptiles. The oral (buccal) cavity contains several sets of salivary glands,[4] including palatine glands (in the roof of the oral cavity), sublingual glands (on the floor of the mouth), and labial glands (around the edge of the mouth). Squamates have salivary glands in the same locations, whereas turtles and crocodilians lack labial glands.[5] The salivary glands of tuatara are poorly developed compared with those of squamates, being essentially simple crypts within the epithelium. The tongue, which is described in the main text, is more richly endowed with lingual glands than are the tongues of squamates.[6]

Towards the rear of the oral cavity, air enters the trachea through a slit-like glottis lying in the floor of the pharynx. The oesophagus lies dorsal to the trachea and is initially wide, but narrows as it approaches the stomach.[7] The stomach (described variously as spindle-shaped or J-shaped) contains fundic and pyloric glands, as expected for a region involved in protein digestion; the fundic glands most closely resemble those of lizards and turtles.[8] The small intestine is longer, narrower and more convoluted than the large intestine, looping back on itself

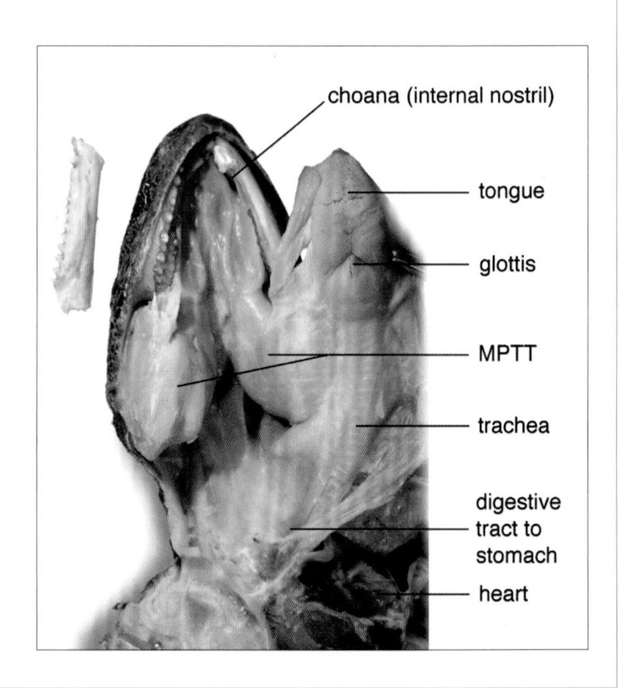

Box 7.4 Fig. 1 Oral cavity of a tuatara from Stephens Island, with the tongue, glottis and trachea held to the side. The ventral skin and part of the lower jaw on the animal's right side (left in photo) have been removed in order to show the distinctive double row of teeth in the upper jaw. Note the large swellings of the M. pterygoideus typicus (MPTT), the largest single muscle of the masticatory muscles. This muscle is active during closing of the mouth, crushing and shearing (Gorniak et al. 1982). Digital montage from a preserved adult female (188 mm snout-vent length, 52 mm head length).

Photo: Alison Cree.

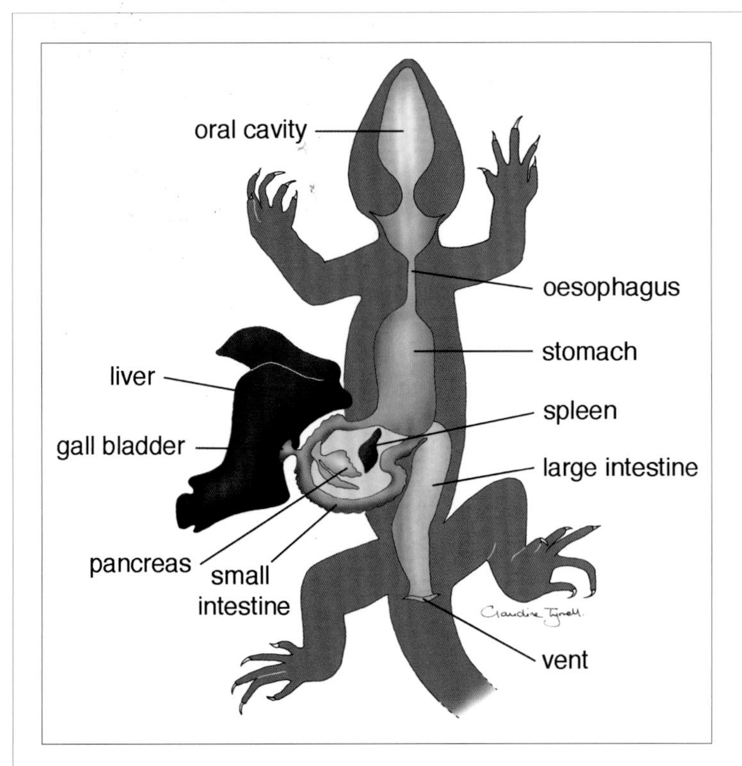

Box 7.4 Fig. 2 Diagram of the digestive system (tongue removed) and nearby structures in the abdominal cavity. The stomach and intestine vary greatly in width depending on the contents.

From an original drawing by Claudine Tyrrell, based on four dissected tuatara from Stephens Island; colouring by Ken Miller.

once or twice.[9] Longitudinal folds line the internal surface of the small intestine,[10] as is typical of a region absorbing foodstuffs. The large intestine or colon is smoother internally and less muscular.[11] The cloaca is divided into three compartments, of which the first, the coprodaeum, stores the faeces before they move through the remaining compartments and exit the vent.

Associated with the alimentary tract are the gall bladder, which stores bile produced by the liver and releases it through the bile duct into the small intestine, and the pancreas, which produces the hormone insulin and secretes digestive fluids into the small intestine. The gall bladder of tuatara is a dark green sac, pea-sized in an adult male, lying close to the transition between the right and left lobes of the liver.[12] The pancreas is pink and diffuse, being divided into dorsal and ventral lobes;[13] histologically, it is similar to that of squamates, containing islets of Langerhans (source of insulin) and exocrine cells in acinar glands.[14]

The liver is the largest abdominal organ in reptiles, playing important roles in nutrient storage as well as in the production of bile. In tuatara, the dark-red liver extends from about the level of the heart into the posterior half of the abdominal cavity. Many have commented on its large size and irregular structure. Albert Günther, in 1867, described the liver as being conspicuously divided into two lateral portions (left and right), each of which is further subdivided 'in a curious and apparently irregular manner'.[15] Similarly, A. K. Newman in 1878 described the liver as 'a most curiously elaborated body'.[16] Nearly 9% of the liver's wet mass is accounted for by lipids, with neutral lipids (mainly triacylglycerols) being major components.[17] A high glycogen content was evident in the hepatocytes (liver cells) of two tuatara examined by Manfred Gabe and Hubert Saint Girons in 1964.

The hepatocytes, like those of other reptiles, produce several types of bile acids. Although tuatara share some bile acids with turtles and crocodilians, only tuatara and lizards produce varanic acid as a common bile acid.[18] Also in common with squamates, the major metabolite of benzoic acid in faeces and urine of tuatara is ornithuric acid, whereas turtles and crocodilians also produce substantial amounts of hippuric acid.[19] Ornithuric acid is probably produced mainly in the liver, although small amounts may be produced by the kidney and other organs.[20]

collection in both summer and winter.[185] These studies suggest that ingested food takes a week or more to pass through the digestive tract of wild adults under cool conditions, but perhaps only a few days under warm conditions. These estimates of gut-passage time are not unusual compared with values ranging from about 1.5–38 days in other reptiles.[186] In recent work in our laboratory, Anne Besson observed that gut passage in juveniles lasted about 8 days at 20°C but apparently ceased at 5°C.[187] We inferred that tuatara might need access to periodic sun for basking during winter to ensure that any food in the digestive tract does not decompose.

Predators

Until the time of human arrival, the predators of adult tuatara were exclusively avian. Many of New Zealand's predatory birds are now extinct and several others no longer overlap in distribution with tuatara. Nonetheless, the evolution of anti-predator strategies in tuatara must be seen in light of these associations. Tuatara tend to spend long periods of time motionless when basking, and often freeze on initial disturbance, strategies that may have been effective for avoiding birds that hunted visually by day.

Among flighted birds that are now extinct, the laughing owl (whēkau, *Sceloglaux albifacies*) was possibly the chief predator of tuatara in pre-human times.[188] Laughing owls consumed tuatara in the small to medium-size range, but not large adults. A smaller, weakly flighted predator, the New Zealand owlet-nightjar (*Aegotheles novaezealandiae*), as well as the omnivorous New Zealand raven (*Corvus moriorum*), may have preyed on small tuatara. Among those that were flightless, adzebills (*Aptornis* spp.) were large enough to have preyed on both juvenile and adult tuatara, probably by digging them from burrows or under logs. The North Island takahē (*Porphyrio mantelli*) and possibly other species assumed to have been primarily herbivorous, such as small moa, may also have eaten small tuatara, at least occasionally.

Among native birds surviving today, few have actually been recorded as predators of tuatara. One that has, on both Cook Strait and northern islands, and probably over a wide range of sizes of tuatara, is the Australasian harrier (kāhu, *Circus approximans*; Figure 7.23).[189] This species may have established in New Zealand only since humans arrived and burnt the forests, creating more open habitat.[190] Australasian harriers frequently caught tuatara on Stephens Island after the initial clearance of forest for lighthouse operations. Many were subsequently shot by lighthouse keepers during the early 1900s.[191]

Other native birds known to prey on tuatara are the sacred kingfisher (kōtare, *Todiramphus sanctus*) and the black-backed gull.[192] Kingfishers, which may have been rare in New Zealand before human settlement,[193] prey only on juvenile tuatara.[194] Black-backed gulls show great cunning, according to a description supplied by a lighthouse keeper on Stephens Island in 1923: 'Sea gulls that are nesting about the Island are destroying a large number of Lizards [*sic*] this year. They actually wait at a hole for hours until a Lizard

Fig. 7.23 The Australasian harrier is today the main predator of adult tuatara on offshore islands.

Photo: Rod Morris.

emerges, grab it and fly to sea and back to where they nest, dropping it every now and then to kill it. Sometimes if the water is rough they loose [*sic*] it and that accounts for the dead Lizards that are picked up on the coast near Kapati [*sic*]'.[195] This behaviour towards tuatara has not subsequently been reported by tuatara biologists.[196]

Several other species of native birds that feed on New Zealand lizards would prey on small tuatara, given the opportunity. These include the New Zealand falcon (kārearea, *Falco novaeseelandiae*), which visits some tuatara islands at least occasionally. Remains of bones in mainland deposits suggest that falcons preyed on tuatara in pre-human times.[197] Other native predators of lizards that are known on at least some offshore islands with tuatara include the banded rail (mohopereru, *Gallirallus philippensis*), morepork (ruru, *Ninox novaeseelandiae*), red-billed gull (tarāpunga, *Larus novaehollandiae*), and pūkeko (swamphen, *Porphyrio porphyrio*).[198] Two further species that eat lizards, the South Island takahē (*Porphyrio hochstetteri*) and the weka (*Gallirallus australis*),[199] no longer overlap naturally in distribution with tuatara. However, South Island takahē are present on Tiritiri Matangi Island (to which tuatara have recently been reintroduced), and weka are present at sites to which reintroduction may occur in future, so predation by these species may soon resume.

Fig. 7.24 The giant centipede (*Cormocephalus rubriceps*) is equipped with poison fangs and reaches up to 250 mm in length on the Poor Knights Islands. The photographer witnessed the prey shown here (a member of the common gecko species complex), reduced to a skeleton within 3 hours. Although predation on tuatara has not been reported, juveniles must surely be vulnerable to this awesome invertebrate.

Photo: Rod Morris.

During the 1980s, the Antipodes Island parakeet (*Cyanoramphus unicolor*) was unwisely introduced for a short time to Stephens Island. On three subsequent occasions parakeets were seen playing with small, dead, juvenile tuatara, tearing them to pieces;[200] the parakeets were soon afterwards removed from the island. Other species of *Cyanoramphus* (kākāriki) that naturally co-occur with tuatara, as well as saddlebacks (tīeke, *Philesturnus* spp.) and kiwi (*Apteryx* spp.), might also prey at least occasionally on small tuatara.

Other potential native predators of small tuatara, known to eat lizards, include reptiles, invertebrates and even fish. This list includes cannibalistic adult tuatara (see section on diets above), as well as large geckos and skinks, the giant centipede (weri, *Cormocephalus rubriceps*; Figure 7.24), large carabid beetles and large spiders[201] (beetle larvae are also predators on, or scavengers of, tuatara eggs[202]). Lizards have been discovered in the digestive tracts of marine fish,[203] so a small tuatara unlucky enough to be washed to sea might also be so eaten.

Among introduced mammals, cats (*Felis catus*) and pigs (*Sus scrofa*) once preyed on tuatara (and house mice, *Mus musculus*, possibly once did as well), but none of these species now remains on any tuatara island. A wealth of evidence implicates rats (*Rattus* spp.) as significant predators of tuatara and their eggs, but most populations of rats on islands with tuatara have now been eradicated (see Chapters 6 and 10). Introduced birds known to eat lizards, and present at least occasionally on offshore islands with tuatara, include the blackbird (*Turdus merula*), European starling (*Sturnus vulgaris)*, Indian or common myna (*Acridotheres tristis*), little owl (*Athene noctua*) and Australian magpie (*Gymnorhina tibicen*).[204] Blackbirds were reported by a lighthouse keeper on Stephens Island to dig for and eat eggs of tuatara that were laid in the keeper's garden.[205]

In summary, most native birds from the pre-human period that might have preyed on medium-sized or large tuatara are now extinct (e.g. laughing owls and adzebills) or much reduced in distribution and hence impacts (e.g. New Zealand falcon). Adult tuatara in open habitats are vulnerable by day to Australasian harriers, but those in forest or that have emerged at night are largely immune from predation. Tuatara are today considered as top predators in their island ecosystems.

Small tuatara were exposed to a much wider range of predatory birds, some of which are now extinct (e.g. New Zealand owlet-nightjar), though many survive today (e.g. kingfishers, red-billed gulls and moreporks). Other potential predators of small tuatara include large native reptiles and invertebrates, as well as several species of introduced birds; however, direct evidence is lacking. The potential impacts of introduced predatory mammals (including rodents that might reinvade) are the most significant concern to conservation.

Parasites

Wild tuatara are hosts to about eight species of parasites.[206] These include three ectoparasites (a tick and two species of mites) and about five species of endoparasites (one blood parasite and probably four species of intestinal worms). Protozoan parasites have been detected in the skeletal muscle of captive tuatara and, in two cases, resulted in death by microsporidiosis (*Plistophora* sp. and possibly *Microsporidium* sp.),[207] but these parasites have not been reported in wild tuatara. The protozoan *Trichomonas* spp. was recently detected in faecal samples from captive-reared juvenile tuatara but, surprisingly, neither these tuatara nor a sample of wild individuals showed any evidence of the bacteria *Salmonella* spp.[208] Species of *Salmonella* are common in the intestinal tracts of reptiles overseas and lizards in New Zealand, and have zoonotic potential (the ability to transfer disease to humans).

The tuatara tick, *Amblyomma* (previously *Aponomma*) *sphenodonti*,[209] was first described in 1943 from tuatara on Stephens Island.[210] This is the largest ectoparasite of tuatara, reaching up to about 6 millimetres in diameter.[211] In colour (olive-brown) and shape (resembling subcutaneous fat nodules), it blends well with the skin, so is easily missed at first glance (Figures 7.25 and 7.26). Common attachment sites include the midline of the back, and the flanks, hind limbs (especially in or around the folds of skin where the limbs meet the body) and tail.[212]

Fig. 7.25 A tuatara tick (*Amblyomma sphenodonti*) attached to the back of a tuatara on Stephens Island.

Photo: Michael Schneider.

The tuatara tick is endemic to New Zealand (related species are found on reptiles in Australia). It is more restricted in distribution than its host, being absent from populations on the Poor Knights Islands,[213] Hen and Chickens Islands[214] and islands in the Bay of Plenty.[215] This pattern of distribution suggests that the ticks are more likely to die out when tuatara are at low density.[216] The tuatara tick shows the same geographic separation into genetically distinct groups from northern New Zealand and Cook Strait as does its host,[217] and, although not considered as a threatened species in the Department of Conservation's Threat Classification System, it is considered 'at risk'.[218] Despite its name, the tick is not entirely exclusive to tuatara. One was found attached to a Duvaucel's gecko (*Hoplodactylus duvaucelii*) on Middle Trio Island, and a specimen of what was probably the same species was found attached to a marbled skink (*Oligosoma alani*), on Atiu.[219] An unlucky biologist had a tick attached to his or her skin for several hours after coming off Stephens Island.[220]

Tick loads on individual tuatara are highly variable, ranging from 0 to 37 per tuatara on Atiu Island,[221] and from 0 to 584 on Stephens Island.[222] Almost all tuatara sampled on Stephens had ticks, with the mean load per tuatara being higher in spring and autumn than in summer,[223] and higher in males than in females.[224] Three life-history stages of ticks attach: larvae, nymphs and adults. Each stage detaches before infesting the next host (which may be the same tuatara), and female ticks detach to lay eggs. The fate of a tick when a tuatara moults is not known. Observations on captive ticks suggest a generation time of 2–3 years (detached adult ticks can survive more than 200 days).[225]

The tuatara tick feeds on blood and is the probable vector for the haemogregarine blood parasite described below,[226] but its overall impacts on tuatara are unclear. In one study on Stephens Island, the mean index of body condition (mass adjusted for snout–vent length) of tuatara with six or more ticks was not significantly different from that for tuatara with five or fewer ticks, but those tuatara with more ticks had a higher count of total leucocytes (number of white blood cells per millilitre of blood).[227] In a later study, tuatara sampled in spring (September) showed a negative relationship between tick load and body-condition index: those tuatara with more ticks weighed less for their length, and these tuatara also tended to decline more in body condition over the next 2 months.[228] On Atiu Island, on the other hand, the body-condition index was significantly higher in tuatara with ticks than in those without.[229] In the Australian sleepy lizard *Tiliqua rugosa*, infestation with a different species of tick caused reductions in home range, daily movements, sprint speed and endurance,[230] but the possibility of such effects in tuatara has yet to be explored.

Two species of red chigger mites (*Neotrombicula naultini* and *N. sphenodonti*) have been described from tuatara on Stephens Island.[231] These mites also parasitise some New Zealand lizards. Red chigger mites attach to the skin, where they feed on lymph fluids and digested cells, but only the larval stage is parasitic.[232] In heavy infestations, the mites form large, vivid clusters of orange-red spots in or around the neck, axilla (armpit) and groin of tuatara, especially in summer and early autumn (Figure 7.13).[233] By extrapolating from the number of mites covering a specified area to the total infested area of the body surface, Stephanie Godfey and colleagues estimated that mite loads could reach 6000 per tuatara;[234] however, mite loads in March (the peak infestation period) were not correlated with body condition of tuatara.[235] The researchers also noted that mite loads did not differ significantly between tuatara living in the pasture and the forest, whereas tick loads were significantly

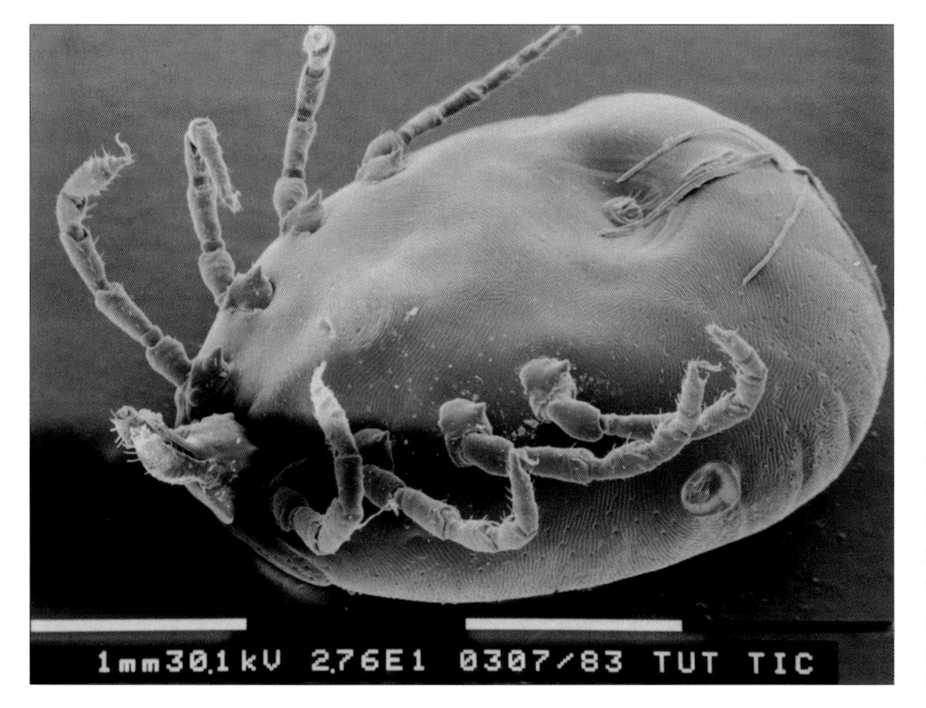

Fig. 7.26 A scanning electron micrograph of the ventral surface of a tuatara tick. The tick feeds on blood using its mouthparts (bottom left), and in so doing may transmit a blood parasite to tuatara. Scale bars = 1 mm.

Photo: Alison Cree and staff of the Electron Microscopy Unit, Victoria University of Wellington.

higher for pasture-dwelling tuatara,[236] a pattern that could have a variety of possible causes.

The blood parasite *Hepatozoon* (previously *Haemogregarina*) *tuatarae* has been detected in 19–23% of tuatara from Stephens Island[237] and in one of four tuatara from Middle Trio Island.[238] However, the infections were very light and may have passed undetected in some individuals. This species parasitises the erythrocytes (red blood cells) and is apparently unique to tuatara. Different species of blood parasites occur in New Zealand geckos and skinks.[239] Although infection with haemogregarines is associated in lizards overseas with reductions in haemoglobin concentration, resting oxygen consumption and mean sprint speed,[240] effects of blood parasites on tuatara have not been examined. There is some indication that smaller tuatara are more likely to be infected, perhaps because they are more likely to ingest the ticks that are the probable vectors.[241]

Tuatara are the sole known hosts of the intestinal nematode *Hatterianema hollandei* (the source population of tuatara is not stated).[242] This nematode and a related nematode in kiwi are the only members of the Kiwinematidae, a family that may have a Gondwanan origin.[243] The digenean trematode (fluke) *Dolichosaccus* (*Lecithopyge*) *leiolopismae* occurs in the small intestine of tuatara from Stephens Island, as well as in several species of New Zealand skinks, and a related species is found in endemic frogs.[244]

Other internal parasites of tuatara on Stephens Island include unidentified hookworms and threadworms, recovered from stomach flushings and scats.[245] Among 222 tuatara that were stomach-flushed for dietary studies, 67% yielded hookworms or threadworms. Prevalence (frequency of occurrence) came close to varying significantly throughout the four seasons of the year,[246] but the seasonal patterns were not reported.

Injuries and abnormalities

Wild tuatara appear remarkably hardy. Missing toes and other injuries are not infrequent, but serious incapacitation is rare. Tail loss is common in some populations and not an abnormality in itself, although the process of regeneration sometimes results in unusually bulbous or forked tails (Figure 7.27). Like many lizards, tuatara show morphological asymmetry in their toes (the fourth toe on the right side has more subdigital lamellae than on the left side, at least in museum specimens), but there is no evidence that 'right-footed' and 'left-footed' tuatara differ in their frequency of tail loss.[247] Many male tuatara have scarred heads and sometimes jaw wounds as well, some of which result from fighting with other males.[248] Occasional tuatara are seen with apparently broken and healed ribs, or with a missing foot or even a missing lower limb.[249] Eye abnormalities, including cataracts, missing eyes or sunken eyes with closed eyelids, have been observed in tuatara on several islands, including Stephens, Kawhitu or Stanley,[250] and Motunau (Plate).[251] On Lady Alice Island about 19% of adult females had eye abnormalities,[252] and on Coppermine Island 6%

Fig. 7.27 A tuatara on Stephens Island with a bulbous forked tail. Unusually, both sides of the fork appear to contain regenerated tissue.

Photo: Michael Schneider, reproduced from Daugherty, C., Cree, A. and Schneider, M. 1990. Tuatara: a survivor from the dinosaur age. *New Zealand Geographic* 6: 66–86.

of tuatara appeared to be blind in one eye; at least some of these injuries are attributed to aggression from seabirds.[253] An open sore with a fungal infection was observed in a tuatara on Green Island.[254] A tuatara from Hauturu with a drooping head was diagnosed with a fungal infection of the bone; by the time of euthanasia, the fungus had spread to the brain, liver and spleen.[255] In general, however, infections seem uncommon.[256] On Tawhiti Rahi Island in 1989, I observed three adult tuatara with grossly swollen limbs and/or feet (Figures 7.28 and 7.29). This undiagnosed abnormality is suggestive of articular gout[257] – a condition in which crystals of uric acid accumulate in and around the joints, possibly in response to high protein levels in the diet and/or dehydration. Symptoms have not been reported on other islands.

Fig. 7.28 An adult female tuatara with swollen limbs and feet on Tawhiti Rahi Island. All four limbs were affected by this condition (possibly articular gout).

Fig. 7.29 (below) Underside of the right rear foot of the tuatara at left

Photos: Alison Cree.

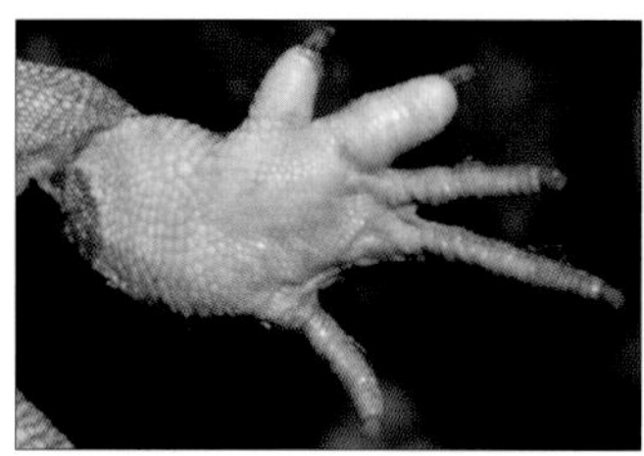

Concluding remarks and ideas for future research

Far from being animals that 'do nothing', tuatara simply go about the business of living on a timescale slower than most humans find visually interesting. In the right circumstances, tuatara are extremely successful animals, attaining high population densities and the highest biomass reported for any lepidosaurian reptile.

Conditions favouring high densities of tuatara include the presence of burrowing seabirds and the absence of rats and other introduced mammals. In addition, tuatara live at low air (and hence body) temperatures; thus, their metabolic demands are low and high densities can be supported by the available prey base. Indeed, tuatara may be considered the ultimate exemplar of the 'low-energy lifestyle' that characterises reptiles and other ectotherms in comparison with endothermic mammals and birds.[258]

The behaviour of tuatara is more complex than is sometimes appreciated. Other reptiles have at times been considered in terms of simple dichotomies, such as nocturnal versus diurnal behaviour, or ambush versus active foraging. These dichotomies are really ends of a spectrum, as is well illustrated by tuatara, which show elements of both ends. In addition, some aspects of behaviour or performance attributed in tuatara to phylogenetic inertia, such as low running speed[259] and a tendency to freeze when disturbed, need to be reconsidered. Such behaviours may reflect specialisations, such as for burrowing, or for avoidance of diurnal predatory birds.

Among topics deserving further research, the ecology of juvenile tuatara, including hatchlings, stands out (Figure 7.30). Juveniles are more diurnal and secretive in their habits, more restricted in dietary breadth and more vulnerable to predators (including cannibalistic adults) than are larger tuatara. Adult males are the largest, least cryptic and probably the most widely foraging of all tuatara, so it is not surprising that much of our ecological knowledge stems from observations of this group. In contrast, we know rather little of the ecology of juveniles and of adult females and small adult males, including how the lifestyles of these groups might be constrained by those of the largest males.

The significance of chemosensory detection in tuatara, and whether olfaction (in its strict sense), vomerolfaction or both are involved, deserve further study. Tuatara do not tongue-flick and their vomeronasal organs lack the vomeronasal duct that in squamates provides a direct opening to the mouth, but this does not mean vomerolfaction does not occur.[260] More information on this and on foraging mode in tuatara would be helpful to our understanding of the evolution of lepidosaurs. Current knowledge suggests that early lepidosaurs were ambush foragers that did not tongue-flick or use chemosensory detection by vomeronasal organs to detect prey, relying instead on tongue prehension (rather than the jaws) as the primary means of prey capture.[261] Although tuatara show (retain?) elements of these characteristics, they appear not to depend on them exclusively.

Fig. 7.30 A juvenile tuatara, perhaps about 2–3 years old, found under a wooden railway sleeper in pasture on Stephens Island. Juveniles are more secretive than adults and little is known about their ecology.

Photo: Alison Cree.

Many recent ecological studies have explored variation in a simple index of body condition (mass relative to snout–vent length) among tuatara on different islands or over time; for example, in relation to the presence/absence of rats, vegetation changes and parasite load. New techniques are needed to establish how well this index reflects underlying variation in fat and/or body water content. In some cases, it should be biologically possible to examine putative influences on body condition (such as the effects of parasites) using an experimental approach, which would move our understanding beyond inference.

Although not a strictly ecological topic, it is worth noting that nutrition in tuatara is generally poorly understood. Further research into the role of the liver in carbohydrate and lipid metabolism is relevant, given the large size of this organ, the lack of abdominal fat bodies, the ability of tuatara to go for long periods without food and, in females, the extended time over which the yolk-protein precursor is produced at the liver and incorporated into oocytes (vitellogenesis).[262] Biochemical analyses of plasma are available,[263] but these should be interpreted cautiously, as many are likely to stem from captive tuatara.[264] Tuatara in captivity have often been overfed and tend to become obese,[265] and their diets differ from those in the wild.[266] Reptiles in captivity that are overfed and under-exercised are vulnerable to a condition known as fatty liver, or steatosis.[267] This condition may occur in captive tuatara, but specific studies are lacking.

The relationship between tuatara and their parasites has been an area of recent growth. Given that the evolution of a parasite is intimately related to the evolution of its host, further questions arise: were the rhynchocephalian ancestors of tuatara already parasitised with the ancestors of current parasites prior to the fragmentation of Gondwana? Do parasites alter the behaviour of tuatara in ways that favour the parasite's transmission to new hosts? In what ways do parasites reduce fitness of tuatara, and do they help to structure ecosystems? (The fact that some tuatara populations appear naturally to lack ticks suggests a geographic starting point for such research.) How should conservation managers balance the demands of maintaining the biodiversity of both tuatara and their parasites? At present, tuatara hatched in captivity and translocated to new islands appear to lack ectoparasites,[268] and even if tuatara are naturally infested with parasites when translocated, their parasites may die out if the released tuatara are at low density.[269] No specific policy exists to translocate ectoparasites, although these may themselves be species at risk of extinction.

Other ecological topics of management value include techniques for estimating the density of tuatara on heavily burrowed islands (important for monitoring population stability), changes in density with island reforestation, and studies of orientation and homing. The consequence of reintroducing tuatara to habitats that currently have few if any seabirds is also of conservation interest.

Endnotes to Chapter 7

See the bibliography for full references to articles on tuatara, which are cited here at first mention by surname and year only. References on other topics are cited here in full at first mention, and then using 'op. cit.'. Except where required for clarity, the convention of parentheses around years has been dispensed with for economy of space.

1 Dawbin 1949, p. 91.
2 For example: Dieffenbach, quoted by Gray 1843; Günther 1867; Buller 1877, 1879; Newman 1878; Thomas 1890.
3 For accounts of the bite of tuatara, see Anon. 1882b; Reischek 1882; Hutton and Drummond 1904, p. 358; Schmidt 1953; Webber 1953; Wojtusiak 1973; Chorlton 1977; Daugherty et al. 1990a.
4 Walls 1978; Newman 1982c, 1987a; Markwell 1997, 1998.
5 Newman 1987a.
6 Markwell 1997.
7 Markwell 1998.
8 Walls 1978; Newman 1987a.
9 McIntyre, M. E. 1988. *Management implications of habitat use and dispersal of juvenile tuatara,* Sphenodon punctatus, *on Stephens Island.* Unpublished report. Victoria University of Wellington, Wellington; East et al. 1995.
10 Newman 1987a; McIntyre 1988, op. cit.
11 Gaston and Scofield 1995.
12 Reischek 1882, 1886; Pierce 2002.
13 Newman 1982a, p. 292.
14 Reischek 1882, 1886.
15 Tony Whitaker, Whitaker Consultants Ltd, Motueka, pers. comm. 2 August 2005.
16 Newman 1982c; Walls 1983.
17 Werner and Whitaker 1978.
18 Gillingham and Miller 1991.
19 Gillingham and Miller 1991.
20 McIntyre 1988, op. cit.
21 Whitworth 2006.
22 Pers. obs. See also Falla 1935.
23 Terezow et al. 2008. See also Whitworth 2006.
24 Wörner 2009; Woerner and Nelson 2010.
25 Goetz and Thomas 1994.
26 Walls 1983.
27 Newman et al. 1979; Walls 1983.
28 Walls 1983.
29 Saint Girons et al. 1980.
30 Walls 1983.
31 Walls 1982, 1983; Cree et al. 1990a, 1992; Thompson and Daugherty 1998.
32 Thoresen 1967.
33 Mair, in Buller 1877; Schmidt 1953.
34 For discussion, see Cassey and Ussher 1999; Tyrrell et al. 2000; Nelson et al. 2002a; and Moore et al. 2010.
35 Older values for capture rates ('number per hour') can be found in Crook 1973a, but with less information on search techniques and habitats.
36 Whitaker, A. H. and Daugherty, C. H. 1991. *Research on the tuatara (*Sphenodon punctatus*) of Little Barrier Island, 5–12 February 1991.* Unpublished report to the Department of Conservation, Wellington; Whitaker, A. H. 1993. *Research on the tuatara (*Sphenodon punctatus*) of Little Barrier Island, 6–20 October 1992.* Unpublished report to the Department of Conservation, Auckland; Cree et al. 1995a.
37 Moore et al. 2010.
38 Garrick (for Moutoki Island) and Heaphy (for Karewa Island) reported capture rates about half those of encounter rates, but on more easily traversed islands such as Stephens, the proportion actually caught (rather than just seen) may be higher. See Garrick, A. 1996. *Preliminary report on the status of the tuatara on Moutoki, Rurima Rocks, Bay of Plenty.* Unpublished report, Department of Conservation, Rotorua. RWL:030; Heaphy, J. 2005. *Karewa tuatara survey, March 2005.* Department of Conservation, Tauranga, TAUAO–2374.
39 Craig 1986; Cassey and Ussher 1999.
40 With more care in reporting, the relationship shown in Figure 7.12 (Pearson correlation coefficient = 0.773; $p = 0.005$) might be improved. For maximum value, future reports should distinguish encounter rate and capture rate, with and without resightings (recaptures).
41 Newman 1987a; Carmichael et al. 1989; Markwell 1997; Moore et al. 2009b.
42 Carmichael et al. 1989; Markwell 1997; Moore 2008.
43 Cassey and Ussher 1999; Tyrrell et al. 2000; Towns et al. 2007.
44 Fleming, in Newman 1982a, p. 291.
45 Turner, F. B. 1977. The dynamics of populations of squamates, crocodilians and rhynchocephalians. In: Gans, C. and Tinkle, D. W. (eds). *Biology of the Reptilia. Vol. 7. Ecology and behaviour A.* Academic Press, London, pp. 157–264.
46 Moore et al. 2009b.
47 Newman 1982c. Working in largely unforested areas with lower density but using a slightly higher estimate of mean body mass than Newman, Crook (1975) calculated a similar biomass of about 250 kg per hectare.
48 Lawless (1994) reported slightly lower estimates (17–45 tonnes of tuatara per km^2), but without indicating their source.
49 Rodda and Dean-Bradley 2002.
50 Hardy, in Newman 1982a, p. 223.
51 Pough, F. H. 1980. The advantages of ectothermy for tetrapods. *American Naturalist* 115: 92–112.
52 Pough, F. H. 1983. Amphibians and reptiles as low-energy systems. In: Aspey, W. P. and Lustick, S. I. (eds). *Behavioral energetics.* Ohio State University Press, Columbus, pp. 141–188. See also Bonnet, X., Pearson, D., Ladyman, M., Lourdais, O. and Bradshaw, D. 2002. 'Heaven' for serpents? A mark-recapture study of tiger snakes (*Notechis scutatus*) on Carnac Island, Western Australia. *Austral Ecology* 27: 442–450.
53 Drummond 1917. For other examples of tuatara surviving for months without eating, see Buller 1871, 1878 and Newman 1878. In an extreme example, a Mr G. G. Norris (presumably Greg Norris, younger brother of Colin Norris, who kept over a hundred tuatara in a colony in Tauranga for years), was reported to have kept one for 15 months without food (Anon. 1908a).
54 Anon. 1979.
55 Rodda and Dean-Bradley 2002, op. cit.
56 Hoare et al. 2006.
57 MacAvoy et al. 2007.
58 Moore et al. 2007.
59 MacAvoy et al. 2007.
60 Moore et al. 2007.

61 Walls 1982, 1983.
62 Carmichael et al. 1989.
63 Mean ± SD 86.7 ± 33.2 per m^2 in pasture, 15.7 ± 19.9 per m^2 in forest (Gillingham et al. 1995; see also Gillingham and Miller 1991).
64 Fine-scale genetic structuring has also been observed among tuatara from different locations on Stephens Island, and this may have been influenced by habitat modification. See Moore et al. 2008a.
65 Wever, E. G. 1978. *The reptile ear: its structure and function*. Princeton University Press, Princeton.
66 Wu, X.-C. 1994. Late Triassic–Early Jurassic sphenodontians from China and the phylogeny of the Sphenodontia. In: Fraser, N. C. and Sues, H.-D. (eds). *In the shadow of the dinosaurs*. Cambridge University Press, Cambridge, pp. 38–69.
67 Gans and Wever 1976.
68 Hutton and Drummond 1904.
69 Anonymous, undated account (early 1900s to 1940s) in AFCM album 6, PA1–f–127, Alexander Turnbull Library, Wellington.
70 Wojtusiak 1973.
71 Sladden and Falla 1928, p. 288. See Buller 1877 and Falla 1935 for similar descriptions.
72 The specimen was made available by R. G. Northcutt.
73 Gans and Wever 1976.
74 von Wettstein 1931.
75 Reischek 1882.
76 Krull 1923; Dawbin 1949.
77 According to W. H. Dawbin, cited in Anon. 1955a.
78 Wojtusiak and Majlert 1973.
79 Gans and Wever 1976.
80 Wojtusiak 1973; Wojtusiak and Majlert 1973.
81 Pers. obs. (see Chapter 8).
82 Pepperell 1982.
83 Reilly et al. 2006.
84 Nelson et al. 2006.
85 Reilly et al. 2006.
86 Pough 1980, op. cit.; Pough 1983, op. cit.
87 Pepperell 1982.
88 Reilly et al. 2006.
89 Nelson et al. 2006.
90 Mean ± SE for naturally incubated = 0.84 ± 0.03 m/s; artificially incubated = 0.67 ± 0.02 m/s; Nelson et al. 2006.
91 See Nelson et al. 2006 for discussion. See also Pepperell 1982.
92 The founder of the manufacturer, US-based Shelby Supercars, is reported to have named the car Tuatara in recognition of what had recently been suggested as a rapid rate of molecular evolution in tuatara (*Otago Daily Times* 20 July 2011; see also Hay et al. 2008). However, this suggestion of rapid evolution had already been contested by other molecular biologists (Miller et al. 2008b).
93 Walls 1981. See also summary in Walls 1982.
94 Fraser 1993.
95 Walls 1981. Given the way that frequency of occurrence was subdivided by Walls within different types of darkling beetles, weevils and reptiles, the percentages that I have calculated here could be underestimates. However, the figure for reptiles would not exceed 6%.
96 Fraser 1993.
97 Carmichael et al. 1989.
98 Fraser 1993.
99 Walls 1981, 1982.
100 Fraser 1993.
101 Walls 1978.
102 Fraser 1993.
103 Markwell 1998. Note that Markwell estimated losses only from nest chambers, not losses of fledglings late in the season.
104 Newman 1978.
105 Walls 1978; Fraser 1993.
106 Fraser 1993.
107 Fraser 1993.
108 Walls 1978, 1981.
109 Walls 1981, 1982; Moller 1985; Fraser 1993.
110 Buller 1877. Though described then as *Naultinus*, the geckos concerned are likely to have been a species of what has until recently been considered as *Hoplodactylus* (Tony Whitaker, pers. comm. 2 August 2005). In a similar account in the same publication, the remaining geckos were described as appearing 'quite paralyzed with fear'.
111 Newman 1977.
112 Mitchell, J. C. 1986. *Cannibalism in reptiles: a worldwide review*. SSAR Herpetological Circular No. 15, Oxford, Ohio.
113 Walls 1981.
114 Daugherty et al. 1990a; Brown [2000].
115 Newman 1987b. Assumed to be total length, though not stated.
116 Fraser 1993.
117 Cree et al. 1992 (see Chapter 8).
118 Gillingham et al. 1995.
119 Walls 1981.
120 Fraser 1993.
121 Walls 1981.
122 Fraser 1993.
123 *Extract from report of Mr B. Bell on visits to Stephen [sic] Island 8.6.59–21.6.59*. ANZ IA 1 46/18/4 Part 2 (copy also in ANZ IA 1 46/14/26).
124 Bredeweg and Nelson 2010.
125 Whitaker, A. H. 1989. *Report on a visit to Moutoki Island, Rurima Rocks Group, Bay of Plenty, 19–21 January 1989*. Unpublished report for the Ngati Awa Trust Board and Tuwharetoa Maori Trust, Victoria University of Wellington, Wellington. See also Thompson et al. 1992.
126 Gaston and Scofield 1995.
127 Buller 1879.
128 Mair 1873; Buller 1877; Reischek 1882, 1886; Sladden 1924; Schmidt 1953.
129 Southey 1985.
130 Ussher 1999a.
131 Ussher 1999a.
132 Southey 1985.
133 Pierce 2002.
134 Dawbin 1982a.
135 Southey 1985.
136 Thoresen 1967.
137 D. J. Grindlay to Under Secretary for Internal Affairs, 17 March 1919, ANZ IA 1 46/18/8 pt 1.
138 Robb 1973.
139 Graham Ussher, Auckland Regional Council, pers. comm. 9 September 2005. Both fur and bones from kiore were present.
140 Pers. obs.
141 Whitaker 1989, op. cit.
142 Blair et al. 2000a.
143 Southey 1985.
144 Whitaker 1989, op. cit.
145 Reischek 1882.
146 Buller 1879.
147 von Wettstein 1931.
148 Southey 1985.

149 Assistant keeper on North Brother Island, reported in *The Dominion*, 20 January 1926. Copy in ANZ M 1 25/611 pt 2, 1923–1928.
150 Bonnet et al. 2002, op. cit.
151 Gorniak et al. 1982.
152 Buller 1879; Falla 1935; Dawbin 1949, 1962a; Walls 1981, 1982; Gorniak et al. 1982. For a detailed technical description, see Schwenk 2000. For descriptions of muscles of the head and neck associated with feeding, see Jones et al. 2009a and references therein. For further information on head musculature and biomechanical models of feeding, see Curtis et al. 2009, 2010a, b, c, and Johnston 2010.
153 Günther 1867; Osawa 1897; Gabe and Saint Girons 1964a; Schwenk 1986.
154 Farlow 1975; Meyer-Rochow and Teh 1991; Gorniak et al. 1982.
155 Walls 1981; Gorniak et al. 1982.
156 Farlow 1975.
157 Gorniak et al. 1982.
158 A propalinal movement in which the power stroke is forwards, as in tuatara, is sometimes referred to as proal (or prooral, e.g. Jones et al. 2009a), as distinct from a backwards or palinal movement.
159 Farlow 1975.
160 Moore and Godfrey 2006.
161 Walls 1981.
162 Newman 1977.
163 Farlow 1975; Walls 1981.
164 Meyer-Rochow 1988; Meyer-Rochow and Teh 1991.
165 Falla 1935; Farlow 1975; Walls 1981; Dawbin 1982a; Refsnider et al. 2008. Krull 1923 described tuatara feeding on pieces of steak. Juvenile tuatara in our laboratory colony fed on fish-based cat food.
166 Schwenk 2000.
167 Besson et al. 2009. See also Cooper et al. 2001.
168 Dawbin 1982a.
169 Gillingham and Miller 1991; pers. obs.
170 Newman 1987b; Cooper, W. E. 1994. Chemical discrimination by tongue-flicking in lizards: a review with hypotheses on its origin and its ecological and phylogenetic relationships. *Journal of Chemical Ecology* 20: 439–487.
171 Newman 1982c.
172 Walls 1978, 1981.
173 Gans 1983; pers. obs.
174 D. J. Grindlay to Under Secretary for Internal Affairs, 17 March 1919, ANZ IA 1 46/18/8 pt 1.
175 See Schwenk 2000 for further discussion of the difficulty of applying these terms as a strict dichotomy.
176 Besson 2009; Besson and Cree 2011.
177 Schaerlaeken et al. 2008.
178 Schaerlaeken et al. 2008.
179 Jones and Lappin 2009.
180 Based on data in Fraser 1993 and Ussher 1999d.
181 Huey, R. B., Pianka, E. R. and Vitt, L. J. 2001. How often do lizards 'run on empty'? *Ecology* 82: 1–7.
182 Fraser 1993.
183 Fraser 1993.
184 Fraser 1993.
185 Fraser 1993; pers. obs.
186 Skoczylas, R. 1978. Physiology of the digestive tract. In: Gans, C. and Gans, K.A. (eds). *Biology of the Reptilia. Vol. 8. Physiology B*. Academic Press, London, pp. 589–717.
187 Besson 2009; Besson and Cree 2011.
188 Worthy and Holdaway 2002, p. 460.
189 Buller 1879; Thomson 1915; *Extract from report of Mr B. Bell on visits to Stephen [sic] Island 8.6.59–21.6.59*. ANZ IA 1 46/18/4 pt 2 (copy also in ANZ IA 1 46/14/26); Kinsky, F. C. and Sibson, R. B. 1959. Notes on the birds of the Poor Knights Islands. *Notornis* 8: 132–142; Campbell, D. J. 1967. *The Trio Islands, Marlborough Sounds; an ecological study of a bird modified island*. Unpublished MSc thesis, Victoria University of Wellington, Wellington; McCallum 1981; Southey 1985.
190 Holdaway, R. N., Worthy, T. H. and Tennyson, A. J. D. 2001. A working list of breeding bird species of the New Zealand region at first human contact. *New Zealand Journal of Zoology* 28: 119–187.
191 Thomson 1915.
192 Thomson 1915; Oliver, W. R. B. 1955. *New Zealand birds*. A. H. and A. W. Reed, Wellington; Whitaker, A. H. 1972. Lizard predators. *Pepeke* 19: 4–5; Dawbin 1974.
193 Worthy and Holdaway 2002, see p. 483.
194 Keepers on Stephens Island who reported kingfishers preying on very small or young tuatara include E. W. Tutt (2 February 1914) and P. R. W. Willers (12 October 1914). In NMNZ MU 000207, Box 4, Item 1, *Miscellaneous files 1913–1915*.
195 F. Woodbury to Secretary for Marine, 23 November 1923. ANZ M 1 25/611 pt 2. Oliver 1955, op. cit., was told directly by Woodbury of these observations, including the identity of the gulls as black-backed gulls.
196 However, similar behaviour by black-backed gulls towards seabird chicks has been seen on Middle Trio Island (see Campbell 1967, op. cit.).
197 Worthy 1997.
198 For evidence of predation on lizards, see Sladden and Falla 1928; Whitaker, A. H. 1968. The lizards of the Poor Knights Islands, New Zealand. *New Zealand Journal of Science* 11: 623–651; Whitaker 1972, op. cit.; Tony Whitaker, pers. comm. 6 January 2005; Atkinson, I. A. E. and Millener, P. R. 1991. An ornithological glimpse into New Zealand's pre-human past. *Acta XX Congressus Internationalis Ornithologici* (20th International Ornithological Congress, Christchurch, New Zealand 2–9 December 1990). New Zealand Ornithological Congress Trust Board, Wellington.
199 Atkinson and Millener 1991, op. cit.; Whitaker, A. H. 1991. *Research on the striped gecko (*Hoplodactylus stephensi*) on Maud Island, Pelorus Sound, Marlborough: 6–12 March 1991*. Unpublished report, Department of Conservation, Nelson–Marlborough Conservancy, Nelson; Tony Whitaker, pers. comm. 6 January 2005.
200 Lindsay Hazley, Southland Museum and Art Gallery, pers. comm. 7 January 2005. Conversely, adult tuatara appear to have preyed on the parakeets: see Brown 2000, op. cit., p. 216.
201 Whitaker 1972, op. cit.; Tony Whitaker, pers. comm. 6 January 2005.
202 Nelson 2001; Nelson et al. 2004a.
203 Tony Whitaker, pers. comm. 6 January 2005.
204 Whitaker 1972, op. cit.; Tony Whitaker, pers. comm. 6 January 2005; Thompson, M. B. 2000. *Oligosoma* spp. (New Zealand skinks): predation. *Herpetological Review* 31: 175.
205 *Extract from report of Mr B. Bell on visits to Stephen [sic] Island 8.6.59–21.6.59*. ANZ IA 1 46/18/4 pt 2 (copy also in ANZ IA 1 46/14/26).
206 Walls 1981; Allison 1982; McKenna 2003.
207 Liu and King 1971; Twentyman 1999; see also Blanchard 2002.
208 Gartrell et al. 2006, 2007. See also Middleton et al. 2011.
209 Barker, S. C. and Murrell, A. 2004. Systematics and evolution

of ticks with a list of valid genus and species names. *Parasitology* 129: S15–S36.

210 Dumbleton 1943; Klompen, H., Dobson, S. J. and Barker, S. C. 2002. A new subfamily, Bothriocrotoninae n. subfam., for the genus *Bothriocroton* Keirans, King & Sharrad, 1994 status amend. (Ixodida: Ixodidae), and the synonymy of *Aponomma* Neumann, 1899 with *Amblyomma* Koch, 1844. *Systematic Parasitology* 53: 101–107.

211 Dawbin 1949.

212 Ruffell 2005; Heath 2006.

213 Pers. obs.; Claudine Tyrrell, University of Otago, pers. comm. 16 March 2005; Miller et al. 2007b.

214 Pers. obs.; Claudine Tyrrell, pers. comm. 16 March 2005; Donald Newman, Department of Conservation, pers. comm. for Chickens; Miller et al. 2007b.

215 J. Heaphy, Department of Conservation, pers. comm. 23 August 2005; Jonathan Ruffell, University of Auckland, pers. comm. 11 June 2005 for Moutoki; Miller et al. 2007b.

216 Miller et al. 2007b.

217 Miller et al. 2007b.

218 Although the tick was recently listed as a threatened species, 'range restricted' (Hitchmough, R., Bull, L. and Cromarty, P. (compilers). 2007. *New Zealand Threat Classification System lists 2005*. Department of Conservation, Wellington), new categories introduced in 2008 make that part of the 'at risk' rather than 'threatened' category.

219 Heath 2006.

220 Nicola Nelson, Victoria University of Wellington, pers. comm. July 2005.

221 Ruffell 2005.

222 Burnham et al. 2006; Godfrey et al. 2008.

223 Godfrey et al. 2008.

224 Godfrey et al. 2010a.

225 Heath 2006.

226 Laird 1950; Herbert et al. 2010. See also Gartrell et al. 2006.

227 Burnham et al. 2006.

228 Godfrey et al. 2010b.

229 Ruffell 2005.

230 Main, A. R. and Bull, C. M. 2000. The impact of tick parasites on the behaviour of the lizard *Tiliqua rugosa*. *Oecologia* 122: 574–581.

231 Dumbleton 1947; Goff et al. 1987.

232 McKenna 2003.

233 Dawbin 1949; Godfrey et al. 2008; pers. obs.

234 Godfrey et al. 2008.

235 Godfrey et al. 2010b.

236 Godfrey et al. 2008.

237 Laird 1950; Godfrey et al. 2011. See also Desser 1978.

238 Laird 1950; Desser 1978.

239 Allison 1982; McKenna 2003.

240 Oppliger, A., Célérier, M. L. and Clobert, J. 1996. Physiological and behaviour changes in common lizards parasitised by haemogregarines. *Parasitology* 113: 433–438.

241 Godfrey et al. 2011. See also Herbert et al. 2010.

242 Chabaud and Dollfus 1966.

243 Inglis, W. G. and Harris, E. A. 1990. Kiwinematidae n. fam. (Nematoda) for *Kiwinema* n. g. and *Hatterianema* Chabaud & Dollfus, 1966: heterakoids of native New Zealand vertebrates. *Systematic Parasitology* 15: 75–79.

244 Allison and Blair 1987; McKenna 2003.

245 Walls 1981; Fraser 1993.

246 $p = 0.062$; Fraser 1993.

247 Seligmann et al. 2003. However, see Seligmann et al. 2008 for further study of related questions.

248 Pers. obs.; Jones and Lappin 2009. See also Jones et al. 2009a for pathologies of jaw bones.

249 Pers. obs.

250 Pers. obs.

251 Jansen, P. W. 1988. *Plate Island survey 17–20 March 1988*. Unpublished report, Department of Conservation, Bay of Plenty.

252 Tyrrell et al. 2000.

253 Tony Whitaker, pers. comm. 2 August 2005.

254 Thoresen 1967.

255 Lee 2010.

256 Infections and diseases observed in captivity are summarised by Boardman and Blanchard 2006. For recent information on immune function, see Burnham et al. 2005 and La Flamme et al. 2010.

257 Brett Gartrell, Massey University, pers. comm. 19 August 2005. Systemic gout has been reported in captive tuatara: see Boardman and Blanchard 2006.

258 Pough 1980, op cit.; Pough 1983, op. cit. See also Chapter 9.

259 Reilly et al. 2006.

260 Schwenk 1986.

261 Cooper 1994, op. cit.; Vitt, L. J. and Pianka, E. R. 2006. The scaly ones. *Natural History Magazine* July–August 115 (6): 28–35.

262 Cree et al. 1992. See also Chapter 8.

263 For example, Desser 1978; Boardman and Sibley 1991; Blanchard 2002; Boardman and Blanchard 2006.

264 But see Chapter 8 for some plasma analyses related to reproduction in wild females.

265 Newman et al. 1979; Cree and Daugherty 1990b.

266 Cartland-Shaw et al. 1998; Cree et al. 1999.

267 Schaffner, F. 1998. The liver. In: Gans, C. and Gaunt, A. S. (eds). *Biology of the Reptilia. Vol. 19. Morphology G. Visceral organs.* Society for the Study of Amphibians and Reptiles, Ithaca, New York, pp. 485–531.

268 For example, no ectoparasites were detected on juvenile tuatara raised in captivity (from eggs or hatchlings from Stephens Island) and translocated to Wakaterepapanui Island (Gartrell et al. 2006). In this case, adult tuatara were also translocated directly from Stephens Island, so the juveniles may subsequently have become infested.

269 Miller et al. 2007b.

Endnotes to boxed text

For explanation of citation format in box endnotes, see endnotes to the main text.

Box 7.1

1 Reischek 1882.
2 Reischek 1971, p. 98.
3 Anon. 1882a; Buller 1888, p. 218.
4 Thomas 1890, p. 153.
5 Sizes of seabird species in this paragraph are from: Marchant, S. and Higgins, P. J. (co-ordinators). 1990. *Handbook of Australian, New Zealand and Antarctic birds. Vol. 1. Part A. Ratites to petrels*. Oxford University Press, Melbourne.
6 Sladden and Falla 1928; McFadden, I. 1986. *Aldermen Islands: 3–6 March 1986*. Unpublished report to the Department of Conservation, Auckland.
7 Falla, R. A. 1924. Discovery of a breeding place of Buller's shearwater, Poor Knights Island, N.Z. *Emu* 24: 37–43.
8 Sutherland, J. H. 1952. Birds at The Brothers. *Notornis* 5: 26–27; Wright 1961; Walls 1981; Fraser 1993; Gaston and Scofield 1995; Markwell 1998.
9 Dunnet, G. M. 1985. Pycroft's petrel in the breeding season at Hen and Chickens Islands. *Notornis* 32: 5–21; Bartle, J. A. 1968. Observations on the breeding habits of Pycroft's petrel. *Notornis* 15: 70–99.
10 Pierce 2002.
11 Walls 1981.
12 Chambers, B. S. 1956. Birds of the Poor Knights Island. *Tane* 7: 66–67; Harper 1983.
13 On Karewa Island, inferred from Reischek 1886 (as *Puffinus brevicaudus*), Thomas 1890 (as *P. tenuirostris*) and Schmidt 1953. See also Dawbin 1949, and Discusssion in Newman 1982a, for evidence of burrow-sharing with *P. carneipes*.
14 Sladden and Falla 1928.
15 Dawbin 1949.
16 Mair 1873; Gaston and Scofield 1995.
17 Harper 1983, p. 313.
18 Kinsky, F. C. and Sibson, R. B. 1959. Notes on the birds of the Poor Knights Islands. *Notornis* 8: 132–142. See p. 135 for quote.
19 See also Newman, in Newman 1982a, p. 289, 292.
20 Walls 1978.
21 Fukami, T., Wardle, D. A., Bellingham, P. J., Mulder, C. P. H., Towns, D. R., Yeates, G. W., Bonner, K. I., Durrett, M. S., Grant-Hoffman, M. N. and Williamson, W. M. 2006. Above- and below-ground impacts of introduced predators in seabird-dominated island ecosystems. *Ecology Letters* 9: 1299–1307.
22 Schmidt 1953; Crook 1970, 1974, 1976; Mulder, C. P. H. and Keall, S. N. 2001. Burrowing seabirds and reptiles: impacts on seeds, seedlings and soils in an island forest in New Zealand. *Oecologia* 127: 350–360.
23 Dawbin 1962a; Markwell, T. J. and Daugherty, C. H. 2002. Invertebrate and lizard abundance is greater on seabird-inhabited islands than on seabird-free islands in the Marlborough Sounds, New Zealand. *Ecoscience* 9: 293–299; Markwell, T. J. and Daugherty, C. H. 2003. Variability in ∂^{15}N, ∂^{13}C and Kjeldahl nitrogen of soils from islands with and without seabirds in the Marlborough Sounds, New Zealand. *New Zealand Journal of Ecology* 27: 25–30.
24 Dawbin 1962a; Gillingham and Miller 1991.
25 Tyrrell 2000. See also Ramstad et al. 2007a; Corkery et al. 2010.
26 Newman, in Newman 1982a, p. 293.
27 Worthy and Holdaway 1994; 2002, p. 461.
28 Crook 1970, 1974, 1975. See also Markwell 1998.
29 Crook 1974, 1975, 1976; Newman 1978.

Box 7.2

1 Wilmshurst, J. M., Anderson, A. J., Higham, T. F. G. and Worthy, T. H. 2008. Dating the late prehistoric dispersal of Polynesians to New Zealand using the commensal Pacific rat. *Proceedings of the National Academy of Sciences* 105: 7676–7680.
2 For example, Pacific rats were described as the 'native rat' (Wodzicki, K. A. 1950. *Introduced mammals of New Zealand: an ecological and economic survey*. Department of Scientific and Industrial Research, Wellington), as 'this native species' (K. Wodzicki, Animal Ecology Section DSIR to Secretary for Internal Affairs: 3 December 1954; ANZ IA 1 46/18/8 pt 1), as 'the harmless native species' (Secretary for Internal Affairs to Secretary for Marine, 17 November 1953; ANZ IA 1 46/18/8 pt 1), and as 'the harmless Polynesian rat' by the future director of the Wildlife Service (Williams, G. R. 1962. Extinction and the land and freshwater-inhabiting birds of New Zealand. *Notornis* 10: 15–31).
3 Atkinson, I. A. E. 1978. Evidence for effects of rodents on the vertebrate wildlife of New Zealand islands. In: Dingwall, P. R., Atkinson, I. A. E. and Hay, C. (eds). *The ecology and control of rodents in New Zealand nature reserves*. New Zealand Department of Lands and Survey Information Series 4, Wellington, pp. 7–30.
4 Worthy and Holdaway 2002, pp. 533–537.
5 For example, Whitaker, A. H. 1973. Lizard populations on islands with and without Polynesian rats, *Rattus exulans* (Peale). *Proceedings of the New Zealand Ecological Society* 20: 121–130; Cree and Butler 1993, p. 37; Tyrrell et al. 2000, p. 6; Towns, D. R., Atkinson, I. A. E. and Daugherty, C. H. 2006. Have the harmful effects of introduced rats on islands been exaggerated? *Biological Invasions* 8: 863–891.
6 H. J. Dusting to Secretary of Internal Affairs, 16 November 1914, NMNZ MU 000207 Box 4, Item 1.
7 G. H. Brown to Secretary of Marine, 15 February 1924, in ANZ IA 1 46/18/8 pt 1, also in ANZ M 1 25/611 pt 2. Another keeper had suggested the previous year that another kind of rat ('not native ones') was becoming plentiful on the island, but no species other than the Pacific rat has ever been confirmed on Cuvier Island.
8 Bell, L. 1950. *Report of visit to offshore islands of the Coromandel Peninsula 20.2.51–3.3.51.* ANZ IA 46/18/4 pt 2). (The date of 17/4/50 on the report conflicts with the dates in the title, and appears to be in error.)
9 Atkinson, I. A. E. 1964. The flora, vegetation, and soils of Middle and Green Islands, Mercury Islands Group. *New Zealand Journal of Botany* 2: 385–402. See also Atkinson 1978, op cit.
10 Crook 1973a. See also Crook 1970, 1973b and 1975, and Whitaker 1978.
11 See Chapter 3.
12 See Appendix 3.
13 Crook 1973a; Whitaker 1978; Cree et al. 1995a. A fourth island (Araara, in the Chickens Islands) is sometimes mentioned in this context, but whether it has been searched adequately for tuatara and the amount of suitable habitat is uncertain (Tony Whitaker, Whitaker Consultants Ltd, Motueka, pers. comm. 2 August 2005).

14 Bell 1951, op. cit.; Crook 1973a; Whitaker 1978; Whitaker, A. H. 1993. *Research on the tuatara* (Sphenodon punctatus) *of Little Barrier Island, 6–20 October 1992*. Unpublished report to the Department of Conservation, Auckland; Cree et al. 1995a; Tennyson and Pierce 1995; Tyrrell et al. 2000. The mean capture rates and densities shown in the box figure differ significantly between rat-inhabited and rat-free islands ($p < 0.05$ by t-test or Mann-Whitney test).

15 Crook 1973a; Whitaker 1993, op. cit.; Cree and Butler 1993; Cree et al. 1995a; Tyrrell et al. 2000.

16 Newman and McFadden 1990b; Cree et al. 1995a.

17 Cree et al. 1995a; Cassey and Ussher 1999. See Table 7.1.

18 Brook, F. J. 1999. Changes in the landsnail fauna of Lady Alice Island, northeastern New Zealand. *Journal of the Royal Society of New Zealand* 29: 135–157.

19 Crook 1973a; Whitaker 1978; Craig 1986; Newman and McFadden 1990; Towns 1991; Cree et al. 1995a.

20 Whitaker 1978; see also Craig 1986.

21 Craig 1986; Craig, J. L. 2002. Eradicating kiore. *Forest and Bird* No. 306 (November): 3. See also Veltman, C. J. 1996. Investigating causes of population decline in New Zealand plants and animals: introduction to symposium. *New Zealand Journal of Ecology* 20: 1–5; Towns et al. 2006, op. cit.

22 Atkinson, I. A. E. and Towns, D. R. 2005. Kiore. In: King, C. M. (ed.). *The handbook of New Zealand mammals*. Oxford University Press, Auckland, pp. 159–174. The voracious habits of Pacific rats were brought home to me during a field trip to Lady Alice Island, when rats ate a hole through our field party's polystyrene chillibin (cooler) to eat the meat inside. Even more eye-opening is a report from Kawhitu, where Pacific rats 'chewed on everything we left exposed including two bars of soap which they completely consumed' (Thoresen 1967: 189).

23 Newman 1988.

24 For example, Whitaker 1978; Lettink, M. L. and Cree, A. 2006. Predation, by the feral house mouse *(Mus musculus)*, of McCann's skinks (*Oligosoma maccanni*) constrained in pitfall traps. *Herpetofauna* 36: 61–62.

25 Newman 1986.

26 Ashley 1902.

27 Newman et al. 1994.

28 Tyrrell 2000; Tyrrell et al. 2000.

29 Towns 1991.

30 Towns, D. R. and Broome, K. G. 2003. From small Maria to massive Campbell: forty years of rat eradication from New Zealand islands. *New Zealand Journal of Zoology* 30: 377–398; Atkinson and Towns 2005, op. cit.

31 Towns and Broome 2003, op. cit.; Atkinson and Towns 2005, op. cit.

32 Towns et al. 2007.

33 Towns et al. 2007.

34 Towns, D. R., West, C. J. and Broome, K. G. 2012. Purposes, outcomes and challenges of eradicating invasive mammals from New Zealand islands: an historical perspective. *Wildlife Research* doi.org/10.1071/WR12064.

35 Although no sign of rats has been seen since 2011 on Taranga, definitive confirmation of rat-free status will await surveys in 2013 (Keith Hawkins, Department of Conservation, pers. comm. 18 September 2012).

36 Tennyson and Pierce 1995.

Box 7.3

1 Cartland-Shaw et al. 1998.

2 Cartland-Shaw et al. 1998.

3 Fraser 1993.

4 Cartland-Shaw et al. 1998.

5 Cartland-Shaw et al. 1998.

6 Cartland et al. 1994; Cartland-Shaw 1996; Cartland-Shaw et al. 1998.

7 Blair et al. 2000a.

8 Blair et al. 2000a.

9 The standard error is 16%; Cree et al. 1999.

10 For discussion and references, see Cartland-Shaw et al. 1998.

11 Blair et al. 2000b.

12 For discussion, see Cartland-Shaw et al. 1998.

Box 7.4

1 For general anatomy of the digestive tract, see Günther 1867 and Osawa 1897. Examples of other studies include those of the buccal cavity (Busch 1898; Lakjer 1927), intestinal relief (Jacobshagen 1920), mesenteries (Klaatsch 1892) and blood supply (Hochstetter 1892). For an early review, see von Wettstein 1931. For description of the cloaca, see Gadow 1887.

2 For histological studies on specimens from Stephens Island: Gabe and Saint Girons 1964a, 1969, 1972.

3 Schwenk 2000 provided a detailed review of feeding anatomy and behaviour. See also Parsons, T. S. and Cameron, J. E. 1977. Internal relief of the digestive tract. In: Gans, C. and Parsons, T. S. (eds). *Biology of the Reptilia. Vol. 6. Morphology E*. Academic Press, London, pp. 159–223; Kochva, E. 1978. Oral glands of the Reptilia. In: Gans, C. and Gans, K. A. (eds). *Biology of the Reptilia. Vol. 8. Physiology B*. Academic Press, London, pp. 43–161; Schaffner, F. 1998. The liver. In: Gans, C. and Gaunt, A. S. (eds). *Biology of the Reptilia. Vol. 19. Morphology G. Visceral organs*. Society for the Study of Amphibians and Reptiles, Ithaca, New York, pp. 485–531.

4 Günther 1867; Gabe and Saint Girons 1969.

5 Kochva 1978, op cit.

6 Gabe and Saint Girons 1969.

7 von Wettstein 1931.

8 Gabe and Saint Girons 1964a, 1972; Luppa, H. 1977. Histology of the digestive tract. In: Gans, C. and Parsons, T. S. (eds). *Biology of the Reptilia. Vol. 6. Morphology E*. Academic Press, London, pp. 225–314.

9 For example, Günther 1867.

10 Günther 1867; Jacobshagen 1920; Parsons and Cameron 1977, op. cit.

11 Gabe and Saint Girons 1964a.

12 Günther 1867.

13 von Wettstein 1931.

14 Gabe and Saint Girons 1964a.

15 Günther 1867, p. 621.

16 Newman 1878, p. 233. See also Osawa 1897; Engelbert, pers. comm. in Newman et al. 1979.

17 Body and Newman 1989.

18 Schaffner 1998, op. cit., p. 513; Hagey, L. R., Vidal, N., Hofmann, A. F. and Krasowski, M. D. 2010. Evolutionary diversity of bile salts in reptiles and mammals, including analysis of ancient human and extinct giant ground sloth coprolites. *BMC Evolutionary Biology 10*: 133 (23 pp).

19 Jordan et al. 1980.

20 Bill (W.) Jordan, Victoria University of Wellington, pers. comm. 10 December 2004.

CHAPTER EIGHT

Reproduction and life history

In 1931 a review concluded that reproduction in tuatara was 'comparatively slight' and that it might be 'one of the reasons for the rapid regression and dying-out of the species.'[1] Numerous discoveries since that time have forced us to rethink this idea. The topic of reproduction in tuatara has been one of the most fruitful areas of modern research, and, in the right circumstances, the process itself is clearly successful.

Introduction

By the early 1850s and probably much earlier, tuatara were known among Māori to be an oviparous or egg-laying reptile.[2] However, formal studies on reproduction essentially began in 1867 with the observation by Günther that male tuatara lacked any intromittent organs.[3] This anomaly, unique among reptiles, fuelled considerable interest in the late 19th century. A New Zealand researcher, A. K. Newman, suggested that the absence of a penis might explain what was perceived as a lack of sexual dimorphism, or differences in appearance between the sexes, and that males had 'probably small sexual passions, and but little rivalry'.[4] Writing in German, von Wettstein compiled these and other observations in a major review in 1931.[5] His and others' inferences of failing reproduction were wrong in many ways, but it has taken nearly 150 years of research, and considerable patience, to obtain an accurate understanding of reproductive processes in tuatara.

This chapter begins with an overview of the seasonal reproductive cycle of tuatara on Stephens Island (Takapourewa). Details of reproductive anatomy, followed by cycles in gonadal activity, hormone production and reproductive behaviour, are then described. The events involved in reproduction are followed in a broadly developmental sequence through the production of gametes (sperm and eggs), mating, ovulation, eggshell formation, nesting and embryonic development. Growth, longevity and other aspects of life history are then explored. Information for islands other than Stephens Island is included throughout, where it exists. Topics of broad general interest, including sexual dimorphism, reproduction in captivity and temperature-dependent sex determination, are addressed in boxes. Suggestions for future research are noted in the relevant sections, and summarised in concluding remarks at the chapter's end.

Overview of the seasonal reproductive cycle on Stephens Island

Like other reptiles from temperate zones, tuatara are seasonal breeders. This and the following paragraph provide a brief summary of the seasonal cycle in animals on Stephens Island, with supporting references added in the detailed sections that follow. Mating occurs in late summer and autumn (January–March), at or just after the annual peak in air temperature (Figure 8.1). Although temperature is probably an important environmental cue for stimulating seasonal reproduction, as in other temperate-zone reptiles,[6] experimental studies addressing this are lacking for tuatara.

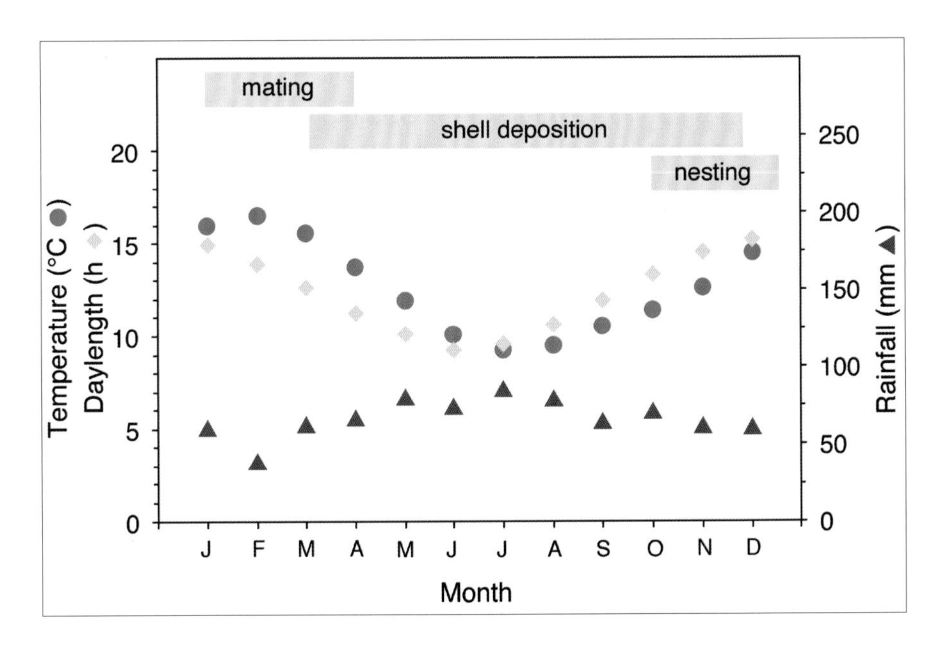

Fig. 8.1 The seasonal cycle of reproduction in tuatara on Stephens Island (Takapourewa) in relation to environmental conditions. Male tuatara have an annual reproductive cycle, with mating occurring in late summer when conditions are warm and relatively dry. Females ovulate in late summer and then begin adding the shell layers to their eggs. Nesting occurs in late spring when temperatures are once again rising. Note that the complete cycle of egg production (including yolking of oocytes) in females takes about 4–5 years on average and only a subset of females ovulate and nest in each year.

For sources on reproductive activity, see text. Mean monthly air temperatures for 1981–2010 and mean monthly rainfall for 1961–1990 from the NIWA National Climate Database (cliflo.niwa.co.nz). Daylength from New Zealand Nautical Almanac 2012–2013 (http://www.linz.govt.nz).

BOX 8.1

Telling the difference between males and females

Adult tuatara have obvious sexual dimorphism, yet distinguishing males from females has been the source of a surprising amount of confusion. The error seems to have begun with Walter Buller, a prominent lawyer-ornithologist in New Zealand who otherwise had a leading knowledge of tuatara during the late 19th century. In 1870 Buller wrote of his pair of tuatara from Karewa Island that the 'male is considerably smaller than the female'.[1] After obtaining further specimens from Karewa, he noted that 'In the adult state, the female is always larger than the male';[2] he made a similar comment for tuatara from North Brother Island.[3] A 'female' from Karewa Island was described as being stouter in the body than any other tuatara Buller had seen, with heavy nuchal folds and a head that was heavily scarred, apparently from fighting. A tuatara that he sent to Canterbury Museum (a medium-sized 'male') was said to be 'remarkable for the extreme smallness of the dorsal spines, which were reduced to a line of mere points along the back';[4] another specimen sent to the same museum, which Buller repeatedly referred to as 'he', has since been identified as a female.[5] From these descriptions, it seems that Buller completely reversed the characters separating males from females.

A medically trained contemporary of Buller's in Wellington, A. K. Newman, also had captive tuatara. Newman provided accurate descriptions of the internal organs of several dissected females. Yet, surprisingly, he concluded that male tuatara 'have no special strongly-marked tints, no special personal attractions; and, unlike the males of several other species of lizards [*sic*], are not much, if at all, bigger than the females ... The males are so like the females that they have not yet been distinguished with certainty'. [6]

During the next three decades, at least two researchers independently corrected these errors. The first was Professor A. P. W. Thomas, of Auckland University College, who went to Karewa Island in 1885 and collected 29 tuatara for breeding in captivity. Thomas later wrote, in an article published in both British and New Zealand journals, that 'We had been assured by those familiar with the tuatara that there was no difference in the external characters of the sexes'.[7] But a year later no eggs had been laid, so Thomas went back to Karewa. There he made an important discovery: 'On dissecting and carefully comparing a number of tuataras, I found that the current statements were not correct. There need seldom be any difficulty in distinguishing the sexes ... .'[8] He went on to note that males were much larger, had a more robust build and more strongly developed crests on the neck and back, and were 'more pugnacious'. He also discovered the reason for his initial lack of success in captive breeding: all 29 of the tuatara he had collected the previous year were male! He attributed this to his desire to collect the largest and most vigorous specimens, as well as the tendency of females to conceal themselves more than males.[9]

A few years later, during the summer of 1896/97, Professor Hugo Schauinsland from Germany visited Stephens Island and the Trio Islands.[10] Apparently unaware of Thomas's research, Schauinsland dissected tuatara to collect eggs for studies of embryonic development. He accurately reported (in German) that females were smaller than males, and had a rounder head, a less well-developed crest (often almost completely lacking on the back), and usually a more rounded body shape. He also observed that males appeared much more numerous than females, although (as also inferred by Thomas) he noted that this might have resulted from females being more inclined to remain in their burrows.[11]

Another German researcher, Dr Georg Thilenius, spent time on Karewa Island and Stephens Island during the late 1890s. He too was apparently unaware of earlier work, and dissected tuatara for studies on egg development. Although Thilenius did not comment on the features he used to sex animals, his observations on sex ratios imply that he could tell the difference.[12]

Despite these advances, two decades later it was still being reported in a newspaper account that females were larger than males – and capable of eating them![13] Government officials seemed to have lost all knowledge of earlier work by Thomas, Schauinsland and Thilenius (and Buller). When replying to a request for information on sexing from a lighthouse keeper, the director of the Dominion Museum replied in 1915 that he knew of no published account of the sex differences in tuatara.[14] Similarly, in 1935, a later director of the Dominion Museum was apparently 'unaware of any characters specially distinguishing the male tuatara from the female'.[15] Lighthouse keepers were thus forced to note with regret, when collecting live specimens for researchers, that they were unable to meet requests for animals of known sex.[16]

Shortly thereafter, William (Bill) Dawbin began field studies on Stephens Island. Initially, he reported difficulty in sexing, noting in 1949 that 'the external difference between the sexes must be very small, as no foolproof method for sexing them has yet been devised – excepting by the tuatara'.[17] It was only after he obtained permission to dissect tuatara during the 1950s that 'a number of criteria for sexing them in the field became apparent'.[18]

Box 8.1 Fig. 1 Sexual dimorphism in crest development, head shape and throat colour becomes apparent as juvenile tuatara approach the size of sexual maturity. At the top is a male of 154 mm snout–vent length, and on the bottom, a female of 160 mm snout–vent length (Stephens Island, January 1993). Sexes were confirmed by laparoscopy.

Photo: Alison Cree.

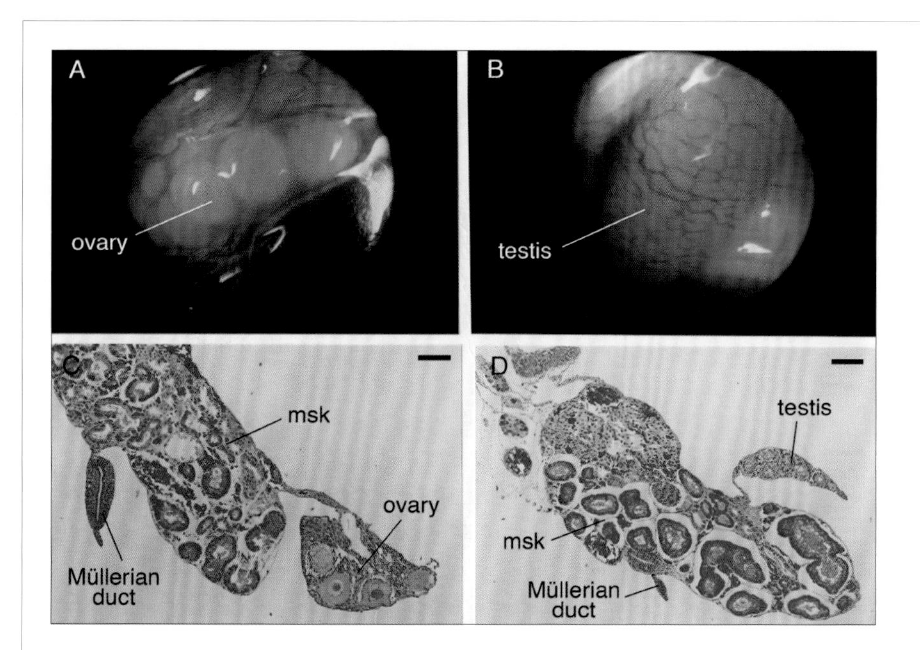

Box 8.1 Fig. 2 The internal reproductive organs of juvenile tuatara are clearly differentiated by hatching, although external differences in sex take some years to become obvious. The top photos show views of an ovary (A) and a testis (B) obtained during laparoscopy of live juveniles (136 and 156 mm snout–vent length, respectively). The bottom photos show histological sections of the gonads and ducts of smaller juveniles that died in a zoo (C, female, 65 mm and D, male, 58 mm snout–vent length). In C, note developing ovarian follicles and Müllerian duct differentiating into an oviduct. In D, note seminiferous tubules in the testis and regressing Müllerian duct. msk = mesonephric kidney. Scale bars in C and D are 100 micrometres (or 0.1 mm). Colour reproduction, relabelled, of image previously reproduced in black and white in Nelson, N. J., Cree, A., Thompson, M. B., Keall, S. N. and Daugherty, C. H. 2004a. Temperature-dependent sex determination in tuatara. In: Valenzuela, N. and Lance, V. (eds). *Temperature-dependent sex determination in vertebrates*. Smithsonian, Washington, pp. 53–58.

Photos: Alison Cree.

In a remarkable turn-around, Dawbin eventually concluded that sexual dimorphism in tuatara was 'marked'[19] and 'amongst the most striking of all reptiles'.[20] He quantified differences in snout–vent length and mass (both much greater in males), noting also that males had a crest (comprising nuchal and back regions) in which the spines were better developed than in females. Males were additionally said to have a proportionately narrower abdomen and larger head.[21]

These differences have since been widely accepted and applied by biologists to distinguish the sexes of adults,[22] and differences in head dimensions have been quantified.[23] Occasionally, there has been misidentification of sex around the time of sexual maturity. Laparoscopies performed on small 'adult females' collected for X-raying on Stephens Island revealed that a few were small males[24] (but not a high enough proportion to interfere with the conclusion that females do not breed each year).[25] On northern islands, some very large females have spines with a rather masculine appearance, but their abdomen width is typical of females.[26]

Laparoscopy has revolutionised the sexing of small tuatara, including juveniles.[27] In this minor surgical

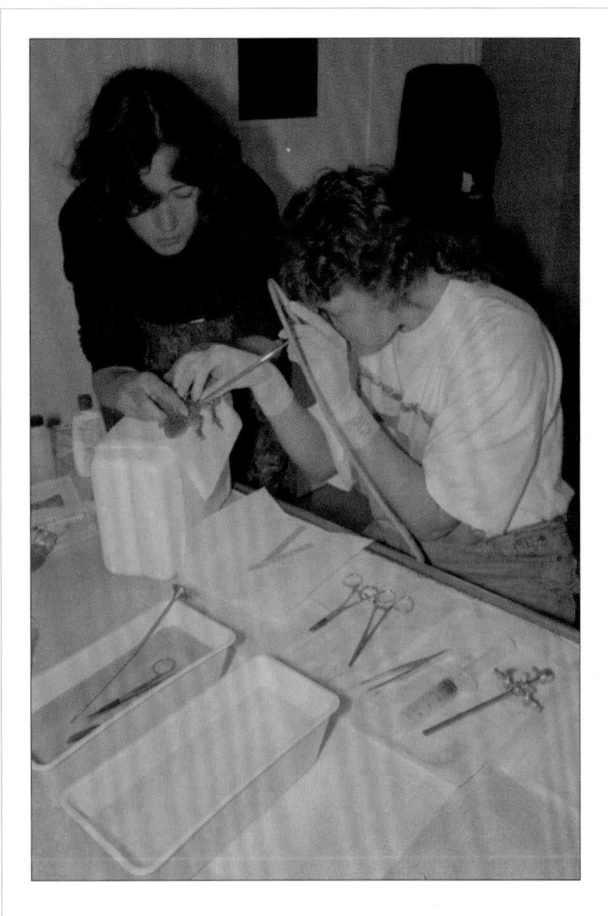

Box 8.1 Fig. 3 With the assistance of the author, Claudine Tyrrell performs a laparoscopy to sex a juvenile tuatara.

Photo: collection of Alison Cree.

procedure, a medical telescope (2–5 millimetres in diameter, depending on the size of the tuatara) is inserted into the abdominal cavity under local anaesthesia. With experience, the procedure requires just a few minutes to distinguish the developing gonads as ovaries or testes (oviducts in females can also often be seen). The laparoscope is then withdrawn and the wound closed with sutures that eventually disappear. Tuatara tolerate the procedure well and recover with no lasting ill-effects. Laparoscopy has been used by Nicola Nelson to identify the sex of hundreds of juveniles within about the first year after hatching, enabling the pattern of temperature-dependent sex determination to be confirmed years in advance of sex becoming apparent externally.[28]

Using laparoscopy to confirm gonadal sex, Jolene Oldman, Claudine Tyrrell and I have begun to examine whether features of sexual dimorphism begin to appear in juveniles on or from Stephens Island before they reach maturity. Some of the same features that distinguish adult males, including darker throat colour and longer neck spines, can also be used to sex at least some half-grown juveniles, and subtle differences in head shape appear to exist in even smaller juveniles.[29] Some of these differences are likely to be enhanced by the rise in plasma testosterone concentration that becomes evident in juvenile males as they approach maturity.[30]

Cycles of gonadal activity in both males and females are extended affairs. Males reproduce annually, their testes reaching peak activity at the time of mating; however, the early stages of sperm development extend throughout the year. Female tuatara take at least 2 and typically about 4–5 years to produce a single clutch of eggs. In the year that a female nests, eggs are laid during late spring or early summer (October–December). Both mating and nesting involve conspicuous patterns of behaviour. Adult males and females are also sexually dimorphic, although this took a surprisingly long time to be accurately recognised.

Reproductive anatomy of males

The paired testes of tuatara are located in the abdominal cavity, slightly closer to the hindquarters than the forelimbs (Figure 8.2).[7] Each testis, its corresponding epididymis (the coiled upper part of the duct system that transports sperm from the testis to the cloaca) and adrenal gland are suspended from the dorsal body wall by a thin membrane known as the mesorchium. In life, the testes are creamy-white with occasional spots of black pigmentation, and blood vessels branch over the surface (Figure 8.3). Each testis is typically an ovoid or elongate disc, flattened dorso-ventrally to about 1–5 millimetres in depth, and tapered at its margin to an edge that is sometimes lobulated. The two testes lie alongside each other on either side of the spine (one testis may be a few millimetres more anterior or posterior than the other), and are similar in size. Among preserved adults from Stephens Island that I examined, the right testis averaged 257 milligrams (equivalent to 0.07% of body mass).[8] Mean dimensions of the right testis were 16.8 x 8.4 millimetres (length x width); those for the left testis were 19.4 x 8.8 millimetres.[9] Shape varied both within and among individuals; some testes were much more elongate than others, but there appears to be little if any seasonal variation in size.[10] A similar range of sizes and shapes was apparent among testes from 13 preserved adults that I examined from other or unknown islands.

In internal structure, the testes consist largely of an outer layer of connective tissue (tunica albuginea) surrounding an inner mass of looped seminiferous tubules (Figure 8.4).[11] Developing sperm cells and their supporting Sertoli cells are found in the seminiferous epithelium (the epithelial, or lining, layer of the seminiferous tubules). Leydig cells (also known as interstitial cells, and assumed to be the main source of testosterone) are seen occasionally in the interstitial tissue between the seminiferous tubules.[12]

Spermatids (a late stage in the development of sperm) and spermatozoa (mature sperm) from the testes of tuatara from Stephens Island have been described using

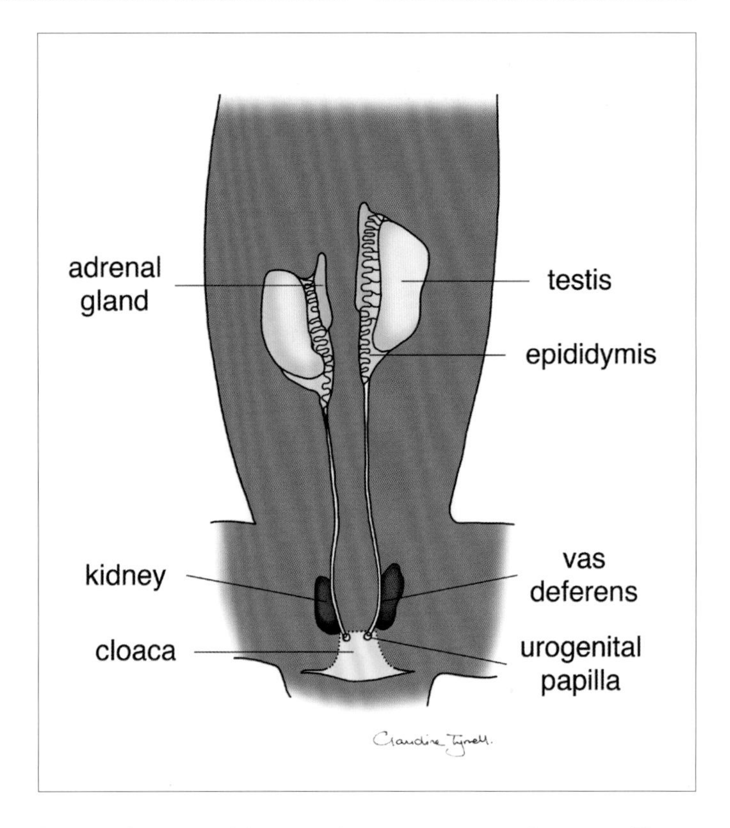

Fig. 8.2 Diagram of the reproductive system in male tuatara. The base of the cloaca is shown partly opened to reveal the urogenital papillae inside.

Based on dissections and a drawing by Claudine Tyrrell, with supporting information from Günther 1867 and Osawa 1897.

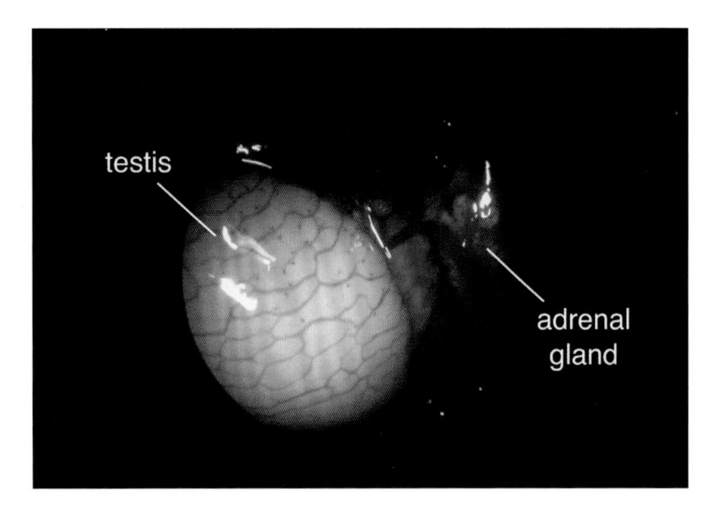

Fig. 8.3 Laparoscopic view of a testis and adrenal gland in a live male tuatara (a captive animal of Stephens Island stock).

Photo: Alison Cree.

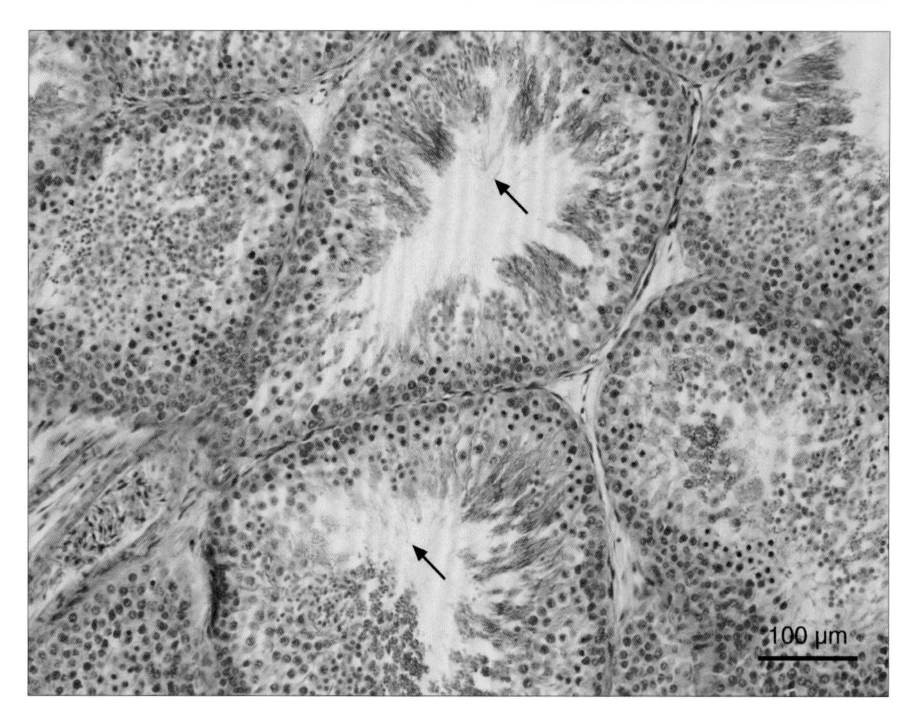

Fig. 8.4 Photomicrograph of the testis of an adult male tuatara in summer (cross-section of a specimen collected 31 December 1955, Stephens Island). Mature sperm (arrows) are being shed into the lumen of the seminiferous tubules, and a variety of earlier stages of sperm production are present in the epithelial layer (scale bar is 100 micrometres or 0.1 mm).

Photo: Alison Cree.

transmission electron microscopy.[13] Testicular sperm are about 140–144 micrometres (0.140–0.144 millimetres) in length with a filiform (thread-like) shape (Figure 8.5). They have the usual components of sperm: a head region containing an acrosome and an elongate, helical nucleus; a mid-piece containing mitochondria around an elongate centriole; a long principal piece containing an axoneme surrounded by a fibrous sheath; and a short end-piece. In their ultrastructure, the sperm of tuatara appear more similar to those of turtles and crocodilians than to those of their squamate relatives (lizards and snakes). For example, tuatara, turtles and a caiman exhibit one or more canals within the nucleus, whereas squamates have apparently lost any endonuclear canals.[14] Whether such similarities reflect comparable functional needs or constraints is unknown.

Sperm ducts lead from the testes to the cloaca. Mature sperm from each testis pass through fine ducts (recently renamed as tubules of the rete testis[15]) to reach the epididymis. The epididymis contains a convoluted mass of efferent tubules or ductuli efferentes,[16] distinguished in part by their ciliated cells[17] and leading into a central epididymal duct or ductus epididymis. The efferent tubules extend posteriorly to the pelvis, lying alongside the anterior third of the kidney,[18] before the epididymal duct straightens and narrows to become the relatively short vas deferens, also known as the ductus deferens. The epididymides and vasa deferentia are largely white, but patches of black pigmentation are evident on their surfaces[19] and in histological sections.[20] The epididymal duct contains a pseudostratified, secretory epithelium[21] and probably functions not just as the main region for sperm storage but also for further processing and maturation of sperm. Sperm from the epididymis show ultrastructural changes compared with those from the testis, including a sheath of flocculent material around the principal piece.[22]

In male squamates, a specialised region of the kidney (the renal sex segment) has evolved. This region becomes seasonally enlarged, producing secretions that may function as sperm nutrients or activators.[23] Male tuatara, however, have no renal sex segment,[24] and in this they resemble crocodilians and turtles. As in some other male reptiles,[25] adult male tuatara sometimes retain a pair of Müllerian ducts as vestigial structures.[26] These open-ended ducts, each one of which lies dorsally to an epididymis and vas deferens, are the remains of embryonic structures that are present initially in both sexes and become the functional oviducts of females. The vestigial Müllerian ducts in male tuatara extend posteriorly from the anterior abdomen[27] to the cloaca, where each ends blindly.[28]

The cloaca of tuatara is divided by folds into three chambers.[29] From the anterior end these are the coprodaeum, the urodaeum and finally the proctodaeum, from which the transverse vent opens to the outside.[30] The transverse vent is a feature that helps unite tuatara, lizards and snakes, distinguishing them from crocodilians and turtles. In male tuatara, the vas deferens and ureter of each side open onto a urogenital papilla in the wall of the urodaeum.[31]

In both sexes of tuatara, two pairs of glands occur in the cloacal region (Figure 8.6). These are the tubular glands in the posterior urodaeum, and the sebaceous glands in the ventro-lateral wall of the proctodaeum.[32] The

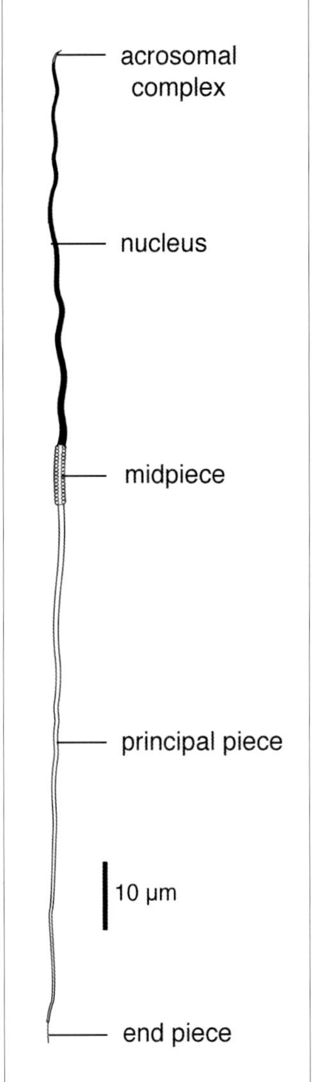

Fig. 8.5 Diagram of sperm structure in tuatara from Stephens Island. The scale bar (10 micrometres) represents one-hundredth of a millimetre.

Modified from Healy, J. M. and Jamieson, B. G. M. 1992. Ultrastructure of the spermatozoon of the tuatara (*Sphenodon punctatus*) and its relevance to the relationships of the Sphenodontida. *Philosophical Transactions of the Royal Society of London B* 335: 193–205.

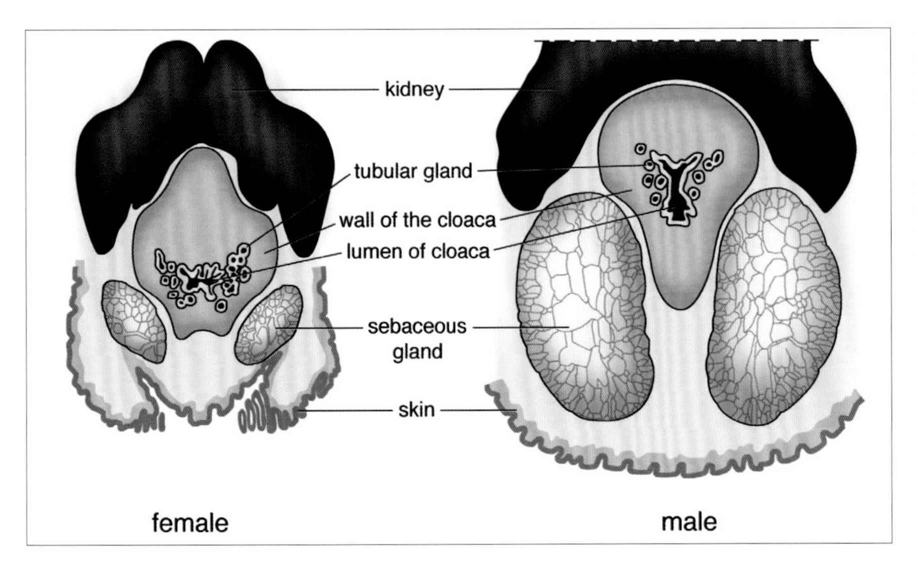

Fig. 8.6 Diagram of cross-sections through the cloacal region of a male and a female tuatara from Stephens Island. Tubular glands are present in the cloacal wall at the level of the posterior urodaeum to proctodaeum (female) and proctodaeum (male). The large, paired sebaceous glands are also visible.

Modified, with colour added, from Gabe, M. and Saint Girons, H. 1965a. Contribution à la morphologie comparée du cloaque et des glandes épidermoïdes de la région cloacale chez les lépidosauriens. *Mémoires du Muséum National d'Histoire Naturelle, Paris* Série A Zoologie 33: 149–292 + 15 plates. See also Gabe and Saint Girons 1964a.

Fig. 8.7 White secretions ooze from the sebaceous glands within the cloaca of a captive male tuatara in late summer.

Photo: courtesy of Lindsay Hazley, Collection: Southland Museum & Art Gallery.

sebaceous glands, which are visible through the opening of the vent, have attracted more attention from scientists. The mass of these large, rounded structures ('double the size of a common pea', in one early researcher's view)[33] averaged 173 milligrams in 12 preserved males from Stephens Island that I examined.[34] They were notably larger in a male than in a female examined by Gabe and Saint Girons,[35] but the male also had a larger body size, and proportional sizes remain unexamined.

Each sebaceous gland within the cloaca is divided by connective tissue into compartments, within which there are cells containing small fat droplets.[36] The glands open via about eight small openings into the proctodaeum.[37] A thick, creamy-white secretion oozes from the ducts during the late summer mating season (Figure 8.7),[38] but no seasonal changes in histological appearance of the glands have been detected.[39] The main component of the secretion has recently been identified as a glycoprotein. The constituent fatty acids include at least two that are thought to be novel to nature, one of which has been termed 'tuataric acid'.[40] Gadow described the glands in the living animal as having 'a strong, rather agreeable smell of musk and violets',[41] but more importantly we do not know how tuatara react to the smell. The glands could have a pheromonal function similar to the femoral and pre-anal glands of some male squamates.[42] The secretions are reported to vary in composition in stable ways among individual tuatara, and could possibly enable individuals to distinguish each other.[43] However, whether the composition varies in consistent ways between the sexes is not known, and studies examining function are needed.

As Günther noted in 1867, the urogenital system of male tuatara is broadly similar to that of lizards – the major exception being the absence of an intromittent organ. In male tuatara, neither a penis (as in turtles or crocodilians) nor paired hemipenes (as in squamates) are present.[44] Various researchers have suggested that features of the cloaca might somehow compensate for this lack. In 1887 Gadow suggested that the circular fold between the proctodaeum and urodaeum might be 'protruded by inward pressure of the cloaca in order to secure conception'[45] and that 'the inner walls of the cloaca form by protrusion and evagination a temporary intromittent organ'.[46] Arnold reported in 1984 that two shallow out-pouchings on the posterior wall of the cloaca might allow the opening of the sebaceous glands to be everted and its secretions released; alternatively or additionally, Arnold suggested that these out-pouchings might allow shallow insertion of the male's cloaca into the female's cloaca at copulation.[47] Subsequent observations confirm that the swollen cloaca is partly everted in male tuatara during courtship, just prior to copulation,[48] but the extent to which the everted tissue is actually inserted into the female's cloaca is unknown. There are no reports of grooves on the cloacal out-pouchings (equivalent to those that transport sperm along the surface of squamate hemipenes), so it may be premature to consider the out-pouchings as rudimentary hemipenes.

Seasonal reproductive cycle of males

Histology of the testes and sperm ducts (epididymides and vasa deferentia) has been examined in healthy adults collected from Stephens Island at known times of the year.[49] These studies confirm that reproductive activity in males is annual. However, the cycle of sperm production (spermatogenesis) is unusual among temperate-zone reptiles in that there is no period of complete regression of the testes and epididymides. Testicular mass shows little if any seasonal variation, and primary spermatocytes are present throughout the year. However, spermatocytogenesis (the production and development of spermatocytes, the cells that eventually become spermatids) is probably slow over winter, with some evidence of abnormal cell division such as the presence of isolated chromosomes. Spermiogenesis (the maturation of spermatids into spermatozoa, or 'sperm' for short), begins in December and continues until March, by which time a new wave of spermatocytogenesis appears imminent.[50] Mating occurs between January and March,[51] and the summer–early autumn spermiogenesis can thus be classified as pre-nuptial. The epididymal ducts are enlarged by the beginning of summer and contain abundant sperm from December until March, but are never completely regressed.[52]

Although continuous spermatocytogenesis is unusual in reptiles generally, the phenomenon seems likely in at least some New Zealand lizards.[53] It probably represents an adaptation to a maritime environment where summers remain relatively cool and winters relatively warm by continental standards, allowing some activity of the reproductive organs throughout the year.[54] A pattern of late summer or autumn spermiogenesis and mating is also characteristic of some New Zealand lizards.[55]

As in other male reptiles, the main androgen in tuatara is testosterone.[56]

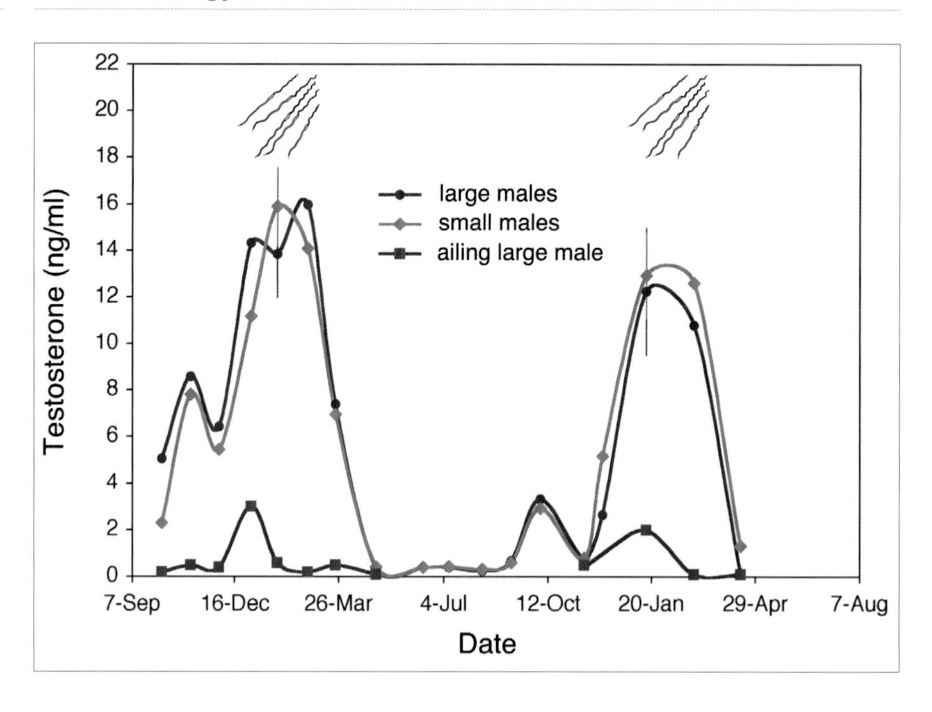

Fig. 8.8 The seasonal cycle of plasma testosterone concentration in male tuatara on Stephens Island from October 1986 to April 1988. In healthy animals, similar peaks occur between December and March in the mean concentrations for large males (241–261 mm snout–vent length, dark blue circles) and in the mean concentrations for small males (187–225 mm snout–vent length, teal diamonds). Males in both size classes have mature sperm in their testes between mid- and late summer (wavy lines at top of figure), but only large males are likely to be seen mating. Immature stages of sperm production are present in the testes of healthy males throughout the year. Standard errors are shown for the January means only. One large, malnourished and apparently aged male had relatively low concentrations of testosterone throughout (red squares).

Modified from Cree et al. 1992.

The seasonal pattern of secretion in males on Stephens Island broadly matches but is somewhat more extended than that for spermiogenesis.[57] Testosterone begins to rise in concentration in the blood plasma (the liquid component of the blood)[58] by mid-spring (October) and peaks between January and March, at the same time that spermiogenesis is at its peak and mating is occurring (Figure 8.8). The concentration then drops rapidly and remains low over winter. A short-term dip in testosterone concentration during late spring–early summer was seen in two consecutive years on Stephens Island, and was possibly weather related. There was no difference in testosterone cycles, including summer peaks, between small adult males (187–225 millimetres snout–vent length) and large adult males (241–261 millimetres), despite the fact that small males are almost never seen mating.[59] However, a large male that was malnourished and probably aged showed little or no summer elevation in plasma testosterone concentration, and spermatogenesis was abnormal in other emaciated males.[60] Plasma testosterone concentrations in healthy males show no difference between day and night in spring or summer, but, as is typical for male reptiles, concentrations fall within 24 hours when tuatara are captured and confined within a cloth bag.[61]

Although spermatogenic cycles have not been studied in tuatara from other islands, some information is available on the plasma concentrations of testosterone in late summer and early autumn. Concentrations on North Brother Island and several Bay of Plenty islands in January, and on more northerly islands during February and March, are broadly comparable with those on Stephens Island in the same months (Figure 8.9).[62] These similarities suggest that reproductive activity is typically high on all islands, not just Stephens Island, during the late-summer period. Males on Kawhitu or Stanley Island in the late 1980s were a possible exception; although mean plasma testosterone concentration in late summer was not significantly lower than on nearby northern islands at the same time, there was a suggestion that reproductive activity (as well as juvenile recruitment) might be reduced in the presence of introduced Pacific rats (*Rattus exulans*).[63] This relict population of aging adults was subsequently transferred to captivity as part of a restoration programme, but reproductive performance has been poor compared with captive groups from nearby islands.[64]

Nothing is known of the hormones that exert higher-level control over the reproductive system in male or female tuatara. For example, the structure of the gonadotrophin-releasing hormone(s) or GnRHs from the hypothalamus in the brain is undescribed.[65] Whether the pituitary gland secretes two gonadotrophic hormones (resembling follicle-stimulating hormone or FSH, and luteinising hormone or LH, as in crocodilians and turtles) or one that seems to be neither (as in squamates)[66] is also unknown, though of obvious phylogenetic interest.

Fig. 8.9 Plasma testosterone concentrations in late summer for male tuatara from northern and Cook Strait islands. Locations are listed from north to south for islands or island groups with at least five tuatara sampled in 1989 (except North Brother Island, sampled in 1988).

Mean concentrations are from Cree et al. 1995a and unpublished observations (vertical lines indicate standard errors, and sample sizes are given at base of bars). For comparison with Stephens Island in the same months in 1987 and 1988, see Figure 8.8.

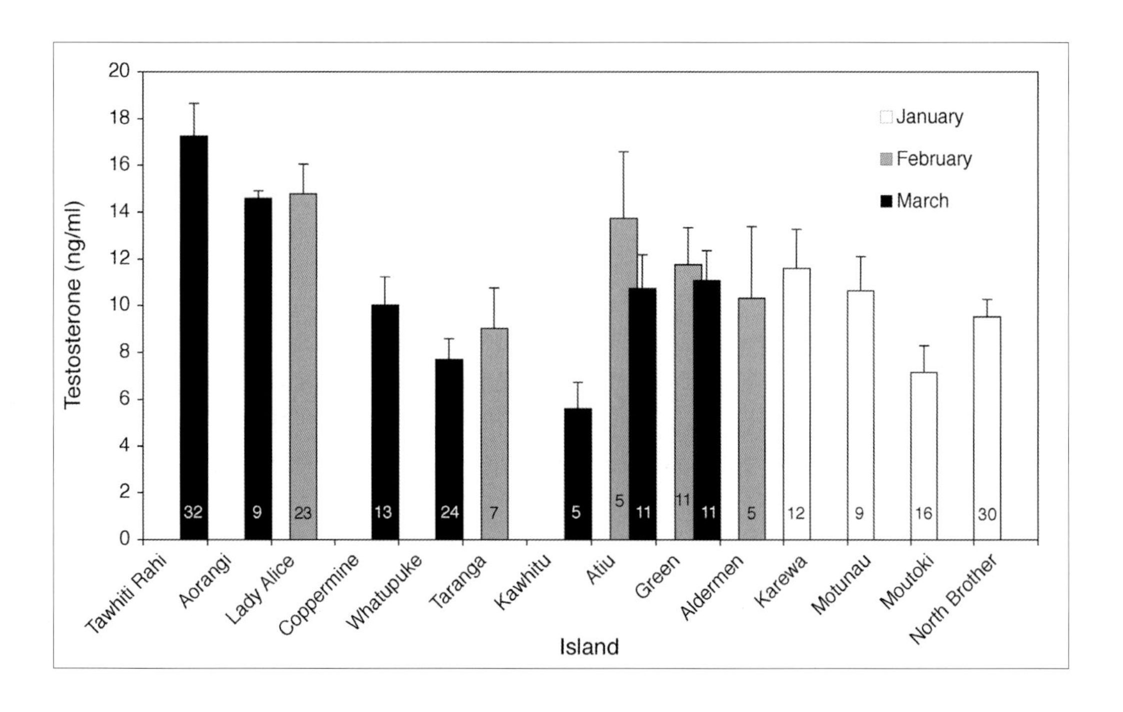

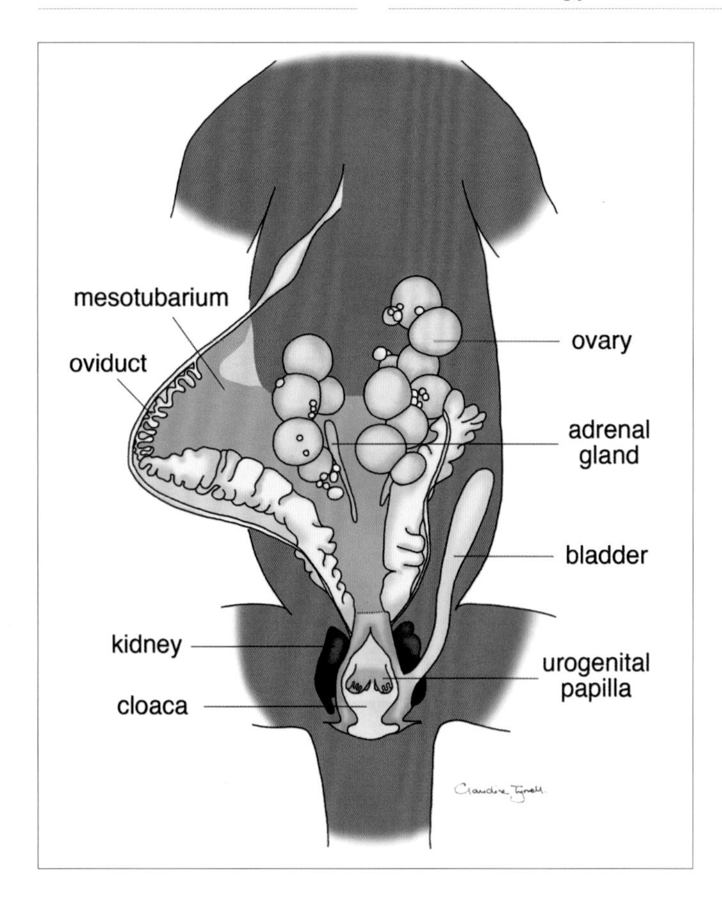

Fig. 8.10 Diagram of the reproductive system in female tuatara. Each of the two ovaries contains numerous ovarian follicles, ranging at this stage of the reproductive cycle from tiny white spheres to large, yolk-filled spheres. The bladder and the right oviduct (as viewed dorsally) have been reflected to the side (the anterior end of the left oviduct is concealed). The cloaca is cut open to reveal the urogenital papillae inside.

Based on dissections and a drawing by Claudine Tyrrell, with supporting information from Osawa 1898b.

Reproductive anatomy of females

Female tuatara have paired ovaries and oviducts (Figure 8.10).[67] The ovaries lie in the posterior abdomen, one on each side of the spine, with one ovary sometimes beginning a few millimetres anterior to the other.[68] The ovaries are longitudinally oriented and taper slightly towards the caudal end.[69] Each is connected to the body wall by a thin membrane (the mesovarium), which also supports the adrenal glands. It is unclear why the ovaries of tuatara have sometimes been described as 'solid',[70] as the text and illustration provided by Osawa in 1898[71] indicate that they are saccular with central fluid-filled cavities (Figure 8.11), the norm for most reptiles.[72] The internal lymph spaces are penetrated by strands of connective tissue, which contain blood vessels and smooth muscle fibres. Enlarging follicles move into the interior of the ovary as they grow, but remain connected by strands of connective tissue with the ovarian epithelium.[73]

Osawa was apparently unable to locate a discrete germinal bed or beds, as immature stages of oocytes (egg cells) were scattered over the ovary on the dorsal and to a lesser extent the ventral surfaces.[74] Although it has been suggested that oogonial division and development into oocytes might be completed during embryogenesis in tuatara,[75] oogonia were later reported in an adult female examined by Gabe and Saint Girons.[76] The granulosa (follicular epithelium) that surrounds the developing oocytes has not been described in detail, but it lacks the pyriform cells that fulfil a 'nurse cell' function in squamate reptiles (Figure 8.12).[77] The granulosa of tuatara thus resembles that of crocodilians and turtles, rather than the multi-layered, polymorphic granulosa of squamates.[78]

Among five mature females from Stephens Island that I dissected for other studies,[79] the fresh mass of both ovaries combined was no more than 1.3% of body mass in three females with relatively inactive ovaries.[80] However, in a female collected in March, with enlarged vitellogenic (yolk-containing) follicles that were close to ovulation, the combined mass of the two ovaries reached 10.6% of body mass (Figure 8.13). Ovaries from these females and seven others that were preserved contained numerous small, translucent follicles less than 2 millimetres (mm) in diameter, as well as a smaller number of creamy-white follicles about 2–4 mm in diameter. Vitellogenesis became visually evident (yellow colouration) in follicles of about 3–4 mm in diameter. Vitellogenic follicles reached a maximum diameter of 19 mm and were well

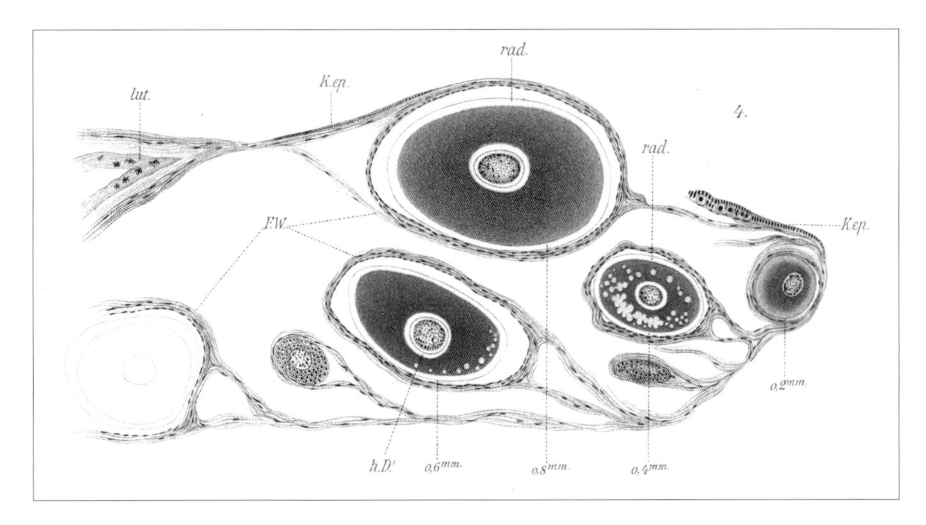

Fig. 8.11 Ovarian structure in a female tuatara. The diameter of several follicles is indicated in millimetres. F.W. = follicle wall; h.D. = inner layer of yolk; K. ep. = germinal epithelium; lut. = scar of corpus luteum; rad. = zona radiata.

Reproduced from Osawa, G. 1898b. Nachtrag zur Lehre von den Eingeweiden der *Hatteria punctata*. Die weiblichen Geschlechtsorgane. *Archiv fur Mikroskopische Anatomie und Entwicklungsmechanik* 51: 764–794 + Tafeln XXIII–XXV.

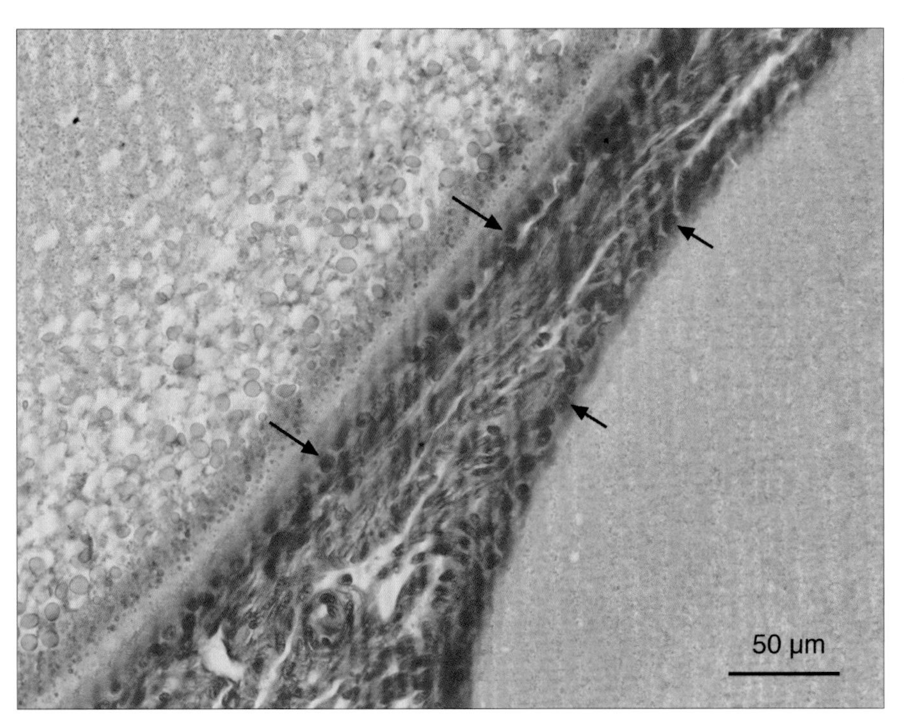

Fig. 8.12 Photomicrograph of two adjacent ovarian follicles from a tuatara. The follicle to the left has begun vitellogenesis, whereas the follicle to the right is in a more immature, pre-vitellogenic stage. Note that the granulosa layer of the pre-vitellogenic follicle (short arrows) has a similar structure to that of the vitellogenic follicle (long arrows). Being only about one cell deep, and lacking the specialised pyriform cells of squamates, the granulosa of tuatara resembles that of turtles and crocodilians (scale bar is 50 micrometres or 0.05 mm).

Photo: Alison Cree.

endowed with blood vessels in the female from March in a pre-ovulatory condition (Figure 8.14). Atretic follicles (those that have aborted the process of vitellogenesis and will be resorbed) are sometimes present in tuatara in vitellogenic condition.[81] Non-ovulated follicles that are presumably becoming atretic have sometimes been seen in gravid females (those carrying eggs in their oviducts),[82] but atresia appears neither extensive in nor universal among females.

Corpora lutea develop from ovulated follicles and remain conspicuous in the ovary until the shelled eggs are laid (Figure 8.15).[83] Small scars of about 1–3 millimetres in diameter and with an orange-brown-black colouration are often seen in the ovaries of mature females. These are probably corpora

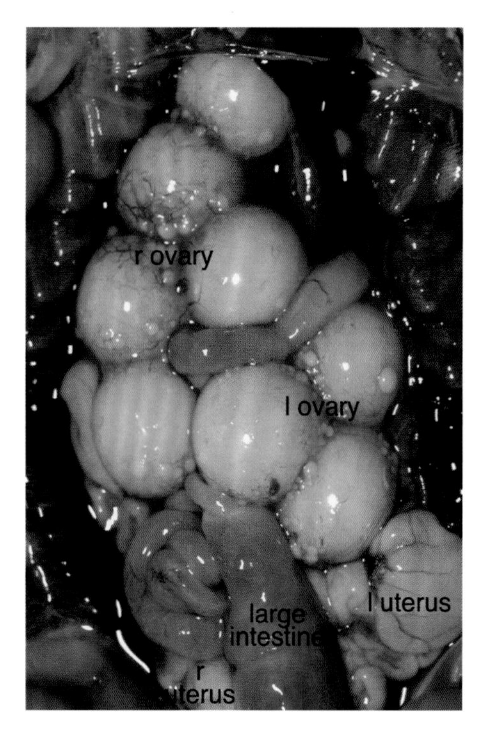

Fig. 8.13 Enlarged ovaries within the abdomen of a female tuatara from Stephens Island in pre-ovulatory condition (liver removed). This female was seen mating on 4 March 1988, collected immediately afterwards, and dissected on 18 March 1988. The right ovary (viewed dorsally) was anterior-most and contained five enlarged, yolk-filled follicles 18–19 mm in diameter. The left ovary contained three enlarged follicles, each 19 mm in diameter. Note the hypertrophied uteri, with small patches of black pigmentation. r = right; l = left.

Photo: Alison Cree.

albicantia (scars of corpora lutea). The luteal scars may number 50 or more in each ovary of large females,[84] and possibly persist for life as in some squamates.[85] The morphology and role of the corpora lutea are discussed later, in the section on the physiology of gravidity (the period when eggs are receiving their shells in the oviducts).

The paired oviducts of tuatara are lengthy tubes for egg processing (Figure 8.16). Their length when removed from the abdomen averaged 79% of snout–vent length among four females that I examined from Stephens Island.[86] Each oviduct begins anteriorly at a location just posterior to the axilla (armpit) and extends back to the cloaca. The oviducts are attached to the body wall on the medial edge by a thin membrane known as the mesotubarium. The posterior mesotubarium is often black, as also noted for tuatara from North Brother Island.[87] Small patches of black pigmentation also sometimes appear on the otherwise white oviduct (Figure 8.13). Each oviduct is extensively folded, and constrained along its lateral edge by a band of connective tissue and muscle known as the cordon ligament, or as the German *Haltband*.[88]

As in other reptiles,[89] the oviducts can be divided longitudinally into several regions, of which four can be conveniently recognised in tuatara.[90] The terminology used here follows that presented for reptiles by Girling in 2002.[91] The infundibulum or anterior-most region begins with a cleft-like ostium abdominale (the opening to the abdominal cavity, through which ovulated oocytes will enter). The infundibulum[92] is thin-walled, translucent and initially flat, but becomes finely pleated at its posterior end. The uterine tube that follows is thicker and more convoluted.[93] There is some discrepancy in the literature about whether either of these regions is glandular in tuatara. Robb stated, without giving a source, that '[g]lands in the upper part of the oviduct produce albumen, with which each ovum is surrounded'.[94] However, neither Osawa[95] nor Gabe and Saint Girons[96] reported glands in these regions, and there is very little albumen in freshly laid eggs (see later section on composition of eggs). Mucus-secreting cells and ciliated cells are present in the epithelium of the uterine tube.[97]

The uterus, which follows the uterine tube, is the widest and longest region of the oviduct. Among four mature females that I dissected (one in pre-ovulatory condition and three that were gravid), the uterus accounted for 52–70% of the total length of the oviduct.[98] The uterus is the region in which shell formation occurs, and it becomes especially thick-walled and folded during late vitellogenesis. Numerous glands are present in the lamina propria of the uterus (the layer of connective tissue immediately below the luminal epithelium; Figure 8.17);[99] these glands open into the lumen and almost certainly secrete components of the eggshell, as in other egg-laying reptiles.[100] The eggs of gravid females greatly distend the uterine walls and fill much of the abdominal cavity (Figure 8.18).

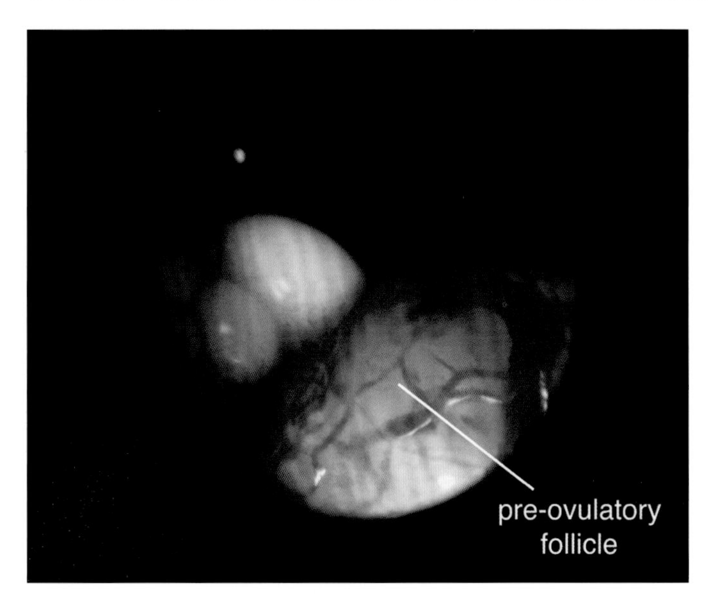

Fig. 8.14 Laparoscopic view of an ovary shown in the preceding figure. Laparoscopy was performed the day before dissection. Note the hierarchy of ovarian follicles, ranging from a small, translucent follicle less than 2 mm in diameter (left) through to a yolk-filled, well-vascularised pre-ovulatory follicle 19 mm in diameter (right, labelled). The intermediate-sized, creamy-white follicle (about 4 mm in diameter) would probably have contributed to the next clutch of vitellogenic follicles. The small brown scar (about 2 mm in diameter) on the surface of the pre-ovulatory follicle is probably a corpus albicans (regressed corpus luteum).

Photo: Alison Cree.

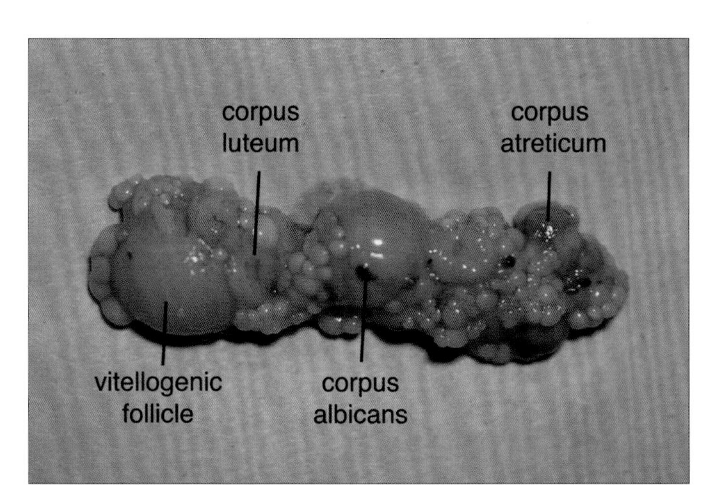

Fig. 8.15 Ovary of a gravid female tuatara collected from Stephens Island on 17 July 1988 and dissected on 4 August 1988. The ovary contained several large vitellogenic but non-ovulated follicles up to 11 mm in diameter, as well as several smaller (5–6 mm), partially yolked follicles that were obviously atretic (corpora atretica). Small, white, pre-vitellogenic follicles are numerous. Each corpus luteum has a prominent ovulation aperture (opening through which the oocyte had passed at ovulation). Small brown scars are probably corpora albicantia.

Photo: Alison Cree.

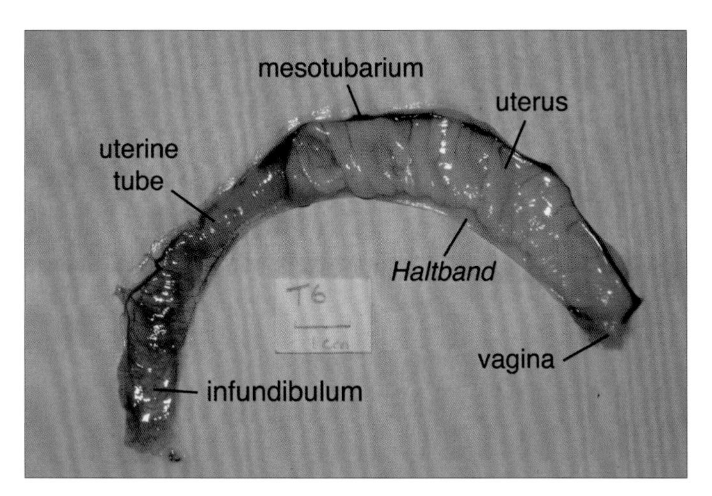

Fig. 8.16 Oviduct of the female tuatara in pre-ovulatory condition shown in Figure 8.13. The hypertrophied oviduct has four main regions. From anterior to posterior these are the infundibulum, which has a cleft-like opening at the anterior end, the uterine tube, the uterus and the vagina. The convoluted oviduct is suspended on its medial edge by a black mesotubarium (cut edge visible), and on its lateral edge is constrained by a strip of connective tissue called the *Haltband*. The heightened vascularisation of the uterine tube may be an artifact in this specimen.

Photo: Alison Cree.

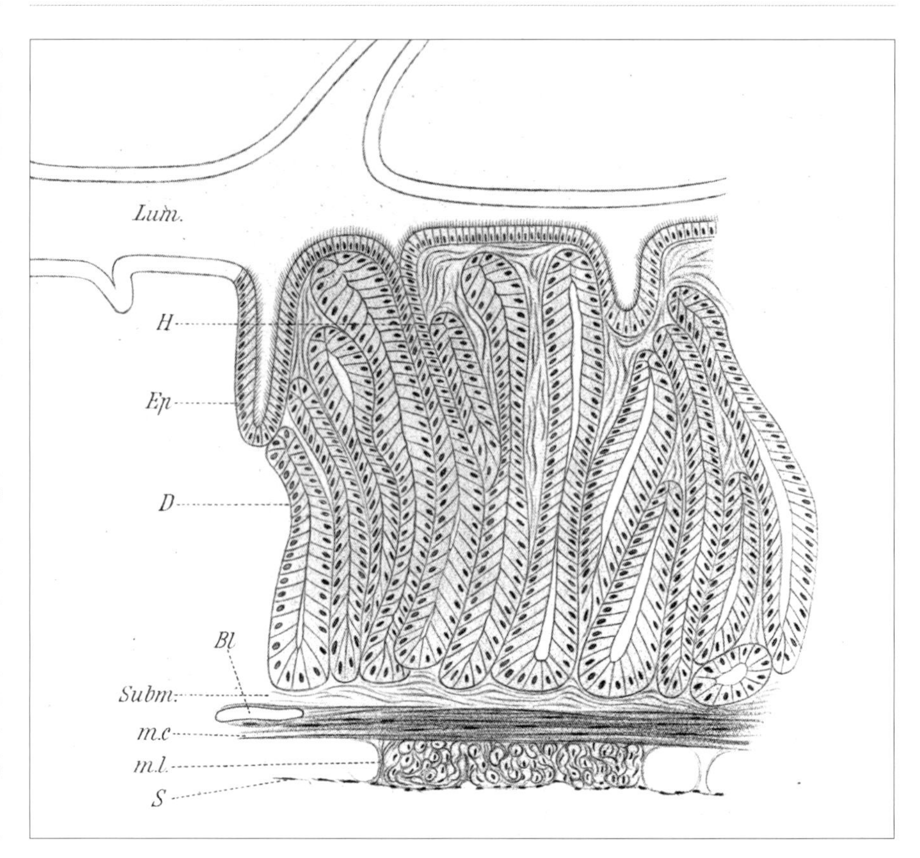

Fig 8.17 Cross-section of the uterine region of an oviduct from a female tuatara with enlarged follicles, presumed to be in mid-vitellogenesis. The luminal (inner) surface is lined with a ciliated epithelium, the lamina propria is densely packed with uterine glands, and below this lies the connective tissue of the submucosa, circular muscle, longitudinal muscle and an outer serosal epithelium. Bl = blood vessel; D = gland; Ep = epithelium; H = head of gland; Lum. = lumen; m.c. = circular muscle; m.l. = longitudinal muscle; S = serosa; Subm. = submucosa.

Reproduced from Osawa, G. 1898b. Nachtrag zur Lehre von den Eingeweiden der *Hatteria punctata*. Die weiblichen Geschlechtsorgane. *Archiv fur Mikroskopische Anatomie und Entwicklungsmechanik* 51: 764–794 + Tafeln XXIII–XXV.

The uterus leads into a very short (about 4 millimetres long), aglandular and muscular vagina, which opens into the cloaca. As in males, the cloaca of female tuatara is divided into three chambers. The vagina and corresponding ureter from each side of the body enter the middle chamber, the urodaeum, alongside each other at a urogenital papilla.[101] Again as in males, tubular glands are present in the urodaeum, and sebaceous glands alongside the proctodaeum.[102] No sperm receptacles or spermatozoa have been described from the reproductive tract of female tuatara.[103]

In a female tuatara in pre-ovulatory condition, the enlarged and well vascularised (but empty) oviducts had a combined mass accounting for 2.8% of body mass.[104] In four females whose oviducts contained eggs, the combined mass of both oviducts (including eggs) ranged between 7% and 18% of body mass.[105] The oviducal eggs lie against the ventral body wall and may receive support and protection from the gastralia (the so-called 'abdominal ribs'; see Chapter 1 for the evolutionary significance of gastralia).[106]

Overall, the reproductive tract of female tuatara is broadly similar to that reported for other egg-laying reptiles.[107] One notable difference in anatomy with possible implications for reproduction is that tuatara of both sexes lack abdominal fat bodies.[108] In squamates, these often-bulky fat stores in the pelvis can be drawn on seasonally to help support reproductive activity (including vitellogenesis) and/or winter dormancy.[109] Tuatara continue to feed intermittently over winter[110] and gonadal cycles are gradual and protracted, so there may be little need for specific fat-storage organs. Alternatively, lipid deposits in other locations in tuatara, for example in the tail,[111] in the large liver[112] or in corpora adiposa reported in the abdominal mesentery of females,[113] might substitute. Leptin, which is produced by the fat cells of vertebrates, might be involved with vitellogenesis; however, this recently discovered hormone has so far received little attention in reptiles,[114] and none in tuatara.

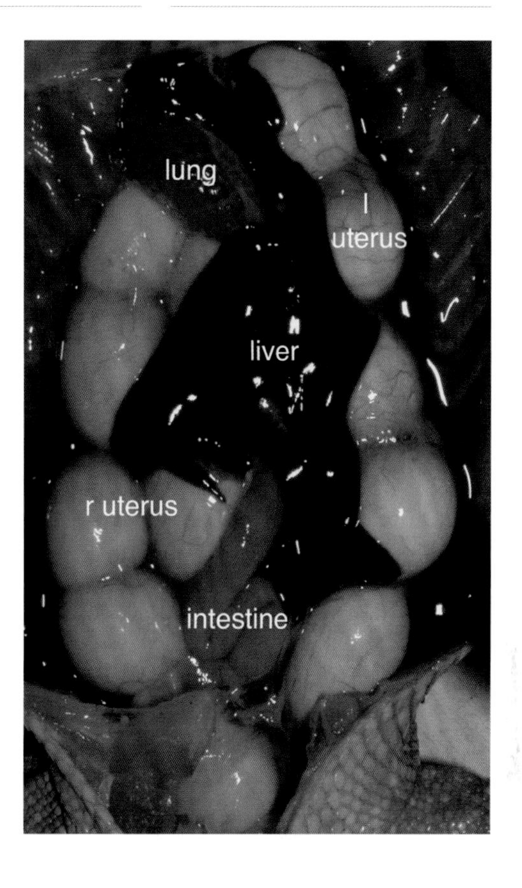

Fig. 8.18 Abdominal contents of the gravid female tuatara whose ovary was shown in Figure 8.15. The right (r) uterus contained seven eggs and the left (l) uterus five eggs. The ovaries are hidden from view. This female was known from radiography to have also been gravid 2 years previously.

Photo: Alison Cree.

Multi-year reproductive cycle of females

The reproductive cycle of female tuatara and other egg-laying reptiles begins with vitellogenesis, the process of egg yolk production. The yolk protein precursor, vitellogenin, is a protein–lipid complex that binds calcium, magnesium, phosphate and other ions. In egg-laying vertebrates, vitellogenin is made by the liver, circulates in the blood and is absorbed by the ovarian oocytes.[115] Within the oocytes, the protein component of vitellogenin is cleaved into yolk proteins. Vitellogenin thus provides the embryos of reptiles with a valuable source of nutrients including amino acids, lipids, phosphate and metal ions. Other nutrients, many derived from the liver, are also deposited into the oocyte of reptiles during vitellogenesis.[116]

The seasonal reproductive cycle of female tuatara has been studied in detail only on Stephens Island (Figure 8.19). There, female tuatara have a remarkably protracted period of vitellogenesis spread over several years, a pattern seen in few if any other reptiles. In the first year after nesting, vitellogenesis resumes in most female tuatara. This is evident from both laparoscopy[117] and semi-quantitative analysis of plasma vitellogenin levels.[118] Plasma concentrations of oestradiol (a specific form of oestrogenic steroid) are detectable during this period,[119] which is expected since this ovarian hormone stimulates vitellogenesis in tuatara,[120] as in other reptiles (see next section). Increased production of oestradiol probably also stimulates the oviduct walls to thicken, although this is unconfirmed in tuatara.

In a 'typical' ovarian cycle, vitellogenesis continues for about 3–4 years before nesting.[121] Growth of ovarian follicles occurs gradually over this period. Oestradiol and vitellogenin remain detectable in the plasma, with the latter showing a peak during mid–late vitellogenesis. However, vitellogenin concentration (as a proportion of total plasma protein) remains lower than reported for other reptiles.[122] Plasma concentrations of calcium, total protein and inorganic phosphate, which provide markers for vitellogenesis in some female reptiles, show less obvious fluctuations in tuatara. This is consistent with the more gradual growth of oocytes.[123]

Plasma concentrations of the steroid hormones testosterone and progesterone are generally low in female tuatara during the first 1–2 years of

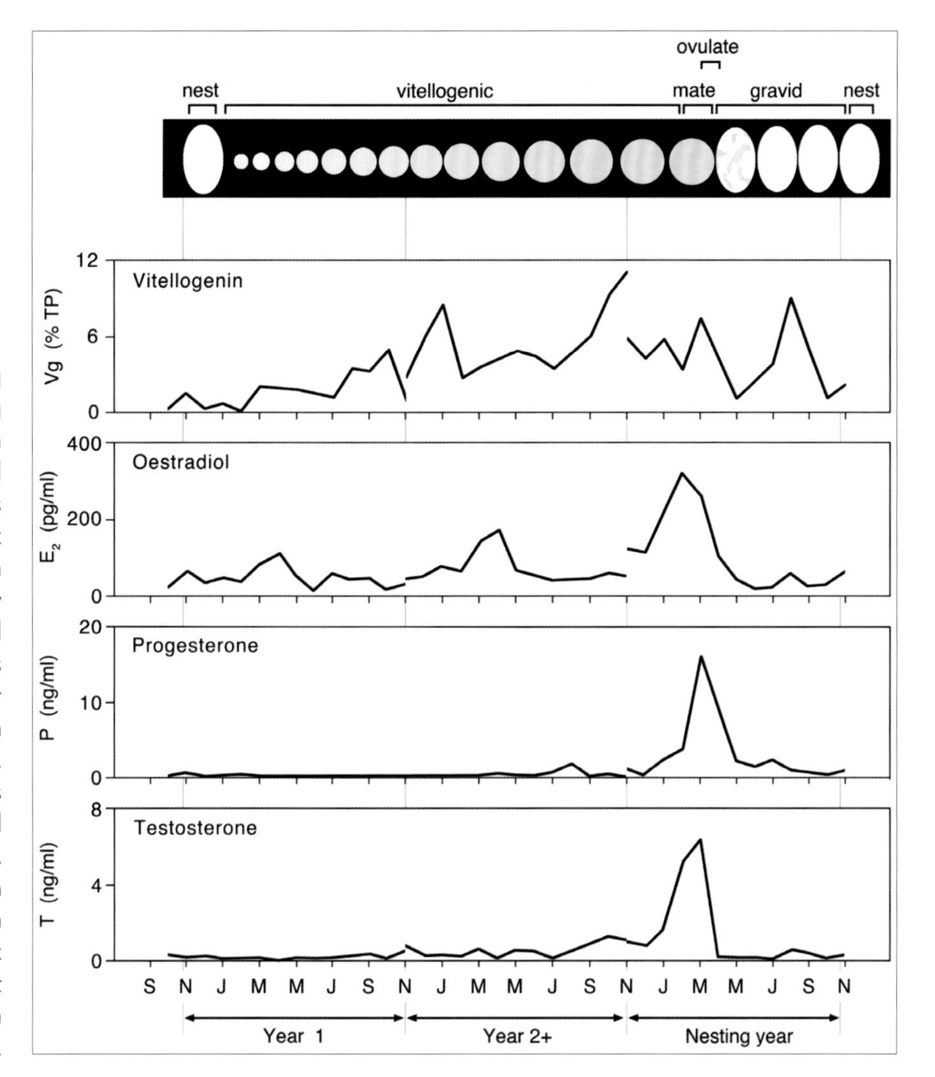

Fig. 8.19 Diagram summarising a typical 4–5-year cycle of egg production in female tuatara on Stephens Island. The top panel shows growth of ovarian follicles (yellow circles) as the yolk protein precursor vitellogenin is incorporated, followed by shell deposition on oviducal eggs (ovals) once ovulation has occurred. Note that the major period of vitellogenic growth (Year 2+) can last several years. Mean plasma vitellogenin is shown as a percentage of total protein (from Brown et al. 1991). Note the gradual rise in plasma oestradiol and vitellogenin during vitellogenesis, the peak in oestradiol and testosterone at about mating, and the peak in progesterone at about ovulation.

vitellogenesis. Testosterone increases during mid–late vitellogenesis, a pattern seen in many other female reptiles. This may simply reflect the fact that testosterone is a precursor in the synthesis of oestradiol, but it is possible that testosterone directly stimulates aspects of reproduction such as growth of the oviducts.[124]

In females, the most dramatic changes in plasma concentrations of sex steroids occur in late summer and early autumn of the nesting year, around the time of ovulation. Plasma concentrations of oestradiol and testosterone reach a peak during February–March, at about the time of mating, when females are in pre-ovulatory condition.[125] The concentration of progesterone peaks soon afterwards, during March–April,[126] and ovulation occurs by April. The surges in testosterone and progesterone within the period of mating and ovulation during late summer and early autumn are very pronounced, allowing females that are destined to nest that year to be distinguished from those that are not by measurement of these hormones.[127]

In other female reptiles, oestradiol stimulates female receptivity, and progesterone may enhance this effect.[128] Given that female tuatara show a pronounced surge in these hormones at about the time of mating only during a nesting year, and that male courtship attempts often fail because of female non-receptivity (see section on courtship and mating below),[129] it seems likely that females are receptive only in the nesting year. However, during the season in which they are receptive, females are known to mate more than once.[130]

Following ovulation, plasma concentrations of oestradiol and testosterone fall rapidly in females and remain low for most of the time that eggs are in the oviducts. Progesterone falls somewhat, but remains detectable throughout the gravid period.[131] Vitellogenin remains detectable in the plasma of many gravid females.[132]

Following nesting, most females on Stephens Island resume the next vitellogenic cycle. However, a few females can be found in which the ovaries remain regressed or contain atretic follicles (but apparently not healthy, vitellogenic follicles).[133] In these females, plasma concentrations of oestradiol, testosterone and progesterone all remain very low[134] and vitellogenin is undetectable.[135] At the other extreme, a few female tuatara on Stephens Island ovulate clutches two years apart. In these females, the entire vitellogenic cycle is compressed into just over one year.[136] No wild female on Stephens Island is known to have laid clutches of eggs in successive years (see later section on nesting frequency).

Some information on parameters of vitellogenesis is available for females from 11 northern islands. Plasma samples collected during February–March (just before or during the probable mating period) were analysed for sex steroids and vitellogenin.[137] Plasma vitellogenin concentrations were highly variable, ranging from the minimum detectable concentration of 67 nanograms per millilitre (seen in 20% of females) up to 2864 micrograms per millilitre,* and were not significantly correlated with plasma oestradiol or testosterone concentrations. The low concentrations support a pattern of gradual and protracted vitellogenesis, as on Stephens Island.[138]

Experimental induction of vitellogenesis

As in other reptiles, the production of vitellogenin (the yolk precursor protein) is stimulated by oestradiol, and both males and females are responsive to treatment.[139] This indicates that the gene for vitellogenin is present in both sexes, although normally only females produce sufficient oestradiol to stimulate the natural process.[140]

In both sexes of tuatara, the vitellogenic response to oestradiol treatment is remarkably slow compared with other reptiles.[141] Tuatara on Stephens Island given oestradiol implants showed a marked increase in plasma oestradiol (exceeding the known natural concentrations) within 2 days, but plasma vitellogenin showed only a small increase by 11 days. By 4–9 months after

* A microgram is one-thousandth of a milligram. A nanogram is one-millionth of a milligram.

oestradiol was first administered, plasma levels of vitellogenin and concentrations of calcium, total protein and inorganic phosphate were markedly elevated above those of control animals. Although the concentrations attained were similar to those seen in other species of oestrogen-treated reptiles, the response took weeks to months rather than days to weeks. The slow response is consistent with the protracted natural cycle of vitellogenesis. Tuatara may not possess the physiological or biochemical capacity to rapidly synthesise large quantities of vitellogenin,[142] perhaps because their body temperatures are often relatively low.[143]

Vitellogenin has been purified from tuatara.[144] The apparent molecular mass is about 218 kilodaltons, similar to that of vitellogenin subunits from other reptiles including turtles, lizards, some snakes and an alligator. Amino acid composition is also similar to that for other egg-laying vertebrates.[145]

Courtship and mating

Stephens Island is the only population of tuatara for which the full sequence of courtship and mating behaviour has been described. Contrary to A. K. Newman's suggestion in 1877 that male tuatara, lacking intromittent organs, would have 'probably small sexual passions, and but little rivalry',[146] recent studies reveal that male tuatara have a complex repertoire of courtship and territorial displays. Furthermore, territorial disputes can involve considerable

Fig. 8.20 A male tuatara on Stephens Island displays near its burrow with an erect crest during the mating season (March 1990). Photo: Michael Schneider.

BOX 8.2

Latex rubber models as aids to understanding reproductive behaviour

During the mating season, male tuatara respond to females with a stereotyped courtship display, and to other males with aggressive territorial displays. But how do they tell the difference in sex? James Gillingham and colleagues suspected that the primary cues were visual, and the approach they used to test this was both simple and effective.[1]

The researchers constructed models of tuatara from latex rubber, and had them appear and 'behave' in ways typical of either males or females. Each model had an erect, detachable crest that could be removed to transform a displaying 'male' into a relatively crest-less 'female'. The model was placed on a motorised base (a camouflaged, radio-controlled vehicle) so that it could be brought into position, without human disturbance, within a real male's territory. The head of the model could be moved up and down by remote control to mimic the slow, subtle

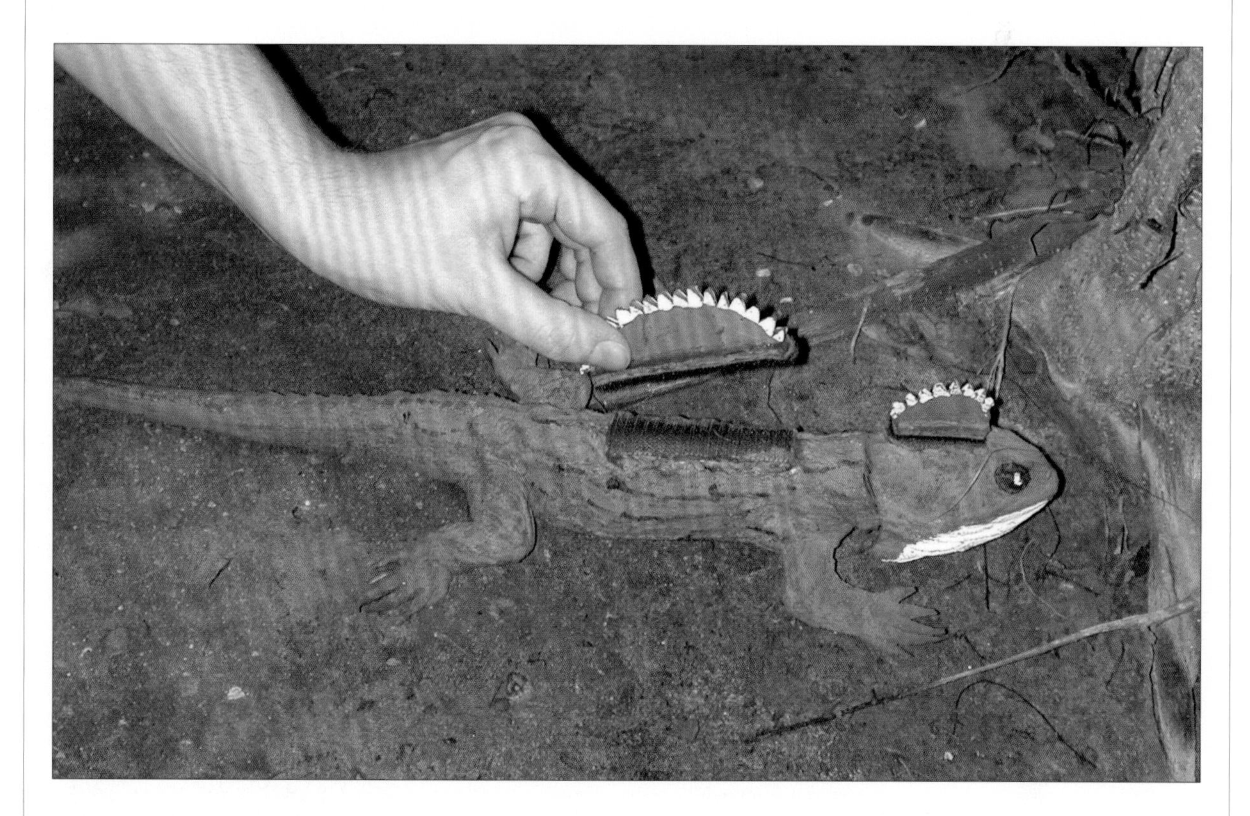

Box 8.2 Fig. 1 Life-size latex rubber models of a tuatara were used by Gillingham and colleagues to investigate social behaviour. Erect nuchal and dorsal crests were attached using Velcro patches so that the model superficially resembled a territorial male. A model with crests, but no head nods, was often attacked but never courted by male tuatara. Without the crests, the model more closely resembled a female. When a slow head nod was added, the crest-less model was courted by most males, and most males also attempted to copulate with it.

Photo: collection of James Gillingham.

head nods that female tuatara make during various social encounters. Real males have never been seen to make these head nods.

Models with typical 'male' or 'female' appearance that also displayed the appropriate sex-specific behaviour were very convincing to male tuatara. With an erect crest attached and no head-nods, the 'male-like' model was aggressively recognised by 8 of 11 territorial males (one even grasped the model and tossed it in the air). The 'female-like' model, which had no crest but which performed head nods, was treated to courtship displays by 9 of 12 males. The males performed a typical *stoltzer Gang* (courtship walk), climbed the vehicle, and even attempted to copulate with the model! The other two trials offered combinations of appearance and behaviour that were less sex-specific and presumably more confusing to male tuatara. When the model had an erect crest but a female-like head-nod, seven of 10 males initially responded with a few *stoltzer Gang*, but moved off without resuming courtship. When the model had no crest and no head-nod, 11 males showed no observable response.

These results confirm that male tuatara respond in very specific ways depending on the visual cues that other 'tuatara' present. A tuatara exhibiting a male-like crest but no head nod indicates a rival male, whereas a crest-less tuatara with a head nod is treated as a female. The head nod may even over-ride the aggressive stimulus provided by an erect crest. Gillingham and colleagues suggest that the head nod is essential to initiation of courtship but may also help to protect females from attacks by males.[2] Females do erect their crest in some situations, such as when displaying their territorial status to other females, so a head-nod in this case could discourage male aggression.

These results do not rule out possible communication by other means, such as sound and smell, during social interactions in tuatara. However, given that the rubber model did not provide such cues, it seems that these are not essential to initiate courtship or territorial defence by male tuatara. The use of similar models to investigate the responses of female tuatara in various social situations (including nesting) would be an interesting future development.

aggression. These improvements in understanding have come about not only through patient observation of wild and captive animals, but also with the use of equipment such as elevated viewing platforms, night-vision equipment and rubber models.[147]

As noted above, mating occurs on Stephens Island between January and March. Males vigorously defend their entire home ranges from intruding males during this period. The following description of territorial defence and courtship is based largely on the detailed accounts of James Gillingham and colleagues.[148] Resident males typically respond to the presence of an intruder by approaching to within about 1 metre, then display in a manner that has the effect of exaggerating body size. This involves elevating at least the anterior trunk off the substrate, inflating the trunk and gular (throat) region, and erecting the crest in the nuchal (back of the neck) and dorsal regions (Figures 8.20, 8.21). The resident often then walks slowly toward the intruder, sometimes swaying his head and neck side to side. Usually, this is sufficient for the intruder to leave the territory. Sometimes, however, the two males may move closer together (0.3–0.5 metres) in a 'face-off' display. One or other male slowly opens his mouth and then rapidly snaps it shut, a behaviour then

Fig. 8.21 A male tuatara displays with its inflated body held high in forest on Stephens Island in November 2005. A second male's head is partially concealed under foliage at the bottom left of the image.

Photo: Anne Besson.

Fig. 8.22a and 22b Two captive males of Stephens Island origin fight at the Southland Museum and Art Gallery. During the mating season, males are capable of propelling themselves through the air to attack other males. Severe wounds can result in the confines of captivity; one male died at a zoo from blood loss when its rival bit and severed an abdominal artery (Blanchard and the Tuatara Recovery Group 2002). In these images, the nictitating membrane sometimes covers the eye, giving it a purplish-white cast.

Photos: courtesy of Lindsay Hazley, Collection: Southland Museum & Art Gallery.

Fig. 8.23 A male tuatara courting a female on Stephens Island, on a track within pasture, in February 1987. Courting males have erect crests and an extended gular region. They move towards and around receptive females with a jerky, lunging walk known as a *stoltzer Gang*. Note the partly everted cloaca of the male.

Photo: Alison Cree.

performed by the other male in an alternating sequence. Eventually, one or other male violently whips his tail and begins a short (up to 2–3-metre) chase. Biting attempts are directed at the head, shoulders, trunk or tail (Figure 8.22) and are often accompanied by a croaking vocalisation. In one instance, a wild male bitten on the lower jaw was thrown into a complete body roll. Another male bitten on its tail dropped the tail tip (promptly eaten by a nearby female) before leaving the territory. Wounds and scars, including broken and healed lower-jaw bones, are reasonably common among male tuatara on Stephens Island, offering indirect evidence of aggression.[149]

Aggressive interactions between male tuatara in late summer are frequently followed by an attempt at courtship.[150] Although courtship occurs during the daytime and evening as well as at night in forest,[151] it has been seen only at night in the more open habitat of pasture (Figure 8.23).[152] Courtship is initiated by males when within 1–2 metres of a female. It initially involves erection of the crest and inflation of the body and gular region, much as in male–male territorial disputes. The skin above the shoulder and around the eyes also darkens. The male, with head elevated at an angle, then approaches the female in an ostentatious walk known by the German term *stoltzer Gang* ('proud walk').[153] The *stoltzer Gang* involves about 5–10 seconds of slow walking with the anterior trunk and head held high and the forelegs making exaggerated steps. The male then pauses for about 30–90 seconds before performing another *stoltzer Gang*. If the male's path is essentially unimpeded, the *stoltzer Gang* occurs in concentric circles around the female – I once saw a male initially pause and then climb about 150 millimetres over a low sloping tree trunk in order to complete the circle. If the male's path is more seriously impeded, the circuit is modified into an incomplete circle or a parade back

Fig. 8.24 Copulation on Stephens Island in March 1990. In some instances, the female seems almost flattened by the heavier male. Old, unsloughed skin has contributed to the dark skin patches on this male.

Photo: Michael Schneider.

and forth in front of the female, slowly moving closer to her.[154] Quiet grunting from the male has sometimes been heard during the *stoltzer Gang*.[155]

When courtship is successful (see below for unsuccessful attempts), the male eventually makes physical contact with the female and mounts her.[156] The male uses his front legs to grasp the female's shoulders; a lighthouse keeper's observation indicates that the male may also bite her shoulder.[157] The male simultaneously grasps at her pelvic region with his rear legs in an attempt to turn her cloacal region to the side. The male's cloaca is noticeably swollen and slightly protruded at this time. Copulation on Stephens Island can last up to 90 minutes (Figure 8.24).[158] The female remains quiescent during this time, as does the male with the exception of periodic squirms and shudders of the front half of his body, during which he may pull back on the female with his forelimbs.[159] Copulation ends when the male draws his cloaca away from the female and moves away.[160]

Observations of courtship and mating for tuatara from Stephens Island in captivity are broadly consistent with descriptions for wild tuatara.[161] In one instance, a courting male showed increased melanism (darkening) of the posterior portion of the nuchal crest as well as of the anterior shoulder region, and during the course of a *stoltzer Gang*, he bit the female on the neck at repeated intervals.[162] On this occasion, courtship lasted 54 minutes, mating for 15 seconds, and the male remained mounted for 15 minutes afterward. Mating in captivity has lasted up to 2 hours,[163] and one mounted male appeared to be performing 'pelvic thrusts'.[164] In one case of male–male aggression, males performed a 'face-off' during which they repeatedly bit each other in reciprocal fashion, with a distinct croak as they did so.[165] Other males in captivity have inflicted severe and occasionally lethal injuries on one another.

BOX 8.3

Reproducing in captivity: problems, successes and opportunities

Attempts in New Zealand to breed tuatara in captivity date back to at least 1885, when Professor A. P. W. Thomas of Auckland University College collected animals from Karewa Island. After resolving difficulties with sex identification (Box 8.1), Thomas apparently had some success with egg incubation and rearing of juveniles,[1] although his procedures and findings were never published in detail. From then into the early 1900s, eggs and young were occasionally produced in captive colonies, which were seemingly maintained out of individual curiosity[2] or occasionally for commerce.[3] However, little was learnt from the experience, and the attempts themselves were quickly forgotten.

Throughout the 1900s isolated attempts at maintenance continued within the zoo community. Despite improvements in husbandry and record-sharing within New Zealand[4] and overseas,[5] survival of captive tuatara was much lower than desirable[6] and breeding success remained poor.[7] During the 1980s the question of how to manage tuatara in captivity within New Zealand for long-term survival and successful reproduction acquired a more rigorous focus. Field observations by biologists[8] suggested that the following problems might contribute:

- inappropriate physical conditions (enclosures were often very small – sometimes less than 1 square metre – or lacked suitable nesting conditions; temperatures were sometimes lethally high; indoor enclosures often lacked ultraviolet light)
- inappropriate social structure (policy at the time allowed only two animals to be collected each year for captive breeding, but this did not allow for the courtship stimulus provided by male–male aggression, the fact that males attempt to breed annually whereas most females are probably receptive only every few years, and the possibility that some paired tuatara were simply incompatible)
- a tendency for captive animals to become overweight or obese
- high juvenile mortality, for a variety of possible reasons.

Box 8.3 Fig. 1 Nutritional secondary hyperparathyroidism is a well-known form of metabolic bone disease in captive reptiles. It arises when calcium levels are low, stimulating increased production of parathyroid hormone. The disease has been observed in tuatara kept indoors with insufficient ultraviolet light or insufficient dietary supplements to maintain the levels of vitamin D3 and calcium needed for normal bone calcification. Signs in this juvenile include conditions known informally as 'rubber jaw' and 'goggle eye' (technically, fibrous osteodystrophy of the lower jaw, and exophthalmia, respectively; see Boardman and Blanchard 2006). In severe cases, paralysis and death can result. Photo: Alison Cree.

In addition to biological problems, management issues were apparent. In the absence of a coordinated captive-management programme, sharing of information within the captive-breeding community was fragmented, and there was little or no integration with conservation efforts in the wild.[9]

The appointment by 1991[10] of a captive-breeding coordinator (Barbara Blanchard of Wellington Zoo), who was also appointed a member of the Department of Conservation's Tuatara Recovery Group, has done much to alleviate these issues. Indeed, the captive management of tuatara has now undergone a radical overhaul with the explicit goal of integrating in situ and ex situ management, in accordance with international philosophy such as that of the IUCN (World Conservation Union). New research findings on topics such as patterns of genetic variation in the wild,[11] techniques for successful egg incubation[12] and the existence of temperature-dependent sex determination[13] now have a much faster uptake within the captive-breeding community.[14] A comprehensive captive-management plan and a husbandry manual have been produced,[15] and veterinary knowledge of tuatara is improving.[16]

As of July 2010, over 380 tuatara were in captivity, all but about 30 in New Zealand.[17] Most captive tuatara originate from Stephens Island stock, and are maintained for advocacy, educational and research purposes. However, some animals, including those from relict island populations, are involved directly in conservation initiatives, which include rearing for eventual release to new wild habitats.[18] Since 1995, captive-reared juveniles have contributed to the establishment of four new populations of tuatara on islands,[19] and captive-bred offspring have helped supplement four other populations in a relict state on islands.[20] Progress with these conservation initiatives, which hinges on close cooperation between the Department of Conservation, community ecosanctuaries, university researchers, Māori iwi (tribes) and the zoos[21] themselves, is discussed in more detail in Chapter 10.

Given that tuatara of Stephens Island stock are relatively abundant in captivity, basic research that does not conflict with conservation objectives should be fostered. Captive colonies will not be suitable for every research question, but could provide opportunities to study some poorly understood aspects of behaviour, such as the role of vocalisation and olfactory communication in breeding, and the role of soil temperature and other factors in nest site selection. As examples of past cooperation, surplus eggs from

Box 8.3 Fig. 2 Obesity has occurred frequently in captive tuatara, reflecting overfeeding and a lack of activity in confined situations. This adult male reached a mass of 1210 g in captivity, whereas a wild male of the same length on Stephens Island would be expected to weigh about 750–850 g. Obesity in tuatara is suggested to cause physical difficulties during mating as a result of large fat deposits at the base of the tail.[22] It is also associated with lipid imbalances, including high plasma cholesterol concentrations, which could affect sperm and egg quality.[23]

Photo: courtesy of Lindsay Hazley, Collection: Southland Museum & Art Gallery.

a productive colony at Southland Museum and Art Gallery have been used at the University of Otago to help explore several research topics. These include the effects of warm temperatures on egg incubation, sex of offspring and performance;[24] olfactory communication in juveniles;[25] the development of sexual dimorphism;[26] and aspects of thermoregulation and growth that might help guide reintroduction of tuatara to the southern ecosanctuaries.[27] Further research is also needed with captive tuatara to address specific husbandry concerns. These include the amount of ultraviolet light and dietary calcium necessary indoors to avoid onset of metabolic bone diseases;[28] and the causes of possible instances of spontaneous ovarian haemorrhage,[29] of shelled eggs becoming fused within the uteri,[30] of failure to lay eggs in properly constructed nests,[31] of low viability of some egg clutches[32] (including possible dietary effects on yolk composition),[33] and of sudden mortality in second-generation young in captivity.

Members of the public have limited opportunities to see wild tuatara, so captive displays are important for education and to help instil a conservation ethic. Provided conditions are appropriate, captive tuatara can survive for many decades as advocates for their species and ecosystems. In 2011, one adult alive in New Zealand had survived for 41 years in captivity, and two others for 31 years.[34] Another adult survived in captivity at Portobello Marine Laboratory, New Zealand for 42 years,[35] and others have survived at zoos or universities in Europe for 25–36 years.[36] Tuatara held in captivity also provide an opportunity for Māori to demonstrate kaitiakitanga or guardianship, in some cases leading to educational tours or 'road shows' on a national scale.[37]

Places where tuatara were on public display in New Zealand in 2011 included the following: Ti Point Reptile Park (Warkworth), Auckland Zoo, Hamilton Zoo, National Aquarium (Napier), Otorohanga Kiwi House, Rainbow Springs Kiwi Wildlife Park (Rotorua), Pukaha Mt Bruce, Nga Manu Nature Reserve (Waikanae), Wellington Zoo, Victoria University of Wellington, Natureland Zoo (Nelson), Eco World Aquarium and Terrarium (Picton), National Kiwi Centre (Hokitika), Orana Park (Christchurch), Southern Encounter Aquarium and Kiwi House (Christchurch), Willowbank Wildlife Reserve (Christchurch), Queenstown Kiwi and Birdlife Park, and Southland Museum and Art Gallery (Invercargill). In addition, a few tuatara are held off-display at other locations. Overseas, tuatara are held at Taronga Zoo in Australia, at four zoos in the US (Toledo, San Diego, Dallas and St Louis), at Chester Zoo in England and at Berlin Zoo in Germany.

Box 8.3 Fig. 3 Southland Museum and Art Gallery in Invercargill has achieved annual reproduction in some females within its captive colony of tuatara,[38] and simultaneously provides an extremely popular public display.[39] Ample space, a group situation allowing controlled social interactions, warm temperatures and regular feeding (within limits, to avoid obesity) may contribute to reproductive success.

Photo: Alison Cree, with permission of Southland Museum & Art Gallery.

Body size has great implications for mating success in wild male tuatara. Although the forest territories of large males on Stephens Island are not greater in area than those of small males, they do provide better access to females.[166] Large males almost inevitably win agonistic encounters with other males, and break up courtship attempts by smaller males whose territories overlap with their own. Thus, multiple paternity of egg clutches in wild tuatara is rare.[167] Although small males greater than 180 millimetres in snout–vent length appear physiologically mature, only males with a snout–vent length greater than 213 millimetres have been seen mating.[168] No genetic differences have been found between males of different snout–vent lengths,[169] although longer males clearly have greater bite forces.[170] Whether all males have an equal chance of eventually reaching the large body size needed for mating success is unclear.

Females seem unlikely to mate in every season. Of 27 courtship attempts observed on Stephens Island by Gillingham and colleagues, only two resulted in mating.[171] In the other 25 attempts, the females appeared non-receptive, and retreated within minutes to a point several metres away from the male or down a burrow. Males persisted in attempting to court females that remained on the surface, and followed other females down their burrows, only to emerge alone a few minutes later. As noted above, the plasma hormone profiles in females that will ovulate and nest that year are very different during the late-summer mating period from those of females that will not, suggesting that non-ovulating females might be non-receptive (see earlier section on the multi-year cycle of females).[172] Consistent with this, females have not been seen to mate in successive years, although within the year that they do mate, they may mate up to five times (usually with the same partner).[173]

In summary, from studies on Stephens Island, it is now clear that reproducing tuatara are not animals that 'do nothing'. Territorial and courtship responses (to other tuatara and to rubber models) are vigorous, with visual communication being extremely important.[174] Although there is no evidence that olfactory communication plays a role in social behaviour (for example, tuatara do not exhibit the tongue-flicking or tongue-contact seen in many squamates),[175] tuatara have recently been shown capable of detecting odours from food.[176] Thus, the role of secretions from the sebaceous glands in reproduction deserves closer study. There is also uncertainty about the significance of vocalisations. Gillingham and colleagues suggested that the croaking heard during male–male aggression is the result of an involuntary compression of the trunk region,[177] but there are also reports of soft, deliberate vocalisations during courtship.

Currently, little is known about the timing of courtship and mating in northern tuatara. Males on northern islands have high concentrations of plasma testosterone and erect crests during late summer and early autumn,[178] as on Stephens Island, suggesting that the time of mating may be similar (Figure 8.25). Unexpectedly, however, northern females seem to nest as late as or later (in spring or summer) than females in Cook Strait (see section

Fig. 8.25 A male tuatara displays with its crest erect on Tawhiti Rahi Island in the Poor Knights Islands, March 1989. By sometimes displaying on rocks, males probably make themselves more visible to other tuatara. The white mark on the snout is typist's correction fluid, applied by researchers to avoid recapture of the tuatara during the field trip.

Photo: Mary McIntyre.

below on nesting behaviour). This raises questions about whether the time of mating is correspondingly adjusted, and about the overall completion of the reproductive cycle for a given population within a calendar year.

Egg-processing during gravidity

As discussed earlier, females that contain eggs in their oviducts are referred to as gravid. On Stephens Island, the gravid period lasts for 6–8 months over winter, at least three times as long as reported for most egg-laying reptiles.[179] Although evidence was lacking at the time, Dawbin was correct in suggesting in 1962 that fertilisation might occur many months before eggs were laid.[180] An alternative inference, that female tuatara store sperm throughout winter,[181] is clearly not the case on Stephens Island.

At ovulation in early autumn, the enlarged oocytes rupture from their follicles and make their way into the anterior oviduct. The remnants of each ovarian follicle form a corpus luteum in the ovary.[182] In early gravidity (May, 1–2 months post-ovulation), the corpora lutea are large, pinkish-white, flattened discs about 7 millimetres in diameter with a visible ovulation aperture (the opening through which the oocyte passes at ovulation; Figure 8.15). The corpora lutea remain similar in size and histological appearance throughout gravidity. As in alligators (but not squamates or turtles), the ovulation aperture never closes and the luteal cell mass never completely fills the central cavity, but reasons for these patterns are unclear.[183] Plasma progesterone concentrations peak around ovulation in March (mean: 16.4 nanograms per millilitre; ng/ml) and then fall rapidly, but remain detectable (up to 2.9 ng/ml) during mid–late gravidity.[184] Histological evidence suggests that corpora lutea probably continue to secrete progesterone during this time, and thus they may help to prevent premature egg expulsion. Within 1–2 months of egg-laying (oviposition), the corpora lutea have shrunk and show signs of regression, and progesterone has become undetectable in the plasma.

Following ovulation, the oocytes move into the oviducts, where fertilisation and shell formation occur (Figure 8.26). The manner in which the anterior end of the oviduct engulfs the oocytes at ovulation is unknown. The process of fertilisation is also undescribed, although it probably happens in the upper reaches of the oviduct, above the uterus where shell deposition occurs.

By the end of the gravid period in late spring, the fully developed, flexible eggshell consists of several layers. Scanning electron microscopy shows that these comprise an innermost inner boundary (a thin, sometimes amorphous layer surrounding the oocyte), a thicker shell membrane containing overlapping, multi-layered, protein fibres, and an outermost calcified layer containing calcium carbonate in the form of calcite.[185] As was first recognised by Mary (Kathy) Packard and colleagues in shells from hatched eggs,[186] calcareous columns penetrate deeply into the fibrous layer and enmesh the protein fibres. Tuatara eggshells thus differ in structure from those of most egg-laying reptiles, in which a clear demarcation exists between the fibrous shell membrane

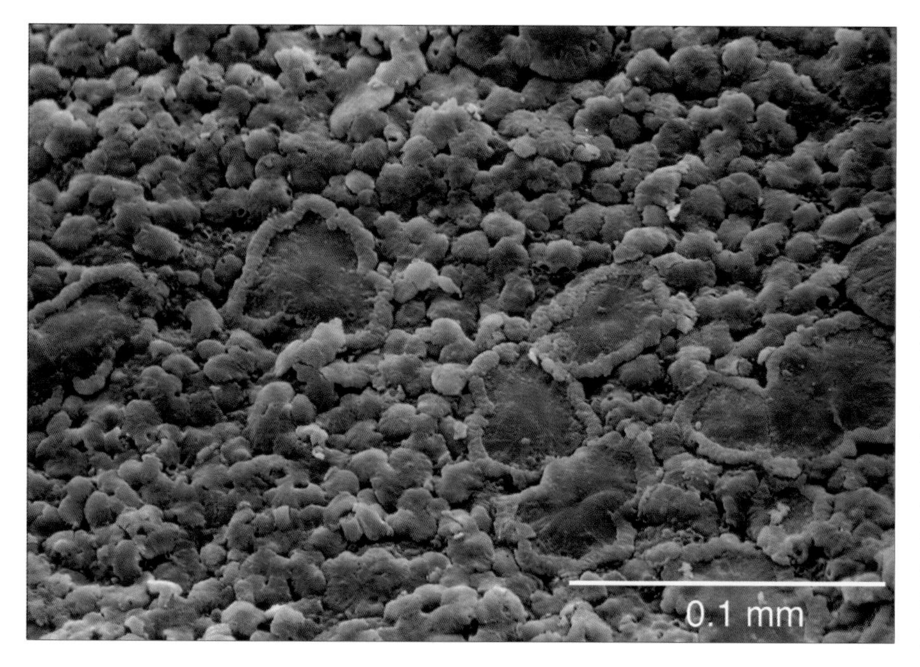

Fig. 8.26 Scanning electron micrograph of the outer surface of an egg from Stephens Island. Removed from the oviduct of a gravid female in early December, this egg appears ready for laying. The outer surface comprises caps and rosettes of calcium carbonate, and overlies a fibrous shell membrane through which calcareous columns penetrate.

Photo: Alison Cree.

and the outermost calcified layer. The unusual structure of the eggshells of tuatara seems to combine features of the flexible and rigid-shelled eggs of other reptiles.[187] A similar structure has been described for eggs from an agamid lizard,[188] but the functional significance remains unclear.

Several colleagues and I compared shell formation in tuatara from Stephens Island between eggs obtained from the oviducts in early gravidity (May), mid-gravidity (August) and late gravidity (December).[189] By May, the eggs were located within the uterine region of the oviducts and had already attained a size and shell thickness (0.2–0.3 millimetres) similar to that at the end of gravidity. The shell membrane was well formed and penetrated by calcareous columns to at least half its depth, but calcification over the outer surface was incomplete. Calcification of the outer surface was inferred to continue gradually over the remaining 6–7 months of gravidity. By the end of gravidity, the outer layer is covered by calcified units in the form of 'caps' and 'rosettes'.[190]

On Stephens Island, Michael Thompson and others have shown that females will lay eggs from October in response to injections of oxytocin, a mammalian neurohormone stored in the pituitary gland.[191] In reptiles, as in mammals, oxytocin stimulates uterine contractions. However, eggs induced at this time are often small with incomplete shell layers, and have relatively low hatching success.[192] In one study, eggs induced in October that subsequently failed averaged only 3.16 grams in mass.[193] In the same study in December, at the end of the natural nesting season, the mean mass of induced eggs that subsequently hatched (5.06 grams) was significantly higher than in October.[194] This and other information on the mass of eggs in the oviducts or at induction[195] suggest that water uptake continues during the final 1–2 months that eggs are in the oviducts, and that this absorbed water contributes to hatching success.

Uterine eggs take up considerable space in the abdominal cavity of tuatara, and for a prolonged time. However, the implications for ecology and behaviour of females are unstudied. Stomach flushing and faecal production confirm that gravid females continue to feed,[196] but whether food intake is limited to a smaller amount than normal is not known. Other female reptiles may become relatively thermophilic (heat-seeking) when gravid with eggs.[197] Such behaviour could help stimulate the completion of egg processing, but whether it occurs in tuatara is unknown.

Egg yolk composition

Eggs of tuatara contain very little albumen ('egg white') at oviposition,[198] a feature typical also of squamate eggs.[199] A crudely punctured egg of a tuatara will typically release what initially appears to be only copious yellow yolk. But by using a fine probe to pierce the poles of several freshly laid (oxytocin-induced) eggs from Stephens Island tuatara in November, I was able to obtain from some eggs up to about 0.5 millilitres of clear fluid, possibly including some albumen.

BOX 8.4

How can gravid females be identified?

In female reptiles that do not nest each year, a way of identifying gravid females (those with eggs in the oviducts) is valuable to understanding patterns of nesting. Gravid female turtles can readily be identified by radiography (X-raying), and this approach has since been used with success for tuatara.[1] Radiography is especially valuable as it yields information on clutch size as well as nesting frequency. The technique relies on the fact that the calcium-containing eggs in a female's oviducts are radio-opaque to X-rays prior to nesting in spring. I found that eggs inside the oviducts of tuatara on Stephens Island were sufficiently radio-opaque to be detected by July, about 3–5 months before nesting, but how rapidly they become detectable after ovulation in March or April is not known.[2]

Given the risks of X-rays to embryos that are rapidly developing and differentiating in pregnant women, it is appropriate to ask whether the procedure is harmful for reptiles. In tuatara, there is no evidence that exposure to the X-ray dose used is risky for either embryos or mothers. Eggs from X-rayed females produced hatchlings that showed no deformities and did not differ in hatching success,

Box 8.4 Fig. 1 Claudine Tyrrell prepares to X-ray three female tuatara on Green Island in October 1996.
Photo: Alison Cree.

incubation period or size scores (based on linear and mass measures) from eggs that had not been X-rayed.[3] Females that have been repeatedly X-rayed every 1–2 years in captivity have continued to produce viable eggs, and females from these eggs have themselves matured and produced viable eggs.[4] The very early and probably arrested development of embryos at the time of X-raying may contribute to the apparent safety of the procedure.

Laparoscopy is another technique used in tuatara to identify gravid females.[5] This procedure involves minor surgery and is thus more invasive and technically challenging than radiography. However, it has the advantage of providing information on reproductive condition in animals that would simply be classed as 'non-gravid' by X-rays. Using laparoscopy, my colleagues and I were able to show that 'non-gravid females' collected for X-raying in early spring are indeed mostly females, with most having ovaries in various stages of vitellogenesis, and only a few in atretic condition. However, a few small males were erroneously included.[6] On the other hand, because it is difficult during surgery to see complete oviducts on both sides of a gravid female, laparoscopy has not provided reliable information on clutch size.

Less invasive techniques for identifying gravid females have been explored for several populations. One might expect that females carrying oviducal eggs would be relatively heavy and plump compared with non-gravid females, and indeed they are – on average. On Stephens Island, Michael Thompson and colleagues showed that gravid females increase more rapidly in abdominal girth in relation to snout–vent length than do non-gravid females.[7] This pattern means that females with eggs in their oviducts (especially those that are relatively long) tend to be wider around the abdomen than those that are not. However, the amount of overlap between groups meant that the reproductive condition of individual females could not be reliably identified. Mass (corrected for both tail loss and snout–vent length) also showed no significant difference between gravid and non-gravid females on four northern islands, and on a fifth (Coppermine Island), where gravid females were significantly heavier than non-gravid females,

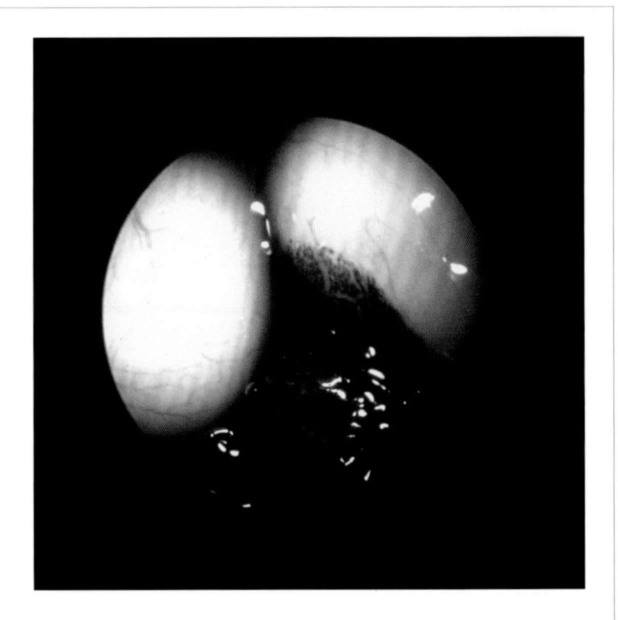

Box 8.4 Fig. 2 Laparoscopic view of eggs in the oviduct of a gravid female.

Photo: Alison Cree.

there was still much overlap.[8] This overlap probably results from the fact that the vitellogenic cycle of female tuatara can be spread over several years. In particular, females that are in late vitellogenesis (but not ready for nesting that year) already have large ovaries that are probably similar in mass to the egg-containing oviducts of gravid females.

As with humans, ultrasound has provided valuable information on reproductive status in species of reptiles. However, in tuatara, attempts in New Zealand using a variety of ultrasound probes have not yielded the quality of images expected from studies on other reptiles, possibly in part because of acoustic shadowing from gastralia, the so-called 'abdominal ribs'.[9] Further exploration of ultrasound would be worthwhile to determine whether gravid and non-gravid females can be discriminated, but at this stage reliable information on clutch size seems unlikely.[10]

Manual palpation for eggs is cost- and equipment-free, and thus an attractive technique that could potentially discriminate gravid from non-gravid reptiles. However, initial reports for tuatara suggested that the presence of gastralia was once again the cause of unreliable results.[11] For example, on Stephens Island, Louis Guillette and I found that palpation of females in the pelvic region did not identify reproductive condition with certainty, mainly because many females palpated as non-gravid turned out to be gravid when X-rayed.[12] However, Nicola Nelson successfully selected gravid females for hormonal induction of egg-laying using palpation, suggesting that there may be fewer false positives than false negatives.[13] On the other hand, the success rate for selecting gravid females on North Brother Island was apparently not as high,[14] leaving uncertainty about the accuracy of the procedure.

Further assessment of palpation and ultrasound at different stages of gravidity, and among females of varying clutch size, would be valuable. Preliminary analysis of blood for hormones or other biochemical analyses that might distinguish gravid from non-gravid females has so far been uninformative,[15] but further approaches using different biochemical markers are worth exploring. The ideal technique for future identification of reproductive condition of tuatara, especially on islands for which no information is currently available on nesting frequency, would be harmless, non-invasive, inexpensive, require no electricity or refrigeration of components, and yield accurate results for every female (gravid or non-gravid) within minutes. Where information on clutch size is required, radiography, hormonal induction of egg laying, or direct observations of nesting seem likely to remain necessary.

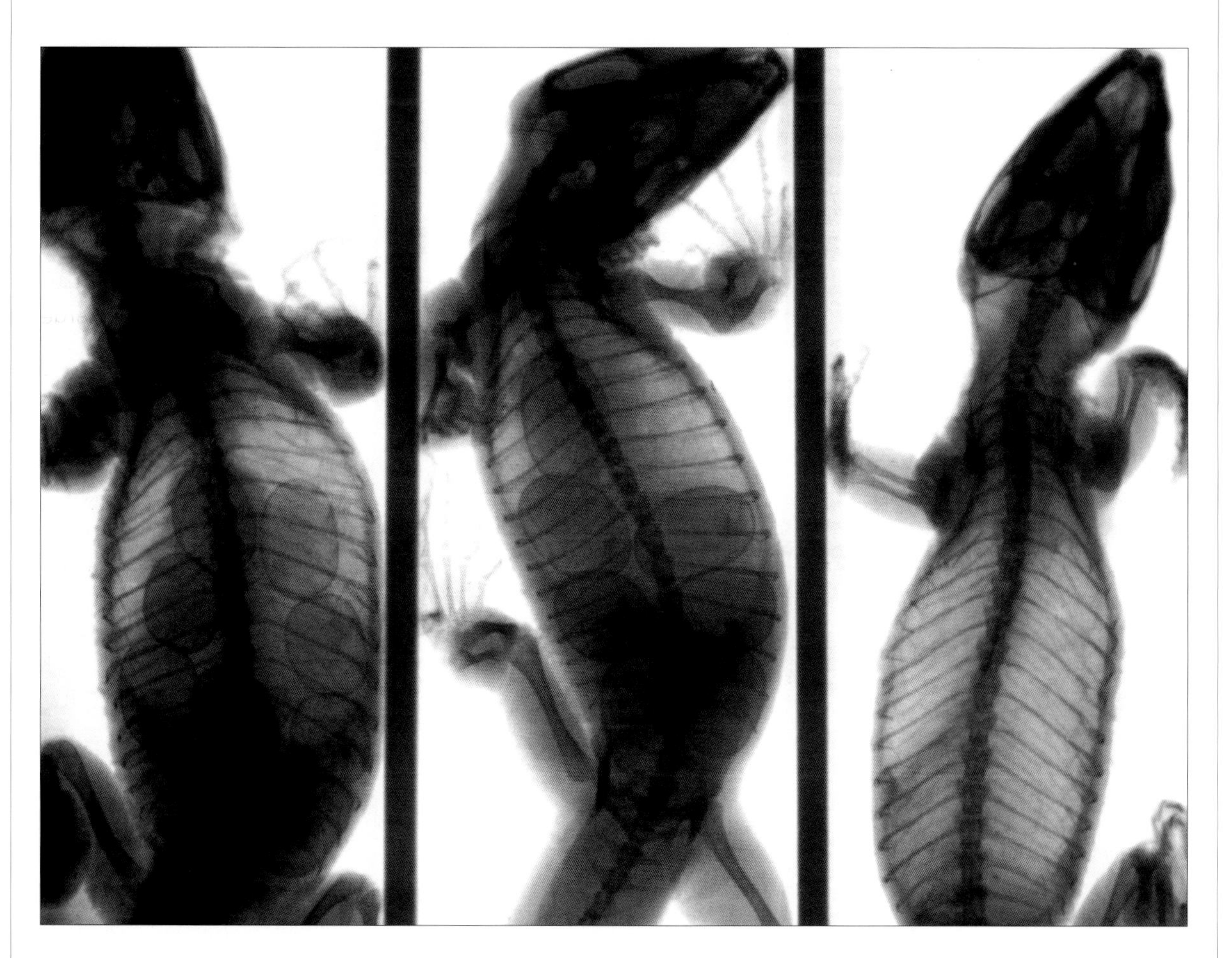

Box 8.4 Fig. 3 Polaroid radiograph of three female tuatara on Coppermine Island in October 1996. The two females on the left are gravid, but the third is not.

Photo: Claudine Tyrrell.

Slightly more information is available on yolk structure and composition in tuatara. Yolk bodies were described from 'fully developed eggs' obtained from a female that died in a New Zealand zoo in November.[200] The yolk bodies were flattened (occasionally elliptical) vesicles consisting of a superficial membrane, a small amount of protein fluid and a large, crystalline and more-or-less rectangular platelet (occasionally two or more platelets) within. The crystalline structure of the yolk platelets[201] and the size range of the vesicles[202] are similar to those in other vertebrates.

The lipid composition of one newly laid egg from Stephens Island has been analysed.[203] The egg ('yolk and albumen') contained lipid and protein. The proportions of the major lipid fractions, namely triacylglycerols (81.6%) and phospholipids (13.8%), were within the ranges reported for the yolk of other reptiles.[204] A variety of fatty acids, including the essential fatty acids linoleic acid and arachidonic acid, were present. The phospholipid fraction contained a high proportion of linoleic acid (22.6%), similar to the values reported for four squamate species and higher than observed in two turtles and a crocodilian.[205]

In addition to vitellogenin, the yolk of reptiles includes another source of yolk lipid (a very-low-density lipoprotein known as VLDL*y*), as well as vitamin-binding proteins, vitamin precursors, vitamins and steroid hormones.[206] These components are likely to be important, if not critical, to embryonic development, but concentrations have yet to be reported in the eggs of tuatara.

Nesting frequency

Like many other long-lived reptiles, including sea turtles and alligators, wild tuatara do not nest each year.[207] On Stephens Island this was evident from dissections dating from the 1890s.[208] However, just how infrequently females reproduced did not emerge until the radiographic surveys initiated by Donald Newman and Peter Watson in the 1980s.

Surveys by Newman and Watson were conducted yearly in spring from 1982 until 1987 on Stephens Island and from 1983 until 1987 on Lady Alice Island.[209] The gravidity rate (the percentage of mature-sized females that were gravid) in spring ranged from 8.2% to 29.6% on Stephens Island and from 22.2% to 45.5% on Lady Alice Island.[210] Variation among years was significant on Stephens Island only, being especially low in 1982 (8.2%) following a severe El Niño event,[211] during which the weather was cooler and windier than normal. When data from 1983 until 1987 were combined for comparison among islands, the proportion of females that were gravid on Stephens Island (23.4%) did not differ significantly from that on Lady Alice Island (28.8%).[212]

In a more recent survey among five northern islands sampled between 1996 and 1998, gravidity rates ranged from 19.0% on Green Island to 48.2% on Aorangi Island (Figure 8.27).[213] Variation among islands was significant and negatively correlated with latitude, but not with island size, presence of

Pacific rats[214] or estimated density of tuatara. The number of islands was small so these patterns should be considered tentative.[215]

Some estimates of gravidity rate have also been made for tuatara on North Brother Island. In 1989 only 19.5% of females caught over three nights oviposited in response to an injection of oxytocin, implying that nesting frequency is low on this island also.[216] In a more extensive study of two nesting seasons (in 2000 and 2001), Nicola Mitchell and colleagues calculated that only 9% of mature-sized females on North Brother Island were gravid. They also concluded that a tendency for gravid females to be seen on more nights than non-gravid females might have inflated the estimate from the earlier study.[217] Certainly, nesting females emerge on cool, windy nights in conditions that would strongly deter other females.[218] Variable emergence rates due to weather conditions might therefore have affected the gravidity rates shown in Figure 8.27 for other islands also. Ideally, comparisons of gravidity rate among islands would involve sampling all populations in the same year, using the same method, across a large period of time to correct for variable patterns of emergence but before nesting had begun, and across a large enough area to avoid a bias toward nesting grounds where gravid females aggregate. However, this would pose extreme logistical challenges.

Laying histories for wild females radiographed in successive years on Stephens, Lady Alice and Coppermine Islands reveal considerable variation in nesting frequency among individuals. On each of these islands, some females produced clutches 2 years apart, but none reproduced in successive years.[219] On Stephens and Lady Alice Islands, two females (one from each island) did not produce eggs within four successive years;[220] for another female on Stephens Island, 5 years elapsed between nestings.[221] Although variation in gravidity rate among size classes was not significant, there was a suggestion for both Stephens and Lady Alice Islands that the largest females might have the lowest gravidity rate.[222] It seems

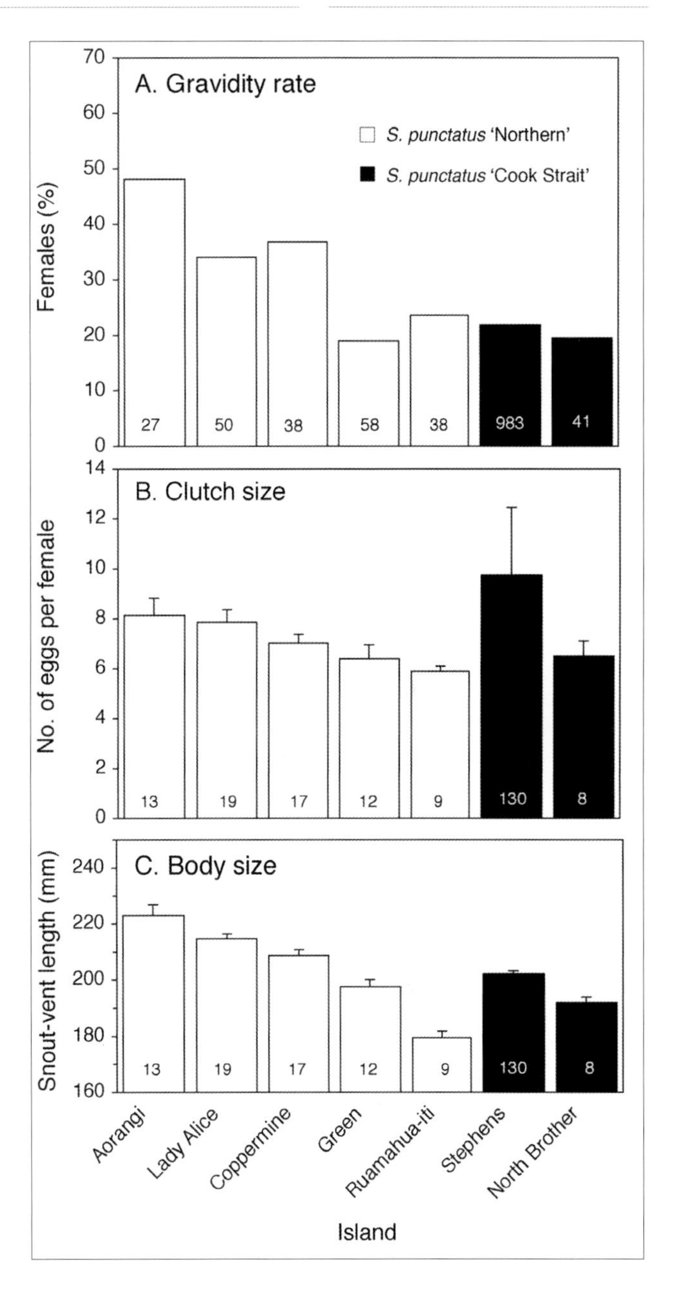

Fig. 8.27 Gravidity rate (the proportion of mature-sized females carrying eggs with shells), clutch size and female body size (snout–vent length) for tuatara on seven islands (listed from north to south). Note a tendency for gravidity rate to decline with increasing latitude, and that clutch size is unexpectedly large on Stephens Island given the intermediate body size of tuatara there.

Radiographic data for five northern islands were collected during 1996–1998 (Tyrrell et al. 2000). Radiographic data for Stephens Island were collected during 1982–1987 (Newman et al. 1994). Data for North Brother Island were obtained from response to oxytocin in November 1989 (Cree et al. 1991d). On each island, gravidity rate was estimated for a sample collected over about 3–14 days, and is not corrected for a possible difference in encounter rate between gravid and non-gravid females. Data for clutch size and body size are mean ± SE.

that as a female tuatara grows, her clutch size increases (see next section) but reproductive frequency may not change much, except perhaps in the largest females. Using laparoscopy, I once observed that a large[223] and apparently old female on Stephens Island contained oviducal eggs in February, 2 months after the usual end of the nesting season. Thus, reproductive timing may be disturbed in the oldest and/or largest females.

Overall, the current information suggests that females on Stephens Island nest on average about every 4–5 years,[224] and that the frequency on North Brother Island is similar or lower.[225] Nesting frequency on the northern-most (warmest) islands may be slightly higher than on Stephens Island, perhaps as frequently as once every 2 years on Aorangi Island.[226] Plasma concentrations of sex steroids in late summer–early autumn (using criteria indicative of incipient ovulation for Stephens Island tuatara) also suggest that females on the Poor Knights, the Hen and Chickens and the Mercury Islands do not nest each year. About one-third of females on these islands met the criteria for incipient ovulation, and on the Hen and Chickens Islands and the Mercury Islands, the proportion was not affected by the presence of Pacific rats.[227] Radiographic data also revealed no significant change in gravidity rate on Lady Alice Island 4 years after the eradication of Pacific rats.[228] Thus, the presence of Pacific rats on Lady Alice Island appears not to have affected nesting frequency, despite inhibiting recruitment of juveniles into the population.[229]

Overall, wild tuatara are unlikely to nest each year, but factors that determine reproductive frequency remain poorly understood. The suggestion that nesting frequency might be influenced by El Niño events via direct effects on weather or indirect effects on food supply (through the transfer of marine nutrients to land via seabirds)[230] deserves closer investigation. Reports that female tuatara from Stephens Island and Hauturu/Little Barrier Island have reproduced annually in captivity[231] also suggest that environmental constraints such as temperature and/or food supply may limit reproductive frequency in the wild.

Clutch size and induced oviposition

Clutch size, or the number of eggs produced by a female tuatara at a single nesting event, is relatively low in tuatara for a medium-sized reptile that reproduces so infrequently.[232] The most reliable data on clutch size come from radiography. In radiographic surveys on Stephens Island between 1982 and 1987, clutch size averaged 9.4 eggs (range 1–18).[233] As is common in egg-laying reptiles, clutch size varies with female body size: larger females produce more eggs (Figure 8.28).[234]

Another way of estimating clutch size is to inject hormones that stimulate the oviducts to contract and expel eggs (Figure 8.29). Oxytocin is cheaper than the native reptilian equivalent, arginine vasotocin or AVT,[235] although both have been used with success. On Stephens Island, oxytocin injection during October–December 1985 yielded a mean clutch size of 10.6 eggs.[236]

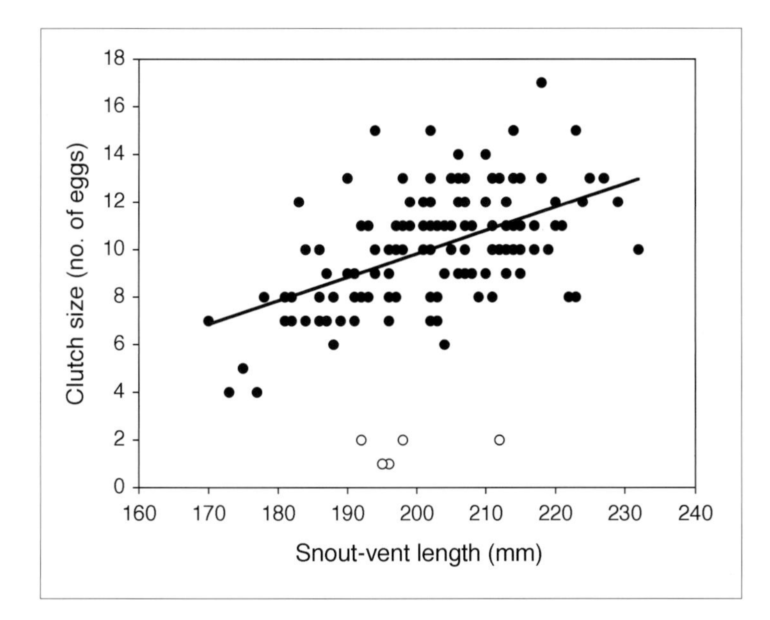

Fig. 8.28 Relationship between female body size and clutch size for tuatara on Stephens Island during 1986–1988. Each symbol represents a different female. The regression for those shown with black dots is significant ($p < 0.001$). Five small clutches that may have been partially laid at the time of X-raying (open circles) have been excluded from the regression line.

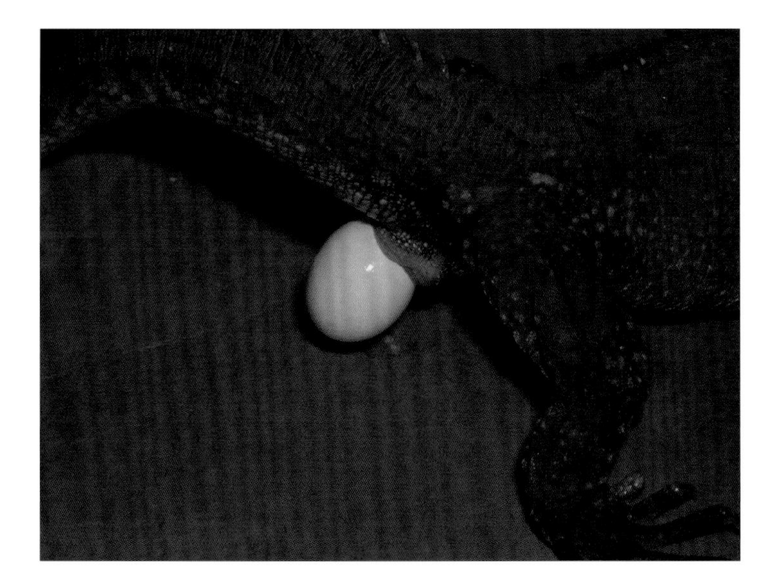

Fig. 8.29 A female tuatara induced to oviposit following an injection of oxytocin on Stephens Island in December 1986. The ovoid egg is extruded end-on through the partially everted cloaca.

Photo: Alison Cree.

This was not significantly different from the mean of 10.7 eggs obtained for the same population in the same year by radiography, suggesting that most responding females laid complete clutches. However, the speed of the response varied, and the following year some females required a second dose after 12–36 hours to recommence laying.[237] An AVT injection of 20 nanograms/gram body mass induced complete oviposition in three females (as determined by prior radiography, in November), whereas a dose 10-fold lower resulted in partial or complete clutches from five females.[238] Neither dose of AVT induced oviposition in July, 4 months earlier in the gravid period.[239] The local-signalling molecule prostaglandin $PGF_{2\alpha}$ also induced oviposition in November but not in July.[240] Thus, the oviducts of tuatara possibly acquire receptors to these contractile agents only late in the gravid period.

Clutch size can also be estimated from counts of eggs in freshly laid nests, prior to any nest disturbance. This involves assumptions about nesting behaviour that are probably valid in most cases (that females oviposit their entire clutch within one nest, and that clutches from different females are not laid within the same nest site without physical separation; see next section for discussion). Clutch size in completed nests on Stephens Island averaged 8.6 in 1986 (range: 1–17 for 26 clutches),[241] a value similar to that from radiography in the same year (9.2; range: 1–17).[242] Historical reports of clutch size (apparently based on nest excavations as well as dissections) gave values within these ranges for the late 1890s,[243] suggesting that clutch size has not changed markedly over the last century on Stephens Island.

On North Brother Island, clutch size is smaller than on Stephens Island. This difference is not surprising given the smaller body size of tuatara there. Clutch size estimated from two studies using induced oviposition and/or radiography averaged 6 eggs (range: 3–10).[244] When corrected for snout–vent length, clutch sizes on North Brother fell within the 95% confidence limits for Stephens Island tuatara of the same size,[245] and overlapped with the distribution for northern tuatara of the same snout–vent length.[246]

Radiographic surveys on Lady Alice Island between 1983 and 1987 revealed a smaller clutch size than on Stephens Island, both in the absolute mean (7.9; range: 5–13) and with female snout–vent length as a covariate.[247] Why should female tuatara on Lady Alice Island, which are if anything larger than those on Stephens Island, produce smaller clutches of eggs? It

Fig. 8.30 During the past century, female tuatara on Stephens Island aggregated in grazed pasture to nest. One such area, known to researchers as the Winch-house Rookery (formerly the Bull Paddock Rookery), is shown here in November 1998. During the 1980s researchers discovered many nest diggings in the grassy bank between the forest patches. When these were marked with reflective stakes, the hillside was lit up at night like a street with road works.

Photo: Alison Cree.

Fig. 8.31 (left) Early stage of a nest digging on Stephens Island, November 1988.

Fig. 8.32 (right) On Stephens Island the same nest sites are sometimes repeatedly used by different females, both within years and in successive years. Sites between and under rocks are popular, as in this example from November 1988. Note the ejected egg at bottom right.

Photos: Alison Cree.

was plausibly suggested that Lady Alice tuatara might produce fewer eggs because of reduced food supplies in the presence of Pacific rats.[248] However, surveys by Claudine Tyrrell and colleagues, involving five northern islands, showed no difference in clutch size (corrected for snout–vent length) with respect to rat status.[249] Further analysis showed that, regardless of the presence of rats, tuatara on northern islands have significantly smaller clutch sizes (corrected for snout–vent length) than on Stephens Island, and clutch size on Lady Alice Island had not changed by 3–4 years after rat eradication.[250] The reason for this difference in reproductive output between northern islands and Stephens Island is unclear; perhaps there is a small trade-off with more frequent reproduction, but other factors such as differences in egg size need investigation (see section below on the size and shape of freshly laid eggs).[251]

Only limited information is available for clutch size on other islands. For the northern population of East Island (Whangaokeno Island), lighthouse keepers reported one nest containing a dozen eggs and another, 17 eggs.[252] These seem surprisingly large clutches for tuatara from such a small island (mean snout–vent length is generally low on small islands; see Chapter 6), but mean body size of tuatara from this now-extinct population is unknown, and the possibility that two females laid in the same site without the keeper realising cannot be discounted (see next section). Clutch sizes for three female tuatara from Karewa Island held in captivity for 3 years in Auckland in the late

1880s were 9, 10 and 12.[253] Clutch sizes on Red Mercury Island (Whakau) for three females that were induced to oviposit at capture using oxytocin were between 5 and 10 eggs.[254] Clutches subsequently produced in captivity by a few females from Red Mercury, Cuvier (Repanga), Kawhitu and Hauturu Islands as part of recent recovery programmes have had ranges of 6–14, 2–19, 11–19 and 2–15, respectively.[255] Relative to snout–vent length, some of these values appear large compared with wild northern females, but egg quality in captivity has sometimes been compromised, resulting in variable incubation success.[256]

Nesting behaviour

One of the most remarkable discoveries of the last 120 years of field research on tuatara is the extended and conspicuous nesting behaviour that occurs on Stephens Island. Female tuatara live at a high density in remnant forest, but have never been observed to nest there. Instead, between mid-October and mid-December (mid-spring to early summer), forest- and pasture-dwelling females move up to 200 metres, and perhaps further, from their home burrows to nest in sunny, open areas.[257] Nesting sites include gardens and other clearings within the forest, as well as pathways, cliff tops and the pasture that was grazed until recently by livestock (Figure 8.30). Nests are constructed in soil or soil-shingle, occasionally beneath rocks, bushes or tussocks (Figures 8.31, 8.32).[258] Some of the earliest observers of nesting on Stephens Island commented on the use of open sites resulting from lighthouse activities, suggesting that females rapidly take advantage of new open areas.[259]

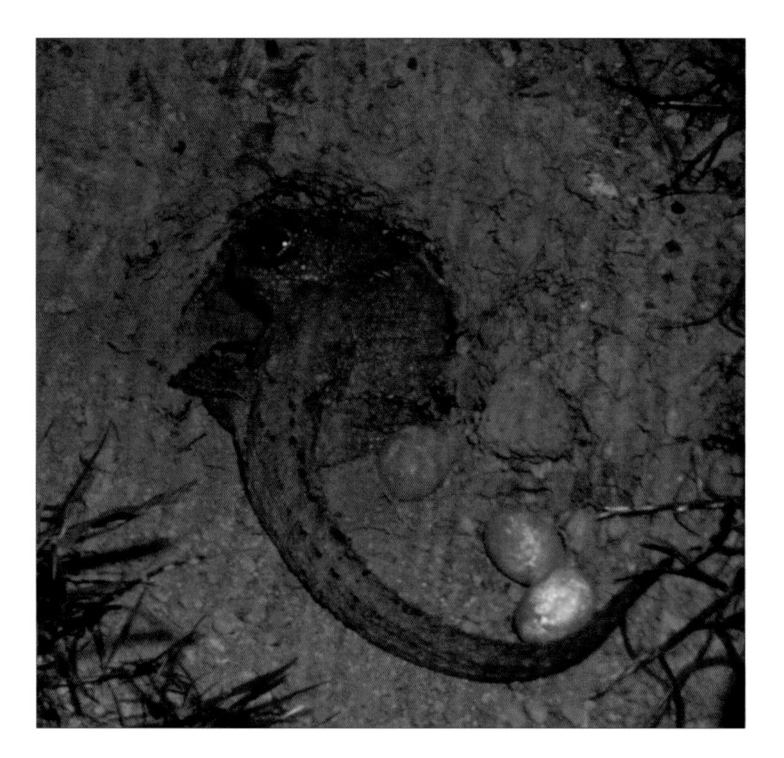

Fig. 8.33 Digging a nest appears to be energetically demanding for female tuatara. Unsurprisingly, females on Stephens Island tend to pick sites where digging is easiest, including the nest sites of other females. This female seems to have been caught in the act of excavating the eggs of another female.

Photo: Alison Cree.

Nesting females often aggregate in specific areas referred to as rookeries.[260] Within rookeries, competition for nest sites occurs. Individual females spend a week or more digging at night (Figure 8.33), sometimes abandoning their initial attempts or taking over diggings begun by other females.[261] Gravid females often return to their digging each night from a temporary retreat nearby, but once their excavations are deep enough, the females may remain within them by day.[262] Sometimes, several females gather at the same nest digging (on one occasion, Nicola Nelson saw seven females with their heads down the same hole!).[263] In such instances, females often have their crests erect and sometimes one will have fresh wounds to the head (Figure 8.34).[264] Tail loss has also been observed

Fig. 8.34 Females on Stephens Island compete at nest sites. In this example, two females had their heads inside the same nest digging in November 1998. Upon disturbance, one female emerged with a fresh wound to her neck. To the right of her crest, the tail of the second female is visible.

Photo: Alison Cree.

Fig. 8.35 A nest site on Stephens Island in which a clutch, probably laid the previous year, has been disturbed by another female (November 1988). Note the advanced embryo in the right foreground, which would have been nearly ready to hatch. Such excavated clutches are doomed to failure.

Photo: Alison Cree.

on rookeries,[265] again suggesting that nesting females are aggressive towards each other. Aggression between females at nest sites is by no means unique to tuatara: for example, many species of iguanid lizards fight among each other and defend their nest sites.[266]

A major consequence of competition among nest sites for tuatara is that, once a female has laid her eggs, other females may take over the same nest site and eject her eggs. This excavation can happen either later in the same season when the first-laid eggs are still early in development, or in the next season, 10–12 months later, when advanced embryos are present (Figure 8.35).[267] The fate of exposed eggs is certain death by desiccation, overheating or predation (including occasional predation by adult male tuatara).[268]

Fig. 8.36 Freshly laid eggs glisten in a nest on Stephens Island in November 1987.

Photo: Alison Cree.

Once the nest chamber is completely dug, the female lays her eggs (Figure 8.36). Females are rarely observed during oviposition.[269] They usually have their posterior end within the nest chamber and their head facing toward the entrance,[270] but I once saw an ovipositing female facing inwards, depositing her eggs at the entrance. In 1899 Thilenius reported digging up a nest in which the female was in the passage with her head in the egg chamber and an egg in her mouth, suggesting to him that females might position the eggs within the chamber in this way.[271] This and a related suggestion of egg movement by mouth from outside the nest warrant further investigation.[272]

Undisturbed females usually complete oviposition within one night,[273] and each female seems likely to lay her entire clutch within one nest. The now-spent females then spend several more nights covering the eggs with soil, grass and other material.[274] During and after this time (for up to two weeks following oviposition), the female remains at her nest by night and sometimes even by day (Figure 8.37),[275] and is assumed to defend her nest from other females.[276] I once saw a nest-guarding female back on her nest, six nights after she was first seen on the completed nest and the night after it had been partially excavated by another female. The nest was by then re-filled, presumably by the nest-guarding female, but three nights later it had again been dug up by the second female. Spent females eventually leave the nest to return to their residence burrow, and there is no further maternal care.

Recently, Jeanine Refsnider and colleagues made detailed studies of nesting behaviour on seven rookeries over four or five nesting seasons on Stephens Island. Most females (93%) that re-nested within this period used the same rookery.[277] Within rookeries, about one-quarter of females nested in sites that contained evidence of previous nesting within the same or

Fig. 8.37 Spent females often guard their completed nests at night for a few days after eggs are laid. Guarding by day, as seen here on Stephens Island in November 1988, is less frequently seen. The plug of grass and soil that fills the nest entrance is in front of the stake and below the female's head.

Photo: Alison Cree.

the previous year. Females might re-use the nesting sites of other females because the digging is easier, or because they independently select the same microhabitat features that are advantageous to egg incubation. To explore these possibilities, Refsnider and colleagues constructed experimental sites with tilled or untilled soil, with or without surface vegetation and with or without overhead shade cloth.[278] Nesting females selected the sites with tilled soil and those that were bare of surface vegetation, supporting inferences from earlier observations in the late 1800s. More surprisingly, females showed no preference between shaded or unshaded sites, despite unshaded sites apparently being warmer by about 2°C.[279] Further experimental work is needed to fully understand the cues that females use to select nest sites, including the role of temperature, independent of shading. However, these observations do confirm that females pick sites where digging will be easier, with or without cues from other females. Whether females usually return to the same rookery from which they themselves hatched remains unknown.

Refsnider and colleagues also explored the potential value of nest-guarding behaviour.[280] About half the females guarded their nests (on average for 3 days). If a nest take-over occurred, the eggs laid by the previous female were almost always excavated. The difference in frequency of take-overs between nests that were guarded (14% of nests) and those that were not (34%) was of borderline statistical significance ($p = 0.053$). Guarding behaviour therefore probably reduces the risk that a female will entirely lose her investment in eggs developed over the previous 2–4 years or more. However, guarding is unlikely ever to be completely effective, given that the nest-guarding period is short compared with the overall period that eggs are vulnerable to excavation (both within the same nesting season, and the following year).[281] This helps explain why nest guarding is not universal among females.

Fig. 8.38 Two clutches discovered by researchers in the same nest site on Stephens Island in November 2005. Most of the eggs on the left are white and freshly laid (the dimpled egg is possibly an older egg that has failed). The swollen and soil-stained eggs on the right are about 1 year old, and one has already hatched. Here, re-use of the same site by a second female a year later (and probably at a shallower depth) did not result in the destruction of the first-laid clutch.

Photo: Anne Besson.

Occasionally, females nest in the same site as another female but without disturbing the eggs of the previous female. This possibility was first raised by Schauinsland, who observed that, over the course of a few days in the late 1890s, the number of eggs (unspecified) in a nest doubled.[282] Recent observations on Stephens Island indicate that clutches can be laid in successive years in the same site without the first-laid eggs always being excavated, perhaps because the second female did not dig as deeply as the first (Figure 8.38).

Overall, nesting females on Stephens Island engage in extraordinary physical activity. Over a period of several weeks, they migrate from their home territory, dig, oviposit, fill and sometimes defend their nests (occasionally in direct sunlight), before departing and re-establishing a home range, often within their original habitat. These activities must involve substantial physiological demands, at a time when females seem unlikely to feed. The metabolic implications of nesting behaviour call for investigation. Another unresolved question is the response of nesting females to reforestation and the growth in rank grass that has occurred following the removal of livestock from Stephens Island.

On North Brother Island, nesting activity has been observed from late October until December, overlapping as expected with that on Stephens Island. Nest diggings are made in sunny, open (un-vegetated) patches of soil or soil-gravel, including tracks and clearings resulting from lighthouse

Fig. 8.39 (left) Nest scrapes (early nest diggings) on a track on North Brother Island in November 1989.

Photo: Alison Cree.

Fig. 8.40 (below) A spent female lies over the plug of a completed nest on Lady Alice Island in December 1997. The female, previously marked with silver pen on her flanks for identification, had taken advantage of a human-made clearing within the forest for her nest.

Photo: Alison Cree.

operations (Figure 8.39).[283] Females often arrive at their nest scrape (an early nest digging) shortly before dark (from burrows up to 20 metres away) and spend 3–15 nights constructing their nests.[284]

Nesting behaviour on northern islands has been reported for only a few populations. On the now-extinct population on East Island, a lighthouse keeper observed repeated opening and closing of the same nest site in his garden (a dozen eggs were eventually found inside), but the time of year and whether more than one female was involved are unknown.[285] Surprisingly, given the warmer conditions, several observations indicate that nesting occurs on more northerly islands at the same time or even later in summer than on Stephens Island. In 1881 Reischek found several tuatara on Lady Alice Island in December 'in holes only half finished',[286] a possible indication of nesting behaviour,[287] and also suggested after dissecting a female carrying 'full-grown eggs' in January that nesting might occur as late as February.[288] Claudine Tyrrell observed that a nest was laid on Lady Alice Island between 7 and 9 December in 1997. A spent female was on the nest on 10 December (Figure 8.40), and within 2 metres of it on 13 December, implying nest attendance for the 5–6 days following oviposition.[289] Several other females oviposited on Lady Alice Island between early November and early December, but their nests were not located.[290] Two other nests were found on Lady Alice Island in November 2004, also in clearings within the forest. Most of the eggs (in clutches of eight and seven) appeared to have hatched or to have advanced embryos, and the signs of disturbance that were reported[291] could indicate the activities of other females. The western dunes and slopes of Grave Bay are likely to form a prime nesting area on Lady Alice, given the abundance of young juveniles seen in this area following eradication of Pacific rats.[292]

Females from several relict populations of northern tuatara held captive in outdoor enclosures (on Hauturu Island), or on the mainland at similar latitude (populations from Cuvier, Red Mercury and Kawhitu Islands), nested between November and January.[293] These observations are consistent with the time of nesting on Lady Alice Island. Although the timing is unexpected, later nesting in warmer climates than on Stephens Island could be adaptive in ensuring that hatchlings continue to emerge in the favourable conditions of spring or summer, after a winter in the nest.[294]

Fig. 8.41 (opposite) Plasma concentrations of the hormones oestradiol, progesterone, testosterone, AVT, PGF, PGE_2 and corticosterone in female tuatara engaged in nesting behaviour on Stephens Island. The columns show mean concentrations for the sample sizes shown at the base of the bars (vertical bars indicate standard errors). Columns with no superscript letters in common differ significantly by post-hoc test ($p < 0.05$).

Based on data in Guillette et al. 1990, 1991a, Cree et al. 1992 and Cree and Tyrrell 2001.

Endocrinology of nesting

In addition to having highly conspicuous nesting behaviour, female tuatara on Stephens Island can tolerate blood sampling without abandoning their nests. These circumstances have enabled Louis Guillette, Claudine Tyrrell and I to study, with others, the fluctuations in blood hormone concentrations associated with nesting. Among other reptiles, perhaps only sea turtles have been examined as closely in this regard.

Our studies of tuatara have focused on three nesting phases: digging, oviposition and nest-guarding (Figure 8.41). Plasma concentrations of oestradiol

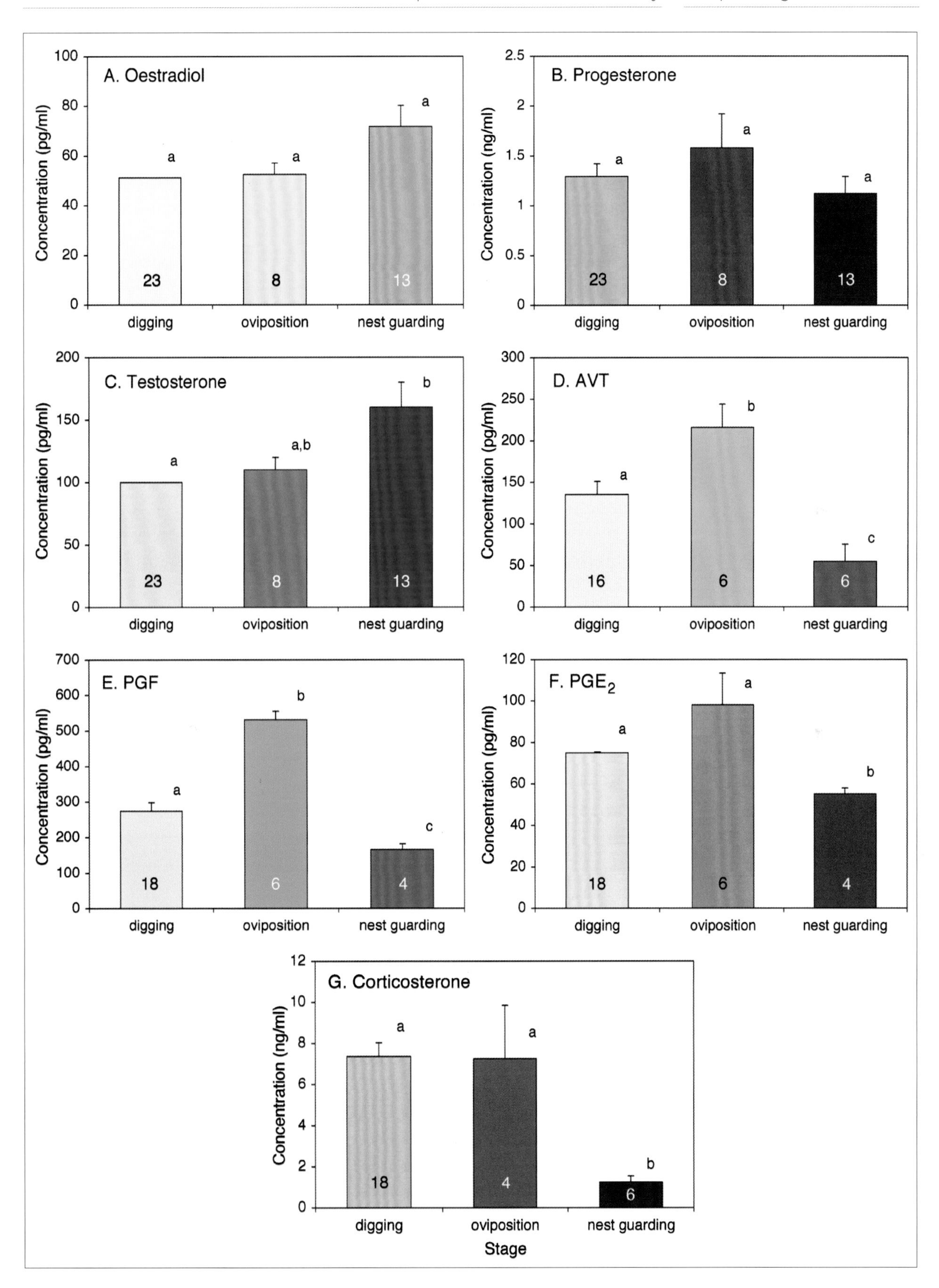
A. Oestradiol
Concentration (pg/ml)
digging
oviposition
nest guarding
B. Progesterone
Concentration (ng/ml)
C. Testosterone
Concentration (pg/ml)
D. AVT
Concentration (pg/ml)
E. PGF
Concentration (pg/ml)
F. PGE2
Concentration (pg/ml)
G. Corticosterone
Concentration (ng/ml)
Stage

and progesterone are low throughout all three phases and show little if any change.[295] These patterns are unsurprising given the minimal ovarian activity at this time. Plasma testosterone is at or below the minimum detectable concentration (100 picograms/millilitre; pg/ml*) during digging and oviposition, but becomes detectable at low concentrations during guarding (100–400 pg/ml). Whether this small rise in testosterone helps stimulate the nest-guarding behaviour seen in some females is not known.[296] Endogenous AVT (AVT made within the body) rises in the plasma of nesting females between digging and oviposition, but falls to low concentrations during guarding.[297] Endogenous PGF[298] shows the same pattern as AVT, with a peak at oviposition and a subsequent fall.[299] Given these patterns, and that both substances stimulate egg laying when administered by injection, a physiological role for AVT and $PGF_{2\alpha}$ in natural oviposition is likely. Plasma concentrations of the prostaglandin PGE_2 do not vary significantly between digging and oviposition, but do fall by the onset of guarding.[300] Although the effects of injected PGE_2 have not been examined in tuatara, this chemical messenger perhaps contributes to relaxation of the posterior reproductive tract in late gravidity.[301] Overall for tuatara, the lack of change in oestradiol and progesterone, and the peak in AVT and PGF at oviposition, are consistent with patterns in nesting loggerhead sea turtles (*Caretta caretta*).[302]

As is typical of vertebrates, reptiles secrete a glucocorticosteroid hormone from the paired adrenal glands in their abdomens (in reptiles, the native form is corticosterone). Although glucocorticoids are often considered as 'stress hormones', this term is somewhat simplistic given that such hormones are essential for survival. Tuatara of both sexes are typical of reptiles in showing an elevated concentration of corticosterone in plasma within about 1 hour of capture.[303] Interestingly, baseline plasma concentrations of corticosterone in females (sampled immediately upon capture, before a response to capture could have arisen) also show natural variation related to nesting. Tuatara in late gravidity (November) that are not engaged in overt nesting behaviour have elevated concentrations of corticosterone at capture compared with females in all other reproductive stages.[304] Plasma concentrations at capture remain elevated during digging and oviposition but, contrary to our prediction, fell dramatically by nest-guarding time despite the continued interactions with other females at this time.[305] A similar pattern of high levels during nest digging followed by a precipitous drop following oviposition has since been observed in Galápagos marine iguanas (*Amblyrhynchus cristatus*).[306] In tuatara, the relatively high concentrations during nest digging and oviposition are perhaps explained by the increased metabolic activity resulting from migration and nest construction.[307] Whether injected corticosterone would stimulate earlier oviposition, as reported for at least one species of lizard,[308] is not known.

* A picogram is one-thousand-millionth of a milligram.

Nest characteristics

Nests of tuatara generally consist of a tunnel up to about 900 millimetres long,[309] slightly expanded into a chamber at the end within which the eggs are laid. The nest is plugged with loose earth, vegetation, and other nearby material. Although descriptions of nests on Stephens Island date from the 1890s,[310] the most detailed information on physical characteristics comes from studies by Michael Thompson and Nicola Nelson since the 1980s.

In the 1986 season, nests consisted of a downward-sloping tunnel ending in an egg chamber. Tunnels varied in length from 110 millimetres (mm) to 500 mm and in width from 50 mm to 90 mm.[311] The egg chamber at the base was sometimes slightly expanded laterally and towards the roof. Eggs were deposited in one to three layers in (typically) an air-filled chamber. Measured from the top of the top-most egg, nests were located 30–155 mm below the soil surface.[312] Nest tunnels were plugged with soil, grass, faeces from sheep and cattle, and other material from near the nest entrance. Observations from the 1998 and 2002 seasons extend the mean depth from the top egg to the soil surface to 230 mm.[313] Mean depth did not differ significantly between the 1998 and 2002 seasons, and mean depth in 2002 was not correlated with date of laying.[314]

Nests of tuatara have only occasionally been found and investigated on other islands. On North Brother Island, nests were 50–250 mm deep.[315] A completed nest on Lady Alice Island had an entrance width of 80 mm, with the eggs located between 25 mm and 90 mm from the entrance. The five eggs were arranged in a single layer, with a 15 mm gap, partly filled with loose soil, between the top of the eggs and the hard ceiling of the egg chamber.[316]

Size and shape of freshly laid eggs

Tuatara lay soft-shelled (flexible-shelled or parchment-shelled) eggs with a blunt-ended ovoid shape. Eggs that I observed within the oviducts of a female about to nest in December averaged 27.1 millimetres in length and 18.7 millimetres in width.[317] Schauinsland reported similar dimensions for eggs in the late 1890s,[318] but slightly larger values reported elsewhere for eggs[319] probably reflect water uptake from the nest substrate prior to measurement. At laying, the shell is soft, and glossy with mucus, but it hardens to a chalky white within a few hours.[320]

Despite differences in clutch size and female body size, the mass of fresh eggs on North Brother Island appears similar to that on Stephens Island. Mean mass of induced eggs in November on North Brother Island (4.9 grams) did not differ from that on Stephens Island (4.4 grams; many of the eggs from both islands were subsequently shown to be viable).[321] In a later study, egg mass on North Brother Island (including induced eggs) averaged 4.1 grams.[322]

Little or no information is available on the dimensions of freshly laid eggs from wild tuatara on northern islands. However, dimensions measured from radiographs for eggs within oviducts suggest that there is variation in egg size among several northern populations. Specifically, tuatara on Lady Alice had oviducal eggs that were wider, with a tendency towards a larger estimated volume, than on three other islands.[323] Further study is needed to determine whether egg size and shape at oviposition vary significantly among northern populations in the manner suggested by radiography. If they do, this could result in differences in hatchling size and potential fitness (see section below on effects of the temperature and other factors on development and growth in the laboratory).

Embryonic development in natural nests

Embryonic development has been examined only in eggs from Stephens Island. Eggs are laid at a very early stage of development,[324] similar to that in turtles.[325] No visible development is apparent (under a binocular microscope) in eggs fixed within 24 hours of oviposition,[326] when embryos are at the gastrula stage.[327] This negligible development is remarkable given the lengthy period of 6–7 months that eggs have been within the oviducts, and implies a prolonged period of pre-ovipositional arrest. The factors that prevent the more advanced development seen within the oviducts of some squamates, and to a lesser extent within crocodilians,[328] are unknown but may involve low oxygen availability.[329]

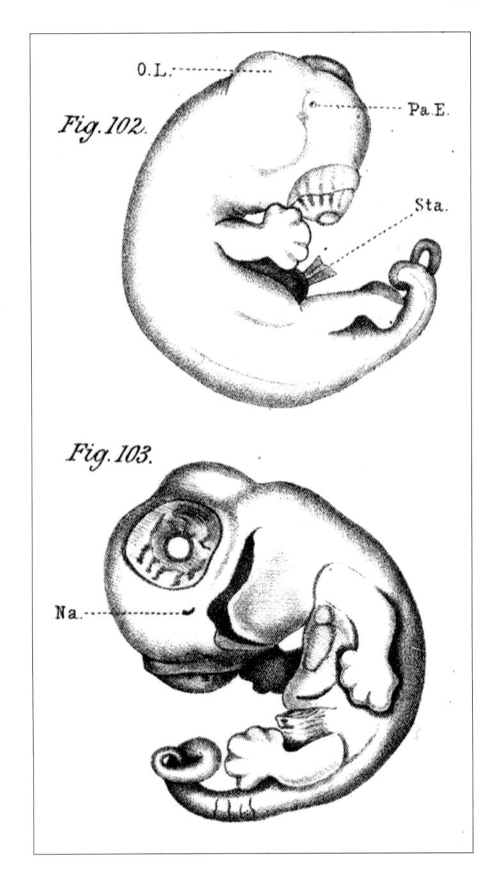

Fig. 8.42 Embryo of a tuatara from Stephens Island at Stage Q. Na. = nasal pit; O. L. = optic lobe, Pa. E. = parietal eye; Sta. = stalk connecting embryo with yolk sac.

Reproduced from Dendy, A. 1899. Outlines of the development of the tuatara, *Sphenodon (Hatteria) punctatus*. *Quarterly Journal of Microscopical Science* 42: 1–87 + plates 1–10. Permission from Company of Biologists conveyed through Copyright Clearance Center Inc.

Embryonic development was described and a staging scheme presented in Dendy's classic study from 1899.[330] Dendy obtained eggs at intervals from the lighthouse keeper on Stephens Island, and examined them upon receipt at his laboratory in Christchurch. Other early studies on development were made by Schauinsland[331] and Thilenius,[332] both of whom collected eggs during spring or summer on Stephens Island and may have had eggs sent to them at other times in development.

Dendy's staging scheme extends from Stage C to Stage S. By beginning at Stage C, Dendy made allowance for earlier stages that might subsequently be described, although none have. By Stage C, reached 8 or more days after laying, the blastoderm appears to have spread completely around the yolk. A distinctive structure known as the posterior amniotic canal (a tubular extension of the amniotic cavity) is present from Stage F until about Stage K. Such a duct also exists in some turtles, and a similar structure may exist in an agamid lizard, so its phylogenetic significance is unclear.[333] The parietal eye (a structure in the top of the head that is probably light sensitive[334]) makes its first appearance at Stage K, the allantois (a membranous sac, attached to the embryo, that stores liquid waste) at Stage L, limb buds at Stage M, and digits at Stage Q (Figure 8.42). The appearance of the limb buds after the allantois suggests an affinity with squamates rather than with

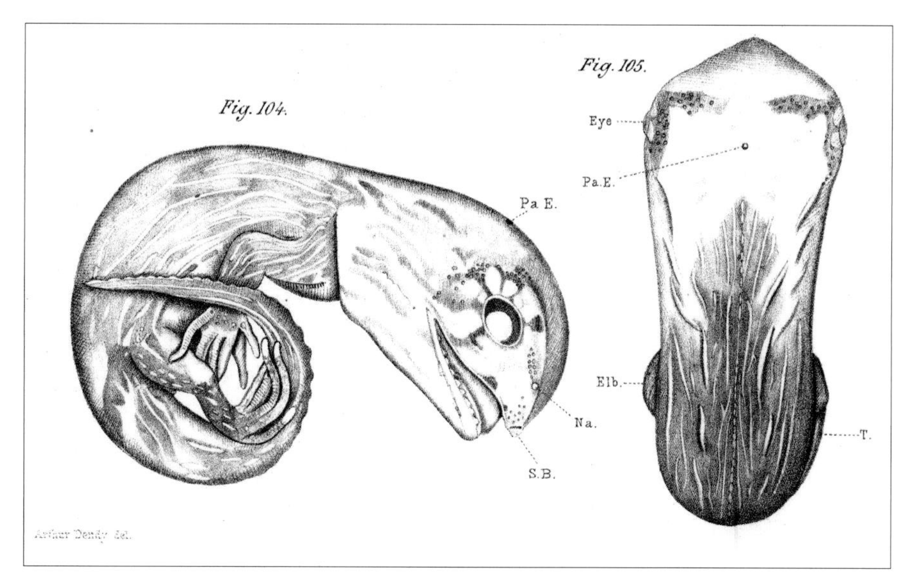

Fig. 8.43 Embryo of a tuatara from Stephens Island at Stage R. This stage is reached by March, and embryos remain in this stage over winter. Stage R embryos have distinct stripes and banding, including radiating bands of pigmentation around each eye. Elb. = elbow; S. B. = shell-breaker; T. = tail. For other abbreviations, see Figure 8.42.

Reproduced from Dendy, A. 1899. Outlines of the development of the tuatara, *Sphenodon (Hatteria) punctatus*. *Quarterly Journal of Microscopical Science* 42: 1–87 + plates 1–10. Permission from Company of Biologists conveyed through Copyright Clearance Center Inc.

turtles or a crocodilian, in which the limb buds precede or appear simultaneously with the allantois; however, more species must be studied to determine whether this is a consistent phylogenetic pattern.[335]

Tuatara embryos reach Stage R by February–March on Stephens Island (Figure 8.43), and stay at this stage over winter, with only slow development. Although this is the penultimate stage before hatching, the small body length given for a Stage R embryo[336] and the presence of much unused yolk in the yolk sac[337] indicate that substantial growth has yet to occur before hatching. The skin of Stage R embryos is distinctively striped and banded.[338] Embryos at Stage R also have a conspicuous 'shell-breaker' formed from cornified epidermal cells on the snout. This structure thus resembles the horny shell-breakers of turtles and crocodilians, rather than the dentinal egg tooth (or teeth) of squamates.[339]

Stage S, the final stage before hatching, was reached in January, about 13 months after oviposition, in eggs examined by Dendy.[340] Embryos at Stage S have scales and claws much as in adults, and rudimentary nuchal and caudal (tail) spines. The parietal eye is no longer visible externally but its probable position is indicated by a median scale. Longitudinal striping has disappeared except under the throat and chin, where it remains distinct, although the white stripes are now narrower than the grey. Transverse banding is still prominent on the back and tail.[341] Both Schauinsland[342] and Thilenius commented on the distinctive striation shown by advanced embryos; in addition, the red body tinges and a light greenish triangle forward of the eyes in freshly hatched tuatara produced what was described by Thilenius, in translation, as a 'multicoloured impression'.[343]

Tuatara eggs have no air chamber[344] but swell noticeably during development. This reflects the absorption of water[345] and its accumulation in the form of a copious, semi-gelatinous liquid in the allantois (Figure 8.44).[346] Swelling of eggs occurs mainly during the last third of development, with some eggs

Fig. 8.44 Candling of a laboratory-incubated egg reveals the darkened body of an embryo developing within, and a lighter region of accumulated fluid.

Photo: Brett Robertson/Victoria University of Wellington.

becoming almost spherical prior to hatching.[347] After 12 months, swollen eggs in the natural nests studied by Michael Thompson and colleagues filled much of the nest chamber, and were grossly distorted in shape from contact with each other and the surrounding soil.[348] Mean mass of the remaining viable eggs ranged from 5.63 grams to 11.41 grams.[349] The largest egg measured by Dendy at Stage S attained dimensions of 35 x 27.5 millimetres,[350] although this egg may have benefited from moisture provided in the laboratory.[351]

Shells incubated in soil often become brown-stained by the end of the incubation. Scanning electron microscopy shows that hatched shells have a more granular, weathered appearance than when laid,[352] but whether resorption from the shell is a major source of calcium for embryonic development, as in turtles and crocodilians,[353] is unknown. Clearly, the unusual structure of the eggshells does not prevent expansion and water uptake during incubation. Little is known about the biochemistry of yolk metabolism during development. In tuatara, the microscopic yolk bodies appear to aggregate during embryonic development.[354] Analyses also suggest that lipid and fatty acid composition may change,[355] although the small sample size (one developing egg and one newly laid egg) limits the interpretation.

Incubation time on Stephens Island is 10–16 months.[356] Nest temperatures vary with location, and eggs in the warmest nests hatch first.[357] However, a suggestion by Thilenius that some eggs might hatch in the same summer as

laid has not been confirmed.[358] Although incubation time in tuatara appears to be the longest known for any reptile,[359] the general features of early development seem similar to those of other vertebrates with heavily yolked eggs.[360]

In moist conditions in the laboratory, eggs become taut and turgid by the end of incubation,[361] and a very small incision causes the shell to split open with, as Dendy put it, 'almost explosive violence'.[362] The shell is then often split cleanly, as if by a razor, along the long axis.[363] The released fluid perhaps allows the surrounding, compacted soil to be more easily penetrated by the emerging hatchling (Figure 8.45).[364] It was initially thought that hatching might be simultaneous for all embryos in a nest,[365] but recent observations indicate asynchrony of up to 25 days.[366] This could reflect the effects of temperature gradients within the nest on embryonic development (see next section). Hatchlings may delay emergence from the nest for up to a week after hatching.[367]

In captivity, hatchlings from eggs incubated in soil made sporadic, convulsive movements of the head and body, slowly working their way over about 8–9 hours through about 130 millimetres of soil to the surface without using their limbs. Only after the snout had emerged were the forelimbs used to help bring the body to the surface.[368] Emergence from the nest is rarely observed in the field (Figure 8.46), but the few sightings made suggest a similar, snout-first 'corkscrewing' behaviour.[369] Hatchlings have a small

Fig. 8.45 A hatchling tuatara emerges from its egg, following laboratory incubation in moist vermiculite. Fluid that has accumulated within the shell is visible just below the hind leg.

Photo: Brett Robertson/Victoria University of Wellington.

Fig. 8.46 A rarely seen sight: a hatchling tuatara emerges from soil above a natural nest on Stephens Island.

Photo: Stu Bisset, courtesy of Mary McIntyre.

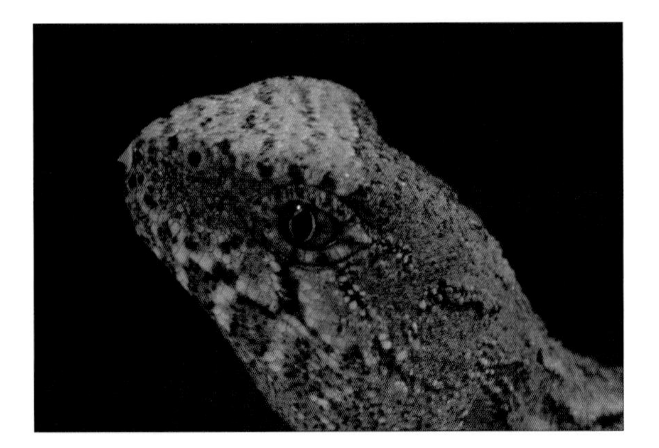

Fig. 8.47 A laboratory hatchling of Stephens Island stock, photographed within 4 weeks of hatching. The horny shell-breaker (not a true tooth) is still attached to the snout.

Photo: collection of Alison Cree.

umbilicus, occasionally with a small remnant of yolk sac,[370] which dries and falls off within a few days.[371] In our laboratory, viable but very small (1.6–1.9 grams) twins once emerged from a normal-sized egg;[372] these apparently had shared a common yolk sac.[373] Although the horny shell-breaker often disappears within about 1–2 weeks of hatching,[374] it can remain attached for at least 6 months under laboratory conditions[375] (Figures 8.47 and 8.48). Hatchlings in captivity are remarkably active, indeed frisky at times. In the field, the emerged hatchlings soon find a nearby cavity or cover. Although some may remain close to the nest site for the first year of life,[376] others migrate toward areas with more cover provided by plants.[377] During the first few months of life, hatchlings appear mainly diurnal in behaviour.[378]

Little information exists on embryonic development in nature for other populations of tuatara. On North Brother Island, one nest excavated in January contained live eggs, plus two shells that contained what appeared to be the faeces of an invertebrate predator or scavenger.[379] In one nest observed at intervals on Lady Alice Island, three of five eggs apparently failed (shells were empty and crumpled, with signs of entry by beetle larvae). The remaining two eggs appeared to have hatched between October and early December, giving an incubation length of between 310 days and 362 days.[380] In the 1880s Thomas once obtained eggs from a captive female from Karewa Island that died, apparently egg-bound.[381] He reported that 'From these eggs were obtained a number of embryos at various stages of development, from a stage equal to a two days' chick up to a stage shortly before hatching.' This confusing description seems to have led others to infer that various stages were simultaneously present at dissection.[382] What Thomas almost certainly meant was that he incubated the eggs for various periods to obtain embryos at different stages.

The pioneering studies by Dendy, Schauinsland and Thilenius on tuatara from Stephens Island were landmarks in the study of embryonic development. Dendy's collections, in particular, enabled numerous additional studies on the development of specific tissues and organs (Chapter 5). However, little recent work has occurred on the anatomy of embryonic development in tuatara. Uncertainties remain about the formation of

Fig. 8.48 Hatchling tuatara are cryptically coloured in shades of earth-brown (often tinged with pink or red). Darker transverse bands are present on the trunk and tail. There is a distinct, lighter-coloured triangle on the head, but minimal signs of a crest. Hatchlings are quick and agile. The two small, 6-day-old hatchlings on the left emerged from the same egg (compare these with the normal-sized hatchling on the right, from the same captive-bred clutch of Stephens Island stock).

Photo: collection of Alison Cree.

the germ layers, about the sequence of events relative to squamates, and about variability in the sequence among different embryos.[383] For example, whether tuatara form an isolated yolk mass during the development of extra-embryonic membranes, a feature otherwise unique among reptiles to squamates, is unknown.[384] Dendy's 17-point staging scheme for tuatara is imprecise, and difficult to match against the 40-point staging scheme developed by Dufaure and Hubert in 1961 for *Lacerta vivipara* and widely used since in other squamates.[385] Some histological work is now underway to record the sequence of changes in embryos, focusing on the middle third of incubation.[386] However, no complete reference series of preserved embryos incubated under standard conditions exists. Given the major advances recently made in understanding the developmental genetics of other vertebrates, modern techniques hold the promise of valuable new discoveries once applied to tuatara.[387]

Temperature and moisture in natural nests

In field studies led by Michael Thompson, nest temperatures and water potentials in natural nests on Stephens Island were monitored about every 2 weeks for a year.[388] During each 24-hour monitoring period, recordings were made every 4 hours. As expected, nest temperatures showed considerable diel (day–night) and seasonal variation, with less variation related to position within the nest.

BOX 8.5

Why do female tuatara not nest in the forest on Stephens Island?

Tuatara live at high density in the forest on Stephens Island (Takapourewa), but have never been observed to nest there. Forest soil is extremely friable, which should make for easy nest construction. The soil is also favourably moist, with water potential being well within the limits for successful incubation.[1] However, soil temperatures under the canopy are much lower than recorded in natural nests, and rarely rise above 15°C. These temperatures are below the laboratory threshold for successful development and hatching,[2] suggesting that soil in the forest is simply too cool for successful incubation.[3]

To test this hypothesis, I conducted a small experiment with several colleagues, using viable eggs that were surplus to other studies examining effects of hormones on laying.[4] We buried half of the freshly laid eggs in five artificial (human-made) 'nests' in the forest, and compared their development with those placed in five artificial 'nests' in the pasture.[5] The artificial 'nests' were constructed to resemble natural nests in depth and other characteristics. Each nest contained eight eggs.

'Nest' temperature recordings over the next 15 months revealed that forest 'nests' were always cooler than those in the pasture. Forest 'nests' were almost always below 15°C (and always below 19°C), whereas pasture nests reached 21–25°C during summer afternoons. After 12 months, the eggs were checked to assess the progress of incubation. Four 'nests' in the pasture still contained at least some turgid and apparently viable eggs, although other eggs had failed and desiccated. One egg from each 'nest' was opened to measure and stage the embryo within. The embryos in the pasture 'nests' were all much larger, more advanced in development and had less yolk remaining in their yolk sacs than those in the

Box 8.5 Fig. 1 Michael Thompson and Jennifer Hay burying eggs in artificial nests in forest on Stephens Island.

Photo: Alison Cree.

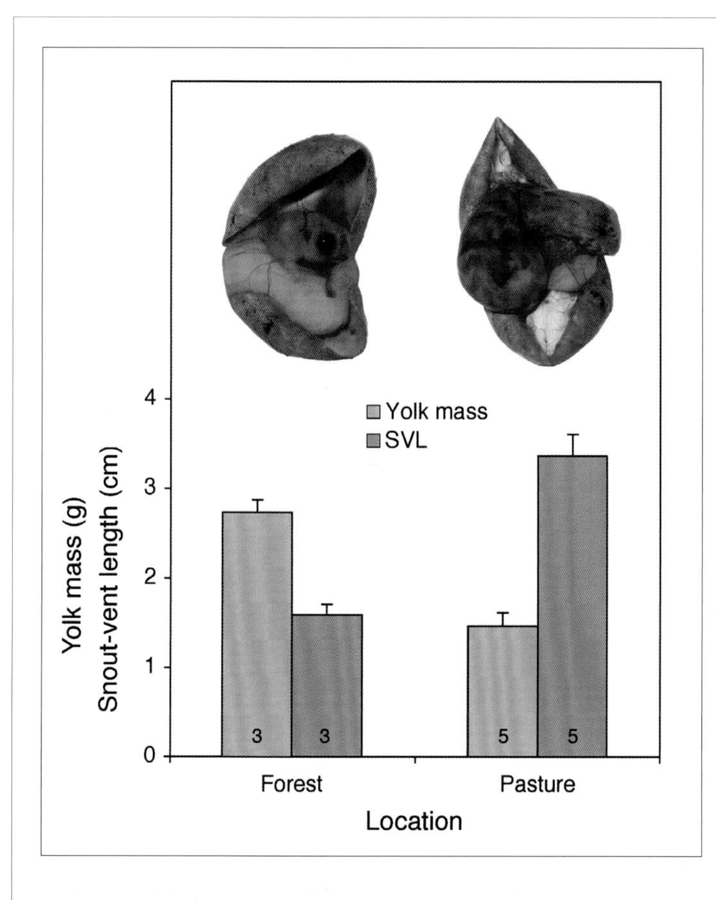

Box 8.5 Fig. 2 Mean yolk mass, mean snout–vent length (SVL) and appearance of embryos from eggs buried for 12 months in the forest or pasture on Stephens Island. Embryos that survived in human-made 'nests' in the forest (left) were only half the length of those in pasture (right), and had twice as much yolk still to absorb. Forest embryos were also less developed (Dendy's Stage Q) and averaged only 20% of the mass of pasture embryos (Dendy Stage R or S). Although some embryos from the pasture hatched about 3 months later, none is known to have hatched from the forest, where soil temperatures were lower. Differences in yolk mass and snout–vent length between forest and pasture were significant ($p < 0.01$; standard errors and sample sizes are shown).

Photos: Alison Cree.

forest.[6] However, some development had occurred in the forest eggs, and most forest 'nests' still contained some viable eggs.

The remaining eggs were checked again at 15 months after burial. Two of the pasture eggs had hatched, and a third 'nest' contained six live hatchlings. All of the remaining eggs in the pasture 'nests' had failed, with desiccation a probable cause. Although there was no evidence of hatching in the forest, eight eggs were still incubating at 15 months. When checked again at 21 months, four were still incubating. Only one egg was still incubating (but with a trace of mould on the surface) when checked finally at 27 months, and it probably never hatched.

These results confirm that tuatara embryos can develop in the cooler conditions that prevail in the forest, and are probably less prone to desiccation there. But despite development continuing for a remarkable 2 years or more in one case, there is no evidence that hatching was ever successful. As in laboratory studies,[7] cool temperatures permit slow development to an advanced stage, but for some reason embryos from such conditions fail to hatch. Not surprisingly, given the temperature differences (see Box 8.6), three sexable embryos from forest 'nests' were all female, whereas five sexable embryos from pasture 'nests' were all male.[8] Hatching success overall in the pasture was a modest 16.4%,[9] which suggests that our pasture 'nests' were not as well sited or constructed as those made by tuatara themselves.

The results of this and other studies raise questions about the future responses of tuatara to forest restoration on Stephens Island, and to climate change on all islands. Will tuatara nest in a forested site if the soil temperature is sufficiently warm for hatching? If so, will they do so only if the soil is warm enough for the production of both sexes? Will embryos from cold sites hatch if they get a short pulse of warm temperature late in development, or are they physically incapable of breaking their shells regardless of the temperature at the end of incubation? Only further research will tell.

Diel fluctuations in summer reached up to 16°C. Temperatures in successful nests (those from which hatching resulted) reached up to 32.5°C (but 38.4°C in a later study).[389] Temperatures recorded by Thompson were typically up to about 1°C warmer by day, and up to about 1°C cooler at night, at the top of the nest than at the bottom.[390] Midday temperatures from early summer until mid-autumn (December until April) were generally above 15°C, which laboratory experiments indicate is a threshold for successful development and hatching (see next section). Nest temperatures fell below 15°C for about 4 months over winter, before rising again in spring. This corresponds with the winter period of developmental arrest inferred by Dendy in 1899.[391]

Soil water potentials in the same study varied greatly among nests in summer.[392] In 'wet' nests, water potential measured in the morning[393] remained at or close to saturation (0 kilopascals, or kPa), whereas in 'dry' nests, water potentials fell to below –2000 kPa. All nests remained at high water potentials over winter, but some began drying again in spring. Egg survival varied from 0% to 100% among nests, but was about 48% overall. Water potential was a better predictor of survival than temperature: embryos survived better in wet nests than dry ones.[394]

Desiccation has long been considered a factor contributing to egg mortality on Stephens Island.[395] Thilenius inferred that desiccation might cause failure of entire clutches where egg chambers were shallowly situated, or laid in shingle with little earth. In deeper nests with dead embryos at an advanced stage of development, he suggested that pressure from swollen eggs within the confines of the nest chamber might have contributed to mortality,[396] an interesting but unconfirmed idea. Thilenius also noted occasional abnormalities of development, and considered that overall only about one-third to one-half of eggs contained viable embryos when excavated in the year following nesting.[397] Failed eggs are sometimes punctured by holes, suggesting that soil-dwelling insect larvae contribute to egg mortality,[398] although whether the larvae kill embryos or simply enter eggs that have already failed is unclear.[399] Recent work by Nicola Nelson confirms that nests less than 50 millimetres deep are less likely to be successful.[400]

On Stephens Island, nest temperatures are related to sex ratios at hatching,[401] as expected for a species that exhibits temperature-dependent sex determination under laboratory conditions. In a recent experimental translocation, eggs were shifted from Stephens Island to a colder climate further south at Orokonui Ecosanctuary near Dunedin, to explore the implications for embryonic development and hatchling sex ratios[402] (Orokonui lies within the past latitudinal range of tuatara in the South Island; see Chapter 2, Fig. 2.8). A few eggs survived 12 months of incubation at Orokonui, during which time soil temperature did not rise above 22°C. Embryos were by this time at Stage R (beyond the point of sex determination), and surviving hatchlings (which seemed of normal size, despite the low incubation success) are known or likely to have been female. However, further measurements in a later year and at different sites showed that at least some shallow (50-millimetre-deep) sites at Orokonui are warm enough potentially to produce both sexes.[403] At a similar latitude but further inland, at Alexandra and Wanaka (also within the past geographic range of tuatara), climates are hotter in summer and colder in winter than at Orokonui. Temperatures at 100 millimetres of depth in Alexandra and Wanaka are likely to support embryonic development and yield hatchlings with a high proportion of males, if eggs have the same characteristics as those from Stephens Island.[404] Ultimately, the risk of desiccation from incubation at shallow depths (the only sites warm enough for successful development of both sexes at cooler sites) may have influenced the southern limits for the distribution of tuatara. Whether egg incubation could be successful if extended over more than the maximum of 16 months seen on Stephens Island (for example, over two winters) is unknown.

Effects of temperature and other factors on development and growth in the laboratory

Controlled studies with eggs from Stephens Island confirm that incubation temperature influences both the success and the rate of development.[405] Eggs of good quality held under moist conditions experienced high hatching success (88% or more) at temperatures between 18°C and 22°C.[406] The lower limit for successful hatching lies between 15°C and 18°C, since embryos incubated at 15°C developed but did not hatch.[407] The upper limit for successful incubation at a constant temperature is uncertain, but must be above 24°C since eggs incubated at this temperature in our laboratory had high hatching success (89%).[408] At 18°C, the incubation time of about 280 days is about twice as long as at 24°C (Figure 8.49). A similar effect of temperature on incubation success and incubation time is apparent for eggs from North Brother Island.[409] Overall, temperatures permitting successful incubation in tuatara are lower than for most other egg-laying reptiles.[410]

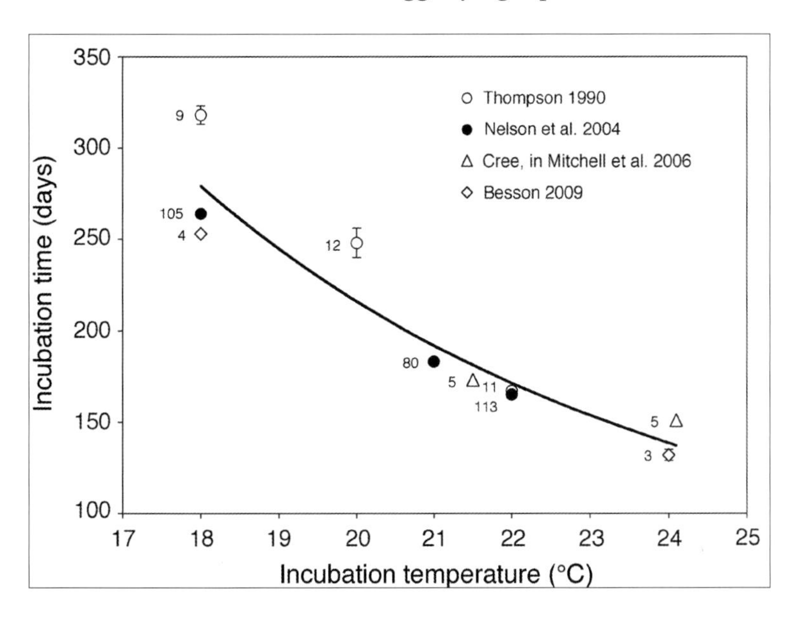

Fig. 8.49 Effect of temperature on incubation time in eggs of Stephens Island origin. Eggs were incubated half-buried in moist vermiculite (water potential about −90 to −170 kPa) and under a range of constant temperatures for which incubation success is generally high. Incubation time includes transit from field sites and up to 4 weeks in an incubator at 18°C. Sample sizes are shown to left or below each mean value (vertical lines indicate standard errors). The curve has been fitted without weighting for sample sizes.

BOX 8.6

Soil as an oven or a fridge: how sex is determined in tuatara

In 1992 I was conducting laparoscopies with my students for studies in which we needed to know the sex of juvenile tuatara in zoos.[1] Intrigued, I noticed that offspring from several zoo-incubated clutches were all one sex or the other,[2] including one clutch known to have experienced high temperatures early in development[3] in which all seven surviving offspring were males. Could it be that tuatara possessed a form of sex determination in which incubation temperature, not genes on chromosomes, decided whether hatchlings turned out to be males or females?

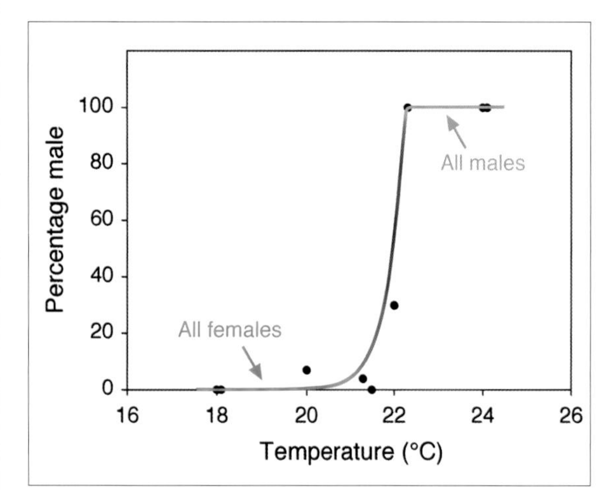

Box 8.6 Fig. 1 At constant egg-incubation temperatures in the laboratory, cool conditions (pink line, below about 21.2°C) produce essentially only female tuatara whereas warm temperatures (blue line, above about 22.2°C) produce only males. At intermediate or transitional temperatures (lilac line), both sexes are produced in varying proportions (at the pivotal temperature of 22.0°C, the expected ratio is 50:50). The fitted line takes into account the precision of temperature control in different studies. These results come from experiments with eggs from Stephens Island, but less extensive data for North Brother Island and Hauturu/Little Barrier Island suggest a broadly similar pattern (see Cree et al. 1995b; Nelson et al. 2004a; Mitchell et al. 2006).

Redrawn from Mitchell et al. 2006, with additional data from Besson 2009.

In collaboration with Michael Thompson and Charles Daugherty, I sexed further juveniles that originated from known-temperature incubations.[4] The results confirmed unequivocally that offspring sex was indeed a function of egg incubation temperature.[5] At low and constant incubation temperatures in the laboratory, only females were produced, and at high incubation temperatures, only males. Tuatara therefore have a pattern of temperature-dependent sex determination, or TSD, that can be described as FM (or female-male).[6] This pattern has since also been observed by Nicola Nelson and colleagues in natural nests on Stephens Island (Takapourewa).[7] Intermediate incubator or nest temperatures yield both sexes, but not individuals in which sex is ambiguous. As is typical of other TSD reptiles, sex-specific chromosomes have never been found in tuatara.[8]

Although many egg-laying reptiles have TSD, the FM pattern is a rare one. In the more common MF pattern,[9] seen in many turtles, high temperatures produce females. Also common is an FMF pattern,[10] in which both low and high temperatures produce females, with males arising in greatest abundance from intermediate temperatures. In some lizards and crocodilians, what was initially thought to be an FM pattern turned out to be an FMF pattern when higher incubation temperatures were tested.[11] Further work with tuatara is needed to establish whether high-temperature as well as low-temperature females can be produced.

The discovery of TSD in tuatara raises many questions, both theoretical and applied. A 'how' question – how does the mechanism work? – is not yet possible to answer with certainty. Extensive research on other reptiles suggests that something about temperature affects the first steps in a complex sequence of genes

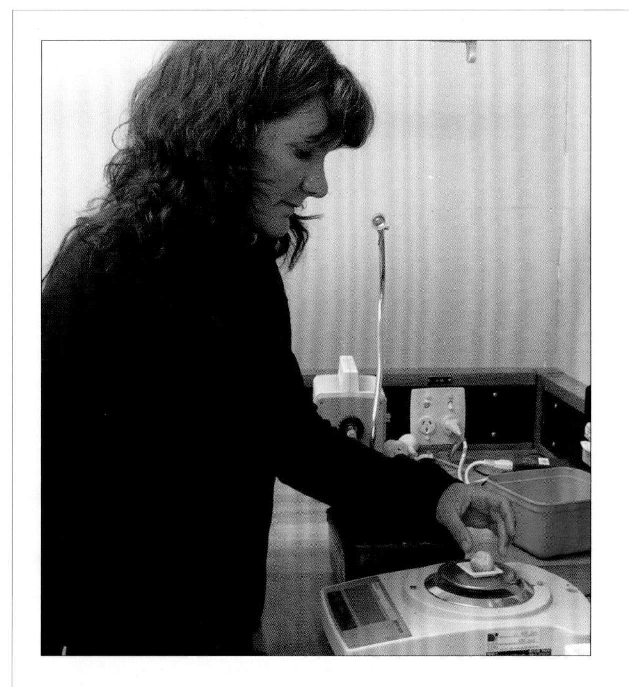

Box 8.6 Fig. 2 Nicola Nelson weighs an egg before its return to an incubator. By shifting eggs between incubators set at different temperatures, Nelson and colleagues have established the period when embryo sex is sensitive to temperature. Eggs that were transferred from a warm, male-producing temperature of 23°C to a cool and potentially female-producing temperature of 20°C remained male if the shift occurred later than week seven of incubation. However, if the shift occurred before week seven, the offspring were influenced by the new temperature to become female. This tells us that the temperature-sensitive period ends at 23°C by about the middle third of incubation, a similar time to that in other TSD reptiles.[19]

Photo: Susan Keall.

routinely involved, in all vertebrate species, in producing a male or a female outcome. The trigger, or 'master switch', may have early effects on aromatase, an enzyme responsible for converting testosterone into oestradiol. This is suggested by experiments in TSD turtles in which manipulations of aromatase activity, or oestradiol levels, over-rode the normal effects of temperature, affecting the outcome of sex determination.[12]

The answer to the 'why' question – why does TSD evolve? – is also not known with certainty. In some respects, TSD would seem to be disadvantageous, especially in short-lived species where a strong bias in a population's sex ratio might quickly lead to extinction. Being a long-lived species, tuatara are less at risk in this respect, and perhaps TSD brings benefits. Research with an Australian lizard suggests that each sex is produced at the temperatures that will bring individuals the greatest reproductive success.[13] That is, TSD may be a way of linking sex with the patterns of hatching time, growth and reproductive performance that arise from different incubation conditions. If, for example, certain incubation temperatures favour rapid post-hatching growth, then it may be advantageous at that temperature to produce the sex that has the most to gain by being large. Male tuatara grow to a larger size than females, and their reproductive success is clearly influenced by size.[14] In tuatara, we can say that the FM pattern links maleness with rapid embryonic development and early hatching,[15] although whether males have more to gain from these conditions than females remains speculative.

Questions of a 'what if' nature can also be asked. What if soil temperatures rise as global climates become warmer? Will the sex ratio of the population change, and could local extinction result? In principle, yes, especially on small islands where there is relatively little variation in available nest temperatures.[16] Whether tuatara on larger, forested islands will change their behaviour and nest more deeply, or in shade, or at a different time of year, is unknown.[17] However, TSD is only one of a number of factors that put tuatara at risk from climate change. Even without TSD, populations on small islands (which are already vulnerable because of small population size and sometimes low genetic variation) will be vulnerable to physical changes in the habitat. The predicted risks include rising sea levels, increased coastal erosion, rising temperatures and in some cases drought,[18] the latter two of which could lead, for example, to reductions in egg incubation success and juvenile survival.

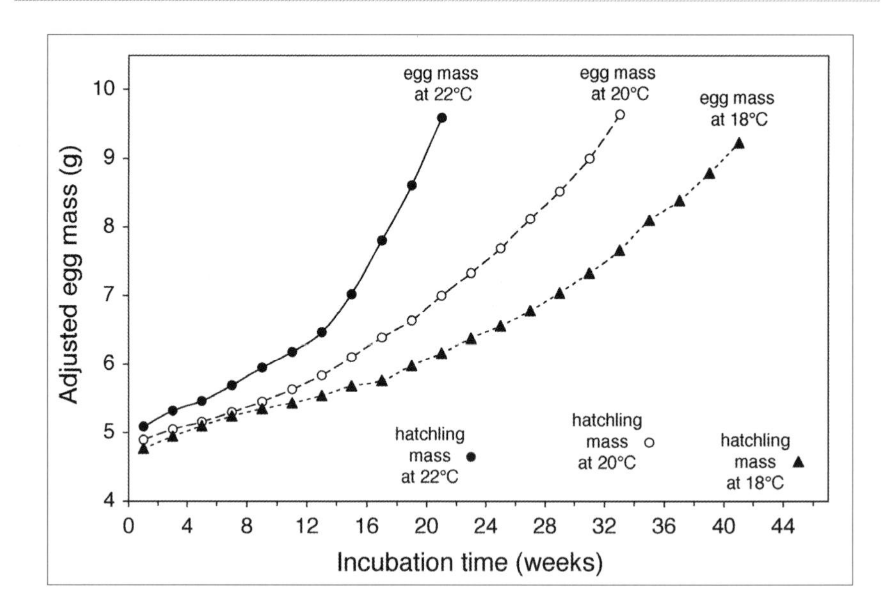

Fig. 8.50 Pattern of water uptake in eggs of Stephens Island origin incubated at three temperatures. Eggs were incubated half-buried at a water potential of −90 kPa. Mass is adjusted for variation in initial egg mass. Mean hatchling mass at the mean time of hatching is also shown for each temperature. Eggs absorbed water more quickly and hatched more quickly at warmer temperatures, but temperature did not significantly affect the adjusted mass of hatchlings.

Drawn from data in Thompson 1990.

A study by Michael Thompson also examined the effect of substrate water potential on incubation. Water uptake may be essential for development, as the wet mass of hatchlings and their shells was always greater than the fresh mass of the eggs.[411] The rate of water absorption (indicated by the rate of change in egg mass, adjusted for variation in initial egg mass) was faster at warmer temperatures, as expected given the shorter incubation time (Figure 8.50). At each temperature, lower water potential (a drier substrate) resulted in a slower rate of mass gain (water uptake) but had little effect on incubation time. There were no consistent effects of temperature, water potential or egg position (surface or fully buried) on hatchling size (mass or linear dimensions).[412] After adjustment for initial egg mass (which affects the rate of water uptake), means for hatchling mass from half-buried eggs incubated at 18–22°C and −90 kPa were 4.6–4.9 grams, and means for snout–vent length were 48.6–49.4 millimetres.[413] In other research, initial egg mass had a marked effect on hatchling size: all else being equal, larger eggs produce larger hatchlings.[414]

In an extension of studies exploring temperature effects, oxygen consumption was examined in embryos of Stephens Island origin.[415] At 20°C and −230 kPa, the rate of oxygen consumption increased for the first 70–80% of incubation and thereafter slowed to approach an asymptote, a pattern known in some other reptiles. There was no evidence of hatching synchrony in tuatara incubated under these conditions: hatching was spread over 58 days.[416] The peak rate of oxygen consumption varied with temperature across the range 18–22°C, but total energy expenditure remained similar.[417]

In a recent study by Nicola Nelson and colleagues, hatchlings from eggs incubated under natural (field) conditions on Stephens Island were compared with hatchlings from eggs from the same island incubated under constant conditions in the laboratory. Incubation conditions did not greatly affect size at hatching, but they did appear to influence post-hatching growth and behaviour. After maintenance for 10 months under comparable conditions, hatchlings from natural nests were smaller and had a greater incidence of tail loss, but were faster for their size and possibly more aggressive than those incubated under constant temperatures.[418] Given the differences in hatching time and sex that arise from differing incubation conditions, it was not possible to fully randomise the way that hatchlings were grouped during the post-hatching period, and hatchlings also did not have access to basking (heat) lamps. Nonetheless, these interesting results argue for further exploration of the consequences of embryonic conditions for post-hatching life. Constant incubation temperatures, which are easy to provide and yield high hatching success, have been used extensively for captive rearing (including the production of offspring for release to new islands, as described in Chapters 6 and 10). However, embryos in natural nests clearly experience variation in temperature and moisture, and the implications for lifetime performance are important to understand.

Growth and longevity in wild populations

Although popular accounts have sometimes proposed a life span of 300 years or more[419] for tuatara, scientists have long recognised such guesses as unfounded.[420] At least some of these high values apparently stem from a misinterpretation of a statement in 1876 that a tuatara was kept in a kūmara pit on Motiti Island for 'over three generations'.[421] Unpublished correspondence from the observer, Captain Gilbert Mair, indicates that the tuatara might have reached a total age of about 100 years, including some later decades in captivity (if indeed it was the same animal throughout, which is possible but unproven).[422]

The only certain information on longevity in the wild stems from recaptures of animals marked from the 1940s onwards. William (Bill) Dawbin toe-clipped 750 tuatara on Stephens Island between 1949 and 1955, many of which he recaptured in visits up until 1980.[423] He used the von Bertalanffy model to fit growth curves for each sex, combining these with the size of sexual maturity (which he inferred from preserved specimens as 180 millimetres snout–vent length in females and 200 millimetres in males; but see below for evidence that the value for males is an overestimate). Using this approach, Dawbin estimated that sexual maturity was reached at about 20 years, that maximum adult size was reached at about 50–60 years, and that life expectancy might exceed 100 years.[424] However, the length at which adults stopped growing varied markedly among individual tuatara,[425] and newer information challenges the assumption that body size and age are strongly correlated in reptiles and amphibians.[426] These observations raise the possibility that some individual tuatara might grow quickly and reach a large maximum size, whereas others might never attain the same dimensions and perhaps end up adopting different behaviours and reproductive strategies (perhaps in part as a consequence of differing egg size and incubation conditions).

An alternative technique, skeletochronology, has been used for estimating growth rates and minimum individual age.[427] This technique involves analysis of long bones to detect rings that take up the dye haematoxylin. These 'lines of arrested growth' are assumed to correspond to a period of reduced growth, in much the same way that rings in tree trunks illustrate seasonal changes in growth. In tuatara, it is assumed, but not tested, that such rings accrue on an annual basis and that adequate compensation can be made for possible resorption of bone. Castanet and colleagues used skeletochronology to examine growth rings in phalanges (segments from digits, either a front or rear toe), which were removed from Stephens Island and Lady Alice Island tuatara of a range of sizes and sexes, as well as in the femurs of seven adults from Stephens Island killed for other purposes.[428]

Based on their results for Stephens Island, Castanet and colleagues questioned the validity of the growth model calculated by Dawbin. The bone rings they counted suggested that tuatara on Stephens Island reached the sizes of sexual maturity reported by Dawbin (180 millimetres snout–vent length for females and 200 millimetres for males) in 11–13 years. On Lady Alice, the same sizes were reached at ages about 2–3 years younger than on Stephens, suggesting faster growth rate in the warmer climate. There were limits to the use of the technique in estimating adult age. Phalangeal growth appeared to cease earliest, yielding no more than about 23 rings, even in the largest individuals. A larger number of femoral rings could be counted, from which it was estimated that on Stephens Island, females stop growing at about 20–25 years and males at 25–35 years. The largest number of femoral rings (35), added to the known period of adult life without growth (30 years, from Dawbin's work)[429] suggested a lifespan of at least 60 years.[430] This investigation also suggested that adult mortality was likely to be low in tuatara.[431]

More research is required to determine which of these approaches for estimating age at maturity is the most accurate. Regardless of method, Dawbin's estimates of size at maturity on Stephens Island are likely to be near the upper limit for females and an overestimate for males. Male tuatara that are 182 millimetres (mm) or more in snout–vent length already have plasma testosterone concentrations and spermatogenic activity indicative of sexual maturity.[432] Although some females of about 180 mm snout–vent length appear from ovarian condition not to have reproduced,[433] others are known to have been gravid at 170 mm snout–vent length,[434] with vitellogenesis beginning below this size.[435] Interpolation from Dawbin's growth curve suggests that these lower estimated sizes at maturity (about 170 mm in females and 180 mm in males) might be reached at about 17 years in females and 13 years in males.[436]

Several female tuatara marked on Stephens Island by Dawbin have subsequently been recaptured by other researchers. One reproducing female that I captured in 1988[437] had been first marked in 1949. Eight reproducing females captured by Newman in the 1980s had been marked over 30 years previously, at which time they were already fully grown.[438] Using the conservative assumption that growth in female tuatara on Stephens Island stops at about 20–25 years of age,[439] these females were at least 55–60 years of age.[440] From more recent recaptures by Nicola Nelson's research group, longevity of up to 91 years has been estimated.[441]

Some observations of growth and longevity are available for North Brother Island, where 21 tuatara were toe-clipped by Richard Barwick between 1957 and 1959. One animal recaptured in 1988 was a relatively large adult female that had not changed in snout–vent length (197 millimetres) since first being marked. Using the assumption that tuatara stop growing at about 25–35 years of age,[442] this female was estimated as at least 60 years old at recapture.[443] Another female recaptured in 2000 or 2001 was already an adult when first marked at least 43 years previously.[444] These and other recaptures over shorter periods enabled Nicola Mitchell and colleagues to construct von Bertalanffy growth curves, from which it was estimated that males and females grow at similar rates until maturity is reached (at about 23 years of age in females). Growth rate then declines more rapidly in females than in males. Adult mortality was low (4.4% per year for females and 5.4% for males).[445]

Taking into account all the available information, including longevity in captivity (see Box 8.3), tuatara can probably survive for 100 years or more in nature, as previously estimated by Schauinsland and Dawbin.[446] However, it remains doubtful, or at least unproven, that any tuatara surviving today could have been alive, for example, when Captain Cook first sailed around New Zealand in 1769–1770. Future studies need to consider the likelihood that individual tuatara within a population might be on different trajectories in terms of growth rate and maximum size. It is also probable that growth patterns, and perhaps longevity, will vary among tuatara from different islands, alongside known or suspected differences in food supply, density and maximum body size (see Chapters 6 and 7). Responses of tuatara translocated to new islands have also raised the interesting possibility that growth in not only mass but also snout–vent length can resume when adults are transferred to new habitats at low density.[447] These observations need to be reconciled with the anatomy of the long bones, including whether growth can continue once the epiphyseal cartilage has turned into bone.[448]

As tuatara are such long-lived animals, it is crucial that any marking scheme is managed on a time-frame that extends beyond the working lives of biologists. Electronic records of all tuatara toe-clipped on Stephens Island from Dawbin's studies onwards have now been compiled[449] and lodged in the Department of Conservation's national database ('Herpetofauna'). PIT-tags (passive integrated transponder-tags; small bar-coded tags about the size of a rice grain, implanted under the skin) are now in use in tuatara on Stephens Island[450] and, although more expensive than toe-clipping, are considered more ethically acceptable for adults and will expand the number of individuals that can be uniquely marked. Future recaptures of these animals, and of marked animals that have since been translocated to new islands, will continue to provide information on growth and longevity in the wild.

Concluding remarks and ideas for further research

Tuatara have, for reptiles, an unusual combination of reproductive and life history characteristics. In many respects, life history resembles that of turtles, almost to the extent that tuatara could be considered as 'turtles without shells'. Features shared with turtles and crocodilians include sperm structure in males, lack of pyriform cells in the ovarian granulosa of females, egg-laying, presence of a horny shell-breaker in hatchlings, delayed maturity and great longevity. However, there are other features of morphology and life history in which tuatara most closely resemble squamates. These include the structure of the cloaca with its transverse vent (and possible hemipenial precursors), as well as the structure and physiology of eggs (including the structure of the shell, the scant amount of albumen, the need for water uptake during embryonic development and the sequence in which the limb buds and allantois appear). During their courtship and mating, tuatara also exhibit a behavioural repertoire and social structure as complex as those in many squamates.[451]

Future studies exploring the phylogenetic significance of these patterns need to consider the possible effects of body size and life history (like turtles and crocodilians, tuatara are relatively large-bodied and long-lived), as well as the cool temperatures that tuatara currently experience. Recent studies of New Zealand lizards reveal that many features of life history once considered 'primitive' in tuatara, such as the prolonged period of embryonic development,[452] are more likely to be adaptations to a cool climate. For example, geckos from southern New Zealand (though live-bearing, unlike tuatara) also have prolonged periods of embryonic development (7–14 months)[453] with less-than-annual reproduction in females of at least one species.[454] Male geckos from New Zealand exhibit prolonged spermatogenic cycles (resembling that of tuatara),[455] and nocturnal species also show great longevity of 40 years or more in the wild,[456] which is remarkable for such small lizards. Prolonged gonadal activity across a year or more, extended periods of embryonic development, low annual reproductive output and great longevity may be common responses of New Zealand reptiles to climates that overall are remarkably cool by the standards of many reptiles studied overseas.[457]

Research gaps that have been identified in this chapter (in the main text and in boxes) are summarised below in the form of questions under several headings.

Functional anatomy of the reproductive tract and gametes

What is the functional significance of differences in sperm structure between tuatara and squamates; does the granulosa of ovarian follicles show any features that could be considered precursors towards the development of the pyriform cells characteristic of squamates; how do the sebaceous glands compare between males and females; what is the functional significance of the secretions of the sebaceous glands; what is the function of cloacal eversion by male tuatara prior to mating; and to what extent are hemipenial precursors present, either anatomically or functionally?

Endocrinology and metabolism

How does the chemical structure of the gonadotrophic-releasing hormone(s) compare with that of other reptiles; is there a single gonadotrophin (as in squamates) or two (as in other reptiles); what role do possible fat stores (in the tail, liver and other sites) play in supporting the protracted vitellogenesis; what is the role and temperature-dependency of leptin secretion during vitellogenesis and other seasonal cycles; and what are the metabolic implications of nesting, including the functions of elevated corticosterone?

Reproductive cycles and nesting behaviour

What is the relative importance of temperature, food supply and age on reproductive frequency in females; how do the physiology and behaviour of females change during late gravidity in spring prior to nesting (including the significance of water uptake by eggs within the oviducts); what role do substrate temperature and moisture play, independent of shading, in the selection of nest sites; how are nesting females on Stephens Island responding to reforestation and growth in rank grass; and what are the seasonal times of mating, nesting and, especially, hatchling emergence on more northerly (and warmer) islands than Stephens Island?

Eggs and embryonic development

What are the cause(s) of pre-ovipositional arrest in embryonic development; what is the detailed sequence of morphological events and gene activities (especially in events and genes of phylogenetic significance) during embryonic development; how does diet influence egg quality; does egg size vary among wild populations; what are the detailed relationships among substrate temperature, moisture and incubation success in natural nests, especially those on small northern islands prone to drought; what are the limits to tolerance of high temperatures and low substrate water potential at different stages of embryonic development (especially in northern populations); and are high-temperature females ever produced during egg incubation?

Growth and life history

Can Dawbin's records and more recent recaptures on Stephens Island be used to produce a detailed analysis of growth patterns, to clarify in particular the age at maturity and age at the time growth ceases; can growth in snout–vent length resume in adults and what is the relationship to epiphyseal calcification; what influence do physical conditions of the substrate in embryonic life have on lifetime growth and reproductive success; what factors control egg size and thus hatchling size; and is egg size the major determinant of maximum adult size and/or lifetime reproductive success?

The above list includes questions that may have phylogenetic significance in terms of understanding the evolution of lepidosaurs. Answers to these questions may help provide clues as to the extraordinary success of squamates relative to tuatara and other living reptiles. Other questions might be said more simply to satisfy our curiosity about the biology of a remarkably long-lived reptile. Some questions may prove intractable, given the length of time required for answers, the availability of funding, and a local regulatory environment that favours, at most, minimally invasive research. Increasingly, attention is likely to turn to applied questions concerned with climate change. Answers to some of the questions posed above will be especially important in helping predict the effects on embryonic development from summer droughts, which may occur with increasing frequency as climates warm, especially on small northern islands.

Endnotes to Chapter 8

See the bibliography for full references to articles on tuatara, which are cited here at first mention by surname and year only. References on other topics are cited here in full at first mention, and then using 'op. cit.'. Except where required for clarity, the convention of parentheses around years has been dispensed with for economy of space.

1 von Wettstein 1931, p. 222, in translation.
2 A Party of Officers of the 58th Regiment 1852.
3 Günther 1867.
4 Newman 1878, p. 225.
5 von Wettstein 1931, who had access to anatomical studies and observations in captivity but very few field studies.
6 For discussion, see Jones, S. M. 2011. Hormonal regulation of ovarian function in reptiles (pp. 89–115) and Kumar, S., Roy, B. and Rai, U. 2011. Hormonal regulation of testicular functions in reptiles (pp. 63–88). Both in: Norris, D. O. and Lopez, K. H. (eds). 2011. *Hormones and reproduction of vertebrates. Vol. 3. Reptiles*. Academic Press, London.
7 Descriptions of gross reproductive anatomy in male tuatara were made by Günther (1867), Gadow (1887) and Osawa (1897), but source populations were not stated and only one or two animals were examined in each report. Observations from these early studies were summarised by von Wettstein (1931) and used in later reviews. This paragraph is based on these sources plus my own observations.
8 Preserved males were ≥ 180 mm snout–vent length and collected at various times of year (not always known). Mass of the right testis was measured for 13 specimens (standard error: 40 mg). The left testis was not weighed.
9 Testis lengths and widths were measured for 11 specimens. Standard errors for dimensions of the right testis were 1.1 mm (length) and 0.4 mm (width); those for the left testis were 1.0 mm and 0.9 mm, respectively (unpubl. data).
10 However, mass (relative to body mass) was known for only seven males from specific months of the year (Cree et al. 1992).
11 Osawa 1897; Hogben 1921; Gabe and Saint Girons 1964a; Saint Girons and Newman 1987.
12 Saint Girons and Newman 1987.
13 Healy and Jamieson 1992, 1994.
14 Jamieson and Healy 1992; Healy and Jamieson 1992, 1994.
15 For conventions on names of ducts, see: Sever, D. M. 2010. Ultrastructure of the reproductive system of the black swamp snake (*Seminatrix pygaea*). VI. Anterior testicular ducts and their nomenclature. *Journal of Morphology* 271: 104–115. The tubules of the rete testis of tuatara were until recently known as efferent tubules or ductuli efferentes (Gabe and Saint Girons 1964a; Saint Girons and Newman 1987). See also Rheubert et al. 2012.
16 Previously known as epididymal tubules or ductuli epididymides.
17 Rheubert et al. 2012.
18 Saint Girons and Newman 1987.
19 Pers. obs.
20 Osawa 1897.
21 Saint Girons and Newman 1987.
22 Healy and Jamieson 1994.
23 Fox, H. 1977. The urogenital system of reptiles. In: Gans, C. and Parsons, T. S. (eds). *Biology of the Reptilia. Vol. 6. Morphology E*. Academic Press, London, pp. 1–157 + addendum pp. 463–464.
24 Saint Girons and Newman 1987; Rheubert et al. 2012.
25 Fox 1977, op. cit. reviewed evidence for the presence of vestigial Müllerian ducts in some male turtles and snakes. I have seen them in a half-grown juvenile male of a gecko from the *Hoplodactylus maculatus* species complex.
26 Osawa 1897; pers. obs.
27 Specifically, the parietal mesentery (Osawa 1897).
28 Osawa 1897.
29 Gadow 1887; Gabe and Saint Girons 1964a.
30 Osawa (1897, 1898b) could not clearly distinguish a urodaeum from the coprodaeum and preferred to recognise just two main chambers.
31 Günther 1867; Osawa 1897; Fox 1977, op. cit.
32 Günther 1867; Osawa 1897, 1898b; Gabe and Saint Girons 1964a; Saint Girons and Newman 1987. Günther referred to the sebaceous glands as 'anal glands'. More recently, Flachsbarth et al. (2009) referred to them as 'cloacal gland[s]', an ambiguous term given that there are two pairs of cloacal glands that differ in nature.
33 Gadow 1887.
34 The standard error was 25 mg.
35 Gabe and Saint Girons 1964a.
36 Osawa 1897.
37 Günther 1867.
38 Saint Girons and Newman 1987; pers. obs.
39 Saint Girons and Newman 1987.
40 Flachsbarth et al. 2009. See also Weldon, P. J., Flachsbarth, B. and Schulz, S. 2008. Natural products from the integument of reptiles. *Natural Product Reports* 25: 738–756.
41 Gadow 1887, p. 19.
42 See Weldon et al. 2008, op. cit. for brief review.
43 Flachsbarth et al. 2009.
44 As a corollary of this absence, the urodaeum of males is developed to a considerable extent, similar to that of females (Gabe and Saint Girons 1964a).
45 Gadow 1887, p. 19.
46 Gadow 1887, p. 31.
47 Arnold 1984.
48 Gillingham et al. 1995.
49 Gabe and Saint Girons 1964a; Saint Girons and Newman 1987; Cree et al. 1992.
50 Cree et al. 1992.
51 Dawbin 1962a; Newman et al. 1979; Cree et al. 1992; Gillingham et al. 1995; Moore et al. 2009a, b.
52 Saint Girons and Newman 1987; Cree et al. 1992.
53 Robinson, M. 1985. *Sexual cycles of New Zealand lizards, with particular reference to the gecko* Hoplodactylus maculatus *(Boulenger)*. Unpublished MSc thesis, Victoria University of Wellington.
54 Saint Girons and Newman 1987; Cree et al. 1992.
55 Robinson 1985, op. cit.; Rock, J., Andrews, R. M. and Cree, A. 2000. Effects of reproductive condition, season, and site on selected temperatures of a viviparous gecko. *Physiological and Biochemical Zoology* 73: 344–355; Cree, A., Tyrrell, C. L., Preest, M. R., Thorburn, D. and Guillette Jr, L. J. 2003. Protecting

embryos from stress: corticosterone effects and the corticosterone response to capture and confinement during pregnancy in a live-bearing lizard (*Hoplodactylus maculatus*). *General and Comparative Endocrinology* 134: 316–329.

56 Rather than dihydrotestosterone (Bradshaw et al. 1988).
57 Cree et al. 1992.
58 After an anti-coagulant has been added to a blood sample, the straw-coloured plasma can be separated from the blood cells by centrifugation.
59 Cree et al. 1992; Moore et al. 2009a.
60 Cree et al. 1992.
61 Cree et al. 1990a, b.
62 Cree et al. 1995a; Tyrrell 2001.
63 Cree et al. 1995a.
64 Tyrrell 2000; Keall et al. 2010. See Chapter 6 for further discussion.
65 Some analysis of the gonadotrophins is underway (John Cockrem, Massey University, pers. comm. 8 February 2011).
66 Jones 2011, op cit.; Kumar et al. 2011, op. cit.
67 Original published descriptions are limited to the following sources: Newman (1878), who made brief observations on the ovaries and oviducts of tuatara from North Brother Island; Gadow (1887), who compared cloacal structure in a female (source population unknown) with that of a male; Osawa (1898b), who described gross morphology and histology of ovaries and oviducts in two females (obtained from Parker and thus likely to be from Karewa Island: see Chapter 5); and Gabe and Saint Girons (1964a), who examined the histology and histochemistry of ovaries and oviducts for one female from Stephens Island. Most of these studies were apparently used as sources for the reviews of von Wettstein (1931), Fox (1977, op. cit.) and Robb (1977). The original sources provide valuable information on comparative morphology, but little or nothing about the seasonality of reproduction in the wild.
68 Osawa 1898b indicated that the right ovary began anterior to the left, but this seems to depend on the number of enlarged follicles in each ovary (pers. obs.).
69 Osawa 1898b.
70 For example, Robb 1977.
71 Osawa 1898b.
72 Wake, M. H. 1979. The comparative anatomy of the urogenital system. In: Wake, M. H. (ed.). *Hyman's comparative vertebrate anatomy.* 3rd edn. University of Chicago Press, Chicago, pp. 555–614. See also Jones 2011, op. cit.
73 Osawa 1898b. An arrangement such as that seen in crocodilians, in which the ovaries have a well-developed stroma with an extensive system of lacunae and smooth muscle bundles around the follicles, has not been reported in tuatara. See Uribe, M. C. A. and Guillette Jr, L. J. 2000. Oogenesis and ovarian histology of the American alligator *Alligator mississippiensis*. *Journal of Morphology* 245: 225–240.
74 Osawa 1898b.
75 Tribe and Brambell 1932.
76 Gabe and Saint Girons 1964a.
77 Osawa 1898b; Gabe and Saint Girons 1964a.
78 Guraya, S. S. 1989. *Ovarian follicles in reptiles and birds.* Springer-Verlag, Berlin. See also Jones 2011, op. cit.
79 Cree et al. 1996; Guillette and Cree 1997.
80 Two of these females were gravid and one was spent, i.e. it had laid eggs recently.
81 Gabe and Saint Girons 1964a.
82 Pers. obs.
83 Guillette and Cree 1997.
84 Pers. obs.
85 Fox 1977, op. cit.
86 The standard error was 3%.
87 Newman 1878.
88 Osawa 1898b; Fox 1977, op. cit.
89 Girling, J. E. 2002. The reptilian oviduct: a review of structure and function and directions for future research. *Journal of Experimental Zoology* 293: 141–170.
90 Osawa 1898b; Gabe and Saint Girons 1964a; pers. obs.
91 Girling 2002, op. cit.
92 The infundibulum is described (in translation) as the anterior or proximal part of the oviduct above the uterus by Osawa (1898b), and in French as the 'pavillon' by Gabe and Saint Girons (1964a).
93 The uterine tube is described (in translation) as the posterior or distal part of the oviduct above the uterus by Osawa (1898b), and in French by the 'trompe' of Gabe and Saint Girons (1964a).
94 Robb 1977, p. 25.
95 Osawa 1898b.
96 Gabe and Saint Girons 1964a.
97 Gabe and Saint Girons 1964a.
98 In absolute length, the oviducts averaged 165 mm (standard error: 3 mm).
99 Osawa 1898b; Gabe and Saint Girons 1964a.
100 Girling 2002, op. cit.
101 Osawa 1898b; Claudine Tyrrell, University of Otago, pers. comm. 27 September 2007.
102 Osawa 1898b; Gabe and Saint Girons 1964a.
103 Osawa 1898b; Gabe and Saint Girons 1964a. However, what appeared to be sperm were detected by scanning electron microscopy in the uterine lumen in March of a female in pre-ovulatory condition known to have mated 2 weeks previously (pers. obs.).
104 Pers. obs.
105 Pers. obs.
106 Newman 1878.
107 Fox 1977, op. cit.; Girling 2002, op. cit.
108 Greer 1986; pers. obs.
109 Derickson, W. K. 1976. Lipid storage and utilization in reptiles. *American Zoologist* 16: 711–723. See also Fox 1977, op. cit.; Greer 1986.
110 Fraser 1993. See Chapter 7 for further discussion.
111 Ali 1941.
112 See Chapter 7, Box 7.4 for further discussion of the liver.
113 Newman 1878.
114 Jones 2011, op. cit.
115 Jones 2011, op. cit.
116 Thompson, M. B. and Speake, B. K. 2004. Egg morphology and composition. In: Deeming, D. C. (ed.). *Reptilian incubation: environment, evolution and behaviour*. Nottingham University Press, Nottingham, pp. 45–74.
117 Cree et al. 1991a, 1992.
118 Brown et al. 1991b. Plasma from a vitellogenic female tuatara (along with that of several other vertebrates) has since been shown to cross-react with a monoclonal antibody to vitellogenin from a species of trout, which increases the possibilities for future analysis. See: Heppell, S. A., Denslow, N. D., Folmar, L. C. and Sullivan, C. V. 2005. Universal assay of vitellogenin as a biomarker for environmental estrogens. *Environmental Health Perspectives* 103 (Supplement 7): 9–15.
119 Brown et al. 1991b; Cree et al. 1992.
120 Brown et al. 1989; Cree et al. 1991b.

121 Cree et al. 1991a.
122 Brown et al. 1991b.
123 Brown et al. 1991b; Cree et al. 1991a.
124 Cree et al. 1992. For discussion, see also Jones 2011, op. cit.
125 Cree et al. 1991a, 1992.
126 Brown et al. 1991b; Cree et al. 1992.
127 Cree et al. 1991c.
128 McNicol, D. Jr and Crews, D. 1979. Estrogen/progesterone synergy in the control of female sexual receptivity in the lizard, *Anolis carolinensis*. *General and Comparative Endocrinology* 38: 68–74.
129 Gillingham et al. 1995.
130 Moore et al. 2009a.
131 Cree et al. 1992; Guillette and Cree 1997.
132 Brown et al. 1991b.
133 As judged by laparoscopy of one ovary (Cree et al. 1991a).
134 Cree et al. 1992.
135 Brown et al. 1991b.
136 Cree et al. 1992.
137 Brown et al. 1994.
138 Brown et al. 1994.
139 Brown et al. 1989, 1997a; Cree et al. 1991b.
140 Vitellogenin is normally undetectable in the plasma of males (Brown et al. 1991b).
141 Cree et al. 1991b.
142 Brown et al. 1989, 1991b.
143 Cree et al. 1991c.
144 Brown et al. 1994, 1997a.
145 Brown et al. 1997a.
146 Newman 1878.
147 Gans et al. 1984; Gillingham and Miller 1991; Gillingham et al. 1995; Moore et al. 2009a, b.
148 Gillingham and Miller 1991; Gillingham et al. 1995. Note, however, that Linklater (2011) has recently asked whether the term 'territorial' has been applied prematurely in tuatara; he considered that more evidence is needed for defence by males of resources such as space, as distinct from defence of females.
149 Gillingham et al. 1995; pers. obs.
150 Gillingham and Miller 1991. See also Jones et al. 2009a for jawbone pathologies.
151 Gillingham et al. (1995) and Moore et al. (2009a) observed copulation during afternoons. Mating has also been observed in forest during heavy mist in the morning (at 1130 hours; Michael Schneider Photographics Ltd, Auckland, pers. comm. 10 March 1990).
152 Gillingham et al. 1995.
153 Gans et al. 1984.
154 Gillingham and Miller 1991.
155 In one instance, I heard a male make two quiet, low grunts ('uh-uh') about half a second apart, repeated 3 or 4 times during a *stoltzer Gang*. In an instance reported to me, when a mating pair was disturbed and the female ran down a burrow, the male high-stepped towards her with a similar, quiet vocalisation (Michael Schneider, pers. comm. 10 March 1990).
156 Gillingham et al. 1995.
157 'The male catches her on the right shoulder with his teeth'. W. Tutt to unidentified recipient, *Copulation of the tuatara lizard*, 2 June 1913. NMNZ MU000207 Box 4, Item 1.
158 Gillingham et al. 1995; Moore et al. 2009a.
159 Gillingham et al. 1995; pers. obs., including observations of video footage taken by Margie Grant-Caplan, Department of Conservation, February 2011.
160 Gillingham et al. 1995.
161 Dawbin 1982a; Gans et al. 1984; Blanchard and the Tuatara Recovery Group 2002.
162 Gans et al. 1984.
163 Blanchard and the Tuatara Recovery Group 2002.
164 Nicola Nelson, Victoria University of Wellington, pers. comm. 27 April 2004.
165 Gans et al. 1984.
166 Moore et al. 2009b.
167 Hay and Lambert 2008; Moore et al. 2009a. See Moore et al. 2008b for study of a captive population on Hauturu/Little Barrier Island.
168 Cree et al. 1992; Moore et al. 2009a. Although observations in forest plots on Stephens Island led to the conclusion that 'only 29% of males … were ever successful [at mating]', (see also H. C. Miller et al. 2009), note that observations did not cover the entire period of mating opportunity (either across the season or across the diel period).
169 H. C. Miller et al. 2009; Moore et al. 2009b.
170 Herrel et al. 2010.
171 Gillingham et al. 1995.
172 Cree et al. 1991d, 1992.
173 Moore et al. 2009a. In study plots observed during three successive mating seasons, only one female was observed mating in more than one year (it mated 2 years apart). But note that it has not yet been possible to obtain continuous observations of females 24 hours a day for the potential 3-month mating period.
174 Gans et al. 1984; Gillingham and Miller 1991; Gillingham et al. 1995.
175 Gans et al. 1984; Gillingham et al. 1995.
176 Besson et al. 2009.
177 Gillingham et al. 1995.
178 Cree et al. 1995a.
179 Cree et al. 1992. Another egg-laying species, the turtle *Deirochelys reticularia*, sometimes carries shelled eggs in oviducts for a prolonged period (at least 4–6.5 months) over winter. Unlike tuatara, however, this species is known to be capable of nesting in autumn. See: Buhlmann, K. A., Lynch, T. K., Gibbons, J. W. and Greene, J. L. 1995. Prolonged egg retention in the turtle *Deirochelys reticularia* in South Carolina. *Herpetologica* 51: 457–462.
180 Dawbin 1962a.
181 Robb 1977.
182 Guillette and Cree 1997.
183 Guillette and Cree 1997.
184 Cree et al. 1992.
185 Packard et al. 1988; Cree et al. 1996.
186 Packard et al. 1982.
187 Packard et al. 1988.
188 Packard, M. J., Hirsch, K. F., Packard, G. C., Miller, J. D. and Jones, M. E. 1991. Structure of shells from eggs of the Australian lizard *Amphibolurus barbatus*. *Canadian Journal of Zoology* 69: 303–310.
189 Cree et al. 1996.
190 Cree et al. 1996; Packard et al. 1988.
191 Specifically, in the neurohypophysis, a neurohaemal region of the pituitary gland.
192 Thompson 1990; Packard et al. 1988; Cree et al. 1996.
193 Thompson 1990 (standard deviation: 0.49 g).
194 Thompson 1990 (standard deviation: 0.96 g).
195 Cree et al. 1996; Nelson 2001.
196 Fraser 1993; pers. obs.
197 Shine, R. 2006. Is increased maternal basking an adaptation or

a pre-adaptation to viviparity in lizards? *Journal of Experimental Zoology* 305A: 524–535.
198 Thilenius 1899.
199 Packard et al. 1988.
200 Grodziński 1980. The source population (not stated) is likely to have been Stephens Island. Prolonged storage and shipping to the northern hemisphere affected the yolk quality, limiting the observations that could be made.
201 Lange and Kilarski 1986.
202 Grodziński 1980.
203 Body 1985.
204 Thompson and Speake 2004, op. cit.
205 Thompson and Speake 2004, op. cit.
206 Thompson and Speake 2004, op. cit.
207 Cree 1994.
208 Schauinsland 1898b, 1899.
209 Newman and Watson 1985; Newman et al. 1994.
210 Newman et al. 1994. Mature size was considered as at least 170 mm snout–vent length.
211 A climatic pattern that recurs every few years as part of the southern oscillation in the Pacific Ocean.
212 Newman et al. 1994.
213 Tyrrell 2000; Tyrrell et al. 2000. Note that not all islands were sampled in all years.
214 Categories were rat-free, rat-removed or rat-inhabited.
215 Tyrrell et al. 2000.
216 Cree et al. 1991d.
217 Mitchell et al. 2010.
218 Pers. obs. Consistent with this, Newman et al. (1994) reported a high proportion of gravid females (45.5%) in a survey on Lady Alice Island in 1986, when the weather was cool. The total number caught was also low (11).
219 Newman et al. 1994; Tyrrell et al. 2000. The appearance of females on nesting rookeries also indicates a minimum of 2 years between successive nesting events on Stephens Island (Refsnider et al. 2010).
220 Newman et al. 1994.
221 Cree 1994.
222 Newman et al. 1994.
223 Snout–vent length: 235 mm.
224 Cree 1994.
225 Cree et al. 1991b; Cree 1994; Mitchell et al. 2010.
226 Tyrrell 2000.
227 Brown et al. 1994; Cree et al. 1995a.
228 Tyrrell 2000.
229 Confirmation followed the removal of rats (see Towns et al. 2007).
230 Newman et al. 1994.
231 Hazley 1995; Blanchard and the Tuatara Recovery Group 2002; Moore et al. 2008b.
232 Cree 1994.
233 Newman et al. 1994. Mean clutch size in different years ranged between 8.1 and 10.8 eggs, a variation that was apparently close to statistical significance.
234 The equation for the regression line in Figure 8.28 is $y = 0.098x - 9.848$ ($r^2 = 0.294$; $p < 0.001$). See also Newman et al. (1994) for an overlapping dataset.
235 Thompson et al. 1991.
236 Thompson et al. 1991. The dosage was usually 10 International Units (IU)/kg (occasionally 15 IU/kg). The standard deviation was 2.8 eggs.
237 Thompson et al. 1991.
238 L. J. Guillette Jr and A. Cree, unpubl. data.
239 Pers. obs.
240 Cree and Guillette 1988; unpubl. data.
241 Thompson et al. 1996.
242 Newman et al. 1994.
243 Dendy 1899b; Schauinsland 1899; Thilenius 1899.
244 Cree et al. 1991b; Mitchell et al. 2010. A clutch of 10 eggs was also reported for a female from North Brother Island dissected in about the 1870s (Newman 1878).
245 For eight induced clutches (Cree et al. 1991b).
246 Tyrrell 2000.
247 Newman et al. 1994.
248 Newman et al. 1994.
249 Tyrrell et al. 2000.
250 Tyrrell 2000.
251 Tyrrell 2000.
252 Thomson 1915.
253 Thomas 1890.
254 Pers. obs.
255 Tyrrell 2000; Moore et al. 2008b; Keall et al. 2010.
256 Tyrrell 2000; Keall et al. 2010.
257 Schauinsland 1898b, 1899; Thilenius 1899; Cree and Thompson 1988; Thompson and Daugherty 1992; Cree and Butler 1993; Thompson et al. 1996.
258 Thilenius 1899; Thompson et al. 1996.
259 Dendy 1899b; Thilenius 1899.
260 Cree and Thompson 1988; Guillette et al. 1990; Thompson and Daugherty 1992; Nelson et al. 2004a.
261 Guillette et al. 1991a; pers. obs.
262 Thilenius 1899; pers. obs.
263 Nicola Nelson, pers. comm. 27 April 2004.
264 Refsnider et al. 2009.
265 Nelson et al. 2004b.
266 Rubenstein, D. R. and Wikelski, M. 2005. Steroid hormones and aggression in female Galápagos marine iguanas. *Hormones and Behavior* 48: 329–341.
267 Thilenius 1899; Thompson et al. 1996.
268 Nicola Nelson, pers. obs. in Refsnider et al. 2009.
269 Thilenius 1899.
270 Thompson and Daugherty 1992; pers. obs.
271 Thilenius 1899.
272 Dendy's correspondent Henaghan suggested that some females might lay their eggs outside the nest chamber and carry them in using their mouth or claws (Dendy 1899b). Henaghan was apparently speculating based on the tight fit for the eggs within the nest chamber; whether he was also aware of Thilenius's observation is unknown.
273 Dawbin 1982b; Thompson and Daugherty 1992.
274 Cree and Thompson 1988; Thompson and Daugherty 1992.
275 Cree and Tyrrell 2001; Refsnider et al. 2009; Nicola Nelson, pers. comm. 27 April 2004.
276 Refsnider et al. 2009. In captivity, a female that had recently nested made biting lunges at the feet of her human keepers within the enclosure, a behaviour interpreted as nest guarding (Blanchard 2002).
277 Refsnider et al. 2010.
278 Refsnider et al. 2010.
279 Refsnider et al. 2010. But note pseudo-replication of temperature measurements, and that there were only two loggers per treatment.
280 Refsnider et al. 2009.
281 Refsnider et al. 2009.
282 Schauinsland 1898b, 1899. In 1913 a lighthouse keeper indicated that nests sometimes contained eggs in 'two benches

in the same hole', and implied that these could be at different developmental stages (W. Tutt to unnamed recipient, 2 June 1913, NMNZ MU000207 Box 4, Item 1).

283 Cree et al. 1991d; Thompson et al. 1992; Mitchell et al. 2008, 2010.

284 Mitchell et al. 2010.

285 Thomson 1915.

286 Reischek 1882, p. 276.

287 Claudine Tyrrell, University of Otago, pers. comm. 2 February 2004.

288 From the presence in January of 'full-grown eggs', Reischek inferred that egg-laying occurred in February (Reischek 1882). Presumably, he was referring to eggs in the oviducts (as opposed to ovarian oocytes), although the statement is ambiguous and whether the eggs (if oviducal) were completely shelled is unknown.

289 Tyrrell 2001.

290 Tyrrell et al. 2000. Radiography confirmed that the females had oviposited during this time.

291 R. Parrish, Department of Conservation, pers. comm. 27 July 2005, based on unpublished observations of Victor Shortland.

292 Towns et al. 2007.

293 C. Smuts-Kennedy, pers. comm. to C. L. Tyrrell, in Tyrrell 2000; Keall et al. 2010.

294 Mitchell et al. 2008.

295 Guillette et al. 1990; Cree et al. 1992.

296 Cree et al. 1992.

297 Guillette et al. 1991a.

298 The assay detected both $PGF_{2\alpha}$ and $PGF_{1\alpha}$ though the former is likely to be the major natural form.

299 Guillette et al. 1990.

300 Guillette et al. 1990.

301 Guillette et al. 1990.

302 Guillette et al. 1991b.

303 Tyrrell and Cree 1998; Tyrrell et al. 2000; Cree and Tyrrell 2001; pers. obs. The speed of the response appears slower in tuatara than in some other reptiles, apparently taking between about 10 and 60 minutes to induce, but has not been fully characterised. For reptiles generally, see Tokarz, R. R. and Summers, C. H. 2011. Stress and reproduction in reptiles. In: Norris, D. O. and Lopez, K. H. (eds). *Hormones and reproduction of vertebrates. Vol. 3. Reptiles*. Academic Press, London, pp. 169–213.

304 Tyrrell and Cree 1998.

305 Cree and Tyrrell 2001.

306 Rubenstein and Wikelski 2005, op. cit.

307 Cree and Tyrrell 2001.

308 Discussed in Tokarz and Summers 2011, op. cit.

309 Dendy 1899b.

310 Schauinsland 1898b, 1899; Thilenius 1899; Dendy 1899b.

311 Thompson et al. 1996. Range for 6–9 nests.

312 Thompson et al. 1996. Range for 26 nests.

313 Nelson 2001; Nelson et al. 2004a, d.

314 Nelson et al. 2004d.

315 Mitchell et al. 2008, 2010.

316 Tyrrell 2001.

317 Cree et al. 1996. Standard errors were 0.3 mm (length) and 0.2 mm (width) for nine eggs.

318 Schauinsland 1898a, 1899.

319 For example, Dendy 1899a; Dawbin 1962a; Moffat 1985.

320 Thilenius 1899; pers. obs.

321 Cree et al. 1991d. The standard error was 0.2 g for both populations.

322 Mitchell et al. 2010. Range: 2.2–5.9 g.

323 Tyrrell 2000.

324 Thilenius 1899.

325 Packard, G. C. and Packard, M. J. 1988. The physiological ecology of reptilian eggs and embryos. In: Gans, C. and Huey, R. B. (eds). *Biology of the Reptilia. Vol. 16*. Alan R. Liss, New York, pp. 523–605.

326 Thompson 1990.

327 Schauinsland 1898a, 1899.

328 Packard and Packard 1988, op. cit.

329 Andrews, R. M. 2004. Patterns of embryonic development. In: Deeming, D. C. (ed.). *Reptilian incubation: environment, evolution and behaviour*. Nottingham University Press, Nottingham, pp. 75–102.

330 Dendy 1899b.

331 Schauinsland 1898a, 1899, 1903a, b, c.

332 Thilenius 1899.

333 Thilenius 1899.

334 See Chapter 9, Box 9.2 for more on the parietal eye.

335 Andrews 2004, op. cit.

336 Dendy 1899b (in both snout–vent length and total length).

337 Dendy 1899d.

338 Dendy 1899b.

339 De Beer 1949. The horny shell-breaker has sometimes been referred to as a caruncle. However, this term has been considered inappropriate, presumably because the structure is horny rather than fleshy. See: Morafka, D. J., Spangenberg, E. K. and Lance, V. A. 2000. Neonatology of reptiles. *Herpetological Monographs* 14: 353–370.

340 Dendy 1899b.

341 Dendy 1899b.

342 Schauinsland 1898a, 1899.

343 Thilenius 1899, p. 255.

344 Dendy 1899b.

345 Thompson 1990.

346 Dendy 1899b.

347 Thilenius 1899.

348 Thompson et al. 1996.

349 Thompson et al. 1996.

350 Dendy 1899a.

351 Moffat 1985.

352 Packard et al. 1982, 1988.

353 Packard and Packard 1988, op. cit.; Andrews 2004, op. cit.

354 Dendy 1899b.

355 Body 1985.

356 Thompson and Daugherty 1992.

357 Thompson et al. 1996.

358 Thilenius 1899.

359 Some egg-laying chameleons have incubation periods of up to a year. Unlike tuatara, however, these species have an obligatory period of embryonic diapause around gastrulation, regardless of temperature. This form of developmental arrest may have evolved in association with dry and highly seasonal climates, contrasting with the cold torpor seen in tuatara. See Andrews, R. M. and Karsten, K. B. 2010. Evolutionary innovations of squamate reproductive and developmental biology in the family Chamaeleonidae. *Biological Journal of the Linnean Society* 100: 656–668.

360 Dendy 1899d.

361 Dendy 1899a.

362 Dendy 1899d, p. 251.

363 Howes 1899. See Howes and Swinnerton 1901 for illustration.

364 Dawbin 1962a, 1982a.
365 Thilenius 1899.
366 Thompson et al. 1996.
367 McIntyre, M. E. 1988. Management implications of habitat use and dispersal of juvenile tuatara, *Sphenodon punctatus*, on Stephens Island. Unpublished report. Victoria University of Wellington, Wellington.
368 Dawbin 1962a, 1982a.
369 McIntyre 1988, op. cit.; Michael Thompson, University of Sydney, pers. comm. April 2004.
370 Thilenius 1899.
371 Dawbin 1962a, 1982a.
372 The egg with twin embryos weighed 4.3 g at oviposition. One of the twins died in an accident about 20 months later. By age 6.5 years, the surviving twin was similar in length and mass to its clutch-mates.
373 Elaine Whitworth, University of Otago, pers. comm. 11 April 2004.
374 Howes 1899; Dawbin 1962a, 1982a.
375 Pers. obs.
376 Thilenius 1899.
377 McIntyre 1988, op. cit.
378 McIntyre 1988, op. cit. See Chapter 7 for further discussion.
379 Thompson et al. 1992.
380 Tyrrell 2001.
381 Thomas 1890, p. 155; Thomas 1891, p. 30
382 Schauinsland 1898a, 1899; Moffat 1985.
383 Moffat 1985.
384 Stewart, J. R. 1993. Yolk sac placentation in reptiles: structural innovation in a fundamental vertebrate fetal nutritional system. *Journal of Experimental Zoology* 266: 431–449.
385 For instance, Moffat (1985) considered that Dendy's stage Q corresponded to four stages (34–37) and stage R to two stages (38–39) in *Lacerta vivipara*, and that it was impossible to match stages earlier than stage H. A slightly different correspondence for stage Q (stages 34–35), stage R (stages 36–38) and stage S (stages 39–40) was suggested by Thompson et al. (1996).
386 Nicola Nelson, pers. comm. 20 January 2011.
387 For an initial study, see Di-Poï et al. 2010 for information on *Hox* gene structure in tuatara compared with squamates and turtles.
388 Thompson et al. 1996.
389 Nelson et al. 2004d.
390 Thompson et al. 1996.
391 Dendy 1899b.
392 Thompson et al. 1996.
393 At 8 am.
394 Thompson et al. 1996. On the other hand, flooding could conceivably produce conditions that are too wet for successful development (see Besson et al. 2012 for a mainland site).
395 Dendy 1899b; Thilenius 1899.
396 Thilenius 1899.
397 Thilenius 1899.
398 Thompson et al. 1996.
399 Nelson et al. 2004a.
400 Nelson et al. 2004d.
401 Nelson 2001; Nelson et al. 2004a; Mitchell et al. 2006.
402 Besson 2009.
403 Besson et al. 2012.
404 Besson 2009.
405 Thompson 1990; Nelson et al. 2004b.
406 Nelson et al. 2004b.
407 These means are for eggs exposed on the surface (half-buried) at a water potential of –90 kPa. Standard deviations were 15, 8 and 5 days for means at 18°, 20° and 22°C, respectively (Thompson 1990).
408 Eight out of nine eggs from captive-laid or wild sources hatched when incubated at 24–24.1°C (Mitchell et al. 2006; Besson 2009; Besson et al. 2012). Most of a zoo-laid clutch hatched successfully when incubated between 23°C and 29°C in the early stages of incubation (Tintinger 1987), and late-stage embryos apparently survived for at least 2 months at about 25°C before hatching successfully (Howes 1899; Howes and Swinnerton 1901). Although Thompson (1990) reported low hatching success at 25°C, variable-quality eggs and a brief over-heating in the incubator may have influenced this result.
409 Cree et al. 1991f; Mitchell et al. 2006.
410 Thompson 1990. However, they are not the lowest. Some eggs of the gecko *Lepidodactylus lugubris* hatched following incubation at constant temperatures of 14°C and 16°C, a remarkable finding for a lizard from a subtropical environment. See: Ota, H. 1994. Female reproductive cycles in the northernmost populations of the two gekkonid lizards, *Hemidactylus frenatus* and *Lepidodactylus lugubris*. *Ecological Research* 9: 121–130.
411 Thompson 1990.
412 Thompson 1990.
413 Thompson 1990.
414 Nelson et al. 2004b.
415 Thompson 1989; Booth and Thompson 1991.
416 Thompson 1989.
417 Booth and Thompson 1991.
418 Nelson 2001; Nelson et al. 2004b.
419 In an extreme example, a newspaper account reported that Norris considered the ages of two tuatara in his possession as 700 to 1000 years (Anon. 1908a).
420 Robb 1977.
421 Buller 1877, quoting Captain [Gilbert] Mair.
422 See Chapter 5, Box 5.3, for further discussion and sources.
423 Dawbin 1982a.
424 Dawbin 1982a.
425 Among a sample of 32 males in which growth had ceased for 4 years or more, the mean snout–vent length was 264.7 mm (standard deviation: 12.6); in an equivalent sample of 24 females, snout–vent length averaged 213.8 mm (standard deviation: 14.1) (Dawbin 1982a).
426 Halliday, T. R. and Verrell, P. A. 1988. Body size and age in amphibians and reptiles. *Journal of Herpetology* 22: 253–265.
427 Castanet et al. 1988; Castanet 1994.
428 Castanet et al. 1988.
429 Dawbin 1982a.
430 Castanet et al. 1988; see also Castanet 1994.
431 Castanet et al. 1988.
432 Cree et al. 1992.
433 Pers. obs.
434 Newman et al. 1994.
435 Vitellogenesis can begin in females as small as 161 mm snout–vent length (pers. obs.), although it could be several years before the first clutch is ovulated.
436 Pers. obs. (but inferences from skeletochronology suggest values of about 10–11 years for both sexes).
437 Cree et al. 1994.
438 Newman et al. 1994.
439 Castanet et al. 1988.
440 Newman et al. 1994. Another method that I used opportunistically to estimate age for a female on Stephens

Island provided a similar estimate. The female was found squashed by a rock that had fallen during heavy rain (more happily, with a freshly captured wētā in her mouth). Each ovary contained vitellogenic and pre-vitellogenic follicles (indicating that she was still reproductively active) but also an estimated 50–70 corpora albicantia, suggesting that between her two ovaries this female had ovulated at least 100 eggs over her lifetime. If she produced a clutch of 9.4 eggs every 4.3 years on average (mean values for the population), she would have been reproductively active for about 45 years, giving a minimum estimated age of about 56–62 years at death.

441 Nicola Nelson, pers. comm. in Moore et al. 2007.

442 From Castanet et al. 1988, for Stephens Island.

443 Thompson et al. 1992.

444 Mitchell et al. 2010.

445 Mitchell et al. 2010.

446 Schauinsland 1898b; Dawbin 1982a.

447 Nelson et al. 2002b; K. A. Miller et al. 2010. For example, several adults, previously thought to have stopped growing on North Brother Island, showed increases in mass and snout–vent length following release on Titi Island, leading Nelson et al. (2002b) to question whether growth was determinate, as previously reported.

448 Haines 1939.

449 Christmas et al. 1996.

450 For example, Moore et al. 2009a; Refsnider et al. 2009.

451 Gillingham et al. 1995.

452 Dendy 1899b; Crook 1975.

453 Cree 1994; Wilson, J. L. and Cree, A. 2003. Extended gestation with late-autumn births in a cool-climate viviparous gecko from southern New Zealand (Reptilia: *Naultinus gemmeus*). *Austral Ecology* 28: 339–348.

454 Cree 1994.

455 Robinson 1985, op. cit.; Saint Girons and Newman 1987.

456 Lettink, M. and Whitaker, T. [A. H.] 2006. *Hoplodactylus maculatus* (common gecko). Longevity. *Herpetological Review* 37: 223–224.

457 Dawbin 1962a; Saint Girons and Newman 1987; Cree and Daugherty 1990a; Thompson and Daugherty 1992; Cree 1994.

Endnotes to boxed text

For explanation of citation format in box endnotes, see endnotes to the main text.

Box 8.1

1 Buller 1871, p. 9.
2 Buller 1877, p. 324.
3 'The female is considerably larger than the male' (Buller 1877, p. 324).
4 Buller 1878, p. 320.
5 Freeman and Freeman 1995.
6 Newman 1878, p. 225.
7 Thomas 1890, p. 154 (repeated in Thomas 1891, pp. 28–29).
8 This and the preceding quote from Thomas 1890, p. 154 (repeated in Thomas 1891, p. 29).
9 Thomas 1890, 1891.
10 Schauinsland 1898b.
11 Schauinsland 1898b.
12 Thilenius concluded that although sex ratios could vary widely depending on season and time of day, sex ratios overall were likely to be about equal (Thilenius 1899).
13 Goodlet c. 1922.
14 J. A. Thomson to Undersecretary of Internal Affairs, 15 February 1915. NMNZ MU000207 Box 4, Item 1. J. A. Thomson Miscellaneous files 1913–1915.
15 J. Heenan to the Secretary of Marine, 7 October 1935. ANZ M1 25/611 pt 3.
16 Principal Keeper [Cockler?] to Secretary of Marine, 19 December 1936, ANZ M1 25/611 pt 3.
17 Dawbin 1949, p. 92.
18 W. Dawbin to Secretary of Internal Affairs, 5 March 1959. ANZ IA 46/18/59.
19 Dawbin 1962a, p. 18.
20 Dawbin 1982a, p. 237.
21 Dawbin 1982a.
22 Cree et al. 1991b, 1995a; Nelson et al. 2002a; Thompson and Daugherty 1992; Hoare et al. 2006; Moore et al. 2007.
23 Herrel et al. (2009) reported smaller head height, relative to snout–vent length, in a small sample of preserved female tuatara (island sources and method of sex identification unstated). Live tuatara sampled in larger numbers on Stephens Island differed in relative head height and head length between sexes (Herrel et al. 2010).
24 Cree et al. 1991a, b.
25 Cree et al. 1991a.
26 Claudine Tyrrell, University of Otago, pers. comm. 2 February 2004.
27 Cree et al. 1991a; Tyrrell and Cree 1994; Nelson et al. 2004a.
28 Nelson et al. 2004a, b, 2010.
29 Oldman 2008; Oldman and Cree 2010; Oldman, Nelson and Cree, unpubl. data.
30 Pers. obs.

Box 8.2

1 Gillingham et al. 1995.
2 Gillingham et al. 1995.

Box 8.3

1 Thomas 1890, 1891; Anon. 1955b.
2 For example, a pair maintained in the early 1900s by Donne in a hutch in his garden. Hatchlings were produced but soon eaten by his dog (Donne 1942). See Chapter 5, Box 5.2, for more examples of tuatara as pets in the late 19th and early 20th centuries.
3 For example, Norris's colony in Tauranga, from which many eggs were said to have been produced, with one young reared (Anon. 1908a). See Chapter 10 for more on Norris's attempts to sell the tuatara after legal protection began in 1895.
4 For example, Wood 1967; Hazley 1982; Tintinger 1987.
5 For example, Krull 1923; Keck 1925; Dowling 1961; Austin 1962; Hartley 1965; Hoessle 1969; Goellner 1984; Tytle 1988.
6 For example, among 63 adult tuatara taken into captivity between 1952 and 1995, mean survival was only 7.51 years (Blanchard 2002).
7 Newman et al. 1979; Cree and Daugherty 1990b.
8 For example, Newman et al. 1979; Newman 1982c; Blanchard 1988; Cree and Daugherty 1990b; Gillingham and Miller 1991; Cree et al. 1994; Cartland-Shaw et al. 1998.
9 Cree and Daugherty 1990b.
10 Blanchard 1991.
11 Daugherty et al. 1990b; Hay et al. 2010.
12 Thompson 1990; Nelson et al. 2004b.
13 Cree et al. 1995b; Nelson et al. 2004a; Mitchell et al. 2006.
14 Cree et al. 1994.
15 Blanchard and the Tuatara Recovery Group 2002.
16 For example, Boardman and Sibley 1991; Boardman and Blanchard 2006; Gartrell et al. 2006; Burgess et al. 2009.
17 Barbara Blanchard, Wellington Zoo, pers. comm. 2 February 2011.
18 Cree et al. 1994.
19 On Titi, Matiu/Somes, Wakaterepapanui and Long Islands (e.g. Nelson et al. 2002b, 2008). These programmes involved captive rearing rather than captive breeding: juveniles were hatched from wild-collected eggs and 'head-started' before release to the wild.
20 On Red Mercury (Whakau), Cuvier (Repanga), Kawhitu and Hauturu/Little Barrier islands (Keall et al. 2010).
21 'Zoos' is used here in a loose sense, and can also include wildlife parks, museums, etc.
22 Newman et al. 1979.
23 Cartland-Shaw et al. 1998.
24 Mitchell et al. 2006; Whitworth 2006; this publication.
25 Besson et al. 2009.
26 Oldman 2008.
27 Besson 2009; Besson and Cree 2010. See also Mello, R. S. R. 2010. *Adjustment of juvenile tuatara (*Sphenodon punctatus*) to the Orokonui Ecosanctuary climate: operative temperatures, emergence behaviour and growth rate.* Unpublished PGDipSci report, University of Otago; Mello et al. 2011.
28 Burgess et al. 2009.
29 Boardman and Blanchard 2006.
30 Blanchard and the Tuatara Recovery Group 2002; pers. obs.
31 Blanchard and the Tuatara Recovery Group 2002; pers. obs.
32 Keall et al. 2010; pers. obs. (for example, a captive-laid clutch failed entirely in our laboratory: embryos developed to a late stage, but had spinal deformities and were incapable of pipping and emerging from the eggshell; see Mitchell et al. 2006).
33 Cartland-Shaw et al. 1998.
34 Based on information supplied by Barbara Blanchard, Wellington Zoo, 16 November 2003. These animals were still alive in 2013

(Lindsay Hazley, Southland Museum and Art Gallery, pers. comm. 12 June 2013).

35 Barbara Blanchard, Wellington Zoo, pers. comm. 16 November 2003.
36 Flower 1937; Toft 1999.
37 Ramstad et al. 2009; see also Chapter 10.
38 Hazley 1995.
39 See http://www.youtube.com/watch?v=Im_VVVZxvDM (Nightlife in the Tuatarium) and Braun-Elwert 2009. An associated short film *Love in cold blood* (2009), produced by Carla Braun-Elwert and Jane Adcroft, demonstrates the intense media interest associated with reproduction in captive tuatara, an interest that sometimes runs away with the facts.

Box 8.4

1 Newman and Watson 1985; Cree et al. 1991a; Cree 1994; Newman et al. 1994; Tyrrell 2000; Tyrrell et al. 2000; Mitchell et al. 2010.
2 Cree et al. 1991a.
3 Thompson et al. 1998.
4 Lindsay Hazley, Southland Museum and Art Gallery, pers. comm. 26 June 2013.
5 Cree et al. 1991a.
6 Cree et al. 1991a.
7 Thompson et al. 1991.
8 Tyrrell 2000.
9 Attempts within New Zealand have employed 3.5, 5 and 7.5 MHz linear array probes (Kuchling, G. 1994. *Assessment of the reproductive condition of female tuatara of critically endangered populations by ultra-sound scanning*. Unpublished report of a study done during March 1994 at the Victoria University of Wellington, Wellington Zoo, and Auckland Zoo; Tyrrell 2000; pers. obs.). A brief report from a European zoo suggested that, as with Kuchling's study, ovarian follicles could be detected with ultrasound (Fritsch et al. 2002), but technical details were not included.
10 Tyrrell 2000.
11 Cree et al. 1991a; Thompson et al. 1998.
12 Pers. obs.
13 Nelson 2001.
14 Not all females palpated as gravid in December 2000 produced eggs when treated with oxytocin (Mitchell et al. 2008).
15 Cree et al. 1991a.

Box 8.5

1 Thompson et al. 1996.
2 Thompson 1990.
3 Thompson et al. 1996.
4 Cree and Guillette 1988.
5 Cree et al. 1989.
6 Five embryos were measured from pasture 'nests' (including one embryo that had died only recently). Three embryos were measured from forest 'nests' (another 'nest' contained only long-dead eggs, and the fifth 'nest' could not be found and was probably disturbed by seabirds).
7 Thompson 1990.
8 These totals include several that were found recently dead.
9 That is, 12 of 73 eggs (excluding seven that were damaged during excavation, or dissected).

Box 8.6

1 Tyrrell 1993; Cartland et al. 1994; Tyrrell and Cree 1994; Cartland-Shaw 1996.
2 Cree et al. 1995b.
3 Tintinger 1987.
4 Thompson 1990.
5 Cree et al. 1995b.
6 An FM pattern is also known, less intuitively, as pattern 1b. For review, see Warner, D. 2011. Sex determination in reptiles. In: Norris, D. O. and Lopez, K. H. (eds). *Hormones and reproduction of vertebrates. Vol. 3. Reptiles*. Academic Press, London, pp. 1–38.
7 Nelson et al. 2004a, d; Mitchell et al. 2006.
8 Wylie et al. 1968; Norris et al. 2004; Norris 2007; O'Meally et al. 2009.
9 Also known as pattern 1a.
10 Also known as pattern 2.
11 Mitchell et al. 2006.
12 For review, see Warner 2011, op. cit.
13 Warner, D. A. and Shine, R. 2008. The adaptive significance of temperature-dependent sex determination in a reptile. *Nature* 451: 566–569.
14 Cree et al. 1992; Moore et al. 2009a, b; H. C. Miller et al. 2009.
15 Mitchell et al. 2006.
16 Nelson et al. 2004d; Mitchell et al. 2008.
17 It is also unknown whether a higher pivotal temperature could evolve, although this is unlikely to happen quickly. Currently, there is little variation in pivotal temperature among females within the Stephens Island population, or between Cook Strait islands and Hauturu/Little Barrier Island to the north (Mitchell et al. 2006).
18 Discussed further in Chapter 10, Box 10.7.
19 Nelson et al. 2010. The temperature-sensitive period ends slightly later in relative terms at 20°C (by about 39% of the full incubation period).

CHAPTER NINE

Environmental relations: temperature, oxygen, water and light

During the first known attempt to measure the body temperature of a tuatara, in 1851, the tuatara bit the thermometer in its mouth into two parts.[1] Fortunately, safer and more successful ways of recording body temperature have been developed, enlightening us about this and other aspects of body function in an unusually cold-adapted reptile.

Introduction

The ability to thrive at cool temperatures is one of the most well known features of tuatara. This tolerance, indeed requirement, for cool conditions influences many physiological processes. The present chapter begins by discussing thermal biology. The anatomy of the respiratory and circulatory systems is then described, along with the influence of temperature on gas exchange and metabolic rate. Water balance, including the related topics of skin structure and nitrogen excretion, is then examined, followed by a section on the anatomy and function of the visual system. Two text boxes are included, one explaining the ectothermic biology of tuatara, and one describing the pineal complex (an outgrowth of the brain that includes a presumed sensory structure known as the parietal eye). As before, most information in this chapter is assumed or known to come from tuatara from Stephens Island (Takapourewa), although not all publications have stated the source population.

Daytime body temperatures and basking behaviour

Like other reptiles that bask, tuatara are able to raise their body temperatures to higher levels than the surrounding air. However, tuatara on Stephens Island do not always take full advantage of the solar radiation available.[2] Basking tuatara are not always fully exposed to the sun;[3] frequently, only the head or part of the abdomen is in sunlight (Figures 9.1 and 9.2; see also Chapter 7,

Fig. 9.1 A tuatara basks in a patch of sunlight on Stephens Island (Takapourewa). Like many tuatara in the forest, this animal appears not to be making full use of the sunlight available.

Photo: Marcus Simons.

Fig. 9.2 A male tuatara rests outside its burrow in forest on Stephens Island. Despite poses that can seem very relaxed, basking tuatara retreat rapidly to their burrows if disturbed.

Photo: Alison Cree.

Figures 7.6–7.8). Although some forest-dwelling tuatara move from time to time during the day following patches of sunlight, others remain in the same position and posture throughout.[4] Adult tuatara living on forest edges often emerge fully into sunlight during the morning,[5] but remain close to cover in the event that they need to make a rapid retreat. Those in pasture tend to bask in a more concealed position at the burrow entrance, often with just the head and sometimes part of the trunk in the sun.[6] Although I have sometimes seen tuatara in pasture that were fully emerged, these were usually large males (probably the least vulnerable of tuatara to attack by predatory birds) and they were always within about 1 metre of their burrow entrance (Figure 9.3).

BOX 9.1

Tuatara, 'cold-bloodedness' and ectothermy

Like other terrestrial reptiles, tuatara are ectotherms – they depend on sources of environmental heat to raise their body temperature above that of the surrounding air. Usually, the heat source is the sun, although there are records of wild tuatara being attracted to campfires on islands[1] and to a fireplace in a lighthouse keeper's cottage.[2] By basking, tuatara are able to raise the temperature of their body tissues during the day to as warm as 30.1°C (potentially 9°C or more above the surrounding air). At night, their body tissues continue to operate, albeit much more slowly, at temperatures of 7°C or below, temperatures similar to those of the surrounding air.

Just as other body tissues do, a tuatara's blood fluctuates in its temperature with environmental conditions. Thus, it is true that tuatara are, at times, 'cold-blooded'. However, this term is of limited validity because at other times, a tuatara's blood can be rather warm. In part for this reason, the terms 'cold-blooded' and 'warm-blooded' are now avoided by physiologists. More fundamentally, these terms do not capture the difference in energy use between ectotherms like

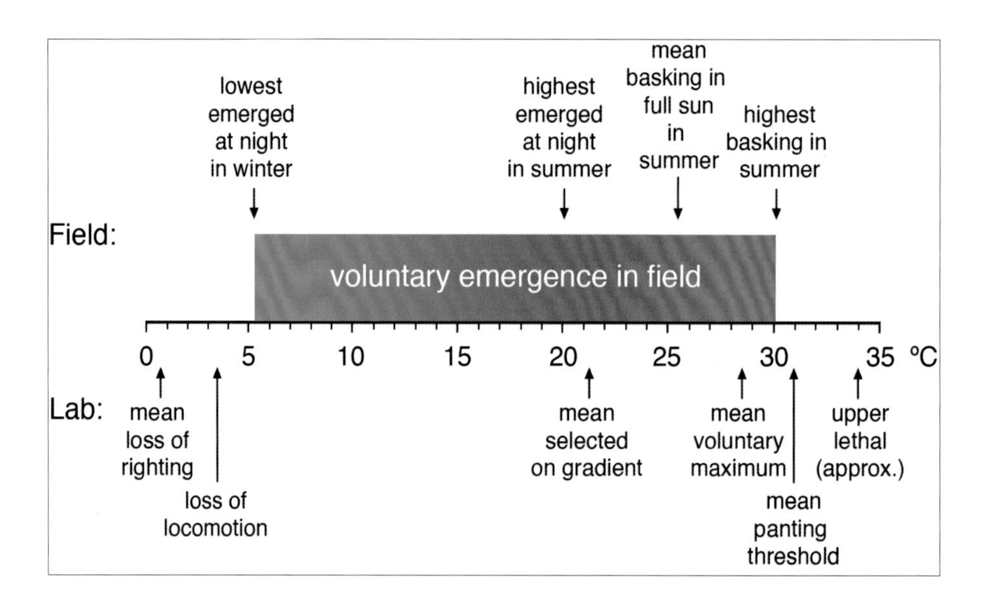

Box 9.1 Fig. 1 Summary of the body temperatures tolerated by tuatara of Stephens Island origin. Juveniles and adults tolerate a wide range of body temperatures from just above freezing to the low 30s (°C). Responses recorded above the horizontal temperature scale are from adults in field studies. Responses recorded below the scale are from adults or juveniles in laboratory situations. Some responses are mean values for several animals, whereas others are reported for individuals. The upper lethal temperature in adults is only approximate, and probably varies with exposure time as well as factors such as relative humidity. Voluntary maximum is the point at which tuatara exposed to radiant heat began to struggle and turned their head to the side. Loss of righting is the loss of ability of a tuatara placed on its back to turn itself upright. For sources, see main text. The upper thermal tolerance in embryos (inferred from transient extremes of soil temperatures for successful nests; see Chapter 8) appears to be higher than shown here for juveniles and adults.

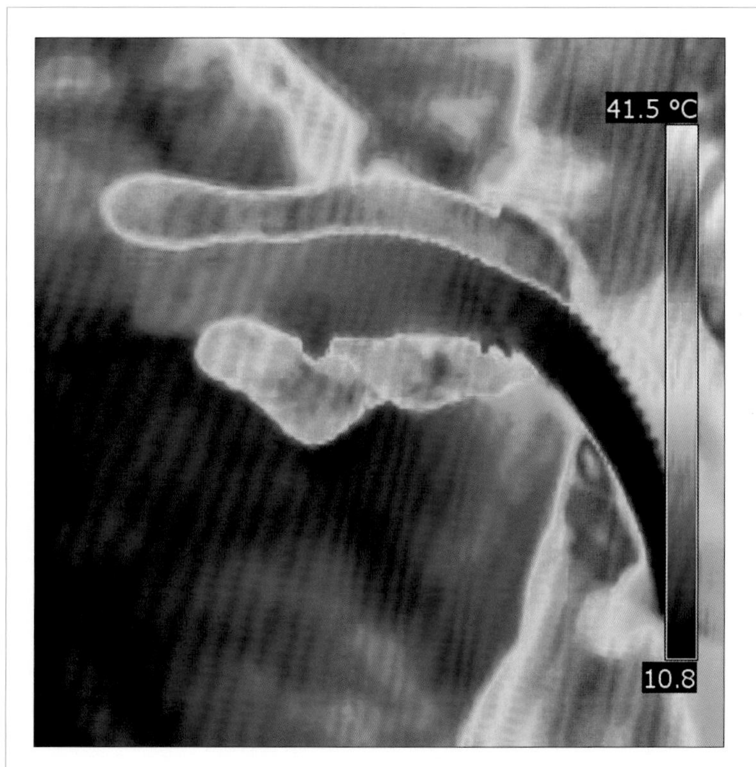

Box 9.1 Fig. 2 Infra-red image of a juvenile tuatara retrieved from its burrow. The surface temperature of the tuatara ranges from 11°C at the tail end (dark blue) to 19°C at the head end (light blue). The gradient arises because the head end was closer to sunlight at the burrow entrance. The human hand, in contrast, ranges from 30°C to 33°C (yellow-to-red areas).

Photo: Alison Cree.

tuatara, which use the sun's energy to help drive metabolic processes, and endothermic mammals and birds, which are constantly warm and metabolically active, fuelling that activity with a very high intake of food. The 'low-energy' lifestyles of reptiles place much lower food demands on their ecosystems (for example, tuatara in cool conditions have been observed to survive for months without food).[3] In fact, studies of field metabolic rates suggest that, on an annual basis, the energy demands of ectothermic reptiles may be only one-thirtieth those of similar-sized endotherms that stay active year-round.[4]

Although adult and juvenile tuatara operate over a wide range of body temperatures, they seem to prefer temperatures of about 22°C, slightly above the mid-range of what can be tolerated. Activities such as locomotion, and processes such as growth, probably occur at a maximum rate at about the preferred (selected) temperature, although this has yet to be tested. It is also likely that thermal preference varies with factors including age, sex and reproductive condition, and that the optimal temperatures vary for different processes.

Fig. 9.3 An adult male tuatara close to its boulder-covered burrow in pasture on Stephens Island. Most tuatara in the pasture are vulnerable to birds of prey, and rarely bask as openly as this large specimen.

Photo: Alison Cree.

The daytime body temperatures that tuatara experience while basking on Stephens Island have been reported by several workers. In a study by Saint Girons and colleagues in December (early summer), tuatara basking in full sun at the forest edge reached body temperatures of 25–26°C, whereas those in forest with only part of the body in sun had body temperatures of about 19–23°C.[7] Those that remained in shade in the forest were at 17–19°C, only slightly (generally 1–2°C) above the temperature of the burrow.[8] Tuatara that were concealed under sheets of corrugated roofing iron lying in forest clearings had body temperatures (21–26°C) that were similar to those in partial or full sun.[9] By basking, tuatara on Stephens Island are able to reach body temperatures at least 6°C warmer than air temperature.[10] Although a study of mine reported daytime body temperatures in January (summer) that appear higher on average for females (23.9°C)[11] than for males (17.3°C),[12] these were from different years. To confirm differences between the sexes, matched comparisons on days of similar basking opportunity are needed. However, tuatara on warm days are challenging to capture – most disappear down burrows before their body temperatures can be recorded.

A study that eliminated this difficulty was performed on Stephens Island by Richard (Dick) Barwick between 1968 and 1977 (Figure 9.4).[13] For this radio-telemetric study, a temperature-sensitive probe was first inserted deeply through the cloaca of a small number of tuatara (the cloaca is the cavity that follows the digestive tract; it also receives the products of the urinary and reproductive systems, just prior to the exit of these materials from the body). A radio-transmitter was attached to the body, allowing temperatures in the cloaca of each tuatara to be recorded by an observer using a radio receiver. Tuatara living in forest in January (summer) emerged from their burrows in the morning and showed a rapid rise (6–9°C) in body temperature during the next 30 minutes (other studies show that tuatara can heat faster than they cool, probably because of an increase in heart rate during heating[14]). Thereafter, some tuatara retreated intermittently to their burrows, only to re-emerge and bask again. Not surprisingly, daytime body temperatures were not especially stable, varying over at least several degrees within a day. Maximal body temperatures varied between days, ranging between 19.7°C and 26.0°C for an adult male on 6 days that it basked, but reaching only 17.3°C on a cooler day when it did not bask. The highest temperature recorded for a free-ranging tuatara (sex not specified) was 30.1°C. The same study suggested that basking behaviour during warm, dry periods is constrained by the need to reduce water loss.[15]

In a more recent telemetric study, on Lady Alice Island to the north, body temperatures of four female tuatara were monitored during daylight hours for 3–4 days in October (spring).[16] Body temperatures ranged from 12.7°C to 27.8°C. The highest value was for a gravid female (one with eggs in the oviducts) in forest, where air temperature in the shade never exceeded 19°C. As sometimes seen on Stephens Island, basking tuatara shifted their position to optimise the amount of sunlight falling on their bodies.

In a recent study, Anne Besson and I examined the responses of captive juveniles (of Stephens Island stock) to changes in basking opportunity.[17] Juveniles that were provided with a heat lamp for 8 hours per day (about the same period of time available for basking on Stephens Island in the autumn) often basked during the first 2 hours, but thereafter used the heat lamp less frequently. When juveniles had access to the heat lamp for only 5 hours or even 3 hours per day, they used the heat lamp more often. In consequence, when the heat lamps were on, tuatara under the regime with least basking

Fig. 9.4 Fluctuations in body temperature of an adult female tuatara over three 24-hour periods on Stephens Island during summer (January 1968). Body temperature was measured using a temperature-sensitive radio-transmitter. Note the steep rise in temperature that follows the time of first emergence, which occurred between about 7:00 and 9:40 a.m. The female thereafter basked intermittently (with periodic retreats into the burrow) until about 3:00 p.m., when direct sun was lost from the site. At night, the female emerged for brief periods, during which time body temperature dropped below burrow temperatures (12–13°C). The black horizontal bars indicate periods of darkness.

Modified from Barwick 1982.

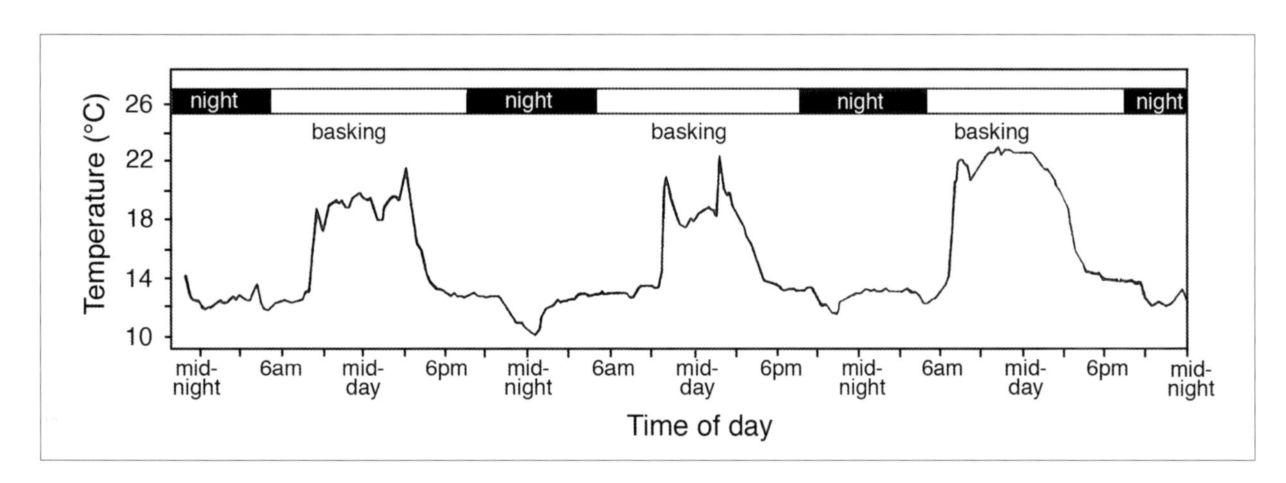

opportunity maintained their body temperature[18] closer to their preferred range (as separately determined on a thermal gradient) than when they had the opportunity to bask for 8 hours per day. We inferred that, if juveniles were shifted from Stephens Island to the southern coast of the South Island (in an attempt to re-establish a southern population, within a predator-controlled ecosanctuary), thermoregulation would still be possible, and it would probably be carried out more effectively.[19] Further work suggests that juveniles are unlikely to bask if potential body temperatures (as measured by calibrated models in basking positions) remain below 11°C.[20] In addition, intermittent basking may be important for continued digestion if food has been consumed.[21]

Body temperatures during nocturnal emergence

Tuatara emerge and roam most widely from their burrows at night. Ambient air temperature has a large effect on emergence: on Stephens Island, few if any tuatara are seen when air temperature is below about 7°C, but emergence increases as air temperatures warm towards 20°C.[22] Given that these air temperatures are low by the standards of many reptiles elsewhere in the world, researchers have been intrigued to record the body temperatures of tuatara emerged at night.

Reports from a variety of studies confirm that nocturnal body temperatures can be as low as 5.2–6.2°C,[23] although in my experience on Stephens Island, the only tuatara likely to be visible with such low temperatures are large, slow-moving adult males that remain partly within their burrows. More typical temperatures of tuatara that are fully emerged and active at night are 10–18°C.[24] Body temperature can be higher than air temperature in tuatara caught soon after sunset.[25] However, as the night progresses, body temperature declines to values similar to air temperature, and can fall slightly below air temperature in damp animals under breezy conditions (when evaporative cooling occurs).[26] Body temperatures of tuatara emerged at night can also fall below burrow temperatures.[27]

Records for body temperatures of tuatara on other islands are limited, but suggest patterns similar to those on Stephens Island. On North Brother Island, body temperatures of emerged tuatara and the gecko *Hoplodactylus duvaucelii* were measured at night in January (summer). Body temperatures (reported jointly for the two species) were within the range 9.5–18.2°C, and usually within 1°C of ambient air temperature.[28] Five northern islands were visited by Claudine Tyrrell during September to December (spring to early summer). Body temperatures during the first hour after nightfall ranged from minima of 10.3–13.2°C to maxima of 14.9–20.6°C,[29] and were similar to those of the surrounding air. For three islands (Lady Alice, Coppermine and Green) on which gravid females, non-gravid females and males were captured, there were no differences in the relationship between night-time body temperature and air temperature among the reproductive groups. Additionally, telemetric studies on Lady Alice Island revealed two females with unexpectedly high body temperatures of 21–22°C in the early daylight period, before there was opportunity for basking. Tyrrell suggested that these high temperatures resulted from sharing burrows with seabirds,[30] a possibility receiving further study on Stephens Island.[31]

Temperatures selected in a thermal gradient

The temperatures that tuatara prefer when given a choice in a thermal gradient – referred to as selected temperatures or preferred body temperatures – have been examined in several studies (Table 9.1). Despite variation in age of the tuatara and prior holding conditions, mean preferences generally converge at about 20–22°C. In one study in our laboratory, there was no significant difference in preference at about 7 months of age between juveniles arising from different egg incubation temperatures (animals that were probably of different sexes);[32] yet, there was significant variation between times of the 24-hour period, with lowest mean values at night (Figure 9.5).[33] However, in older juveniles and adult males sampled less frequently, no influence of time of day was detected.[34] Although mean selected temperatures of 20–22°C in tuatara are low by comparison with many lizard species overseas,[35] they are not significantly lower than for (non-pregnant) individuals of two New Zealand geckos and a skink under similar conditions.[36]

TABLE 9.1

Mean selected temperature of tuatara placed in a thermal gradient (± standard error)[1]

Animal stocks were of Stephens Island origin. Selected temperature (T_{sel}, also known as preferred body temperature) was measured in the cloaca.

Animal source	Season and/or prior thermal history	Sex and life-history stage	Sample size	Mean T_{sel} ± SE (°C)	Range (°C)	Reference and comment
Indoor captive colony	Constant 20 ± 2°C for 2 years	Not stated	7	21.3 ± 0.6	10–28 (over this and other trials)	Sanderson 2009[2]
Indoor captive colony	Constant 25 ± 2°C for 1–26 weeks	Not stated	3	22.6 ± 0.5	10–28 (over this and other trials)	Sanderson 2009
Indoor captive colony	Spring-summer conditions (12–18°C with basking opportunity)	Unsexed juveniles about 7 months old	10	20.9 ± 0.6	12.8–28.1	Hare et al. 2007 (see also Fig. 9.5 and Whitworth 2006 for significant time-of-day effect, but no effect of egg incubation temperature)
Free-roaming population outdoors in ecosanctuary	Autumn	Adult males	10	21.3 ± 0.4	Not stated	Besson and Cree 2010 (no time-of-day effect)
Indoor captive colony	Spring–summer conditions (18°C with basking opportunity)	Unsexed juveniles	8	21.3 ± 0.9	Not stated	Besson and Cree 2010 (no time-of-day effect) (see also Besson 2009; Besson and Cree 2011)

Notes

1 Additionally, Stebbins (1953) indicated a mean selected temperature of 18.0°C (range 3.5–27.9) for two adult males (prior to shielding of the parietal eye), but the records were collected over a 2-month period in more complex housing pens, and one animal suffered inflammations that may have influenced the results.

2 Mean selected temperatures across recording periods were essentially unchanged for four animals following transfer to 15°C (n = 3) or 10°C (n = 1) for 1–26 weeks.

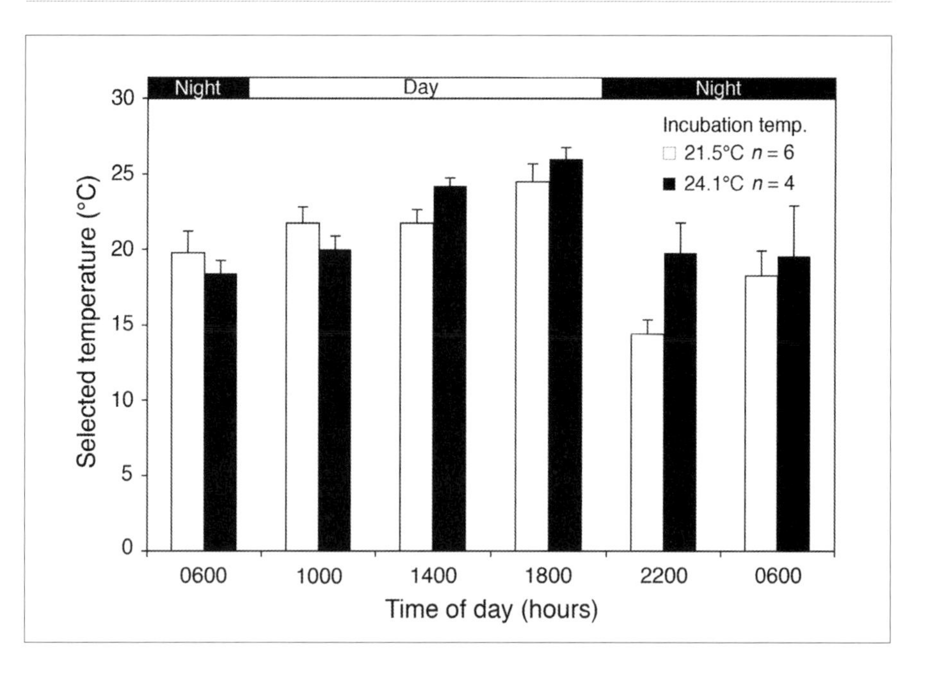

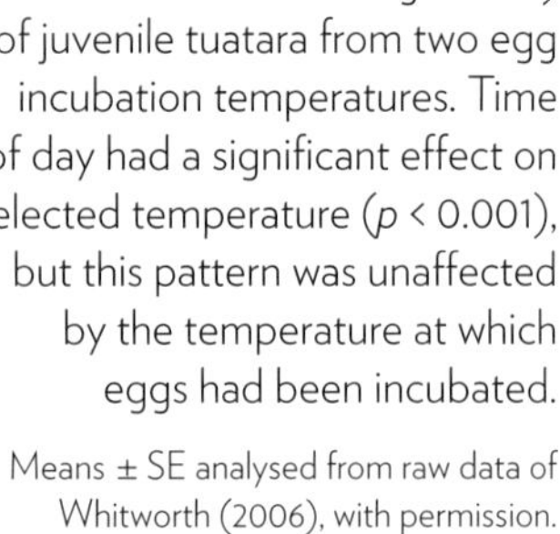
Fig. 9.5 Selected body temperature (temperature chosen on a thermal gradient) of juvenile tuatara from two egg incubation temperatures. Time of day had a significant effect on selected temperature ($p < 0.001$), but this pattern was unaffected by the temperature at which eggs had been incubated.

Means ± SE analysed from raw data of Whitworth (2006), with permission.

It should be noted, however, that most studies on thermal preference of tuatara have been for animals housed indoors under relatively benign conditions. No studies have examined thermal preferences in field-fresh tuatara in spring or summer, when females producing eggs, for example, might be expected to prefer higher temperatures (recall that body temperatures for basking tuatara on Stephens Island have reached up to 30.1°C[37]). Unpublished observations on laboratory-incubated hatchlings and young juveniles also indicate high preferences in some individuals of 30–32°C, apparently maintained for several hours.[38] Given the many factors that can influence selected temperature within lizard species, including digestive state, season and reproductive condition,[39] it would be unwise to assume a single, tightly regulated value for thermal preference in any wild population of tuatara.

Limits to thermal tolerance

Embryos of tuatara appear remarkably tolerant of warm temperatures, in that successful hatching of eggs can occur from nests in which the surrounding soil has reached up to 38.4°C for brief periods.[40] Juvenile and adult tuatara, however, are unable to tolerate such warm temperatures. In the only quantitative study of upper limits to tolerance, restrained animals on Stephens Island (adults of both sexes, plus juveniles) were exposed to radiant heat from a heat lamp.[41] Initial deep body temperatures (within the intestine) were 19.1–26.0°C and the increase in deep body temperature was controlled to a rate not exceeding 1°C per minute. The animals were initially alert and usually motionless, with the head temperature (measured inside the mouth, against the posterior palate) increasing at a faster rate than the body. At a point designated as the voluntary maximum, the tuatara suddenly struggled, lowering and twisting their heads so that the side of the head faced upward. Mean temperatures at this point were significantly higher for the head (30.8°C) than for the body (28.3°C).[42] Above the voluntary maximum, the heating rate of the head declined to the same rate as that of the body, presumably in part because of the reduced exposure of the head, but possibly supported by changes in blood circulation and breathing patterns. Soon afterwards (when head temperature averaged 33.0°C and body temperature 31.0°C), the tuatara began to pant (that is, they opened their mouths while breathing continued). Panting, which was often accompanied by salivation and urination, reduced the rate at which head temperature increased. Thereafter, all animals were allowed to recover and were later returned to the field, in apparently healthy condition.[43]

Other reports indicate some risk from allowing body temperatures to become so warm. On Stephens Island during March, one adult male (in which the skin over the parietal eye was painted black for experimental reasons) was heated to the point where it panted (body temperature of 33.8°C), but it did not survive cooling and died a few hours later, although others survived.[44] In another study, a tuatara heated to a body temperature of 34.5°C (apparently with forced activity) developed paralysis of the posterior half of the body and died within 24 hours.[45] Although captive tuatara in Sydney, Australia, apparently survived air temperatures of 40°C for at least several hours,[46] this report should be interpreted with caution.[47] Body temperatures of animals in burrows would have been at least several degrees lower, and some acclimatisation to local conditions may have occurred.

Responses to and tolerance of low temperatures have received limited attention in tuatara. A captive adult male was observed digging intermittently at body temperatures of about 7–9°C and could still move forward laboriously at 4.5°C. At a body temperature of 3.5°C, it could still right itself but was unable to crawl.[48] Captive juveniles with a body temperature of 5°C were capable of coordinated movement and sometimes ate food that was offered (albeit more slowly than at warmer temperatures, and without evidence of digestion). However, at a mean body temperature of 0.7°C they lost the ability to right themselves.[49] This value (defined as the critical thermal minimum) was similar to those for two species of New Zealand geckos in the same study, but lower than that for a strictly diurnal skink. A captive tuatara (an adult female) in an outdoor enclosure in Christchurch survived periods during which air temperature fell below freezing on several occasions,[50] but its body temperature was not reported. Whether any life-history stage of tuatara can survive body temperatures below 0°C, either by supercooling (a situation in which body fluids remain unfrozen) or by tolerating freezing, as in the European lizard *Lacerta vivipara* and some North American turtles,[51] is unknown.

Overall, the body temperatures that tuatara experience during field activity, at the upper voluntary maximum, at the panting threshold and at the critical thermal minimum are much lower than for many reptiles studied elsewhere in the world.[52] There was once speculation that the tolerance, indeed requirement, of tuatara for cold temperatures was an ancestral feature retained from rhynchocephalians of the Mesozoic Era.[53] However, ample evidence now exists that the same cool-temperature tolerance is typical of several New Zealand geckos, at least.[54] (Furthermore, temperatures of the Mesozoic appear in general to have been warmer, rather than colder, than today.[55]) Thus, the ability of tuatara to survive and reproduce at cool temperatures (as with other New Zealand reptiles) is more reasonably interpreted as an adaptation to current or recent cool climates,[56] than as a feature of distant rhynchocephalian ancestors.

Metabolic rate

Metabolic rate, the rate at which energy is being used, is inferred in animals from the rate at which oxygen is consumed or carbon dioxide produced. In laboratory studies on tuatara of Stephens Island origin, oxygen consumption (measured as microlitres consumed per gram of body mass per hour) varied with body mass, activity and temperature.[57] As expected, juveniles have higher mass-specific metabolic rates than adults, and metabolic rate increases with temperature (Figure 9.6). Metabolic rate in juveniles also varies with time of day, being lowest in the morning and highest during the late afternoon and evening,[58] when activity levels in captive animals are also generally higher.[59]

The precise shape of the metabolic rate–temperature curve in tuatara has been debated. An early study with two adult tuatara suggested that metabolic rate increased exponentially with temperature over the range 7–35°C (thus indicating that the relationship between the logarithm of metabolic rate and body temperature was linear).[60] However, later studies, often having larger sample sizes, indicate a stepped relationship, with a reduced increase (a region of relative thermal independence, as shown by a low value for Q_{10}, the temperature coefficient) within the range 12–20°C.[61] For example, in a study by Cartland and Grimmond, oxygen consumption in juveniles of 2–3 years of age showed a Q_{10} value of 7.1 between 5.0°C and 12.5°C, but a value of 1.3 between 12.5°C and 20.0°C.[62]

Although older studies have often asserted that tuatara have a low metabolic rate compared with lizards,[63] such statements are limited by small sample sizes, poor control with respect to body size and/or a limited range of temperatures for comparison. A more rigorous study, by Michael Thompson and Charles Daugherty, examined oxygen consumption in tuatara ranging from small juveniles to large adults (6.8–746 grams) on Stephens Island. The authors concluded that mass-specific metabolic rates in adult tuatara at 13°C were about the same as expected for lizards at this temperature, and that rates in juvenile tuatara were higher (not lower) than expected for lizards of similar mass.[64] Thus, the crucial difference between tuatara and other species is not that mass-specific metabolic rate is low at a given temperature. Rather, tuatara have low activity temperatures. Thus, tuatara are active with

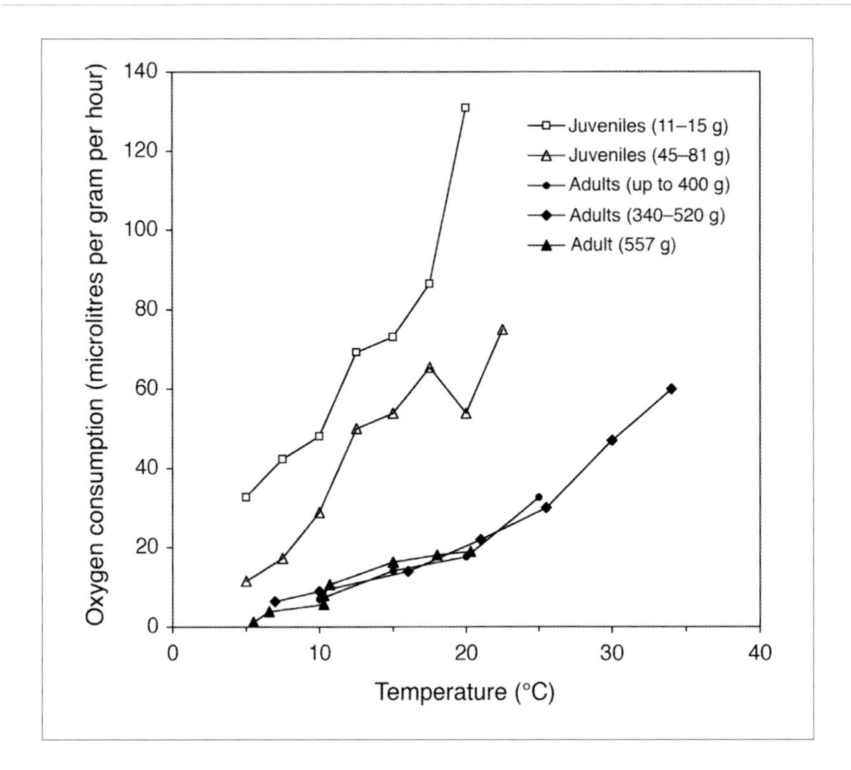

Fig. 9.6 Effect of temperature on mass-specific oxygen consumption in tuatara known or assumed to be inactive. Rates were obtained under laboratory conditions for tuatara of Stephens Island origin. Mean resting metabolic rate ('during the active part of the daily cycle', although time of day was not stated) was measured for five juveniles, in each of two size and age groups (spanning 1–3 years of age), by Cartland and Grimmond (1994). Mean oxygen consumption in adults up to 400 grams (g; sample size and time of day not stated) was measured by Hill (1982). Standard metabolic rate (i.e. during the inactive daytime period) was averaged for two adults of 340 g and 520 g by Wilson and Lee (1970). Carbon dioxide production was measured by day in one adult of 557 g by Wells et al. (1990), and here converted to oxygen consumption assuming an arbitrary respiratory quotient of 0.75 (depending on diet, this value could change slightly). Standard errors are available for juveniles in Cartland and Grimmond 1994 but not for other sources.

lower metabolic rates than many overseas lizards, whose activity temperatures are often between 20°C and 40°C.[65] In a consistent vein, Linda Cartland and Nicola Grimmond concluded that metabolic rates of juvenile tuatara, at temperatures between 5°C to 15°C, were comparable with those of some New Zealand lizards of similar mass. These included a nocturnal gecko with similar activity temperatures.[66]

Field metabolic rate, the amount of energy used by wild animals going about their activities at the temperatures they naturally experience, has been examined by Claudine Tyrrell and colleagues in two populations of northern tuatara. The rates were derived from the turnover of doubly labelled water

(water enriched with specific, measurable isotopes of hydrogen and oxygen) during spring to early summer.[67] The mean value for field metabolic rate differed significantly between non-gravid females on Lady Alice Island (0.049 millilitres of carbon dioxide per gram body mass per hour) and those on Coppermine Island (0.036 millilitres of carbon dioxide per gram body mass per hour).[68] Whether the lower value on Coppermine Island resulted from differences in weather during the longer period for measurement, or from other possible causes such as constraints on activity and food intake arising from the more recent presence of Pacific rats (*Rattus exulans*), is unclear. These field metabolic rates in tuatara have been compared with predicted values derived from 55 species of reptiles overseas during their active periods.[69] Not surprisingly, given the low activity temperatures of tuatara, field metabolic rates were lower than (about 20–30% of,[70] and below the 95% confidence interval for) the value predicted from the overall rate for other species. As pointed out by Tyrrell, although values similar to those in active tuatara have been seen in Australian frillneck lizards (*Chlamydosaurus kingii*), these values were only seen during the dry season when the lizards were inactive and losing mass.[71]

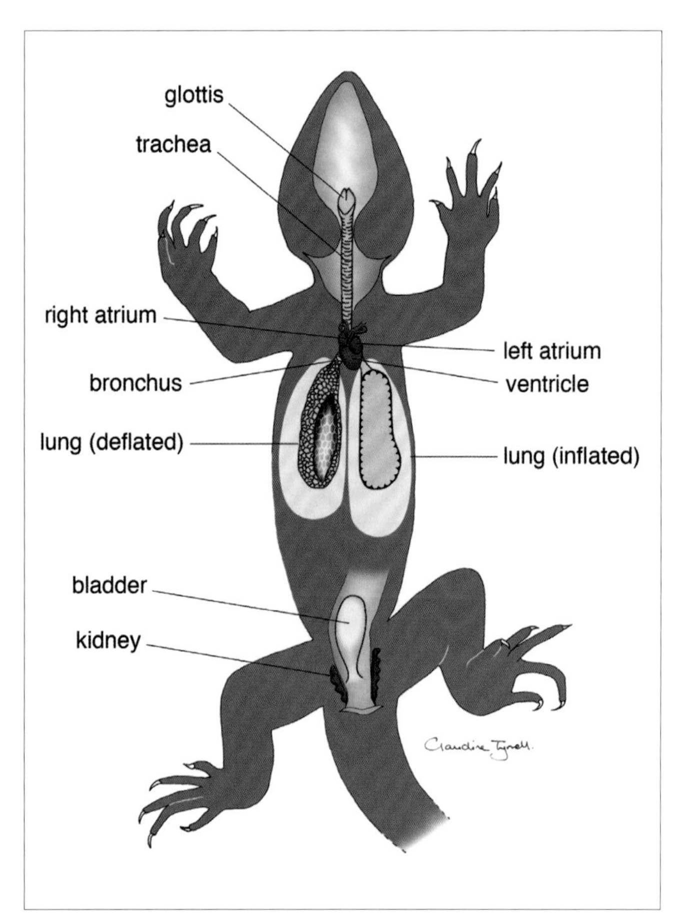

Fig. 9.7 Diagram of the respiratory system, heart and excretory system of tuatara, as viewed from the ventral surface. The cobble-stoned external appearance of the lung (as drawn on the left of the diagram) reflects the presence of internal ridges of tissue, or septa, as seen in the cut-away view. The sectional view of the lung (as drawn on the right) shows how the shallow septa increase, to a small extent, the surface area for gas exchange. Although the internal complexity is remarkably modest for a reptile of this size, it is consistent with the low metabolic demands of tuatara. Capacity of the inflated lungs may not always vary between sides in the manner shown here.

Diagram based on dissections and a drawing by Claudine Tyrrell, with further detail and colouring by Ken Miller. Additional details of lung and heart structure are based on Milani 1894, Mienertz 1966, Grigg and Simons 1972 and Willnow and Willnow 1976.

Respiratory anatomy and breathing patterns

In its general features, the respiratory system of tuatara is typical of reptiles (Figure 9.7). Air enters through external nostrils (nares) on the snout, and then passes through internal nostrils (choanae) into the mouth. The passage of air continues through a slit-like glottis on the floor of the throat (pharynx), then through the larynx, before entering a trachea ('windpipe') lined with cartilaginous rings. The trachea divides into two very short bronchi,[72] each leading to a large and sac-like lung. Although some differences between the two lungs in dimensions or capacity have been reported,[73] whether these are in a consistent direction is unclear and other sources indicate a general similarity of size.[74] Internally, the surface of the lung has septa (dividing walls) that divide the lung incompletely into air spaces

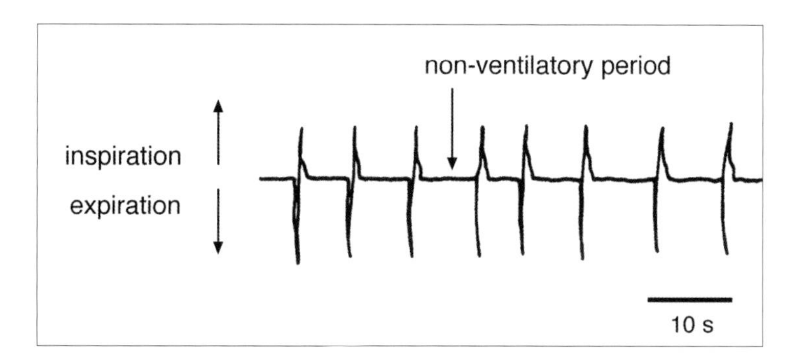

Fig. 9.8 Pattern of inspiration and expiration in an adult tuatara of Stephens Island origin. The trace shows a strain gauge recording for a resting, unrestrained individual at 21°C.

Modified from Wells et al. 1990.

known as infundibuli. However, the level of internal complexity (and hence respiratory surface area) is exceptionally low for reptiles, especially given the moderately large body size of tuatara.[75] Technically, the lungs are single-chambered (unicameral), resembling those seen in some lizards, but differing from the single-chambered lung or lungs of some snakes, the transitional lungs seen in some lizards, and the multi-chambered lungs seen in other squamates, turtles and crocodilians (see later section entitled 'Heart and lung function in a low-temperature specialist' for the possible evolutionary significance of differences in lung structure).[76] Although some histological information on lung structure is available for tuatara,[77] its comparative significance is unclear and ultrastructural details have not been reported.

Tuatara show a biphasic breathing cycle similar to that of many lizards (Figure 9.8).[78] During inspiration, the rib cage and lungs expand. This is followed by a small and incomplete expiration (possibly caused by movement of abdominal organs), then a period without breathing (the non-ventilatory period). Finally, there is an active expiration involving muscles of the rib cage, followed by another inspiration in quick succession. Although the length of the ventilatory period is relatively independent of temperature (over the range 5–30°C), the length of the non-ventilatory period is strongly and inversely affected by temperature.[79] In consequence, the frequency of breaths

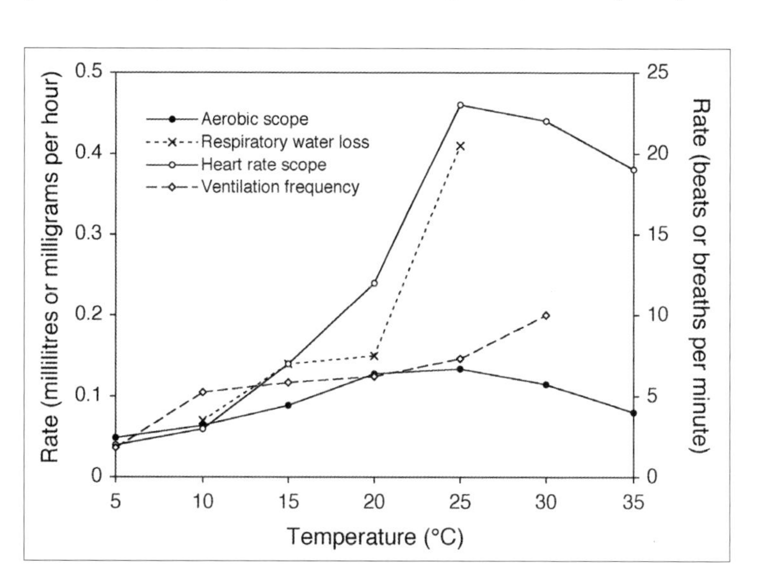

Fig. 9.9 Effects of body temperature on aspects of metabolic function in adult tuatara of Stephens Island origin. Aerobic scope (in millilitres O_2 per gram per hour) and respiratory water loss (in milligrams H_2O per gram per hour) are shown on the left vertical axis. Heart rate scope (in beats per minute), and ventilation frequency in inactive tuatara (in breaths per minute), are shown on the right vertical axis. Note the sharp increase in respiratory water loss between 20°C and 25°C, the increase in ventilation frequency between 20°C and 30°C, and the decrease in aerobic scope and heart rate scope above 25°C. These patterns suggest that adult tuatara are increasingly physiologically challenged by body temperatures above 25°C, especially if they are active.

Data for aerobic scope and heart rate scope are means for two adults from Wilson and Lee 1970; ventilation frequency is for one adult from Wells et al. 1990; respiratory water loss at 0% relative humidity is for an unknown number of adults from Hill 1982.

(ventilatory frequency) increases markedly with temperature. At 5°C, tuatara average only about two breaths per minute, compared with about 10 breaths per minute when lightly restrained at 30°C (Figure 9.9).[80] The frequency may be much higher during activity, as one early study reported that 26 breaths could occur in a minute 'during the heat of the day, when the animal was running about'.[81] At low temperatures, there can be prolonged periods of breath holding (apnoea). Rufus Wells and colleagues noted that tuatara at body temperatures below 7°C could hold their breath for up to 8 minutes,[82] and famously, R. R. D. Milligan reported in 1924 that a tuatara at 9°C (body temperature, presumably) was 'closely watched for one hour, during which time no respirations occurred'.[83] Tuatara also have a disconcerting ability to remain under water for prolonged periods (hours, according to a publication from 1877,[84] although temperatures were not reported).

Heart structure and circulation

Among other reptiles, the heart consists externally of four chambers: a sinus venosus, two atria and a ventricle.[85] Crocodilians are notable for having a ventricle divided internally into two separate chambers by a solid, ventricular septum. In the incompletely divided ventricles of turtles and squamates, various other septa nonetheless operate to produce a degree of separation between communicating chambers, and thus between oxygenated and deoxygenated blood.[86] Under normal conditions, blood flowing to the lungs (pulmonary flow) is separated from that flowing to the head and body (systemic flow). The system is also flexible, allowing cardiac shunting to occur. This means not only that deoxygenated blood can be recirculated back into the systemic circulation, bypassing the lungs, but also that oxygenated blood can be recirculated back into the pulmonary circulation, bypassing the head and body. This system may be an evolutionary specialisation that offers advantages during periods of breath holding, exercise and thermoregulation (heating and cooling), although the exact physiological significance of different shunts continues to be debated.[87]

Given these patterns in other reptiles, the structure of the heart and circulatory system of tuatara is of obvious interest. Externally, the heart has four chambers,[88] with an overall structure most similar to that of lizard hearts.[89] The ventricle itself contains an atrioventricular valve, a vertical septum and a muscular ridge (the *Muskelleiste*), thus possessing the main components needed to achieve a subdivision into three communicating chambers.[90] Although the septa are not highly developed, there is nonetheless considerable separation of the pulmonary and systemic blood flow, similar to that in other reptiles. This was shown by the injection of radio-labelled technetium particles into a peripheral vein of an adult male: more than 92% of the particles lodged, after passing through the heart, in the area corresponding on a radiograph to the lungs.[91] Although cardiac shunting has not been investigated in tuatara, it could be part of the mechanism by which heating occurs more rapidly than cooling.[92] Overall, the arterial and venous systems of tuatara are similar to those of lizards.[93]

Heart and lung function in a low-temperature specialist

As expected, heart rate is strongly affected by temperature in tuatara.[94] In a study on restrained tuatara that were described as inactive, heart rate increased from about 9 beats per minute at 7°C to about 80 beats per minute at 32°C.[95] However, in a later study, values were often lower, suggesting that restraint in the earlier study may have led to an elevation of heart rate above resting levels.[96]

Several measures of respiratory or cardiac function suggest that tuatara are increasingly challenged (especially if forced into activity) as body temperatures rise above about 25°C. Oxygen consumption in an active adult tuatara peaked between about 25°C and 30°C, but reached a plateau or declined slightly at temperatures above this.[97] Thus, aerobic scope (the difference between oxygen consumption when active versus when resting, representing the extent to which energy can be generated aerobically) reached a maximum at 25°C, and declined slightly at higher temperatures. Heart-rate scope also reached a maximum at 25°C in the same study.[98] Ventilatory frequency in adults increased markedly between 25°C and 30°C.[99]

As in other reptiles, the oxygen-carrying capacity of the blood is enhanced in tuatara by respiratory pigments (haemoglobins).[100] However, the oxygen-dissociation curve for tuatara shows unusual properties compared with those of many vertebrates.[101] The curve indicates that the blood has high affinity for oxygen, but with low co-operativity (meaning that the binding of the first oxygen molecule has little positive effect on binding of further oxygen molecules to the same haemoglobin molecule).[102] The Bohr effect (the shift of the curve to the right with raised carbon dioxide levels or reduced pH) is also small. These features seem consistent with a low demand for oxygen and limited scope for aerobic activity.[103] The amino-acid structure of the α- and β-globin chains contributing to different forms of haemoglobin in tuatara have been described[104] and compared with those for other reptiles.[105] Overall, the haemoglobin chains of vertebrates have apparently evolved in ways that do not always correspond with conventional phylogenies (for example, the α-globins of tuatara show greater similarity with those of turtles and birds than with those of squamates). Some haemoglobin chains may have evolved more rapidly within squamates than within other reptiles.[106]

Many of the unusual metabolic features of tuatara, including the large size of the blood cells,[107] aspects of the oxygen-dissociation curve[108] and low metabolic rate,[109] have been considered as 'primitive'. However, the putatively ancestral nature of such features[110] should be re-evaluated in light of low oxygen demand in an animal specialised to low body temperatures.[111] As a group, reptiles, like amphibians, are much more efficient than endothermic mammals and birds at converting energy obtained from food into body tissue.

Reptiles are able to achieve this efficiency because they do not depend on metabolic heat production to raise their body temperatures above ambient. Thus, reptiles have been able to exploit low-energy lifestyles, leading to the evolution of small and/or elongate body forms and the ability to tolerate periods with low supplies of oxygen and/or food.[112] Rather than retaining physiologically ancestral features, tuatara may have taken the reptilian strategy to an extreme: in other words, their exceptionally low-energy lifestyle may be a highly specialised feature.

Lung structure provides an example of changing views about 'primitiveness'. Although the lungs of tuatara have often been considered 'frog-like',[113] it does not follow that the single-chambered lung is necessarily an ancestral condition for reptiles. In discussing this issue, Steven Perry pointed out in 1998 that evolution of lung structure should be considered in connection with lifestyle, and with other possible non-respiratory functions and constraints such as those associated with locomotion.[114] For example, single-chambered lungs are well suited for lepidosaurs (lizards, snakes and tuatara) with long-term, low-to-medium levels of oxygen demand. Further, in elongate species (such as tuatara), deformability of the lungs may help smooth the body contour and thus help in maintaining contact with the substratum during locomotion. In reptiles whose bodies swing from side to side during movement (with axial bending or lateral undulation, as in tuatara[115]), the short length or lack of bronchi could be a derived condition that enables movement of air between the two lungs. Overall, Perry proposed that the first reptiles may have had unspecialised, multi-chambered lungs, and that single-chambered lungs may have evolved on several occasions among lepidosaurs.[116] If so, at least some features of the lungs of tuatara may be derived rather than ancestral.

Water balance, skin structure and nitrogen excretion

If tuatara are to remain in water balance, then amounts of water gained and lost must be equivalent over time. However, only one detailed study has been made of water balance, involving a laboratory colony of Stephens Island origin studied by Lance Hill in Sydney, Australia.[117] Total amounts of water gain (influx) and loss (efflux) increased markedly with temperature, but were reasonably balanced at each temperature (Figure 9.10).

The only measurable routes for water gain in Hill's animals were free water taken in by drinking, pre-formed water ingested in the diet (of chopped heart, mealworms and snails), and, to a small extent, water produced internally as a result of metabolism (metabolic water). There was no evidence of water absorption through the skin. The relative importance of different routes depended on factors including temperature and activity level (Figure 9.10A).[118] Drinking was always the main route of gain, increasing from 63% of total water intake at 10°C to 95% at 25°C.[119] Water in food was the next most important route, but the relative importance declined with increasing

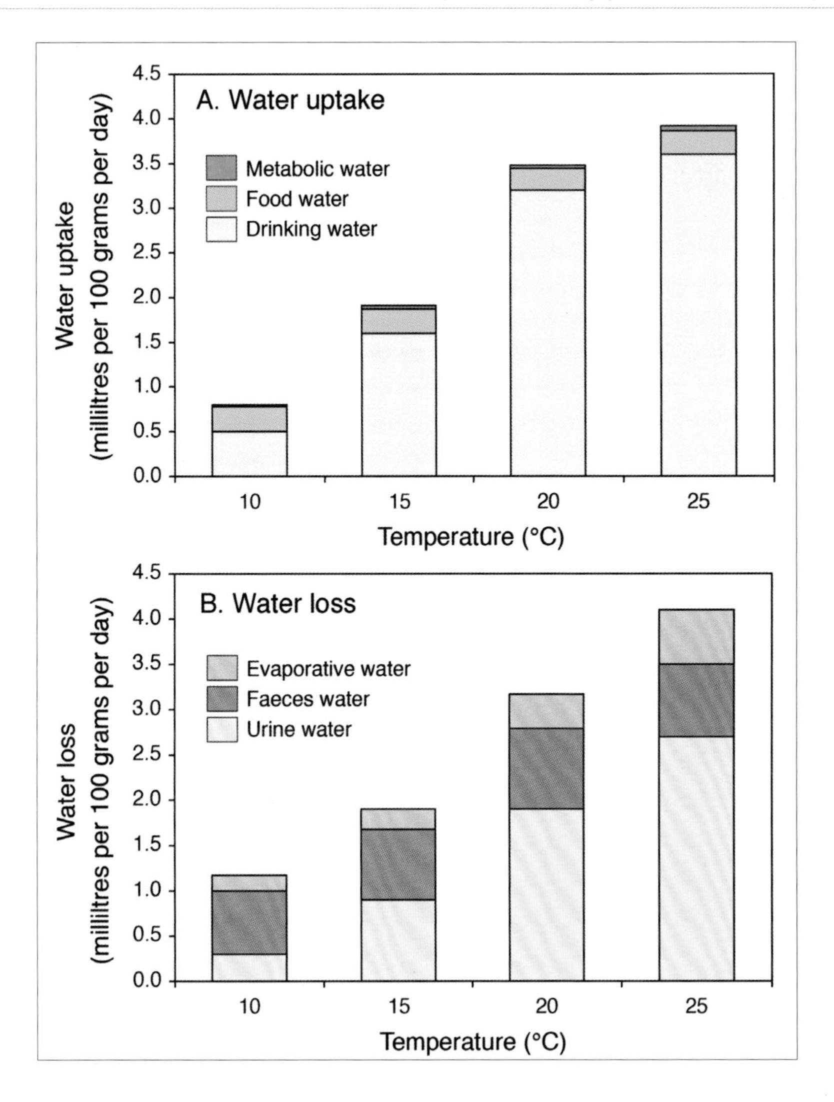

Fig. 9.10 Water fluxes in tuatara (a combined sample of adults and juveniles of Stephens Island origin) under laboratory conditions. A: water uptake. B: water loss.

Modified from Hill 1982.

temperature. Metabolic water accounted for less than 4% of total water intake at all temperatures tested.

In terms of water loss, the measurable routes were evaporation (through the skin and from the respiratory tract), urination and faecal production (Figure 9.10B). (Unlike some reptiles, tuatara have no salt glands, so water loss by this route is non-existent.) When results from a range of conditions were pooled, temperature had a marked effect on total water loss.[120] Faecal water loss was relatively high at 10°C but declined at warmer temperatures, whereas urinary and evaporative losses became increasingly important at warm temperatures. In partitioning experiments in dry air (0% relative humidity), evaporative loss through the skin exceeded evaporative loss from the respiratory system at temperatures at or below 20°C. However, at 25°C, evaporative loss from the respiratory system increased substantially to become double that through the skin. This increase in evaporative water loss from the respiratory system is probably due to the effect of temperature on metabolic rate and breathing rate, as discussed in earlier sections.

Fig. 9.11 Detail of moult skin being shed from a tuatara on Stephens Island. The old skin is shed in fragments, giving the animals a tatty appearance.

Photo: Alison Cree.

Fig. 9.12 Following moulting, the skin of tuatara has a much fresher, more vivid appearance. The last vestiges of old skin are still visible on this male tuatara on Stephens Island.

Photo: Alison Cree.

Using the same colony of animals, Hill also investigated the changes in urine production that occur with dehydration, which resulted from lack of access to free water for drinking.[121] Once tuatara had lost 10% of body weight by dehydration, urine flow was reduced, reflecting a reduction in glomerular filtration rate (the rate at which blood was filtered by the kidneys) and an increase in tubular reabsorption (the rate of water reabsorption by the kidney tubules). However, urine leaving the kidneys in the ureters did not become hyperosmotic to the plasma (in other words, like other reptiles,[122] tuatara are not able to concentrate osmotically active substances such as salts to higher levels than in the blood). Dehydrated tuatara reduced the volume of urine that they spontaneously voided, which probably also resulted from an increase in resorption of water from the bladder and/or cloaca. Overall, Hill's detailed study of water balance of tuatara under laboratory conditions revealed no major differences in water balance compared with other reptiles from equivalent habitats.[123]

The structure of the skin is relevant to water loss. Like lizards and snakes, tuatara have scaly skin with a derived manner of skin shedding (sloughing or ecdysis) not seen in turtles or crocodilians.[124] Sloughing in lepidosaurs involves the production of a specialised shedding complex that separates an 'outer epidermal generation' (the more mature, outermost layers of epidermis) from an 'inner epidermal generation' (the innermost layers of epidermis). However, some features of the process in tuatara may be ancestral to those in squamates.[125] In wild tuatara, the shedding of the outer epidermis in ragged pieces, usually in late summer, briefly gives the animals a tatty appearance (Figures 9.11, 9.12).[126] Nonetheless, evaporative water loss (presumably measured on a whole-animal basis) was apparently unaffected by shedding.[127] I observed that when juvenile tuatara were held in still air under conditions that were relatively warm and dry,[128] the rate of water loss through the skin averaged 1.4 milligrams per gram body weight per hour.[129] This rate was relatively high compared with diurnal New Zealand lizards or lizards elsewhere, of similar body mass, from mesic or arid environments and held under similar conditions. In the same study, cutaneous water loss from isolated pieces of dorsal trunk skin was increased about five-fold when lipids (fats) were removed, indicating that, as in other reptiles, lipids provide a major barrier to permeability of the skin to water. High rates of cutaneous and/or evaporative water loss appear to be a general phenomenon among nocturnal New Zealand lepidosaurs,[130] but the ultrastructural basis is unclear.

The kidneys of adult tuatara are paired, irregularly lobed structures located within the pelvis, dorsal to the cloaca.[131] In their internal structure, the kidneys appear similar to those of other reptiles.[132] The short duct or ureter that transports urine from each kidney opens onto a urogenital papilla in the wall of the cloaca.[133] The ureteral opening on the papilla is close to the opening from the reproductive duct on the same side. A capacious, long-necked urinary bladder lies ventral to the cloaca and opens into it.[134] The cloaca is itself a potential site of water exchange, whereas the contribution of the bladder in this regard is unclear.[135] Overall, the structure of the urinary system of tuatara is similar to that of lizards.

Tuatara produce a liquid urine containing various dissolved solutes plus semi-solid concretions. Although tuatara are predominantly uricotelic (meaning that their major nitrogenous waste is uric acid, as a semi-solid concretion), they also excrete significant amounts of urea, the relative proportions varying with diet and hydration state.[136] On a low-protein diet, uric acid accounted for 80.8%, urea 10.9% and ammonia 3.7% of total nitrogenous wastes.[137] On a high-protein diet, the uric acid component decreased to 64.5%, urea increased to 27.3%, and ammonia was essentially unchanged at 4.1% (small amounts of substances including creatinine, creatine and amino acids made up the balance). The increase in urea excretion is considered a metabolically efficient way of excreting excess nitrogen under conditions of free access to water. In another study by the same research group (protein level of diet not specified, but probably intermediate), loss of 10% of body weight by dehydration led to an increase in the uric acid component to about 80% and a decrease in the urea component to about 10% of total nitrogenous wastes.[138] Uric acid is less toxic and much less soluble than urea, so increased production of uric acid serves to conserve water.[139]

How do the findings from laboratory studies relate to the water balance experienced by tuatara in the wild? The first and most obvious point is that free water for drinking is often not available in nature (Chapter 6). Most islands on which tuatara survive today have little or no free-standing water. Thus, most wild tuatara probably have access to water only in food, in intermittent dew or in puddles that form after heavy rain. During summer droughts on Stephens Island, tuatara readily make use of trays of water provided by humans, and when rain finally falls tuatara often lie in and drink from puddles

Fig. 9.13 Viewed from a window in the field station, a tuatara on Stephens Island drinks from a puddle following the end of a summer dry spell.

Photo: Alison Cree.

(Figure 9.13). A lighthouse keeper on Cuvier Island (Repanga Island) implied that tuatara living near a creek used it for drinking as well as bathing, noting that when the creek was almost dry the tuatara were 'obliged to omit their regular swim'.[140] In captivity, tuatara also drink readily and immerse themselves when water is available.[141] For example, when T. E. Donne provided a pan of water to his seven captives, they 'promptly piled themselves [into the pan] and remained there night and day.'[142]

A second point about water balance in nature is that burrowing reduces the potential for water loss. On a summer's day at 25°C and 70% relative humidity, a 400 gram tuatara that emerges is estimated to lose water at nearly three times the rate experienced if it remained in a burrow, where air temperature is cooler (estimated at around 15°C) and humidity higher (estimated as near saturation).[143] This may explain why tuatara reduce their basking and remain in burrows during prolonged periods of warm dry weather.[144] Samples of urine collected from the bladder or cloaca of tuatara on Stephens Island following a dry period of several weeks apparently showed a similar change in the ratio of uric acid to urea to those dehydrated in the laboratory, implying that tuatara do become dehydrated in the wild.[145]

A final point about water balance in field situations is that there is no evidence that wild tuatara make deliberate use of seawater. Tuatara can certainly swim, at least for short periods until they tire: an early account reported specimens swimming freely in captivity with nostrils either above or below the water,[146] and more recently it was noted that a captive tuatara 'swam well though it disliked diving'.[147] However, there appear to be no records of wild tuatara voluntarily entering seawater. Dead or dying tuatara have occasionally been found at sea,[148] but these probably represent animals swept from their shoreline habitats in storms. Even if tuatara washed to sea could avoid drowning, they would probably die eventually from dehydration and salt loading.

Water flux in wild tuatara has been reported for non-gravid females from two populations in the Hen and Chickens Islands (on Lady Alice and Coppermine Islands).[149] Water flux was measured, in combination with field metabolic rates, by examining the turnover of doubly labelled water. The study was carried out during mild spring weather, with brief showers falling on about 15% of days. Estimates of total body water ranged among individuals from 69.5% to 82.3%, with no significant difference in mean values between islands.[150] Mean rates of water influx and water efflux were 20.7–22.6 millilitres per kilogram per day, values that were similar to those in the literature for reptiles from semi-arid or arid zones, but lower than those of reptiles from tropical zones.[151] These field values are somewhat lower than the turnover of 37.5 millilitres per kilogram per day estimated for tuatara in a captive colony where drinking water was presumably available.[152]

Light, vision and the lateral eyes

Tuatara are visually responsive animals. Their paired, lateral eyes have a similar structure to those of other reptiles, especially lizards[153] (Figure 9.14). (Tuatara, like many lizards, also have a well developed, median parietal eye, but this is not considered to be image-forming and is discussed separately.) Each lateral eye has a scale-clad upper and lower eyelid;[154] the lower lid, which covers most of the eye, is stiffened by a tarsal plate (an internal cartilaginous disc). A nictitating membrane ('third eyelid') can be drawn across the cornea from the corner nearest the snout to clean and lubricate the eye (see Chapter 7, Figure 7.19 for illustration). Secretions from the well-developed Harderian gland provide lubrication but, as in some lizards, a lachrymal gland ('tear gland') is lacking.

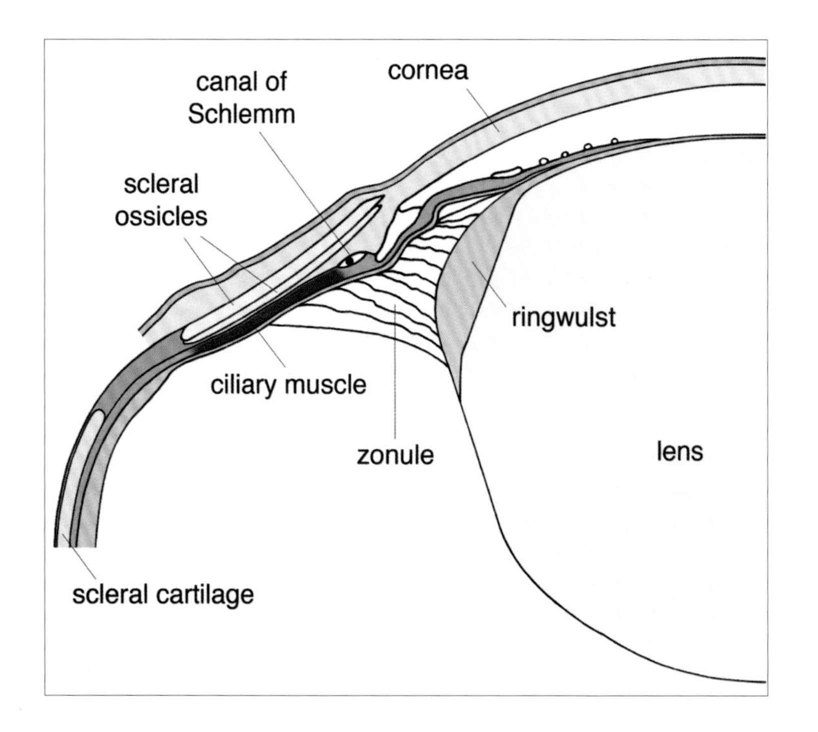

Fig. 9.14 Structure of the anterior segment of the lateral eye of a tuatara.

Modified, with arbitrary colouring, from Walls 1942 with additional labelling from Osawa 1898c.

BOX 9.2

The parietal eye: a device to help regulate basking and to orient using the sun?

On top of the head of a tuatara, in the mid-line under the skin, lies a structure that has generated enormous interest. This parietal eye[1] is often referred to in the popular literature as a third eye, although the term is somewhat misleading given that the parietal eye is not an image-forming structure. It is, however, potentially sensitive to light.

The parietal eye is one of two main components of a neuroendocrine structure called the pineal complex.[2] The parietal eye is the more superficial part, lying above the brain, beneath a gap in the roof of the skull known as the parietal foramen. The other main component of the pineal complex, the pineal gland, is a deeper structure.[3] Elements of a pineal complex are present to varying degrees in representatives of all vertebrate classes. Among early fossil reptiles, many species possessed a parietal foramen (implying also a parietal eye),[4] but the parietal eye has been lost from modern turtles, crocodilians and snakes. However, many lizards retain a parietal eye that in some species is at least as, if not better, developed than in tuatara. Thus, the well-developed state of the parietal eye in tuatara is not unique among reptiles, and its evolutionary significance has often been overrated.[5] Most other reptiles (and mammals, including humans) also retain a pineal gland, although this structure has been lost from crocodilians.[6]

One theory to explain the evolution of the pineal complex proposes that early vertebrates may have possessed more than one pair of photo-sensory structures. According to this theory, what we now call the lateral eyes became specialised over time for vision, whereas an ancillary pair became modified into a mid-line, asymmetrical pineal complex.[7] Of the two main components of the complex in modern lepidosaurs, the parietal eye has retained (or evolved) a more eye-like structure than the pineal gland.

The existence of the parietal eye in tuatara was established in 1886 by W. Baldwin Spencer.[8] In the late 1890s, Hugo Schauinsland reported on the development of the complex in embryos.[9] However, it is the exhaustive studies of Arthur Dendy from 1898, on the pineal complex of embryonic and adult tuatara, that have deservedly received the greatest attention.[10] The parietal eye arises in embryos as an outgrowth of the roof of the forebrain. It first appears slightly to the left of the mid-line, but by late in development is in a more or less mid-line position, appearing on the surface of the head as a white spot surrounded by a small black circle. The white spot indicates the presence of a lens (formed from the dorsal wall of the outgrowth) and the black border indicates the pigmented margin of the retina (formed from the lateral and ventral walls). The retina is differentiated into two layers, with the inner layer being the most pigmented.[11] A histological study in adults indicates that the pigment is probably melanin.[12] Late in embryonic development, a parietal nerve develops from the outer layer of the retina of the parietal eye,[13] extending to the brain.[14]

By hatching, the parietal eye of a tuatara sometimes remains visible externally as a dark spot beneath translucent skin covering the parietal foramen.[15] In other specimens, the parietal eye may be essentially invisible, although its position remains evident from a small rounded scale (or sometimes a weakly scaled region) lying a short distance in front of the first nuchal spine.[16] As juveniles develop, the pigmentation in the overlying skin increases, but to a variable degree. In adults, the skin over the parietal eye remains translucent, despite being covered with small and irregular scales.[17] Although there may be a slight depression in the skin over the parietal foramen, the scale pattern does not indicate the

Box 9.2 Fig. 1 The head of a juvenile tuatara of Stephens Island origin within about a month of hatching. The arrow points to a weakly scaled and poorly pigmented region, below which lies the parietal eye. This region often lies within a partial rosette of small scales.

Photo: collection of Alison Cree.

Box 9.2 Fig. 2 The head of a 7-year-old juvenile of Stephens Island origin, showing the weakly scaled depression (arrow) overlying the parietal eye. As tuatara age, the area becomes less obvious and may be hard to distinguish from patches of scar tissue that accumulate on the head.

Photo: Alison Cree.

underlying presence of the parietal eye to the extent seen in some lizards.[18] Beneath the parietal foramen, the parietal eye of adults remains highly organised. The foramen is filled with a plug of connective tissue that is gelatinous-like in life, and more transparent than the surrounding connective tissue.[19]

The pineal organ[20] of tuatara also arises during embryonic development as an outgrowth of the roof of the forebrain, appearing at a later stage than the parietal eye, and in a median position.[21] Although no lens thickening is formed, the histological features of the pineal organ are initially identical with those of the parietal eye, including the deposition of pigment in the inner layer of the wall.[22] In adult tuatara, bundles of nerve fibres from the wall of the pineal organ apparently converge with the nerve from the parietal eye.[23] Overall, the histological features of the pineal organ in adult tuatara correspond well with those of lizards and turtles.[24]

What can be said about the function in tuatara of this curious complex? Both the parietal eye and pineal organ contain what appear to be photoreceptive cells.[25] The parietal eye also contains a lens but, given the overlying skin and connective tissue, the parietal eye cannot form images in the way that the paired lateral eyes do.[26] Nevertheless, it is possible that the parietal eye, pineal organ or both are able to sense and respond to varying intensities of light. According to Dendy, light is able to pass through the skin and the parietal plug of tuatara.[27] However, limited attempts to explore the role of the parietal eye, including the effects of shielding, have yielded inconclusive results. In 1958, Robert Stebbins reported that shielding the parietal eye of an adult had no obvious effect on exposure to light, locomotory activity or use of the warm end of a thermal gradient.[28] In our laboratory, Elaine Whitworth observed no consistent effect of shielding the parietal eye on substrate temperatures selected by juveniles on a thermal gradient, and no effect on preference for light with or without a UVA component.[29]

Despite the limited investigations to date, the most likely scenario is that the pineal complex of tuatara functions in a similar way to that of lizards with an equivalent structure. Experiments on lizards show that both the parietal eye and the pineal organ are sensitive to light, and that overall the complex acts as a 'radiation dosimeter' (in effect, it helps to

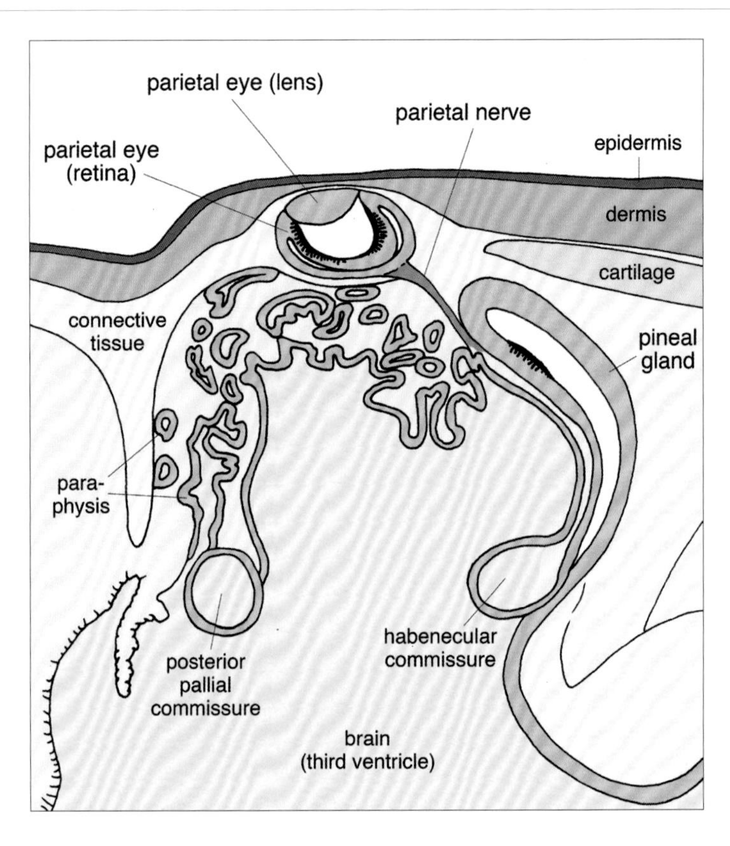

Box 9.2 Fig. 3 Diagram showing the location and components of the pineal complex in a late-stage embryo of Stephens Island origin. Modified, based on a sagittal view (in the vertical plane) for an embryo at stage R in Dendy 1899c, with additional labelling from Quay 1979. The black areas in the parietal eye and pineal gland are indicative of melanin; other colours are arbitrary.

regulate basking behaviour).[30] For example, when the pineal complex of green iguanas (*Iguana iguana*) is surgically removed, iguanas select lower body temperatures by day and higher temperatures at night.[31] Both components of the pineal complex of lizards produce melatonin, a vertebrate hormone known as the 'chemical expression of darkness',[32] although in lizards the main source of the melatonin that enters the bloodstream appears to be the pineal organ.[33] Secretion of melatonin, which is sensitive to both light and heat, helps to regulate daily cycles in selected body temperature as well as seasonal cycles of reproduction in lizards.[34] In some species, including the sleepy lizard of Australia (*Tiliqua rugosa*), the parietal eye also plays a role in homing behaviour, possibly by analysing patterns of polarised light in the sky.[35]

Tuatara are known to secrete melatonin (which is assumed to come from the pineal complex), and blood concentrations are highest at night, as in lizards.[36] However, in other respects, our knowledge of the function of the pineal complex in tuatara is meagre. Modern techniques, including electrophysiological studies as well as surgical and hormonal manipulations, could be fruitfully applied to resolve questions that have existed for over a hundred years. Studies on the parietal eye seem likely to be most rewarding with young juveniles, in which the parietal eye is less concealed by overlying structures than in adults, and in which activity is more diurnal.[37]

Structural features indicate that the curvature of the lens can be sharpened, implying good powers of accommodation (the ability to focus on objects at different distances, an ability typical of diurnal species), despite the ciliary muscle being rather weakly developed. Examples of such structural features are the presence of scleral ossicles, or thin overlapping plates of bone that form a circlet around the lens, and a well-developed ringwulst, or annular (ring-shaped) pad of the lens. The presence of scleral ossicles is an ancestral feature of amniotes also retained in turtles and lizards, whereas the ossicles have been lost from snakes and crocodilians.[155] The presence of a fovea (an area of the retina with an increased density of photoreceptor cells, giving enhanced visual acuity) is consistent with diurnal origins.[156] The vasculature, innervation and musculature of the eye are generally similar to those of lizards, although tuatara retain two processes to a muscle that attaches to the eyeball, whereas lizards have only one.[157]

Overall, the lateral eyes have many features suggesting that tuatara have evolved from diurnal ancestors.[158] However, some modifications for nocturnality are present. These include: the large size of the eyeball, with its relatively large corneal diameter (allowing for a large pupil);[159] the shallowness of the fovea;[160] and the colourless nature of the oil droplets in the visual cells.[161]

Tuatara exhibit eye-shine at night, which some field biologists have taken to imply the existence of a tapetum lucidum (literally 'shining carpet', a layer within or behind the retina that reflects light back through the retinal photoreceptors).[162] This inference is consistent with an early description of the fundus oculi (the concave interior of the eye, including the retina, choroid and other components) in which a tapetal layer is mentioned.[163] However, there are later reports that tuatara lack a tapetum,[164] so the structural basis for eye-shine requires clarification. Each lateral eye also possesses a circular pupil (Figure 9.15) that closes to a near-vertical slit in bright light,[165] a design considered well suited to nocturnal reptiles that also bask in the sun.[166] The nature of the visual receptors in the retina remained unresolved for many years, with different authors recognising various types of cones, or cones that have become 'physiological rods'.[167] A recent ultrastructural study confirms the presence of at least three types of cone-like photoreceptors, with some rod-like features considered advantageous to nocturnality.[168]

The most recent studies on the visual system of tuatara have focused on performance in the

Fig. 9.15 The large, circular pupil makes the eye of this adult male tuatara on Stephens Island appear almost black at night. By day, the pupil reduces to a near-vertical slit.

Photo: Marcus Simons.

laboratory. As in other reptiles, tuatara track moving objects primarily by head and eye movements rather than whole body movements.[169] The lateral eyes can accommodate by a surprisingly large amount (eight dioptres), and are able to focus independently,[170] but whether tuatara have colour vision is unclear. Visual acuity during prey capture is considerable, even under low light intensity,[171] but a juvenile tuatara showed no response to prey in total darkness.[172] Tuatara can be trained to discriminate visual stimuli, and do so up to a critical flicker-frequency (the frequency of an intermittent light pulse) similar to that in other reptiles (above this frequency, the light is presumably perceived as being continuous).[173]

As an aside, several selective pressures have been proposed for the evolution of nocturnality in tuatara. These include the availability of nocturnal invertebrates as prey,[174] the avoidance of predatory diurnal raptors[175] and the avoidance of high daytime temperatures.[176] It is possible that these and other factors, such as the need to maintain water balance, have acted in combination.[177] Ultimately, however, it is not possible to make definitive statements about the timing and reasons for the evolution of nocturnal behaviour in tuatara.

Concluding remarks and ideas for further research

Tuatara provide an exceptional example of the difficulty of classifying some reptiles using a strict dichotomy of diurnal versus nocturnal. Furthermore, although tuatara are in part diurnal, they bask less avidly than opportunity provides, even at the southern (coolest) end of their current latitudinal range. It remains possible that heat-seeking behaviour was more effective in the colder habitats of southern New Zealand where tuatara once existed. The lower temperature limits for survival, including whether or not freezing can be tolerated, remain unknown.

The ways in which thermoregulatory behaviour of tuatara might vary among different life-history stages have not been explored. It is conceivable, for example, that hatchlings (which have often experienced transiently high temperatures as embryos) and gravid females have higher thermal preferences than adult males. In terms of voluntary exposure to warm temperature, there is also some inconsistency between studies. The highest body temperature reported for a wild tuatara in the field is 30.1°C, compatible with values in other studies for thermal preference in some small juveniles on a thermal gradient. However, in another study, the mean body temperature at the voluntary maximum was only 28.3°C, with the mean panting threshold at 31.0°C. Whether these apparent discrepancies arise from genuine biological variation (e.g. from acclimation, or differences among life-history stages) or technical limitations (e.g. inaccurately calibrated equipment) requires resolution.

Although activity temperatures and thermal preference in tuatara are low by the standards of many overseas reptiles, they are not unique among New Zealand reptiles. In particular, activity temperatures in tuatara are similar to those for New Zealand geckos with nocturnal habits, which should not be a surprise given that these species often live in the same habitats and emerge in similar weather conditions. Thus, and contrary to a view from the 1950s, studies of tuatara do not necessarily inform us about the thermal biology of ancient rhynchocephalians. Similarly, inferences that the relative simplicity of lung structure in tuatara might be characteristic of ancestral lepidosaurs should be treated with caution.

Laboratory studies dating from the 1970s to the 1990s have provided a broad indication of the effects of temperature on oxygen consumption, ventilatory frequency and heart rate in tuatara. In general, rates of these processes seem to reach a maximum between body temperatures of about 25°C and 30°C. However, little is known about how acute changes in temperature affect the physiological performance of unrestrained tuatara. For example, optimum temperatures for locomotion (which have been widely studied in lizards) are unreported for tuatara. The constraints that activity and dehydration place on respiratory and cardiac performance at warm temperatures are also unclear. There is already some indication that rates of water loss through the skin may be relatively high in tuatara compared with diurnal lizards, and that dehydration can occur in natural environments.

Studies to date on whole-animal physiology in tuatara have often focused independently on topics such as thermoregulation, cardiac function, lung function or water balance. In addition, many studies have been performed in laboratory situations, with very few animals, often only adults. There is a need for integrated studies that consider possible trade-offs in performance among different aspects of body function (for example, between lung function and locomotory activity, including both running and burrowing). The ability to carry out such studies in field situations has been revolutionised for other reptiles over recent decades by the development of miniaturised, computer-assisted recording devices, and by other approaches of relatively low invasiveness (for example, the doubly-labelled water technique for measuring field metabolic rates and water turnover). These approaches deserve further application in tuatara. Such studies will, however, be incomplete from an evolutionary perspective if they neglect juveniles, given that the selective advantages of physiological features may vary with life-history stage.

In addition to furthering our understanding of diversity among reptiles, continued studies of physiological function in tuatara will address a pressing conservation imperative. New Zealand is not immune to global climate change. Mainland areas of northern New Zealand that are close to islands inhabited by tuatara are predicted to become warmer over coming decades, with changing rainfall patterns contributing to possible increases in the incidence and severity of droughts (see Chapter 10, Box 10.7). The significance of such changing environments will be important to explore, not only for adults and embryos but also for hatchlings and juveniles. Small, young tuatara have a high surface-area-to-volume ratio and thus a relatively high risk of dehydration for at least the first few years of life.

Despite speculation for over a hundred years, the role of the parietal eye in tuatara remains clouded. The most likely scenario is that this structure is light sensitive, and operates, as in more closely studied lizards, to help regulate basking and to orient using polarised light from the sky. The structure is perhaps of greatest importance to hatchling tuatara, in which pigmentation in the overlying skin is incomplete and in which activity is more strictly diurnal than in adults. Insofar as the lateral eyes are concerned, research examining the structural basis for eye-shine, the molecular evolution of visual pigments, and the significance of ultraviolet light for communication will be of interest. Given the low temperature preference of tuatara, and lack of annual reproduction in females, studies of the interactions between cycles in light and temperature with secretion of melatonin from the pineal gland, and with reproduction, also hold potential interest.

Endnotes to Chapter 9

See the bibliography for full references to articles on tuatara, which are cited here at first mention by surname and year only. References on other topics are cited here in full at first mention, and then using 'op. cit.'. Except where required for clarity, the convention of parentheses around years has been dispensed with for economy of space.

1 A Party of Officers of the 58th Regiment 1852.
2 Werner and Whitaker 1978; Carmichael and Gillingham 2004.
3 Werner and Whitaker 1978.
4 Werner and Whitaker 1978. See also Carmichael and Gillingham 2004.
5 Saint Girons et al. 1980.
6 Saint Girons et al. 1980; Carmichael and Gillingham 2004.
7 Saint Girons et al. 1980.
8 Depth within the burrow at which temperature was measured was not specified.
9 Saint Girons et al. 1980; see also Saint Girons 1980.
10 Werner and Whitaker 1978.
11 Standard error: 1.0°C.
12 Standard error: 0.5°C; Cree et al. 1990a.
13 Barwick 1982.
14 Wilson and Lee 1970.
15 Barwick 1982.
16 Tyrrell 2000.
17 Besson and Cree 2010. See also Besson 2009.
18 Measured as dorsal skin temperature over the posterior abdomen with a non-contact, infrared thermometer; in small juveniles, this temperature closely matches cloacal temperature measured with a thermocouple (Hare et al. 2007; Besson and Cree 2010). However, in juveniles that have been basking at the entrance to their burrows, the snout may sometimes be slightly warmer (Mello, R. S. R. 2010. Adjustment of juvenile tuatara (*Sphenodon punctatus*) to the Orokonui Ecosanctuary climate: operative temperatures, emergence behaviour and growth rate. Unpublished Postgraduate Diploma of Science dissertation, University of Otago, Dunedin).
19 Besson and Cree 2010.
20 Mello 2010, op. cit.
21 Besson and Cree 2011.
22 Newman et al. 1979; Walls 1983. See also Newman 1977.
23 Bogert 1953a, b (see also Dawbin 1949); Thompson and Daugherty 1998.
24 Werner and Whitaker 1978; Saint Girons 1980; Saint Girons et al. 1980; pers. obs. (raw data from Cree et al. 1990a).
25 Werner and Whitaker 1978 for Stephens Island; Tyrrell 2000 for northern islands.
26 Werner and Whitaker 1978.
27 Barwick 1982.
28 Thompson et al. 1992.
29 Tyrrell 2000.
30 Tyrrell 2000.
31 Corkery et al. 2010.
32 Juveniles from 24.1°C are likely or known to have been male; those from 21.5°C, female (unpubl. obs.; see also Chapter 8, Box 8.6).
33 Figure 9.5 shows an analysis of raw data from Whitworth 2006. Cloacal temperatures were measured with a thermocouple for juveniles (with no parietal patch) from a single clutch of eggs. Repeated-measures ANOVA confirmed a significant effect of time of day ($F_{5,40} = 8.819, p < 0.001$), but no interaction between time of day and egg incubation temperature ($p = 0.158$). The same conclusions were drawn from analysis of skin temperatures measured with an infrared thermometer (Whitworth 2006).
34 Besson and Cree 2010.
35 See, for example, Besson 2009 for data on geckos and skinks.
36 Besson and Cree 2011.
37 Barwick 1982.
38 Unpubl. data; Nicola Nelson, Victoria University of Wellington, pers. comm. 28 April 2004.
39 Underwood, H. 1992. Endogenous rhythms. In: Gans, C. and Crews, D. (eds). *Biology of the Reptilia. Vol. 18. Physiology E. Hormones, brain and behavior.* University of Chicago Press, Chicago, pp. 229–297; Rock, J., Andrews, R. M. and Cree, A. 2000. Effects of reproductive condition, season, and site on selected temperatures of a viviparous gecko. *Physiological and Biochemical Zoology* 73: 344–355.
40 Nelson et al. 2004d.
41 Heatwole 1982.
42 $p < 0.01$; standard errors were 0.34°C for both mean temperatures.
43 Heatwole 1982.
44 Body temperature was measured in the cloaca. Bruce Firth, University of Adelaide, pers. comm. 6 January 2004.
45 Wilson and Lee 1970. Body temperature was measured with a thermocouple inserted either 7 cm through the cloaca or 11 cm down the oesophagus.
46 Dawbin 1962a; see also Dawbin 1974.
47 Newman et al. 1979.
48 Stebbins 1958. On the other hand, Hill (1982, p. 186) reported that captive animals remained 'coordinated and capable of locomotion even at 2°C', but whether this value was for body or air temperature is unclear.
49 Besson and Cree 2011.
50 Desser 1979.
51 Voituron, Y., Storey, J. M., Grenot, C. and Storey, K. B. 2002. Freezing survival, body ice content and blood composition of the freeze-tolerant European common lizard, *Lacerta vivipara*. *Journal of Comparative Physiology B* 172: 71–76; Dinkelacker, S. A., Costanzo, J. P. and Lee, R. E. Jr. 2005. Anoxia tolerance and freeze tolerance in hatchling turtles. *Journal of Comparative Physiology B* 175: 209–217.
52 For comparisons of voluntary thermal maxima and panting thresholds, see Heatwole 1982; for field body temperatures and critical thermal minima in diurnal lizards, see: Huey, R. B., Deutsch, C. A., Tewksbury, J. J., Vitt, L. J., Hertz, P. E., Pérez, H. J. A. and Garland, T. Jr. 2009. Why tropical forest lizards are vulnerable to climate warming. *Proceedings of the Royal Society B* 276: 1939–1948.
53 Bogert 1953b.
54 Werner and Whitaker 1978; Cree 1994; Rock et al. 2000, op. cit.; Besson and Cree 2011.
55 Price, G. 2009. Mesozoic climates. In: Gornitz V. (ed.) *Encyclopedia of paleoclimatology and ancient environments.* SpringerReference (http://www.springerreference.com).

Springer-Verlag Berlin Heidelberg. doi: 10.1007/SpringerReference_77402 2011-01-31 23:00:00 UTC.
56 Dawbin 1962a; Crook 1975; Heatwole 1982; Gans 1983; Cree and Daugherty 1990a; Cree 1994.
57 Wilson and Lee 1970; Cartland and Grimmond 1994. However, metabolic rate was unaffected by a fish-oil supplement in the diet (Blair et al. 2000b).
58 Birchard et al. 2006.
59 Terezow et al. 2008.
60 Wilson and Lee 1970.
61 Hill 1982; Wells et al. 1990; Cartland and Grimmond 1994. The Q_{10} value, which indicates the rate of change over a 10°C interval, is often between 2 and 3 for physiological processes that are dependent on enzymatic reactions.
62 Cartland and Grimmond 1994.
63 Milligan 1924; Wilson and Lee 1970.
64 Thompson and Daugherty 1998.
65 Thompson and Daugherty 1998.
66 Cartland and Grimmond 1994.
67 Tyrrell 2000; Tyrrell et al. 2000.
68 $p < 0.05$; standard errors were 0.04 and 0.03, respectively (Tyrrell 2000).
69 The predicted rate, derived from data for 55 species of reptiles, is from: Nagy, K. A., Girard, I. A. and Brown, T. K. 1999. Energetics of free-ranging mammals, reptiles, and birds. *Annual Review of Nutrition* 19: 247–277.
70 Tyrrell et al. 2000.
71 Christian, K. and Green, B. 1994. Seasonal energetics and water turnover of the frillneck lizard, *Chlamydosaurus kingii*, in the wet-dry tropics of Australia. *Herpetologica* 50: 274–281.
72 The review of Perry (1998) suggested that tuatara lack bronchi. However, original sources describe the presence of short bronchi; see Günther 1867; Newman 1878; Osawa 1897; Gabe and Saint Girons 1964a; Willnow and Willnow 1976.
73 Milani 1894; Willnow and Willnow 1976. See also Grigg and Simons 1972.
74 Günther 1867; Newman 1878.
75 Willnow and Willnow 1976.
76 Perry 1998.
77 Osawa 1897; Gabe and Saint Girons 1964a.
78 Gans and Clark 1978; Wells et al. 1990.
79 Wells et al. 1990.
80 Wells et al. 1990.
81 A Party of Officers of the 58th Regiment 1852.
82 Wells et al. 1990.
83 Milligan 1924.
84 Newman 1878 (see also A Party of Officers of the 58th Regiment 1852, and Buller 1879, for indications of unspecified periods under water).
85 Farrell et al. 1998.
86 See also Webb et al. 1974 for squamates.
87 Farrell et al. 1998; Hicks, J. W. 1998. Cardiac shunting in reptiles: mechanisms, regulation, and physiological functions. In Gans, C. and Gaunt, A. S. (eds). *Biology of the Reptilia. Vol. 19. Morphology G.* Society for the Study of Amphibians and Reptiles. Contributions to Herpetology Vol. 14, Ithaca, New York, pp. 425–483.
88 Günther 1867; Newman 1878; O'Donoghue 1920. The heart also possesses a distinct conus arteriosus (= bulbus cordis; O'Donoghue 1920; Simons 1965; Hart 1969), a region that is well developed in amphibians but apparently absorbed within the ventricle during development in most reptiles, although exceptions exist among turtles and lizards (Farrell et al. 1998).
89 White 1959; Simons 1965.
90 White 1959; Simons 1965; Meinertz 1966, reviewed by Webb et al. 1974; Farrell et al. 1998.
91 Grigg and Simons 1972.
92 Wilson and Lee 1970. The difference between rates of heating and cooling is known as thermal hysteresis.
93 O'Donoghue 1920; Adams 1953.
94 Wilson and Lee 1970; McDonald and Heath 1971.
95 Wilson and Lee 1970.
96 McDonald and Heath 1971.
97 Wilson and Lee 1970.
98 Wilson and Lee 1970.
99 Wells et al. 1990.
100 Brittain 1988.
101 Wells et al. 1983; Tetens et al. 1984; Weber et al. 1989.
102 Low co-operativity is apparent from the hyperbolic (rather than sigmoidal) shape.
103 Wells et al. 1983; Brittain 1988.
104 Abbasi et al. 1988.
105 Gorr et al. 1998.
106 Gorr et al. 1998.
107 Desser 1978. For information on ultrastructure, see Desser and Weller 1979a, b.
108 Wells et al. 1983.
109 Wells et al. 1990.
110 Thompson and Daugherty 1998.
111 Gans 1983.
112 Pough, F. H. 1980. The advantages of ectothermy for tetrapods. *American Naturalist* 115: 92–112.
113 Günther 1867; Milani 1894; see also Willnow and Willnow 1976; Dawbin 1982b; Perry 1998.
114 Perry 1998.
115 Newman 1878; Dawbin 1974; Pepperell 1982.
116 Perry 1998.
117 Hill 1982.
118 Hill 1982.
119 Hill 1982.
120 Hill 1982. Humidity or saturation deficit was unspecified.
121 Hill 1982.
122 Schmidt-Nielsen, K. 1997. *Animal physiology: adaptation and environment.* 5th edn. Cambridge University Press, Cambridge.
123 Hill 1982.
124 Gabe and Saint Girons 1964a; Maderson 1968; Peterson 1984; additional references in Dawbin 1982b.
125 Maderson 1968; Alibardi 2003; Alibardi and Maderson 2003a, b. See also Alibardi and Toni 2006a, b; Alibardi and Gill 2007.
126 Pers. obs. For skin shedding in captivity, see Mitteilungen 1971.
127 Hill 1982.
128 Temperature was 22 ± 1.5°C and relative humidity 25 ± 7%.
129 The standard error for the mean rate of water loss was 0.2. The 3-year-old juveniles had a mean mass of 40.7 g. Cree, A. and Daugherty, C. H. 1991. *High rates of cutaneous water loss in nocturnal New Zealand reptiles.* Unpublished report to the Department of Conservation, Wellington. 27 p.
130 Cree and Daugherty 1991, op. cit.; Neilson, K. 2002. Evaporative water loss as a restriction on habitat use in endangered New Zealand endemic skinks. *Journal of Herpetology* 36: 342–348.
131 Günther 1867; Osawa 1897. Development of the embryonic kidneys is reviewed by: Fox, H. 1977. The urogenital system

of reptiles. In: Gans, C. and Parsons, T. S. (eds). *Biology of the Reptilia. Vol. 6. Morphology E.* Academic Press, London, pp. 1–157 + addendum pp. 463–464.

132 Osawa 1897; Gabe and Saint Girons 1964a; Hill 1982.

133 Specifically, in the urodaeum, the middle chamber of the cloaca.

134 Specifically, the urinary bladder opens into the coprodaeum, the first chamber of the cloaca from the anterior end (Günther 1867; Gadow 1887; Osawa 1897; Gabe and Saint Girons 1964a).

135 Schmidt-Nielsen and Schmidt 1973; Hill 1982.

136 Dawbin and Hill 1969; Hill and Dawbin 1969.

137 Hill and Dawbin 1969.

138 Hill 1982.

139 Another study to examine renal function in tuatara (a single adult) was that of Schmidt-Nielsen and Schmidt (1973). Their tuatara was maintained in an incubator at 15°C, in conditions that were described as 'normal' (held overnight in a pan with water in the base), 'dehydrated' (same conditions but without water for 24 h) or 'waterloaded' (receiving aliquots of water by stomach tube before and during the experiment). Some results from the study are puzzling and differ from those of Hill (1982). For instance, there was no change in glomerular filtration rate with dehydration or water loading, tubular water reabsorption declined during both dehydration and water loading, and there was no evidence of bladder water reabsorption. Given the mild holding conditions and the absence of data showing loss of body weight in the 'dehydrated' situation, it seems likely that their animal was relatively well hydrated under all treatments.

140 D. J. Grindlay to Under Secretary for Internal Affairs, 17 March 1919, ANZ IA 1 46/18/8 pt 1.

141 Buller 1877, 1879; Newman 1878.

142 Donne 1942.

143 Hill 1982.

144 Pers. obs.; see also Barwick 1982.

145 Hill 1982.

146 Newman 1878.

147 Wojtusiak 1973, p. 98. These tuatara were presumably in fresh water, although when captive animals were offered a trough filled with seawater, Buller (1879) noted that 'they have taken to it just as readily as when the bath was of fresh water'.

148 Anon. 1996; Nicola Nelson, Victoria University of Wellington, pers. comm. 15 March 2010.

149 Tyrrell 2000; Tyrrell et al. 2000.

150 Tyrrell 2000.

151 See Tyrrell 2000 for discussion.

152 Hill 1982.

153 For original descriptions of the lateral eyes, see Günther 1867; Osawa 1898c, d; Virchow 1901; Bage 1912; Walls 1942; Vilter 1951a, b, c. For reviews, see Underwood 1970; Robb 1977; Dawbin 1982b. There is passing mention of tuatara in chapters by Repérant, J., Rio, J.-P., Ward, R., Hergueta, S., Miceli, D. and Lemire, M. Comparative analysis of the primary visual system of reptiles (pp. 175–240) and Ulinski, P. S., Dacey, D. M. and Sereno, P. S. Optic tectum (pp. 241–366). Both in: Gans, C. and Ulinski, P. S. (eds). 1992. *Biology of the Reptilia. Vol. 17. Neurology C. Sensorimotor integration.* University of Chicago Press, Chicago.

154 Mann 1932.

155 Tuatara and lizards also have no ciliary processes alongside the lens, unlike turtles and crocodilians, but the loss of these structures in lepidosaurs is of obscure significance (Underwood 1970).

156 Walls 1942; Hall 2008.

157 The retractor bulbi: Walls 1942; Underwood 1970.

158 Walls 1942; Underwood 1970.

159 Hall 2008.

160 Walls 1942; Underwood 1970.

161 Walls 1942. See also: Peterson, E. H. 1992. Retinal structure. In: Gans, C. and Ulinski, P. S. (eds). 1992. *Biology of the Reptilia. Vol. 17. Neurology C. Sensorimotor integration.* University of Chicago Press, Chicago, pp. 1–135.

162 Robb 1977; Dawbin 1982b.

163 Mann 1933.

164 For statements that, among reptiles, 'only the crocodile and the alligator have a tapetum', and that some vertebrates can have a 'bright reflection' without a conventional tapetum, see: Schwab, I. R., Yuen, C. K., Buyukmihci, N. C., Blankenship, T. N. and Fitzgerald, P. G. 2002. Evolution of the tapetum. *Transactions of the American Ophthalmological Society* 100: 187–200. For a report that tuatara 'do not have a tapetum, whatsoever', see: Schwab, I. R. 2005. Mirror, mirror, on the wall … *British Journal of Ophthalmology* 89: 528. No source reference was provided.

165 Johnson 1927; Mann 1932.

166 Walls 1942.

167 Bage 1912; Mann 1933; Walls 1942; Vilter 1951a, b, c; Underwood 1970; Peterson 1992, op. cit.; Meyer-Rochow 1988; Meyer-Rochow and Teh 1991.

168 Meyer-Rochow et al. 2005.

169 Ireland and Gans 1977.

170 Schmid et al. 1992. Given the position of the eyes, Mann (1933) considered that binocular vision was not possible.

171 Meyer-Rochow and Teh 1991.

172 Meyer-Rochow 1988.

173 Woo et al. 2009.

174 Hazley 1982.

175 Saint Girons 1985.

176 Underwood 1970.

177 Meyer-Rochow and Teh 1991; see also Gans 1983.

Endnotes to boxed text

For explanation of citation format in box endnotes, see endnotes to the main text.

Box 9.1

1 A Party of Officers of the 58th Regiment 1852.
2 W. Marshall to the Secretary of Internal Affairs, 1 July 1928. ANZ M 1 25/611 pt 2.
3 Drummond 1917.
4 Nagy, K. A. 2005. Field metabolic rate and body size. *Journal of Experimental Biology* 208: 1621–1625.

Box 9.2.

1 Also known as the parapineal organ, or in older literature as the 'pineal eye'.
2 Tosini, G. 1997. The pineal complex of reptiles: physiological and behavioral roles. *Ethology Ecology and Evolution* 9: 313–333.
3 Also known as the pineal organ or epiphysis cerebri.
4 Edinger, T. 1955. The size of the parietal foramen and organ in reptiles. A rectification. *Bulletin of the Museum of Comparative Zoology at Harvard College* 114: 1–34.
5 Bogert 1953b.
6 Quay 1979; Tosini 1997, op. cit.
7 Dendy 1911; Walls 1942, p. 338; Bogert 1953b.
8 Spencer 1886a, b, c.
9 Schauinsland 1898a, 1899. See also Schauinsland 1903a.
10 Dendy 1898, 1899b, c, d, 1911. For a summary, see Dendy 1910.
11 In this respect, the parietal eye contrasts with the lateral eyes, reflecting the different ways in which the retinas are formed. See also Ung and Molteno 2004; Schwab and O'Connor 2004, 2005.
12 Gabe and Saint Girons 1964a.
13 Dendy 1899c.
14 Dendy 1910.
15 Thomas 1890. Schauinsland (1898a, 1899, 1903a) also considered the parietal eye still visible in hatchlings. For another colour photograph, see Daugherty et al. 1990a.
16 Dendy 1899b; Dawbin 1962a.
17 Dendy 1911. Spencer (1886a, c) wrote of a 'general transparency' or 'absence of pigment' in the skin above the parietal eye, but the age of his specimens is unclear.
18 Spencer 1886a, b, c.
19 Dendy 1911.
20 Termed 'parietal stalk' in Dendy's earlier papers, and 'pineal sac' in his later ones.
21 The two structures may form from the same outgrowth (Schauinsland 1898a, 1903a; Dendy 1910; Quay 1979).
22 Dendy 1899c. However, Ung and Molteno (2004) observed no discernable pigment in the pineal gland of a late-stage embryo or neonate.
23 Dendy 1910.
24 Gabe and Saint Girons 1964a.
25 Dendy 1910, 1911; Quay 1979.
26 Spencer 1886b; Dendy 1910.
27 Dendy 1910, 1911.
28 Stebbins 1958. Unexpectedly, a control (sham-shielded) animal did show some change in temperature selection but was also affected by mild disease, exacerbating the difficulty of drawing conclusions from so few animals. Another study reported no sign of orientation by an adult to a source of infrared radiation at night, but without comparing responses to a parietal-shielded situation (Wojtusiak 1973).
29 Whitworth 2006. The possibility that the pineal complex is involved in calcium metabolism of tuatara has also been raised (Hazley 1998).
30 Underwood, H. 1992. Endogenous rhythms. In: Gans, C. and Crews, D. (eds). *Biology of the Reptilia. Vol. 18. Physiology E. Hormones, brain and behavior*. University of Chicago Press, Chicago, pp. 229–297; Tosini 1997, op. cit.
31 Tosini, G. and Menaker, M. 1996. The pineal complex and melatonin affect the expression of the daily rhythm of behavioral thermoregulation in the green iguana. *Journal of Comparative Physiology* 179: 135–142.
32 Reiter, R. J. 1991. Melatonin: the chemical expression of darkness. *Molecular and Cellular Endocrinology* 79: C153–158.
33 Underwood 1992, op. cit.; Tosini 1997, op. cit.
34 Tosini 1997, op. cit.; Lovern, M. B. 2011. Hormones and reproductive cycles in lizards. In: Norris, D. O. and Lopez, K. H. (eds). *Hormones and reproduction of vertebrates. Vol. 3. Reptiles*. Academic Press, London, pp. 321–353.
35 Freake, M. J. 2001. Homing behaviour in the sleepy lizard (*Tiliqua rugosa*): the role of visual cues and the parietal eye. *Behavioral Ecology and Sociobiology* 50: 563–569.
36 Cockrem et al. 1988; Firth et al. 1988, 1989.
37 Whitworth 2006.

PART THREE

FUTURE SURVIVAL

CHAPTER TEN

Conservation: past, present and future

During the 19th century, the disappearance of tuatara from all wild habitats seemed imminent to many biologists. Today, however, and despite past mistakes, New Zealanders are passionately committed to the survival of tuatara, seeing the species as an important animal in its own right and as a symbol of a healthy native ecosystem.

Introduction

When Albert Günther described the tuatara as a rhynchocephalian species in 1867, extinction of this 'new' reptilian curiosity seemed almost inevitable. Fossil evidence revealed that tuatara had once existed on the North and South Islands; Māori tradition recorded that tuatara had also been eaten there. But live specimens could reliably be found only on Karewa Island, a tiny (5-hectare) island in the Bay of Plenty free of introduced mammals. Would the species survive further onslaught? Günther was not optimistic, concluding that tuatara might 'one day be enumerated among the forms of animal life which have become extinct within the memory of man'.[1]

Fortunately, tuatara are hardy creatures. Thanks to a long lifespan, a vigorous recovery programme and an element of luck, tuatara are found today on about 37 offshore islands, and their continued existence – with provisos relating to the unsettling issue of climate change – seems probable. This chapter traces the history of conservation efforts for tuatara, from the mid-1800s through to 2011.[2] For developments since 2012, see Chapter 11.

Mounting pressure from collection: 1867–1885

For the first few decades after 1831, following J. E. Gray's scientific description of the tuatara as a lizard, there was a modest, and manageable, interest in collection (Chapter 5). However, the announcement from Günther that the tuatara should instead be considered the sole living representative of a separate order of reptiles changed that essentially overnight. Suddenly, everyone wanted a tuatara. Live specimens were sought for display and observation, and preserved specimens were essential raw material for anatomical studies.

Initial collecting for New Zealand scientists was by locals, apparently without financial reward. Walter Buller, for instance, obtained specimens from his brothers-in-law, the Mairs, and indirectly via the newly established lighthouse tender.[3] Many locals also kept tuatara as pets. E. C. Gold-Smith, the District Surveyor for Tauranga, brought 20 home after surveying Karewa Island in 1884 and turned them out in his garden, and Gilbert Mair also kept one for many years in Tauranga[4] (see Chapter 5 for further examples). Scientists and politicians together ensured a steady supply of tuatara during the 1860s and 1870s to British institutions, including the British Museum (later the Natural History Museum), the Zoological Society of London and Kew Gardens.[5] These early collections came mainly, if not exclusively, from Karewa, Moutoki and North Brother Islands.

Within a few years, opportunities for financial gain from collection were being realised. Enterprising locals included some Māori, who reportedly sold a dozen large 'green lizards' (thought to be tuatara) from East Island (Whangaokeno Island) to the Sydney Museum; the stuffed specimens reached four shillings each, according to a newspaper quoted by Buller in 1877. A boatman, Colin Norris, maintained a large collection of live tuatara in Tauranga from the 1880s, some of which were sold or made available to local and visiting scientists[6] (discussed further below in the section on management of collection).

Professional natural-history collectors were also soon on the scene. Among these was Henry Augustus Ward, a collector from New York and former university professor, who owned what may have been the most substantial taxidermy firm of the time.[7] Ward spent several days travelling by boat in the Bay of Plenty during 1881. After collecting three tuatara from Karewa Island and about two dozen from Moutoki Island, he wrote with satisfaction in his company's natural science bulletin in 1882: 'No longer will our catalogues, which aim at systematic fullness, have an odious blank after "Order III, Rhyncocephalina" [*sic*]'.[8] A later catalogue for Ward's Natural Science Establishment listed prices for tuatara skins (US$15 to $18) and mounted specimens (US$25 to $30) that were at least double those for other reptiles (Figure 10.1).[9] In England, specimens of tuatara were said to be reaching £5 by 1882.[10]

The collector who drew the most attention and eventual opprobrium, however, was the Austrian Andreas Reischek, resident in New Zealand from 1877 to 1889. Between periods of taxidermy for local museums, Reischek amassed a large collection of ethnographic material and specimens of natural history that he hoped to sell in Europe. He published useful observations on the natural history of wild tuatara,[11] but his exporting ultimately drew veiled censure from New Zealand biologists. The full scale of his collecting cannot be established from current museum holdings, which merely confirm that he exported at least some dozens of tuatara from Taranga Island, the

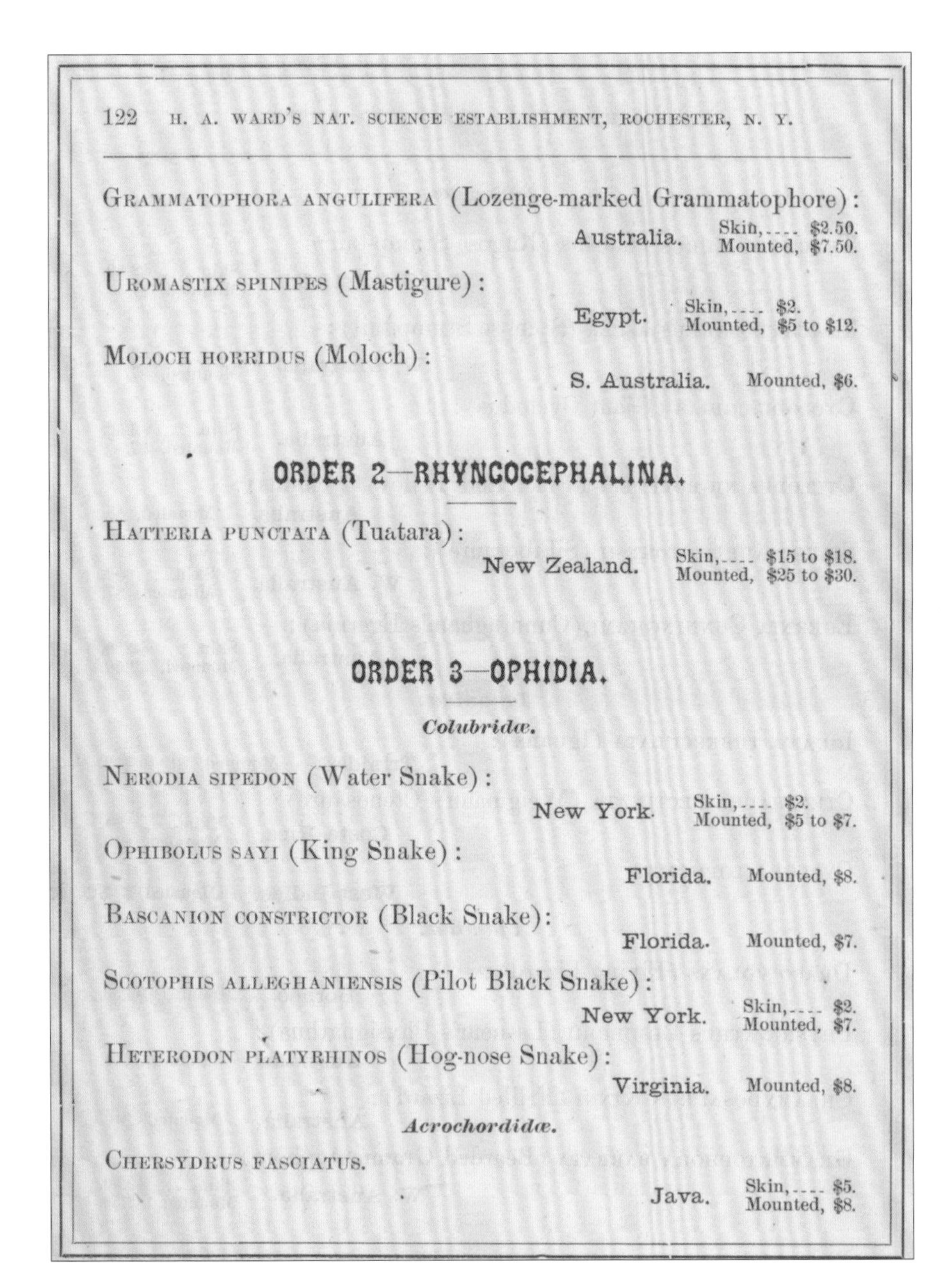
122 H. A. WARD'S NAT. SCIENCE ESTABLISHMENT, ROCHESTER, N. Y.

GRAMMATOPHORA ANGULIFERA (Lozenge-marked Grammatophore):
Australia. Skin, ---- $2.50. Mounted, $7.50.

UROMASTIX SPINIPES (Mastigure):
Egypt. Skin, ---- $2. Mounted, $5 to $12.

MOLOCH HORRIDUS (Moloch):
S. Australia. Mounted, $6.

ORDER 2—RHYNCOCEPHALINA.

HATTERIA PUNCTATA (Tuatara):
New Zealand. Skin, ---- $15 to $18. Mounted, $25 to $30.

ORDER 3—OPHIDIA.

Colubridæ.

NERODIA SIPEDON (Water Snake):
New York. Skin, ---- $2. Mounted, $5 to $7.

OPHIBOLUS SAYI (King Snake):
Florida. Mounted, $8.

BASCANION CONSTRICTOR (Black Snake):
Florida. Mounted, $7.

SCOTOPHIS ALLEGHANIENSIS (Pilot Black Snake):
New York. Skin, ---- $2. Mounted, $7.

HETERODON PLATYRHINOS (Hog-nose Snake):
Virginia. Mounted, $8.

Acrochordidæ.

CHERSYDRUS FASCIATUS.
Java. Skin, ---- $5. Mounted, $8.

Fig. 10.1 An early advertisement for the sale of tuatara skins and mounts, from H. A. Ward's Natural Science Establishment, New York. Ward collected specimens from Moutoki and Karewa Islands in 1881. This advertisement, using the outdated genus name *Hatteria*, dates from sometime between then and 1913. A German catalogue from about the same period similarly listed specimens of tuatara for at least double the price of lizards. By 1980 the advertised price for a preserved tuatara had reached US$7500 in an Indian catalogue (ANZ IA 1 46/18/4 pt 1).

Photo: Archives New Zealand Te Rua Mahara o te Kāwanatanga, Wellington Office, ref. IA 1 46/18/4 pt 1.

Marotere (Chicken) Islands, Hauturu/Little Barrier Island and Karewa Island.[12] However, Reischek may have been involved with a reported export of about 120 live tuatara in 1885 (see next section). He was also said by Buller in 1888 to have collected some 30–40 tuatara 'for the market' from the Chickens, and he may have been the person to whom Buller referred when he wrote in 1895 that tuatara had been 'exposed to the persecution of travelling natural-history collectors, one of whom is said to have forwarded at one time to Europe no less than three hundred specimens preserved in spirits'.[13] If Reischek's collecting was on the scale of hundreds, as seems likely, this was clearly enough to justify concern.

Initial attempts to secure legal protection: 1885

Several notable efforts were made as early as 1885 to draw government and public attention to the need for conservation management of tuatara. The first came in a paper presented to the Nelson Philosophical Society by Hugh Martin in February, on the topic of 'The protection of native birds'.[14] After arguing cogently for the importance of island reserves that could serve as refuges for birds from introduced predators, Martin digressed to 'say a few words on behalf of the Tuatara'. He noted that the tuatara survived only on certain islands and that 'being a rare and singular lizard [*sic*] it is well worth preserving'. Although no specific recommendations were made, the implication was that islands secure from introduced predators would be highly desirable for tuatara as well.

The prelude to a more specific recommendation came in the same month when two university college scientists, Professor T. Jeffery Parker from Otago (Figure 10.2) and his colleague Professor A. P. W. Thomas from Auckland, travelled to Karewa Island with plans to study the embryonic development of tuatara. The trip was made in company with Reischek, who was familiar with offshore islands and tuatara. As no nests were found, Parker and Thomas instead collected live tuatara for attempts at captive breeding in their home cities.[15]

This field trip probably helped increase awareness of the potential for collecting tuatara (a bird-dealer from Dunedin advertised the next month in the *Bay of Plenty Times* for '100 Tuatara Lizards, alive or dead').[16] However, it also led to the first serious attempt to persuade the colonial government to confront the issue of over-collection. Parker was then the secretary of the Otago Institute (later a branch of the Royal Society of New Zealand); his friend G. M. [George] Thomson was also a member. Like many men of his generation, Thomson was at one and the same time a supporter of the naturalisation of certain exotic mammals (some of which would later prove to be problematic) and an early advocate for the preservation of native flora and fauna. Thomson had already spoken to the Otago Acclimatisation Society in April that year of his concern that native animals were 'rapidly being exterminated and that steps should be taken to preserve them'.[17]

An opportunity for Thomson and Parker to address their concerns for tuatara came soon afterwards. At a meeting of the Otago Institute on 9 June 1885, 4 months after Parker's visit to Karewa, the issue of over-collection was raised by Thomson. The meeting went on to record that 'Very recently, a shipment had been sent Home of something like 120 living specimens ...'[18] Parker added to the discussion, noting that he and a companion had recently visited Karewa Island and had 'brought away a number of specimens, but had, of course, been careful to run no risk of depopulating the island'. However, if such shipments as the one mentioned were repeated two or three times, then, Parker noted, 'the island would quickly be cleared, and it would amount almost to a crime to allow this'.

Fig. 10.2 T. Jeffery Parker (1850–1897), DSc, FRS, Professor of Biology at the University of Otago and curator of the Otago Museum (left), and his colleague G. M. Thomson (1848–1933), FLS, a Dunedin science teacher and fellow member of the Otago Institute (right). Parker and Thomson were early leaders in recommending, via the Otago Institute and the Australasian Association for the Advancement of Science, that tuatara be legally protected to prevent over-collection. Parker died in 1897, aged 47. Thomson was a member of parliament 1908–14, allowing him to support other legal initiatives, including the creation of island sanctuaries for tuatara.

Photos: Thomas Jeffery Parker, Hocken Collections Uare Taoka o Hākena, University of Otago, ref. c/n F193/12; George Malcolm Thomson c. 1918, Alexander Turnbull Library, Wellington, New Zealand, ref. 35mm-00091-b-F, General Assembly Library Collection PAColl-0838.

The following motion was proposed by Thomson, seconded by Parker, and passed unanimously:

> That the Institute draw the attention of the Government to the recent wholesale deportation of Tuatara lizards which has taken place from this colony, and respectfully suggest that steps be taken to preserve these animals in the localities in which they still occur.[19]

Two days later, Parker forwarded the resolution to the Colonial Secretary, the government minister responsible for internal administration of the colony.[20] Parker's letter noted that the institute had explored the possibility of legal protection of tuatara. No laws then existed in New Zealand for the specific benefit of native animals; the only potential protection was under the Animals Protection Act of 1880, which allowed harvesting seasons to be set for introduced species of game. However, the Otago Institute had obtained a legal opinion that this act could offer no immediate protection to tuatara because it protected only specified animals and birds, 'animals' then being interpreted in a legal sense as mammals only. The institute therefore called for an amendment to the legislation, hinting that to be effective, this would need to protect tuatara on both Crown-owned and Māori-owned land.

BOX 10.1

Tuatara in popular culture: a national icon

Māori have for centuries recognised the tuatara as one of New Zealand's most distinctive native animals, so much so that their rich traditions regarding New Zealand reptiles are the subject of an entire chapter (Chapter 4). Today, most New Zealanders are highly protective of tuatara, and united in viewing the genus as an important symbol of national identity. Indeed, tuatara are the only non-avian land animals in New Zealand, and one of few reptiles world-wide, accorded this level of popular attention.

Early evidence of European pride and interest in tuatara comes from public exhibitions in Victorian times. Following the 1851 exhibition at the Crystal Palace in London, such extravaganzas became popular ways of showcasing the 'best' features of British colonies. An early diorama featuring tuatara and seabirds in their island habitat was included in the New Zealand and South Seas Exhibition held in Dunedin during 1889–1890.[1] The display was organised by Professor T. Jeffery Parker of the Otago

Box 10.1 Fig. 1 An early diorama showing a male and female tuatara, seabirds and lizards, which featured at the Natural History Court of the New Zealand and South Seas Exhibition of 1889–1890.

Photo: Burton Brothers image no. 4602 [1890], Pearson Album, Alexander Turnbull Library, Wellington, New Zealand, ref. PA1-f-048-22-2.

Box 10.1 Fig. 2 Tuatara first appeared on a New Zealand stamp in the 1935 pictorial series. Early images of tuatara often showed rather frog-like or overweight animals, perpetuating a misleading view of the appearance of a healthy tuatara.

Artist L. C. Mitchell. Photo: ©New Zealand Post, reproduced with consent.

Box 10.1 Fig. 3 A series of four New Zealand stamps released in 1991 in conjunction with the World Wide Fund for Nature. This series, by artist Pauline Morse, offered a variety of ecologically realistic images: a juvenile (top left), a female (top right), a tuatara in its burrow (bottom left) and a male with reddish-brown colour variation (bottom right). Three of these images resemble published photographs of wild tuatara (see Crook 1975) although the fourth (the female) seems overweight.

Photo: ©New Zealand Post, reproduced with consent.

University Museum, who had visited Karewa Island to collect tuatara in 1885. Live tuatara were displayed at the international exhibition held in Hagley Park, Christchurch, during 1906–1907.[2] The exhibition had nearly 2 million attendances, more than twice the population of New Zealand at the time.

Tuatara first appeared on a New Zealand postage stamp, the 8d brown, in the 1935 pictorial series. Like many others of the time, the artist appears to have been unfamiliar with wild tuatara. A series of four 40c stamps released in 1991 were far more realistic,[3] reflecting the greater accessibility of wild specimens in recent years to photographers and artists.[4]

From 1967, when decimal currency was introduced, New Zealanders could have a form of 'close encounter' with tuatara by fingering the 5c coins in their pockets. Tuatara are the only reptiles to have graced any of the 16 coins and 18 bank notes issued by New Zealand's Reserve Bank to 2006 (in contrast, birds have appeared on six coins and 15 bank notes issued since 1933). However, after nearly 40 years in circulation, the 5c coin was withdrawn from 31 July 2006 because inflation had greatly reduced its value; from 1 October 2006 it was no longer legal tender. Although no firm commitment has been made by the Reserve Bank, submissions indicate

Box 10.1 Fig. 4 The New Zealand 5c coin (inset), introduced in 1967, featured a tuatara on a coastal rock, with a gull in the background. By 2006 the coin had been reduced by inflation to less than one-tenth of its original value. Having been withdrawn from circulation, it is now effectively 'extinct'. Members of the public donated the soon-to-be withdrawn coins to several conservation groups. Here, a trustee of the Orokonui Ecosanctuary (the late Diane Campbell-Hunt, centre) holds a poster of a tuatara, created with 5c coins by members of the Dunedin Hospital Early Childhood Education Centre (the fund-raising yielded over $4000 to assist the reintroduction of tuatara).

Main photo: *The Star*, Dunedin. Inset: Ken Miller.

that there is public support for retaining an image of tuatara on local currency[5] – so perhaps a new coin or banknote featuring tuatara will eventually appear.[6]

More-lasting recognition comes from geographic place names that recognise tuatara. There is a Tuatara Drive in Whangarei, a Tuataranui (meaning 'great tuatara') Stream near Kerikeri and a Tuatara Bay on Ruamahuaiti Island in the Aldermen Islands.[7] Surely an extreme, even for such a cryophilic reptile, is the existence of a Mount Tuatara in Antarctica near the Byrd Glacier. It was named by a New Zealand geological expedition during 1960/61 in recognition of the long and spiny summit ridge.[8]

Graced with such a distinctive crest, tuatara (especially males) are a graphic designer's delight. In the early 1950s the tuatara was one of several native animals considered for the new logo of the Wildlife Division of Internal Affairs, although the ultimate choice was a bird, the white heron (kōtuku, *Ardea modesta*).[9] In recent times tuatara have featured in several logos, including those for the magazine *New Zealand Geographic* and for Environment Canterbury's environmental awards. A tuatara and an Australian sleepy lizard (*Tiliqua rugosa*) combine in trans-Tasman unity to form the emblem of the Neurosurgical Society of Australasia; this choice recognises the fascination that neurosurgeons have for the well-developed parietal eye in these species.[10]

Box 10.1 Fig. 5 (above left) A poster from 1985 advertising 'Tuatara', a Flying Nun compilation featuring '12 great bands on 1 LP', designed by Lesley MacLean.

Photo: ©Flying Nun, photolithograph from 1985, Alexander Turnbull Library, Wellington, New Zealand, ref Eph-D-PHONO-1985-02.

Box 10.1 Fig. 6 (above right) The robust shape of tuatara (with a little artistic licence to keep the spines flat) lends itself to children's playground equipment. This example, from a busy pedestrian thoroughfare in central Wellington, was made by the visual-effects company Weta Workshop.

Photo: Alison Cree.

Box 10.1 Fig. 7 (left) The name or image of tuatara features in many commercial products, including wine and beer.

Photo: Ken Miller, collection of Alison Cree.

For an animal widely associated in the public mind with dinosaurs and lack of adaptation, tuatara have been surprisingly popular symbols in the commercial world. Older followers of alternative rock music will recall 'Tuatara' as the first Flying Nun compilation, released with its distinctive 'Dunedin sound' in 1986 and described nostalgically as 'a party-starter for student flats the world over'.[11] Today, the word 'tuatara' is part of the name of several commercial companies in New Zealand, including a craft shop, a backpackers' lodge, a trekking-tour company and a software design company, and images of tuatara appear on numerous products ranging from silk ties to cast-metal doorhandles. Tuatara have achieved what has been considered for snakes in Australia as the 'ultimate symbol of acceptance into ... mainstream culture'– appearance on a beer bottle label[12] – but they also grace the classier end of the beverage market – including the wine label 'Tuatara Bay'.[13]

Tuatara are popular models for sculpture, ranging from miniatures in the form of exquisite jewellery to gargantuan, climb-aboard statues. A friendly-looking specimen eyes up thousands of city shoppers every day from a playground in Wellington's busy Cuba Mall. Models of this size or larger are found in other parts of the country, including a 2.5-metre bronze sculpture in the grounds of Invercargill's Southland Museum and Art Gallery, a 5-metre-long concrete sculpture at the entrance to Pyes Pa School near Tauranga, and a record-breaking 12-metre concrete tuatara built by children at Kimbolton School in Manawatu with assistance from local engineers.

The Pyes Pa sculpture is just one among many roadside images of tuatara or other reptiles that the observant traveller will see in the Bay of Plenty–East Cape region. Here, traditions of reptiles as local identities or guardians feature strongly in Māori-related artwork. Examples include carvings that grace the entrances to marae (open areas in front of meeting houses) and schools, as well as a prominent, carved pole featuring tuatara and other New Zealand animals in the town centre of Opotiki.

Tuatara have been a stimulus to creative writing at many levels, and not just among children's books.[14] 'Tuatara' was the name of the scientific journal published by the Biological Society of Victoria University of Wellington between 1947 and 1993 (tuatara biologist W. H. Dawbin was the first editor),[15] and the journal's namesake inspired visiting scientists to poetry on at least two occasions. Unsurprisingly for the times, the poems emphasised a perceived evolutionary stasis. 'Resisting evolution's mighty flow ... How, but for you, should we the Mesozoic know?' wrote Karl Schmidt of the Chicago Natural History

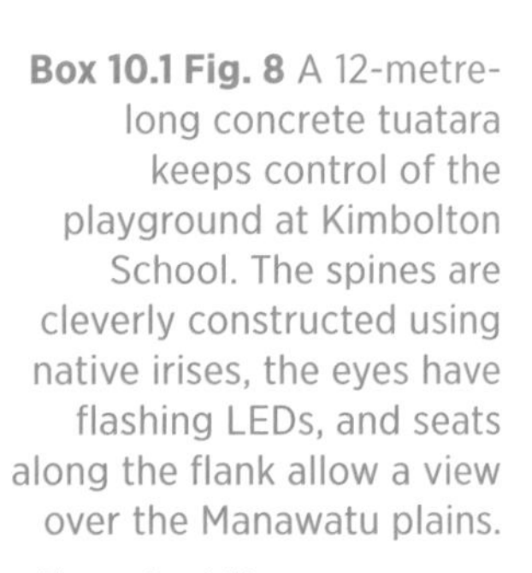

Box 10.1 Fig. 8 A 12-metre-long concrete tuatara keeps control of the playground at Kimbolton School. The spines are cleverly constructed using native irises, the eyes have flashing LEDs, and seats along the flank allow a view over the Manawatu plains.

Photo: David Lupton, courtesy of Horizons Regional Council.

Box 10.1 Fig. 9 A green tuatara (guardian of the forest of Tāne) sits on the shoulder of a warrior in this Māori carving over the entranceway to Torere School, Bay of Plenty. Another tuatara-like reptile in brown (guardian of Earth or Papatūānuku) is found between the two warriors. The carving was made by a team supervised by Wikuki Kingi senior of Ngāti Ngāi Tai. According to tribal tradition, these reptiles remind humankind of the need to protect our land, forest, sea, shore, sun and moon, and all that live within these realms.[21]

Photo: Marcus Simons.

Museum in 1949.[16] A seven-verse poem written in 1967 by Ralph Lewin of the University of California is delightfully humorous, despite its questionable emphasis on biological 'shortcomings':

> ... So pity tuataras
> Who give the triple wink,
> At such a pace, their very race
> Is verging on the brink.[17]

A more positive view of survival came in a recent poem by Nora Borrell, speaking in the voice of a tuatara reintroduced to Matiu/Somes Island ('Keep your distance ... I've learnt a thing or two over 200 million years ... quit drooling over me').[18] In 1989 acclaimed author Barbara Anderson published a short story about the pleasure brought by the hatching of tuatara to a university scientist facing challenges in his private life,[19] and in 1998 composer Gareth Farr produced a work entitled 'Tuatara' for percussion and piano; the piece premiered in a concert featuring the New Zealand forest as a unifying theme.[20]

In summary, the tuatara, like kiwi (*Apteryx* spp.) and the silver fern (ponga, *Cyathea dealbata*), is firmly entrenched in New Zealand culture as a symbol that helps define national identity. Moreover, images of tuatara contribute to spiritual wealth in ways that trace shifts in scientific understanding. Whereas the species was once often viewed in European popular culture as an anachronism – an example of essentially failed evolution – the tuatara is today more likely to be portrayed as a survivor. In a refreshing combination of modern evolutionary perspective and a Māori tradition of 'reptiles as guardians', tuatara now fill a valuable role as symbols of a sustainable future with humans.

Parker's letter was passed to James Hector, the senior government scientist. Hector was first and foremost a geologist, and, as Director of the Geological Survey and Colonial Museum, and Manager of the New Zealand Institute, undoubtedly overworked. Hector ignored the recommendation, responding to Parker with a rather testy 'How many Tuataras did you and Thomas take into captivity and what has become of them?' Parker replied immediately by telegraph on 17 June 1885: 'I have about eighteen still alive Thomas probably about twenty'.[21] No further response from Hector is recorded.

The Otago Institute's suggestion must have received some publicity, however, because over the following two months, a correspondent from Wellington, Edward S. Dodgson, wrote twice to the government expressing enthusiasm for the legal protection of tuatara.[22] In his first letter, forwarded to Hector, Dodgson wrote: 'Permit me to ask you not to let the proposed Tuatara Protection Act drop. This Colony has so few antiquities or historical associations, and so little local art to make it interesting, that it ought to be careful to keep in existence as long as possible its unique things in the way of natural history …' His second letter was addressed directly to Hector: 'Will you allow me to suggest that you should urge the Government to bring in a bill to preserve and protect the Tuatara lizards which are among the chief objects of interest in this Colony, being the largest lizards in it and the rarest in the world?' These letters reveal an emerging view that tuatara and other distinctive native animals could contribute to a sense of national identity. Nonetheless, Dodgson's recommendation, for the meantime, drew no response.

Action at last: legal protection of tuatara in 1895

In 1888 the Australasian Association for the Advancement of Science (AAAS) was formed, uniting New Zealand and Australian scientists and providing a forum for resolutions that could influence the New Zealand colonial government, at least, from beyond its border. Over the course of several meetings between 1891 and 1895, G. M. Thomson took an influential role in moves by AAAS that led to the protection of several offshore islands as sanctuaries for native animals, including one for the specific benefit of tuatara.

The first two such sanctuaries were established with birds in mind, although tuatara were present on one. At the third AAAS meeting in Christchurch in 1891, Thomson had joined forces with Professor A. P. W. Thomas to move a resolution that the New Zealand government should establish reserves for the protection of native flora and fauna.[23] A recommendation that Resolution Island (in Fiordland) be made a reserve was implemented by the government the same year, and a similar recommendation for Hauturu Island was actioned in 1894. These were world-leading developments in the conservation of island faunas (though achieved on Hauturu at the expense of some Māori owners, following legislation to enable compulsory purchase by the government).[24] Tuatara were known from Reischek's work in the 1880s to exist in low numbers on Hauturu, but seem to have received little if any official recognition in discussions about the island's reservation.[25]

But Thomson and the AAAS had not lost interest in tuatara. From the time of the third meeting in Christchurch in 1891, Thomson had been a member of a committee formed to report upon the protection of native fauna.[26] The committee's report was presented and accepted at the fifth AAAS meeting in Adelaide in 1893. It noted approval of what had been achieved for Resolution and Hauturu islands, but added a recommendation to the New Zealand government that 'islands should be set apart for the preservation of the Tuatara lizard'.[27] Ironically, it was Hector, a vice-president of AAAS, who moved the motion to approve the report.[28] However, no government action was taken at this point.

Undeterred, members of Section D (Biology) of AAAS made a more specific resolution at the sixth meeting, held in Brisbane in January 1895: 'that Stephens Island, Cook Strait, may be proclaimed as a reserve for the Tuatara lizard, in order to prevent the extinction of this rare and interesting creature'.[29] The resolution was put forward by the section's president, Professor Arthur Dendy, of Canterbury College (Dendy would soon take a leading role in the study of embryonic development in tuatara); however, Thomson very likely had a hand in preparing the resolution. Additionally, Parker had been a local (New Zealand) secretary for AAAS from 1891 to 1895 and would have lent support.

This time, the Colonial Secretary's office was favourably inclined. Hector's minute on the proposal noted that 'The tuatara is one of the most interesting animals now surviving – & its last home is Stephens Island & there they are being wantonly exterminated'.[30] A lighthouse had been constructed on Stephens Island (Takapourewa) between 1892 and 1894, bringing knowledge of its populations of rare birds, and its abundant population of tuatara, to public attention. Attention quickly turned to dismay, however, at the impact that cats, introduced during or soon after construction, were having on these species.[31]

Ironically, the government did not follow the recommendation from AAAS to make Stephens Island, which was already a lighthouse reserve, a reserve for tuatara. Presumably, the island could have been given the same reserve status as Hauturu, which might have helped stem further habitat destruction and the loss of other native species. Instead, a decision was made to protect tuatara throughout their range.[32] This was admittedly a more effective way of addressing the problem of over-collection, although it came too late to influence Reischek, who had left the country 6 years previously.

The restriction on collection was achieved by declaring tuatara to be 'native game' under the Animals Protection Act 1880 – exactly what Thomson, Parker and the Otago Institute had argued for, 10 years earlier. Species of native game were protected from 'hunting, shooting, pursuing, taking, or killing' in any but a specified season (which could be nil, if publicly notified – effectively the case for tuatara). The protection of tuatara, published in the

Tuatara Lizards protected under the Animals Protection Acts.

GLASGOW, Governor.

IN exercise of the powers vested in me by "The Animals Protection Act, 1880," and the Acts amending the same, I, David, Earl of Glasgow, the Governor of the Colony of New Zealand, do hereby declare that, from and after the date hereof, the lizard known as tuatara (*Sphenodon punctatum*) shall come within the operation of the said Acts as fully and effectually as if it had been included in the Fourth Schedule to the said "Animals Protection Act, 1880."

As witness the hand of His Excellency the Governor, this first day of April, one thousand eight hundred and ninety-five.

P. A. BUCKLEY.

Tuatara Lizards and their Eggs protected under the Animals Protection Acts.

RANFURLY, Governor.

IN exercise of the powers vested in me by "The Animals Protection Act, 1880," I, Uchter John Mark, Earl of Ranfurly, the Governor of the Colony of New Zealand, do hereby declare that from and after the publication hereof the lizard known as tuatara (*Sphenodon punctatum*) shall come within the operation of the said Act, and do hereby declare that the eggs of such lizard are hereby absolutely protected.

As witness the hand of His Excellency the Governor, this second day of September, one thousand eight hundred and ninety-eight.

J. CARROLL.

Fig. 10.3 Publications in the *New Zealand Gazette* announcing the legal status of tuatara as native game (4 April 1895), and the absolute protection of eggs (8 September 1898).

New Zealand Gazette on 4 April 1895 (Figure 10.3), was among the first in New Zealand for any native animal. Intriguingly, the announcement refers only to '*Sphenodon punctatum*', suggesting that Buller's identification in 1877 of tuatara from North Brother Island as a separate species, *S. guntheri*, had already fallen into disuse.[33]

Further legal protection of tuatara and their eggs: 1898–1922

Almost immediately, some local wildlife managers and scientists raised concerns that the matter of legal protection was not settled. By 1898 Dendy's research on embryonic development was underway at Canterbury University College; the Canterbury Acclimatisation Society was also aware that two German researchers (Hugo Schauinsland and Georg Thilenius) wished to obtain eggs for the same purpose. The society therefore questioned the government as to whether the eggs of tuatara were protected. Dendy and the Philosophical Institute of Canterbury went further, recommending protection of the eggs; the Philosophical Institute also suggested increasing the penalty for collection of tuatara without a permit from £5 to at least £20.[34] In response, on 8 September 1898, the eggs of '*Sphenodon punctatum*' were gazetted as 'absolutely protected'.

Both the previous declaration of tuatara as 'native game', and the new declaration regarding eggs, gave officials private concern for some years. The term 'native game' was open to misinterpretation and effectively required a nil season to be announced each year. This ambiguity was rectified, under the provisions of the Animals Protection Act 1908, by declaring tuatara to be 'protected absolutely' (i.e. have no open season) on 13 November 1913.[35] Officials were also concerned that a prosecution for unauthorised collection of eggs might fail because the strict wording of the Animals Protection Act 1880 (and its successor, the Animals Protection Act 1908) allowed the eggs of only birds, not reptiles, to be protected.[36] This situation was apparently not resolved until the passing of the Animals Protection and Game Act 1921–22, which specifically protected the eggs of reptiles, as well as birds, that were absolutely protected.[37]

Management of collection: 1895–1915

Following legal protection in 1895, collection of tuatara without a permit from the Colonial Secretary became an offence. This does not mean that wild tuatara were unobtainable. Authorised collection continued, including of specimens to be used in research and displays, both locally and abroad. Undoubtedly, there was also unauthorised harvesting, the reported collection of tuatara from the Mokohinau Islands in 1890, in part for the private collection of a millionaire, being a possible example.[38]

One application for the legal collection of six tuatara, 'from Stephens Island or elsewhere', was made on behalf of Andrew Johnson only 2 months after tuatara had been protected. Johnson was a former curator of the Canterbury Acclimatisation Society and ran a fish hatchery, aquarium and gardens known as Troutdale Farm in Christchurch. The initial response within the Colonial Secretary's office was negative, reflecting concern that if permission were given to Johnson, others would be encouraged to make similar applications.[39] Johnson must have been persuasive, however, as live tuatara were on display at his property by 1903.[40]

On Stephens Island, tuatara continued to be removed under authorisation. Up to six adults and 100 eggs were collected personally by Schauinsland and Thilenius, who visited the island between 1896 and 1898.[41] From 1899 to 1908, the lighthouse keepers collected and forwarded at least 52 adult tuatara and 74 eggs, and probably more that were not recorded.[42] Payment was received by the keepers for at least one collection (£15 for 50 eggs), which compared very favourably with a keeper's annual salary of £100.[43] One surprisingly large request (for 200 tuatara) emanated from the Governor of New Zealand in 1901, but apparently only 12 were forwarded. On one occasion in 1906, the principal keeper received a request for a number of tuatara (unstated) that he considered excessive; he recommended supplying only half the number.[44]

In 1907 New Zealand made the transition from colony to dominion, and the Colonial Secretary's Office was renamed the Department of Internal Affairs. Six tuatara were among a collection of native animals presented that year by the Dominion Government to the Emperor of Austria.[45] The following year, a meeting of the Canterbury Acclimatisation Society resolved to draw to the attention of the Minister that tuatara were apparently 'very plentiful' in London and Hamburg; dealers in those cities were said to be 'selling tuataras from ten shillings upwards', implying that tuatara were also being exported illegally.[46] Such concern may at least partly explain a subsequent reduction in the number of tuatara authorised for collection from Stephens Island,[47] although awareness of the harmful effects of cats there was probably also a factor. Local interest in specimens for research was dwindling in the face of economic and wartime pressures, and such interest as remained was to some extent met with specimens from Moutoki Island. In total, the years from 1896 until 1912 saw permits issued for the export of 76 tuatara and for collection, for local purposes, of a further 55 tuatara from various islands.[48]

Apparently, a single pet tuatara could still be publicly carried about without question as late as 1909 (14 years after legal protection),[49] but a large private colony in Tauranga raised more substantial concerns. Georg Thilenius had been authorised in 1897 to export six tuatara from Karewa Island or Cuvier Island (Repanga Island). A Tauranga resident soon became concerned that Thilenius had greatly exceeded his collection permit, also reporting to officials that there was 'a man in Tauranga who does quite a business in selling tuataras'.[50] Upon investigation by a Tauranga policeman, Thilenius confirmed that six tuatara had been exported, and that others had been collected from Karewa for observation, in collaboration with local boatman Colin Norris (about 60 were in Norris's colony in Tauranga at the time). An assurance was given by Thilenius that 'when finished with they will be liberated either on Karewa Island or at any place the Government may wish'. Thilenius left New Zealand by about 1899, his one known publication on tuatara making no reference to the captive colony.[51]

Six years later, however, Norris announced that he had about 100 tuatara for sale, claiming that 125 had already been in his possession before tuatara became legally protected and that others, left with him by Thilenius 'for certain observations which I have completed ... are now my property'.[52]

BOX 10.2

Early translocations of tuatara

Translocation – the deliberate shifting of organisms from one place to another – has long been part of the repertoire of the New Zealand conservation manager. Most early translocations for conservation purposes involved birds, but a few little-known cases featured tuatara. These provide instructive contrasts with recent translocations over the past decade (described elsewhere in this chapter).

Several translocations of tuatara from offshore islands were made during the late 1800s, with releases occurring onto the mainland or other islands. Most appear to have been performed out of curiosity, or for personal benefit, rather than to assist the conservation of tuatara. One example was the probable introduction of tuatara to Kawau Island in the late 1870s, for the private menagerie of Governor George Grey.[1] A better-documented release was that made by Walter Buller to his country estate at Papaitonga, near Wellington. Buller had already placed kiwi (*Apteryx* spp.) on an island in his private lake, and in 1893 he boasted about how he had used tuatara to place a tapu (restriction) on the island:

> ... to prevent any chance of Maori depredations in the breeding-season I have also placed on the island three large live tuataras, kindly supplied to me by Captain Fairchild [Master of the government steamer servicing the lighthouses]. The fame of these lizards [*sic*], of which the Maoris have a most unaccountable dread, has spread far and wide ... and the Kiwis could not have three better guardians, for with this dread of the *ngarara* no Maori will ever willingly set foot on the island.[2]

The fate of the tuatara and kiwi was never reported; probably they succumbed to predatory mammals.

During 1913, and seemingly unaware of the experience of Buller and others, government officials hatched plans to shift eggs of tuatara from one island to another. This proposal, initiated by the lighthouse keeper E. W. Tutt, was one of a range of actions intended to assist the survival of tuatara.[3] Possible destinations were discussed by lighthouse keepers, scientists and others.[4] The Minister of Internal Affairs gave instructions for eggs to be shifted from Stephens Island (Takapourewa) and conveyed to 'various other islands on which the tuatara is scarce'.[5] The keeper on Stephens was asked to collect the eggs, and Captain Bollons, by then the Master of the government steamer, undertook to distribute the eggs to 'the Brothers, Cuvier, Little Barrier and Tiri Tiri'.[6] The last of these, Tiritiri Matangi Island, is not known to have had tuatara at the time, but the other islands did. Tutt supplied information on how to maintain the eggs during transport and bury them at their destination. Frustratingly, there are no published details of when and how many eggs were shifted to each island, or whether any hatchlings resulted.[7]

In the same year (1913), several other islands were proposed as possible sites for stocking with tuatara. Among these were Tokomapuna and Motungarara islands, two islets off Kapiti Island not known to have resident tuatara.[8] Yet more destinations (the Mokohinau Islands, Hauturu/Little Barrier Island, the Poor Knights Islands, the Hen and Chickens Islands, the Mercury Islands and Manawatawhi/the Three Kings Islands) were recommended by an anonymous newspaper correspondent (probably H. H. Travers, a well-known natural-history dealer), apparently without the author realising that tuatara already existed on all except perhaps the last-named group.[9]

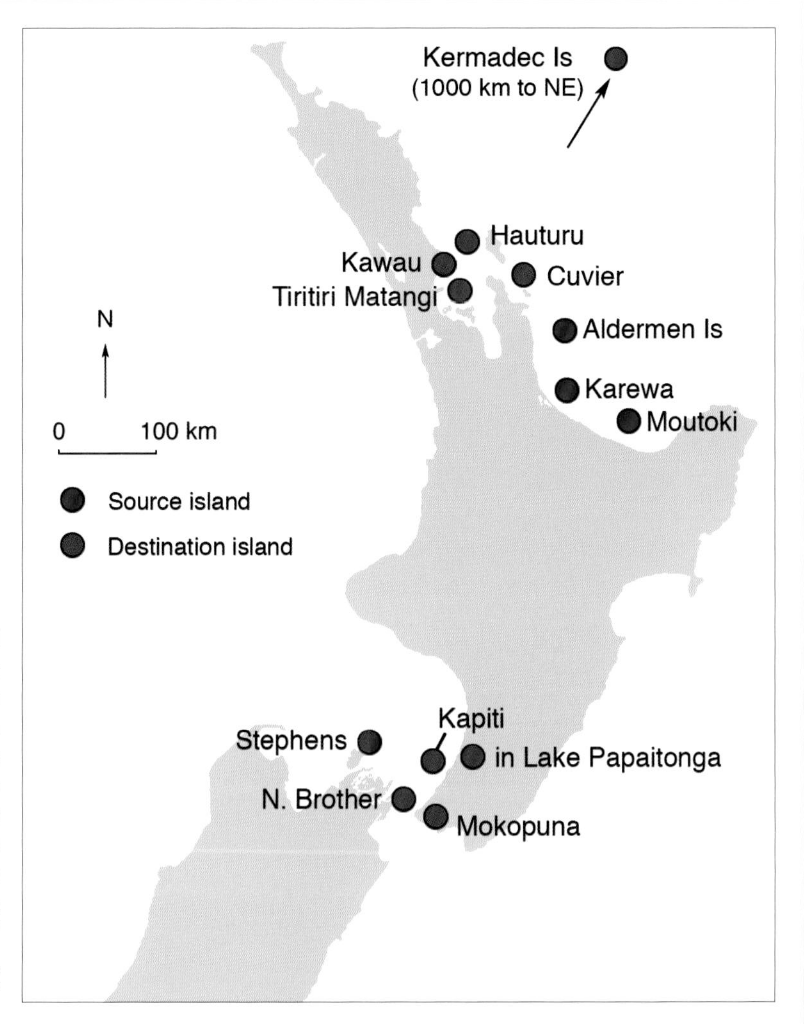

Box 10.2 Fig. 1 Reported translocations of tuatara (including eggs) to islands between 1870 and 1951. Translocations to Mokopuna and Kapiti Islands definitely occurred, but details of other reported translocations (including source populations) are less certain. Whether such translocations contributed to the gene pool of current populations of tuatara is unknown. For references, see text.[26]

These suggestions received some support,[10] but there is no evidence that any were taken up. On the other hand, tuatara are said to have been released on one of the subtropical Kermadec Islands, although details are scant and no subsequent sightings were made.[11] A newspaper article from about the 1920s reported that 150 tuatara were collected from Karewa Island, the Aldermen Islands and 'Ruarima' (presumably Moutoki Island) by Mr Norris and 'a German naturalist on the staff of the Strasburg Museum, who came to New Zealand more than 30 years ago'.[12] It continued, 'some of these are known to have been placed on Stephens Island', which, if true, would be a disconcerting mix of existing stocks, but no confirmation is known.

A later, official translocation, from Moutoki Island in 1920, is somewhat better documented. Along with the other Bay of Plenty islands, Moutoki (a tiny (0.8-hectare), Māori-owned island then without sanctuary status) was considered by government officials to be vulnerable to disturbance from picnickers, fishermen and Māori,[13] and so a transfer was performed. A government ranger collected 'some 23 or 24 Tuatara lizards of various sizes',[14] and the tuatara were released on Mokopuna Island,[15] a rocky islet (less than 1 hectare) about 80 metres north of Matiu/Somes Island. The lighthouse keepers on Matiu were appointed as Special Protectors (see main text); they later wrote of having, on numerous occasions, to dissuade boys and men from landing on the islet.[16] However, despite making several searches in the thick undergrowth at the release site, the keepers were unable to relocate the tuatara by 1924,[17] and their allowance as Special Protectors was discontinued that year. It is unclear whether Mokopuna Island was free of introduced mammals when the release occurred, but rabbits had established by about 1946 and caused considerable damage to the dense coastal scrub. In 1951 a Wildlife Division officer exterminated the rabbits but found no trace of tuatara.[18]

Another officially sanctioned but little-known transfer of tuatara involved their release on Kapiti Island, a large (1970-hectare) bird sanctuary.[19] Five tuatara were collected from Stephens Island by W. H. (Bill) Dawbin, a tuatara biologist at Victoria University of Wellington, and released in December 1951 near the resident caretaker's house. The caretaker later reported that the tuatara were 'of varied sizes' and that one was seen dead a week after liberation, whereas others were sighted for some months afterwards. Despite searching for the animals, the caretaker was not aware of further sightings from 1952 until the head of a dead animal was reported in 1954.[20] By 1955 it seemed likely that all the tuatara had died.[21] Dawbin himself later wrote that 'several specimens are known to have been killed by wekas a very short time after they were liberated'; he was therefore opposed to further releases of tuatara on islands where weka were present.[22] Curiously, Dawbin made no mention of Norway (*Rattus rattus*) and Pacific rats (*Rattus exulans*), both of which were also present on Kapiti;[23] Norway rats seem the most likely predators of adult tuatara.[24]

In summary, these early translocations, while often well intentioned, teach us many things about how not to do a successful translocation for conservation. First, no thought seems to have been given to ensuring that likely threats to tuatara from introduced mammals at the release site were identified, controlled or removed (for example, rats were present on Kapiti, Tiritiri Matangi, Cuvier and Hauturu Islands, and cats were present on the latter two).[25] We know little about the numbers and sizes of the translocated tuatara and nothing about their sex ratios; indeed, given historical difficulties in sexing tuatara and lack of knowledge at the time of temperature-dependent sex determination, it is not impossible that all the translocated animals or eggs were of the same sex. Potential effects on the fauna of the host island were not considered (including genetic mixing with, or transfer of parasites or disease to, existing populations of tuatara). There is little evidence of formal monitoring, a casual approach aptly described by today's conservation workers as 'dump and run'. Limited reporting meant a low awareness of, and opportunity to learn from, the experience of previous translocations.[26] There was also little if any consultation with private landowners and local communities, including Māori. Today, these and other such matters must be taken into account before translocations can proceed.

Upon further challenge by officials, he brazenly offered to sell the tuatara to the Crown at £9 per dozen.[53] This offer was initially declined, with officials uneasy about the legality of sale. Nonetheless, some tuatara were bought the following year by Augustus Hamilton, Director of the Dominion Museum.[54] In 1908 Norris still retained a 'large collection' in Tauranga, including two that were publicly displayed in Dunedin.[55]

In 1913 Hamilton recommended to the Minister of Internal Affairs that the government 'acquire the collection which was offered to them a year or so ago and which is held by a resident of Tauranga as it is a constant source of leakage'.[56] Hamilton's recommendation was approved, but the exact fate of the tuatara is unknown.[57] Norris apparently still had some in his possession until January 1915, when he wrote to Internal Affairs that he had not seen any that summer and was afraid that the 'sharp snap of cold we experienced last winter has done for them'.[58]

For years afterwards, escaped specimens were occasionally seen about the cliffs at Tauranga.[59] A woman found one in her garden; after 'giving fright to various people on account of its large size … [the tuatara] was ultimately killed by a dog'.[60] A newspaper article from about the late 1920s suggested that tuatara from Norris's colony were even released on Stephens Island – a quite astonishing mixing of populations, if it happened.[61]

Doomed to extinction through evolutionary stagnation?

At the same time as ideas for the protection of tuatara were gathering force, there was doubt in some minds as to whether tuatara were even capable of survival. Perhaps this strange reptile was in the process of disappearing regardless, and Europeans had arrived just in time to see the final stages of extinction? This idea of 'genetic senescence' was not held uniquely for tuatara – several influential ornithologists of the early to mid 1900s argued that moa (a group of large flightless birds) had become extinct for similar reasons.[62] It was also a process widely believed by influential men such as Buller in the late 1800s to apply to the Māori people themselves.[63] But the tuatara, with its extreme level of taxonomic distinctiveness and affinities with Mesozoic fossils, was a leading candidate.[64]

One person who contributed with enthusiastic rhetoric to this view was the journalist James Drummond. Drummond wrote about natural history for the *Lyttelton Times*, and in one of his articles in 1905, he reported on the behaviour of a pet tuatara that he kept in a cage in his garden, probably in Christchurch (a similar article appeared in the *Boy's Own Annual* in 1910).[65] The newspaper article's title encapsulated the over-riding theme: 'A New Zealander who would not progress. Must end his days soon'. Drummond went on to write about tuatara: 'He is not of the busy, bustling today, but of the old, old long ago, the past and forgotten yesterday … As his [Jurassic] ancestors were, so is he. What was good enough for them is good enough for him … He is dying, slowly but surely. The world has no time for him now'.

The coup de grâce was a moral one: 'He has lived too long in the Castle of Indolence. Having rebelled against nature and committed the sin of inaction, he must earn the wages of sin, which is death'!

Fig. 10.4 H. G. (Harry) Ell (1862–1934), a leading political campaigner for tuatara in the early 1900s. Ell pushed effectively for an increased bounty for lighthouse keepers to kill cats on Stephens Island (Takapourewa), and together with others, including the Reverend J. Crewes of Wellington, promoted the establishment of legal sanctuaries for tuatara.

Photo: Henry George Ell, photographed c. 1902 by William Henshaw Clarke. Alexander Turnbull Library, Wellington, New Zealand ref. 35mm-00188-c-F, General Assembly Library Collection PAColl-0838.

Belief that tuatara were maladapted, and their extinction inevitable, continued to pervade popular and to some degree scientific understanding for decades. Bernard Sladden, a local authority on tuatara in the Bay of Plenty in the 1930s, and an honorary ranger for Karewa Island, explained it thus to the Minister of Internal Affairs: 'Sphenoden [*sic*] (tuatara) is the survival [*sic*] of a long extinct group of reptiles, and the race itself is naturally a decadent one'.[66] Even in the latter half of the 1900s, as government scientist Ian Crook noted in 1970, '[s]ome authorities describe the tuatara as an "obsolete" animal in the evolutionary sense', an example of an animal whose 'final extinction ... is only a matter of time and the persistent advance of evolutionary innovation'.[67] Crook himself seemed uncertain about the extent to which he rejected this argument. It was only during the final two decades of the 20th century that this view of inevitable extinction became widely discredited.[68]

Legal protection is not sufficient: other initiatives during 1904–1917

By the early 1900s it was clear that legislation forbidding unauthorised collection was not sufficient to guarantee the survival of wild tuatara. The effects of feral cats were now being keenly felt on Stephens Island, and several other island populations were experiencing disturbance through human activities, including mutton-birding (harvesting the chicks of certain burrow-nesting procellariiform seabirds), fires and sightseeing. Such events were widely reported in newspapers, with one writer making the (unlikely) estimate in 1913 that perhaps only a thousand tuatara remained.[69] These reports stimulated a vigorous flurry of interest in the management of tuatara.

A leading figure in this debate was the politician Henry ('Harry') Ell (Figure 10.4). Ell is more widely remembered today for his role in preserving the forest remnants and walkways of Christchurch's Summit Road. Nonetheless, the impression from newspaper accounts and government files of the period is that, when it came to concern for tuatara, Ell was like a terrier on the heels of the Minister of Internal Affairs, F. H. D. Bell. One of Ell's early successes was to get the bounty for cats increased on Stephens Island in 1906; by October 1910, 511 cats had been killed on the island and none had been seen for 10 months.[70] Reverend John Crewes, the president of the Wellington Zoological Society and a prominent conservation-minded citizen, also took a strong interest in tuatara at this time. In 1912 he led deputations to Ell and to the Minister for Internal Affairs, urging more protection for tuatara and steeper fines, even imprisonment, for illegal collection.[71]

Bell responded to the pressure from Ell, Crewes and others by requesting that the Director of the Dominion Museum look into the situation. Hamilton's interim response in February 1913 was rather negative, suggesting that tuatara were

> … probably doomed to partial extinction, if not quite, from the fact –
>
> 1. That the habitats are becoming more and more accessible.
> 2. From the ravages of cats which are imported into several places, especially where lighthouse keepers are resident.
> 3. From unauthorised visits by scientists and others. It is well known that a few years ago a German doctor went to one of the islands solely for the purpose of killing Tuataras with the object of getting their eggs in various states. He destroyed a great number of specimens and left them lying on the ground.
> 4. The reproduction is extremely slow and it is almost impossible to find young specimens.[72]

Among those who corresponded with Hamilton was Professor William Benham of the University of Otago. Tuatara, noted Benham in 1912, were quite unlike any other living reptile and thus of great international interest. Was there an island where they would 'be free from molestation by visitors and vermin'? Based in Dunedin, Benham felt insufficiently familiar with tuatara islands to make specific suggestions, but wasted no words in emphasising the importance of the task: 'this much I can say and most emphatically – that New Zealand will be regarded by Naturalists the world over as little better than barbaric if, with all the experience in other countries and all the information accessible to those who look for it, – something is not done to preserve the Tuatara'.[73]

Hamilton accordingly sought information on the distribution and status of tuatara populations from those familiar with offshore islands, including the botanist Thomas Cheeseman (curator of the Auckland Museum), Captain Gilbert Mair, and the lighthouse keepers. These sources confirmed the recent existence of tuatara on at least 20 islands, but sightings on several (including at least seven where tuatara subsequently became extinct) were scarce.[74] Lighthouse keepers considered cats and rats as known or potential predators of tuatara;[75] their observations on the numbers and behaviour of tuatara were compiled in 1915 and presented in the annual report to Parliament by the Director of the Dominion Museum.[76]

Several leading figures, including Ell, Crewes and Mair, were convinced that the government could do more to protect the habitats of tuatara from disturbance. Independently, they applied pressure on the Minister for Internal Affairs from 1912 to 1914 to acquire islands as reserves.[77] In 1914 the government responded, introducing an amendment to the Animals Protection Act to allow the state to acquire land for sanctuaries. Karewa

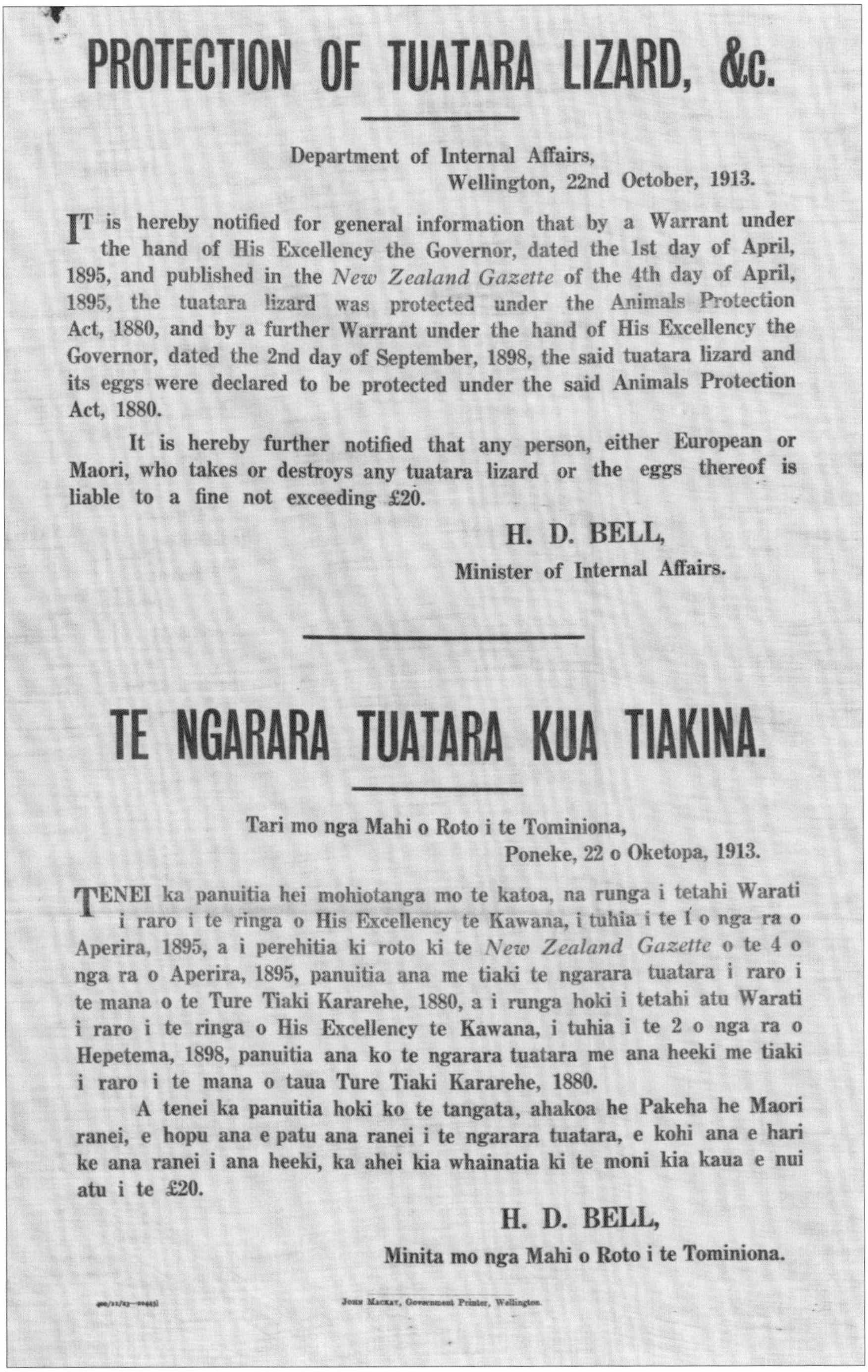

PROTECTION OF TUATARA LIZARD, &c.

Department of Internal Affairs,
Wellington, 22nd October, 1913.

IT is hereby notified for general information that by a Warrant under the hand of His Excellency the Governor, dated the 1st day of April, 1895, and published in the *New Zealand Gazette* of the 4th day of April, 1895, the tuatara lizard was protected under the Animals Protection Act, 1880, and by a further Warrant under the hand of His Excellency the Governor, dated the 2nd day of September, 1898, the said tuatara lizard and its eggs were declared to be protected under the said Animals Protection Act, 1880.

It is hereby further notified that any person, either European or Maori, who takes or destroys any tuatara lizard or the eggs thereof is liable to a fine not exceeding £20.

H. D. BELL,
Minister of Internal Affairs.

TE NGARARA TUATARA KUA TIAKINA.

Tari mo nga Mahi o Roto i te Tominiona,
Poneke, 22 o Oketopa, 1913.

TENEI ka panuitia hei mohiotanga mo te katoa, na runga i tetahi Warati i raro i te ringa o His Excellency te Kawana, i tuhia i te 1 o nga ra o Aperira, 1895, a i perehitia ki roto ki te *New Zealand Gazette* o te 4 o nga ra o Aperira, 1895, panuitia ana me tiaki te ngarara tuatara i raro i te mana o te Ture Tiaki Kararehe, 1880, a i runga hoki i tetahi atu Warati i raro i te ringa o His Excellency te Kawana, i tuhia i te 2 o nga ra o Hepetema, 1898, panuitia ana ko te ngarara tuatara me ana heeki me tiaki i raro i te mana o taua Ture Tiaki Kararehe, 1880.

A tenei ka panuitia hoki ko te tangata, ahakoa he Pakeha he Maori ranei, e hopu ana e patu ana ranei i te ngarara tuatara, e kohi ana e hari ke ana ranei i ana heeki, ka ahei kia whainatia ki te moni kia kaua e nui atu i te £20.

H. D. BELL,
Minita mo nga Mahi o Roto i te Tominiona.

John Mackay, Government Printer, Wellington.

Fig. 10.5 A notice to publicise the protected status of tuatara, printed in English and Māori on linen. The notice was distributed in 1913 to acclimatisation societies, post offices and police stations and to a schoolteacher on Motiti Island.

Photo: Archives New Zealand Te Rua Mahara o te Kāwanatanga, Wellington Office, ref. IA 1 46/18/4 pt 1.

Island was purchased from its Māori owners and gazetted as a sanctuary in 1917, among the first for a native species.

Another initiative taken in this period was to publicise the protected status of tuatara in ways intended to reach both Māori as well as Europeans. In August 1913 Crewes was informed, by a teacher from Motiti Island in the Bay of Plenty, that local Māori were unaware of the penalty for killing tuatara; they were said to be killing tuatara on nearby islands 'out of pure mischief' while collecting muttonbirds. The teacher recommended that a notice be printed, in Māori, to explain the protected status of tuatara.[78] This suggestion was pursued by Internal Affairs; notices in Māori and English were printed on linen (Figure 10.5) and distributed to the Motiti Island teacher, as well as to acclimatisation societies, post offices and police stations in northern New Zealand, with a request that they be displayed where they would be seen by Māori.[79] A further announcement was placed in the *Gazette,* and instructions were issued for publication in *Kahiti*, the Māori Gazette, to reiterate the status of tuatara as an absolutely protected species.[80]

Given the shortcomings of government efforts during this period, conservation-minded individuals recognised a need for a public lobby group. In 1914 Harry Ell, with support from G. M. Thomson, James Drummond and others, initiated the first such group, the New Zealand Forest and Bird Protection Society. Although it survived only five years, this group was a forerunner to what became a later group under the same banner, and which today is New Zealand's leading public organisation for conservation.

Appointment of Special Protectors: 1913–1932

Two other initiatives in 1913 involved the lighthouse service. One was to instruct the captain of the government steamer to shift tuatara eggs among islands. Another was to appoint the lighthouse keepers (employees of the Marine Department) on Stephens, Brothers and Cuvier Islands, and the caretaker (an employee of the Department of Tourist and Health Resorts) on Hauturu, as 'special protectors of the tuatara lizard'. The recommended duties of these men, for which they received a salary of £5 per annum,[81] were as follows:

1. To protect the lizards in every way.
2. To see that they [the tuatara] are not interfered with by children or visitors.
3. To kill <u>all</u> cats on the islands.
4. To report from time to time generally as to the number of lizards seen and whether they think they are increasing or decreasing.[82]

This was undoubtedly a 'feel-good' scheme, but its success and value to conservation are debatable. The greatest achievement of the lighthouse keepers was on Stephens Island, where cats were eradicated by about 1925 (but note that, with the incentive of a bounty, eradication was virtually complete by the time that the Special Protector scheme was initiated). Cats never reached the Brothers Islands, and on both Stephens and the Brothers, hawks were for a time also killed under a bounty scheme (Chapter 3).

On Cuvier Island, however, the commitment of the lighthouse keepers was erratic. Attempts to control rats around the lighthouse station, underway by 1924 with reported benefits for tuatara (Chapter 7, Box 7.2), continued into the 1930s.[83] Control of cats, however, was more controversial. Although the Marine Department had banned the keeping of cats on all light stations with tuatara in 1922,[84] some keepers on Cuvier Island continued to bring their pets, and it was not until 1970 that cats (domestic and feral) were completely eradicated. Feral goats were also present; these eventually became an obvious source of damage to the vegetation, and thus to the habitat of tuatara. Although some were shot, more goats were introduced by the keepers, and eradication (eventually completed in 1961) was resisted for decades, in part because of the value of goats as dog food, indicating that dogs were also present.[85]

On Hauturu Island, the Tourist Department caretaker, Robert Nelson, assiduously reported to head office of Internal Affairs from 1913 to 1932. He made regular searches for tuatara, but saw very few (Chapter 6). Apart from receiving encouragement to keep killing cats (a hopeless mission for one man on a mountainous island of 3083 hectares), he seems to have been given little direction.[86]

Further examples of negligent management, by today's standards, of the Special Protector scheme have been identified with respect to Cuvier Island,[87] and many of the problems applied elsewhere. Each keeper (there were several on each island) was made a Special Protector and, in the initial years at least, wrote his own report, with consequent duplication of effort. There were few guidelines from Internal Affairs on what to put in the report, and requests from keepers for information on the biology of tuatara were not always met. Although some of the early observations of the keepers were published in newspapers, and reports were forwarded to biologists at the Dominion Museum, most of the information was essentially filed and forgotten. Some of the reports had useful observations, but the opinions on tuatara numbers were highly variable and clearly influenced by the keepers' skills and interests.

The involvement of two government departments in the scheme also created friction and tremendous bureaucratic inefficiency. All correspondence between Internal Affairs and the keepers (in either direction) had to pass through the Secretary of Marine (one keeper found it necessary to apologise to the Secretary after approaching Internal Affairs directly for more ammunition to kill hawks[88]). There was also confusion about which government department was to pay for the killing of cats and hawks, whether the Special Protector allowance covered the costs of ammunition, and how inequities in salary introduced by the bounty for killing hawks (in which the Principal Keeper was not allowed to participate) would be resolved.[89] Issues concerning limits on livestock numbers and the fencing of remnant forest on Stephens Island,[90] as well as the management of cats on Cuvier Island, indicate that the Marine Department held the upper hand.

Meanwhile, vulnerable populations of tuatara continued to dwindle and disappear. With the intention of protecting a vulnerable Bay of Plenty population, officials organised a transfer of tuatara to Mokopuna Island, an islet off Matiu/Somes Island, in 1920. The lighthouse keepers on Matiu/Somes Island were made Special Protectors, but the tuatara soon disappeared from Mokopuna Island with little comment. Despite public pleas dating from 1895 for the Marine Department not to allow cats on islands with lighthouses,[91] cats were able to establish on East Island (Whangaokeno Island) when the lighthouse was constructed between 1898 and 1900. Inexplicably, the lighthouse keepers there and on Burgess Island (Pokohinu) in the Mokohinau Islands (which had no cats but whose vegetation was heavily affected by sheep, cattle, pigs and feral goats) were never made Special Protectors, despite the remnant populations of tuatara there – all now extinct (Chapter 3, Table 3.1). If the European mammals introduced during lighthouse construction and operation had been controlled or removed, tuatara might have survived a few more decades until Pacific rats (*Rattus exulans*) could be eradicated from these islands.

The stipend for Special Protectors ceased in 1932. Internal Affairs considered that tuatara were no longer threatened, but the financial pressure of the Great Depression clearly played a role in the decision.[92] One keeper tried, unsuccessfully, to have the allowance reinstated in 1936.[93] Another keeper who wrote to the Secretary of Marine in 1937, asking him 'just what the keeper's duties are in respect to the Tuataras', was told that there were no specified duties.[94] The bounty on hawks had also finally ended (in 1929), although keepers on Stephens and the Brothers Islands, at least, were still killing hawks in 1941–1942.[95] Despite the official end of the Special Protector scheme, from time to time Internal Affairs continued to request reports from the keepers on the 'Tuatara Lizards [*sic*] … and opinion as to whether or not the lizards are increasing'.[96]

Development of a national wildlife service: 1920s–1970s

In 1922 an expedition to Stephens Island by staff of the Dominion Museum found that tuatara 'were plentiful and apparently increasing'.[97] This report, following soon after the translocation of tuatara to Mokopuna Island, may have contributed to a sense of reassurance among officials that the survival of tuatara was no longer in doubt. Not long afterwards, Internal Affairs was criticised by a public conservation group for the possible mixing of different types of birds in island transfers. Subsequently, the department largely withdrew for many years from an active role in native species management.[98]

The decades from 1930–1970 thus illustrate a period of 'benign neglect' in the conservation of tuatara.[99] Although a Wildlife Branch within Internal Affairs became officially recognised by 1946, resources for native species conservation were few and fieldwork fell effectively to one man, field officer Logan Bell. It was Bell who made the first official visits by Internal Affairs to many offshore islands of wildlife interest. Bell was also the first to recognise that tuatara were less abundant where Pacific rats were present, but this observation received little if any attention and did not lead to changes in management (see Chapter 7, Box 7.2).

In 1951 the Wildlife Branch was renamed the Wildlife Division. New legislation soon followed; the Wildlife Act 1953 consolidated earlier amendments to the Animals Protection and Game Act 1921–22 but brought few if any changes to the management of tuatara. Studies of wild tuatara during this and the following decade were limited to those carried out by university biologists. The leading figure was W. H. [Bill] Dawbin, a researcher at Victoria University of Wellington who studied life history and general biology, mainly on Stephens Island. Members of the Auckland University Field Club made expeditions to many northern offshore islands, providing reports on the abundance (or otherwise) of tuatara there.[100]

Officials had long held a conservative, protectionist attitude toward the collection of wild tuatara. In 1930 it was reported that Stephens Island was closed to the 'taking of tuatara lizards or their eggs',[101] and in 1938 that there was 'a long standing decision, not to approve of the taking of live specimens of absolutely protected [species, including tuatara] for sending overseas'.[102] In the face of considerable international interest, officials were concerned by the precedent that would be set by approvals to collect, and were disinclined to allow collection without definite evidence that populations were increasing.[103] By 1970 the policy had apparently relaxed to the point of allowing the collection of up to 12 tuatara each year from Stephens Island. Tuatara could be collected only for the purposes of public exhibition (preferably in New Zealand zoos), breeding in captivity (without exhibition), or research (with a detailed research plan if more than one animal was required).[104] This quota was probably rarely if ever reached, and the policy eventually drew criticism from overseas for the limitations it placed on research that could have aided management.[105]

The first serious government research on tuatara began in 1970, following the appointment of scientist Ian Crook to the Wildlife Division. Crook began studies on the interactions between Pacific rats, seabirds and tuatara, and provided the first published evidence, in 1973, that recruitment of tuatara was negligible on islands where Pacific rats were present. Following Crook's departure, research on the status of tuatara populations, interactions with seabirds and nesting frequency in females continued in the Wildlife Division (renamed the Wildlife Service in 1974) under the guidance of Donald Newman.[106] During the 1980s a new wave of university scientists expanded this fundamental research into many aspects of reproductive biology and genetic variation.[107] Collectively, these developments helped set the scene for more effective management of wild populations and for improvements in captive husbandry.

At the time of Crook's work in 1973, the likelihood that rats could be eliminated from offshore islands with tuatara seemed remote. Management of tuatara thus continued to remain much more passive than the recovery operations for endangered birds like the black robin, for which the Wildlife Service was winning local and international acclaim.[108] It was only during the late 1970s, when rats were discovered to have been accidentally eliminated on several small islands (following localised rat control around seabird colonies) that thinking began to change.[109] Attempts with increasingly larger islands confirmed that deliberate eradication of rats was not only feasible; it could bring enormous benefits for natural ecosystems. By the early 1990s, tuatara would be among the beneficiaries.

Creation of a Department of Conservation in 1987: new opportunities for management

For many decades, management of conservation in New Zealand had been acknowledged as fragmented. At the same time as the Wildlife Service was attempting to protect rare species, the habitats of these species were generally being managed separately by other sections of government, such as the Department of Lands and Survey. Calls were also being made, for economic reasons, to separate the commercial, production-oriented aspects of resource management practised by some government departments from the protection or conservation-oriented aspects.

These pressures, combined with a growing international movement towards sustainable development, culminated in 1987 with a radical shake-up of government departments and the establishment of a Department of Conservation. Known rather aptly in New Zealand with the physician's title of 'DOC', the department is charged under the Conservation Act 1987 with managing natural and historical resources for conservation purposes, promoting the benefits (to present and future generations) of conservation, and allowing, within limits, for the use of natural and historic resources for recreation and tourism.

BOX 10.3

Is the tuatara a threatened species?

The IUCN Red List of Threatened Species™, published by the World Conservation Union, is the internationally recognised scheme for ranking the global risk of species extinctions. The scheme includes three categories of 'Threatened': in increasing order of concern, these are 'Vulnerable', 'Endangered' and 'Critically Endangered'. The 2011 Red List[1] assessed tuatara using taxonomy dating from 1990, according to which two species of tuatara were recognised.[2] Of these, the more common species, *Sphenodon punctatus,* was not considered Threatened (it was placed within a lower-level category of 'Lower Risk/least concern'. The Brothers tuatara, *S. guntheri,* was listed as Vulnerable (VU D1+2), the lowest of the three categories for threatened species (it was not considered Endangered). This ranking reflected a moderate risk of extinction in the wild, given that the total population of *S. guntheri* was estimated to number fewer than 1000 mature individuals (criterion D1) and that the species occurred in fewer than five locations, making it vulnerable to stochastic events (criterion D2).

The IUCN ratings are, however, in urgent need of updating. Importantly, they do not take into account the recent conclusion, published by Jennifer Hay and colleagues in 2010, that the genus *Sphenodon* 'is best described as a single species'.[3] This conclusion was

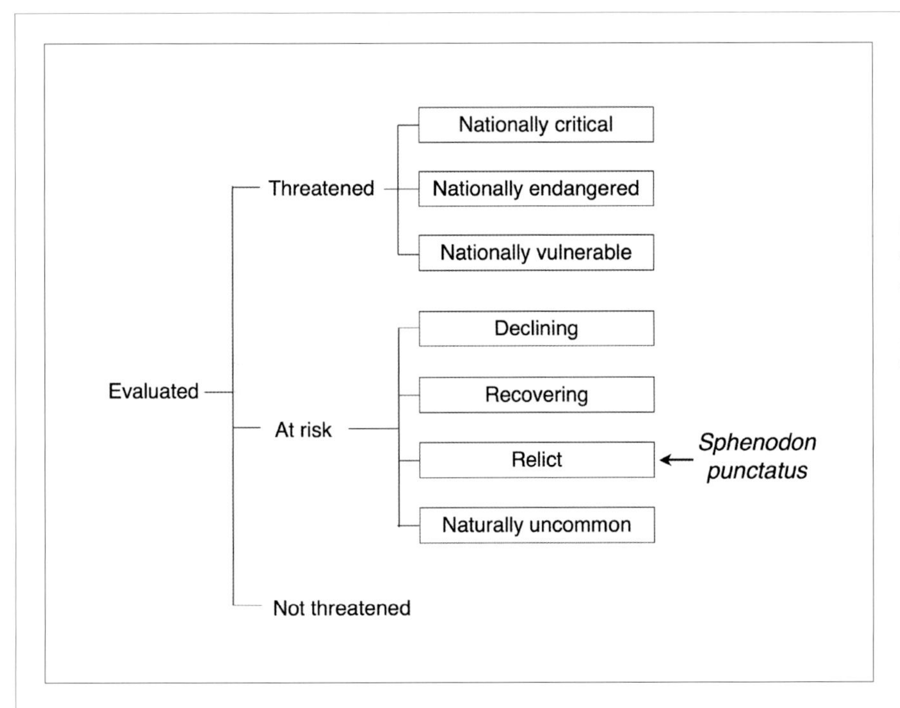

Box 10.3 Fig. 1 Categories of threatened species according to the current New Zealand Threat Classification System. The status of *S. punctatus* (representing all living populations of tuatara) is rated as 'Relict' within the category 'At risk'.

Modified from Townsend, A. J., de Lange, P. J., Duffy, C. A. J., Miskelly, C. M., Molloy, J. and Norton, D. A. 2008. *New Zealand Threat Classification Manual*. Department of Conservation, Wellington.

reached from the authors' own analyses of variation among populations in microsatellite DNA, as well as from a number of earlier genetic analyses examining variation in mitochondrial and nuclear DNA.[4] In other words, the Brothers tuatara is no longer recognised as a separate species (one study in 2007 went so far as to suggest that the population is naturally inbred as a result of being founded by few individuals).[5]

Within New Zealand, the Department of Conservation has its own set of criteria for ranking the national threat of extinction.[6] Under this system, which helps guide conservation actions and spending, the three subcategories within 'Threatened' are very similar to those of the IUCN (though prefaced by the word 'Nationally'). At a lower risk of extinction, there is a category of 'At risk'. This is subdivided into four subcategories considered appropriate for New Zealand conditions, given that many animals and plants have populations that are naturally small in size and geographic range but that are not at immediate risk of extinction. The New Zealand classifications were updated during 2009; all living populations of tuatara were recognised as *S. punctatus* and placed within the sub-category 'Relict' of the category 'At risk'.[7]

To summarise, the sole living species of tuatara, *S. punctatus*, cannot be considered by local or global criteria as Threatened; it is, in fact, relatively secure compared with many New Zealand animals and plants. It is, nonetheless, vastly reduced in total population size and in area of occupancy compared with its pre-human distribution, and it therefore remains a legitimate subject for recovery actions.

In addition to the above threatened-species rankings, tuatara are also affected by an international agreement known as the Convention on International Trade in Endangered Species of Wild Fauna and Flora (CITES). This convention, ratified in 1989 in New Zealand as the Trade in Endangered Species Act, lists tuatara as a 'species endangered by trade' (Appendix 1 of CITES). This uses the term 'endangered' in a different way from threatened-species rankings, in essence recognising that tuatara are of sufficient interest that unregulated export from New Zealand could lead to an endangered status. Export of tuatara therefore requires permission from the Director-General of Conservation, who will seek reassurance that each export is not detrimental to the survival of the species.

BOX 10.4

Remember the Treaty: Māori and the conservation of tuatara

Shortly before New Zealand became a colony of the British Empire in 1840, a brief written agreement, in Māori and English versions, was reached between the British Crown and Māori chiefs. Proposed with the mutual benefit of Māori residents and British settlers in mind,[1] the Treaty of Waitangi has come to be viewed by New Zealanders as a founding document of nationhood.

Nonetheless, the Treaty was for more than a century largely overlooked by those in powerful positions. As a result, many Māori have grown up with histories of injustice, and many non-Māori have grown up with a level of ignorance, regarding their country's colonial history. Since the 1980s, the establishment of the Waitangi Tribunal has provided a mechanism for Māori to seek acknowledgement and redress of past injustices. This settling of grievances continues today, amid much discussion and controversy.

One major development since the late 1970s is that many pieces of legislation, including the Conservation Act 1987, now require public organisations to give effect to the principles of the Treaty of Waitangi. This is not always simple in practice, in part because of uncertainty about the intended meaning of the Treaty. Under Article 1, Māori ceded 'sovereignty' (English version) or 'kawanatanga' (governorship or governance, Māori version) to the British Crown, but for Māori these were not necessarily the same thing. Under Article 2, Māori were to retain 'full exclusive and undisturbed possession of their Lands and Estates Forests Fisheries and other properties ...' (English version), or 'te tino rangatiratanga o o ratou wenua o ratou kainga me o ratou taonga katoa' (the unqualified exercise of their chieftainship over their lands, villages and all their treasures; Māori version), so long as it was their wish and desire to retain these in their possession.[2] Regardless of text or translation, it is clear, when one looks at the history of conservation management, that Māori in 1840 were promised much greater control over natural resources than they received. This extends to tuatara.[3]

In current practice, the Treaty principles have been interpreted by the courts to mean, among other things, that 'the Government has the right to govern and make laws ... Māori have the right to control their own resources ... the relationship between Māori and the Crown is akin to a partnership ... the Crown has an obligation actively to protect Māori interests ... [and] to consult with Māori to be properly informed'.[4] With respect to tuatara, consultation occurs between the Department of Conservation (DOC) and Māori at various levels. In recent years Māori representatives have often attended, and contributed valued perspectives to, meetings of the DOC-led Tuatara Recovery Group. Specific actions, such as proposed translocations of tuatara or the eradication of pest species from islands, only proceed after discussion with the appropriate iwi; Māori have themselves sometimes initiated and implemented island restoration projects, with practical support from DOC. An agreement was reached between Ngāti Koata and DOC over co-management of Takapourewa Nature Reserve, which encompasses virtually all of Stephens Island (Takapourewa),[5] and discussion with iwi over co-management of other islands continues.

Researchers, too, now engage in direct consultation with iwi as a result of memoranda of understanding reached between universities or government science agencies and Māori. This heightened level of consultation between the Crown, iwi and other interest groups has led to developments to help foster the interests and abilities of Māori. A leading example of 'capacity-building' involves the training of young members of Te Ātiawa by Victoria University of Wellington researchers, with the support of the Royal Society of New Zealand, to lead educational

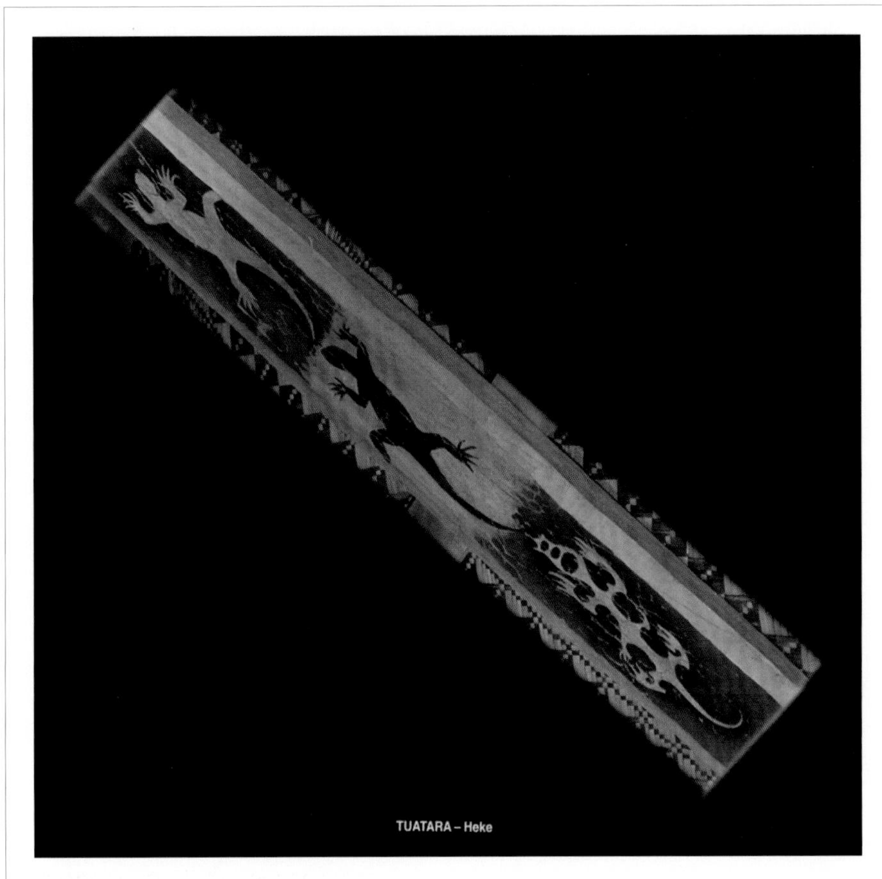

Box 10.4 Fig. 1 Māori today retain a sense of traditional knowledge about reptiles, and view tuatara as a taonga (treasure). This strong interest in and concern for tuatara is revealed in various aspects of culture. In this contemporary artwork, created by Daniel Ormsby of Ngāti Maniapoto at Te Wānanga o Aotearoa, tuatara are seen delicately climbing a heke (rafter) of a carved storehouse. In striving to reach an attainable yet perpetually advancing pinnacle, tuatara are illustrating a message for humans: 'the point being, although the pinnacle is very important, enjoy the process where you are'.[14]

Photo: Te Wānanga o Aotearoa.

'tuatara road shows' around New Zealand.[6] As a result, Te Ātiawa have now moved to display tuatara in a captive-breeding centre within their rohe (tribal area). Another such centre, Natureland, in Nelson, holds captive tuatara under a licence that recognises a special relationship with Ngāti Koata, and Ngāti Koata is also involved with the Spinyback Tuatara Education and Conservation Charitable Trust, which aims to educate schoolchildren and others to care for tuatara.[7] These developments are consistent with the second Tuatara Recovery Plan, which specifically notes that full use should be made of the knowledge and skills held by iwi and other special interest groups when tuatara are managed for advocacy purposes.[8]

From a researcher's perspective, this new environment has been both intellectually stimulating and immensely challenging. It requires those trained in Western science to think in non-traditional ways, to listen to different viewpoints on nature and its management, to explain their science in effective terms, to convey that their interest in research originates from a deep respect for animals, and to take the hard knocks arising from unresolved differences between the Crown and Māori when they come. It requires sensitivity over the use of Māori names for islands, over the confused history of island 'purchases', and of relationships between different iwi. It means accepting that each iwi acts with autonomy (there is no national 'Māori' policy on tuatara), and that many iwi have a non-transparent decision-making structure and a time-frame for responding to research or management proposals that may not suit Western pressures of grant applications, conferences and publications.

One area of specific tension concerns the accommodation of Māori spiritual values. Māori culture places great emphasis on whanaungatanga (kinship), whakapapa (genealogy, family history) and mauri (life force); sometimes these values appear threatened by proposals to separate parts of a tuatara's body (for example, to take blood, to clip toes from hatchlings[9] or to dissect dead specimens). Genetic studies pose a particular challenge, in part because of the perception that different life forms will be combined (e.g. incorporation of tuatara genes into bacterial genomes for analysis). Tribal mana (authority, prestige) is also important, and kaumātua (Māori elders) often wish to accompany tuatara (which may be viewed almost as family members) when the latter are shifted away from their tribal area. Such shifts may be relatively straightforward when the destination is close, but if tuatara or their tissues are to be accompanied to a zoo or research institution in the northern hemisphere, then the financial costs can become problematic. There is also apprehension among Māori that animals sent overseas may yield commercial gains for others, with no return for Māori.

For some Māori, recent moves toward greater consultation are not enough. A continued sense of injustice among six iwi led in 1991 to a challenge to Crown management of natural resources under a Treaty of Waitangi claim known as Wai 262.[10] This claim, sometimes known as the 'flora and fauna claim', is complex and has broader concerns for cultural and intellectual property than the informal title implies. Among other things, with respect to tuatara, it states that the claimants have been prevented from exercising te tino rangatiratanga (sovereignty,

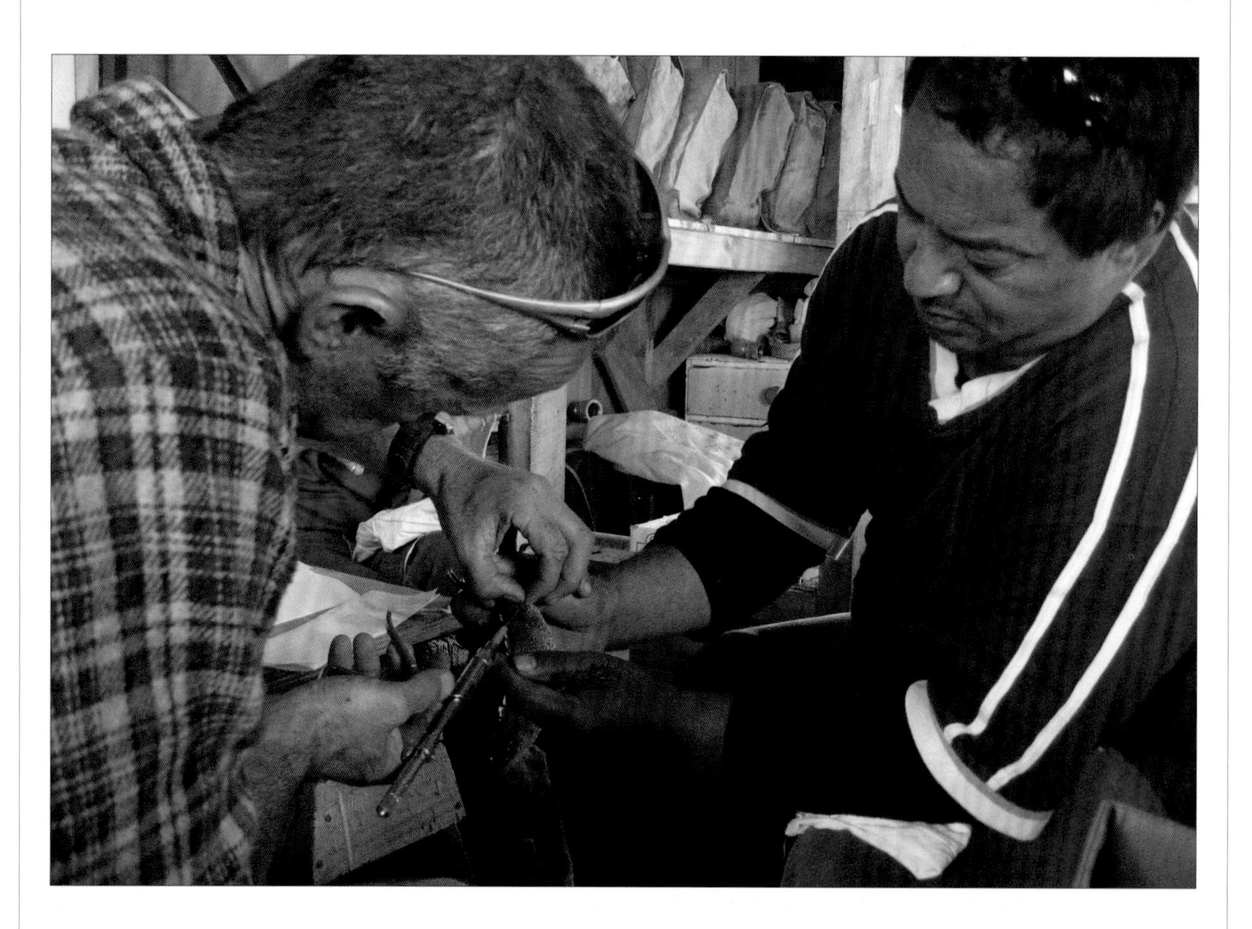

Box 10.4 Fig. 2 Anaru Paul of Ngāti Koata (right) assists Peter Gaze of DOC (left) with the implantation of PIT-tags (passive integrated transponder tags) into a tuatara. The tags provide unique identification for tuatara translocated from Stephens Island to Wakaterepapanui Island in 2003. An agreement with DOC for co-management of Takapourewa Nature Reserve identifies Ngāti Koata as kaitiaki (guardians) of the island, and hence of the tuatara there.

Photo: Lee Pagni.

Box 10.4 Fig. 3 Rachael Burgess of Te Ātiawa shows a captive-reared tuatara to schoolchildren from New Plymouth as part of a road show for tuatara in May 2005.

Photo: Susan Keall, Victoria University of Wellington.

or unqualified exercise of chieftainship), including kaitiakitanga (stewardship or guardianship), as a result of the alienation of iwi from islands that were subsequently made reserves or sanctuaries, and as a result of the legal protection of tuatara (including export for scientific or diplomatic purposes) without negotiation or input from Māori. The Crown is also considered to have acted in breach of the Treaty by, and without iwi consent, facilitating the introduction of exotic pest species and by permitting the destruction or substantial modification of the habitats of indigenous flora and fauna. Essentially, the claimants have sought the ability to make decisions about (not just to be consulted over) the future conservation and control of natural resources, including tuatara. Hearings began in 1998 and the Waitangi Tribunal released its lengthy and wide-ranging report in late 2011.[11] In its findings, the Tribunal emphasises that the ability of Māori to exercise the responsibilities of kaitiakitanga is essential to the preservation of Māori cultural identity. It makes a variety of recommendations to government about how this might be achieved, emphasising a forward-looking partnership that has the potential in some respects to be world leading.

In summary, these recent developments emphasise that management of natural resources today is as much about people as it is about non-human animals and plants. Many biologists consider the support and involvement of local human communities to be crucial to successful conservation; the Tuatara Recovery Plan specifically sees the increased involvement of iwi as essential to the plan's success.[12] My own experience is that inputs from a diverse range of humans have the potential to bring new and valuable perspectives on the conservation of tuatara, as well as a more unified sense of purpose and shared responsibility. Although Western-style conservation has historically leant toward protectionism and Māori tradition favours resource use,[13] this does not necessarily mean that conservation goals need be in conflict. Existing grievances held by some Māori must be resolved, but overall I am optimistic that the national commitment to conservation of tuatara will be strengthened by these recent developments.

Some mourned the loss of the Wildlife Service and its replacement with a larger and more bureaucratic department; it seemingly spelled the end of a 'can-do' attitude that had been very successful in managing endangered birds in remote locations. On the other hand, it was now possible to conserve wild animals and their habitats on Crown-managed land in a coordinated manner. Importantly for tuatara, neglected non-avian species were now given explicit attention. In another fundamental shift in policy, DOC was required to give effect to the principles of the Treaty of Waitangi, setting the scene for a much greater involvement of Māori in the management of tuatara and other native species. This process of bicultural engagement continues to evolve today.

The first Tuatara Recovery Plan

Following international trends, the new department chose to manage threatened species, including tuatara, under a species recovery plan. Recovery plans produced by DOC are not legally binding, but provide a general statement of intent, subject to available resources. Discussions among DOC staff, university scientists, zoo curators and others led to a draft recovery plan for tuatara in 1990;[110] the final version was eventually published in 1993 following a period of wider consultation.[111]

Until this time, tuatara had been managed as one taxon (the species *Sphenodon punctatus*), present on about 30 islands. However, genetic research led by Charles Daugherty and published in 1990 had indicated that tuatara consisted of two species, one of which comprised two probable subspecies.[112] The plan therefore recognised three genetic stocks, each with its own management needs: *S. punctatus punctatus* (northern tuatara, on about 25 islands), an unnamed subspecies of *S. punctatus* (Cook Strait tuatara, on four islands), and rarest of all, *S. guntheri*, the Brothers tuatara, on one island only (but note that *S. guntheri* is no longer recognised as a separate species[113]). The plan pointed out that, if present trends continued, between four and seven populations of tuatara (of the total of 30 then remaining) were likely to become extinct within the next 50 years.

Key aims identified for the 5-year life of the recovery plan were to raise the security of existing populations (mainly by minimising the risk of new arrivals of rodents), to establish new populations by captive breeding and translocations of at-risk genetic stocks, and to restore several islands (by removing rats and, where present, rabbits) to allow the eventual return of tuatara, housed in temporary captivity elsewhere, to those islands. The plan also advocated for controlled public access to two of the newly established populations, so that wild tuatara and an island restoration programme could be viewed first-hand. This would provide enhanced opportunities to educate the public and advocate for conservation, while also providing a visible return on conservation spending for the New Zealand taxpayer and treasury officials.

In another major paradigm shift, the recovery plan recognised that proposed recovery actions for relict populations would require involvement from the captive-management community for success.[114] Until this time, tuatara in captivity had effectively been managed in isolation according to the wishes of the holding institution, with no specific mandate for conservation. To help integrate zoo activities with the recovery plan's goals, a scientific workshop on captive management was held in 1990 and a captive-management coordinator, Barbara Blanchard of Wellington Zoo, was appointed.[115] The plan also actively encouraged research to assist conservation, including topics such as the effects of rat eradication, the restoration of wild tuatara populations and captive breeding.

A recovery group was appointed to provide national oversight of the plan's implementation. The group included DOC staff, scientists from universities and a professional herpetological society,[116] and the captive-management coordinator. On a day-to-day basis, local activities were implemented by the seven conservancies (geographic management areas within DOC) with responsibility for wild populations of tuatara. Significant financial contributions toward aspects of the plan coordinated by Victoria University of Wellington came from overseas sources, mainly the Zoological Society of San Diego.

TABLE 10.1

Steps taken from 1991 to 2011 to restore populations of tuatara depleted in the presence of Pacific rats (kiore, *Rattus exulans*)

Island	Area (ha)	Population status prior to restoration	Year of rat eradication	Management of tuatara population[1]	Results[2]
Lady Alice	155	Low density; adults and juveniles	1994	Recovery in situ	Dramatic increase in recruitment (in 2001, 41.2% were juveniles)
Whatupuke	102	Low density; adults only	1993	Recovery in situ	Dramatic increase in recruitment (in 2002, 43% were juveniles); increased body mass of adults
Coppermine	80	Low density; adults only	1997 (following an unsuccessful attempt in 1992)	Recovery in situ	Increased recruitment (in 2002, 13.2% were juveniles); increased body mass of adults
Hauturu/Little Barrier	3083	Exceptionally low density (< 10 adults known)	2004	Captive breeding since 1991 with 8 adults in rat-proof enclosure on island; eggs incubated off-site	131 captive-reared offspring (aged up to 12 years) released 2006–2010; 1 of original adults released 2008 (7 adults, 7 offspring and 31 eggs remained in captivity)
Cuvier (Repanga)	170	Exceptionally low density (< 10 adults known)	1993	Captive breeding on North Island with 6 adults	39 captive-bred juveniles released 2001–2010 (4 adults and 7 juveniles remained in captivity)
Red Mercury (Whakau)	225	Exceptionally low density (< 20 adults known)	1992	Captive breeding on North Island with 11 adults	11 adults released 1996–1998; 24 captive-bred juveniles released 1996–2001
Kawhitu or Stanley	100	Exceptionally low density (< 20 adults known)	1991	Captive breeding on North Island with 15 adults	17 captive-bred juveniles released 2003–2010 (7 adults, 10 juveniles and 4 eggs remained in captivity)

Notes

1 Off-site captive management has involved Auckland, Hamilton and Wellington zoos, Otorohanga Kiwi House and Victoria University of Wellington.

2 Results to December 2010. Sources: Cree et al. 1991e, 1994; Keall and Daugherty 1997; Towns et al. 2001, 2007; Moore et al. 2008b; minutes of Tuatara Recovery Group meetings; Barbara Blanchard, Wellington Zoo, pers. comm. 25 July 2006, 28 October 2009 and 28 January 2011; Leigh Marshall, Department of Conservation, pers. comm. 29 August 2006; Susan Keall, Victoria University of Wellington, pers. comm. 23 October 2009 and 8 March 2011; Shane McInnes, Department of Conservation, pers. comm., 20 December 2010 and 21 February 2011; Rob Chappell, Department of Conservation, pers. comm. 22 December 2010. See also Chapter 6.

BOX 10.5

Playing the pied piper: eliminating rats from islands that support tuatara

Eradication of introduced rats from offshore islands has been central to the recent conservation of tuatara. One obvious benefit is that recruitment of tuatara has resumed on islands where populations were previously dwindling in the presence of rats (see Chapter 7, Box 7.2). In addition, several rat-free populations of tuatara are now less at risk from the possible arrival of rats from neighbouring islands. Finally, by 2011 tuatara had been reintroduced to seven islands within their former range, as part of ecological restoration programmes (see Box 10.6).

The first deliberate attempts to remove rats completely from offshore islands in New Zealand occurred in the early 1980s, following the accidental demonstration that eradication using anti-coagulant poisons was possible.[1] Early attempts at eradication involved labour-intensive, ground-based distribution of bait, but recent operations have used helicopters equipped with global positioning systems to ensure an accurate and even spread of bait. The toxin used, brodifacoum, is a second-generation anticoagulant related to warfarin; it inhibits the synthesis of vitamin K-dependent clotting factors in the liver, and causes haemorrhagic death in rats within about a week of consumption. The bait is spread in late winter, when alternative food sources for rats are in short supply and activity of non-target species is low. Brodifacoum is insoluble in water and breaks down in soil within a few months of dispersal. It appears to have no direct effects on many invertebrates, and although it is poisonous if consumed by mammals, birds and some reptiles,[2] harmful side effects on island ecosystems have been minor, short-lived and manageable.[3]

By 1997 Pacific rats (kiore, *Rattus exulans*) had been eradicated from six islands with resident free-roaming tuatara. Listed by the year in which bait was aerially distributed, these are Kawhitu or Stanley Island (1991), Red Mercury Island (Whakau) (1992), Cuvier Island (Repanga Island) (1993), Whatupuke Island (1993), Lady Alice Island (1994) and Coppermine Island (1997, following an unsuccessful ground-based distribution in 1992). On the first three, most tuatara were removed to temporary captivity prior to spreading the bait, but subsequent operations on the three Marotere (Chicken) Islands left the tuatara in residence. Tuatara are attracted primarily to moving food, so deliberate consumption of bait is unlikely and no harmful effects of bait distribution on islands have been observed. Once 2 years had passed without sign of rats, the islands were declared rat-free. A dramatic increase in recruitment of juvenile tuatara has since been observed on the Chickens Islands.[4] The numbers of tuatara on Kawhitu, Red Mercury and Cuvier Islands are as yet too low to detect such a response, but similar results are expected in time.

As of 2011, the most recent confirmed eradication of rats from an island with tuatara involved Hauturu/Little Barrier Island in 2004. At 3083 hectares, this is by far the largest of any island with tuatara to be treated.[5] It also presented more political difficulty than any previous eradication of rats, and the project came to fruition only after at least 7 years of consultation led by the New Zealand Department of Conservation (DOC).

The main objection to eradication of Pacific rats from Hauturu arose from within the Māori community. Hauturu was inhabited by Māori for several centuries before being forcibly acquired by the Crown as a nature reserve in the early 1890s. Today, the island is within the rohe (tribal area) of Ngāti Manuhiri, Ngāti Rehua and Ngātiwai. The first two hapū (sub-tribes) were supportive of the proposal to eradicate rats, but individuals within the Ngātiwai Resource Management Unit, on behalf of the Ngātiwai Trust Board, made strenuous objections on

Box 10.5 Fig. 1 (top) The eradication of rats from Hauturu/Little Barrier Island in 2004 was an industrial-scale operation. Here, Department of Conservation (DOC) workers store the poison bait on the island in preparation for distribution.

Box 10.5 Fig. 2 (far left) Richard Griffiths, DOC programme manager for island biodiversity, holding pellets baited with brodifacoum.

Box 10.5 Fig. 3 (left) The bait was accurately distributed over Hauturu by a helicopter fitted with a global positioning system.

Photos: Department of Conservation.

Box 10.5 Fig. 4 A temporary sign warning of the presence of poison bait.

Photo: Department of Conservation.

several grounds. Pacific rats, known as kiore, are seen as a taonga (treasure) by Māori: they have shared the New Zealand landscape with humans for over 700 years, become part of cultural traditions and were once a valued item of food. The Ngātiwai Resource Management Unit thus proposed that a pre-European condition (with kiore) rather than a pre-human condition (without kiore) should be the restoration goal for the island; the unit also sought an agreement with DOC for future co-management of the island prior to eradication taking place. After protracted negotiation, members of Ngātiwai were assisted to remove 179 kiore from the island to be established in another location or maintained in captivity,[6] and the eradication went ahead. Representatives of Ngātiwai have been invited to participate in developing a restoration plan for Hauturu following the eradication.

The operation itself was a feat of military-style planning.[7] Some 55 tonnes of bait were barged to the island and stored while awaiting suitable weather. Three helicopters spread the first load in June 2004 at a rate of 8 kilograms per hectare in overlapping swaths; a second application was made a month later. To minimise negative impacts, the enclosures in which the remaining few tuatara had been held since 1991 for a captive-breeding programme (see Table 10.1) were covered with tarpaulins to prevent entry of bait, and later with insect-proof mesh for a month to prevent the potential entry of affected prey. The island was declared rat-free in June 2006, by which time the nesting success of a threatened seabird, Cook's petrel (tītī, *Pterodroma cookii*) had already increased dramatically.[8] In November 2006 around 70 people gathered on the island to celebrate the release of captive-bred tuatara: 68 offspring, 8–12 years in age, were released then or soon afterwards into an area of former habitat for tuatara. By 2010 about 63 more had been released.[9]

Tension remains within New Zealand over how best to manage the conflicts between the ecological benefits that would come from rat removal, versus the need to acknowledge iwi interest in maintaining kiore on some islands. These challenges are legally complex: on the one hand, DOC has an obligation under the Reserves Act 1977 to remove pest species and under the Conservation Act 1987 to preserve and protect natural resources; on the other, the Conservation Act requires the department to administer both acts so as to 'give effect to the principles of the Treaty of Waitangi'.[10] In practice, this includes the need to consult with Māori and to actively protect Māori interests. Some members of Ngātiwai had suggested that rats could be controlled, rather than completely removed, on islands such as Hauturu, but this is considered by DOC staff to be logistically impractical, financially costly and of uncertain effectiveness compared with benefits that eradication is known to bring.[11] The focus in Northland has thus moved towards the possible maintenance of kiore in other, less vulnerable locations.

Elsewhere, not all iwi organisations have been opposed to eradication of kiore. The Te Whanau a Tauwhao ki Tuhua Trust Board, for example, actively supported eradication from Mayor Island (Tuhua), which was subsequently a release site for tuatara.[12] Wherever an eradication of kiore is proposed, it is essential that such agreements are reached, for it would be all too easy for conservation actions to be blocked if new grievances are created.

The only two islands with tuatara on which kiore remained in 2011 were Taranga and Mauitaha Islands in the Hen and Chickens group. With a confirmed sighting of only one tuatara, in 1994,[13] the 'population' of tuatara on the small island of Mauitaha (26 hectares) is doomed. However, the population of about 50 adult tuatara[14] on the much larger Taranga (500 hectares) is anticipated to recover, now that a rat-eradication programme agreed with Ngātiwai is underway.[15] The eradication attempt began in May 2011 and, if successful, will be one of the most significant events of the past decade for conservation of tuatara. Because of its large size and varied altitude, Taranga is likely to be less vulnerable than most other northern islands with tuatara to some of the predicted impacts of climate change. The agreement has come just in time, as the surviving tuatara will soon be nearing the end of their lifespan.

Box 10.5 Fig. 5 The release of tuatara on Hauturu began in 2006, two years after rat bait was distributed. Here, Whetu McGregor (right), a kuia (female elder) of Ngāti Rehua whose family has a long association with tuatara, and Susan Keall (left), who assisted with egg incubation at Victoria University of Wellington, admire a feisty young tuatara just prior to its release. The photo is taken inside the captive-breeding enclosure on the island, within which the few remaining adult tuatara have been held since 1991 to provide protection from rats.

Photo: Alison Cree.

TABLE 10.2

Islands on which tuatara were released between 1995 and 2011

All except Wakaterepapanui, Titi and Long Islands have some evidence for the past presence of tuatara (see Chapter 3). Although it is too early to claim that self-sustaining populations have been established, all reintroductions monitored for at least 5 years have met preliminary targets for successful translocation.

Destination is. (hectares)	Rats eradicated (by year)[1]	Other introduced mammals removed	Date of release	Tuatara translocated[3]	Iwi involved	Source island & genetic stock	Follow-up assessments[2]: Survival?	Follow-up assessments[2]: Mating, nesting?	Follow-up assessments[2]: Juveniles hatched on island?	Questions researched experimentally
Titi (32)	Norway (1975)	–	Nov. 1995	18 wild adults (7 M, 11 F) and 50 captive-reared juveniles aged 3–5 y	Te Ātiawa, Ngāti Kuia	North Brother (Cook Strait)	61% of adults and 56% of those released as juveniles recaptured over 5 y; 44% of adults and 28% of those released as juveniles recaptured at 11 y after release	One failed nest seen at 27 mo	1 island-hatched juvenile found at 11 y after the release	Effect of egg incubation history on translocation success for captive-reared juveniles
Moutohora (Whale) (143)	Norway (1986)	Cattle, sheep, goats, cats, rabbits	Oct. 1996 & Oct. 2008	32 wild adults (12 M, 20 F) in 1996; 30 large juveniles or small adults (14 M, 16 F) in 2008	Ngāti Awa	Moutoki (northern)	First release cohort: 88% recaptured over first 9 mo; 25% recaptured over 5 y; 33% re-sighted at 9 y after release	First release cohort: courtship seen during first 9 mo	First release cohort: 7 island-hatched young (some already adult size) seen at 9 y after the release	First release cohort: use of artificial refuges
Matiu/Somes (25)	Ship (1990)	Agricultural & quarantined species	Oct. 1998	20 wild adults (10 M, 10 F) and 35 captive-reared juveniles aged 7 y	Te Ātiawa	North Brother (Cook Strait)	45% of adults and 26% of juveniles re-sighted at 5 y after release; 40% of adults and 31% released as juveniles recaptured at 8 y after release	Mating seen; nest found at 9 y	2 from island nest (found at 9 y after release) later released after hatching off-island	Effect of egg incubation history on translocation success for captive-reared juveniles
Tiritiri Matangi (197)	Pacific (1993)	Rabbits, cats	Oct. 2003	60 wild adults (28 M, 32 F)	Ngāti Paoa, Ngāti Horowhenua ki Hauraki, Kawerau a Maki, Ngāti Karaua, Ngāti Whaunaunga, Ngāti Hako	Atiu (northern)	45% of adults recaptured at 5 y after release	Courtship seen; 2 nests (1 failed) seen at 3 mo after release	3 half-grown juveniles, and several hatchlings in nest, seen at 6 y after release	Use of artificial refuges
Wakaterepapanui (61)	Pacific and Norway (1999)	Sheep	Oct. 2003 & Oct. 2004	89 wild animals (37 M and 22 F adults, 30 juveniles) in 2003); 343 captive-reared juveniles aged 5 y in 2004	Ngāti Koata	Stephens Island (Takapourewa) (Cook Strait)	20% of first cohort recaptured 3 y after release; 6% of second cohort found 2 y after release	Not reported	Not reported	Effect of egg incubation history on translocation success for captive-reared juveniles
Mayor (Tuhua) (1277)	Pacific & Norway (2003)	Pigs, cats	Oct. 2007	30 wild adults (15 M, 15 F)	Ngāi Te Rangi	Karewa (northern)	27% recaptured at 4 mo after release; 23% recaptured at 14 mo after release	Not reported	Not reported	–
Long (142)	Pacific (1997)	–	Nov. 2007	53 captive-reared juveniles aged 7 y	Te Ātiawa, Rangitāne	North Brother (Cook Strait)	Several individuals seen within 3 y of release	Not reported	Not reported	–

The second Tuatara Recovery Plan

The directions set, and met so successfully, by the first Tuatara Recovery Plan were broadened in the second Tuatara Recovery Plan, published in 2001. The plan's long-term goal, to maintain the genetic diversity of tuatara, is addressed by attempting the return of 'all existing populations to their natural levels and [by] establishing new wild populations of tuatara throughout their pre-human range as components of healthy ecosystems'.[117] Listed below are the specific objectives set for the 10-year life of the second plan, the actions taken to achieve these objectives, and the successes to date for the two recovery plans as a whole. Both recovery plans also seek to obtain new knowledge of tuatara through research.

1. To preserve the genetic diversity of all existing stocks of tuatara

The major action is to continue to eradicate rats wherever they co-exist with tuatara. Over the life of the two recovery plans, this has been achieved on seven islands to date (Table 10.1). Additional actions are to maintain the (mammalian) predator-free status of existing habitats, to develop a security system to discourage illegal visitors, to monitor trends in island populations that may warrant management intervention, to survey islands likely to hold unknown relict populations, and to enhance the recovery of relict populations using captive breeding.

2. To reinstate tuatara as components of healthy ecosystems throughout their pre-human range

The major action is to reintroduce tuatara, where possible, to islands and other wild habitats where they are thought, or known, to have once occurred. Between 1995 and 2011 tuatara have been reintroduced to seven islands and a mainland sanctuary. Translocations to establish new populations are such a major component of the recovery programme that they are addressed in more detail in the following section.

3. To promote public awareness of tuatara and related conservation issues through accessibility to captive animals and certain wild populations of tuatara

The main actions are to promote controlled public access to certain wild populations of tuatara, and to promote conservation through the use of captive tuatara, high-quality interpretive materials, and the knowledge and skills of iwi and other special-interest groups. Achievements include the reintroduction of tuatara to three islands and one mainland sanctuary where public access is allowed (see next section), and the involvement of Māori in translocations, educational road shows and captive displays involving live tuatara.

Notes for Table 10.2 (opposite)

1 Scientific names for introduced mammals: Norway rats (*Rattus norvegicus*); ship rats (*Rattus rattus*); Pacific rats (*Rattus exulans*); cattle (*Bos taurus*); sheep (*Ovis aries*); goats (*Capra hircus*); cats (*Felis catus*); rabbits (*Oryctolagus cuniculus*); pigs (*Sus scrofa*).

2 To 2010; y = year(s), mo = month(s). Sources: Titi Island: Nelson and Daugherty 1997; Nelson 1998, Nelson et al. 2002b; Gaze, P. and Cash, B. 2008. *A history of wildlife translocations in the Marlborough Sounds*. Occasional Publication No. 2, Department of Conservation (DOC), Nelson; K. A. Miller et al. 2010; K. Miller, Victoria University of Wellington, pers. comm. 23 September 2009. Moutohora Island: Smale, S. and Owen, K. 1990. Motuhora: a whale of an island. In Towns, D. R., Daugherty, C. H. and Atkinson, I. A. E. (eds). *Ecological restoration of New Zealand islands*. Conservation Sciences Publication No. 2. Department of Conservation, Wellington, pp. 109–112; Owen 1998; Ussher, G. 1997. *Translocation success of tuatara (*Sphenodon punctatus*) to Red Mercury (Whakau) and Moutohora (Whale) Islands*. Unpublished report, University of Auckland, Auckland; Ussher, G. 2002. *Reintroduction of tuatara (*Sphenodon p. punctatus*) to Moutohora (Whale Island): population status after 5 years*. Unpublished report to the Bay of Plenty Department of Conservation, Rotorua, 47 pp.; G. Ussher, Auckland Regional Council, pers. comm. 6 March 2006, Matiu/Somes Island: K. A. Miller et al. 2010. K. Miller, pers. comm. 23 September 2009, and Box 10.6. Tiritiri Matangi Island: Ruffell 2005; G. Ussher pers. comm. 20 July 2006; van Winkel, D. and Habgood, M. 2009. *Tuatara survey on Tiritiri Matangi Island: five years post-translocation*. Unpublished report to the Auckland Conservancy, Department of Conservation; D. van Winkel, Bioresearches Group Ltd, Auckland, pers. comm. 28 September 2009, Wakaterepapanui: Gaze and Cash 2008. op. cit.; Nelson et al. 2008; K. Miller, pers. comm. 23 September 2009. Mayor: DOC press releases. Long: DOC press releases; K. A. Miller et al. 2010. All islands: reports to Tuatara Recovery Group. For dates of rodent eradications, see chapters on kiore by I. A. E. Atkinson and D. R. Towns (pp. 159–174) and on the Norway rat by J. G. Innes (pp. 174–187) in King, C. M. (ed.). 2005. *The handbook of New Zealand mammals*. Oxford University Press, Auckland.

3 M = male; F = female.

BOX 10.6

Homeward bound: reintroduction of tuatara to offshore islands

Translocations involving the return of organisms to parts of their former distribution are known as reintroductions or repatriations. Between 1995 and 2011, tuatara have been reintroduced to seven islands where they are known, or assumed with a high level of confidence, to have once occurred. Three islands, namely Matiu/Somes, Moutohora and Tiritiri Matangi, have strong evidence for the past presence of tuatara (Chapter 3); these islands are also significant because, for the first time in many decades, members of the public are allowed to visit them and see wild tuatara. The remaining four islands, Titi, Wakaterepapanui,

Box 10.6 Fig. 1 Aerial photo of Matiu/Somes Island (right) and Mokopuna Island (left) in Wellington Harbour (February 1996).

Photo: Lloyd Homer, GNS Science, http://www.gns.cri.nz.

Box 10.6 Fig. 2 Reintroduction of tuatara to Matiu/Somes Island in 1998: members of Te Ātiawa, Victoria University of Wellington and Department of Conservation accompany the tuatara ashore. Visible in the background are Mokopuna Island (the green hump near top left), and the North Island.

Photo: Brett Robertson/Victoria University of Wellington.

Mayor (Tuhua) and Long, are also likely to have had natural populations of tuatara,[1] but historical information is limited.

Matiu/Somes Island (25 hectares) provides a particularly exciting reintroduction close to an urban centre. The island, known to Māori as Matiu after a female ancestor, rises as an emerald-and-bronze landmark in the glistening waters of Wellington Harbour (Port Nicholson), just 8 kilometres from the capital city. Sixty metres to the north, a cluster of rocky islets, Nga Mokopuna (the grandchildren), also have a historical connection with tuatara, following a failed attempt to reintroduce tuatara to the largest islet in 1920 (see Box 10.2).

Human occupation of Matiu/Somes Island has a long and complex history. For centuries the island was a strategic refuge for Māori. Following settlement of nearby Petone by Europeans in 1839, the island was renamed Somes Island[2] and became a destination for picnics and farming. A tuatara was pulled from a burrow in 1842 by a child searching for her pet rabbit; the animal was examined locally and then sent to the British Museum,[3] where it remains today as a stuffed specimen.[4] What may have been a different specimen was sent by a New Zealand Company representative (by 1845), to Richard Owen, Britain's leading anatomist at the time.[5] Forest cover on the island was probably negligible by then. A lighthouse station was established in 1866 and automated in 1924. In 1872 the island became a quarantine station for immigrants and, officially from 1889, for animals as well. One immigrant, later to become the Director of the Dominion Museum, saw a tuatara there in 1877 – the last known sighting.[6] During the two World

Wars, the island housed an internment camp; gun emplacements were built in 1942 but never used in war. Following World War II, the island continued to function primarily as an animal quarantine station. Ship rats (present by 1961) were eradicated by 1990.[7] Re-vegetation began in 1981 with the involvement of community groups; in 1995, after much public debate, the island became a scientific and historic reserve managed by the Department of Conservation, with a resident caretaker.

Ecological restoration of Matiu/Somes Island, as it was renamed in 1997, is now well underway. Reintroduction of tuatara is consistent with restoration goals for the island. In total, 55 tuatara from North Brother Island were released (all but one in October 1998); 20 adults (10 males and 10 females) were transferred directly from the wild, and 35 juveniles, hatched at Victoria University of Wellington from wild-collected eggs, were reared at Nga Manu Nature Reserve in Waikanae for 7 years until release.[8] Juveniles were inferred, from the temperatures of egg incubation, to have an even sex ratio. Adults and juveniles were released at different parts of the island to minimise the risk of negative interactions, including cannibalism.

During monitoring by university student Katrina Merrifield over the next 2 years, 90% of the released adults were re-sighted; many had dispersed (up to 300 metres from the release site) and gained weight.[9] Although only 17% of the released juveniles were re-sighted, this may reflect the inherent difficulty of finding juveniles, without radio-transmitters, in dense cliff-side vegetation. In a subsequent survey 5 years after the release, 45% of adults and 26% of those that were juveniles at the time of release were confirmed alive (some juveniles had reached adult size, but growth rates were highly variable).[10] Further recaptures were made at about 8 years after release,[11] and in 2008 a nest was found from which two hatchlings were obtained.[12]

Matiu/Somes Island is only a 25-minute ferry ride from Wellington and receives about 15,000 visitors a year.[13] Local people, school children and overseas tourists are among those who come to absorb the island's history. Although a sighting of tuatara cannot be guaranteed, it frequently happens. For conservation managers, the rewards of providing such experiences must be balanced against the risks of adverse events, including accidental introduction of rodents. To help minimise the risk, visitors must check their bags before arrival, and again in a special 'whare kiore' (quarantine building) upon arrival.

Recent reviews of animal translocations worldwide suggest that the support and involvement of local communities contribute to success. Reintroductions of tuatara to islands such as Matiu/Somes have certainly benefited from widespread community interest. Moreover, they provide common ground, bringing together diverse groups of people, including local Māori, community conservation groups and the general public, as well as zoo and wildlife-park curators, government conservation officials and university students and scientists. Just as the community may aid the success of the reintroduction, the experience of contributing to an environment-positive project may bring social cohesion to the community.

Recent translocations to form new island populations

Between 1995 and 2011, tuatara have been translocated to seven 'new' islands (i.e. islands not already inhabited by tuatara; Table 10.2). Tuatara were once present on at least three of these (Tiritiri Matangi in the Hauraki Gulf, Moutohora in the Bay of Plenty and Matiu/Somes in Wellington Harbour), but disappeared following the introduction of mammals (see Chapter 3, Table 3.1). Translocations to these three islands can therefore also be described as reintroductions. Although there is no certain evidence for the past presence of tuatara populations on the remaining four islands (Mayor (Tuhua) in the Bay of Plenty, Titi and Wakaterepapanui in Cook Strait, and Long in the Marlborough Sounds), it is probable that tuatara once lived in these places as well.

The first step in preparing islands for the release of tuatara was the removal of introduced mammals, including rats. Tuatara were later released following the preparation of a formal translocation proposal, in consultation with affected iwi (Figure 10.6). Translocations have been made in spring, a season favourable for tuatara activity and for the transfer of females gravid with

Fig. 10.6 An adult male tuatara, held by John Heaphy of the Department of Conservation, is admired by members of Ngāi Te Rangi at Otawhiwhi Marae in 2007. Thirty tuatara from Karewa Island were subsequently translocated to Mayor Island.

Photo: Dean Evans, Department of Conservation.

Fig. 10.7 All dressed up and somewhere to go? A tuatara from Stephens Island, held inside a cardboard postage tube by Nicola Nelson, ready for translocation to Wakaterepapanui Island in October 2003.

Photo: Lee Pagni, courtesy of Peter Gaze.

eggs, thus increasing the effective genetic size of the founding population. Transfers in spring also allow for establishment of territories before the driest part of summer and before the mating season in late summer–autumn. The number of tuatara translocated to each island has ranged from 30 to 427. Founders have included wild adults from nearby islands (Figure 10.7), as well as captive-reared juveniles of about 3–7 years of age originating from wild eggs incubated in the laboratory (Table 10.2). 'Head-starting', which involves raising eggs and the resulting juveniles in captivity until release, has the potential to reduce the impact on the source population, but the possible disadvantages of rearing tuatara in relatively benign environments in captivity require further study. When adults and juveniles have been released on the same island, releases were made in different locations to minimise the risk of cannibalism.

Ultimately, the ecological goal of translocation is to establish populations that are self-sustaining in perpetuity, thus increasing the overall security of tuatara from extinction in the wild. If these translocations prove successful, the distribution of tuatara will have increased from 30 to 37 islands, and the island area available for occupation from 5003 hectares to 6880 hectares (a

38% increase). Nevertheless, confirmation that populations are self-sustaining will take decades or even centuries. The following, staged criteria have been developed by the Tuatara Recovery Group for provisionally assessing the success of a translocation in the early years.[118]

- After 5 years, over a 12-month period, 50% of the adults released should be recaptured and evidence of breeding should be obtained (e.g. capture of at least 2 or 3 unmarked, young animals).
- After 10 years, from a sample of 30–35 animals, the percentage of small tuatara (up to 180 millimetres snout–vent length, roughly equating to juveniles and subadults) in the sample should be at least 10%. A high percentage of small animals in the sample (20% or more) would indicate that the population has established, whereas a moderate percentage (10–19%) would indicate that the population has not established but that prospects for establishment appear good. A small percentage of juveniles in the total sample (less than 10%) would be cause for concern.

According to these criteria, the five earliest translocations to islands are showing promising signs of success. Recaptures of released animals have been frequent, considering the inherent difficulty of finding small numbers of tuatara in dense vegetation on relatively large islands (Figure 10.8). Island-hatched juveniles found on one island (Moutohora) accounted for 47% of the

Fig. 10.8 A tuatara recaptured in fine condition in 2002, 6 years after release on Moutohora Island, held by Ropata Kopae of Ngāti Awa.

Photo: Joanne Peace, courtesy of Graham Ussher.

tuatara detected at 9 years after release,[119] and nesting has clearly occurred on three others.

On some islands the released tuatara have been closely monitored, often with the aid of radio-telemetry (Figure 10.9).[120] These field studies, undertaken mostly by university students, have yielded many insights about establishment.[121] Artificial burrows, provided on some islands, appear to have been little used by tuatara (Figure 10.10). Dispersal has been variable: some adults have remained within a few tens of metres of the release site, whereas others have been found up to about 300 metres away.

Especially interesting is the marked increase in the body condition index (mass relative to snout–vent length) observed in animals on some source or destination islands following translocations. Research by Nicola Nelson and colleagues has shown that some adults, which had apparently stopped growing while on the source island of North Brother, resumed growth in both mass and snout–vent length when transferred to Titi Island,[122] and a similar response has since been seen in tuatara released on Matiu/Somes Island.[123] A detailed study by Jonathon Ruffell has examined recovery from harvest on the small source island of Moutoki (from which about 25% of the population was removed for release on Moutohora Island).[124] Animals remaining on Moutoki Island showed a significant increase in body condition, and the apparent relative abundance of juveniles (representing about 25% of the population prior to harvest) had increased to 51% by 8 years post-harvest. This is more than can be accounted for by the removal of adults, and suggests that recruitment has increased, enabling the population to return to about its pre-harvest size. Taken together, these responses suggest that tuatara on small source islands may be limited in body size by competition with other tuatara. Translocation seems to provide a release from these constraints, at least in the short term.

Fig. 10.9 Radio-telemetry has been a valuable tool for tracking the movements of tuatara released onto islands. In studies by Graham Ussher on Moutohora Island and Red Mercury Island (Whakau), transmitters were attached using a shoulder harness.

Photo: Alex Eagles, courtesy of Graham Ussher.

Fig. 10.10 Artificial burrows made of drainage coil have sometimes been provided for translocated tuatara, but seem to have been little used. Here, a tuatara, repatriated to Red Mercury Island after several years in captivity for breeding while rats were removed, is radio-tracked to an artificial burrow by Graham Ussher.

Photo: Len Doel.

Given that the number of tuatara translocated is often only a few dozen, how secure genetically and demographically are the new populations likely to be? These issues have recently been explored by Kimberley Miller and colleagues. The genetic diversity in founders translocated to two islands and a mainland site (Zealandia, formerly known as Karori Wildlife Sanctuary) was compared with that in the source populations on Stephens Island or North Brother Island.[125] Values for natural-history parameters were then entered into population viability analyses, which allowed the amount of genetic diversity that might be lost over the subsequent 10 generations (about 400 years) to be modelled. As one would expect, a large founder group of 200 adults (as released at Zealandia, for instance) is expected to lose diversity less rapidly, relative to the Stephens Island source, than a translocation of 30 adults. A translocation of 30 juveniles is expected to lose more diversity than the same number of adults, because some juveniles would die before breeding.[126] Overall, Miller and colleagues concluded that translocations of 30 or more adults are likely to retain about 94% of their original genetic heterozygosity (a measure of genetic variation)[127] after 10 generations, and to have a negligible probability of extinction. A population with high genetic diversity (such as that on Stephens Island) is also preferred as a source to one with low genetic diversity (such as that on North Brother Island).[128]

Translocations have been undertaken in part for ecological goals (to establish self-sustaining populations), but advocacy and education provide additional reasons. By 2011, members of the public had the opportunity to see tuatara in the wild on three islands undergoing restoration. The number of visitors is substantial on Tiritiri Matangi Island (about 30,000 visitors per year) and Matiu/Somes (about 15,000 visitors per year), with a modest

Fig. 10.11 The discovery in May 1997 of four tuatara in a box, on a roadside between Timaru and Waimate in the South Island, led to a successful second prosecution of New Zealand's most notorious wildlife smuggler, Frederick (Freddie) Angell, for offences involving tuatara. Angell's first tuatara-related conviction, in 1992, yielded a sentence of 21 months' jail for the theft of a tuatara from Stephens Island and two tuatara from Southland Museum and Art Gallery. The collection of the four tuatara shown here, from Stephens Island in 1997, combined with an attempt to harvest more and evidence of conspiracy to trade, resulted in a further jail sentence of 3 years. Angell died in 2003.

Photo: Colin Hitchcock.

number visiting Moutohora (about 700 per year).[129] On both Tiritiri and Matiu/Somes Islands, several tuatara have established themselves in conspicuous locations and are often seen.

Various issues confront DOC and its iwi partners in co-managing these reintroduced populations. Biological issues include the potential impacts of reintroduced tuatara on rare invertebrates and lizards, and conversely the impacts of reintroduced birds that possibly prey on tuatara, such as the South Island takahē (*Porphyrio hochstetteri*) and saddlebacks (tīeke, *Philesturnus* spp.). More pressing are the steps that must be taken to prevent rodent reinvasion (currently, visitors are encouraged to check their bags before landing and/or again in a special 'whare kiore' immediately upon landing; bait stations or gnaw sticks are also placed to help indicate the absence of rodents). Visitors must be managed to minimise disturbance in sensitive areas such as wāhi tapu (restricted or sacred sites) and urupā (burial grounds), to minimise wastes and to prevent fires (smoking is banned).

Illegal collection of tuatara is an ever-present risk on all tuatara islands. Currently, natural populations and colonies in captivity seem to provide the greatest attraction for such attempts – such as the well-publicised thefts of tuatara from Stephens Island and Southland Museum in the 1990s (Figure 10.11).[130] Rangers from DOC are resident on two islands with reintroduced populations (Tiritiri Matangi and Matiu/Somes Islands) as well as on Stephens Island. Their presence, together with heightened local vigilance and the activities of a Wildlife Enforcement Group (including officers from

DOC, the New Zealand Customs Service and the Ministry of Agriculture and Forestry), seems to have provided an effective deterrent to illegal collection, judging by the small number of cases leading to criminal charges.[131]

To summarise this section, although the value of translocations to conservation of reptiles and amphibians has been viewed with some scepticism internationally,[132] recent experience in New Zealand with tuatara (and many lizards) has been extremely positive.[133] For tuatara, further translocations are proposed, including to Korapuki Island and Ohinau Island (in or near the Mercury Islands), Middle Island (in the Aldermen Islands) and East Island (Whangaokeno Island, near East Cape). Releases to secure mainland sanctuaries in the North and South Islands are also in discussion or underway. Given recent successes, there has been some question as to whether a tuatara recovery group is even now needed (Figure 10.12). However, the large number of plans in development, and concerns about the possible impacts of climate change, argue for some level of national oversight to continue.

Looking back, looking forward

In the preface to a recent international review of biodiversity issues confronting New Zealand, the comment was made that 'since European settlement, catastrophes outnumber successes in New Zealand wildlife conservation'.[134] The conservation of tuatara provides us with one of the more remarkable success stories to date, but only after a period of near-annihilation when distribution is estimated to have fallen to less than 0.1% of the former (pre-human) range.[135]

Fig. 10.12 An endangered species-recovery group? A meeting of the Tuatara Recovery Group and interested parties in Wellington, April 2005. Present, from left foreground in a clockwise direction: Barbara Blanchard (Wellington Zoo, captive-breeding coordinator), Leigh Marshall (Department of Conservation, DOC), Alison Cree (University of Otago), Donald Newman (DOC), Rob Stone (DOC), Peter Gaze (DOC, recovery group leader), John Heaphy (DOC), Peter Barrow (DOC), Chris Smuts-Kennedy (Maungatautari Ecological Island Trust) and Nicola Nelson (Victoria University of Wellington).

Photo: collection of Alison Cree.

BOX 10.7

Reintroduction to the mainland: insurance against global climate change?

Until about a thousand years ago, tuatara were widely distributed over the North and South Islands, from Cape Reinga in the north to near Invercargill in the south (see Chapter 2, Figure 2.8). The 'reintroduction' of tuatara to outdoor habitats on the mainland has, until recently, been possible only in relatively small enclosures in zoos and wildlife parks, where only the largest mammals such as pigs, cats and mustelids have been definitively excluded. However, the development of new fences that also prevent entry of at least some rodents has increased the prospects for reintroducing tuatara to more substantial areas of the mainland.

The first public display of tuatara in a large mainland area, free of all introduced mammals except mice, was established at Zealandia™ (formerly the Karori Wildlife Sanctuary) in Wellington in 2005. Of the 70 tuatara transferred from Stephens Island (Takapourewa), 60 were placed within a small mouse-proof area (1 hectare) within the larger sanctuary. The remaining 10 adults were released into the outer sanctuary (225 hectares), and were monitored using radio-transmitters to assess the possible impacts of mice before further releases were made.[1] In 2007, following observations of high survival, a further 130 adult tuatara were transferred from Stephens Island into the outer sanctuary. Mating has been seen, at least one nest has been found, and in 2009 the first hatchling was reported.[2] In the longer term there are plans to make the fenced perimeter an effective barrier to mice.

Other proposed sanctuaries in the North and South Islands involve an improved fence design that was initially promoted as being mouse-proof under normal operation. Among these is Maungatautari Ecological Island, an ambitious (3400-hectare) sanctuary proposed for the central North Island near Hamilton. This inland site encompasses a small mountain, rising to a higher altitude (797 metres above sea level) than any current tuatara island. Another

Box 10.7 Fig. 1 The recent development of mesh fences with capped tops, preventing the entry of all pest mammals except perhaps small rodents, has enabled the reintroduction of tuatara to mainland sites such as Zealandia™, the former Karori Wildlife Sanctuary.
Photo: Alison Cree.

exciting proposal, far to the south, involves the Orokonui Ecosanctuary (307 hectares, 45°S) near Dunedin.[3] Sites such as Maungatautari and especially Orokonui might initially seem sub-optimal; after all, tuatara may grow and reproduce more slowly here than they do on their source islands.[4] But higher altitudes, cooler latitudes and greater geographic protection from extreme weather events may provide important 'insurance' to help protect tuatara from the possible impacts of global climate change.

Many scientists consider the changes forecast in the Earth's climate over the next century, largely resulting from increased greenhouse gas emissions, to be the biggest challenge to global biodiversity within human existence. New Zealand is somewhat buffered by the surrounding ocean from predicted increases in mean global temperature. Nonetheless, if current emission patterns continue (producing a 'high carbon world'), New Zealand is likely to experience mean annual temperatures by 2080–2099 that are about 2.6°C warmer than a century earlier.[5] Even under a 'rapidly decarbonising world', which assumes rapid and concerted action on a global scale to reduce emissions, a rise of about 1.1°C seems unavoidable. How will warmer temperatures affect tuatara?

One question of obvious interest, studied recently on North Brother Island, is the effect that warmer soils will have on sex ratios arising from egg incubation. Tuatara have a pattern of temperature-dependent sex determination in which cool nests produce females and warm nests, males.[6] If nesting on North Brother continues at current soil depths and seasons, and temperatures rise to the high end of current predictions, the population will become strongly skewed towards males.[7] On some islands, nesting females might be able to adjust their behaviour in response to changing

Box 10.7 Fig. 2 At Zealandia™ in Karori, members of the public are invited to assist monitoring by writing their sightings on a whiteboard. Some of the released tuatara have been individually identified with coloured beads threaded through the skin of the nuchal crest.

Main photo: Alison Cree. Inset: Tom Lynch.

Box 10.7 Fig. 3 The capacity for eggs of Cook Strait tuatara to develop at more southerly soil temperatures was investigated by Anne Besson, as part of a recent site assessment for Orokonui Ecosanctuary.[12] Several hatchlings from this site in the southern South Island were obtained after undergoing most of their embryonic development in artificial outdoor nests. During incubation, the eggs were protected from introduced mammals that were still present using mesh cages.

Photo: Alison Cree.

temperatures or sex ratios; if they nested at greater depths or in forest, for example, they might continue to produce females. However, these options are lacking on small, rocky, forestless islands such as North Brother. On that island, management options include providing artificial shade, or hatching eggs off-island under controlled temperatures to produce females.[8] However, the low genetic diversity on North Brother[9] and the likelihood of extra environmental challenges for tuatara on small islands suggest that management effort might be better focused in other directions.

The additional challenges posed by climate change for current islands include altered rainfall. Parts of New Zealand (including the northeast) are predicted to become somewhat drier, in winter or spring at least.[10] In addition, sea level is very likely to rise by at least 0.5 metres by 2100, though the rise could be much greater if polar ice flows change, and the rise will continue for centuries.[11] This will lead to increased coastal erosion and reduced shoreline habitats. Once cannot be optimistic about the combined effects of increased temperature, drought, sea-level rise and extreme weather events on the 13 populations of tuatara currently on islands of 5 hectares or less.

The wider effects of climate change on island ecosystems also need attention. For example, if sea and wind currents change, and if a greater concentration of

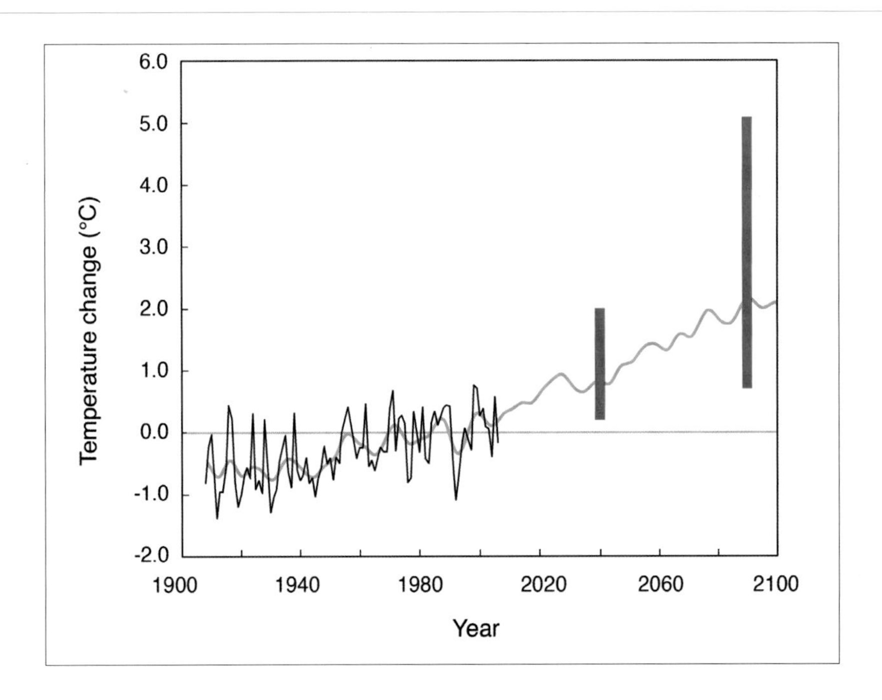

Box 10.7 Fig. 4 Rising concentrations of greenhouse gases are predicted to cause increases in air temperature in New Zealand over the coming century. The figure shows annual mean temperatures in New Zealand since the early 1900s (black line) as a deviation in degrees Celsius from records for 1980–1999 (horizontal grey line). The blue line is a smoothed curve showing decadal variation in the observations, continuing forwards as a projection from one of the Fourth Assessment models of the Intergovernmental Panel on Climate Change (IPCC). The vertical red bars show the full range of warming in New Zealand predicted by the IPCC for six greenhouse-gas emissions scenarios (centred for the years 2030–2049 and 2080–2099).

Figure modified with permission from Ministry for the Environment 2008. *Climate change effects and impacts assessment: a guidance manual for local government in New Zealand.* 2nd edn. Prepared by Mullan, B., Wratt, D., Dean, S., Hollis, M., Allan S., Williams, T., Kenny, G. and MfE staff. Ministry for the Environment, Wellington. Crown Copyright Reserved; data sourced from NIWA.

carbon dioxide in the atmosphere leads to increased acidity of the oceans, then marine food chains could be disrupted and the productivity that seabirds currently bring to islands could decline. In such an event, the density of tuatara populations is likely to fall. In addition, climate change increases the risk that new pest species will establish in northern New Zealand; at the same time, increased social and economic pressures may well affect the willingness and ability of human communities to engage in conservation actions. The overall impacts, direct and indirect, of climate change on tuatara are thus impossible to predict with certainty, but are of serious concern.

Tuatara belong to a lineage that has survived many environmental challenges over the millennia. Rapid international action to reduce greenhouse gas emissions, combined with vigilant monitoring of existing populations and environments, protection and enhancement (e.g. through rat removal) of habitats on large islands, continued research into the broad-scale impacts of climate change, and insurance in the form of a wider (and partly restored) southern distribution for tuatara are all important. Together, these offer our best hope that the human-induced challenge of climate change will not be the final downfall for the world's last rhynchocephalian.

In looking back, what general lessons about conservation history may we draw from the example of tuatara? The following points offer a summary of this chapter.

- The first written accounts to advocate strongly for the protection of tuatara date from 1885. However, uncertainty remains about when efforts to conserve tuatara began. Māori of the early 1800s were clearly aware of the disappearance of tuatara from the North Island; whether earlier attempts were made to avert the decline is unclear. Although it is difficult to know what further evidence might be gathered, this is an area in which new interpretations may come to light.
- The legal protection of tuatara in 1895 is rightly hailed as being an early example of a conservation initiative undertaken by a colonial government. It is less widely appreciated that this achievement required a decade of effort by committed individuals and institutions facing a high level of bureaucratic indifference.
- The protectionist legislation of 1895 came down heavily against those whose actions created a sense of righteous indignation (especially foreigners who wished to export tuatara from New Zealand). However, legal protection of tuatara (and subsequently, to some extent, their habitats) did not acknowledge or address the most serious agents of decline, namely the introduction of mammals. Legal protection was thus insufficient to prevent the disappearance of at least seven and probably 10 or more island populations of tuatara during the 1900s. This ongoing decline has been described as a 'sinking lid' for the species.[136]
- Some well-intentioned and, in principle, innovative steps (including translocations) were taken in the early 1900s to enhance the security of tuatara. Given that the fundamental agents of decline had not been identified and removed, these efforts were doomed to failure. By today's standards, they were also poorly conceived, monitored and documented.
- A strongly protectionist environment prevailed from the early 1900s until the 1970s. Although several field studies on islands with tuatara had been conducted during the 1880s and 1890s, visits were subsequently discouraged and no professional biologist was appointed by the government specifically to study wild tuatara until 1969. 'Management' was for many years largely in the hands of bureaucrats and on-site amateurs (lighthouse keepers), and, when the crunch came, economic imperatives came first. This gives the impression of an 'anti-people' rather than a 'pro-wildlife' style of management.[137]
- In the early years of study, the actions of scientists may have contributed to the perceived need for a protectionist environment. One visiting scientist possibly exceeded the terms of, or behaviour expected under, a permit for study, leading to concern that such activity was contributing to declining abundance.

BOX 10.8

Managing a (no-longer threatened) member of the 'charismatic megafauna'

Biologists around the world have repeatedly highlighted the disproportionate attention, or 'taxonomic bias', given to large, 'fuzzy' animals (essentially birds and mammals) in global conservation.[1] In New Zealand, where conservation history has been undeniably avian-centred, tuatara have enjoyed special status as honorary 'birds'. But does current attention on tuatara detract from the needs of rarer but less well-known creatures – such as the 17 species or proposed species of endemic lizards in New Zealand that are threatened with extinction?[2]

Perhaps the relative emphasis is extreme, but the attention has not been without wider benefits. Tuatara serve as a 'flagship species' for numerous types of animals – including small seabirds, lizards and invertebrates, many much rarer than tuatara – that benefit from the removal of rats from islands. From an administrative viewpoint, the effective cooperation within the recovery group for tuatara, including the manner in which captive management has been integrated with conservation in wild habitats,[3] has been a useful model for other species recovery groups. Additionally, tuatara are 'taonga species' (of special significance to Māori) and generate much international attention, meaning that there are also cultural and political reasons to support a strong and visible management programme.

Nevertheless, species recovery programmes in New Zealand operate within an environment in which there is a strong imperative on managers to obtain more conservation gains for every dollar spent. This places pressure on 'successful' recovery groups to dissolve when targets have been reached, and on managers to focus on site-based, rather than species-based, conservation programmes. These developments, as well as staffing cuts in the Department of Conservation (DOC) during 2011, have left the Tuatara Recovery Group with an uncertain future. As a single species, tuatara are no longer threatened with extinction, and future reintroductions may be largely managed by local communities rather than by DOC. Nonetheless, it is hard to imagine a time when national leadership in conservation management of tuatara would be unnecessary.

An authoritative national perspective is needed to ensure that local actions within conservancies are appropriate, to ensure that captive management is consistent with management objectives, to evaluate and coordinate research proposals (of which there is a steady stream from both within and outside New Zealand), and to speak on behalf of management of tuatara to the international community. One possible scenario is that the recovery group will change into a taxon advisory group, thus keeping a 'TAG' on tuatara for the foreseeable future.

- The creation in 1987 of the Department of Conservation – a new government agency for conservation of species and habitats – led a shift towards active management of tuatara. This new, interventionist approach involved a formal recovery-planning process, with the specific goal of reversing the decline in abundance and distribution of tuatara by eliminating its immediate causes. It required managers to brave risks, setting aside the fear of political fall-out should the outcomes have been adverse for this high-profile genus. It benefited from the active involvement of scientists conducting publicly defendable, conservation-oriented research.
- The new directions set by the recovery plans of 1993 and 2001 involve a much wider community of interest: in other words, conservation – in both its activities and the financing of those activities – has come to be seen as a communal exercise.[138] As part of that shift, the captive-breeding community has gradually changed its focus from one of husbandry for display to one involving husbandry for conservation (including advocacy).
- By 2011, as a result of the directions set by the recovery plans, seven populations of tuatara have been made more secure by the removal of introduced mammals, and seven new island populations have been founded. This has increased the total area of island habitat potentially occupied by tuatara by 38%, and the area of rodent-free island habitat by 1028%. Central to this success has been the development of methods for eliminating rats and performing effective translocations. Several populations have also become more accessible to the public, allowing greater community involvement with conservation.
- Much of the history of conservation management of tuatara reveals little attempt to involve Māori or to consider Māori interests and perspectives. Indeed, it could be said that early steps to protect tuatara and their habitats were at some expense to the indigenous people. Since the late 1980s, efforts to address grievances and engage in partnerships between Crown agencies, other research agencies and Māori have been ongoing. Māori support has been instrumental to a recent initative to remove rats from most of the final 526 hectares of tuatara habitat still shared in 2011 with rats.[139] The future is likely to see much greater involvement of Māori in the conservation of tuatara and their habitats, on both Māori-owned islands and in other locations.
- The future is also likely to see increased involvement of community groups generally, with the reintroduction of tuatara to mainland sanctuaries. In 2005 tuatara were reintroduced to a North Island sanctuary virtually free of introduced mammals and with public access, and as of 2011 plans for reintroduction to other mainland sanctuaries, including one in the South Island, are well advanced. These steps are small (on an area basis) but significant for the gradual return of tuatara to their areas of former occupation, and for allowing the public to view free-roaming tuatara. The

prospect that tuatara might one day be freely released within the North and South Islands is an enticing vision with which to challenge future generations of herpetologists.

Regardless of recent progress in conservation, there will always be a role for further research. Questions of current interest include methods for assessing and monitoring changes in abundance of tuatara on islands without causing extensive damage to seabird burrows; genetic viability of small populations (both natural and translocated); relationships between island size and density, body size and condition of tuatara following translocation; population viability analyses for harvested and translocated populations; the desirability or otherwise of mixing different genetic stocks in future translocations; factors influencing translocation success (including the relative success of captive-raised and wild-collected animals); survival of tuatara in the presence of mice in mainland sanctuaries; new methods for detecting tuatara (e.g. using trained dogs); long-term effects of and possible alternatives to current methods of marking for permanent identification; management of Māori aspirations for seabird harvest while ensuring the safety of tuatara; and the responses of tuatara to climate change. A reassessment of management priorities is also needed following the conclusion in 2009 that living populations should be viewed as one species of tuatara rather than two.[140]

Fig. 10.13 Captive-bred juvenile tuatara of Stephens Island stock: ambassadors for the survival of their species.

Photo: Aude Thierry, collection of Alison Cree.

In closing, it is sobering to remember the astonishingly short history of humans within New Zealand compared with that of tuatara. Within about 800 years, two main waves of human impact, first Polynesian/Māori and then European/Pākehā, wreaked a reduction of about 99.9% in the area occupied by tuatara. This reduction, and the resulting near-extinction of the genus, spanned only about one ten-thousandth of the time that the tuatara lineage has been evolving within New Zealand. After several generations of human presence in the landscape, a transition from a 'frontier' mindset to one of stewardship has developed. Research-based conservation has produced, over the last two decades, the first known increases in the distribution of tuatara for hundreds of years. The security of tuatara has now increased to the point where alternative or additional approaches to management can be considered in carefully chosen situations. This is exciting and healthy. No doubt we are still making mistakes for which future generations will berate us; the looming issue of global climate change seems certain to be one of these. But, at least on a local scale, there is reason to feel hopeful that the number and severity of mistakes are diminishing (Figure 10.13).

Endnotes to Chapter 10

See the bibliography for full references to articles on tuatara, which are cited here at first mention by surname and year only. References on other topics are cited here in full at first mention, and then using 'op. cit.'. Except where required for clarity, the convention of parentheses around years has been dispensed with for economy of space.

1 Günther 1867, p. 596.
2 Whether conservation of tuatara can be said to have begun during the pre-European Māori period is unclear. The impacts of harvesting would perhaps have been reduced by the tapu status of tuatara, including a practice (at least among Tūhoe) of restricting harvesting to men only (Chapter 4). However, it is unclear whether such restrictions were introduced for conservation purposes, or of their overall effect.
3 Buller 1871, 1877.
4 Gold-Smith 1885; G. Mair to H. D. Bell, Minister for Internal Affairs, 8 June 1913, ANZ IA 1 46/18/4 pt 1.
5 Sclater 1870, 1871; Buller 1871, 1877; Anon. 1895; Mair to Bell 1913, op. cit. For an example of the value of tuatara in museum exchanges, see Gill 2010b.
6 C. Norris to Superintendent of Tourist Dept, 12 February 1906, ANZ IA 1 46/18/61.
7 Sheets-Pyenson, S. 1988. *Cathedrals of science: the development of colonial natural history museums during the late nineteenth century*. McGill-Queen's University Press, Kingston. See also Jordan, T. 1989. Visit of an American naturalist to the Bay of Plenty, 1881. *Historical Review* 37: 23–32.
8 Ward 1882, p. 14 A tally of the figures in Ward's account suggests that he collected about 17–28 tuatara from Moutoki Island.
9 Page 122 from catalogue, H. A. Ward's Natural Science Establishment, Rochester, New York; date not stated, but filed between items from 1913 (ANZ IA 1 46/18/4 pt 1).
10 Anon. 1882b.
11 Reischek 1882, 1886, 1971. See also Chapter 5.
12 Based on catalogue information supplied by F. Tiedemann (curator) and K. Krey (visitor), at least 23 and perhaps up to 99 specimens of tuatara (plus six eggs) in the Natural History Museum of Vienna (Naturhistorisches Museum Wien) were collected by Reischek. Two specimens catalogued in 1922 by the Natural History Museum in London, collected from Hauturu/Little Barrier Island and received from the 'Vienna Museum', are probably also from Reischek's collection (see also Chapter 5).
13 Buller 1888, see p. 213. See also Buller 1895.
14 Martin, H. 1886. The protection of native birds. *Transactions and Proceedings of the New Zealand Institute* 18: 112–117.
15 Thomas 1890. See also Chapter 5 and O'Sullivan, M. J. 1968. *Algernon Phillips Withiel Thomas, 1856-1937: educationalist, scientist, horticulturist*. University of Auckland, Auckland.
16 Anon. 1885b.
17 Galbreath, R. 2002. *Scholars and gentlemen both: G.M. & Allan Thomson in New Zealand science and education*. Royal Society of New Zealand, Wellington; see p. 82 and sources therein.
18 For quotes in this paragraph, see Otago Institute, *Otago Daily Times*, 10 June 1885, p. 3. Further information on the institute's meeting on 9 June 1885 is found in *Otago Institute Minute Book 1883–1897* (MS 128/B, Hocken Library, Dunedin); *New Zealand Journal of Science* 1885 2: 503–504 and p. 584; *Transactions and Proceedings of the New Zealand Institute* 1886 18: 427 (the latter account records the date of the meeting, apparently incorrectly, as 10 June 1885). See also Chapters 5 and 6 for further information regarding the collection of tuatara by Parker, Thomas and Reischek from Karewa Island.
19 *Transactions and Proceedings of the New Zealand Institute* 1886, 18: 427.
20 T. J. Parker to Colonial Secretary, 12 June 1885, NMNZ Series No. MU000147 Location R–1M03–130E Box 6, Item 118 (other number: 85/2010).
21 For the wording of Hector's telegraph and Parker's response, see NMNZ Series No. MU000147 Location R–1M03–130E Box 6, Item 118 (other number: 85/2010).
22 E. S. Dodgson to Mr Stout, 18 July 1885, E. S. Dodgson to Dr Hector, 28 July 1885, both in NMNZ Series No. MU000147 Location R–1M03–130E Box 6, Item 118 file 85/2010.
23 *Report of the third meeting of the Australasian Association for the Advancement of Science, held at Christchurch, New Zealand, in January, 1891*, p. xxiii. See also Thomas, A. P. W. 1891. (1). The preservation of the native fauna and flora of New Zealand. *New Zealand Journal of Science (new series)* 1: 93, and Galbreath 2002, op. cit., p. 84.
24 Discussed by Young, D. 2004. *Our islands, our selves: a history of conservation in New Zealand*. University of Otago Press, Dunedin, pp. 88–95.
25 For example, discussion about Hauturu as a sanctuary centred on its value for birds and plants, with no reference to the presence of tuatara, in the following: Buller, W. L. 1895. Notes on the ornithology of New Zealand; with an exhibition of rare specimens. *Transactions and Proceedings of the New Zealand Institute* 27: 104–126; Kirk, T. 1896. The displacement of species in New Zealand. *Transactions and Proceedings of the New Zealand Institute* 28: 1–27. See also Galbreath, R. 1989. *Walter Buller: the reluctant conservationist*. GP Books, Wellington, p. 165.
26 For membership of the committee, see *Report of the third meeting of the Australasian Association for the Advancement of Science, held at Christchurch, New Zealand, in January, 1891*, p. xxii.
27 Report of Committee appointed to make recommendations for the protection of native fauna (as amended and approved by Council), in *Report of the fifth meeting of the Australasian Association for the Advancement of Science, held at Adelaide, South Australia, September, 1893,* pp. 241–242.
28 *Report of the fifth meeting of the Australasian Association for the Advancement of Science, held at Adelaide, South Australia, September, 1893*, p. xxii.
29 Australasian Association for the Advancement of Science to Colonial Secretary, 24 January 1895, ANZ IA 1 46/18/4 pt 1. See also *Report of the sixth meeting of the Australasian Association for the Advancement of Science, held at Brisbane, Queensland, January, 1895*, p. xix.
30 Hector's minute on Colonial Secretary memorandum, 8 February 1895, ANZ IA 1 46/18/4 pt 1.
31 Superintendent of Tourist and Health Resorts to Secretary

of Marine, 7 April 1902; Secretary of New Zealand Institute to Colonial Secretary, 26 January 1904; both in ANZ M 1 25/611 pt 1. See also a newspaper report of a recent meeting of the New Zealand Institute, *New Zealand Herald*, 28 January 1904, copy in ANZ IA 1 46/18/61.

32 The distinction was apparently not appreciated at the next AAAS meeting in 1898, at which it was reported that Stephens Island (Takapourewa) had been set aside as a reserve for tuatara. See *Report of the seventh meeting of the Australasian Association for the Advancement of Science, held at Sydney, 1898*, p. 11.

33 Except by Buller himself (Buller 1895).

34 A. Dendy to J. Hector, 15 August 1898, NMNZ MU000095 Box 2, Item 115 1885/200; President of the Philosophical Institute of Canterbury to the Colonial Secretary, 29 August 1898, ANZ IA 1 46/18/4 pt 1.

35 J. Hislop, memo for Minister of Internal Affairs, 15 October 1913, ANZ IA 1 46/18/4 pt 1.

36 For the protection of eggs under the Animals Protection Act 1880, see L. G. Reid, memorandum in Colonial Secretary's office, 31 August 1898, ANZ IA 1 46/18/4 pt 1: 'I have added a paragraph protecting the eggs of the tuatara, but strictly speaking I do not think that the statute authorises this, tho' it is manifestly desirable'. For the situation in 1913, see J. Hislop, memo for Minister of Internal Affairs, 15 October 1913, ANZ IA 1 46/18/4 pt 1: 'At present … a prosecution for taking Tuatara eggs would probably fail on the ground that the Governor had no power to protect them.'

37 Section 40 of the Animals Protection and Game Act 1921–22.

38 Eight or nine tuatara are said to have been collected from the principal island of the Mokohinau Group, including some for the private collection of a millionaire. Two apparently found their way to the London Zoo. See *The Dominion*, 11 August 1931, copy in ANZ M 1 25/611 pt 3. Within about 40 years (if not sooner), tuatara were effectively extinct on the Mokohinau Islands.

39 Colonial Secretary's office, memo 14 June 1895 re application from A. M. Johnson per C. A. Baker, Opawa, ANZ IA 46/18/4 pt 1.

40 Newspaper report cited by Hutton and Drummond 1904, p. 358. For more on Johnson's tuatara, see Chapter 5.

41 Schauinsland received at least two permits for Stephens Island, one in December 1896 to catch or take six tuatara, and one in October 1898 to receive up to 50 eggs. Thilenius received at least two permits for Stephens Island (October and December 1898), each for up to 50 eggs (see *List of permits to take or export tuatara lizards from 1896 to 1912*, NMNZ MU 000207 Box 4, Item 1; MU000001 Box 23, Item 14). Thilenius's second instalment of 50 eggs was apparently forwarded by a lighthouse keeper. The government received some public censure at the time for 'the extraordinary latitude and license allowed … to foreign scientists [for obtaining eggs] compared with the scanty facilities vouchsafed to our own investigators' (Anon. 1899b; see also Benham 1899).

42 The totals were tallied from the lighthouse keepers' records by Schmidt 1952. Six of these tuatara were forwarded from Stephens Island under permit to 'Mr Thomas Edward Dunne' [*sic*] on 1 Feb 1900. Donne was Superintendent of the Department of Tourist and Health Resorts from 1901–1909; two of the tuatara were initially housed as pets in his office, and later taken with him to London (Donne 1942; see also Chapter 5).

43 Brown [2000], pp. 185–186.

44 Schmidt 1952.

45 Anon. 1907.

46 Newspaper report in the *Auckland Star*, 22 October 1908, copy in ANZ IA 1 46/18/61. Also reported in the *Bay of Plenty Times*, 23 October 1908, p 2.

47 Brown [2000], p. 186.

48 *List of permits to take or export tuatara lizards from 1896 to 1912*, NMNZ MU000207 Box 4, Item 1. The total includes the specimens previously noted for Stephens Island.

49 Anon. 1909.

50 Untitled summary of events regarding collections made during 1896–1898 by Thilenius, Schauinsland and Dendy. NMNZ MU 000001 Box 23, Item 14, File No. 14/6/9. See also articles during 1898 in the *Bay of Plenty Times and Thames Valley Warden*: 18 May, p 2; 29 May, p 2; 16 September, p. 2.

51 Thilenius 1899.

52 C. Norris to E. S. Corlett, Department of Tourist and Health Resorts, 23 February 1905, ANZ IA 1 46/18/61. A 'Mr G. G. Norris' was also reported in 1908 as stating 'that he had kept tuataras ever since he could remember, and had had as many as 200 in his possession at one time.' (Anon. 1908a).

53 C. Norris to T. E. Donne, Superintendent of Tourist Department, 12 February 1906, ANZ IA 1 46/18/61.

54 C. Norris to Hamilton, 10 July 1906, Director [of the Dominion Museum] to C. Norris, 24 July 1906, both in NMNZ MU000207 Box 4, Item 1.

55 The two in Dunedin were said to be 'the property of Mr G. G. Norris…and brought from his home in Tauranga, where he has a large collection….[and that he] had had as many as 200 in his possession at one time…' (Anon. 1908a, p. 18).

56 Hamilton to Minister of Internal Affairs, 5 June 1912, ANZ IA 1 46/18/4 pt 1. The pencilled marginal note reads 'Appro. Take necessary steps. GWR 14/6/12'.

57 The collection was referred to in a newspaper report in the *Lyttelton Times*, 30 June 1913, copy in ANZ IA 46/18/4 pt 1: 'a resident of Tauranga has kept tuataras in his garden for a long time'.

58 Extract of letter to Mr Hislop from Colin Norris, 27 January 1915, NMNZ MU000207 Box 4, Item 1.

59 B. Sladden to Under Secretary for Internal Affairs, 7 November 1935, ANZ IA 1 52/30 pt 1.

60 B. Sladden to Director of the Dominion Museum, 9 April 1929, ANZ IA 1 52/30 pt 1.

61 Goodlet [c. 1922]. William Goodlet was the janitor of the Otago Museum from 1913 until about 1930. In his article, which is detailed and knowledgeable, Goodlet indicated that he had been in direct communication with Norris.

62 Worthy and Holdaway 2002, pp. 531–537.

63 Galbreath 1989, op. cit., pp. 137–140.

64 Not all contemporary biologists held this view of inevitable extinction through genetic senescence. Kirk (1896, op. cit.), for instance, more realistically attributed the decline of tuatara on the mainland to wild pigs, cats and rats.

65 Drummond 1905; see also Drummond 1911.

66 B. Sladden to Under-Secretary of the Department of Internal Affairs, 7 November 1935, ANZ IA 1 52/30 pt 1.

67 Crook 1970, p. 16.

68 On islands where the impacts of introduced mammals have been controlled or prevented, tuatara are often exceedingly abundant and clearly well adapted to their current environment (Chapter 7). See also Gans 1983; Cree and Daugherty 1990a; Thompson and Daugherty 1992.

69 *The Dominion*, 16 July 1913, ANZ IA 1 46/18/4 pt 1.

70 H. G. Ell to Colonial Secretary, 20 July 1904; H. G. Ell to W. Hall-Jones, 29 March 1906; H. G. Ell to J. Millar, 21 October 1907; H. G. Ell to Minister of Marine, 4 October 1910; all in ANZ M 125/611 pt 1. See also: H. G. Ell to F. H. Bell, 1 November 1912; newspaper report in the *Lyttelton Times*, 30 June 1913; both in ANZ IA 1 46/18/4 pt 1. See also: clipping from unidentified newspaper, 11 June 1913, ANZ ABWN 6095 W5021 Box 688 23/856 pt 1.

71 Interview between Messrs Crewes and others of the Wellington Zoological Society and H. G. Ell, Postmaster-General, for the consideration of the Minister of Internal Affairs, 24 April 1912; memo re deputation consisting of Rev. J. Crewes and others of the Wellington Zoological Society to the Minister of Internal Affairs, 2 May 1912; *The Dominion*, 16 July 1913; all in ANZ IA 1 46/18/4 pt 1.

72 A. Hamilton to Minister of Internal Affairs, 4 February 1913, ANZ IA 1 46/18/4 pt 1. The 'German doctor' was presumably Thilenius (or Schauinsland, but this seems less likely).

73 Benham to Hamilton, May 1912, ANZ IA 1 46/18/4 pt 1; the same letter was published in an unidentified newspaper on 11 June 1913 (copy in ANZ ABWN 6095 W5021 Box 688 23/856 pt 1).

74 Islands where tuatara were reported in the early 1900s but are now extinct include the Moturoa Islands, at least three in the Mokohinau Islands, Shoe Island (Motuhoa), Slipper Island (Whakahau) and East Island (Whangaokeno Island). During the late 1800s, tuatara were also reported from Moutohora Island and the Chetwode Islands, and during the 1920s they were reported on Manawatawhi/Three Kings Islands (see Chapter 3, Table 3.1 for sources).

75 For effects of cats on Cuvier (Repanga), East and Stephens Islands, see Chapter 3, including Box 3.1. For concern about the effects of rats on Cuvier Island, see Chapter 7, Box 7.2.

76 Thomson 1915. In an interesting family connection, the author was Allan Thomson, G. M. Thomson's son. Allan Thomson succeeded Hamilton as Director of the Dominion Museum after the latter's sudden death in 1913.

77 H. G. Ell to F. H. D. Bell, 1 November 1912; Gilbert Mair to H. D. Bell, 8 June 1913; Rev. J. Crewes to Minister for Internal Affairs, 21 October 1914; all in ANZ IA 1 46/18/4 pt 1.

78 C. M. Clench to Rev J. Crewes, 18 August 1913, ANZ IA 1 46/18/4 pt 1.

79 J. Hislop [to Clench?] 10 January 1914; J. Hislop to Secretary of Post and Telegraph, 10 January 1914; J. Hislop to Commissioner of Police, 10 January 1914; J. Hislop to secretaries of acclimatisation societies, 12 January 1914; all in ANZ IA 1 46/18/4 pt 1.

80 *New Zealand Gazette*, 30 October 1913. For instruction to publish the information in *Kahiti*, see J. Hislop to J. Castle, 31 October 1913, ANZ IA 46/18/4 pt 1.

81 Salary reported for R. Nelson, Caretaker on Little Barrier Island, ANZ IA 1 46/18/61.

82 Duties of Special Protectors as recommended by J. Hislop to Minister of Internal Affairs, 12 February 1913, ANZ IA 1 46/18/4 pt 1. There subsequently seem to have been some differences of opinion as to whether keepers were required to kill all cats on the island (the expectation of Internal Affairs, as noted above, and reiterated by J. Hislop to Secretary for Marine, 21 April 1920, ANZ M1 25/611 pt 1), or just wild cats (as reported by the Director of the Dominion Museum; see Thomson 1915). By about 1966–1967, the Marine Department was retrospectively reporting that the duty of the keepers had simply been 'to kill cats', dodging the issue of whether domestic cats were outlawed (undated press release c. 1966, Department of Marine, *Marine Department's work in protection of tuataras*, source unknown).

83 F. Young (Principal Keeper on Cuvier) to Secretary of Marine, 1 July 1930, ANZ M 1 25/611 pt 3: 'It is not known if rats trouble the lizards but an extermination campaign is also being carried out against them'. For reported benefits to tuatara, see Chapter 7.

84 Secretary of Marine to Principal Keepers on Brothers, Stephens and Cuvier, 2 December 1922, in response to a request from the Under Secretary of Internal Affairs to the Secretary of Marine, 28 November 1922; both in ANZ M 1 25/611 pt 1.

85 Report on visit to Cuvier Island [from 3–8 August 1953?], ANZ IA 1 46/18/8 pt 1.

86 Insofar as can be judged from the contents of ANZ IA 1 46/14/17.

87 Wilkes, O. 1996. *Provisional history of human activity on Cuvier (Repanga) Island*. Unpublished manuscript, Department of Conservation, Waikato Conservancy, Hamilton, p. 17.

88 W. J. Ross to Secretary of Marine, 28 November 1925, ANZ M 1 25/611 pt 2: 'I am sorry for infringing rules by writing direct to another department without first putting the correspondence through your Department, but as you informed me when I visited your office that the hawks had nothing to do with your Department I did not think I was doing wrong …'.

89 Various correspondence in ANZ M 1 25/611 pt 1.

90 Blanchard 1935.

91 Letter published in the Canterbury *Press*, quoted (author unnamed) by Buller 1896, p. 342.

92 P. J. Kelleher, Permanent Head to Secretary of Public Service Commissioner, 29 March 1932; Secretary of Marine to Principal Keepers on Cuvier, Stephens and the Brothers, 1 April 1932; both in ANZ M 1 25/611 pt 2.

93 J. Cocker (Principal Keeper on Stephens) to Secretary of Marine, 24 December 1936, ANZ M 1 25/611 pt 3.

94 H. B. Jamieson to Secretary of Marine, 30 January 1937, ANZ M 1 25/611 pt 3.

95 S. F. Marriott to Secretary of Marine, 7 September 1944, ANZ M 1 25/611 pt 3.

96 For example, in 1944 (ANZ M 1 25/611 pt 3).

97 Hislop, J. 1922. Annual report of the Department of Internal Affairs for the year ended 31st March 1922. *Appendix to the Journals of the House of Representatives* H-22: 1–12.

98 Galbreath, R. 1993. *Working for wildlife: a history of the New Zealand Wildlife Service*. Bridget Williams Books and Historical Branch, Department of Internal Affairs, p. 15.

99 Daugherty et al. 1992 considered the management history of tuatara to be 'characterised as much by benign neglect as concerned vigilance'. This is especially true of the period 1930–1970.

100 For an overview, see McCallum, J. 1983. A review of field club research on the northern offshore islands. *Tane* 29: 223–245.

101 Under Secretary of Internal Affairs to Secretary of Marine, 22 August 1930, ANZ M 1 25/611 pt 3: 'It was decided to close Stephens Island against the taking of tuatara lizards or their eggs after the 31st March, 1929'.

102 Under Secretary of Internal Affairs to Minister of Internal Affairs, 21 December 1938, ANZ IA 1 46/18/4 pt 1.

103 But a few collections were made. In 1952 three tuatara were sent to zoos in the US, and a fourth went to London Zoo (Oliver 1953).
104 *Tuataras in captivity*, WIL 32/3/11, in ANZ ABWN 6095 W5021 Box 688 23/856 pt 1.
105 Gans 1991. See also Daugherty et al. 1992.
106 For example, Newman 1978, 1982a, b. In practice, however, the 'wildlife' of interest to the Wildlife Service continued largely to mean birds and fish, as acknowledged by the director in 1979. See Galbreath 1993, op. cit., p. 218.
107 For example, Cree and Thompson 1988; Cree and Daugherty 1990a; Daugherty et al. 1990a, b; Thompson and Daugherty 1992.
108 Galbreath 1993, op. cit., p. 190.
109 Towns, D. R. and Broome, K. G. 2003. From small Maria to massive Campbell: forty years of rat eradication from New Zealand islands. *New Zealand Journal of Zoology* 30: 377–398.
110 Discussed by Daugherty et al. 1992.
111 Cree and Butler 1993.
112 Daugherty et al. 1990.
113 Hay et al. 2010.
114 Cree and Daugherty 1990b; Daugherty et al. 1992; Cree et al. 1994; Towns et al. 2001.
115 Blanchard 1991.
116 SRARNZ, the Society for Research on Amphibians and Reptiles in New Zealand.
117 Gaze 2001, p. 13.
118 Newman, D. G. 1998. *Minutes of the Tuatara Recovery Group meeting held on Matiu/Somes Island, 4–5 March 1998*. Unpublished report, Department of Conservation, Wellington.
119 G. Ussher, Auckland Regional Council, pers. comm. 6 March 2006.
120 Ussher 1999b, c.
121 Ussher 1999b, c, 2000; Ussher, G. 2002. *Reintroduction of tuatara (*Sphenodon p. punctatus*) to Moutohora (Whale Island): population status after 5 years*. Unpublished report to the Bay of Plenty Department of Conservation, Rotorua, 47 pp.; Merrifield 2001a, b; Nelson et al. 2002b; Ruffell 2005.
122 Nelson et al. 2002b; Moore et al. 2007; K. A. Miller et al. 2010.
123 Nelson, N. 2004. *Matiu/Somes Island tuatara – 5 years post-translocation survey, 2–5 February 2004*. Unpublished report, Victoria University of Wellington; K. A. Miller et al. 2010.
124 Ruffell 2005.
125 K. A. Miller et al. 2009. See also Miller 2009.
126 Additionally, if competition for matings leaves few males able to breed (in other words, if reproductive skew for males is high) then genetic diversity would be lost more rapidly; however, given the low density upon release of translocated tuatara, it seems reasonable to assume that the skew will initially be low. See K. A. Miller et al. 2009 for further discussion.
127 Genetic heterozygosity in a population refers to the frequency of individuals that have different forms (alleles) of a given gene at a particular chromosomal position (locus). Such individuals are said to be heterozygous at that locus.
128 K. A. Miller et al. 2009. See also K. A. Miller et al. 2010.
129 Visitor numbers for Tiritiri Matangi: pers. comm. from B. Walter to G. Ussher in Ussher, G. 2003. *Proposal to translocate northern tuatara (*Sphenodon punctatus punctatus*) from Middle Island, Mercury Island Group to Tiritiri Matangi Island, North Auckland in mid-October 2003*. Unpublished report, Department of Conservation, Auckland. Visitor numbers for Matiu/Somes from Department of Conservation. 2000. *Matiu/Somes Island: a plan for conservation management*. Department of Conservation, Wellington. Visitor numbers for Moutohora from Department of Conservation. 1999. *Moutohora (Whale) Island Conservation Management Plan 1999–2009*. Bay of Plenty Conservation Management Planning Series No. 7. Department of Conservation, Rotorua.
130 Sources for the activities of Frederick Angell: *Nelson Mail* 7 July 1997; New Zealand Press Association 11 September 1997; Bodnar 1997; *Otago Daily Times*, 11 April 2003; Colin Hitchcock Wildlife Enforcement Group, pers. comm. 24 November 2006. Prior to his death in a car accident in 2003, Angell was also convicted for several wildlife-related offences involving species other than tuatara.
131 In another instance of poaching, two men were fined $40 each in about 1967 for taking two tuatara, later named 'Darby' and 'Joan', from the Aldermen Islands (photo in *Evening Post*, June 1967, Alexander Turnbull Library Pictorials Collection, Zoology Tuatara 17/4). More recently (March 2005), an unsuccessful attempt was made to break in to a display enclosure housing tuatara at Victoria University of Wellington (*The Press*, 12 March 2005).
132 Dodd, C.K. and Siegel, R.A. 1991. Relocation, repatriation, and translocation of amphibians and reptiles: are they conservation strategies that work? *Herpetologica* 47: 336–350.
133 See also Sherley et al. 2010.
134 Duffey, E. 2001. Introduced pest species and biodiversity conservation in New Zealand. *Biological Conservation* 99: 1.
135 Cree and Butler 1993.
136 Daugherty et al. 1992; Towns et al. 2001.
137 Daugherty et al. 1992.
138 Daugherty et al. 1992.
139 Eradication of rats on Taranga began in 2011 (Towns, D. R., West, C. J. and Broome, K. G. 2012. Purposes, outcomes and challenges of eradicating invasive mammals from New Zealand islands: an historical perspective. *Wildlife Research* doi.org/10.1071/WR12064). Confirmation of rat-free status is expected during 2013 (Keith Hawkins, Department of Conservation, pers. comm. 18 September 2012).
140 See Hay et al. 2010.

Endnotes to boxed text

For explanation of citation format in box endnotes, see endnotes to the main text.

Box 10.1

1 Parker 1891. A similar early diorama was exhibited from 1886 at what is now the Auckland Museum (Gill, B. J. 1999. History of the land vertebrates collection at Auckland Museum, New Zealand, 1852–1996. *Records of the Auckland Museum* 36: 59–93).
2 Lamb, R. C. 1964. *Birds, beasts and fishes: the first 100 years of the North Canterbury Acclimatisation Society*. North Canterbury Acclimatisation Society, Christchurch. See p. 123.
3 The pre-release advertisement sent around the world for the 1991 stamp series was unfortunately not so realistic: in a saurian mix-up, images of a green iguana were substituted for tuatara. The brochure was reprinted and redistributed with the correct images.
4 Tuatara have also featured on stamps from other countries, including a 9d Jersey illustration of the 'tuatara lizard' in 1972.
5 New Zealand wildlife (including birds, fish and tuatara) was identified as a design theme that consumers would like to see on banknotes by 10% of survey respondents. See *Bank notes survey 2010 consumer*. Report prepared for the Reserve Bank by AC Nielson. http://www.rbnz.govt.nz/notes_and_coins/banknote_upgrade/4436446.pdf (accessed 6 September 2013).
6 In December 2006, New Zealand Post announced the issue of two limited-edition coins featuring tuatara (a 'silver proof' coin for $79 and a 'brilliant uncirculated' coin for $29). These are sold as collectors' items.
7 http://www.linz.govt.nz (accessed 2 June 2006).
8 http://aadc-maps.aad.gov.au (accessed 7 June 2006).
9 Galbreath, R. 1993. *Working for wildlife: a history of the New Zealand Wildlife Service*. Bridget Williams Books and Historical Branch, Department of Internal Affairs. See p. 93.
10 http://www.nsa.org.au (accessed 6 June 2006).
11 http://www.flyingnun.co.nz (accessed 6 June 2006). There is also an experimental band named 'Tuatara' in Seattle, US.
12 Shine, R. 1998. *Australian snakes: a natural history.* Revised edition. Reed Books, Sydney. See p. 206. Tuatara beer is bottled by the Tuatara Brewery Company.
13 Bottled by Mt Olympus Wines, Marlborough.
14 A classic example, first published in 1955 and reprinted several times is: Acres, A. 2002. *Hutu and Kawa meet tuatara*. Reed, Auckland.
15 Dawbin 1962b.
16 Schmidt 1949, p. 90.
17 Lewin 1987, p. 32. The poem was written 20 years prior to publication.
18 Borrell, N. 2004. Tuatara. In: Morris, P., Ricketts, H. and Grimshaw, M. (eds). *Spirit abroad: a second selection of New Zealand spiritual verse*. Godwit, Auckland, p. 216.
19 Anderson, B. 1989. Tuataras. In: *I think we should go into the jungle*. Victoria University Press, Wellington, pp. 96–109.
20 *Tuatara (1998) for percussion and piano, programme note*. Promethean Editions, 1997–2006. http://www.promethean-editions.com (accessed 23 August 2006).
21 Communicated by Wikuki Kingi Senior via Te Rangi Martrell, 9 December 2006.

Box 10.2

1 R. Brassey, Auckland Regional Council, pers. comm. 18 September 2003, citing the diary of T. A. Harris for 27 December 1878 (NZ MSS 640, Auckland Public Library). See also Brassey, R. 1999. *Mansion House gardens, Kawau Island: a conservation plan and resource document*. 2nd edn. Department of Conservation, Auckland. Brassey considered that the 'Lizzards' introduced to Kawau Island via the 'Henumoa' (*Hinemoa*?) were likely to have been tuatara.
2 Buller 1893, p. 87. For a detailed account of Buller's activities, see Park, G. 1995. *Nga uruora: the groves of life.* Victoria University Press, Wellington. See pp. 168–197.
3 For the suggestion, and discussion of how to transport eggs: E. W. Tutt to Secretary of Internal Affairs, 12 March 1913, NMNZ MU000207 Box 4, Item 1. Tutt had apparently been on Stephens Island during egg collections by Schauinsland and Thilenius, and for Dendy.
4 Destination islands: Benham to Hamilton, May 1912, G. Mair to H. D. Bell, 8 June 1913, both in ANZ IA 1 46/18/4 pt 1; H. Harvey [lighthouse keeper on Cuvier Island] to Under Secretary for Internal Affairs, 4 April 1913, ANZ IA 1 46/18/8 pt 1.
5 H. D. Bell to G. Mair, 1 July 1913, ANZ IA 1 46/18/4 pt 1.
6 Assistant Under Secretary of Internal Affairs to Secretary of Marine, 5 May 1913; Secretary of Marine to Assistant Under Secretary of Internal Affairs, 28 May 1913; Bollons to Secretary of Marine, 23 May 1913 (all in ANZ M 1 25/611 pt 1); Assistant Under Secretary of Internal Affairs to E. W. Tutt, 31 May 1913, ANZ IA 1 46/18/4 pt 1.
7 I found no details of the releases in files from Internal Affairs on tuatara held at Archives New Zealand. It is possible that day journals kept by the keepers might record receipt of the eggs.
8 Secretary of the New Zealand Institute to J. Hislop, 14 July 1913, ANZ IA 1 46/18/4 pt 1. The name 'Motungarara' ('island of reptiles', or 'island of creeping things') suggests that tuatara might have been known on at least one of these islands during pre-European times. The islands' current names are Tokomapuna Island (Aeroplane Island) and Motungarara Island (Fishermans Island).
9 The Collector 1913. For the likely identity of 'The Collector' as H. H. Travers, see: Galbreath, R. and Brown, D. 2004. The tale of the lighthouse-keeper's cat: discovery and extinction of the Stephens Island wren (*Traversia lyalli*). *Notornis* 51: 193–200.
10 *Evening Post* 13 June 1913, p. 6.
11 Bacon, A. 1938. Appendix D in *Kermadec Islands: report of Aeradio Committtee*. Unpublished report, Alexander Turnbull Library, Wellington. See Appendix D, p. 10: 'Some tuatara lizards were released on one island about the time I was here'. Bacon first resided on the main island (Raoul) of the Kermadec Islands during 1889–1891, and returned briefly in 1927 and again in 1935. He was present during the Aeradio Expedition of 1937, when no trace of tuatara was seen. The government steamer occasionally visited the Kermadec Islands (for example, Captain Fairchild visited in 1887 to declare the islands part of New Zealand). This provides a possible link with tuatara, as Fairchild and his successor, Captain Bollons, were known to collect tuatara from other islands. (With thanks to John Parkes for drawing Bacon's account to the

attention of Charles Daugherty, who passed it on to me.)

12 Goodlet c. 1922, p. 17. From 1913 until c. 1930, William Goodlet was the janitor of the Otago Museum; he wrote knowledgeably about the live tuatara kept there. The German naturalist in question would have been Georg Thilenius.

13 The Under-Secretary of Internal Affairs referred to 'senseless raids by picnic parties and Natives' on Bay of Plenty islands; see Hislop, J. 1920. Annual report of the Department of Internal Affairs for the year ended 31st March, 1920. *Appendix to the Journals of the House of Representatives* H-22: 1–5; see also W. Hill to General Manager of the Government Tourist Bureau, Department of Tourist and Health Resorts, 24 May 1919, ANZ IA 1 46/18/61.

14 W. Hill to General Manager of the Government Tourist Bureau, Department of Tourist and Health Resorts, 13 February 1920, ANZ IA 46/18/61.

15 Known also as Leper Island during the 1900s. The proposal to transfer tuatara to this island may have come from the lighthouse keepers on Matiu/Somes Island; it was supported by the Acting Director of the Dominion Museum (letter to Secretary of Marine, ANZ M 1 25/611 pt 1). See also Thomson 1920.

16 A. V. Pearce to Secretary of Marine, 15 November 1920, ANZ M 1 25/611 pt 1.

17 W. Cleverly, Principal Keeper on Somes, to Secretary of Marine, 22 January 1924, ANZ M 1 25/611 pt 2. A search for tuatara was also planned for September 1921 by staff of the Department of Agriculture and Dominion Museum (A. H. Cockayne, Biology Section of the Department of Agriculture, to Secretary of Marine, 16 September 1921, ANZ M 1 25/611 pt 1), but results of such a survey have not been found.

18 Wildlife Branch of the Internal Affairs Department. 1951. Rabbits on Leper Island (Mokopuna). *Forest and Bird* 102 (November): 11.

19 W. H. Dawbin to Under Secretary for Internal Affairs, 19 December 1951, ANZ IA 1 46/18/4 pt 2.

20 G. F. Fox to Commissioner of Crown Lands, 17 July 1955; Chairman's Report of the Fauna Protection Advisory Council, Tenth Meeting, 1956; both in ANZ IA 1 46/18/4 pt 2.

21 Memorandum from Secretary of Internal Affairs to Commissioner of Crown Lands, 28 June 1955, ANZ IA 46/18/4 pt 2.

22 W. H. Dawbin to Acting Secretary of Internal Affairs 28 May 1959, ANZ IA 1 46/14/17.

23 See chapters on Pacific rats by I. A. E. Atkinson and D. R. Towns (pp. 159–174) and on the Norway rat by J. G. Innes (pp. 174–187) in: King, C. M. (ed.). 2005. *The handbook of New Zealand mammals*. Oxford University Press, Auckland.

24 No tuatara population is known to co-exist with Norway rats. The arrival of Norway rats on Whenuakura Island near Whangamata in the 1980s was followed by the disappearance of the tuatara there, and tuatara may have died out on other islands on which Norway rats were present (see Chapter 3).

25 Later proposed translocations of tuatara to the inner of the two Chetwode Islands and to Hauturu/Little Barrier Island also lacked explicit recognition that introduced mammals were present. The proposal to transfer tuatara to Inner Chetwode Island, made by staff of the Dominion Museum in 1929, was dropped when Captain Bollons argued that it was too easily accessible by motor launch to remain free from (human) interference (Secretary of Marine to Under Secretary of Internal Affairs, 4 March 1929, ANZ M 1 25/611 pt 2). A proposal in 1960 (source unknown) to transfer tuatara to Hauturu was turned down by the Fauna Protection Advisory Council because 'liberation on this island would serve no useful purpose' (Fauna Protection Advisory Council Minutes of Meeting 18–19 May 1960, ANZ IA 46/18/4 pt 2). Pacific rats were almost certainly already present on Inner Chetwode Island, and Pacific rats and cats were present on Hauturu.

26 Additionally, the transfer, in about 1963, of an unknown number of tuatara from the Trio Islands to Rangitoto ke te Tonga (D'Urville Island) has recently been reported (Sherley et al. 2010). The reason for and outcome of the transfer are unknown.

Box 10.3

1 IUCN 2011. *IUCN red list of threatened species*. Version 2011.2. http://www.iucnredlist.org (accessed 9 January 2012).

2 Daugherty et al. 1990b.

3 See Hay et al. 2010, p. 1063.

4 Hay et al. 2003, 2004; Bell et al. 2004; Miller et al. 2006.

5 MacAvoy et al. 2006.

6 Townsend, A. J., de Lange, P. J., Duffy, C. A. J., Miskelly, C. M., Molloy, J. and Norton, D. A. 2008. *New Zealand threat classification system manual*. Department of Conservation, Wellington.

7 Hitchmough et al. 2010.

Box 10.4

1 For a brief history of the Treaty and its outcome, see Chapter 11 (pp. 151–167) and subsequent chapters of King, M. 2003. *The Penguin history of New Zealand*. Penguin Books, Auckland.

2 Quotes from the Treaty of Waitangi (Te Tiriti o Waitangi) are from http://www.treatyofwaitangi.govt.nz/treaty/ (accessed 10 September 2006).

3 In researching this chapter, I found no evidence that wildlife officials consulted with or involved Māori in the management of tuatara prior to about the 1980s. Consideration of Māori 'interests' seems to have been limited to rather paternalistic efforts to ensure that Māori knew of and understood protective legislation. One early example of an apparent lack of consultation involved the removal of tuatara from Māori-owned Moutoki Island to Mokopuna Island in 1920 (Box 10.2). This translocation was initiated amid concern that the tuatara population on Moutoki was endangered by human visitors (including Māori); the irony is that tuatara thrive today on Moutoki whereas the translocation was unsuccessful.

4 Tipene-Matua, B. and Dawson, J. 2003. The Treaty of Waitangi and research. In: Dawson, J. and Peart, N. (eds). *The law of research: a guide*. University of Otago Press, Dunedin, pp. 61–79.

5 However, the terms of the agreement (including whether or not Ngāti Koata has right of veto over proposed actions, and whether additional iwi should be involved in the agreement), have been the subject of a recent claim to the Waitangi Tribunal; see pp. 1306–1317 of New Zealand Waitangi Tribunal. 2008. *Te Tau Ihu o te Waka a Māui: report on northern South Island claims. Vol. 1.* Legislation Direct, Wellington, New Zealand.

6 Ramstad et al. 2009.

7 The trust was established in 2007 with sponsor and winemaker Waimea Estates.

8 Action 3.4, p. 25 of the *Tuatara Recovery Plan 2001–2011*; see Gaze 2001.

9 Levine 2010.

10 So-named because it is the 262nd claim lodged with the Waitangi Tribunal. The six iwi are Ngāti Kuri, Ngātiwai, Te Rarawa, Ngāti Porou, Ngāti Kahungunu and Ngāti Koata. See *First amended statement of claim, WAI 262, Claim 1.1(a); Second amended*

statement of claim on behalf of Ngati Koata, October 2001, WAI 262, Claim 1.1(f). See also McClean, R. and Smith, T. 2001. *The Crown and flora and fauna: legislation, policies, and practices, 1983–1998*. Waitangi Tribunal, Wellington.

11 New Zealand Waitangi Tribunal. 2011. *Ko Aotearoa tēnei: a report into claims concerning New Zealand law and policy affecting Māori culture and identity*. Wai 262 Waitangi Tribunal Report. Legislation Direct, Wellington.

12 *Tuatara Recovery Plan 2001–2011*, p. 7; see Gaze 2001.

13 Interestingly, although evidence from middens, and from oral traditions recorded in the early colonial period, attest to tuatara once being harvested for food (Chapter 3), such a tradition is contested by some Māori today (Ramstad et al. 2007a, b) and there have been no proposals to resume it. On the other hand, recent applications by Māori to resume the harvest of muttonbirds from Karewa Island, which also has tuatara, have been declined by DOC because of concern about the effects on the island's habitat. An eco-tourism proposal to take paying customers to Stephens Island to see tuatara, made by Tuatara Māori Ltd, a company with links to Ngāti Koata, drew opposition from a national conservation group (*Forest and Bird* No. 320, May 2006, p. 48) and was eventually declined in 2007 by DOC on the grounds of increased risk of disease (*The Press*, 29 August 2007). Among Māori, the issue of what constitutes an appropriate koha (reciprocal gift) for proposals involving the 'use' of tuatara by others (e.g. removal of tuatara or their eggs from islands for research or translocation to other areas outside of that iwi's rohe) has also been raised.

14 Rautangata, K. and Rautangata, M. (eds). 2003. *Te moko tukupu wananga a rua i te wheke rangi*. Te Wānanga o Aotearoa, Te Awamutu. See p. 105.

Box 10.5

1 Towns, D. R. and Broome, K. G. 2003. From small Maria to massive Campbell: forty years of rat eradication from New Zealand islands. *New Zealand Journal of Zoology* 30: 377–398.

2 The deaths of three tuatara at Auckland Zoo in 2002 were attributed to ingestion (directly or indirectly) of rat bait. See Richardson 2002.

3 Griffiths, R. W. and Wilson, G. 2002. *Assessment of environmental effects for the eradication of kiore (*Rattus exulans*) from Little Barrier Island (Hauturu)*. Unpublished report, Department of Conservation, Auckland. See also Towns and Broome 2003, op. cit.

4 Towns et al. 2007.

5 But not the largest island from which rats have been successfully eradicated: that record goes to sub-antarctic Campbell Island (11,300 ha), from which Norway rats were eradicated in 2001.

6 Kiore are not endangered within New Zealand: as of 2003, they survived in Fiordland, Southland and south Westland, and on about 17 offshore islands of at least 5 ha, excluding Hauturu (Atkinson, I. A. E. and Towns, D. R. 2005. Kiore. In: King, C. M. (ed.). *The handbook of New Zealand mammals*. Oxford University Press, Auckland, pp. 159–174).

7 Details from DOC media releases 10 June 2004, 14 July 2004 and 18 April 2005, http://www.doc.govt.nz (accessed 7 October 2005); Griffiths and Wilson 2002, op. cit.

8 Anon. 2006. Cook's petrel's 'hit town'. Forest and Bird No. 322 (November): 7.

9 Shane McInnes, Department of Conservation, pers. comm. 20 December 2010.

10 Donohue, D. 2002. A test of animal conservation and conflicting legal demands: kiore on Little Barrier Island. *Animal Rights Legal Advocacy Network Newsletter* 1: 1–5.

11 Discussed by Towns et al. 2007 in relation to Taranga Island.

12 Towns and Broome 2003, op. cit.

13 Tennyson and Pierce 1995.

14 Towns et al. 2007. See also Chapter 6.

15 Eradication of rats began in 2011 (Towns, D. R., West, C. J. and Broome, K. G. 2012. Purposes, outcomes and challenges of eradicating invasive mammals from New Zealand islands: an historical perspective. *Wildlife Research* doi.org/10.1071/WR12064). Definitive confirmation of rat-free status is expected during 2013 (Keith Hawkins, Department of Conservation, pers. comm. 18 September 2012).

Box 10.6

1 Based on their locations near to other islands that have or once had tuatara, and the availability of suitable habitats. Tuhua also has one historical sighting of a tuatara, although whether it was a natural resident or transferred there is unclear (Chapter 3).

2 After Joseph Somes, Deputy Governor of the New Zealand Company. The human history of the island is recounted in detail by: McGill, D. 2001. *Island of secrets: Matiu/Somes Island in Wellington Harbour*. Steele Roberts & Silver Owl Press, Wellington.

3 Knox 1870. See also Drummond 1917.

4 A specimen from Knox was catalogued by the British Museum in 1844 (BMNH 1844.10.29.11).

5 William Williams to Richard Owen, 1 August 1845, Natural History Museum, Owen Collection, Vol. XXVI. Quoted on p. 400 of Temple, P. 2002. *A sort of conscience: the Wakefields*. Auckland University Press, Auckland. The suggestion that this was 'the first specimen of a tuatara to be sent off for scientific examination' cannot be correct (Gray had described a skull in 1831 and a preserved specimen from Dieffenbach in 1842; see Chapter 5). There is no known evidence from Owen's published work that he received the specimen from Williams.

6 Acting Director of Dominion Museum to Secretary of Marine, 19 March 1919, ANZ M 1 25/611 pt 1: 'the late Mr A. Hamilton, while in quarantine, saw one on Somes Island in 1877' (for a similar report, see also: Best, E. 1918. The land of Tara and they who settled it. Part III. *Journal of the Polynesian Society* 27: 49–71). Augustus Hamilton, who became the Director of the Dominion Museum, arrived in New Zealand on a typhoid-stricken ship in 1875 (hence the quarantine) and died in 1913 (Dell, R. K. 2010. Hamilton, Augustus 1853–1913. Dictionary of New Zealand Biography, updated 1 September 2010, http://www.teara.govt.nz/en/biographies/2h8/1 (accessed 31 July 2012).

7 Veitch, C. R. and Bell, B. D. 1990. Eradication of introduced animals from the islands of New Zealand. In: Towns, D. R., Daugherty, C. H. and Atkinson, I. A. E. (eds). *Ecological restoration of New Zealand islands*. Conservation Sciences Publication No. 2. Department of Conservation, Wellington, pp. 137–146.

8 One juvenile was belatedly released in February 2000.

9 Merrifield 2001a, b.

10 Nicola Nelson, Victoria University of Wellington, pers. comm. October 2005. Mating has been seen (McGill 2001, op. cit., p. 9). Two animals have been taken into captivity for periods of up to 6 months to treat injury or infection.

11 Kimberley Miller, Victoria University of Wellington, pers. comm. 23 September 2009; K. A. Miller et al. 2010; see Table 10.1.

12 Baker, K. 2008. KCC goes to Matiu/Somes. *Forest and Bird* No. 328 (May): 47. The eggs were taken off-island to hatch after the nest was disturbed.
13 Department of Conservation. 2000. *Matiu/Somes Island: a plan for conservation management*. Department of Conservation, Wellington.

Box 10.7

1 McKenzie 2007.
2 *Kereru* (newsletter of the Karori Sanctuary Trust), April 2009 No. 39: 1.
3 Besson 2009; Mello et al. 2011.
4 However, an initial analysis suggests that the mean growth rate at Orokonui, for juveniles within outdoor pens where food is provided, is greater than for similar-sized juveniles on Stephens Island (Mello, R. S. R. 2010. *Adjustment of juvenile tuatara (*Sphenodon punctatus*) to the Orokonui Ecosanctuary climate: operative temperatures, emergence behaviour and growth rate*. Unpublished Postgraduate Diploma of Science dissertation, University of Otago, Dunedin). See also Mello et al. 2011.
5 Reisinger, A., Mullan, B., Manning, M., Wratt, D. and Nottage, R. 2010. Global and local climate change scenarios to support adaptation in New Zealand. In: Nottage, R. A. C., Wratt, D. S., Bornman, J. F. and Jones, K. (eds). *Climate change adaptation in New Zealand: future scenarios and some sectoral perspectives*. New Zealand Climate Change Centre, Wellington, pp. 26–43.
6 Cree et al. 1995b; Nelson et al. 2004a; Mitchell et al. 2006.
7 Mitchell et al. 2008.
8 Mitchell et al. 2010.
9 MacAvoy et al. 2007; Miller et al. 2008a. See also K. A. Miller et al. 2009, 2010.
10 Reisinger et al. 2010, op. cit.; see also Ministry for the Environment. 2008. *Climate change effects and impacts assessment: a guidance manual for local government in New Zealand*. 2nd edn. Prepared by Mullan, B., Wratt, D., Dean, S., Hollis, M., Allan S., Williams, T., Kenny, G. and MfE staff. Ministry for the Environment, Wellington.
11 Reisinger et al. 2010, op. cit.
12 Besson 2009; Besson et al. 2012.

Box 10.8

1 For discussion of taxonomic bias and flagship species, see Seddon, P. J., Soorae, P. S. and Launay, F. 2005. Taxonomic bias in reintroduction projects. *Animal Conservation* 8: 1–8.
2 Hitchmough et al. 2010.
3 Cree et al. 1994; Towns et al. 2001.

CHAPTER ELEVEN

Latest developments

Introduction

As the preceding chapters have illustrated, tuatara are remarkable reptiles for their great economy of movement. On the other hand, the scientific study of tuatara does not sit still. This final chapter updates earlier chapters by reviewing discoveries made since 2012 about the evolution and biology of tuatara, as well as new conservation initiatives and changes in conservation management.

Origins: evolution and early interactions with humans

Discoveries of fossil rhynchocephalians, the Mesozoic relatives of the modern tuatara *Sphenodon punctatus*, continue at a remarkable pace. Newly described rhynchocephalians include *Sphenotitan leyesi*, a large and probably herbivorous sphenodontian from the Late Triassic of Argentina,[1] *Sphenocondor gracilis*, a small sphenodontian from the Middle Jurassic of Argentina,[2] and *Oenosaurus muehlheimensis*, a small sphenodontid from the Late Jurassic of Germany.[3] The latter species had broad tooth plates, possibly suited to eating molluscs. The form of this dentition appears unique among tetrapods, further emphasising the diversity that existed within Mesozoic rhynchocephalians.[4] These three new genera have been described based on the structure of a dentary (main bone of the lower jaw) or partial skull. Thus, as with many fossil forms, the remainder of their skeleton (let alone appearance in the flesh) remains unknown or so far undescribed.

Other recent finds of rhynchocephalian material include partial dentaries from the Middle Triassic of Germany that resemble those of *Diphydontosaurus*. At about 238–240 million years of age, these bones are the oldest known lepidosaurian fossils; their existence indicates that the rhynchocephalian and squamate lineages had already diverged by that date.[5] A partial maxilla (upper jaw bone) and several isolated teeth of a rhynchocephalian have been described from the Late Cretaceous of Argentina.[6] Although similar in structure and size to equivalent structures in tuatara, and being placed in the

Sphenodontinae,[7] the material is too incomplete to be either referred to an existing genus or described as a new taxon. Discovery of this material does, however, further emphasise the importance of South America as a stronghold for rhynchocephalians during the Late Cretaceous.[8]

In a recent article, Marc Jones and I provide a brief overview of the evolutionary significance of tuatara.[9] We emphasise that, as a modern representative of a once-diverse lineage (one of the six major lineages of amniotes), the tuatara remains extremely important, even though it is not helpful to describe it as a 'living fossil'. In evolutionary studies generally, tuatara continue to be used as an out-group for establishing relationships within squamates. Challenging discrepancies have continued to arise between morphological and molecular analyses, particularly in the placement of Iguania. For example, analyses of cranial anatomy involving computed tomography (CT) of fossil and extant lepidosaurians, including tuatara, provide continued support for recognising Scleroglossa as a sister group to Iguania,[10] whereas molecular analyses place Iguania within what were previously recognised as scleroglossan squamates.[11] If the latter interpretation is correct, unresolved questions remain about evolutionary convergence in the morphology of iguanians,[12] including in aspects of feeding-related structures that resemble those of tuatara.

With their unique dentition and jaw movements, tuatara remain significant to our understanding of the biomechanics of feeding in living tetrapods. Recent studies have used computer-assisted approaches. These analyses emphasise the importance of a flexible symphysis at the anterior contact between the lower jaws to allow proal jaw movement,[13] and the contribution of skull sutures to accommodating the strains created during feeding.[14]

A powerful new kind of analysis known as next-generation sequencing is set to expand knowledge of the genetic make-up of tuatara. Hilary Miller and colleagues have recently applied this technique to assemble a partial transcriptome (a reflection of the genes being expressed at that time) from a 9-week-old embryo of Stephens Island (Takapourewa) origin.[15] The mRNA sequences obtained suggest the presence of genes involved in immune function, sexual differentiation and low temperature exposure. Within New Zealand, a major project is underway by university scientists collaborating with the Māori iwi Ngātiwai to sequence the genome (the entire hereditary information) of a tuatara from Lady Alice Island. This project should provide a wealth of new genetic information for comparison with squamates and other reptiles. Challenges in interpretation will remain, however, given that the single male individual examined cannot represent all surviving and extinct populations of tuatara, let alone the full genetic diversity of rhynchocephalians. Progress with the project is described in the blog 'Sequencing the tuatara genome'.[16]

As described in Chapter 2, one of the most profound debates about New Zealand biogeography in recent years has concerned whether Zealandia, the largely drowned continent incorporating today's New Zealand, was completely submerged during the Oligocene, about 25 million years ago. If correct, such a proposition would require that the ancestors of tuatara and all other forms of terrestrial life in New Zealand had arrived more recently, by long-distance dispersal. However, geological evidence for complete inundation is lacking, and molecular analyses increasingly provide support for the continued existence of several lineages in New Zealand (including leiopelmatid frogs, harvestmen and midges) throughout this time.[17] In addition, there are theoretical reasons for concluding that molecular analyses may not preserve full details of a lineage's evolutionary history following a major and prolonged extinction event; this is particularly the case for lineages where there are few surviving species.[18] From a biological perspective, the evidence is strong that New Zealand remained an emergent archipelago throughout the Oligocene, and that for tuatara, an ancestral taxon survived the near drowning in situ. Nonetheless, how and why *Sphenodon* came to be a relict genus with so little contemporary diversity (currently recognised as a single species), in comparison with the diversity seen in New Zealand geckos,[19] for example, is an unresolved question.

Molecular analyses, including that of ancient DNA from bones and eggshells, continue to provide overwhelming evidence for the heavy impact of the first humans in New Zealand on the native fauna, especially on the large and flightless moa.[20] Compared with the attention received by the nine species of moa,[21] the extinct populations of tuatara from mainland New Zealand remain poorly understood. Extensive bone deposits are available and could provide valuable new information on genetic structuring, diets (through stable isotope analysis) and time of disappearance from the North, South and offshore islands.[22] Coprolites (fossilised faeces) and eggshell have yet to be reported for mainland tuatara, but if discovered could, as with moa,[23] yield further insights into the diversity and lifestyles of the once extensive and now-extinct mainland populations.

On the offshore islands of the Poor Knights group, at the northern end of the range of tuatara, evidence from pollen and ancient DNA has revealed the recent reduction or complete loss of certain plants (a palm and podocarps) following human occupation and associated fires. These changes may have enhanced the establishment of burrow-nesting seabirds.[24] Such findings emphasise that even now, nearly 200 years after human habitation ended, offshore islands with tuatara are not completely free of the continuing influence of past human activity.

Biology of tuatara today

The population of tuatara on Stephens Island continues to receive the most attention in field ecological studies. Using temperature-sensitive data loggers attached to the side of the tail near its base, Ilse Corkery confirmed that wild adult tuatara do not take full advantage of the basking opportunities available to them; tuatara often operate with tail surface temperatures below that of their preferred temperature range even when the environment offers opportunities to be warmer.[25] The same study showed that, when data were pooled over several years, mean tail temperatures for adult tuatara sharing burrows with fairy prions (*Pachyptila turtur*) in October were significantly warmer than for those of tuatara not sharing burrows with seabirds. However, given the low absolute value for mean tail temperature (13°C when with seabirds) and the very small elevation (less than 1°C above that when not with seabirds), any benefit from the temperature difference seems likely to be small. A laboratory component of the same study confirmed the low thermal preference of juvenile tuatara, with a small and significant difference in mean body temperatures between daytime (20.3°C) and night-time (19.3°C), similar to that in an earlier study of juveniles.[26] Corkery's study also showed an effect of relative humidity, but no effect of feeding status, on the thermal preference of juveniles.[27]

On Stephens Island, further information about the reproductive behaviour of tuatara has emerged. During the spring nesting season, nesting females move on average 84 metres between their residence area and nesting rookery, and preferentially use the rookery that is nearest to their residence area.[28] During the autumn mating season, females as well as males exhibit aggression toward realistic models (fashioned from moulds of dead tuatara). Females presented with models of female tuatara performed mouth gaping and biting, in addition to the head raising or nodding and body inflation that were also seen in males presented with models of male tuatara.[29]

A recent study on preserved tissues of adult male tuatara provided an updated nomenclature for the sperm-transporting ducts that lead from each testis, consistent with that for squamates and other amniotes.[30] The same study identified an unusual alignment of spermatozoa within the epididymis (in which the heads of sperm point toward the centre of spherical bundles, an arrangement not known in squamates) and made the novel suggestion that this arrangement might aid in sperm transfer to females in the absence of an intromittent organ.

The cellular structure of the skin of tuatara has received further attention. The arrangement of epidermal melanocytes and dermal chromatophores (cells that contain pigment or that reflect light in ways that produce colour, namely melanophores, xanthophores and iridophores in the case of tuatara) has been further described.[31] In addition, immunocytochemical evidence for the location of keratin-associated beta-proteins within the epidermis[32] and of beta-defensin-like peptides within the dermis[33] has been presented.

Developments in conservation: further translocations, and changes in conservation management

The most profound developments concerning tuatara since 2012 are those relating to conservation. Taranga Island, at 500 hectares the second largest of the 27 islands naturally inhabited by northern tuatara, was declared free of Pacific rats (*Rattus exulans*) in June 2013 following 2 years of rat-free monitoring.[34] The recovery of the relict population of adult tuatara there is

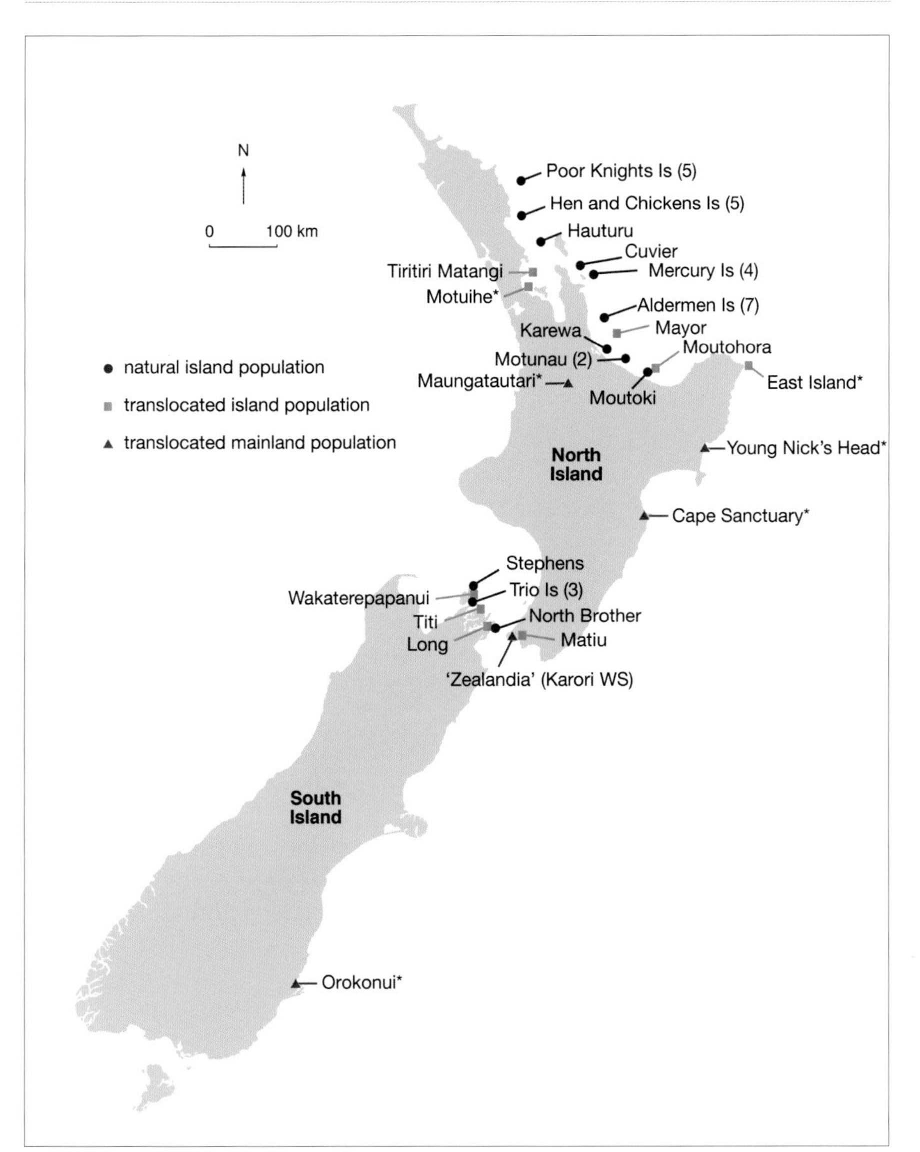

Fig. 11.1 Map of New Zealand showing the current distribution of populations of free-roaming tuatara, including translocations since 2012 (asterisks). Numbers within parentheses are the number of islands on which tuatara are present in an island group.

eagerly awaited. Nearby, on Lady Alice Island, recruitment of juvenile tuatara has already been enhanced by the eradication of Pacific rats.[35] This recovery has made possible the recent translocation of tuatara from Lady Alice to Motuihe Island in the Hauraki Gulf (Figure 11.1, Table 11.1). The proximity of Motuihe Island (even closer than Tiritiri Matangi Island) to the city of Auckland provides a further opportunity for humans to see tuatara on an island undergoing ecological restoration.

In a dramatic development to the south, and in what has been described as 'the largest and most complex tuatara relocation ever undertaken',[36] Cook Strait tuatara from Stephens Island were translocated in October 2012 to five new sites. One site is an island (East Island or Whangaokeno Island) and four are fenced mainland sanctuaries. East Island and three of the mainland sites (Maungatautari Ecological Island, Young Nick's Head Sanctuary and Te Matau a Maui–Cape Sanctuary) fall within the central latitudes of the North

TABLE 11.1

New locations to which tuatara have been translocated since 2012

Islands are listed from north to south (see Fig. 11.1 for map)

Release site (area in hectares)	Nearest evidence for past presence of tuatara[1]	Introduced mammals known to remain at release site	Date of release	Source population and genetic stock
Motuihe Island (179 ha)[2]	c. 6 km to midden site on Motutapu Island	None	March 2012	Lady Alice Island (northern)
East Island (Whangaokeno Island) (13 ha)[4]	Present on East Island until early 1900s	None	October 2012	Stephens Island (Takapourewa) (Cook Strait)
Maungatautari Ecological Island (3400 ha)[5]	c. 40 km to nearest cave deposit	None (mice on main mountain)	October 2012	Stephens Island (Cook Strait)
Young Nick's Head (35 ha)[7]	c. 15 km from natural deposit at Wainui	None except for occasional mouse incursions, quickly controlled	March 2012 (juveniles); October 2012 (adults)	Stephens Island (Cook Strait)
Te Matau a Maui–Cape Sanctuary (1.5 ha mammal-free within 2500 ha)[8]	< 15 km from natural deposits at Ocean Beach and nearby	None except for occasional mouse incursions, quickly controlled	March 2012 (juveniles); October 2012 (adults)	Stephens Island (Cook Strait)
Orokonui Ecosanctuary (307 ha)[9]	c. 7 km to midden site at Long Beach; c. 40 km to natural deposit at Dunback	None except for occasional mouse incursions, quickly controlled	October–December 2012 (juveniles); October 2012 (adults)	Stephens Island (Cook Strait)

Notes

1 See Box 3.1 and Appendices 1, 2 and 3.
2 Motuihe Island: site information from Matt Baber, Tonkin & Taylor Ltd, pers. comm. 8 November 2013; release and recapture information from Lindsay Anderson, Victoria University of Wellington, pers. comm. 18 October and 4 November 2013.
3 By Lindsay Anderson, Victoria University of Wellington.
4 East Island: site information from Sandra Groves, Department of Conservation, pers. comm. 21 November and 28 November 2013; release information from Nicola Nelson, Victoria University of Wellington, pers. comm. 7 October 2013.
5 Maungatautari: site, release and recapture information from Kate Richardson, Maungatautari Ecological Island Trust, pers. comm. 5 November and 26 November 2013.
6 By Stephanie Price, Victoria University of Wellington.
7 Young Nick's Head: release information from Nicola Nelson, Victoria University of Wellington, pers. comm. 7 October 2013; site and recapture information from Steve Sawyer, Ecoworks NZ, pers. comm. 28 November 2013.
8 Te Matau a Maui–Cape Sanctuary: site and recapture information from Tamsin Ward-Smith, Cape Sanctuary, pers. comm. 13 November and 25 November 2013.
9 Orokonui: recapture information from Scott Jarvie, University of Otago, pers. comm. 17 October 2013.

Table 11.1 (cont.)

Tuatara released (M = male; F = female)	Survival and breeding post-release	Questions being researched	Managing body and status of public access	Iwi involved
60 wild adults (20 M,40 F) spread equally between 2 release sites	26 animals (43%) known to have survived at least 6 months; 17 animals (28%) known to have survived at least 12 months	Hormonal stress response to translocation[3]	Motuihe Trust (public access)	Ngātiwai (source); Ngai Tai ki Tamaki, Ngāti Paoa and Ngāti Whātua o Orakei (destination)
44 wild adults (25 M, 19 F)	No details available	No details available	Ngāti Porou (no public access)	Ngāti Koata (source); Ngāti Porou (destination)
50 tuatara in total, comprising 20 wild adults (12 M, 8 F) in a 900-m² 'tuatarium' plus 30 wild adults (18 M, 12 F) on main mountain	19 (95%) in the tuatarium and 18 (60%) on the main mountain known to have survived at least 1–3 months	Hormonal stress response to translocation;[3] phenotypic plasticity[6]	Maungatautari Ecological Island Trust (public access, with fee for tuatarium)	Ngāti Koata (source); Ngāti Koroki Kahukura, Raukawa ki Wharepuhunga and Ngāti Haua (destination)
62 tuatara in total, comprising 20 captive-reared juveniles from Nga Manu Nature Reserve in a 60-m² sub-enclosure plus 42 wild adults (23 M, 19 F)	20 juveniles (100%) known to have survived at least 20 months and many adults seen at 12 months	Phenotypic plasticity[6]	Private land (access limited)	Ngāti Koata (source); Ngai Tamanuhiri (destination)
60 tuatara in total, comprising 20 captive-reared juveniles from Nga Manu Nature Reserve in a 300-m² sub-enclosure plus 40 wild adults (20 M, 20 F) in a 500-m² sub-enclosure	12 juveniles (60%) known to have survived at least 11 months; 20 adults (50%) known to have survived at least 13 months	Hormonal stress response to translocation;[3] phenotypic plasticity[6]	Private land (access limited)	Ngāti Koata (source); Ngāti Mihiroa (destination)
87 tuatara in total, comprising 14 wild juveniles, 43 captive-reared juveniles (28 from Nga Manu Nature Reserve and 15 from Orokonui) plus 30 wild adults (15 M, 15 F)	Adults and juveniles released into separate areas at least 40 m apart; at least 20 (23%) known to have survived at least 10–12 months	Nesting of females;[10] performance of juveniles from different sources;[11] diets of adults and juveniles[12]	Otago Natural History Trust (public access with fee)	Ngāti Koata (source); Kāti Huirapa Rūnaka ki Puketeraki (destination)

Notes (cont.)

10 By Stephen Adolph, Anne Besson, Scott Jarvie, Sophie Penniket and Alison Cree, University of Otago.
11 By Scott Jarvie, University of Otago.
12 From scat analysis by Jenny Kitchin, Barbara Barratt, Scott Jarvie and Alison Cree, University of Otago.

Fig. 11.2 Marlin Elkington (centre) and Noela McGregor (towards left rear) of Ngāti Koata exhibit live adult tuatara at the release ceremony for tuatara translocated to Orokonui Ecosanctuary in October 2012. In Marlin's childhood, tuatara were viewed as kaitiaki (guardians) of the people of Ngāti Koata. With the development of time, the people of Ngāti Koata have become the guardians of tuatara, allowing tuatara to continue playing their original role as protectors of Ngāti Koata (Marlin Elkington, pers. comm. 1 October 2013).

Photo: Ken Miller.

Island, whereas the fourth mainland site (Orokonui Ecosanctuary) is in the southeastern South Island. All of these sites are known to have once had natural populations of tuatara nearby (either at the release site or within about 40 kilometres at similar elevation), so following the latest IUCN terminology,[37] these translocations can be described as reintroductions.[38]

The translocations of wild tuatara from Stephens Island were the fruition of, in some cases, many years of planning and negotiation. Harvesting proceeded with the support and involvement of Ngāti Koata, the Māori guardian iwi of tuatara from Stephens Island, members of whom escorted the tuatara to their respective destinations. The transfers also involved the Department of Conservation (DOC), scientists from Victoria University of Wellington and the national airline, Air New Zealand, with some passengers treated to displays of hand-held tuatara during the flight.[39] The release of tuatara at their destination sites involves the development of new and potentially exciting relationships between Ngāti Koata, local iwi, community trusts and university scientists (Figure 11.2).

In the absence of predatory mammals, the expectation is that tuatara will establish self-sustaining populations at each of the new island and mainland sites. Mice (*Mus musculus*) are periodically detected at low densities at some mainland sites; although the effects of mice on tuatara (including eggs) are uncertain, preliminary indications of survival of the released tuatara

Fig. 11.3 An adult male tuatara, freshly moulted, basks near its release burrow, five months after release at Orokonui Ecosanctuary.

Photo: Alison Cree.

are encouraging (Figure 11.3, Table 11.1). The reintroduction of tuatara to Orokonui Ecosanctuary is of particular significance in the context of climate change. This is the first translocation of tuatara to a site substantially south of, and thus cooler than, the current range of tuatara (though well within the past latitudinal range of the genus[40]). Prior studies show that juvenile tuatara reared in enclosures on-site are capable of surviving and growing in this climate,[41] and that soil temperatures at Orokonui, within the known nest depths used elsewhere by tuatara, are capable of producing both male and female offspring.[42] In a high-carbon world, in less than 100 years (the potential lifespan of a tuatara), Orokonui is predicted to have warmed to about the same mean annual air temperature as Stephens Island experiences now.[43] Thus, temperatures at Orokonui over the next few decades seem likely to remain suitable for the establishment of tuatara.

Several features of these recent translocations of Cook Strait tuatara are favourable in terms of genetic diversity. First, the source population (on Stephens Island) has high genetic diversity relative to other populations.[44] Second, each translocation consisted of between 44 and 87 tuatara, which modelling suggests is a sufficient number for maintaining about 94% of original genetic heterozygosity over 10 generations, with a negligible probability of extinction.[45] Third, given the timing for the transfer of wild adults (October, austral spring), at least some of the females should have been carrying fertilised eggs,[46] adding to the released population's genetic diversity if nesting is successful. There is also potential for enhancing (or minimising the loss of) genetic diversity by the selective transfer of individuals between ecosanctuaries in the future. (The possibility has also been raised that benefits might arise from the mixing of Cook Strait and northern tuatara at sites that fall between the natural distributions of these two stocks;[47] this suggestion deserves further exploration.)

Three of the translocations to mainland ecosanctuaries involved the release of captive-reared juveniles as well as wild-collected adults. At Orokonui, juveniles from three sources have been released: wild juveniles direct from Stephens Island, captive-reared juveniles raised on-site at Orokonui, and captive-reared juveniles raised at an outdoor reserve in the lower North Island. This situation offers the opportunity to compare the growth, behaviour and survival of the three groups, results from which should help guide the sourcing of animals for future translocations. Only the wild tuatara (adults and juveniles) had ectoparasitic ticks (*Amblyomma sphenodonti*) when translocated, and whether the tick species will survive the translocation and establish on captive-reared animals in the process is uncertain.[48]

The translocations described above bring the current distribution of free-roaming tuatara to the following totals: 32 natural populations on islands, nine translocated populations on islands, and five translocated populations in fenced mainland ecosanctuaries. At least four of the translocated island populations and three of the mainland ecosanctuaries allow some level of public access with, over time, the potential for humans to encounter tuatara.

The optimism generated by the developments described above must now be tempered by concern arising from recent changes in conservation management in New Zealand. Staffing reductions and organisational changes within DOC in the past 2 years have seen the disbanding of many species recovery groups, including the Tuatara Recovery Group. Instead, a specialist group to cover all native reptiles and amphibians (over 100 species) is being established.[49] Current indications are that a satellite group attending to tuatara matters will continue in some form, although the precise remit of the group and the extent to which written recovery plans will continue to guide future directions are still under consideration.

Within this new system for natural heritage management, which aspires to combine species and ecosystem management more explicitly, tuatara remain recognised with the national conservation status of 'At risk – relict',[50] a lesser ranking than 'Threatened'. In addition, following a survey of several thousand New Zealanders, tuatara have been identified by DOC as one of about 10 'nationally iconic species' that 'help define New Zealand and New Zealanders'.[51] Recognition of this status involves a commitment to managing tuatara in a manner that ensures that their populations are maintained or restored. In aiming for this, there is now a greater emphasis within DOC on 'partnerships' in conservation with the wider community, including with businesses and volunteer organisations. Whether this new direction for management will provide the same level of national leadership and conservation benefits as under the previous approach remains uncertain.

Internationally, the effects of habitat disruption and harvesting continue to affect many species of reptiles,[52] as they have tuatara in the past. Global climate change is the new and growing threat from which tuatara are not immune. As the world's scientists become ever more certain that climate warming is unequivocal, that it is largely the result of human activity including the release of greenhouse gases, and that most aspects of climate change will be ongoing for centuries even if emissions of carbon dioxide cease,[53] the lack of substantive action to reduce greenhouse gas emissions becomes ever more incomprehensible. Growing social and economic challenges, as well as direct biophysical ones, undoubtedly lie ahead for the conservation of biological diversity. As indirect as it may seem, attending to social factors that are linked with indices of wellbeing for biodiversity, such as indices of socioeconomic equality,[54] may prove to be an investment in the conservation of tuatara and other distinctive organisms that we hold dear.

Endnotes to Chapter 11

See the bibliography for full references to articles on tuatara, which are cited here at first mention by surname and year only. References on other topics are cited here in full at first mention, and then using 'op. cit.'. Except where required for clarity, the convention of parentheses around years has been dispensed with for economy of space.

1 Martínez, R. N., Apaldetti, C., Colombi, C. E., Praderio, A., Fernandez, E., Santi Malnis, P., Correa, G. A., Abelin, D. and Alcober, O. 2013. A new sphenodontian (Lepidosauria: Rhynchocephalia) from the Late Triassic of Argentina and the early origin of the herbivore opisthodontians. *Proceedings of the Royal Society B* 280 (20132057): 1–7.

2 Apesteguía, S., Gómez, R. O. and Rougier, G. W. 2012. A basal sphenodontian (Lepidosauria) from the Jurassic of Patagonia: new insights on the phylogeny and biogeography of Gondwanan rhynchocephalians. *Zoological Journal of the Linnean Society* 166: 342–360.

3 Rauhut, O. W. M., Heyng, A. M., López-Arbarello, A. and Hecker, A. 2012. A new rhynchocephalian from the Late Jurassic of Germany with a dentition that is unique among tetrapods. *PLOS ONE* 7(10): e46839, 1–9.

4 Rauhut et al. 2012, op. cit. See also Meloro and Jones 2012.

5 Jones, M. E. H., Anderson, C. L., Hipsley, C. A., Müller, J., Evans, S. E. and Schoch, R. R. 2013. Integration of molecules and new fossils supports a Triassic origin for Lepidosauria (lizards, snakes, and tuatara). *BMC Evolutionary Biology* 13(208): 1–21.

6 Apesteguía, S. and Jones, M. E. H. 2012. A Late Cretaceous "tuatara" (Lepidosauria: Sphenodontinae) from South America. *Cretaceous Research* 34: 154–160.

7 For biological and cultural reasons, it is inaccurate to describe the taxon as a 'tuatara' as suggested by the authors. The common name tuatara refers specifically to the genus *Sphenodon*, known only from New Zealand and named by Māori in reference to the crest-bearing status of the genus.

8 Apesteguía and Jones 2012, op. cit.

9 Jones and Cree 2012.

10 Gauthier et al. 2012.

11 Pyron, R. A., Burbrink, F. T. and Wiens, J. J. 2013. A phylogeny and revised classification of Squamata, including 4161 species of lizards and snakes. *BMC Evolutionary Biology* 13(93): 1–53.

12 Gauthier et al. 2012.

13 Jones et al. 2012.

14 Curtis et al. 2013.

15 H. C. Miller et al. 2012.

16 http://sciblogs.co.nz/tuataragenome/

17 Sharma, P. P. and Wheeler, W. C. 2013. Revenant clades in historical biogeography: the geology of New Zealand predisposes endemic clades to root age shifts. *Journal of Biogeography* 40: 1609–1618.

18 Sharma and Wheeler 2013, op. cit.

19 Nielsen, S.V., Bauer, A. M., Jackman, T. R., Hitchmough, R. A. and Daugherty, C. H. 2011. New Zealand geckos (Diplodactylidae): cryptic diversity in a post-Gondwanan lineage with trans-Tasman affinities. *Molecular Phylogenetics and Evolution* 59: 1–22.

20 Oskam, C. L., Allentoft, M. E., Walter, R., Scofield, R. P., Haile, J., Holdaway, R. N., Bunce, M. and Jacomb, C. 2012. Ancient DNA analyses of early archaeological sites in New Zealand reveal extreme exploitation of moa (Aves: Dinornithiformes) at all life stages. *Quaternary Science Reviews* 52: 41–48.

21 Allentoft, M. E. and Rawlence, N. J. 2012. Moa's Ark or volant ghosts of Gondwana? Insights from nineteen years of ancient DNA research on the extinct moa (Aves: Dinornithiformes) of New Zealand. *Annals of Anatomy* 194: 36–51.

22 The existence of geographic variation in jaw structure has been reported (Humphries and Jones 2010).

23 Allentoft and Rawlence 2012, op. cit.

24 Wilmshurst, J. M., Moar, N. T., Wood, J. R., Bellingham, P. J., Findlater, A. M., Robinson, J. J. and Stone, C. 2013. Use of pollen and ancient DNA as conservation baselines for offshore islands in New Zealand. *Conservation Biology*: doi: 10.1111/cobi.12150.

25 Corkery 2012. Studies are needed to show how closely the temperatures of data loggers taped to the base of the tail of adults match cloacal body temperatures under a variety of conditions, including when heating (in direct solar radiation) and cooling.

26 Whitworth 2006. See also Chapter 9, Figure 9.5.

27 Corkery 2012.

28 Refsnider et al. 2013.

29 Ramstad et al. 2012.

30 Rheubert et al. 2012.

31 Alibardi 2012a.

32 Alibardi 2012b, 2013a.

33 Alibardi 2013b.

34 Keith Hawkins, Department of Conservation, pers. comm. 2 September 2013.

35 Towns et al. 2007.

36 See http://blog.doc.govt.nz/2012/10/18/tuatara-on-tour/ (accessed 22 November 2013).

37 IUCN/SSC. 2013. *Guidelines for reintroductions and other conservation translocations. Version 1.0.* Gland, Switzerland.

38 Opinions are divided on how to describe a conservation translocation of tuatara to the mainland North Island or South Island in the absence of certain knowledge that the original population of tuatara was of the same species. Some have assumed that the original mainland population was outside the indigenous range (i.e. of a different species), and have therefore described such a translocation (from a Cook Strait island such as Stephens or North Brother) as a conservation introduction (Rout et al. 2013) or, more specifically, assisted colonisation (K. A. Miller et al. 2012). My view is that it is more parsimonious to assume that the mainland populations were of the existing species *Sphenodon punctatus* until proven otherwise (consistent with the wide range of latitude (6°) that the extant species currently spans, comparable with the latitude span from Cook Strait to the southernmost location of fossils in the South Island (6°)). On these grounds, such a translocation is described as a reintroduction (IUCN/SSC 2013, op. cit.). If, on the other hand, it transpired that the mainland fossils were of a different species, a conservation translocation of tuatara from a Cook Strait island to the mainland would, following IUCN guidelines, be best

described as an ecological replacement. This recommendation allows us to save the term 'assisted colonisation' for a conservation introduction that is clearly outside the known range of any population of *Sphenodon* (for example, translocation to a subantarctic island), and motivated by a desire to avoid extinction of the translocated population (IUCN/SSC 2013, op. cit.).

39 See http://blog.doc.govt.nz/2012/10/18/tuatara-on-tour/ (accessed 22 November 2013).

40 See Chapter 2, Figure 2.8, for map of past distribution.

41 Mello et al. 2013.

42 Besson et al. 2012.

43 Reisinger, A., Mullan, B., Manning, M., Wratt, D. and Nottage, R. 2010. Global and local climate change scenarios to support adaptation in New Zealand. In: Nottage, R. A. C., Wratt, D. S., Bornman, J. F. and Jones, K. (eds). *Climate change adaptation in New Zealand: future scenarios and some sectoral perspectives*. New Zealand Climate Change Centre, Wellington, pp. 26–43.

44 MacAvoy et al. 2007; Hay et al. 2010; H. C. Miller et al. 2010.

45 K. A. Miller et al. 2009.

46 Cree et al. 1991a, 1992.

47 Anon. 2009. *Minutes of the Tuatara Recovery Group meeting 23–24 November 2009*. Department of Conservation, Nelson. DOCDM–525029; K. A. Miller et al. 2012. This could have value as an experimental approach, though Māori cultural issues regarding whakapapa (genealogy, familial relationships) would need first to be worked through (K. A. Miller et al. 2012; Towns et al. 2012).

48 Ticks were present on some of the tuatara translocated to Tiritiri Matangi Island in 2003, but appear to have since disappeared (Graham Ussher, Tonkin & Taylor, Auckland, pers. comm. 2 December 2013; Dylan van Winkel, Bioresearches Group, Auckland, pers. comm. 2 December 2013). The possibility that ticks have failed to establish on Tiritiri Matangi Island and at the North Island fenced ecosanctuary known as Zealandia (formerly Karori Wildlife Sanctuary) was briefly discussed by Moir et al. 2012, although in that study information about ticks on Tiritiri Matangi Island was incorrectly attributed to personal communications from myself and Chris Smuts-Kennedy.

49 Oliver Overdyck, Department of Conservation, pers. comm. 4 November 2013.

50 Hitchmough et al. 2013.

51 Department of Conservation. 2013. *Identifying conservation priorities factsheet.* Updated March 2013. http://www.doc.govt.nz/about-doc/policies-and-plans/managing-natural-heritage/identifying-conservation-priorities/ (accessed 4 November 2013).

52 Böhm, M. and more than 200 others. 2013. The conservation status of the world's reptiles. *Biological Conservation* 157: 372–385.

53 IPCC. 2013. Summary for policymakers. In: Stocker, T. F., Qin, D., Plattner, G.-K., Tignor, M., Allen, S. K., Boschung, J., Nauels, A., Xia, Y., Bex, V. and Midgley, P. M. (eds). *Climate change 2013: The physical science basis. Contribution of Working Group I to the fifth assessment report of the Intergovernmental Panel on Climate Change.* Cambridge University Press, Cambridge, UK and New York, NY, USA, pp. 1–27.

54 Holland, T. G., Peterson, G. D. and Gonzalez, A. 2009. A cross-national analysis of how economic inequality predicts biodiversity loss. *Conservation Biology* 23: 1304–1313.

Glossary of scientific terms

acinar gland: a form of gland made up of oval or spherical sacs, such as the exocrine glands seen in the pancreas of tuatara

acrodont implantation: a type of tooth implantation in which the teeth are attached to the crest of the jawbone without sockets; such teeth are generally fused and ankylosed (attached with surrounding bone) to the jawbone, as in tuatara

acrosome: a cap-like structure that develops over the head of a spermatozoon

adipose tissue: body fat

aerobic scope: the difference between oxygen consumption when an animal is active versus that when resting, representing the extent to which energy can be generated aerobically

albumen: the white of an egg

alimentary tract: the digestive tract, from the mouth to the vent

allantois: a membranous sac, attached to the embryo, that stores liquid waste

allele: a variant form of a gene

allozyme electrophoresis: a method that compares the mobility of variant forms of enzymes from blood or other tissues through a gel carrying an electric current

amniote: a vertebrate in which the embryo has a surrounding membrane known as the amnion (in other words, mammals, birds and reptiles, including tuatara)

amphicoelous vertebrae: vertebrae in which the centrum (solid, central part) is concave at both anterior and posterior ends, as in tuatara

anapsid: a skull without arches of bone or fenestrae (openings) in the temporal region; also a reptile (among living species, the turtles) possessing this skull type

androgen: a natural or synthetic compound involved in the development and maintenance of male characteristics (in tuatara, testosterone is the main androgen)

ankylosed: firmly attached by surrounding bone

annular: ring-shaped

anterior: towards the front or head end

apnoea: breath-holding

apomorphy: an evolutionarily derived feature (that is, differing from the condition in ancestors)

archosauromorphs: one of two main branches of diapsid reptiles, comprising the archosaurs and some early reptiles (known only as fossils) such as rhynchosaurs

archosaurs: crocodilians and birds

aromatase: enzyme responsible for converting testosterone into oestradiol

astragalus: one of the proximal bones in the ankle

atretic follicles: ovarian follicles that have aborted the process of vitellogenesis and are being resorbed

atrium: a chamber of the heart, receiving blood from the lungs and body

autotomy septa: fracture zones within the tail vertebrae that enable the tail to be dropped and later regrown

axial bending: a form of movement in which the body swings side to side, seen in tuatara

axilla: armpit

axoneme: the 'skeleton' within whip-like cellular appendages known as cilia and flagella

bi-concave: hollowed at both ends

bifid: having two tips (a condition sometimes seen in the regrown tails of tuatara)

blastoderm: the layer of cells on the surface of the yolk that gives rise to the germinal disk from which the embryo develops

body condition (index): a measure of relative bulkiness of the body, calculated in tuatara as mass adjusted statistically for variation in snout–vent length

brodifacoum: a second-generation anticoagulant poison, used in New Zealand to eradicate introduced rodents from islands

buccal cavity: oral cavity

calcaneum: one of the proximal bones in the ankle

caniniform: canine-like

cardiac shunt: a system of blood circulation, differing from the normal systemic circulation, during which blood flow to and from the lungs can be separated from that to the rest of the body

caudal: relating to the tail

caudal autotomy: the ability to drop the tail at predetermined zones of weakness

Cenozoic Era: the era that extends from 65 million years ago to the present time

centriole: a cell organelle involved in cell division

centrum: the central bony cylinder of a vertebra

choanae: paired internal nostrils

choanal tube: part of the nasal system

cloaca: the chamber that follows and receives products from the digestive, reproductive and urinary systems; the final chamber before the vent

cochlear potential: a measure of the sensitivity of the ear

coprodaeum: the first compartment of the cloaca (from the anterior end), in which the faeces are stored before they move through the remaining compartments and exit through the vent

coronoid (bone): a slender bone forming part of the lower jaw of some vertebrates, including tuatara

corpora albicantia: scars of the corpora lutea

corpora lutea: temporary endocrine structures formed from the remains of ovarian follicles after ovulation

cranial kinesis: flexibility between bones of the skull

critical thermal minimum: the minimum body temperature at which an animal loses some important measure of performance, such as the ability to right itself

deep body temperature: temperature taken within the intestine

dentary (bone): the tooth-bearing bone of the lower jaw

dermis: the inner layer of the skin

DHA: docosahexaenoic acid, an omega-3 fatty acid

diapsid: a skull type with two arches of bone defining the temporal fenestrae; also a reptile possessing this type of skull (among living species, the crocodilians, squamates and tuatara)

diel variation: day–night variation

diurnal: active by day

diurno-nocturnal: active by day and night

doubly labelled water: water enriched with specific, measurable isotopes of hydrogen and oxygen

ecdysis: sloughing or shedding of skin

efferent tubules: fine, convoluted ducts within the epididymis through which mature sperm pass

egg tooth: a toothy projection from the upper jaw of fully developed squamate embryos that aids in hatching (as distinct from the horny shell-breaker seen in tuatara, turtles and crocodilians)

EPA: eicosapentaenoic acid, an omega-3 fatty acid

epidermal generation: an outer portion of skin that is shed in cycles, present in tuatara and squamates

epididymis: the coiled upper part of the duct system that transports sperm from the testis to the cloaca

epiphysis: a secondary ossification centre at the end of a long bone (initially separated from the main bone by a layer of cartilage, but later uniting with it)

epithelium: a tissue that lines the cavities and surfaces of structures throughout the body

exocrine gland: a gland that discharges its secretions through a duct

fenestra(e): small, window-like opening(s), for example in bones

filiform: thread-like in shape

flicker frequency: the frequency of an intermittent light pulse

fovea: an area of the retina with an increased density of photoreceptor cells, giving enhanced visual acuity

fundic gland: gastric gland, found in the stomach

fundus oculi: the concave interior of the eye, including the retina, choroid and other components

gall bladder: a sac that stores bile produced by the liver, releasing it through the bile duct into the small intestine

gametes: sperm and eggs

gastralia: bones that develop within the dermis of the animal's undersurface; although sometimes referred to as 'abdominal ribs', they are not true ribs

gastrula: an early stage of embryonic development

genetic heterozygosity: a measure of genetic variation

glomerular filtration rate: the rate at which blood is filtered by the kidneys

glottis: the vocal cords and the opening between them; part of the larynx

GnRH: gonadotrophin-releasing hormone

Gondwana: a supercontinent of the Mesozoic Era that included most of the present-day landmasses of the Southern Hemisphere

granulosa: a cell layer that surrounds developing oocytes in the ovary

gravid: a term used in connection with egg-laying reptiles such as tuatara to identify individual females containing eggs in their oviducts

gravidity rate: for tuatara, the proportion of a sample of mature-sized females carrying shelled eggs in their oviducts

gular region: the area of the lower jaw between the dentary bones

haemoglobins: respiratory pigments that enhance the oxygen-carrying capacity of the blood

head starting: a conservation practice that involves raising eggs and the resulting juveniles in captivity until later release, at which time the prospects for survival are thought to be improved

hemipenes: the paired, eversible sacs that form the penises of squamates

hepatocytes: liver cells

hyperosmotic: having a higher osmotic pressure than another solution, arising, for example, from a higher concentration of salts or other dissolved materials

hypothalamus: a region of the brain with important control functions, including the secretion of hormones that influence activity of the pituitary gland

iguanians: a lineage of squamates that includes iguanas, chameleons and agamids

inferior: lower

infundibuli: air spaces within the lung

infundibulum: anterior-most region of the oviduct

inner epidermal generation: the innermost layers of epidermis in squamates and tuatara

intromittent organ: male sex organ, such as the hemipenes of squamates

labial gland(s): in tuatara, small gland(s) around the edge of the mouth

lachrymal (lacrimal) gland(s): gland(s) that secrete tears, not present in tuatara

lamina propria: the layer of connective tissue below the luminal epithelium in the uterus

Langerhans cells: cells within the acinar glands of the pancreas that secrete insulin

laparoscopy: a minor surgical procedure used to examine internal organs, such as when sexing small tuatara by inserting a laparoscope (medical telescope) through the flank

lateral undulation: a movement in which the body swings side to side in a wavelike motion

Laurasia: a supercontinent of the Mesozoic Era that included most of the present day landmasses of the Northern Hemisphere

Lepidosauria: a major group of reptiles including lizards, snakes and tuatara (technically, the rhynchocephalians plus the squamates)

lepidosauromorphs: one of two main branches of diapsid reptiles, comprising the surviving lepidosaurs and some related lizard-like reptiles known only as fossils

leptin: a hormone produced by the fat cells of vertebrates

Leydig cells: cells (also known as interstitial cells) within the interstitial tissue between the seminiferous tubules of the testes; the main source of androgens in male vertebrates

lipids: a group of naturally occurring molecules, including fats

loci: specific locations on a chromosome

lumen: the inside space of a tubular structure

maxilla: bone of the upper jaw on which, in tuatara, the outer row of teeth are situated

melanism: darkening

mesorchium: a thin membrane that suspends the testes, epididymides and adrenal glands from the dorsal body wall

mesotubarium: a thin membrane that attaches the oviducts to the body wall

mesovarium: a thin membrane that connects the ovaries to the body wall and supports the adrenal glands

Mesozoic Era: the period of time from 250 to 65 million years ago

metabolic rate: the rate at which energy is being used, inferred in animals from the rate at which oxygen is consumed or carbon dioxide produced

metabolic water: water produced internally as a result of metabolism

metatarsal(s): bone(s) of the foot between the ankle and the toes

midden: an old dump site for domestic waste, often containing animal bones and artifacts

mitochondrial DNA: DNA contained in the chromosome of a mitochondrion (a kind of cellular organelle)

Müllerian ducts: open-ended ducts that become the functional oviducts of females (initially present in embryos of both sexes)

myomeres: muscle blocks

nares: paired external nostrils

nictitating membrane: a third eyelid present as a translucent, full membrane in tuatara and many other vertebrates

nocturnal: active by night

non-ventilatory period: a period without breathing

notochord: a flexible, rod-like structure that is absent in most adult vertebrates, but present in a constricted state in adult tuatara

nuchal spines: the crest of spines that runs along the back of the neck, seen in tuatara

oestradiol: an ovarian steroid hormone that stimulates vitellogenesis in tuatara; a form of oestrogen

oocytes: egg cells; the female gametes

opisthocoelous: vertebrae that typically interlock in a concave–convex arrangement, with the concave surface facing the tail

orbit: eye socket

ornithuric acid: the major metabolite of benzoic acid in faeces and urine of tuatara, probably produced mainly in the liver

ostium abdominale: the entrance to the oviduct from the abdominal cavity, through which ovulated oocytes pass

outer epidermal generation: the more mature, outermost layers of epidermis in the skin of squamates and tuatara

oviducts: ducts through which eggs travel from the ovaries to the outside of the body (in comparative zoology, each may be subdivided into various regions such as the uterine tube, uterus and vagina)

oviparity: the egg-laying parity mode

oviposition: the actual laying of eggs

ovulation aperture: opening in the corpus luteum through which the oocyte passes at ovulation

oxygen-dissociation curve: a curve that relates oxygen saturation to the partial pressure of oxygen in the blood; used to establish how blood carries and releases oxygen

oxytocin: a mammalian neurohormone stored in the pituitary gland, capable of inducing oviposition when injected into tuatara

palatine (bone): a bone of the upper jaw; in tuatara, this bears the inner, shorter row of teeth

palatine gland: salivary gland in the roof of the oral cavity

pancreas: abdominal organ that produces the hormone insulin and secretes digestive fluids into the small intestine

Pangaea: a supercontinent, composed of all the present day continents, that existed during the Palaeozoic and early Mesozoic eras

parietal crest: a bony attachment site for muscles on the head

parietal eye: a structure near the top surface of the head, part of the pineal complex of some reptiles and probably light-sensitive; the so-called 'third eye' of tuatara and some lizards

parietal foramen: an opening in the top of the skull that lies above the parietal eye

pharynx: the region of the alimentary canal between the oral cavity and the oesophagus, present in tuatara as a slit-like glottis on the floor of the throat

pineal complex: a sensory, melatonin-secreting outgrowth of the brain, light-sensitive and including a parietal eye in some reptiles such as tuatara

PIT-tags: passive integrated transponder tags; small bar-coded tags about the size of a rice grain, implanted under the skin for unique identification

(blood) plasma: the liquid component of the blood, containing soluble clotting elements but not cells; a clear-to-pale yellow fluid in tuatara

plesiomorphy: an ancestral feature

pleurodont: a form of tooth attachment in which the teeth are loosely attached to the outside edge of the jaw

posterior: towards the rear or tail end

postorbital area: the region of the skull or head posterior to the eye socket

power of accommodation: the ability to focus the eyes on objects at different distances, an ability typical of diurnal species

preferred body temperature: body temperature preferred by an animal when given a choice in a thermal gradient; also known as selected temperature

premaxilla (bone): a bone at the front of the upper jaw; in tuatara, it bears a pair of overhanging, incisor-like teeth

proal (prooral): a propalinal jaw movement in which the power stroke is forwards, as in tuatara

procoelous vertebrae: vertebrae that typically interlock in a concave–convex arrangement, with the concave surface facing towards the head

proctodaeum: the third, final chamber of the cloaca (from the anterior end)

progesterone: a steroid hormone, produced in females by the corpora lutea; associated in tuatara with mating, ovulation and early gravidity, when eggs enter the oviducts

propalinal: jaw action in which the lower jaw is moved forward or back, relative to the upper jaw

pulmonary flow: blood flowing into the lungs from the body

pyloric gland: gland type found in the pyloric section of the stomach

pyriform cell: a distinctive cell type found in the granulosa layer of the ovarian follicle of female squamates

quadrate (bone): a bone of the skull forming part of the jaw joint

radio-telemetry: a technique in which a radio-transmitter is attached to an animal, allowing information such as location to be established by an observer using a radio receiver

renal sex segment: a secretory region of the kidney present in male squamates but not in tuatara or other reptiles

Reptilia: as traditionally recognised, a class of ectothermic vertebrates including tuatara, lizards, snakes, turtles and crocodiles, and their fossil relatives

Rhynchocephalia: the order of lepidosaurs to which tuatara and related fossil forms belong

rhynchosaurs: a group of extinct herbivorous archosauromorphs

scleral ossicles: thin, overlapping plates of bone that form a circlet around the lens of the eye

scleroglossans: as traditionally recognised based on anatomical features, a lineage of squamates that includes snakes, geckos, skinks and some other lizards but not iguanians (a scleroglossan lineage is not recognised by recent molecular analyses)

sebaceous glands: in tuatara, epithelial glands in the cloaca (specifically, the ventro-lateral wall of the proctodaeum) that possibly have a pheromonal function

seminiferous epithelium: the epithelial layer within the seminiferous tubules of the testes

seminiferous tubules: tubules within the testes in which spermatozoa are produced

septa: dividing walls, for example on the inside surface of the lung

Sertoli cells: cells within the seminiferous epithelium that support developing sperm cells

SINES: short, interspersed, retro-transposable elements of the genome

sinus venosus: a chamber of the heart that precedes the atrium, in terms of blood flow

skeletochronology: a technique for estimating growth rates and minimum individual age by counting the growth rings found in cross sections of bone

snout–vent length: a measure of body length (excluding the tail) in a reptile such as tuatara; the distance from the tip of the snout to the vent

spermatids: a late stage in the development of sperm

spermatocytogenesis: the production and development of spermatocytes, the cells that eventually become spermatids

spermatogenesis: sperm production

spermatozoa: male gametes, mature sperm

spermiogenesis: the maturation of spermatids into spermatozoa

Sphenodontia: a suborder of the more derived rhynchocephalians, including tuatara

Sphenodontidae: the family of sphenodontians that includes tuatara

Sphenodontinae: the sub-family of Sphenodontidae that includes tuatara

Squamata: the lizards and snakes (informally, squamates)

stable carbon isotopes: non-radioactive forms of carbon (^{13}C and ^{12}C)

sternum: a long, flat bone that lies in the middle of the chest

stoltzer Gang: a distinctive walk performed by male tuatara as part of the courtship display

streptostyly: a condition in which the quadrate bone is able to rotate freely where it articulates with the skull; seen in squamates but not tuatara

sublingual gland: salivary gland on the floor of the oral cavity

superior: upper

sutures: immoveable or barely moveable joints between bones, such as the fibrous joints between bones of the skull

synapomorphy: a shared derived character, not present in an ancestor in common with another lineage

synapsids: a group of amniotes that have a single opening on the lower cheek; represented today by mammals

systemic flow: blood flowing from the lungs to the head and body

tapetum lucidum: literally 'shining carpet', a layer within or behind the retina that reflects light back through the retinal photoreceptors

tarsal plate: a cartilaginous disc in the lower lid of the eye

temperature-dependent sex determination (TSD): a form of sex determination in which incubation temperature, rather than genes on chromosomes, determines the sex of offspring

temporal bar: a bony bar extending from the upper jaw to the quadrate

temporal fenestrae: openings in the skull situated behind the eye socket

thecodont: a form of tooth implantation in which the teeth are socketed

thermophilic: heat-seeking

thyroid fenestra: an opening seen in the pelvic girdle of lepidosaurs

torpor: dormancy

trachea: the respiratory tube transporting air from the larynx to the bronchi; the 'windpipe'

transverse vent: a side-to-side orientation of the external opening of the cloaca

trifid: three-tipped (a condition occasionally seen in the regrown tails of tuatara)

tubular reabsorption: water reabsorption by the kidney tubules

tunica albuginea: an outer layer of connective tissue surrounding an inner mass of looped seminiferous tubules in the testes

tympanic membrane: membrane that transmits sound vibrations to the middle ear; eardrum

umbilicus: scar on the abdomen left at the attachment site of the umbilical cord

uncinate processes: small, hook-like structures connected to the ribs that provide attachment for muscles in the chest wall

unicameral: single-chambered (e.g. of lungs)

ureter: the short duct that transports urine from the kidneys of tuatara to a urogenital papilla in the wall of the cloaca

uricotelic: excreting uric acid as the major nitrogenous waste

urodaeum: the second or middle chamber of the cloaca, into which the vas deferens or oviduct and ureter open in tuatara

urogenital papilla: the small projection at which the vas deferens or oviduct and ureter open into the cloaca of tuatara

uterus: the region of the oviduct in which, in egg-laying species such as tuatara, the eggs become shelled

varanic acid: a common bile acid of lepidosaurs, including tuatara

vas deferens: duct that transports sperm from the epididymis to the cloaca

vent: the opening of the cloaca at the base of the tail, from which urine, faeces and eggs or sperm exit

ventricle: the chamber of the heart that pumps blood into the lungs and body

vertebra: a bony or cartilaginous segment of the spinal column (plural vertebrae) vicariance: the physical separation of populations following development of geographical barriers

vitellogenic follicle: an ovarian follicle in which the oocyte contains yolk

vitellogenin: a protein–lipid complex, made by the liver, that is the yolk protein precursor

viviparity: a parity mode in which live offspring develop within the mother and are delivered as such, rather than as eggs

vomeronasal organs: paired organs in the roof of the mouth, which provide a chemical sense akin to smell; also known as Jacobson's organs

Glossary of Māori terms

Macrons indicate long vowels

amo – upright post at side front of house, supporting the barge boards
atua – god or spirit
aurei – mat or cloak pin
hapū – sub-tribe
ika – fish, crocodile
inohi – scale
iwi – tribe, people
kai moana – shellfish
kaitiaki – guardian, guard
karakia – prayer, incantation, chant
kārara – reptile, insect (southern-dialect equivalent of ngārara)
kata – to laugh, laughter
kaumātua – elder
kaurehe – a big, amphibious, lizard-like creature of mythical nature
kaweau or kawekaweau – large tree lizard (sometimes considered a reference to the extinct giant gecko *Hoplodactylus delcourti*, which perhaps inhabited New Zealand)
kiore – the Pacific rat (*Rattus exulans*)
kōauau – a kind of flute
koruru – carved face at front apex of house, gable head
kuia – female elder
kūmara – sweet potato
maihi – barge board of house
mana – status, prestige, authority
Māori – indigenous New Zealander or descendant of such (literally, 'common' or 'ordinary')
mauri – life force
moko – lizard, tattoo
moko-kākāriki – green gecko (one of several species in the genus *Naultinus*)
mokomoko – skink
moko-pāpā – brown gecko (one of several species until recently all placed in the genus *Hoplodactylus*)
moko-tāpiri – see moko-pāpā
motu – island
ngārara – reptile, insect
noa – safe, unrestricted, not tapu
pā – fortified village
Pākehā – New Zealander of European descent
pātaka – storehouse raised upon posts
pou – post, upright support
poupou – side wall posts or slabs of house
poutokoihi – central free-standing post at front of house, supporting ridgepole
poutokomanawa – central post inside house, supporting ridgepole
poutuarongo – central post of rear wall inside house, supporting ridgepole
pūrerehua – sound-producing device or musical instrument known as a 'bull-roarer', made of wood, stone, or bone attached to a long string, and used among other things to lure lizards
rua – hole, burrow
rohe – tribal area, domain
tāhuhu – ridgepole of house
tangata whenua – local or indigenous people, Māori (literally, people of the land)
taniwha – monster of watery places, often with reptilian features (potentially dangerous to the unaware, but sometimes considered a guardian by local people)
taonga – treasured material, possession, property
tapu – under religious restriction, forbidden, feared or sacred
tara – spine, peak
taratara-o-kai – decorative carving style with alternating zigzags and ridges (literally, peaks of food; also a reference to the jagged teeth of Kai (Kae) who stole and ate a pet whale)
taumanu – thwart (of a canoe)
tekoteko – free-standing ancestor figure at the apex of a house's bargeboards
tītī – muttonbird (a chick of certain burrow-nesting petrels)
tohunga (tohuka) – priestly expert
toropakihi or tupakihi – a common grass-dwelling lizard (probably a skink) in the South Island
tua – the back or other side (of a solid body)
tuatara wawata – a brown, crumbly soil suitable for kūmara
urupā – burial ground, cemetery
wāhi tapu – sacred place, such as a burial ground
waka huia – treasure box, feather box
waka taua – war canoe
waka tūpāpaku – bone casket, burial chest
whakairo rākau – wood carving
whakapapa – genealogy, family history
whānau – extended family, family group
whanaungatanga – kinship
whare – house, dwelling
whare rūnanga – tribal house, meeting house
whare whakairo – carved house
wharenui – large house, meeting house

APPENDIX 1 (refer Chapter 2)

Deposits reported to contain tuatara bones and considered likely to be of natural origin

Sites are listed approximately from north to south. Row entries may refer to more than one site in close proximity.

Location	Millener Grid Ref.[1]	Type of Deposit[2]	Reference[3, 4]
North Island			
Spirits Bay – East Loc. 2	0395	dune	Millener 1981
Tom Bowling – West Loc 4, Central Loc 5, East Loc 6	0495	dune	Millener 1981; Hay et al. 2008 for RCD[5]
Waikuku Beach Loc 7–9	0495	dune	Millener 1981; Hay et al. 2008 for RCD
Motuopao Island Loc 12	0194	dune	Millener 1981
Cape Maria van Diemen Loc 13	0194	dune	Millener 1981
Herangi – West Loc 14	0194	dune	Millener 1981
Te Werahi Loc 17	0194	dune	Millener 1981
Spirits Bay – West Loc 19	0394	dune	Millener 1981
Whareana Loc 20	0494	dune	Millener 1981
Henderson Bay Loc 23–25	0691	dune	Millener 1981; Hay et al. 2008 for RCD
Kowhai Beach Loc 26	0690	dune	Millener 1981
Matai Bay Loc 30	0990	dune	Millener 1981
Puheke Beach Loc 32	0889	dune	Millener 1981
Tokerau Beach Loc 33–39	0889	dune	Millener 1981; Hay et al. 2008 for RCD
Loc 39 Tokerau South	0888	dune	Phil Millener, National Museum of N.Z, pers. comm. June 1990
Otangaroa Station Cave		cave	Worthy 1991
Loc 59 Ocean Beach	2078	dune	Phil Millener, pers. comm. June 1990
Loc 62 Waikiekie	1776	cave	Phil Millener, pers. comm. June 1990
Elver Canyon Cave, Paryphanta Passage		cave	Worthy 1991
Loc 80 Skippers, Opito	3267	dune	Phil Millener, pers. comm. June 1990
Loc 128 Templers	2653	cave	Phil Millener, pers. comm. June 1990
Miller's Waterfall Cave Loc 156	2549	cave	Millener 1981
MacDonald's Mud Cave		cave	Trevor Worthy, Palaeofaunal Surveys, Masterton, pers. comm. 26 September 2003
Tapuae Weka Cave Loc 171	2649	cave	Millener 1981
Flowstone Hole Loc 172	2649	cave	Millener 1981; Hay et al. 2008 for RCD
St Benedicts Cavern Loc 183	2548	cave	Millener 1981
Ruakuri A/T Cave Loc 185	2548	cave	Millener 1981; Hay et al. 2008 for RCD
Cave of False Hopes Loc 230	2648	cave	Millener 1981
Loc 184 Whites	2548	cave	Phil Millener, pers. comm. June 1990
Hilltop Cave Loc 243	2547	cave	Millener 1981
Loc 234 Briars	2447	cave	Phil Millener, pers. comm. June 1990
Loc 235 Waitanguru	2447	cave	Phil Millener, pers. comm. June 1990
Loc 239 F1 Cave	2547	cave	Phil Millener, pers. comm. June 1990; Worthy 1984
Loc 241 Papamaru	2547	cave	Phil Millener, pers. comm. June 1990; Trevor Worthy, pers. comm. 26 September 2003
Lucky Strike Cave		cave	Trevor Worthy, pers. comm. 26 September 2003
Loc 247 Little Lost World	2547	cave	Phil Millener, pers. comm. June 1990; Trevor Worthy, pers. comm. 26 September 2003
Cutthroat Cave		cave	Trevor Worthy, pers. comm. 26 September 2003
Dinornis Cave Loc 260	2446	cave	Millener 1981
Rorison's Quarry Loc 287	2545	cave	Millener 1981
Loc 274 Blue Gum	2345	cave	Phil Millener, pers. comm. June 1990
Loc 278 Porthole	2345	cave	Phil Millener, pers. comm. June 1990

Appendix 1 (cont.)

Location	Millener Grid Ref.[1]	Type of Deposit[2]	Reference[3, 4]
Loc 282 Spotlight	2445	cave	Phil Millener, pers. comm. June 1990
Loc 298 Aussie	2344	cave	Phil Millener, pers. comm. June 1990
Loc 313 Wainui	5443	dune	Phil Millener, pers. comm. June 1990
Wheturau Quarry		pitfall –PL	Worthy 2000
Loc 336 Whittles Farm	3936	cave	Phil Millener, pers. comm. June 1990
Deb's Cave		cave/predator	Worthy and Holdaway 2000
Loc 348 Te Waka 1	3936	cave/predator	Phil Millener, pers. comm. June 1990; Worthy and Holdaway 2000; Worthy et al. 2002a
Loc 341 Hukanui 1	3936	cave	Phil Millener, pers. comm. June 1990; Worthy and Holdaway 2000
Hukanui 3		cave	Worthy and Holdaway 2000
Loc 347 Hukanui 7	3936	cave	Phil Millener, pers. comm. June 1990
Loc 364 Bird and Lizard	4035	cave/predator	Phil Millener, pers. comm. June 1990; Worthy and Holdaway 2000
Loc 366 Bushface 1	4035	cave/predator	Phil Millener, pers. comm. June 1990; Worthy and Holdaway 2000
Loc 367 Bushface 2	4035	cave	Phil Millener, pers. comm. June 1990
Loc 377 Dartmoor	4134	cave	Phil Millener, pers. comm. June 1990; Worthy and Holdaway 2000
Loc 384 Napier	4333	alluvium	Phil Millener, pers. comm. June 1990
Loc 396 Matarua	4432	dune	Phil Millener, pers. comm. June 1990
Ocean Beach (sites 7, 7c and 8), near Cape Kidnappers		dune	Trevor Worthy, pers. comm. 26 September 2003, 16 July 2010; Hay et al. 2008 for RCD
Palmers Owl Site		predator	Trevor Worthy, pers. comm. 26 September 2003
Mataikona High Dune		dune	Trevor Worthy, pers. comm. 26 September 2003; Hay et al. 2008 for RCD
Loc 504 Harrison's Hole	2811	cave	Phil Millener, pers. comm. June 1990; Hay et al. 2008 for RCD for Ruakokoputuna
Loc 512 Haurangi 11	2811	cave	Phil Millener, pers. comm. June 1990
Loc 514 Haurangi 12	2811	cave	Phil Millener, pers. comm. June 1990
Loc 517 Haurangi 15	2811	cave	Phil Millener, pers. comm. June 1990
Loc 523 Fissure 2	2811	cave	Phil Millener, pers. comm. June 1990
South Island			
Bullock Bay, D'Urville Island		dune	Trevor Worthy, pers. comm. 26 September 2003, Trevor Worthy, University of New South Wales, 16 July 2010
Aorere	6089	cave	Phil Millener, pers. comm. June 1990; Trevor Worthy, pers. comm. 11 July 2010
Burnt Bush Cave, Upper Aorere		cave	Trevor Worthy, pers. comm. 26 September 2003
Road Cave, Sandhills Creek, Paturau		cave	Trevor Worthy, pers. comm. 26 September 2003
Road Cave, Sandhills Creek, Paturau		cave - PL	Trevor Worthy, pers. comm. 26 September 2003
Gouland Downs	5787	cave	Phil Millener, pers. comm. June 1990
GD101 Cave, Gouland Downs		cave/predator	Worthy 2001
Megamania Cave		cave	Worthy 1998d
Tarakohe Quarry Fissure		quarry	Trevor Worthy, pers. comm. 26 September 2003
Base of limestone cliff, near Takaka		predator	Worthy and Holdaway 1994
Sims Cave – tomo entrance		tomo entrance	Worthy and Holdaway 1994
Kiwi Hole	6286	cave	Phil Millener, pers. comm. June 1990

Appendix 1 (cont.)

Location	Millener Grid Ref.[1]	Type of Deposit[2]	Reference[3, 4]
Little Harwoods	6286	cave/pitfall	Phil Millener, pers. comm. June 1990; Worthy and Holdaway 1994
Commentary Cave		pitfall	Worthy and Holdaway 1994
Takaka Fossil Cave, Takaka Hill		pitfall/predator	Worthy and Roscoe 2003
Kairuru Cave		cave/predator	Trevor Worthy, pers. comm. 26 September 2003
Cave 1, Limestone Bluff, Heaphy River	5684	cave, bottom of a chimney	CM Rep 65 and others; Hay et al. 2008 for RCD for Heaphy River
Honeycomb Hill		cave	Phil Millener, pers. comm. June 1990
Whangamoa River mouth		dune	Trevor Worthy, pers. comm. 26 September 2003, 16 July 2010
Delaware Bay		dune	Trevor Worthy, pers. comm. 26 September 2003
Marfell's Beach, Lake Grassmere		dune	Worthy 1998c; Hay et al. 2008 for RCD
John's Pot, Inangahua	5476	cave	Phil Millener, pers. comm. June 1990
Metro Cave, near Charleston	4973	cave	Phil Millener, pers. comm. June 1990; Worthy and Holdaway 1993
Wharakiri, Charleston	4974	cave	Phil Millener, pers. comm. June 1990
Hermits, Charleston	4974	cave	Phil Millener, pers. comm. June 1990
Babylon Cave, near Punakaiki	4973	cave	Phil Millener, pers. comm. June 1990; Worthy and Holdaway 1993
Te Ana Titi Cave, near Punakaiki		cave	Worthy and Holdaway 1993
Meybille Bay	4873	cave	Phil Millener, pers. comm. June 1990
Cobden Quarry		quarry	Worthy et al. 2002c; Hay et al. 2008 for RCD (as 'Greymouth')
Fyffe Site, Avoca Point, Kaikoura		beach rubble	Trotter 1980; CM Rep 461 and others; Hay et al. 2008 for RCD for Avoca Stream
Holocene Hole, Mt Cookson		pitfall	Worthy and Holdaway 1995; Hay et al. 2008 for RCD for Mt Cookson
Falcon site #2 & Cave Falcon site, Mt Cookson		predator	Worthy and Holdaway 1995; Hay et al. 2008 for RCD for Mt Cookson
Ardenest & Arden Rockshelter 3, near Waikari		predator	Worthy and Holdaway 1996
Falcon Site, near Waikari		predator	Worthy and Holdaway 1996
Rabbit Warren Cave, near Waikari		predator	Worthy and Holdaway 1996
Pyramid Valley Swamp, near Waikari		swamp	Worthy and Holdaway 1996; Holdaway and Worthy 1997
Euan Murchison's Rockshelter, near Waikari		predator	Worthy and Holdaway 1996
The Deans, near Waikari		swamp	Worthy and Holdaway 1996
Bell Hill Vineyard		spring hole in a swamp	Trevor Worthy, pers. comm. 26 September 2003, 16 July 2010
Waipara River	6160	cave	Phil Millener, pers. comm. June 1990
Vincent's Limestone Quarry, Mt Somers		quarry	CM Rep 11; Worthy 1997
Mt Somers, a cave		cave	Worthy 1997
Site near Kimbell, Canterbury		cave/pitfall	Paul Scofield, Canterbury Museum, pers. comm. 14 July 2010
Rocky Ridges, Geraldine		pitfall	Worthy 1997
Glenlea Station		ledge cave	Worthy 1997
Hanging Rock, site 2		predator and vagrant	Worthy 1997
Tengawai River, Albury Park		rockshelter/ predator	Worthy 1997
Braeburn Station, near Cave		predator	Worthy 1997; Hay et al. 2008 for RCD

Appendix 1 (cont.)

Location	Millener Grid Ref.[1]	Type of Deposit[2]	Reference[3, 4]
Sterndale Rd rockshelter, near Cave		rockshelter/ predator	Worthy 1997
Kings Cave, South Canterbury		pothole/tomo	Phil Millener, pers. comm. 1990; Worthy 1997
Gordons Valley Site, South Canterbury		predator	Worthy 1997
Waitaki River	4239	cave	Phil Millener, pers. comm. June 1990
Earthquakes & Duntroon, North Otago		predator	Worthy 1998a; Wilmshurst et al. 2008 for RCD and stable isotopes of C and N
Island Cliffs & Ngapara, North Otago		predator/pitfall	Worthy 1998a
Several sites at Cape Wanbrow, Oamaru		colluvial and interdune–PL	Worthy and Grant-Mackie 2003
Oamaru Location D	4536	dune/loess	Phil Millener, pers. comm. June 1990
Rocklands Owl Site		predator	Worthy 1998a
Site 6, Glendhu Station, Wanaka		pitfall	Worthy 1998a
Station Deposit, near Cromwell		pitfall	Clark et al. 1996
Manuherikia, near Alexandra			Worthy 1998a
SM 93.10, Alexandra			Worthy 1998a; Hay et al. 2008 for RCD for Alexandra
Earnscleugh Cave, near Alexandra		cave	Clark et al. 1996; Worthy 1998a; Hay et al. 2008 for RCD
Knobby Range, near Alexandra		rockshelter	Worthy 1998a
Makaraeo Limeworks, Dunback		limestone cleft	Wood 2009
Castle Rock, Southland		cave	Phil Millener, pers. comm. June 1990; Worthy 1998b
Wakapatu, Southland		dune	Worthy 1998b; Hay et al. 2008 for RCD for Pahia

Notes

1 For explanation of grid system, see Millener 1981.
2 'PL' indicates deposit of Pleistocene age. Most if not all others are of Holocene age.
3 References additional to those in the bibliography:
Trotter, M. M. 1980. Archaeological investigations at Avoca Point, Kaikoura. *Records of the Canterbury Museum* 9: 277–288.
Wilmshurst, J. M., Anderson, A. J., Higham, T. F. G. and Worthy, T. H. 2008. Dating the late prehistoric dispersal of Polynesians to New Zealand using the commensal Pacific rat. *Proceedings of the National Academy of Sciences* 105: 7676–7680.
4 CM = Canterbury Museum.
5 RCD = radiocarbon date.

APPENDIX 2 (refer Chapter 2)

Deposits reported to contain tuatara bones of uncertain origin (natural and/or archaeological)

Sites are listed approximately from north to south.

Location	Millener Grid Ref.[1]	Type of Deposit	Reference or Museum Specimen[2, 3]
North Island region			
Ngahau Bay		unknown	Dickison 1998
Parauwanui Beach, Pataua North		unknown	AIM H2109
South Kaipara Head		unknown	Dickison 1998
Korapuki Island		bones in soil and burrow	Hicks et al. 1975; M. Meads, pers. comm. in Cree et al. 1995; David Towns, Department of Conservation. pers. comm. 3 September 2003
Parker's, Opito	3267	dune/midden	Phil Millener, National Museum of NZ. pers. comm. June 1990
Between Lake Wainamu and Bethells Beach		unknown	AIM H1359
Tairoa [Tairua?], Coromandel Peninsula		dune	CM Rep 53
Port Waikato		unknown	Dickison 1998
Waihora, near Lake Taupo	3243	cave/midden	Phil Millener, pers. comm. June 1990
Whakamoenga, near Lake Taupo	3443	cave/midden	Phil Millener, pers. comm. June 1990
Wainui, near Gisborne		dune/midden	Hutchinson 1898; Trevor Worthy, Palaeofaunal Surveys, Nelson, pers. comm. 4 October 1999
Rocky Hill Rockshelter	3936	cave/midden	Phil Millener, pers. comm. June 1990; Worthy and Holdaway 2000
Rockshelter #2, Whittle's Farm, Puketiri		rockshelter/ midden	Worthy and Holdaway 2000
Farm of R. H. Hartree, next to 'Manaroa', Patoka		rockshelter	CM Rep 345
Ocean Beach, Hawke's Bay	4431	dune/midden	Phil Millener, pers. comm. June 1990; see also CM Rep 505, 506
Blackhead Beach	4125	dune/midden	Phil Millener, pers. comm. June 1990
Akitio, southern Hawke's Bay		unknown	NMNZ RE 2367
Foxton	2722	dune/midden	Phil Millener, pers. comm. June 1990
Levin	2619	dune/midden	Phil Millener, pers. comm. June 1990
Paremata	2414	dune/midden	Phil Millener, pers. comm. June 1990
Arawhata Stream	3312	dune/midden	Phil Millener, pers. comm. June 1990; see also CM Rep 77
Kilbirnie, Wellington		unknown (buried skeleton)	NMNZ RE 1118
Near Lyall Bay, Wellington		unknown	Anon. 1873
White Rock	2908	dune/midden	Phil Millener, pers. comm. June 1990
South Island region			
Bullock Creek, D'Urville Island		dune	NMNZ RE 112
Heaphy Lewis III	5587	cave/midden	Phil Millener, pers. comm. June 1990; see also CM Rep 92 and others
Heaphy Cave 1	5586	cave/midden	Phil Millener, pers. comm. June 1990
Heaphy Fissure 3	5586	cave/midden	Phil Millener, pers. comm. June 1990
Heaphy below Fissure 3	5586	cave/midden	Phil Millener, pers. comm. June 1990
Marfell Beach	7477	dune/midden	Phil Millener, pers. comm. June 1990; see also CM Rep 7 and others

Appendix 2 (cont.)

Location	Millener Grid Ref.[1]	Type of Deposit	Reference or Museum Specimen[2,3]
Mussel Point	7477	dune/midden	Phil Millener, pers. comm. June 1990; see also CM Rep 83
Hareka [Karaka?] Pa site, Cape Campbell, Marlborough		sandhill	CM Rep 81
Lighthouse Beach, Cape Campbell, Marlborough		unknown	CM Rep 48
Whalers Bay, Kaikoura	6969	cave/midden	Phil Millener, pers. comm. June 1990
Andersons Bay, Kaikoura	6969	cave/midden	Phil Millener, pers. comm. June 1990
South Bay, Kaikoura	6968	dune/midden	Phil Millener, pers. comm. June 1990
Weka Pass	6062	cave/midden	Phil Millener, pers. comm. June 1990
Motunau Island		unknown (protruding from bank)	CM Rep 340
Redcliffs	6055	dune/midden	Phil Millener, pers. comm. June 1990
Moa Bone Point Cave, Redcliffs		unknown	CM Rep 69 and others
Paradise Cave	6155	cave/midden	Phil Millener, pers. comm. June 1990
Purau Bay	6154	dune/midden	Phil Millener, pers. comm. June 1990; see also CM Rep 82
Tumbledown Bay	6151	dune/midden	Phil Millener, pers. comm. June 1990
Near Kakahu's [?] Old Lime Kiln, Timaru		unknown	CM Rep 38
Shepherds Creek II	3844	cave/midden	Phil Millener, pers. comm. June 1990; see also CM Rep 141 and others
Near Bluff		dune	Fig. 5 of Crook 1975

Notes

1 For explanation of grid system, see Millener 1981.

2 References additional to those in the bibliography:
Anon, 1873. Wellington Philosophical Society. Ninth Meeting. 25th September, 1872. *Transactions and Proceedings of the New Zealand Institute* 5: 431–432.
Dickison, M. 1998. *Are island tuatara representative?* Unpublished BSc Hons dissertation, Victoria University of Wellington, Wellington.
Hicks, G. R. F., McColl, H. P., Meads, M. J., Hardy, G. S. and Roser, R. J. 1975. An ecological reconnaissance of Korapuki Island, Mercury Islands. *Notornis 22*: 195–220.

3 AIM = Auckland Institute and Museum; CM = Canterbury Museum; NMNZ = Museum of New Zealand.

APPENDIX 3

Midden sites and other archaeological deposits reported to contain bone(s) of tuatara

Sites are listed approximately from north to south.

Location	Other Associated Species	Date of Deposit[1]	Reference or Museum Specimen[2, 3]
North Island region			
Kawau Island, Auckland		c. AD 1600s	Robert Brassey, Department of Conservation, pers. comm. 18 September 2003
Tiritiri Matangi Island, Auckland	moa, other coastal and bush birds, sea lion, dog, fish, shellfish	c. AD 1400s	Robert Brassey, pers. comm. 13 October 1999
Motutapu Island, Auckland (Sunde site)	moa, crow, seal, dog, fish, shellfish	c. AD 1400	Scott 1970; Davidson 1972; Gill 1985
Port Jackson, Coromandel Peninsula		c. AD 1250–1500	Davidson 1979
Sarah's Gully, Coromandel Peninsula	moa, crow, petrels, many other birds, fur seal	c. AD 1250–1500	Davidson 1979; Scarlett 1979; CM Rep 51
Opito Beach, Coromandel Peninsula	moa, crow, petrels, many other birds, fur seal	c. AD 1250–1500	Davidson 1979; Scarlett 1979; CM Rep 50, 68
Hot Water Beach layer 5, Coromandel Peninsula		c. AD 1250–1500	Davidson 1979
Tararu			Trevor Worthy, Palaeofaunal Surveys, Masterton, pers. comm. 26 September 2003
Cooks Cove, Tolaga Bay, East Coast	moa, other forest birds, various seabirds, dog, fur seal, sea lion, Pacific rat, fish, shellfish	c. AD 1391–1420	Walter et al. 2010
Waihora Bay, Taupo	takahe	Classic Phase, possibly 17th–18th century	Hosking and Leahy 1982
Whakamoenga Cave, Taupo	moa and many other forest birds	c. AD 1300s–1400s, and possibly also more recent	Leahy 1976
Kaupokonui, South Taranaki	moa, other forest birds including adzebills, seabirds, dog, sea lion, Pacific rat	c. AD 1350	Foley 1980
Whanganui District			Downes 1937
Paekakariki			NMNZ RE 1129
Plimmerton (Taupo midden)	moa bones and eggshell fragments, dog		Beckett 1955
Mana Island			Miskelly 1999
Paremata, Wellington Coast	Moa, crow, swan, other birds including penguins and petrels, dog, fur seal, Pacific rat	c. AD 1400s	Scarlett 1979
Lyall Bay, Wellington	moa eggshell fragments		McLeod 1919
South Island region			
Nukuwaiata Island			Richard Walle, Department of Conservation, pers. comm. 15 April 2010
Wairau Bar, Marlborough	moa, crow, swan, eagle	AD 1288–1300	Duff 1977; Higham et al. 1999
Lake Grassmere Bar, Marlborough	moa, swan, eagle		Duff 1977

Appendix 3 (cont.)

Location	Other Associated Species	Date of Deposit[1]	Reference or Museum Specimen[2, 3]
Marfells Beach, Lake Grassmere		≤ 1500 y BP (also in natural deposits at same site)	Worthy 1998c; CM Rep 41
Te Hiku o te Waeroa, South Bay, Kaikoura	seabirds, other birds, seal, dog, Pacific rat		Fomison 1963; CM Rep 125
Hamilton's, Redcliffs	moa, other birds, marine mammal, dog, rat	AD 1296–1464	Dawson and Yaldwyn 1975; Challis 1995; CM Rep 52, 63
Paradise Cave, Redcliffs			Challis 1995
Tumbledown Bay, Banks Peninsula	moa, penguin, other birds, seal, dog, fish		Mason and Wilkes 1963; Challis 1995; CM Rep 128, 465
Kakanui			Trevor Worthy, pers. comm. 26 September 2003
Long Beach, Otago	moa, fish	c. AD 1200s–1400s	Hamel 2001
Otago			Phillipps 1997

Notes

1 Date for associated material, not tuatara bone specifically. y BP = years before present, where present = AD 1950.

2 References for sources not listed in the bibliography:

Beckett, P. 1955. Taupo Midden, Plimmerton, 1910–1914. *Journal of the Polynesian Society* 64: 161–62.

Challis, A. J. 1995. *Ka pakihi whakatekateka o Waitaha: the archaeology of Canterbury in Maori times*. Science and Research Series No. 89. Department of Conservation, Wellington.

Dawson, E. W. and Yaldwyn, J. C. 1975. Appendix II. Excavations and faunal remains at Hamilton's and adjacent properties, Redcliffs, 1946–1948. *Records of the Canterbury Museum* 9: 214–218.

Davidson, J. M. 1972. Archaeological investigations on Motutapu Island, New Zealand. *Records of the Auckland Institute and Museum* 9: 1–14.

Davidson, J. M. 1979. Archaic middens of the Coromandel region: a review. In: Anderson, A. (ed.), *Birds of a feather: osteological and archaeological papers from the South Pacific in honour of R. J. Scarlett. New Zealand Archaeological Association Monograph II*. BAR International Series 62, Oxford, pp. 183–202.

Duff, R. 1977. *The moa-hunter period of Maori culture*. 3rd edn. Government Printer, Wellington.

Foley, D. 1980. *Analysis of faunal remains from the Kaupokonui site (N128/3B)*. Unpublished MA thesis, University of Auckland, Auckland.

Fomison, T. 1963. Excavations at South Bay Kaikoura - Site S49/43. *New Zealand Archaeological Association Newsletter* 6: 100–102.

Gill, B. J. 1985. Subfossil bones of a large skink (Reptilia: Lacertilia) from Motutapu Island, New Zealand. *Records of the Auckland Institute and Museum* 22: 69–76.

Hamel, J. 2001. *The archaeology of Otago*. Department of Conservation, Wellington.

Higham, T., Anderson, A. and Jacomb, C. 1999. Dating the first New Zealanders: the chronology of Wairau Bar. *Antiquity* 73: 420–427.

Hosking, T. and Leahy, A. 1982. Investigations at rock shelter N93/5 Waihora Bay, Taupo, New Zealand. *Records of the Auckland Institute & Museum* 19: 81–99.

Leahy, A. 1976. Whakamoenga Cave, Taupo, N94/7: a report on the ecology, economy and stratigraphy. *Records of the Auckland Institute and Museum* 13: 29–75.

Mason, G. M. and Wilkes, O. 1963. Tumbledown Bay – a Banks Peninsula moa-hunter site S94/30. *New Zealand Archaeological Association Newsletter* 6: 98–100.

McLeod, H. N. 1919. Maori occupation of the Wellington District. Notes on some archaeological remains. Part VI of Best, E. The Land of Tara and they who settled it. *Journal of the Polynesian Society* 28: 1–17.

Miskelly, C. 1999. *Mana Island ecological restoration plan*. Department of Conservation, Wellington, p. 72.

Phillips, W. J. 1997. *Maori carving illustrated*. 4th edn. (Revised by Simmons, D. R.) Reed, Auckland.

Scarlett, R. J. 1979. Avifauna and man. In: Anderson, A. (ed.). *Birds of a feather: osteological and archaeological papers from the South Pacific in honour of R. J. Scarlett. New Zealand Archaeological Association Monograph II*. BAR International Series 62, Oxford, pp. 75–90.

Scott, S. D. 1970. Excavations at the 'Sunde site', N38/24, Motutapu Island, New Zealand. *Records of the Auckland Institute and Museum* 7: 13–30.

Walter, R., Jacomb, C. and Brooks, E. 2010. *Final report on archaeological excavations at Cooks Cove Z17/311, Tolaga Bay, East Coast, North Island*. Unpublished report to the Historic Places Trust Poutere Taonga.

3 CM = Canterbury Museum, NMNZ = Museum of New Zealand.

APPENDIX 4 (refer Chapter 7)

Abundance of tuatara on various islands, based on capture rates and density estimates from 1980 onwards

Islands are listed from north to south, and multiple entries for capture rates or densities on each island are listed in chronological order.

Island	Habitat	Introduced Mammals[1]	Capture Rate[2]	Density (no. per ha.)	Sources
Tawhiti Rahi	Forest	None	0.8	–	Cree (pers. obs. for March 1989)
Aorangi	Forest	None	1.0	–	Cree (pers. obs. for March 1989)
Whatupuke	Mainly forest	Pacific rat	1.3; 1.0		Newman 1986; Cree et al. 1995a
Coppermine	Mainly forest	Pacific rat	3.0; 0.6	12–26	Newman 1986; Cree et al. 1995a; Tyrrell et al. 2000
Lady Alice	Mainly forest	Pacific rat	2.4; 1.1	13.6–17.7	Newman 1986; Cree et al. 1995a; Cassey and Ussher 1999
Taranga	Forest	Pacific rat	0.45	1.1	Cree et al. 1995a; Towns et al. 2007
Hauturu/Little Barrier	Forest	Pacific rat	0.01	–	Whitaker and Daugherty 1991[3]; Whitaker 1993[4]; Cree et al. 1995a
Cuvier (Repanga)	Mainly forest	Pacific rat	0.05	–	Cree et al. 1995a
Red Mercury (Whakau)	Forest	Pacific rat	0.09	–	Cree et al. 1995a
Atiu or Middle	Forest/beach or cliff/beach margin	None	1.3	25	Cree et al. 1995a; G. Ussher (pers. comm. 12 September 2005, for 2003 prior to harvest)
	Forest	None	3.6	80	Towns 1991 (and pers. comm. 8 August 2005); G. Ussher (pers. comm. 12 September 2005, for 2003 prior to harvest)
Kawhitu or Stanley	Forest	Pacific rat, European rabbit	0.15	–	Cree et al. 1995a
Green	Forest	None	2.2	–	Cree et al. 1995a
Hongiora	Forest and scrub	None	3.6	–	Newman and McFadden 1990a
Ruamahuaiti	Forest and scrub	None	4.8	220–335	Newman and McFadden 1990a; G. Ussher in Tyrrell et al. 2000
Whenuakura (tuatara now extinct)	Low forest	House mice	5.8	99	Newman 1986
Karewa	Low forest	None	1.4; 0.6	–	Heaphy 2005[5]
Moutoki	Forest	None	4.7[ER]	c. 500; c. 300	Whitaker 1989[6]; Garrick 1996[ER 7]; G. Ussher (pers. comm. 9 September 2005, for 1996 prior to harvest)
Stephens (Takapourewa)	Pasture	Sheep	8.7	415; 1100; 208	Newman 1986; Carmichael et al. 1989; Markwell 1997; Moore 2008
	Mostly unforested	Sheep, cattle?	–	Up to c. 560	Crook 1975, as discussed by Newman 1987a
	Grassland	None	2.7–3.3[ER]	400	East et al. 1995[ER]; Markwell 1997
	Shrubland	None	–	300	Markwell 1997
	Mixed (including grassland and forest)	None	9.8–25.6	674	Moore et al. 2010; N. Nelson (pers. comm. 3 May 2011)
	Forest	None	18.2[ER]	1420–1500; 2015; 1500; 2732	East et al. 1995[ER]; Newman 1987a; Carmichael et al. 1989; Markwell 1997; Moore et al. 2009b
North Brother	Windswept scrub	None		134; 159	Thompson et al. 1992; Nelson et al. 2002a

Notes

1 Introduced mammals are those present during or within a year prior to estimate. All introduced mammals have since been eliminated, except on Taranga Island, where eradication is underway. Scientific names as follows: Pacific rat (*Rattus exulans*), cattle (*Bos taurus*), sheep (*Ovis aries*), rabbit (*Oryctolagus cuniculus*) and house mouse (*Mus musculus*).
2 Number per person per hour. Values are rounded to the nearest 0.1 per person per hour, except for the very low values for Taranga, Hauturu, Cuvier and Red Mercury (reported to the nearest 0.01). Most estimates are based on night surveys, which focus on adults and large juveniles and thus underestimate total abundance or density. ER = encounter rate, which overestimates capture rate, but probably by not more than double. Density estimates are for limited areas and should not be extrapolated over entire islands.
3 Whitaker, A. H. and Daugherty, C. H. 1991. *Research on the tuatara (*Sphenodon punctatus*) of Little Barrier Island, 5–12 February 1991*. Unpublished report to the Department of Conservation, Wellington.
4 Whitaker, A. H. 1993. *Research on the tuatara (*Sphenodon punctatus*) of Little Barrier Island, 6–20 October 1992*. Unpublished report to the Department of Conservation, Auckland.
5 Heaphy, J. 2005 *Karewa tuatara survey – March 2005*. Unpublished report, Department of Conservation, Tauranga, TAUAO-2374 (data for January 1989 and March 2005).
6 Whitaker, A. H. 1989. *Report on a visit to Moutoki Island, Rurima Rocks Group, Bay of Plenty, 19–21 January 1989*. Unpublished report for the Ngati Awa Trust Board and Tuwharetoa Maori Trust, Victoria University of Wellington, Wellington.
7 Garrick, A. 1996. *Preliminary report on the status of the tuatara on Moutoki, Rurima Rocks, Bay of Plenty*. Unpublished report, Department of Conservation, Rotorua. RWL:030.

Scientific bibliography

Note

Single-authored articles are listed in alphabetical order of author, then in chronological order.

Articles with two authors are listed in alphabetical order of first author, then in alphabetical order of second author, then in chronological order.

Articles with three or more authors are treated as "et al." and listed in chronological order, regardless of number or alphabetical order of secondary authors.

Multiple articles by the same author(s) in the same years are listed using unique letter codes (a, b, c, etc.) in order of entry into the bibliographical database. Thus, order within a year may appear arbitrary here.

A Party of Officers of the 58th Regiment. 1852. The ngararas of the Rurimas. *New Zealander*, Auckland, 24 April pp. 3–4. [Reprinted 1982 in *Historical Review*, (Whakatane and District Historical Society) 30: 54–61.]

Abbasi, A., Wells, R. M. G., Brittain, T. and Braunitzer, G. 1988. Primary structure of the hemoglobins from *Sphenodon* (*Sphenodon punctatus*, tuatara, Rhynchocephalia) – evidence for the expression of α^D-gene. *Biological Chemistry* 369: 755–764.

Abbie, A. A. 1933. The blood supply of the lateral geniculate body, with a note on the morphology of the choroidal arteries. *Journal of Anatomy* 67: 491–521.

Adams, W. E. 1953. The carotid arch in lizards with particular reference to the origin of the internal carotid artery. *Journal of Morphology* 92: 115–155.

Aitken, N., Hay, J. M., Sarre, S. D., Lambert, D. M. and Daugherty, C. H. 2001. Microsatellite DNA markers for tuatara (*Sphenodon* spp.). *Conservation Genetics* 2: 183–185.

Ali, S. M. 1941. Studies on the comparative anatomy of the tail in Sauria and Rhynchocephalia. *Proceedings of the Indian Academy of Sciences* 13: 171–193.

Alibardi, L. 1992. Glial cell composition and ultrastructure of the caudal spinal cord of young and adult tuataras, *Sphenodon punctatus*. *Acta Zoologica (Stockholm)* 73: 157–162.

Alibardi, L. 1999. Keratohyalin-like granules in embryonic and regenerating epidermis of lizards and *Sphenodon punctatus* (Reptilia, Lepidosauria). *Amphibia-Reptilia* 20: 11–23.

Alibardi, L. 2003. Adaptation to the land: the skin of reptiles in comparison to that of amphibians and endotherm amniotes. *Journal of Experimental Zoology* 298B: 12–41.

Alibardi, L. 2009. Development, comparative morphology and cornification of reptilian claws in relation to claws evolution in tetrapods. *Contributions to Zoology* 78: 25–42.

Alibardi, L. 2010. *Morphological and cellular aspects of tail and limb regeneration in lizards: a model system with implications for tissue regeneration in mammals*. Springer, Berlin. 112 p.

Alibardi, L. 2011. Cytology and localization of chromatophores in the skin of the tuatara (*Sphenodon punctatus*). *Acta Zoologica* 00: 1–8. doi: 10.1111/j.1463-6395.2011.00506.x.

Alibardi, L. 2012a. Cytology and localization of chromatophores in the skin of the tuatara (*Sphenodon punctaus* [*sic*]). *Acta Zoologica (Stockholm)* 93: 330–337.

Alibardi, L. 2012b. Immunolocalization of keratin-associated beta-proteins (beta-keratins) in scales of the reptiles [*sic*] *Sphenodon punctatus* indicates that different beta-proteins are present in beta- and alpha-layers. *Tissue and Cell* 44: 378–384.

Alibardi, L. 2013a. Cornification in reptilian epidermis occurs through the deposition of keratin-associated beta-proteins (beta-keratins) onto a scaffold of intermediate filament keratins. *Journal of Morphology* 274: 175–193.

Alibardi, L. 2013b. Granulocytes of reptilian sauropsids contain beta-defensin-like peptides: a comparative ultrastructural survey. *Journal of Morphology* 274: 877–886.

Alibardi, L. and Gill, B. J. 2007. Epidermal differentiation in embryos of the tuatara *Sphenodon punctatus* (Reptilia, Sphenodontidae) in comparison with the epidermis of other reptiles. *Journal of Anatomy* 211: 92–103.

Alibardi, L. and Maderson, P. F. A. 2003a. Observations on the histochemistry and ultrastructure of the epidermis of the tuatara, *Sphenodon punctatus* (Sphenodontida, Lepidosauria, Reptilia): a contribution to an understanding of the lepidosaurian epidermal generation and the evolutionary origin of the squamate shedding complex. *Journal of Morphology* 256: 111–133.

Alibardi, L. and Maderson, P. F. A. 2003b. Observations on the histochemistry and ultrastructure of regenerating caudal epidermis of the tuatara *Sphenodon punctatus* (Sphenodontida, Lepidosauria, Reptilia). *Journal of Morphology* 256: 134–145.

Alibardi, L. and Meyer-Rochow, V. B. 1989. Comparative fine structure of the axial skeleton inside the regenerated tail of some lizard species and the tuatara (*Sphenodon punctatus*). *Gegenbaurs Morphologisches Jahrbuch* 135: 705–716.

Alibardi, L. and Meyer-Rochow, V. B. 1990a. Fine structure of regenerating caudal spinal cord in adult tuatara (*Sphenodon punctatus*). *Journal für Hirnforschhung* 31: 613–621.

Alibardi, L. and Meyer-Rochow, V. B. 1990b. Ultrastructural survey of the spinal cord of young tuatara (*Sphenodon punctatus*) with emphasis on the glia. *New Zealand Journal of Zoology* 17: 73–85.

Alibardi, L. and Sawyer, R. H. 2002. Immunocytochemical analysis of beta (β) keratins in the epidermis of chelonians, lepidosaurians, and archosaurians. *Journal of Experimental Zoology* 293: 27–38.

Alibardi, L. and Toni, M. 2006a. Immunological characterization and fine localization of a lizard beta-keratin. *Journal of Experimental Zoology* 306B: 528–538.

Alibardi, L. and Toni, M. 2006b. Distribution and characterization of keratins in the epidermis of the tuatara (*Sphenodon punctatus*; Lepidosauria, Reptilia). *Zoological Science* 23: 801–807.

Allen, O. M. 1939. Museum case in living flesh. *Digest of World Reading* 1 August: 32–34.

Allendorf, F. W. 2001. Genetics and the viability of insular populations of reptiles. In: Abstracts of papers presented at the 9th Conference of the Society for Research on Amphibians and Reptiles in New Zealand, St Arnaud, Nelson Lakes, New Zealand, 2–4 February 2001. *New Zealand Journal of Zoology* 28: 361.

Alley, M. R., Gartrell, B. D. and Morgan, K. J. 2004. Wildlife cases from Massey University October 2003–April 2004. *Kokako* 11: 13.

Allison, B. and Blair, D. 1987. The genus *Dolichosaccus* (Platyhelminthes: Digenea) from amphibians and reptiles in New Zealand, with a description of *Dolichosaccus* (*Lecithopyge*) *leiolopismae* n. sp. *New Zealand Journal of Zoology* 14: 367–374.

Allison, F. R. 1982. Parasites of New Zealand reptiles. In: Newman, D. G. (ed.). *New Zealand Herpetology*. New Zealand Wildlife Service Occasional Publication No. 2, Wellington, pp. 419–422.

Ananjeva, N. B. and Dujsebayeva, T. N. 1997. SEM study of skin sense organs in two *Uromastyx* species (Sauria: Agamidae) and *Sphenodon punctatus* (Rhynchocephalia: Sphenodontidae). *Russian Journal of Herpetology* 4: 46–49.

Anon. 1872. [On a tuatara captured on Ruarimu [*sic*] Rocks by Major Mair]. *Transactions and Proceedings of the New Zealand Institute* 4: 388.

Anon. 1882a. A new case of commensalism. *Nature* 26: 608–609.

Anon. 1882b. [On tuatara collected from Karewa Island by Captain Fairchild, and others kept as pets]. *New Zealand Herald*, Auckland, 9 February, p. 5.

Anon. 1883. Local and general [on tuatara collected by Captain Fairchild from the Brothers Islands near Stewart's Island]. *Otago Witness*, Dunedin, 13 January, p. 9.

Anon. 1885a. Land and water [on Mr Charles Bills taking tuatara 'Home']. *Otago Witness*, Dunedin, 18 April, p. 21.

Anon. 1885b. Tuatara lizards. *Bay of Plenty Times*, Tauranga, 10 March, p. 2.

Anon. 1886. Otago Institute. *Transactions and Proceedings of the New Zealand Institute* 18: 427.

Anon. 1887. Local and general [on tuatara found in a Ponsonby coal cellar]. *Otago Witness*, Dunedin, 11 March, pp. 9–10.

Anon. 1890. [On specimens of *Hatteria* exhibited by G. B. Howes showing the 'pro-atlas' and vomerine teeth]. *Proceedings of the Zoological Society of London* 25: 357–360.

Anon. 1892. [On Captain Fairchild taking 20 tuatara for the Sydney Museum]. *Bay of Plenty Times and Thames Valley Warden*, Tauranga, 13 June, p. 2.

Anon. 1893. [On tuatara escaped from Government House in Auckland]. *Bay of Plenty Times*, Tauranga, 7 April, p. 6.

Anon. 1895. Exhibits. *Transactions and Proceedings of the New Zealand Institute* 27: 659–660.

Anon. 1896. A lizard of sense. *Southlander*, Invercargill, 10 April (page unknown).

Anon. 1898. [On the likely disappearance of the few tuatara that remain on East Island]. *Poverty Bay Herald*, Gisborne, 12 September, p. 2.

Anon. 1899a. Southland news notes [on a tuatara at the Invercargill Athaneum]. *Otago Witness*, Dunedin, 20 July, p. 34.

Anon. 1899b. Foreign scientists in New Zealand. *Press*, Christchurch, 11 July, p. 4.

Anon. 1903. Passing notes [on tuatara at Opawa Fisheries]. *Otago Witness*, Dunedin, 9 December, p. 5.

Anon. 1904. [On Cockayne's observations on the effects of cats on Stephens Island]. *New Zealand Herald*, Auckland, 28 January, p. 4.

Anon. 1907. General news [on the exhibition of six tuatara at the Imperial Zoological Gardens in Schoenbrunn, Austria]. *Bay of Plenty Times*, Tauranga, 4 November, p. 4.

Anon. 1908a. The tuatara lizard. *Otago Witness*, Dunedin, 11 November, p. 18.

Anon. 1908b. Interesting scientific find. *Bay of Plenty Times*, Tauranga, 6 May, p. 2.

Anon. 1909. Local and general [on Percy Isaac and his pet tuatara]. *Bay of Plenty Times*, Tauranga, 25 October, p. 2.

Anon. 1913a. Ancient reptile: the disappearing tuatara: efforts at preservation. *Evening Post*, Wellington, 7 June, p. 9.

Anon. 1913b. The tuatara: more protection: interesting reptiles on islands. *Lyttelton Times*, Christchurch, 30 June [page unknown; copy in ANZ IA 1 46/18/4 pt 1].

Anon. 1913c. Topics of the day. Sanctuaries invaded. *Evening Post*, Wellington, 13 June, p. 6.

Anon. 1915a. [On a tuatara found at the back of a house in Auckland]. In: *Fildes Cuttings #639*. Beaglehole Collection, Victoria University of Wellington Library, Wellington, p. 65.

Anon. 1915b. Habits of the tuatara: many observers' notes. In: *Fildes Cuttings #639*. Beaglehole Collection, Victoria University of Wellington Library, Wellington, p. 61.

Anon. 1922. The tuatara lizard. *Forest Magazine (New Zealand Out-of-Doors)* 1 July: 113.

Anon. 1931. [On tuatara of the Mokohinau Islands]. *Dominion*, Wellington, 11 August [page unknown; copy in ANZ IA 1 M1 25/611 pt 3].

Anon. 1952. Oldest inhabitants. *New Zealand Listener* 28 (No. 701, 12 December): 6.

Anon. 1954. First home-grown tuataras. *Natural History (New York)* 63: 422–423.

Anon. 1955a. Home-reared tuatara in Auckland. *New Zealand Herald*, Auckland, 17 June, p. 10.

Anon. 1955b. First captive-bred tuataras were in Auckland. *New Zealand Herald*, Auckland, 20 June, p. 8.

Anon. 1966. Old three eyes. *New Zealand Listener* 54 (No. 1400, 5 August): 67.

Anon. 1967. The tuataras of New Zealand. *NAC Airline Review* 6: 6–7.

Anon. 1979. Lizard was a puzzle (New Zealand 100 years ago from Herald files). *New Zealand Herald*, Auckland, 20 April, p. 8.

Anon. 1982. Tuataras take passage (New Zealand 100 years ago from Herald files). *New Zealand Herald*, Auckland, 9 February, p. 8.

Anon. 1996. Unusual find off Poor Knights. *Northern Advocate*, Whangarei, 16 January, p. 1.

Anon. 2006. The flight of the tuatara. *Forest and Bird* No. 319: 16.

Apesteguía, S. 2007. La evolucíon de los lepidosaurios. *Investigacíon y Ciencia* April: 54–63.

Arnold, E. N. 1984. Variation in the cloacal and hemipenial muscles of lizards and its bearing on their relationships. *Symposium of the Zoological Society, London* 52: 47–85.

Ashley, J. 1902. Notes on the breeding habits of the tuatara. *Transactions and Proceedings of the New Zealand Institute* 34: 580.

Atkinson, I. A. E. 1968. An ecological reconnaissance of Coppermine Island, Hen and Chickens Group. *New Zealand Journal of Botany* 6: 285–294.

Austin, W. A. 1962. Tuatara (*Sphenodon punctatus*) in captivity. *International Zoo Yearbook* 4: 124–125.

Bage, F. 1912. On the histological structure of the retina of the lateral eyes of *Sphenodon punctatus*, with special reference to the sense-cells. *Quarterly Journal of Microscopical Science* 57: 305–328 + plates 27–29.

Bain, R. 1951. The four immortals. *Scientific Monthly* 73: 274.

Baird, I. L. 1970. The anatomy of the reptilian ear. In: Gans, C. and Parsons, T. S. (eds). *Biology of the Reptilia. Vol. 2. Morphology B*. Academic Press, London, pp. 193–275.

Baker, R. 1991. *Tuatara: a resource for sixth and seventh form biology*. Ministry of Education, Wellington. 60 p.

Barwick, R. E. 1982. Observations on active thermoregulation in the tuatara, *Sphenodon punctatus* (Reptilia: Rhynchocephalia). In: Newman, D. G. (ed.). *New Zealand Herpetology*. New Zealand Wildlife Service Occasional Publication No. 2, Wellington, pp. 225–236.

Batchelor, A. 1913. The tuatara lizard. *Evening Post*, Wellington, 11 June, p. 7.

Baur, G. 1886a. The ribs of *Sphenodon (Hatteria)*. *American Naturalist* 20: 979–981.

Baur, G. 1886b. Die zwei Centralia im Carpus von *Sphenodon (Hatteria)* und die Wirbel von *Sphenodon* und *Gecko verticellatus* Laur. (*G. verus*, Gray). *Zoologischer Anzeiger* 9: 188–190.

Baur, G. 1891. The lower jaw of *Sphenodon*. *American Naturalist* 25: 489–490.

Baur, G. 1896. Das Gebiss von *Sphenodon* (*Hatteria*) und einege Bemerkungen über Prof. Rud. Burckhardt's Arbeit über das Gebiss der Sauropsiden. *Anatomischer Anzeiger* 11: 436–439.

Bayer, F. 1884. Über die Extremitäten einer jungen *Hatteria*. *Sitzungsberichte der kaiserlichen Akadamie der Wissenschaften in Wien* 90: 237–245 + 1 Tafel.

Beattie, J. H. 1994. *Traditional lifeways of the southern Maori: the Otago University Museum Ethnological Project, 1920*. Anderson, A. (ed.). University of Otago Press and Otago Museum, Dunedin. 636 p.

Bell, B. D., Daugherty, C. H., Hay, J. M. and Hitchmough, R. A. 2004. Species identification of allopatric populations with conflicting datasets: a Cook Strait islands case study and round-table discussion. In: Abstracts of papers presented at the 10th Biennial Conference of the Society for Research on Amphibians and Reptiles in New Zealand, Whakatane, New Zealand, 31 January–2 February 2003. *New Zealand Journal of Zoology* 31: 101.

Bellairs, A., d'A. 1949. The anterior brain-case and interorbital septum of Sauropsida, with a consideration of the origin of snakes. *Journal of the Linnean Society of London* 41: 482–512 + plates 9–11.

Bellairs, A., d'A. 1969. *The life of reptiles. Vol. 1 and 2*. Weidenfeld and Nicolson, London. 282 p.

Bellairs, A., d'A. 1984. Closing address: with comments on the organ of Jacobson and the evolution of Squamata, and on the intermandibular connection in Squamata. In: Ferguson, M. W. J. (ed.). *The structure, development and evolution of reptiles. Symposia of the Zoological Society of London No. 52*. Academic Press, London, pp. 665–683.

Bellairs, A., d'A. and Bryant, S. V. 1985. Autotomy and regeneration in reptiles. In: Gans, C. and Billett, F. (eds). *Biology of the Reptilia. Vol. 15. Development B*. John Wiley and Sons, New York, pp. 301–410.

Benham, W. B. 1899. The development of the tuatara. *Nature* 60: 79–80.

Benham [W.B.]. 1918. The tuatara. In: *Fildes Cuttings #639*. Beaglehole Collection, Victoria University of Wellington Library, p. 61.

Bennett, K., Ragg, J. and Childerhouse, S. 1993. *Construction of a database for toe-clipped tuatara (*Sphenodon punctatus*) on Stephens Island*. University of Otago Wildlife Management Report No. 33. University of Otago, Dunedin. 25 p.

Benton, M. J. 1986. The demise of a living fossil? *Nature* 323: 762.

Berg, J. 1894. Zur Kenntnis der Brückenechse. *Zoologischer Garten Frankfurt am Main* 32: 102–105, 146–150.

Besson, A. A. 2009. *Effects of cool temperature on egg incubation, thermoregulation and physiological performance of tuatara (*Sphenodon punctatus*): implications for conservation programmes*. Unpublished PhD thesis, University of Otago, Dunedin.165 p.

Besson, A. A. and Cree, A. 2010. A cold-adapted reptile becomes a more effective thermoregulator in a thermally challenging environment. *Oecologia* 163: 571–581.

Besson, A. A. and Cree, A. 2011. Integrating physiology into conservation: an approach to help guide translocations of a rare reptile in a warming environment. *Animal Conservation* 14: 28–37.

Besson, A. A., Thierry, A., Boros, E., Allen, K., Bradley, S., Norrie, C. and Cree, A. 2009. Evidence of food chemical discrimination in tuatara (O. Rhynchocephalia): comparison with a gekkotan lizard (O. Squamata). *Journal of Herpetology* 43: 124–131.

Besson, A. A., Nelson, N. J., Nottingham, C. M. and Cree, A. 2012. Is cool egg incubation temperature a limiting factor for the translocation of tuatara to southern New Zealand? *New Zealand Journal of Ecology* 36: 90–99.

Best, E. 1909. Maori forest lore: being some account of native forest lore and woodcraft, as also of many myths, rites, customs, and superstitions connected with the flora and fauna of the Tuhoe or Ure-wera District. Part II. *Transactions and Proceedings of the New Zealand Institute* 41: 231–285.

Best, E. 1923. Notes on the occurrence of the lizard in Maori carvings, and various myths and superstitions connected with lizards. *New Zealand Journal of Science and Technology* 5: 321–335.

Birchard, G. F., Nelson, N. J. and Daugherty, C. H. 2006. A circadian rhythm in oxygen consumption rate in juvenile tuatara (*Sphenodon punctatus*). *New Zealand Journal of Zoology* 33: 185–188.

Blair, T. 1998. *The significance of dietary n-3 fatty acids to captive and wild tuatara (*Sphenodon*)*. Unpublished MSc thesis, University of Otago, Dunedin. 92 p.

Blair, T., Cree, A., Grimmond, N. M. and Skeaff, C. M. 1997. Growth, oxygen consumption, and lipid metabolism in captive juvenile tuatara (*Sphenodon punctatus*) on two different diets. In: Abstracts of papers presented at the 7th Conference of the Society for Research on Amphibians and Reptiles in New Zealand, Kaikoura, New Zealand, 31 January–2 February 1997. *New Zealand Journal of Zoology* 24: 324–325.

Blair, T. A., Cree, A. and Skeaff, C. M. 1999. Plasma fatty acid composition and lipid concentrations in tuatara (*Sphenodon punctatus punctatus*) from a rat-free and a rat-inhabited island. In: Abstracts of papers presented at the 8th Conference of the Society for Research on Amphibians and Reptiles in New Zealand, Great Barrier Island, New Zealand, 5–7 February 1999. *New Zealand Journal of Zoology* 26: 256.

Blair, T. A., Cree, A. and Skeaff, C. M. 2000a. Plasma fatty acids, triacylglycerol and cholesterol of the tuatara (*Sphenodon punctatus punctatus*) from islands differing in the presence of rats and the abundance of seabirds. *Journal of Zoology, London* 252: 463–472.

Blair, T. A., Cree, A., Skeaff, C. M. and Grimmond, N. M. 2000b. Physiological effects of a fish oil supplement on captive juvenile tuatara (*Sphenodon punctatus*). *Physiological and Biochemical Zoology* 73: 177–191.

Blanchard, B. 1988. The breeding of tuataras in the wild and in captivity in New Zealand. *Thylacinus* 13: 17–24.

Blanchard, B. 1991. Future plans for the captive breeding of tuatara. In: Abstracts of papers presented at the 3rd Annual Conference of the Society for Research on Amphibians and Reptiles in New Zealand, Whitianga, Coromandel Peninsula, 28–30 November 1990. *New Zealand Journal of Zoology* 18: 343.

Blanchard, B. 2002. *Tuatara (*Sphenodon*) in captivity*. Unpublished MSc thesis, Victoria University of Wellington, Wellington. 141 p.

Blanchard, B. and the Tuatara Recovery Group. 2002. *Tuatara captive management plan and husbandry manual.* Threatened Species Occasional Publication 21. Department of Conservation, Wellington. 75 p.

Blanchard, F. C. 1935. 'Living fossils' walk on well-nigh inaccessible rocky islands off the coast of New Zealand. *National Geographic Magazine* 67: 649–662.

Boardman, W. and Blanchard, B. 2006. Biology, captive management, and medical care of tuatara. In: Mader, D. R. (ed.). *Reptile medicine and surgery.* Saunders Elsevier, St. Louis, pp. 1008–1012.

Boardman, W. S. J. and Sibley, M. D. 1991. The captive management, diseases and veterinary care of tuatara. In: *Proceedings of the American Association of Zoo Veterinarians.* Calgary, pp. 159–166.

Body, D. R. 1985. The egg-lipid composition of the 'living fossil' reptile tuatara (*Sphenodon punctatus*). *Experientia* 41: 1055–1057.

Body, D. R. and Newman, D. G. 1989. The lipid composition of liver, lung and adipose tissues from tuatara (*Sphenodon punctatus*) (Reptilia: Sphenodontia). *Comparative Biochemistry and Physiology* 93B: 223–227.

Bogert, C. M. 1953a. Body temperatures of the tuatara under natural conditions. *Zoologica* 38: 63–64.

Bogert, C. M. 1953b. The tuatara: why is it a lone survivor? *Scientific Monthly* 76: 163–170.

Booth, D. T. and Thompson, M. B. 1991. A comparison of reptilian eggs with those of megapode birds. In: Deeming, D. C. and Ferguson, M. W. J. (eds). *Egg incubation: its effects on embryonic development in birds and reptiles.* Cambridge University Press, Cambridge, UK, pp. 325–344.

Boulenger, G. A. 1891. On British remains of *Homœosaurus*, with remarks on the classification of the Rhynchocephalia. *Proceedings of the Zoological Society of London* 1891: 167–172.

Bradshaw, S. D., Owen, F. J. and Saint Girons, H. 1988. Seasonal changes in plasma sex steroid levels in the male tuatara, *Sphenodon punctatus,* from Stephens Island, New Zealand. *General and Comparative Endocrinology* 70: 460–465.

Brattstrom, B. H. 1965. Body temperatures of reptiles. *American Midland Naturalist* 73: 376–422.

Braun-Elwert, C. 2009. *Tuatara and their living fossil label.* Unpublished MSciComm thesis, University of Otago, Dunedin. 39 p.

Bredeweg, E. M. and Nelson, N. J. 2010. *Sphenodon punctatus* (Tuatara). Frugivory. *Herpetological Review* 41: 211–212.

Brittain, T. 1988. An investigation of the functioning of the two major haemoglobins of the *Sphenodon* using fast reaction kinetic methods. *Biochemical Journal* 251: 771–776.

Brittain, T. 1991. Cooperativity and allosteric regulation in non-mammalian vertebrate haemoglobins. *Comparative Biochemistry and Physiology* 99B: 731–740.

Broom, R. 1906. On the organ of Jacobson in *Sphenodon. Journal of the Linnean Society of London* 29: 414–420 + plates 41–42.

Brown, D. [2000]. *Stephens Island: ark of the light.* Derek Brown, [Havelock]. 248 p.

Brown, M. A., Cree, A., Chambers, G. K. and Newton, J. D. 1989. Techniques for detecting vitellogenesis in the tuatara *Sphenodon punctatus. New Zealand Journal of Zoology* 16: 25–35.

Brown, M. A., Cree, A., Chambers, G. K., Newton, J. D. and Cockrem, J. F. 1991a. Monitoring of plasma constituents during the natural vitellogenic cycle of tuatara, *Sphenodon punctatus.* In: Abstracts of papers presented at the 3rd Annual Conference of the Society for Research on Amphibians and Reptiles in New Zealand, Whitianga, Coromandel Peninsula, 28–30 November 1990. *New Zealand Journal of Zoology* 18: 343.

Brown, M. A., Cree, A., Chambers, G. K., Newton, J. D. and Cockrem, J. F. 1991b. Variation in plasma constituents during the natural vitellogenic cycle of tuatara, *Sphenodon punctatus. Comparative Biochemistry and Physiology* 100B: 705–710.

Brown, M. A., Cree, A., Daugherty, C. H., Dawkins, B. P. and Chambers, G. K. 1994. Plasma concentrations of vitellogenin and sex steroids in female tuatara (*Sphenodon punctatus punctatus*) from northern New Zealand. *General and Comparative Endocrinology* 95: 201–212.

Brown, M. A., Carne, A., Daugherty, C. H. and Chambers, G. K. 1995. Identification of a 130-kDa albumin in tuatara (*Sphenodon*) and detection of a novel albumin polymorphism. *Biochemical Genetics* 33: 189–204.

Brown, M. A., Carne, A. and Chambers, G. K. 1996. Identification and partial characterization of α_2-macroglobulin from the tuatara (*Sphenodon punctatus*). *Comparative Biochemistry and Physiology* 133B: 731–736.

Brown, M. A., Carne, A. and Chambers, G. K. 1997a. Purification, partial characterization and peptide sequences of vitellogenin from a reptile, the tuatara (*Sphenodon punctatus*). *Comparative Biochemistry and Physiology* 117B: 159–168.

Brown, M. A., Chambers, G. K. and Licht, P. 1997b. Purification and partial amino acid sequences of two distinct albumins from turtle plasma. *Comparative Biochemistry and Physiology* 118B: 367–374.

Browne, C. M. 2005. *The use of dogs to detect New Zealand reptile scents.* Unpublished MSc thesis, Massey University, Palmerston North. 153 p.

Buller, W. 1871. A list of the lizards inhabiting New Zealand, with descriptions. *Transactions and Proceedings of the New Zealand Institute* 3: 4–11.

Buller, W. L. 1877. Notes on the tuatara lizard (*Sphenodon punctatum*), with a description of a supposed new species. *Transactions and Proceedings of the New Zealand Institute* 9: 317–325.

Buller, W. L. 1878. Notice of a new variety of tuatara lizard (*Sphenodon*) from East Cape Island. *Transactions and Proceedings of the New Zealand Institute* 10: 220–221.

Buller, W. L. 1879. Further notes on the habits of the tuatara lizard. *Transactions and Proceedings of the New Zealand Institute* 11: 349–351.

Buller, W. L. 1888. *A history of the birds of New Zealand.* [Illustrated by J.G. Keulemans]. *Vol. 1 and 2.* 2nd edn. The author for the subscribers, London. 250 p., 359 p.

Buller, W. L. 1893. Further notes on the birds of New Zealand. *Transactions and Proceedings of the New Zealand Institute* 25: 63–88.

Buller, W. L. 1895. Illustrations of Darwinism; or, the avifauna of New Zealand considered in relation to the fundamental law of descent with modification. *Transactions and Proceedings of the New Zealand Institute* 27: 75–104.

Buller, W. L. 1896. Notes on New Zealand ornithology, with an exhibition of specimens. *Transactions and Proceedings of the New Zealand Institute* 28: 326–367.

Burgess, T. L., Gartrell, B. D. and Blanchard, B. 2009. A survey of the husbandry of captive tuatara (*Sphenodon* spp.) in relation to factors implicated in nutritional secondary hyperparathyroidism. *New Zealand Veterinary Journal* 57: 378–382.

Burnham, D. K., Keall, S. N., Nelson, N. J. and Daugherty, C. H. 2005. T cell function in tuatara (*Sphenodon punctatus*). *Comparative Immunology, Microbiology and Infectious Diseases* 28: 213–222.

Burnham, D. K., Keall, S. N., Nelson, N. J. and Daugherty, C. H.

2006. Effects of sampling date, gender, and tick burden on peripheral blood cells of captive and wild tuatara (*Sphenodon punctatus*). *New Zealand Journal of Zoology* 33: 241–248.

Burrows, C. J., McSaveney, M. J., Scarlett, R. J. and Turnbull, B. 1984. Late Holocene forest horizons and a *Dinornis* moa from an earthflow on North Dean, North Canterbury. *Records of the Canterbury Museum* 10: 1–8.

Burton, M. 1954. *Living fossils.* Thames and Hudson, London. 282 p.

Busch, C. H. 1898. Beitrag zur Kenntniss der Gaumenbildung bei den Reptilien. *Zoologische Jahrbücher, Abteilung für Anatomie und Ontogenie der Tiere* 11: 441–500 + Tafeln 34–40.

Byerly, T. C. 1925a. Note on the partial regeneration of the caudal region of *Sphenodon punctatum*. *Anatomical Record* 30: 61–66.

Byerly, T. C. 1925b. The myology of *Sphenodon punctatum*. *University of Iowa Studies in Natural History* 11: 3–51 + plates I–IV.

Cairney, J. 1926a. The pallial commissures in *Sphenodon punctatum*. *Anatomical Record* 32: 203.

Cairney, J. 1926b. A general survey of the forebrain of *Sphenodon punctatum*. *Journal of Comparative Neurology* 42: 255–348.

Carden, S. M. 2005. The pineal eye of the tuatara. *Survey of Ophthalmology* 50: 291–292.

Carmichael, C. K. and Gillingham, J. C. 2004. Thermoconformers or thermoregulators: is the tuatara (*Sphenodon punctatus*) truly a maladapted living fossil or a result of non-squamate lepidosaurian adaptation to cold climates? *Ohio Journal of Science* 104: A32–A33 (abstract).

Carmichael, C. K., Gillingham, J. C. and Keall, S. N. 1989. Feeding ecology of the tuatara (*Sphenodon punctatus*) on Stephens Island based on niche diversification. In: Abstracts of papers presented at the 2nd Annual Conference of the Society for Research on Amphibians and Reptiles in New Zealand, Middlemarch, Otago, 2–5 December 1988. *New Zealand Journal of Zoology* 16: 269.

Cartland, L. and Grimmond, N. M. 1991. Aspects of growth and changes in oxygen consumption during digestion in juvenile tuatara, *Sphenodon punctatus*. In: Abstracts of papers presented at the 3rd Annual Conference of the Society for Research on Amphibians and Reptiles in New Zealand, Whitianga, Coromandel Peninsula, 28–30 November 1990. *New Zealand Journal of Zoology* 18: 344.

Cartland, L. K. and Grimmond, N. M. 1994. The effect of temperature on the metabolism of juvenile tuatara, *Sphenodon punctatus*. *New Zealand Journal of Zoology* 21: 373–378.

Cartland, L. K., Cree, A., Grimmond, N. M., Sutherland, W. H. F. and Skeaff, C. M. 1993. Plasma cholesterol and triacylglycerol concentrations in wild and captive tuatara, *Sphenodon punctatus*. In: Abstracts of papers presented at the 4th Conference of the Society for Research on Amphibians and Reptiles in New Zealand, Invercargill, New Zealand, 27–29 November 1992. *New Zealand Journal of Zoology* 20: 127.

Cartland, L. K., Cree, A., Sutherland, W. H. F., Grimmond, N. M. and Skeaff, C. M. 1994. Plasma concentrations of total cholesterol and triacylglycerol in wild and captive juvenile tuatara (*Sphenodon punctatus*). *New Zealand Journal of Zoology* 21: 399–406.

Cartland-Shaw, L. K. 1996. *Lipids and fatty acids in the plasma and diets of wild and captive tuatara (Sphenodon punctatus).* Unpublished MSc thesis, University of Otago, Dunedin. 102 p.

Cartland-Shaw, L., Cree, A., Skeaff, C. M. and Grimmond, N. M. 1995. Fatty acid composition of plasma and selected dietary items of captive and wild juvenile tuatara (*Sphenodon punctatus*). *New Zealand Journal of Zoology* 22: 403.

Cartland-Shaw, L. K., Cree, A., Skeaff, C. M. and Grimmond, N. M. 1998. Differences in dietary and plasma fatty acids between wild and captive populations of a rare reptile (tuatara, *Sphenodon punctatus*). *Journal of Comparative Physiology B* 168: 569–580.

Cassey, P. 1997. *Estimating animal abundance: an assessment of distance sampling techniques for New Zealand populations.* Unpublished MSc thesis, University of Auckland, Auckland. 123 p.

Cassey, P. and Ussher, G. T. 1999. Estimating abundance of tuatara. *Biological Conservation* 88: 361–366.

Castanet, J. 1983. La squelettochronologie: une méthode de choix pour déterminer l'âge de *Sphenodon punctatus* (Reptilia, Lepidosauria, Rhynchocephalia). *Société Zoologique de France* 108: 678–679.

Castanet, J. 1994. Age estimation and longevity in reptiles. *Gerontology* 40: 174–192.

Castanet, J., Newman, D. G. and Saint Girons, H. 1988. Skeletochronological data on the growth, age, and population structure of the tuatara, *Sphenodon punctatus*, on Stephens and Lady Alice Islands, New Zealand. *Herpetologica* 44: 25–37.

Cawston, F. G. 1946. The fused teeth of *Sphenodon*. *South African Dental Journal* 20: 344.

Chabaud, A. G. and Dollfus, R. P. 1966. *Hatterianema hollandei* n. g., n. sp., nématode hétérakide parasite de Rhynchocéphale. *Bulletin du Muséum National D'Histoire Naturelle* 37: 1041–1045.

Chambers, G. K. and MacAvoy, E. S. 2000. Microsatellites: consensus and controversy. *Comparative Biochemistry and Physiology B* 126: 455–476.

Chorlton, R. 1977. Tuataras are living fossils. *Forest and Bird* No. 204: 40–43.

Christensen, K. 1927. The morphology of the brain of *Sphenodon*. *University of Iowa Studies in Natural History* 12: 1–29.

Christmas, E., Coddington, E. and Cree, A. 1996. *A database for toe-clipped tuatara (*Sphenodon punctatus*) on Stephens Island (Takapourewa).* University of Otago Wildlife Management Report No. 77. Dunedin. 126 p.

Churton, H. 1845. *Letters from Wanganui, New Zealand.* E. Churton, London. 42 p.

Clark, G. R., Petchey, P., McGlone, M. S. and Bristow, P. 1996. Faunal and floral remains from Earnscleugh Cave, Central Otago, New Zealand. *Journal of the Royal Society of New Zealand* 26: 363–380.

Clemance, M. 1996. Bleeding tuatara and other small creatures. *Kokako* 3(1): 3.

Cockrem, J. F., Firth, B. T., Cree, A. and Eynon, G. R. 1988. Plasma melatonin in the tuatara (*Sphenodon punctatus*): diurnal rhythm and response to light and heat. *Proceedings of the Endocrine Society of Australia*: S10 (abstract).

Colenso, W. 1844. An account of some enormous fossil bones of an unknown species of the Class Aves, lately discovered in New Zealand. *Annals and Magazine of Natural History* 14: 81–96.

Colenso, W. 1846. An account of some enormous fossil bones, of an unknown species of the Class Aves, lately discovered in New Zealand. *Tasmanian Journal of Natural Science* 2: 81–107.

Colenso, W. 1879. Contributions towards a better knowledge of the Maori race. *Transactions and Proceedings of the New Zealand Institute* 11: 77–106.

Colenso, W. 1880. Notes and observations on the animal economy and habits of one of our New Zealand lizards, supposed to be a new species of *Naultinus*. *Transactions and Proceedings of the New Zealand Institute* 12: 251–264.

Colenso, W. 1886. Notes on the bones of a species of *Sphenodon*, (*S. diversum*, Col.,) apparently distinct from the species already known. *Transactions and Proceedings of the New Zealand Institute* 18: 118–123.

Cooper Jr, W. E., Ferguson, G. W. and Habegger, J. J. 2001. Responses to animal and plant chemicals by several iguanian insectivores and the tuatara, *Sphenodon punctatus*. *Journal of Herpetology* 35: 255–263.

Cope, E. D. 1870. On the homologies of some of the cranial bones of the Reptilia and on the systematic arrangement of the class. *Proceedings of the American Association for the Advancement of Science* 19: 194–247.

Corkery, I. 2012. *Interspecific interactions: a case study using the tuatara–fairy prion association*. Unpublished PhD thesis. Victoria University of Wellington, Wellington, 159 p.

Corkery, I., Nelson, N. J. and Bell, B. D. 2010. Does sharing a burrow with a seabird increase a tuatara's fitness? In: Abstracts of papers presented at the Second Meeting of Australasian Societies for Herpetology (Society for Research on Amphibians and Reptiles in New Zealand and the Australia Society for Herpetologists), Massey University, Auckland, New Zealand, 20–22 February 2009. *New Zealand Journal of Zoology* 37: 69.

Cowan, J. 1908. *New Zealand, or Ao-teä-roa (the Long Bright World): its wealth and resources, scenery, travel-routes, spas, and sport*. Government Printer, Wellington. 279 p.

Craig, J. L. 1986. The effects of kiore on other fauna. *New Zealand Department of Lands and Survey Information Series* 16: 75–83.

Craigie, E. H. 1941. Vascularization in the brains of Reptiles. II. The cerebral capillary bed in *Sphenodon punctatum*. *Journal of Morphology* 69: 263–277.

Cree, A. 1994. Low annual reproductive output in female reptiles from New Zealand. *New Zealand Journal of Zoology* 21: 351–372.

Cree, A. 2002. Tuatara. In: Halliday, T. and Adler, K. (eds). *The new encyclopedia of reptiles and amphibians*. Oxford University Press, Oxford, pp. 210–211.

Cree, A. 2005. Discovered by science: an analysis of the first 200 years of research on tuatara (*Sphenodon*). In: Abstracts of papers presented at the 11th Biennial Conference of the Society for Research on Amphibians and Reptiles in New Zealand, Springbrook National Park, south-east Queensland, Australia, 7–11 February 2005. *New Zealand Journal of Zoology* 32: 221.

Cree, A. and Butler, D. 1993. *Tuatara recovery plan (*Sphenodon *spp.)*. Threatened Species Recovery Plan No. 9. Department of Conservation, Wellington. 71 p.

Cree, A. and Daugherty, C. 1990a. Tuatara sheds its fossil image. *New Scientist* No. 1739: 22–26.

Cree, A. and Daugherty, C. H. 1990b. Captive breeding of the New Zealand tuatara: past results and future directions. In: Dresser, B. L., Reece, R. W. and Maruska, E. J. (eds). *Proceedings of the Fifth World Conference on Breeding Endangered Species in Captivity*. Cincinnati Zoo and Botanical Garden, Cincinnati, Ohio, pp. 477–491.

Cree, A. and Guillette Jr, L. J. 1988. Effects of arginine vasotocin and prostaglandin $F_{2\alpha}$ on oviposition or parturition in two New Zealand reptiles: the tuatara and the common gecko. *Proceedings of the Endocrine Society of Australia*: S11 (abstract).

Cree, A. and Thompson, M. B. 1988. Unravelling the mysteries of tuatara reproduction. *Forest and Bird* No. 250: 14–16.

Cree, A. and Tyrrell, C. L. 2001. Patterns of corticosterone secretion in tuatara (*Sphenodon*): comparisons with other reptiles, and applications in conservation management. In: Goos, H. J. T., Rastogi, R. K., Vaudry, H. and Pierantoni, R. (eds). *Perspective in Comparative Endocrinology: Unity and Diversity (14th International Congress of Comparative Endocrinology, Sorrento, Italy, May 26–30, 2001)*. Monduzzi Editore, Bologna, pp. 433–441.

Cree, A., Thompson, M. B., Guillette Jr, L. J., Hay, J. M. and McIntyre, M. E. 1989. Embryonic development of tuatara in forested and open habitats on Stephens Island, New Zealand. In: Abstracts of papers presented at the 2nd Annual Conference of the Society for Research on Amphibians and Reptiles in New Zealand, Middlemarch, Otago, 2–5 December 1988. *New Zealand Journal of Zoology* 16: 270.

Cree, A., Guillette Jr, L. J., Cockrem, J. F., Brown, M. A. and Chambers, G. K. 1990a. Absence of daily cycles in plasma sex steroids in male and female tuatara (*Sphenodon punctatus*), and the effects of acute capture stress on females. *General and Comparative Endocrinology* 79: 103–113.

Cree, A., Guillette Jr, L. J., Cockrem, J. F. and Joss, J. M. P. 1990b. Effects of capture and temperature stresses on plasma steroid concentrations in male tuatara (*Sphenodon punctatus*). *Journal of Experimental Zoology* 253: 38–46.

Cree, A., Cockrem, J. F., Brown, M. A., Watson, P. R., Guillette Jr, L. J., Newman, D. G. and Chambers, G. K. 1991a. Laparoscopy, radiography, and blood analyses as techniques for identifying the reproductive condition of female tuatara. *Herpetologica* 47: 238–249.

Cree, A., Daugherty, C. H., Schafer, S. F. and Brown, D. 1991b. Nesting and clutch size of tuatara (*Sphenodon guntheri*) on North Brother Island, Cook Strait. *Tuatara* 31: 9–16.

Cree, A., Guillette Jr, L. J., Brown, M. A., Chambers, G. K., Cockrem, J. F. and Newton, J. D. 1991c. Slow estradiol-induced vitellogenesis in the tuatara, *Sphenodon punctatus*. *Physiological Zoology* 64: 1234–1251.

Cree, A., Guillette Jr, L. J. and Cockrem, J. F. 1991d. Identification of female tuatara in ovulatory condition using plasma sex steroid concentrations. *New Zealand Journal of Zoology* 18: 421–426.

Cree, A., Hay, J. M. and Daugherty, C. H. 1991e. Rescue operations for tuatara on Stanley, Red Mercury, and Cuvier Islands. In: Abstracts of papers presented at the 3rd Annual Conference of the Society for Research on Amphibians and Reptiles in New Zealand, Whitianga, Coromandel Peninsula, 28–30 November 1990. *New Zealand Journal of Zoology* 18: 344.

Cree, A., Hay, J. M., Daugherty, C. and Keall, S. N. 1991f. Effects of constant and variable incubation temperatures on growth of tuatara (*Sphenodon guntheri*) embryos. In Sinclair, J. D. (ed.). *Proceedings of the Physiological Society of New Zealand. Vol. 11*. Dunedin, p. 59 (abstract).

Cree, A., Cockrem, J. F. and Guillette Jr, L. J. 1992. Reproductive cycles of male and female tuatara (*Sphenodon punctatus*) on Stephens Island, New Zealand. *Journal of Zoology, London* 226: 199–217.

Cree, A., Daugherty, C. H., Towns, D. R. and Blanchard, B. 1994. The contribution of captive management to the conservation of tuatara (*Sphenodon*) in New Zealand. In: Murphy, J. B., Adler, K. and Collins, J. T. (eds). *Captive management and conservation of amphibians and reptiles*. Society for the Study of Amphibians and Reptiles, Ithaca, New York, pp. 377–385.

Cree, A., Daugherty, C. H. and Hay, J. M. 1995a. Reproduction of a rare New Zealand reptile, the tuatara *Sphenodon punctatus*, on rat-free and rat-inhabited islands. *Conservation Biology* 9: 373–383.

Cree, A., Thompson, M. B. and Daugherty, C. H. 1995b. Tuatara sex determination. *Nature* 375: 543.

Cree, A., Fraser, J. R., Cartland-Shaw, L. and Lyon, G. L. 1995c. Contribution of seabirds to the diet of wild tuatara (*Sphenodon punctatus*). In: Abstracts of papers presented at the 6th Conference of the Society for Research on Amphibians and Reptiles in New Zealand, Manaia, Whangarei, New Zealand, 10–12 February 1995. *New Zealand Journal of Zoology* 22: 403.

Cree, A., Guillette Jr, L. J. and Reader, K. 1996. Eggshell formation during prolonged gravidity of the tuatara *Sphenodon punctatus*. *Journal of Morphology* 230: 129–144.

Cree, A., Lyon, G. L., Cartland-Shaw, L. and Tyrrell, C. 1999. Stable carbon isotope ratios as indicators of marine versus terrestrial inputs to the diets of wild and captive tuatara (*Sphenodon punctatus*). *New Zealand Journal of Zoology* 26: 243–253.

Crook, I. G. 1970. Tuatara. *Wildlife: A Review* 2: 16–19.

Crook, I. G. 1973a. The tuatara, *Sphenodon punctatus* Gray, on islands with and without populations of the Polynesian rat, *Rattus exulans* (Peale). *Proceedings of the New Zealand Ecological Society* 20: 115–120.

Crook, I. G. 1973b. Tuatara and the Polynesian rat. *Wildlife: A Review* 4: 47–48.

Crook, I. G. 1974. Are tuataras dependent on petrels? *Wildlife: A Review* 5: 43–46.

Crook, I. G. 1975. The tuatara. In: Kuschel, G. (ed.). *Biogeography and ecology in New Zealand*. Junk, The Hague, pp. 331–352.

Crook, I. G. 1976. Archaic elements and a host of immigrants. *Australian Natural History* 18: 362–365.

Curtis, N., Jones, M. E. H., Evans, S. E., O'Higgins, P. and Fagan, M. J. 2009. Visualising muscle anatomy using three-dimensional computer models – an example using the head and neck muscles of *Sphenodon*. *Palaeontologia Electronica* 12.3.7T: 18 p.

Curtis, N., Jones, M. E. H., Evans, S. E., Shi, J., O'Higgins, P. and Fagan, M. J. 2010a. Predicting muscle activation patterns from motion and anatomy: modelling the skull of *Sphenodon* (Diapsida: Rhynchocephalia). *Journal of The Royal Society Interface* 7: 153–161.

Curtis, N., Jones, M. E. H., Lappin, A. K., O'Higgins, P. O., Evans, S. E. and Fagan, M. J. 2010b. Comparison between *in vivo* and theoretical bite performance: using multi-body modelling to predict muscle and bite forces in a reptile skull. *Journal of Biomechanics* 43: 2804–2809.

Curtis, N., Jones, M. E. H., Evans, S. E., O'Higgins, P. O. and Fagan, M. J. 2010c. Feedback control from the jaw joints during biting: an investigation of the reptile *Sphenodon* using multibody modelling. *Journal of Biomechanics* 43: 3132–3137.

Curtis, N., Jones, M. E. H., Shi, J., O'Higgins, P., Evans, S. E. and Fagan, M. J. 2011. Functional relationships between skull form and feeding mechanics in *Sphenodon*, and implications for diapsid skull development. *PLOS ONE* 6 (e29804): 1–11.

Curtis, N., Jones, M. E. H., Evans, S. E., O'Higgins, P. and Fagan, M. J. 2013. Cranial sutures work collectively to distribute strain throughout the reptile skull. *Journal of the Royal Society Interface* 10 (20130442): 1–6 [Correction at 20130584: 1].

Daiber, M. 1920. Das Bauchrippensystem von *Sphenodon (Hatteria) punctatus* Gray. *Anatomischer Anzeiger* 53: 371–382 + Tafeln 1–2.

Daugherty, C., Cree, A. and Schneider, M. 1990a. Tuatara: a survivor from the dinosaur age. *New Zealand Geographic* 6: 66–86.

Daugherty, C. H., Cree, A., Hay, J. M. and Thompson, M. B. 1990b. Neglected taxonomy and continuing extinctions of tuatara (*Sphenodon*). *Nature* 347: 177–179.

Daugherty, C. H., Patterson, G. B. and Hay, J. M. 1991. Techniques for identification of new species of skinks and tuatara. *New Zealand Journal of Zoology* 18: 344.

Daugherty, C. H., Towns, D. R., Cree, A. and Hay, J. M. 1992. The roles of legal protection versus intervention in conserving the New Zealand tuatara, *Sphenodon*. In: Willison, J. H. M., Bondrup-Nielsen, S., Drysdale, C., Herman, T. B., Munro, N. W. P. and Pollock, T. L. (eds). *Developments in landscape management and urban planning, 7. Science and the management of protected areas*. Elsevier, Amsterdam, pp. 247–259.

Daugherty, C. H., Patterson, G. B. and Hitchmough, R. A. 1994. Taxonomic and conservation review of the New Zealand herpetofauna. *New Zealand Journal of Zoology* 21: 317–323.

Daugherty, C. H., Gaze, P., Keall, S. N. and Nelson, N. J. 1999. Conservation of tuatara in the Cook Strait region. In: Abstracts of papers presented at the 8th Conference of the Society for Research on Amphibians and Reptiles in New Zealand, Great Barrier Island, New Zealand, 5–7 February 1999. *New Zealand Journal of Zoology* 26: 256–257.

Daugherty, C. H., Keall, S. N., Nelson, N. J., Hay, J. M., Petrove, I., Smuts-Kennedy, J. C. and Whitaker, T. H. 2001. Conservation of tuatara on Hauturu (Little Barrier Island). In: Abstracts of papers presented at the 9th Conference of the Society for Research on Amphibians and Reptiles in New Zealand, St Arnaud, Nelson Lakes, New Zealand, 2–4 February 2001. *New Zealand Journal of Zoology* 28: 362–363.

Dawbin, W. H. 1949. The tuatara. *Tuatara* 2: 91–96.

Dawbin, W. H. I. 1953. Fauna of Stephens Island. *Forest and Bird* No. 108: 8–9.

Dawbin, W. H. 1962a. The tuatara in its natural habitat. *Endeavour* 21: 16–24.

Dawbin, W. H. 1962b. The beginnings and early development of Tuatara. *Tuatara* 10: 2–4.

Dawbin, W. H. 1974. Rhynchocephalia. *New Encyclopedia Britannica* 15: 824.

Dawbin, W. H. 1982a. The tuatara *Sphenodon punctatus*: aspects of life history, growth and longevity. In: Newman, D. G. (ed.). *New Zealand Herpetology*. New Zealand Wildlife Service Occasional Publication No. 2, Wellington, pp. 237–250.

Dawbin, W. H. 1982b. The tuatara *Sphenodon punctatus* (Reptilia: Rhynchocephalia): a review. In: Newman, D. G. (ed.). *New Zealand Herpetology*. New Zealand Wildlife Service Occasional Publication No. 2, Wellington, pp. 149–181.

Dawbin, W. H. and Hill, L. 1969. Uric acid and urea excretion in the tuatara *Sphenodon punctatus*. *Nature* 224: 1325–1326.

De Beer, G. R. 1949. Caruncles and egg-teeth: some aspects of the concept of homology. *Proceedings of the Linnean Society of London* 161: 218–224 + plates 13–15.

Dendy, A. 1898. Summary of the principal results obtained in a study of the development of the tuatara (*Sphenodon punctatum*). *Proceedings of the Royal Society of London* 63: 440–443.

Dendy, A. 1899a. The hatching of tuatara eggs. *Nature* 59: 340.

Dendy, A. 1899b. Outlines of the development of the tuatara, *Sphenodon* (*Hatteria*) *punctatus*. *Quarterly Journal of Microscopical Science* 42: 1–87 + plates 1–10.

Dendy, A. 1899c. On the development of the parietal eye and adjacent organs in *Sphenodon* (*Hatteria*). *Quarterly Journal of Microscopical Science* 42: 111–153 + plates 11–13.

Dendy, A. 1899d. The life-history of the tuatara (*Sphenodon punctatum*). *Transactions and Proceedings of the New Zealand Institute* 31: 249–255.

Dendy, A. 1909. The intracranial vascular system of *Sphenodon*. *Philosophical Transactions of the Royal Society of London B* 200: 403–426 + plate 31.

Dendy, A. 1910. On the structure, development, and morphological interpretation of the pineal organs and adjacent parts of the brain in the tuatara (*Sphenodon punctatus*). *Anatomischer Anzeiger* 37: 453–462.

Dendy, A. 1911. On the structure, development and morphological interpretation of the pineal organs and adjacent parts of the brain in the tuatara (*Sphenodon punctatus*). *Philosophical Transactions of the Royal Society of London B* 201: 227–331 + plates 19–31.

Desser, S. S. 1978. Morphological, cytochemical, and biochemical observations on the blood of the tuatara, *Sphenodon punctatus*. *New Zealand Journal of Zoology* 5: 503–508.

Desser, S. S. 1979. Haematological observations on a hibernating tuatara, *Sphenodon punctatus*. *New Zealand Journal of Zoology* 6: 77–78.

Desser, S. S. and Weller, I. 1979a. Ultrastructural observations on the erythrocytes and thrombocytes of the tuatara, *Sphenodon punctatus* (Gray). *Tissue and Cell* 11: 717–726.

Desser, S. S. and Weller, I. 1979b. Ultrastructural observations on the granular leucocytes of the tuatara, *Sphenodon punctatus* (Gray). *Tissue and Cell* 11: 703–715.

Di-Poï, N., Montoya-Burgos, J. I., Miller, H., Pourquié, O., Milinkovitch, M. C. and Duboule, D. 2010. Changes in Hox genes' structure and function during the evolution of the squamate body plan. *Nature* 464: 99–103.

Dieffenbach, E. 1843. *Travels in New Zealand; with contributions to the geography, geology, botany, and natural history of that country. Vol. I and II.* John Murray, London. 431 p., 396 p.

Djorovic, A., Keall, S. N., Mitchell, N. J. and Daugherty, C. H. 2001. Morphometric variability and fluctuating asymmetry: tools for assessing population variability in tuatara. In: Abstracts of papers presented at the 9th Conference of the Society for Research on Amphibians and Reptiles in New Zealand, St Arnaud, Nelson Lakes, New Zealand, 2–4 February 2001. *New Zealand Journal of Zoology* 28: 363–364.

Donne, T. E. 1942. On friendly terms with the tuatara. *Natural History* 50: 100–102.

Dowling, H. 1961. Tuatara (*Sphenodon punctatus*) at New York Zoo. *International Zoo Yearbook* 3: 83–84.

Downes, T. W. 1937. Maori mentality regarding the lizard and *taniwha* in the Whanganui river area. *Journal of the Polynesian Society* 46: 206–224.

Drummond, J. 1905. A very conservative party. A New Zealander who would not progress. Must end his days soon. *Lyttelton Times*, Lyttelton, 9 December, p. 10.

Drummond, J. 1911. A queer catch: or, an aristocrat in a democratic land. *The Boy's Own Paper* 33 (1910–1911): 9.

Drummond, J. 1917. Nature notes. *Auckland Weekly News*, Auckland, 30 August, p. 15.

Drummond, J. and Hutton, F. W. 1902. *Nature in New Zealand.* Whitcombe and Tombs Ltd., Christchurch. 188 p.

Dujsebayeva, T. N., Ananjeva, N. B. and Iohanssen, L. K. 2004. Reduced state of skin sense organs in *Sphenodon punctatus* (Rhynchocephalia: Sphenodontidae) and its phylogenetical value. *Russian Journal of Herpetology* 11: 106–110.

Dumbleton, L. J. 1943. A new tick from the tuatara (*Sphenodon punctatus*). *New Zealand Journal of Science and Technology* 24: 185B–190B.

Durward, A. 1930. The cell masses in the forebrain of *Sphenodon punctatum*. *Journal of Anatomy* 65: 8–44.

East, K. T., East, M. R. and Daugherty, C. H. 1995. Ecological restoration and habitat relationships of reptiles on Stephens Island, New Zealand. *New Zealand Journal of Zoology* 22: 249–261.

Edinger, T. 1930. *Homoeosaurus* und *Sphenodon*. *Bericht Senckenbergische Naturforschende Gesellschaft (Natur und Museum)* 60: 20.

Engelbert, V. E. and Young, A. D. 1970. Erythropoiesis in peripheral blood of tuatara (*Sphenodon punctatus*) and turtle (*Malaclemys terrapin*). *Canadian Journal of Zoology* 48: 209–212.

Evans, S. E. 2003. At the feet of the dinosaurs: the early history and radiation of lizards. *Biological Review* 78: 513–551.

Evans, S. E. 2008. The skull of lizards and tuatara. In: Gans, C., Gaunt, A. S. and Adler, K. (eds). *Biology of the Reptilia. Vol. 20. Morphology H. The skull of Lepidosauria*. Society for the Study of Amphibians and Reptiles, Ithaca, NY. Contributions to Herpetology Vol. 23, pp. 1–347.

Falla, R. A. 1935. The tuatara (*Sphenodon punctatus*). *Bulletin of the Auckland Zoological Society* No. 2. 3–5 + 4 plates.

Falla, R. A. 1937. A 'living fossil' of New Zealand: the crested, pineal-eyed tuatara. *Illustrated London News*, London, pp. 340–341.

Falla, R. A. 1960. The fauna of New Zealand. In: McLintock, A. H. (ed.). *A descriptive atlas of New Zealand.* R.E. Owen, Government Printer, Wellington, pp. 25–27.

Farlow, J. D. 1975. Observations on a captive tuatara (*Sphenodon punctatum*). *Journal of Herpetology* 9: 353–355.

Farrell, A. P., Gamperl, A. K. and Francis, E. T. B. 1998. Comparative aspects of heart morphology. In: Gans, C. and Gaunt, A. S. (eds). *Biology of the Reptilia. Vol. 19. Morphology G. Visceral organs.* Society for the Study of Amphibians and Reptiles. Contributions to Herpetology Vol. 14, Ithaca, New York, pp. 375–424.

Fawcett, J. D. and Smith, H. M. 1970. An overlooked synonym of *Sphenodon punctatus*, the New Zealand tuatara. *Journal of Herpetology* 4: 89–91.

Finch, M. O. and Lambert, D. M. 1996. Kinship and genetic divergence among populations of tuatara *Sphenodon punctatus* as revealed by minisatellite DNA profiling. *Molecular Ecology* 5: 651–658.

Firth, B. T., Cockrem, J. F., Cree, A. and Eynon, G. R. 1988. Plasma melatonin in the tuatara (*Sphenodon punctatus*): diurnal rhythm and response to light and heat. *Chinese Journal of Physiological Science* 4: 237 (abstract).

Firth, B. T., Thompson, M. B., Kennaway, D. J. and Belan, I. 1989. Thermal sensitivity of reptilian melatonin rhythms: 'cold' tuatara vs. 'warm' skink. *American Journal of Physiology* 256: R1160–R1163.

Fisk, A. 1949. Modification of the ectoderm in *Sphenodon*. *Nature* 164: 617–618.

Fisk, A. and Tribe, M. 1949. The development of the amnion and chorion of reptiles. *Proceedings of the Zoological Society of London* 119: 83–114 + plates I–V.

Fisk, A. and Tribe, M. 1951. The glomerular border in the development of the vertebrate kidney. *Nature* 167: 266–268.

Flachsbarth, B., Fritzsche, M., Weldon, P. J. and Schulz, S. 2009. Composition of the cloacal gland secretion of tuatara, *Sphenodon punctatus*. *Chemistry and Biodiversity* 6: 1–37.

Flower, S. S. 1937. Further notes on the duration of life in animals. III. Reptiles. *Proceedings of the Zoological Society Series A* 1: 1–37.

Fraser, J. and Cree, A. 1993. Diets of wild adult and juvenile tuatara (*Sphenodon punctatus*) on Stephens Island. In: Abstracts of papers presented at the 4th Conference of the Society for Research on Amphibians and Reptiles in New Zealand, Invercargill, New Zealand, 27–29 November 1992. *New Zealand Journal of Zoology* 20: 128.

Fraser, J. R. 1993. *Diets of wild tuatara (*Sphenodon punctatus*) on Stephens Island.* Unpublished MSc thesis, University of Otago, Dunedin. 71 p.

Freeman, A. B. and Freeman, A. N. D. 1995. Rediscovery of an original type specimen of *Sphenodon guntheri* Buller in the

Canterbury Museum, New Zealand. *New Zealand Journal of Zoology* 22: 357–359.

Fritsch, G., Jaroffke, D. and Hildebrandt, T. B. 2002. Ultrasonography of a living fossil: the tuatara (*Sphenodon punctatus*). Contributions to the 4th International Symposium on Physiology and Behaviour of Wild and Zoo Animals. *Advances in Ethology* 37: 133 (abstract).

Gabe, M. and Saint Girons, H. 1964a. *Contribution a l'histologie de Sphenodon punctatus Gray*. Éditions du Centre National de la Recherche Scientifique, Paris. 149 p.

Gabe, M. and Saint-Girons, H. 1964b. Histologie. Particularités histologiques de la glande surrénale chez *Sphenodon punctatus* Gray (Reptile Rhynchocéphale). *Comptes Rendus de l'Academie des Sciences Paris* 258: 3559–3562.

Gabe, M. and Saint-Girons, H. 1964c. Histophysiologie. Le troisième type de contact hypothalamo-hypophysaire proximal: l'éminence médiane de *Sphenodon punctatus*. *Comptes Rendus de l'Academie des Sciences Paris* 259: 2136–2139.

Gabe, M. and Saint Girons, H. 1965a. Contribution a la morphologie comparée du cloaque et des glandes épidermoïdes de la région cloacale chez les lépidosauriens. *Mémoires du Muséum National d'Histoire Naturelle, Paris* Série A Zoologie 33: 149–292 + 15 plates.

Gabe, M. and Saint Girons, H. 1965b. Histologie de *Sphenodon punctatus*. *Copeia* 1965: 393–394.

Gabe, M. and Saint Girons, H. 1967. Données histologiques sur le tégument et les glandes épidermoïdes céphaliques des Lépidosauriens. *Acta Anatomica* 67: 571–594.

Gabe, M. and Saint Girons, H. 1969. Données histologiques sur les glandes salivaires des lépidosauriens. *Mémoires du Muséum National d'Histoire Naturelle, Series A* 58: 1–112.

Gabe, M. and Saint Girons, H. 1972. Contribution à l'histologie de l'estomac des *Lépidosauriens* (Reptiles): a contribution to the histological study of the stomach in Lepidosauria (Reptilia). *Zoologische Jahrbücher Abteilung für Anatomie und Ontogenie der Tiere* 89: 579–599.

Gabe, M. and Saint Girons, H. 1973. Contribution à l'histologie des glandes nasales externes de *Sphenodon punctatus* Gray (*Reptilia, Rhynchocephalia*). *Acta Anatomica* 84: 452–464.

Gabe, M. and Saint Girons, H. 1976. Contribution a la morphologie comparée des fosses nasales et de leurs annexes chez les lépidosoriens. *Mémoires du Muséum National D'Histoire Naturelle Série A* 98: 1–87.

Gadow, H. 1887. II. Remarks on the cloaca and on the copulatory organs of the Amniota. *Philosophical Transactions of the Royal Society of London Series B* 178: 5–37 + plates 2–5.

Gans, C. 1965. 'Histologie de *Sphenodon punctatus*' by M. Gabe and H. Saint Girons. *Copeia* 1965: 393–394.

Gans, C. 1984. Apologia. *Journal of Herpetology* 18: 499.

Gans, C. 1991. Survival test. *Nature* 349: 112.

Gans, C. 1983. Is *Sphenodon punctatus* a maladapted relict? In: Rhodin, A. G. J. and Miyata, K. (eds). *Advances in herpetology and evolutionary biology*. Museum of Comparative Zoology, Cambridge, Massachusetts, pp. 613–620.

Gans, C. and Clark, B. D. 1978. Air flow in reptilian ventilation. *Comparative Biochemistry and Physiology* 60A: 453–457.

Gans, C. and Wever, E. G. 1976. Ear and hearing in *Sphenodon punctatus*. *Proceedings of the National Academy of Sciences USA* 73: 4244–4246

Gans, C., Gillingham, J. C. and Clark, D. L. 1984. Courtship, mating and male combat in tuatara, *Sphenodon punctatus*. *Journal of Herpetology* 18: 194–197.

Gartrell, B. D., Jillings, E., Adlington, B. A., Mack, H. and Nelson, N. J. 2006. Health screening for a translocation of captive-reared tuatara (*Sphenodon punctatus*) to an island refuge. *New Zealand Veterinary Journal* 54: 344–349.

Gartrell, B. D., Youl, J. M., King, C. M., Bolotovski, I., McDonald, W. L. and Nelson, N. J. 2007. Failure to detect *Salmonella* species in a population of wild tuatara (*Sphenodon punctatus*). *New Zealand Veterinary Journal* 55: 134–136.

Gaston, A. J. and Scofield, P. 1995. Birds and tuatara on North Brother Island, Cook Strait, New Zealand. *Notornis* 42: 27–41.

Gauthier, J. A., Kearney, M., Maisano, J. A., Rieppel, O. and Behlke, A. D. B. 2012. Assembling the squamate tree of life: perspectives from the phenotype and the fossil record. *Bulletin of the Peabody Museum of Natural History* 53: 3–308.

Gaze, P. 2001. *Tuatara recovery plan 2001–2011*. Threatened Species Recovery Plan No. 47. Department of Conservation, Wellington. 37 p.

Gill, B. 2010a. New Zealand museum collections as a resource for herpetology. In: Abstracts of papers presented at the Second Meeting of Australasian Societies for Herpetology (Society for Research on Amphibians and Reptiles in New Zealand and the Australia Society for Herpetologists), Massey University, Auckland, New Zealand, 20–22 February 2009. *New Zealand Journal of Zoology* 37: 75–76.

Gill, B. J. 2010b. The Cheeseman–Giglioli correspondence, and museum exchanges between Auckland and Florence, 1877–1904. *Archives of Natural History* 37: 131–149.

Gill, B. J. and Coory, R. 1999. Herpetological collections in New Zealand museums. *Herpetological Review* 30: 133–134.

Gillingham, J. C. and Miller, T. J. 1991. Reproductive ethology of the tuatara *Sphenodon punctatus*: applications in captive breeding. *International Zoo Yearbook* 30: 157–164.

Gillingham, J. C., Carmichael, C. and Miller, T. 1995. Social behavior of the tuatara, *Sphenodon punctatus*. *Herpetological Monographs* 9: 5–16.

Gisi, J. 1908. Das Gehirn von *Hatteria punctata*. *Zoologische Jahrbücher* 25: 71–236.

Godfrey, S. S., Bull, C. M. and Nelson, N. J. 2008. Seasonal and spatial dynamics of ectoparasite infestation of a threatened reptile, the tuatara (*Sphenodon punctatus*). *Medical and Veterinary Entomology* 22: 374–385.

Godfrey, S. S., Moore, J. A., Nelson, N. J. and Bull, C. M. 2010a. Social network structure and parasite infection patterns in a territorial reptile, the tuatara (*Sphenodon punctatus*). *International Journal for Parasitology* 40: 1575–1585.

Godfrey, S. S., Moore, J. A., Nelson, N. J. and Bull, C. M. 2010b. Unravelling causality from correlations: revealing the impacts of endemic ectoparasites on a protected species (tuatara). *Parasitology* 137: 275–286.

Godfrey, S. S., Nelson, N. J. and Bull, C. M. 2011. Ecology and dynamics of the blood parasite, *Hepatozoon tuatarae* (Apicomplexa), in tuatara (*Sphenodon punctatus*) on Stephens Island, New Zealand. *Journal of Wildlife Diseases* 47: 126–139.

Goellner, R. 1984. The tuatarium, an off-exhibit, climate-controlled chamber at the St. Louis Zoological Park. In: Bels, V. L. and Van den Sande, A. P. (eds). *Maintenance and reproduction of reptiles in captivity*. Acta Zoologica et Pathologica Antverpiensia No. 78, pp. 319–324.

Goellner, R. 1985. The little dragon of New Zealand. *Animal Kingdom* 88: 38–43.

Goetz, B. G. R. and Thomas, B. W. 1994. Use of annual growth and activity patterns to assess management procedures for captive tuatara (*Sphenodon punctatus*). *New Zealand Journal of Zoology* 21: 473–485.

Goetz, B. G. R. and Thomas, B. W. 1997. The effect of introduced stress on growth rates in captive tuatara (*Sphenodon punctatus*). In: Abstracts of papers presented at the 7th Conference of the Society for Research on Amphibians and Reptiles in New Zealand, Kaikoura, New Zealand, 31 January–2 February 1997. *New Zealand Journal of Zoology* 24: 327.

Goff, M. L., Loomis, R. B. and Ainsworth, R. 1987. Redescription of *Neotrombicula naultini* (Dumbleton, 1947) and descriptions of two new species of chiggers from New Zealand (Acari: Trombiculidae). *New Zealand Journal of Zoology* 14: 385–390.

Gold-Smith, E. C. 1885. Description of Mayor Island. *Transactions and Proceedings of the New Zealand Institute* 17: 417–427.

Goodlet, W. [c. 1922]. A New Zealand reptile: notes on the tuatara. In: *Scrapbook of newspaper clippings relating to New Zealand, c. 1894–1948. Variae 34.* Hocken Library, Dunedin, p. 16.

Goold, M. 1997. Tuatara – Ill-thrift post oxytocin induced egg-laying. *Kokako* 4(3): 3–4.

Goold, M. and Smits, B. 1995a. Ill thrift and wasting in a tuatara. *Kokako* 2(2): 4–5.

Goold, M. and Smits, B. 1995b. Tuatara: some normal haematological and chemistry values. *Kokako* 2(2): 5.

Gorniak, G. C., Rosenberg, H. I. and Gans, C. 1982. Mastication in the tuatara, *Sphenodon punctatus* (Reptilia: Rhynchocephalia): structure and activity of the motor system. *Journal of Morphology* 171: 321–353.

Gorr, T. A., Mable, B. K. and Kleinschmidt, T. 1998. Phylogenetic analysis of reptilian hemoglobins: trees, rates, and divergences. *Journal of Molecular Evolution* 47: 471–485.

Grant-Mackie, J. A. and Scarlett, R. J. 1973. Last inter-glacial sequence, Oamaru. In: *IX INQUA Congress, guidebook for excursion 7.* Christchurch, New Zealand, pp. 87–99.

Gray, J. E. 1831. Note on a peculiar structure in the head of an agama. *Zoological Miscellany* 1: 13–14.

Gray, J. E. 1842. Descriptions of two hitherto unrecorded species of reptiles from New Zealand; presented to the British Museum by Dr. Dieffenbach. *Zoological Miscellany* 4: 72.

Gray, J. E. 1843. Fauna of New Zealand: materials towards a fauna of New Zealand, Auckland Island, and Chatham Islands. In: Dieffenbach, E. (ed.). *Travels in New Zealand; with contributions to the geography, geology, botany, and natural history of that country. Vol. II.* John Murray, London, pp. 177–296.

Gray, J. E. 1845a. Reptiles. In: Richardson, J. and Gray, J. E. (eds). *The zoology of the voyage of H.M.S.* Erebus *&* Terror, *under the command of Captain Sir James Clark Ross, R.N., F.R.S., during the years 1839 to 1843. Vol. II.* E. W. Janson, London, pp. 1–8, + plates 1–4, 8–9, 12–14 and 20.

Gray, J. E. 1845b. *Catalogue of the specimens of lizards in the collection of the British Museum.* Edward Newman, London. 289 p.

Gray, J. E. 1867. *The lizards of Australia and New Zealand in the collection of the British Museum.* Bernard Quaritch, London. 7 p. + 18 plates.

Gray, J. E. 1869. *Sphenodon, Hatteria*, and *Rhynchocephalus*. *Annals and Magazine of Natural History Series* 3: 167–168.

Green, B. 1989. Water and energy turnover of wild tuataras (*Sphenodon punctatus*). In: Halliday, T., Baker, J. and Hosie, L. (eds). *First World Congress of Herpetology 11–19 September 1989.* Open University, Canterbury, UK, p. R5 (abstract).

Greer, A. E. 1986. On the absence of visceral fat bodies within a major lineage of scincid lizards. *Journal of Herpetology* 20: 267–269.

Grigg, G. C. and Simons, J. R. 1972. Preferential distribution of left and right auricular blood into the arterial arches of the tuatara, *Sphenodon punctatus*. *Journal of Zoology, London* 167: 481–486.

Grimmond, N. M. and Cartland, L. 1991. The effect of temperature and body size on the metabolism of juvenile tuatara, *Sphenodon punctatus.* In: Abstracts of papers presented at the 3rd Annual Conference of the Society for Research on Amphibians and Reptiles in New Zealand, Whitianga, Coromandel Peninsula, 28–30 November 1990. *New Zealand Journal of Zoology* 18: 344–345.

Grodziński, Z. 1980. The yolk of the tuatara *Sphenodon punctatus* Gray (*Reptilia*). *Acta Biologica Cracoviensia* 22: 65–71 + plates 9–10.

Guillette Jr, L. J. and Cree, A. 1997. Morphological changes in the corpus luteum of tuatara (*Sphenodon punctatus*) during gravidity. *Journal of Morphology* 232: 79–91.

Guillette Jr, L. J., Cree, A. and Gross, T. S. 1990. Endocrinology of oviposition in the tuatara (*Sphenodon punctatus*): I. Plasma steroids and prostaglandins during natural nesting. *Biology of Reproduction* 43: 285–289.

Guillette Jr, L. J., Propper, C. R., Cree, A. and Dores, R. M. 1991a. Endocrinology of oviposition in the tuatara (*Sphenodon punctatus*): II. Plasma arginine vasotocin concentrations during natural nesting. *Comparative Biochemistry and Physiology* 100A: 819–822.

Guillette Jr, L. J., Dubois, D. H. and Cree, A. 1991b. Prostaglandins, oviducal function, and parturient behavior in nonmammalian vertebrates. *American Journal of Physiology* 260: R854–R861.

Gunther, A. E. 1975. *A century of zoology at the British Museum through the lives of two keepers 1815–1914.* Dawsons, London. 533 p.

Günther, A. 1867. Contribution to the anatomy of *Hatteria* (*Rhynchocephalus*, Owen). *Philosophical Transactions of the Royal Society of London* 157: 595–629 + plates XXVI–XXVIII.

Günther, A. 1875. A list of the saurians of Australia and New Zealand. In: Richardson, J. and Gray, J. E. (eds). *The zoology of the voyage of H.M.S.* Erebus *&* Terror, *under the command of Captain Sir James Clark Ross, R.N., F.R.S., during the years 1839 to 1843. Vol. II.* E. W. Janson, London, pp. p. 9–19, plus plates 5–7, 10, 11 and 15–19.

Haas, G. 1973. Muscles of the jaws and associated structures in the Rhynchocephalia and Squamata. In: Gans, C. and Parsons, T. S. (eds). *Biology of the Reptilia. Vol. 4. Morphology D.* Academic Press, New York, pp. 285–490.

Haines, R. W. 1939. The structure of the epiphyses in *Sphenodon* and the primitive form of secondary centre. *Journal of Anatomy* 74: 80–90.

Hall, M. I. 2008. Comparative analysis of the size and shape of the lizard eye. *Zoology* 111: 62–75.

Hamilton, A. 1898. On rock pictographs in South Canterbury. *Transactions of the New Zealand Institute* 30: 24–29 + plates I–X.

Hard, G. 1954. Notes on Reptilia of islands visited by Field Club 1953–1954. *Tane* 6: 143–146.

Hare, J. R., Whitworth, E. and Cree, A. 2007. Correct orientation of a hand-held infrared thermometer is important for accurate measurement of body temperatures in small lizards and tuatara. *Herpetological Review* 38: 311–315.

Harper, P. C. 1983. Biology of the Buller's Shearwater (*Puffinus bulleri*) at the Poor Knights Islands, New Zealand. *Notornis* 30: 299–318.

Harrison, H. S. 1901a. *Hatteria punctata*, its dentitions and its incubation period. *Anatomischer Anzeiger* 20: 145–158.

Harrison, H. S. 1901b. The development and succession of teeth in *Hatteria punctata*. *Quarterly Journal of Microscopical Science (New Series)* 44: 161–213 + plates 10–12.

Hart, N. H. 1969. The bulbus cordis and its septa in *Sphenodon punctatus*. *Journal of Morphology* 129: 369–374.

Hartley, J. 1965. Notes on tuataras *Sphenodon punctatus* in captivity. *International Zoo Yearbook* 5: 170–171.

Hatteria [pseudonym]. 1942. Out of the Mesozoic: the tuatara (*Sphenodon punctatus*). *Forest and Bird* No. 65: 3.

Hay, J. M. 1998. *A genetic perspective of evolution and biogeography in some New Zealand reptiles*. Unpublished PhD thesis. Pennsylvania State University, Pennsylvania. 180 p.

Hay, J. M. and Lambert, D. M. 2008. Microsatellite DNA loci identify individuals and provide no evidence for multiple paternity in wild tuatara (*Sphenodon*: Reptilia). *Conservation Genetics* 9: 1039–1043.

Hay, J. M., Daugherty, C. H. and Cree, A. 1991. Taxonomic tangles in tuatara (*Sphenodon* spp.). In: Abstracts of papers presented at the 3rd Annual Conference of the Society for Research on Amphibians and Reptiles in New Zealand, Whitianga, Coromandel Peninsula, 28–30 November 1990. *New Zealand Journal of Zoology* 18: 345.

Hay, J. M., Sarre, S., Daugherty, C. H., Cree, A. and Maxson, L. 2001. Molecular phylogeography of tuatara (*Sphenodon*). In: Abstracts of papers presented at the 9th Conference of the Society for Research on Amphibians and Reptiles in New Zealand, St Arnaud, Nelson Lakes, New Zealand, 2–4 February 2001. *New Zealand Journal of Zoology* 28: 365.

Hay, J. M., Daugherty, C. H., Cree, A. and Maxson, L. R. 2003. Low genetic divergence obscures phylogeny among populations of *Sphenodon*, remnant of an ancient reptile lineage. *Molecular Phylogenetics and Evolution* 29: 1–19.

Hay, J. M., Sarre, S. D. and Daugherty, C. H. 2004. Nuclear mitochondrial pseudogenes as molecular outgroups for phylogenetically isolated taxa: a case study in *Sphenodon*. *Heredity* 93: 468–475.

Hay, J. M., Subramanian, S., Millar, C. D., Mohandesan, E. and Lambert, D. M. 2008. Rapid molecular evolution in a living fossil. *Trends in Genetics* 24: 106–109.

Hay, J. M., Sarre, S. D., Lambert, D. M., Allendorf, F. W. and Daugherty, C. H. 2010. Genetic diversity and taxonomy: a reassessment of species designation in tuatara (*Sphenodon*: Reptilia). *Conservation Genetics* 11: 1063–1081.

Hazley, L. C. 1982. *Tuatara*. Southland Museum and Art Gallery, Invercargill. 24 p.

Hazley, L. C. 1995. Annual breeding of captive tuatara (*Sphenodon punctatus*). In: Abstracts of papers presented at the 6th Conference of the Society for Research on Amphibians and Reptiles in New Zealand, Manaia, Whangarei, New Zealand, 10–12 February 1995. *New Zealand Journal of Zoology* 22: 404.

Hazley, L. 1998. Ultraviolet light and life. *New Zealand GP: New Zealand's Medical Newspaper* 44 (Wed 8 April): 16.

Healy, J. M. and Jamieson, B. G. M. 1992. Ultrastructure of the spermatozoon of the tuatara (*Sphenodon punctatus*) and its relevance to the relationships of the Sphenodontida. *Philosophical Transactions of the Royal Society of London B* 335: 193–205.

Healy, J. M. and Jamieson, B. G. M. 1994. The ultrastructure of spermatogenesis and epididymal spermatozoa of the tuatara *Sphenodon punctatus* (Sphenodontida, Amniota). *Philosophical Transactions of the Royal Society of London B* 344: 187–199.

Heath, A. C. G. 2006. A reptile tick, *Aponomma sphenodonti* Dumbleton (Acari: Ixodidae), parasitic on the tuatara, *Sphenodon punctatus* Gray (Reptilia: Rhyncocephalia), in New Zealand: observations on its life history and biology. *Systematic and Applied Acarology* 11: 3–12.

Heatwole, H. 1982. Panting and other responses to high temperature in the tuatara, *Sphenodon punctatus*. In: Newman, D. G. (ed.). *New Zealand Herpetology*. New Zealand Wildlife Service Occasional Publication No. 2, Wellington, pp. 251–269.

Hedges, S. B. and Poling, L. L. 1999. A molecular phylogeny of reptiles. *Science* 283: 998–1001.

Heidsieck, E. 1929. Der Bau der Skeletteile der freien Extremitäten bei den Reptilien. 2. Mitteilung: *Hatteria* (*Sphenodon*) *punctata*. *Gegenbaurs Morphologisches Jahrbuch* 62: 319–354.

Hemming, F. 1957. Opinion 455. Validation under the Plenary Powers (a) of the emendation to '*Sphenodon*' of the generic name '*Sphaenodon*' Gray (J.E.), 1831, and (b) of the family-group name 'Sphenodontidae' Cope, 1870 (Class Reptilia). *Opinions and Declarations Rendered by the International Commission on Zoological Nomenclature* 15: 381–392.

Herbert, J. D. K., Godfrey, S. S., Bull, C. M. and Menz, R. I. 2010. Developmental stages and molecular phylogeny of *Hepatozoon tuatarae*, a parasite infecting the New Zealand tuatara, *Sphenodon punctatus* and the tick, *Amblyomma sphenodonti*. *International Journal of Parasitology* 40: 1311–1315.

Herrel, A., Schaerlaeken, V., Moravec, J. and Ross, C. F. 2009. Sexual shape dimorphism in tuatara. *Copeia* 2009: 727–731.

Herrel, A., Moore, J. A., Bredeweg, E. M. and Nelson, N. J. 2010. Sexual dimorphism, body size, bite force and male mating success in tuatara. *Biological Journal of the Linnean Society* 100: 287–292.

Hill, L. 1982. Water relations and excretion of the tuatara, *Sphenodon punctatus*: an overview. In: Newman, D. G. (ed.). *New Zealand herpetology*. New Zealand Wildlife Service Occasional Publication No. 2, Wellington, pp. 183–203.

Hill, L. and Dawbin, W. H. 1969. Nitrogen excretion in the tuatara, *Sphenodon punctatus*. *Comparative Biochemistry and Physiology* 31: 453–468.

Hill, R. P. 1951. Last of an ancient race. *Discovery* 12: 181–183.

Hindenach, J. C. R. 1931. The cerebellum of *Sphenodon punctatum*. *Journal of Anatomy* 65: 283–318.

Hines, M. 1923. The development of the telencephalon in *Sphenodon punctatum*. *Journal of Comparative Neurology* 35: 483–537.

Hitchmough, R. A., Hoare, J. M., Jamieson, H., Newman, D., Tocher, M. D., Anderson, P. J., Lettink, M. and Whitaker, A. H. 2010. Conservation status of New Zealand reptiles, 2009. *New Zealand Journal of Zoology* 37: 203–224.

Hitchmough, R., Anderson, P., Barr, B., Monks, J., Lettink, M., Reardon, J., Tocher, M. and Whitaker, T. 2013. *Conservation status of New Zealand reptiles, 2012*. New Zealand Threat Classification Series 2. Department of Conservation, Wellington. 16 p.

Hoare, J. M., Pledger, S. A., Keall, S. N., Nelson, N. J., Mitchell, N. J. and Daugherty, C. H. 2004. A long-term trend of decline in body condition of the Brothers Island tuatara, *Sphenodon guntheri*. In: Abstracts of papers presented at the 10th Biennial Conference of the Society for Research on Amphibians and Reptiles in New Zealand, Whakatane, New Zealand, 31 January–2 February 2003. *New Zealand Journal of Zoology* 31: 104.

Hoare, J. M., Pledger, S., Keall, S. N., Nelson, N. J., Mitchell, N. J. and Daugherty, C. H. 2006. Conservation implications of a long-term decline in body condition of the Brothers Island tuatara (*Sphenodon guntheri*). *Animal Conservation* 9: 456–462.

Hochstetter, F. 1892. Über die Arterien des Darmkanals der Saurier. *Morphologisches Jahrbuch* 26: 213–273 + Tafeln V–VII.

Hoessle, C. 1969. Display of tuatara *Sphenodon punctatus* at St Louis Zoological Park. *International Zoo Yearbook* 9: 32–33.

Hoffstetter, R. and Gasc, J.-P. 1969. Vertebrae and ribs of modern reptiles. In: Gans, C., Bellairs, A. d. A. and Parsons, T. S. (eds).

Biology of the Reptilia. Vol. 1. Morphology A. Academic Press, London, pp. 201–310.

Hogben, L. T. 1921. A preliminary account of the spermatogenesis of *Sphenodon*. *Journal of the Royal Microscopical Society*: 341–352.

Holdaway, R. N. and Worthy, T. H. 1997. A reappraisal of the late Quaternary fossil vertebrates of Pyramid Valley Swamp, North Canterbury, New Zealand. *New Zealand Journal of Zoology* 24: 69–121.

Holmes, B. 1995. Second chance for the tuatara. *New Scientist* 148: 8.

Hongi, H. (Stowell, H. M.). 1922. The tuatara: why the Maori dreads it. In: *Varlae 34. [Scrapbook of newspaper clippings relating to New Zealand, c. 1894–1948.]* Hocken Library, Dunedin, pp. 15–16.

Hoppe, G. 1934. Das Geruchsorgan von *Hatteria punctata*. *Zeitschrift für Anatomie und Entwickelungsgeschichte* 102: 434–461.

Howes, G. B. 1899. [On egg incubation and hatchlings]. *Nature* 59: 340–341.

Howes, G. B. and Swinnerton, H. H. 1901. On the development of the skeleton of the tuatara, *Sphenodon punctatus*; with remarks on the egg, on the hatching and on the hatched young. *Transactions of the Zoological Society of London* 16: 1–86 + plates I–VI.

Huey, R. B. and Janzen, F. J. 2008. Climate warming and environmental sex determination in tuatara: the last of the sphenodontians? *Proceedings of the Royal Society B* 275: 2181–2183.

Humphries, E. and Jones, M. E. H. 2010. Geographic variation in the jaws of Holocene *Sphenodon* (Lepidosauria: Rhynchocephalia) demonstrated by landmark analysis. *Journal of Vertebrate Palaeontology* 30 (Suppl. 2): 108A (abstract).

Hunter, S. A. and Alley, M. R. 2010. Colonic impaction in a juvenile tuatara, *Sphenodon punctatus*. *Kokako* 17(2): 41, 44.

Hutchinson, F. J. 1898. On Maori middens at Wainui, Poverty Bay. *Transactions and Proceedings of the New Zealand Institute* 30: 533–536.

Hutton, F. W. 1875. Notice of the Earnscleugh Cave. *Transactions and Proceedings of the New Zealand Institute* 7: 138–141.

Hutton, F. W. and Drummond, J. 1904. *The animals of New Zealand: an account of the colony's air-breathing vertebrates*. Whitcombe and Tombs, Wellington. 381 p.

Hyde, V. 1990. Tuatara by any other name. *New Zealand Science Monthly*: 8–10.

Internal Affairs [various dates]. *Series 1*. Archives New Zealand/Te Rua Mahara o te Kāwanatanga, Wellington Office.

Ireland, L. C. and Gans, C. 1977. Optokinetic behavior of the tuatara, *Sphenodon punctatus*. *Herpetologica* 33: 339–344.

Jacobshagen, E. 1920. Zur Morphologie des Oberflächenreliefs der Rumpfdarmschleimhaut der Reptilien. *Jenaische Zeitschrift fur Naturwissenschaft* 56: 361–430 + Tafeln 17–30.

Jakob-Hoff, R. 1996. Pre-release health evaluations for tuatara. *Kokako* 3(2): 11.

Jakob-Hoff, R. 1997. Shoulder swellings in tuataras. *Kokako* 4(2): 4–6.

Jamieson, B. G. M. and Healy, J. M. 1992. The phylogenetic position of the tuatara, *Sphenodon* (Sphenodontida, Amniota), as indicated by cladistic analysis of the ultrastructure of spermatozoa. *Philosophical Transactions of the Royal Society of London B* 335: 207–219.

Johnson, G. L. 1927. Contributions to the comparative anatomy of the reptilian and the amphibian eye, chiefly based on opthalmological examination. *Philosophical Transactions of the Royal Society of London B* 215: 315–353 + plates 20–25.

Johnston, P. 2010. The constrictor dorsalis musculature and basipterygoid articulation in *Sphenodon*. *Journal of Morphology* 271: 280–292.

Jones, J. 1993. *The tuatara*. Heinemann Education, Auckland. 24 p.

Jones, J. and Daugherty, C. 1995. *Tuatara*. WWF-NZ, Wellington. 8 p.

Jones, M. E. H. 2006a. *Skull evolution and functional morphology in* Sphenodon *and other Rhynchocephalia (Diapsida: Lepidosauria)*. University of London, London. 565 p.

Jones, M. E. H. 2006b. Tooth diversity and function in the Rhynchocephalia (Diapsida: Lepidosauria). In: Barrett, P. M. and Evans, S. E. (eds). *Ninth International Symposium on Mesozoic Terrestrial Ecosystems and Biota*. Natural History Museum, London, pp. 55–58.

Jones, M. E. H. 2007a. Geometric morphometric analysis of variation in *Sphenodon* 'sub-fossil' material. *Palaeontological Association Newsletter* 65: 95–99.

Jones, M. E. H. 2007b. Cranial suture morphology of the lepidosaur *Sphenodon* (Diapsida: Rhynchocephalia) and implications for functional morphology. *Journal of Morphology* 268: 1090–1091.

Jones, M. E. H. 2008. Skull shape and feeding strategy in *Sphenodon* and other Rhynchocephalia (Diapsida: Lepidosauria). *Journal of Morphology* 269: 945–966.

Jones, M. E. H. 2009. Dentary tooth shape in *Sphenodon* and its fossil relatives (Diapsida: Lepidosauria: Rhynchocephalia). In: Koppe, T., Meyer, G. and Alt, K. W. (eds). *Comparative dental morphology. Frontiers in oral biology. Vol. 13*. Karger, Basel, pp. 9–15.

Jones, M. E. H. and Cree, A. 2012. Tuatara. *Current Biology* 22: R986–987.

Jones, M. E. H. and Lappin, A. K. 2009. Bite-force performance of the last rhynchocephalian (Lepidosauria: *Sphenodon*). *Journal of the Royal Society of New Zealand* 39: 71–83.

Jones, M. E. H., Curtis, N., O'Higgins, P., Fagan, M. J. and Evans, S. E. 2009a. The head and neck muscles associated with feeding in *Sphenodon* (Reptilia: Lepidosauria: Rhynchocephalia). *Palaeontologia Electronica* 12.2.7A: 56 p.

Jones, M. E. H., Tennyson, A. J. D., Worthy, J. P., Evans, S. E. and Worthy, T. H. 2009b. A sphenodontine (Rhynchocephalia) from the Miocene of New Zealand and palaeobiogeography of the tuatara (*Sphenodon*). *Proceedings of the Royal Society B* 276: 1385–1390.

Jones, M. E. H., Curtis, N., Fagan, M. J., O'Higgins, P. and Evans, S. E. 2011. Hard tissue anatomy of the cranial joints in *Sphenodon* (Rhynchocephalia): sutures, kinesis, and skull mechanics. *Palaeontologia Electronica* 14.2.17A: 92 p.

Jones, M. E. H., O'Higgins, P., Fagan, M. J., Evans, S. E. and Curtis, N. 2012. Shearing mechanics and the influence of a flexible symphysis during oral food processing in *Sphenodon* (Lepidosauria: Rhynchocephalia). *The Anatomical Record* 295: 1075–1091.

Jones, P. 2002. Tuatara population boom follows rat eradication. *Forest and Bird* No. 304: 10.

Jordan, T. W., Smith, J. N. and Vaughn, L. 1980. Benzoic acid conjugation in tuatara. *Biochemical Systematics and Ecology* 8: 101–103.

Keall, S. N. and Daugherty, C. H. 1997. Captive incubation programmes for conservation of rare tuatara (*Sphenodon*). In: Abstracts of papers presented at the 7th Conference of the Society for Research on Amphibians and Reptiles in New Zealand, Kaikoura, New Zealand, 31 January–2 February 1997. *New Zealand Journal of Zoology* 24: 327–328.

Keall, S. N., Nelson, N. J., Phillpot, P., Pledger, S. and Daugherty, C. H. 2001. Conservation in small places: reptiles on North

Brother Island. In: Abstracts of papers presented at the 9th Conference of the Society for Research on Amphibians and Reptiles in New Zealand, St Arnaud, Nelson Lakes, New Zealand, 2–4 February 2001. *New Zealand Journal of Zoology* 28: 367.

Keall, S. N., Nelson, N. J. and Daugherty, C. H. 2010. Securing the future of threatened tuatara populations with artificial incubation. *Herpetological Conservation and Biology* 5: 555–562.

Keck, W. N. 1925. Some additional observations on *Sphenodon punctatum* in captivity. *Proceedings of the Iowa Academy of Science* 32: 429–430.

Keenan, R. D. 1932. The chromosomes of *Sphenodon punctatum*. *Journal of Anatomy* 67: 1–17 + plates I–IV.

Kieser, J. A., Tkatchenko, T., Dean, C., Jones, M. E. H., Duncan, W. and Nelson, N. J. 2009. Microstructure of dental hard tissues and bone in the tuatara dentary, *Sphenodon punctatus* (Diapsida: Lepidosauria: Rhynchocephalia). In: Koppe, T., Meyer, G. and Alt, K. W. (eds). *Comparative dental morphology. Frontiers of oral biology. Vol. 13.* Karger, Basel, pp. 80–85.

Kieser, J. A., He, L.-H., Dean, M. C., Jones, M. E. H., Duncan, W. J., Swain, M. V. and Nelson, N. J. 2011. Structure and compositional characteristics of caniniform dental enamel in the tuatara *Sphenodon punctatus* (Lepidosauria: Rhynchocephalia). *New Zealand Dental Journal* 107: 44–50.

King, F. W. and Burke, R. L. 1989. *Crocodilian, tuatara, and turtle species of the world: a taxonomic and geographic reference.* Association of Systematics Collections, Washington DC, USA. 216 p.

Kingsley-Smith, C. 1966. The tuatara (ngarara) bearers for the models. *Historical Review (Whakatane and District Historical Society)* 14: 124–125.

Klaatsch, H. 1892. Zur Morphologie der Mesenterialbildungen am Darmkanal der Wirbelthiere. I. Thiel. Amphibien und Reptilien. *Morphologisches Jahrbuch* 18: 385–450 + Tafel XII.

Klutzny, S. 2002. Three-dimensional computerized reconstructions of the embryonic chondrocranium of *Sphenodon punctatus*. *Journal of Vertebrate Paleontology* 22: 75A (abstract).

Knox, F. J. 1870. On the *tuatara* (Hatteria punctata, *Gray*); or great fringed lizard of New Zealand. *Transactions and Proceedings of the New Zealand Institute* 2: 17–20.

Komocki, W. 1936. Nouvelles observations sur la désagrégation physiologique des leucocytes granuleux ainsi que sur les leucocytes du sang de *Sphenodon punctatus* Grey (*Hatteria*). *Bulletin d'Histologie Applique, Physiologie, et Pathologie* 13: 194–201.

Komocki, W. 1938. La formation des érythrocytes dans le sang du *Sphenodon* (*Hatteria*) *punctatus* Gray. *Archives de Biologie* 49: 101–109.

Krull, W. 1923. Observations on *Sphenodon punctatum* in captivity. *Proceedings of the Iowa Academy of Science* 30: 151–155.

La Flamme, A. C. 2010. Toll-like receptor responses in tuatara. *New Zealand Journal of Zoology* 37: 235–242.

Laird, M. 1950. *Haemogregarina tuatarae* sp. n., from the New Zealand rhynchocephalian *Sphenodon punctatus* (Gray). *Proceedings of the Zoological Society* 120: 529–533 + plate 1.

Lakjer, T. 1927. Studien über die Gaumenregion bei Sauriern im Vergleich mit Anamniern und primitiven Sauropsiden. *Zoologische Jahrbücher, Abteilung für Anatomie und Ontogenie der Tiere* 49: 57–356.

Lanfear, R. and Ho, S. Y. W. 2009. Mitochondrial evolution in tuatara. *Mitochondrial DNA* 20: 3.

Lange, R. H. and Kilarski, W. 1986. Similarity in yolk-platelet structure of an ancient bony fish (*Acipenser*) and an ancient reptile (*Sphenodon*). *Tissue and Cell* 18: 117–124.

Lawless, P. 1994. Teeming tuatara. *New Scientist* 142: 52.

Lee, D. 2010. CT scanning the tuatara. *Shadows: The New Zealand Journal of Medical Radiation Technology* 53: 16–17.

Levine, H. B. 2010. Claiming indigenous rights to culture, flora, and fauna: a contemporary case from New Zealand. *PoLAR: Political and Legal Anthropology Review* 33: 36–56.

Lewin, R. A. 1987. To a tuatara. *Tuatara* 29: 32.

Linklater, W. L. 2011. Territorial tuatara? – a hypothesis still to be tested. *New Zealand Journal of Ecology* 35: 308–311.

Liu, S.-K. and King, F. W. 1971. Microsporidiosis in the tuatara. *Journal of the American Veterinary Medical Association* 159: 1578–1582.

Lowe, C. B., Bejerano, G., Salama, S. R. and Haussler, D. 2010. Endangered species hold clues to human evolution. *Journal of Heredity* 101: 437–447.

Lubkin, S. R. 1997. On pattern formation in reptilian dentition. *Journal of Theoretical Biology* 186: 145–157.

Lutz, D. 2006. *Tuatara: a living fossil.* Dimi Press, Salem. 108 p.

MacAvoy, E. S., Sainsbury, J., Wilson, C., Daugherty, C. H. and Chambers, G. K. 2001. An evaluation of microsatellite loci for use in population studies of tuatara. In: Abstracts of papers presented at the 9th Conference of the Society for Research on Amphibians and Reptiles in New Zealand, St Arnaud, Nelson Lakes, New Zealand, 2–4 February 2001. *New Zealand Journal of Zoology* 28: 367.

MacAvoy, E. S., McGibbon, L. M., Sainsbury, J. P., Lawrence, H., Wilson, C. A., Daugherty, C. H. and Chambers, G. K. 2007. Genetic variation in island populations of tuatara (*Sphenodon* spp) inferred from microsatellite markers. *Conservation Genetics* 8: 305–318.

Mackay, R. S. 1956. *A bibliography of the tuatara.* Unpublished report. Alexander Turnbull Library, Wellington. 11 p.

Maderson, P. F. A. 1968. Observations on the epidermis of the tuatara (*Sphenodon punctatus*). *Journal of Anatomy* 103: 311–320.

Mair, W. G. 1873. Notes on Rurima Rocks. *Transactions and Proceedings of the New Zealand Institute* 5: 151–153.

Manley, G. A. 2002. Evolution of structure and function of the hearing organ of lizards. *Journal of Neurobiology* 53: 202–211.

Manley, G. A. and Köppl, C. 1998. Phylogenetic development of the cochlea and its innervation. *Current Opinion in Neurobiology* 8: 468–474.

Mann, I. 1932. A demonstration of the structure of the lateral eyes of the adult *Sphenodon*. *Proceedings of the Royal Society of Medicine* 25: 834–836.

Mann, I. 1933. Notes on the lateral eyes of *Sphenodon* with special reference to the macular region. *British Journal of Ophthalmology* 17: 1–15.

Marchalonis, J. J., Ealey, E. H. M. and Diener, E. 1969. Immune response of the tuatara, *Sphenodon punctatum*. *Australian Journal of Experimental Biology and Medical Science* 47: 367–380.

Markwell, T. J. 1997. Video camera count of burrow-dwelling fairy prions, sooty shearwaters, and tuatara on Takapourewa (Stephens Island), New Zealand. *New Zealand Journal of Zoology* 24: 231–237.

Markwell, T. J. 1998. Relationship between tuatara *Sphenodon punctatus* and fairy prion *Pachyptila turtur* densities in different habitats on Takapourewa (Stephens Island), Cook Strait, New Zealand. *Marine Ornithology* 26: 81–83.

Markwell, T. J. 1999. *Keystone species on New Zealand offshore islands: ecological relationships of seabirds, rats, reptiles and invertebrates on Cook Strait islands.* Unpublished PhD thesis, Victoria University of Wellington, Wellington. 124 p.

May, R. M. 1990. Taxonomy as destiny. *Nature* 347: 129–130.

McCallum, J. 1980. Reptiles of the northern Mokohinau Group. *Tane* 26: 53–59.
McCallum, J. 1981. Reptiles of Tawhiti Rahi Island, Poor Knights Islands, New Zealand. *Tane* 27: 55–58.
McCallum, J. and Harker, F. R. 1981. Reptiles of Cuvier Island. *Tane* 27: 17–22.
McCallum, J. and Harker, F. R. 1982. Reptiles of Little Barrier Island. *Tane* 28: 21–27.
McDonald, H. S. and Heath, J. E. 1971. Electrocardiographic observations on the tuatara, *Sphenodon punctatus*. *Comparative Biochemistry and Physiology* 40A: 881–892.
McGibbon, L. M. 2003. *Genetic variation in tuatara (*Sphenodon*) populations, as inferred from microsatellite DNA markers.* Unpublished MSc thesis, Victoria University of Wellington, Wellington. 119 p.
McIntyre, M. 1997. *Conservation of the tuatara.* Victoria University Press, Wellington. 24 p.
McKenna, P. B. 2003. An annotated checklist of ecto- and endoparasites of New Zealand reptiles. *Surveillance* 30: 18–25.
McKenzie, K. L. 2007. *Returning tuatara (*Sphenodon punctatus*) to the New Zealand mainland.* Unpublished MSc thesis, Victoria University of Wellington, Wellington. 82 p.
McLeod, H. R. 1922. Tuatara at Miramar. *New Zealand Journal of Science and Technology* 5: 186.
Meinertz, T. 1966. Eine Untersuchung über das Herz bei Tuatara, *Sphenodon* (*Hatteria*) *punctatus* Gray. *Morphologisches Jahrbuch* 108: 568–594.
Mello, R. S. R., Fay, V., Smith, E., Hare, K. M. and Cree, A. 2011. Adjustment of juvenile tuatara (*Sphenodon punctatus*) to a cool-climate ecosanctuary. In: Abstracts of papers presented at the 14th Biennial Conference of the Society for Research on Amphibians and Reptiles in New Zealand, Tautuku, Otago, New Zealand, 11–13 February 2011. *New Zealand Journal of Zoology* 37: 272–273.
Mello, R. S. R., Besson, A. A., Hare, K. M., Fay, V., Smith, E. and Cree, A. 2013. Adjustment of juvenile tuatara to a cooler, southern climate: operative temperatures, emergence behaviour and growth rate. *New Zealand Journal of Zoology*: doi org/10.108 0/03014223.2013.775167.
Meloro, C. and Jones, M. E. H. 2012. Tooth and cranial disparity in the fossil relatives of *Sphenodon* (Rhynchocephalia) dispute the persistent 'living fossil' label. *Journal of Evolutionary Biology* 25: 2194–2209.
Merrifield, K. 2001a. Conservation management of the Brothers' tuatara (*Sphenodon guntheri*): monitoring the translocated Matiu/Somes Island population. In: Abstracts of papers presented at the 9th Conference of the Society for Research on Amphibians and Reptiles in New Zealand, St Arnaud, Nelson Lakes, New Zealand, 2–4 February 2001. *New Zealand Journal of Zoology* 28: 367–368.
Merrifield, K. 2001b. *Conservation management of the Brothers Island tuatara (*Sphenodon guntheri*): monitoring the translocated Matiu/Somes Island population.* Unpublished MConSci thesis, Victoria University of Wellington, Wellington. 274 p.
Mertens, R. 1955. Proposed use of the plenary powers for the purpose (a) of validating the currently accepted emendation '*Sphenodon*' of the generic name '*Sphaenodon*' Gray (J.E.), 1831 and (b) of validating the family-group name 'Sphenodontidae' Cope, 1870 (Class Reptilia). *Bulletin of Zoological Nomenclature* 11: 139–141.
Mertens, R. 1958. Eine lebende Tuatera oder Brückenechse. *Natur und Volk* 88: 15–21.
Mertens, R. 1967. Erfolgreiche Behandlung einer Hautkrankheit und einer Augenverletzung bei der Tuatara (*Sphenodon punctatus*). *Salamandra* 3: 7–8.
Mertens, R. and Wermuth, H. 1954. Die rezenten Schildkröten, Krokodile und Brückenechsen. *Zoologische Jahrbücher* 83: 1–413.
Meyer-Rochow, V. B. 1988. Behaviour of young tuatara (*Sphenodon punctatus*) in total darkness. *Tuatara* 30: 36–38.
Meyer-Rochow, V. B. and Teh, K. L. 1991. Visual predation by tuatara (*Sphenodon punctatus*) on the beach beetle (*Chaerodes trachyscelides*) as a selective force in the production of distinct colour morphs. *Tuatara* 31: 1–8.
Meyer-Rochow, V. B., Wohlfahrt, S. and Ahnelt, P. K. 2005. Photoreceptor cell types in the retina of the tuatara (*Sphenodon punctatus*) have cone characteristics. *Micron* 36: 423–428.
Middleton, D. M., La Flamme, A. C., Gartrell, B. D. and Nelson, N. J. 2011. The epidemiology of *Salmonella* in New Zealand island fauna. In: Abstracts of papers presented at the 14th Biennial Conference of the Society for Research on Amphibians and Reptiles in New Zealand, Tautuku, Otago, New Zealand, 11–13 February 2011. *New Zealand Journal of Zoology* 38: 273.
Milani, A. 1894. Beiträge zur Kenntniss der Reptilienlunge. I. Lacertilia. *Zoologische Jahrbücher Abtilung für Anatomie und Ontogenie der Tiere* 7: 545–592 + Tafeln 30–32.
Millener, P. R. 1981. *The Quaternary avifauna of the North Island, New Zealand.* Vol. I and II. Unpublished PhD thesis, University of Auckland, Auckland. 897 p
Miller, H. C. 2006. Cloacal and buccal swabs are a reliable source of DNA for microsatellite genotyping of reptiles. *Conservation Genetics* 7: 1001–1003.
Miller, H. C., Belov, K. and Daugherty, C. H. 2005a. Characterization of MHC class II genes from an ancient reptile lineage, *Sphenodon* (tuatara). *Immunogenetics* 57: 883–891.
Miller, H. C., Belov, K., Edwards, S. V. and Daugherty, C. H. 2005b. Unravelling the reptile MHC. In: Abstracts of papers presented at the 11th Biennial Conference of the Society for Research on Amphibians and Reptiles in New Zealand, Springbrook National Park, south-east Queensland, Australia, 7–11 February 2005. *New Zealand Journal of Zoology* 32: 227.
Miller, H. C., Belov, K. and Daugherty, C. H. 2006. MHC class I genes in the tuatara (*Sphenodon* spp.): evolution of the MHC in an ancient reptilian order. *Molecular Biology and Evolution* 23: 949–956.
Miller, H. C., Andrews-Cookson, M. and Daugherty, C. H. 2007a. Two patterns of variation among MHC class I loci in tuatara (*Sphenodon punctatus*). *Journal of Heredity* 98: 666–677.
Miller, H. C., Conrad, A. M., Barker, S. C. and Daugherty, C. H. 2007b. Distribution and phylogenetic analyses of an endangered tick, *Amblyomma sphenodonti*. *New Zealand Journal of Zoology* 34: 97–105.
Miller, H. C., Miller, K. A. and Daugherty, C. H. 2008a. Reduced MHC variation in a threatened tuatara species. *Animal Conservation* 11: 206–214.
Miller, H. C., Moore, J. A., Allendorf, F. W. and Daugherty, C. H. 2008b. The evolutionary rate of tuatara revisited. *Trends in Genetics* 25: 13–15.
Miller, H. C., Moore, J. A., Nelson, N. J. and Daugherty, C. H. 2009. Influence of major histocompatibility complex genotype on mating success in a free-ranging reptile population. *Proceedings of the Royal Society B* 276: 1695–1704.
Miller, H. C., Allendorf, F. and Daugherty, C. H. 2010. Genetic diversity and differentiation at MHC genes in island populations of tuatara (*Sphenodon* spp.). *Molecular Ecology* 19: 3894–3908.
Miller, H. C., Biggs, P., Voelckel, C. and Nelson, N. J. 2012. De

novo sequence assembly and characterisation of a partial transcriptome for an evolutionarily distinct reptile, the tuatara (*Sphenodon punctatus*). *BMC Genomics* 13(439): 1–12.

Miller, K. A. 2009. *Founding events and the maintenance of genetic diversity in reintroduced populations*. Unpublished PhD thesis, Victoria University of Wellington, Wellington. 125 p.

Miller, K. A., Nelson, N. J., Smith, H. G. and Moore, J. A. 2009. How do reproductive skew and founder group size affect genetic diversity in reintroduced populations? *Molecular Ecology* 18: 3792–3802.

Miller, K. A., Gruber, M. A. M., Keall, S. N., Blanchard, B. and Nelson, N. J. 2010. Changing taxonomy and the need for supplementation in the management of reintroductions of Brothers Island tuatara in Cook Strait, New Zealand. In: Soorae, P. S. (ed.). *Global re-introduction perspectives: additional case-studies from around the globe*. IUCN/SSC Re-introduction Specialist Group, Abu Dhabi, UAE, pp. 93–97.

Miller, K. A., Miller, H. C., Moore, J. A., Mitchell, N. J., Cree, A., Allendorf, F. W., Sarre, S. D., Keall, S. N. and Nelson, N. J. 2012. Securing the demographic and genetic future of tuatara through assisted colonization. *Conservation Biology* 26: 790–798.

Milligan, R. R. D. 1924. The respiration and metabolism of the tuatara. *Australian Association for the Advancement of Science* Report 16: 404–406.

Miskelly, C. M. 2011. In the dragon's keep: herpetological holdings at Museum of New Zealand Te Papa Tongarewa. In: Abstracts of papers presented at the 14th Biennial Conference of the Society for Research on Amphibians and Reptiles in New Zealand, Tautuku, Otago, New Zealand, 11–13 February 2011. *New Zealand Journal of Zoology* 38: 273–274.

Mitchell, N. J. and Janzen, F. J. 2010. Temperature-dependent sex determination and contemporary climate change. *Sexual Development* 4: 129–140.

Mitchell, N. J., Nelson, N. J., Cree, A., Pledger, S., Keall, S. N. and Daugherty, C. H. 2006. Support for a rare pattern of temperature-dependent sex determination in archaic reptiles: evidence from two species of tuatara (*Sphenodon*). *Frontiers in Zoology* 3(9): 1–12.

Mitchell, N. J., Kearney, M. R., Nelson, N. J. and Porter, W. P. 2008. Predicting the fate of a living fossil: how will global warming affect sex determination and hatchling phenology in tuatara? *Proceedings of the Royal Society B* 275: 2185–2193.

Mitchell, N. J., Allendorf, F. W., Keall, S. N., Daugherty, C. H. and Nelson, N. J. 2010. Demographic effects of temperature-dependent sex determination: will tuatara survive global warming? *Global Change Biology* 16: 60–72.

Mitteilungen, K. 1971. Bemerkungen über die Häutung der Brückenechse (*Sphenodon punctatus*) in Gefangenschaft. *Salamandra* 7: 81–82.

Moazen, M., Curtis, N., O'Higgins, P., Evans, S. E. and Fagan, M. J. 2009. Biomechanical assessment of evolutionary changes in the lepidosaurian skull. *Proceedings of the National Academy of Sciences* 106: 8273–8277.

Moffat, L. A. 1985. Embryonic development and aspects of reproductive biology in the tuatara, *Sphenodon punctatus*. In: Gans, C., Billet, F. and Maderson, P. (eds). *Biology of the Reptilia. Vol. 14. Development A*. John Wiley and Sons, New York, pp. 493–521.

Moir, M. L., Vesk, P. A., Brennan, K. E. C., Poulin, R., Hughes, L., Keith, D. A., McCarthy, M. A. and Coates, D. J. 2012. Considering extinction of dependent species during translocation, ex situ conservation, and assisted migration of threatened hosts. *Conservation Biology* 26: 199–207.

Moller, H. 1985. Tree wetas (*Hemideina crassicruris*) (Orthoptera: Stenopelmatidae) of Stephens Island, Cook Strait. *New Zealand Journal of Zoology* 12: 55–69.

Moore, J. A. 2008. *Fitness implications of the mating system and reproductive ecology of tuatara*. Unpublished PhD thesis, Victoria University of Wellington, Wellington. 140 p.

Moore, J. A. and Godfrey, S. S. 2006. *Sphenodon punctatus* (common tuatara): opportunistic predation. *Herpetological Review* 37: 81–82.

Moore, J. A., Hoare, J. M., Daugherty, C. H. and Nelson, N. J. 2007. Waiting reveals waning weight: monitoring over 54 years shows a decline in body condition of a long-lived reptile (tuatara, *Sphenodon punctatus*). *Biological Conservation* 135: 181–188.

Moore, J. A., Miller, H. C., Daugherty, C. H. and Nelson, N. J. 2008a. Fine-scale genetic structure of a long-lived reptile reflects recent habitat modification. *Molecular Ecology* 17: 4630–4641.

Moore, J. A., Nelson, N. J., Keall, S. N. and Daugherty, C. H. 2008b. Implications of social dominance and multiple paternity for the genetic diversity of a captive-bred reptile population (tuatara). *Conservation Genetics* 9: 1243–1251.

Moore, J. A., Daugherty, C. H., Godfrey, S. S. and Nelson, N. J. 2009a. Seasonal monogamy and multiple paternity in a wild population of a territorial reptile (tuatara). *Biological Journal of the Linnean Society* 98: 161–170.

Moore, J. A., Daugherty, C. H. and Nelson, N. J. 2009b. Large male advantage: phenotypic and genetic correlates of territoriality in tuatara. *Journal of Herpetology* 43: 570–578.

Moore, J. A., Grant, T., Brown, D., Keall, S. N. and Nelson, N. J. 2010. Mark-recapture accurately estimates census for tuatara, a burrowing reptile. *Journal of Wildlife Management* 74: 897–901.

Müller, J. 2003. Early loss and multiple return of the lower temporal arcade in diapsid reptiles. *Naturwissenschaften* 90: 473–476.

Murphy, R. W. and Matson, R. H. 1986. Evolution of isozyme characters in the tuatara, *Sphenodon punctatus*. *New Zealand Journal of Zoology* 13: 573–581.

Nelson, N. 1998. *Conservation of Brothers Island tuatara (*Sphenodon guntheri*)*. Unpublished MConSc thesis. Victoria University of Wellington, Wellington. 262 p.

Nelson, N. J. 2001. *Temperature-dependent sex determination and artificial incubation of tuatara,* Sphenodon punctatus. Unpublished PhD thesis, Victoria University of Wellington, Wellington. 123 p.

Nelson, N. J. 2004. Sex determination and incubation of tuatara, *Sphenodon punctatus*. In: Abstracts of papers presented at the 10th Biennial Conference of the Society for Research on Amphibians and Reptiles in New Zealand, Whakatane, New Zealand, 31 January–2 February 2003. *New Zealand Journal of Zoology* 31: 107.

Nelson, N. J. 2005. Conservation-biased research on a New Zealand treasure, the tuatara. In: Abstracts of papers presented at the 11th Biennial Conference of the Society for Research on Amphibians and Reptiles in New Zealand, Springbrook National Park, south-east Queensland, Australia, 7–11 February 2005. *New Zealand Journal of Zoology* 32: 228.

Nelson, N. J. and Daugherty, C. H. 1997. The first experimental translocation of a tuatara population: conservation of *Sphenodon guntheri*. In: Abstracts of papers presented at the 7th Conference of the Society for Research on Amphibians and Reptiles in New Zealand, Kaikoura, New Zealand, 31 January–2 February 1997. *New Zealand Journal of Zoology* 24: 328.

Nelson, N., Pledger, S., Keall, S. and Daugherty, C. 1999. Conservation evaluation of Brothers Island tuatara (*Sphenodon guntheri*). In: Abstracts of papers presented at the 8th Conference of the Society for Research on Amphibians and Reptiles in New Zealand, Great Barrier Island, New Zealand, 5–7 February 1999. *New Zealand Journal of Zoology* 26: 259.

Nelson, N. J., Keall, S. N., Daugherty, C. H. and Thompson, M. B. 2001. Possible fitness consequences of artificial induction of oviposition in tuatara (*Sphenodon punctatus*). In: Abstracts of papers presented at the 9th Conference of the Society for Research on Amphibians and Reptiles in New Zealand, St Arnaud, Nelson Lakes, New Zealand, 2–4 February 2001. *New Zealand Journal of Zoology* 28: 369.

Nelson, N. J., Keall, S. N., Pledger, S. and Daugherty, C. H. 2002a. Male-biased sex ratio in a small tuatara population. *Journal of Biogeography* 29: 633–640.

Nelson, N. J., Keall, S. N., Brown, D. and Daugherty, C. H. 2002b. Establishing a new wild population of tuatara (*Sphenodon guntheri*). *Conservation Biology* 16: 887–894.

Nelson, N. J., Cree, A., Thompson, M. B., Keall, S. N. and Daugherty, C. H. 2004a. Temperature-dependent sex determination in tuatara. In: Valenzuela, N. and Lance, V. (eds). *Temperature-dependent sex determination in vertebrates*. Smithsonian, Washington, pp. 53–58.

Nelson, N. J., Thompson, M. B., Pledger, S., Keall, S. N. and Daugherty, C. H. 2004b. Egg mass determines hatchling size, and incubation temperature influences post-hatching growth, of tuatara *Sphenodon punctatus*. *Journal of Zoology, London* 263: 77–87.

Nelson, N. J., Thompson, M. B., Pledger, S., Keall, S. N. and Daugherty, C. H. 2004c. Induction of oviposition produces smaller eggs in tuatara (*Sphenodon punctatus*). *New Zealand Journal of Zoology* 31: 283–289.

Nelson, N. J., Thompson, M. B., Pledger, S., Keall, S. N. and Daugherty, C. H. 2004d. Do TSD, sex ratios, and nest characteristics influence the vulnerability of tuatara to global warming? *International Congress Series* 1275: 250–257.

Nelson, N. J., Thompson, M. B., Pledger, S., Keall, S. N. and Daugherty, C. H. 2006. Performance of juvenile tuatara depends on age, clutch, and incubation regime. *Journal of Herpetology* 40: 399–403.

Nelson, N., Keall, S., Gaze, P. and Daugherty, C. 2008. Re-introduction of tuatara as part of an ecological restoration project on Wakatere-papanui Island, Marlborough Sounds, New Zealand. In: Soorae, P. S. (ed.). *Global re-introduction perspectives: re-introduction case-studies from around the globe*. IUCN/SSC Re-introduction Specialist Group, Abu Dhabi, UAE, pp. 58–61.

Nelson, N. J., Moore, J. A., Pillai, S. and Keall, S. N. 2010. Thermosensitive period for sex determination in the tuatara. *Herpetological Conservation and Biology* 5: 324–329.

Newman, A. K. 1878. Notes on the physiology and anatomy of the tuatara *(Sphenodon güntheri)*. *Transactions and Proceedings of the New Zealand Institute* 10: 222–239.

Newman, D. G. 1977. Some evidence of the predation of Hamilton's frog (*Leiopelma hamiltoni* (McCulloch)) by tuatara (*Sphenodon punctatus* (Grey)) on Stephens Island. *Proceedings of the New Zealand Ecological Society* 24: 43–47.

Newman, D. G. 1978. Tuataras and petrels. *Wildlife: A Review* 9: 16–23.

Newman, D. G. 1980. Herpetological symposium. *Wildlife: A Review* 11: 39–47.

Newman, D. G. (ed.). 1982a. *New Zealand Herpetology*. New Zealand Wildlife Service Occasional Publication No. 2, Wellington. 495 p.

Newman, D. G. 1982b. Current distribution of the tuatara. In: Newman, D. G. (ed.). *New Zealand herpetology*. New Zealand Wildlife Service Occasional Publication No. 2, Wellington, pp. 145–147.

Newman, D. G. 1982c. Tuatara, *Sphenodon punctatus,* and burrows, Stephens Island. In: Newman, D. G. (ed.). *New Zealand herpetology*. New Zealand Wildlife Service Occasional Publication No. 2, Wellington, pp. 213–221.

Newman, D. G. 1982d. Breeding tuataras, *Sphenodon punctatus,* in captivity. In: Newman, D. G. (ed.). *New Zealand Herpetology*. New Zealand Wildlife Service Occasional Publication No. 2, Wellington, pp. 277–284.

Newman, D. G. 1982e. New Zealand herpetological research – the work of the New Zealand Wildlife Service. *Herpetofauna* 14: 1–10.

Newman, D. G. 1983. Tuatara/kiore relationships. *Wildlife: a review* 12: 60–63.

Newman, D. G. 1986. Can tuatara and mice co-exist? The status of the tuatara, *Sphenodon punctatus* (Reptilia: Rhynchocephalia), on the Whangamata Islands. In: Wright, A. E. and Beever, R. E. (eds). *The offshore islands of northern New Zealand*. New Zealand Department of Lands and Survey Information Series, No. 16, Wellington, pp. 179–185.

Newman, D. G. 1987a. Burrow use and population densities of tuatara (*Sphenodon punctatus*) and how they are influenced by fairy prions (*Pachyptila turtur*) on Stephens Island, New Zealand. *Herpetologica* 43: 336–344.

Newman, D. G. 1987b. *Tuatara*. John McIndoe Ltd, Dunedin. 24 p.

Newman, D. G. 1988. Evidence of predation on a young tuatara, *Sphenodon punctatus,* by kiore, *Rattus exulans,* on Lady Alice Island. *New Zealand Journal of Zoology* 15: 443–446.

Newman, D. G. 1998. Tuatara. In: Cogger, H. G. and Zweifel, R. G. (eds). *Encyclopedia of reptiles and amphibians*. Academic Press, San Diego, pp. 218–223.

Newman, D. G. and McFadden, I. 1990a. Status of the tuatara, *Sphenodon punctatus*, on Hongiora and Ruamahua-iti Islands, Aldermen Group, New Zealand. *New Zealand Journal of Zoology* 17: 153–156.

Newman, D. G. and McFadden, I. 1990b. Seasonal fluctuations of numbers, breeding, and food of kiore (*Rattus exulans*) on Lady Alice Island (Hen and Chickens group), with a consideration of kiore: tuatara (*Sphenodon punctatus*) relationships in New Zealand. *New Zealand Journal of Zoology* 17: 55–63.

Newman, D. G. and Watson, P. R. 1985. The contribution of radiography to the study of the reproductive ecology of the tuatara, *Sphenodon punctatus*. In: Grigg, G., Shine, R. and Ehmann, H. (eds). *Biology of Australasian frogs and reptiles*. Surrey Beatty and Sons, Chipping Norton, New South Wales, pp. 7–10.

Newman, D. G., Crook, I. G. and Moran, L. R. 1979. Some recommendations on the captive maintenance of tuataras *Sphenodon punctatus* based on observations in the field. *International Zoo Yearbook* 19: 68–74.

Newman, D. G., Watson, P. R. and McFadden, I. 1994. Egg production by tuatara on Lady Alice and Stephens Island, New Zealand. *New Zealand Journal of Zoology* 21: 387–398.

Nicholas, J. L. 1817. *Narrative of a voyage to New Zealand, performed in the years 1814 and 1815, in company with the Rev. Samuel Marsden, Principal Chaplain of New South Wales. Vol. 2.* James Black and Son, London. 400 p.

Norris, T. B. 1997. *Chromosomal studies of the New Zealand herpetofauna*. Unpublished MSc thesis, Victoria University of Wellington, Wellington. 206 p.

Norris, T. B. 2007. *Chromosomes, nuclear genes and the phylogenetic placement within the Reptilia of* Sphenodon *(tuatara).*

Unpublished PhD thesis, Victoria University of Wellington, Wellington. 196 p.

Norris, T. B., Rickards, G. K. and Daugherty, C. H. 2004. Chromosomes of tuatara, *Sphenodon*, a chromosome heteromorphism and an archaic reptilian karyotype. *Cytogenetic and Genome Research* 105: 93–99.

Northcutt, R. G. and Heath, J. E. 1973. T-maze behavior of the tuatara (*Sphenodon punctatus*). *Copeia* 1973: 617–620.

Northcutt, R. G., Braford, M. R., Jr. and Landreth, G. E. 1974. Retinal projections in the tuatara *Sphenodon punctatus*: an autoradiographic study. *Anatomical Record* 178: 428 (abstract).

Nutting, C. C. 1926. Work on *Sphenodon*. *Science* (New Series) 63: 210.

Nye, E. R. and Buchanan, H. 1969. Adipose tissue reactivity of *Sphenodon punctatus* and a species of New Zealand skink. *Comparative Biochemistry and Physiology* 28: 483–485.

O'Donoghue, C. H. 1920. The blood vascular system of the tuatara, *Sphenodon punctatus*. *Philosophical Transactions of the Royal Society of London B* 210: 175–252 + plates 6–8.

O'Meally, D., Miller, H., Patel, H. R., Marshall Graves, J. A. and Ezaz, T. 2009. The first cytogenetic map of the tuatara, *Sphenodon punctatus*. *Cytogenetic and Genome Research* 127: 213–223.

Oldman, J. M. 2008. *Non-surgical methods for sexing small juvenile tuatara*. Unpublished MSc thesis, University of Otago, Dunedin. 90 p.

Oldman, J. M. and Cree, A. 2010. Non-invasive techniques for sexing small juvenile tuatara. In: Abstracts of papers presented at the Second Meeting of Australasian Societies for Herpetology (Society for Research on Amphibians and Reptiles in New Zealand and the Australia Society for Herpetologists), Massey University, Auckland, New Zealand, 20–22 February 2009. *New Zealand Journal of Zoology* 37: 93–94.

Oliver, J. A. 1953. The timeless tuatara. *Animal Kingdom* 61: 2–8, 31.

Ombler, K. 2004. A turnaround for tuatara. *Forest and Bird* No. 313: 28–31.

Orbell, M. 1996. *The natural world of the Maori*. Revised edn. David Bateman, Auckland. 128 p.

Osawa, G. 1896. Beiträge zur feineren Struktur des Integumentes der *Hatteria punctata*. *Archiv für Mikroskopische Anatomie und Entwicklungsgeschichte* 47: 570–583.

Osawa, G. 1897. Beiträge zur Lehre von den Eingeweiden der *Hatteria punctata*. *Archiv für Mikroskopische Anatomie und Entwicklungsgeschichte* 49: 113–226 + Tafeln VIII–XIV.

Osawa, G. 1898a. Beiträge zur Anatomie der *Hatteria punctata*. *Archiv für Mikroskopische Anatomie und Entwicklungsgeschichte* 51: 481–691.

Osawa, G. 1898b. Nachtrag zur Lehre von den Eingeweiden der *Hatteria punctata*. Die weiblichen Geschlechtsorgane. *Archiv für Mikroskopische Anatomie und Entwicklungsmechanik* 51: 764–794 + Tafeln XXIII–XXV.

Osawa, G. 1898c. Beiträge zur Lehre von den Sinnesorganen der *Hatteria punctata*. *Archiv für Mikroskopische Anatomie und Entwicklungsgeschichte* 52: 268–366 + Tafeln XVI–XVIII.

Osawa, G. 1898d. Uber die Stellung der *Hatteria punctata* in der Teirreihe. *Verhandlungen der Anatomischen Gesellschaft* 12: 100–106.

Osawa, G. 1899. Ueber die Fovea centralis von *Hatteria punctata*. Eine Erwiderung an Prof. Kallius in Göttingen. *Anatomischer Anzeiger* 15: 226–227.

Osborn, H. F. 1900. Intercentra and hypapophyses in the cervical region of mosasaurs, lizards, and *Sphenodon*. *American Naturalist* 34: 1–7.

Ostrom, J. H. 1962. On the constrictor dorsalis muscles of *Sphenodon*. *Copeia* 1962: 732–735.

Owen, K. 1998. Introduction of northern tuatara to Moutohora Island, Bay of Plenty. *Ecological Management* No. 6: 23–33.

Owen, R. 1845. Report on the reptilian fossils of South Africa. *Transactions of the Geological Society of London* 7: 59–84 + plates III–VI.

Owen, R. 1853. *Descriptive catalogue of the osteological series contained in the museum of the Royal College of Surgeons of England. Vol. 1.* Taylor and Francis, London.

Packard, M. J., Hirsch, K. F. and Meyer-Rochow, V. B. 1982. Structure of the shell from eggs of the tuatara, *Sphenodon punctatus*. *Journal of Morphology* 174: 197–205.

Packard, M. J., Thompson, M. B., Goldie, K. N. and Vos, M. 1988. Aspects of shell formation in eggs of the tuatara, *Sphenodon punctatus*. *Journal of Morphology* 197: 147–157.

Parham, W. T. 1982. Tuatara – the classic survivor. *Historical Review (Whakatane and District Historical Society)* 30: 62–63.

Parker, T. J. 1891. The Otago University Museum. In: Hastings, D. H. (ed.). *Official record of the New Zealand and South Seas Exhibition, held at Dunedin, 1889–1890*. Government Printer, Wellington, pp. 136–143.

Parsons, T. S. 1959a. Studies on the comparative embryology of the reptilian nose. *Bulletin of the Museum of Comparative Zoology* 120: 104–277.

Parsons, T. S. 1959b. Nasal anatomy and the phylogeny of reptiles. *Evolution* 13: 175–187.

Parsons, T. S. 1970. The nose and Jacobson's organ. In: Gans, C. and Parsons, T. S. (eds). *Biology of the Reptilia. Vol. 2. Morphology B.* Academic Press, London, pp. 99–191.

Pepperell, J. G. 1982. Tuatara *Sphenodon punctatus* locomotion: a summary. In: Newman, D. G. (ed.). *New Zealand herpetology*. New Zealand Wildlife Service Occasional Publication No. 2, Wellington, pp. 207–211.

Perrin, A. 1895. Recherches sur les affinités zoologiques de l'Hatteria punctata. *Annales des Sciences Naturelles Zoologie et Paléontologie* 20: 33–102 + plates II–V.

Perry, S. F. 1998. Lungs: comparative anatomy, functional morphology, and evolution. In: Gans, C. and Gaunt, A. S. (eds). *Biology of the Reptilia. Vol. 19. Morphology G. Visceral organs.* Society for the Study of Amphibians and Reptiles. Contributions to Herpetology Vol. 14, Ithaca, New York, pp. 1–92.

Peters, W. 1874. Über die Gerhörknöchelchen und ihre Verhältnisse zu dem ersten Zungenbeinbogen bei *Sphenodon punctatus*. *Monatsberichte der Königlichen Preußischen Akademie der Wissenschaften zu Berlin* 15: 40–45.

Peterson, J. A. 1984. The scale microarchitecture of *Sphenodon punctatus*. *Journal of Herpetology* 18: 40–47.

Pieau, C., Dorizzi, M. and Richard-Mercier, N. 2001. Temperature-dependent sex determination and gonadal differentiation in reptiles. In: Scherer, G. and Schmid, M. (eds). *Genes and mechanisms in vertebrate sex determination*. Birkhäuser Verlag, Basel, pp. 117–141.

Pierce, R. J. 2002. *Kiore (*Rattus exulans*) impact on breeding success of Pycroft's petrels and little shearwaters*. Department of Conservation Science Internal Series 39. Department of Conservation, Wellington. 24 p.

Platel, R. 1989. L'encéphalisation chez le tuatara de Nouvelle-Zélande *Sphenodon punctatus* Gray (Lepidosauria, Sphenodonta). Etude quantifiée des principales subdivisions encéphaliques. *Journal für Hirnforschung* 30: 325–337.

Poglayen-Neuwall, I. 1953. Untersuchungen über die Trigeminusmuskulatur von *Hatteria*. *Zeitschrift für Wissenschaftliche Zoologie* 157: 57–76.

Polack, J. S. 1838. *New Zealand: being a narrative of travels and adventures during a residence in that country between the years 1831 and 1837. Vol. I and II.* Richard Bentley, London. 403 p., 441 p.

Polack, J. S. 1840. *Manners and customs of the New Zealanders; with notes corroborative of their habits, usages, etc., and remarks to intending emigrants, with numerous cuts drawn on wood. Vol. I and II.* James Madden & Co., Piccadilly. Reprint published 1976 by Capper Press, Christchurch. 288 p., 288 p.

Poluhowich, J. J. and Brush, A. H. 1972. An electrophoretic study of *Sphenodon* proteins. *Comparative Biochemistry and Physiology* 41B: 281–285.

Pratt, C. W. M. 1948. The morphology of the ethmoidal region of *Sphenodon* and lizards. *Proceedings of the Zoological Society of London* 118: 171–201.

Prebble, G. K. 1971. *Tuhua – Mayor Island.* Ashford-Kent, Tauranga. 228 p.

Quay, W. B. 1979. The parietal eye-pineal complex. In: Gans, C., Northcutt, R. G. and Ulinski, P. (eds). *Biology of the Reptilia. Vol. 9. Neurology A.* Academic Press, London, pp. 245–406.

Quinn, T. W. and Mindell, D. P. 1996. Mitochondrial gene order adjacent to the control region in crocodile, turtle, and tuatara. *Molecular Phylogenetics and Evolution* 5: 344–351.

Ramstad, K. M., Nelson, N. J., Daugherty, C. H. and Allendorf, F. W. 2004. Integrating traditional Maori and scientific ecological knowledge of the tuatara (*Sphenodon*). In: Abstracts of papers presented at the 10th Biennial Conference of the Society for Research on Amphibians and Reptiles in New Zealand, Whakatane, New Zealand, 31 January–2 February 2003. *New Zealand Journal of Zoology* 31: 107–108.

Ramstad, K. M., Nelson, N. J., Paine, G., Beech, D., Paul, A., Paul, P., Allendorf, F. W. and Daugherty, C. H. 2007a. Species and cultural conservation in New Zealand: Maori traditional ecological knowledge of tuatara. *Conservation Biology* 21: 455–464.

Ramstad, K., Nelson, N., Paine, G., Beech, D., Paul, A., Paul, P., Allendorf, F. and Daugherty, C. 2007b. Tuatara: our living ancient taonga. *Mana* 76 (June–July): 18–21.

Ramstad, K. M., Paine, G., Dunning, D. L., Geary, A. F., Keall, S. N. and Nelson, N. J. 2009. Effective partnerships between universities and indigenous communities: a case study in tuatara conservation in Aotearoa. *Journal of the Royal Society of New Zealand* 39: 229–231.

Ramstad, K. M., Moore, J. A. and Refsnider, J. M. 2012. Intrasexual aggresion in tuatara: males and females respond differently to same-sex intruders. *Herpetological Review* 43: 19–21.

Refsnider, J., Moore, J. and Streby, H. 2008. *Sphenodon punctatus* (common tuatara). Prey detection. *Herpetological Review* 39: 347–348.

Refsnider, J. M., Keall, S. N., Daugherty, C. H. and Nelson, N. J. 2009. Does nest-guarding in female tuatara (*Sphenodon punctatus*) reduce nest destruction by conspecific females? *Journal of Herpetology* 43: 294–299.

Refsnider, J. M., Daugherty, C. H., Keall, S. N. and Nelson, N. J. 2010. Nest-site choice and fidelity in tuatara on Stephens Island, New Zealand. *Journal of Zoology, London* 280: 396–402.

Refsnider, J. M., Daugherty, C. H., Godfrey, S. S., Keall, S. N., Moore, J. A. and Nelson, N. J. 2013. Patterns of nesting migrations in the tuatara (*Sphenodon punctatus*), a colonially nesting island reptile. *Herpetologica* 69: 282–290.

Reilly, S. M., McElroy, E. J., Odum, R. A. and Hornyak, V. A. 2006. Tuataras and salamanders show that walking and running mechanics are ancient features of tetrapod locomotion. *Proceedings of the Royal Society of London B* 273: 1563–1568.

Reiner, A. and Northcutt, R. G. 2000. Succinic dehydrogenase histochemistry reveals the location of the putative primary visual and auditory areas within the dorsal ventricular ridge of *Sphenodon punctatus*. *Brain, Behaviour and Evolution* 55: 26–36.

Reischek, A. 1882. Notes on zoological researches made on the Chicken Island, east coast of the North Island. *Transactions and Proceedings of the New Zealand Institute* 14: 274–277.

Reischek, A. 1886. Observations on *Sphenodon punctatum*, fringe-back lizard (Tuatara). *Transactions and Proceedings of the New Zealand Institute* 18: 108–110.

Reischek, A. 1971. *Yesterdays in Maoriland: New Zealand in the 'eighties.* Wilson and Horton, Auckland (fascimile edition). 312 p.

Renous, S. 1975. Particularités des systèmes musculaire et nerveux du membre antérieur de *Sphenodon punctatus* (Reptiles – rhynchocéphales). *Gegenbaurs Morphologisches Jahrbuch* 121: 230–238.

Renous-Lécuru, S. 1973. Morphologie comparée du carpe chez les lépidosauriens actuels (rhynchocéphales, lacertiliens, amphisbéniens). *Gegenbaurs Morphologisches Jahrbuch* 119: 727–766 + 3 plates.

Rest, J. S., Ast, J. C., Austin, C. C., Waddell, P. J., Tibbetts, E. A., Hay, J. M. and Mindell, D. P. 2003. Molecular systematics of primary reptilian lineages and the tuatara mitochondrial genome. *Molecular Phylogenetics and Evolution* 29: 289–297.

Rheubert, J. L., Cree, A., Downes, M. and Sever, D. M. 2012. Reproductive morphology of the male tuatara, *Sphenodon punctatus*. *Acta Zoologica* 00: 1–8. doi: 10.1111/j.1463-6395.2012.00574.x.

Ribbing, L. 1911. Kleinere Muskelstudien. 1. Die Vorderarm- und Handmuskulatur von *Sphenodon*. *Acta Universitatis Lundensis, Nova Series (Lunds Universitets Årsskrift, N. Folge) [Fysiografiska Sällskapets Handlingar N. Folge, 21(8)]* 6: 1–9.

Richardson, A. 2002. Poison kills four tuatara at zoo. *Sunday-Star Times*, Auckland, 1 September, p. 5.

Richardson, S. J., Bradley, A. J., Duan, W., Wettenhall, R. E. H., Harms, P. J., Babon, J. J., Southwell, B. R., Nicol, S., Donnellan, S. C. and Screiber, G. 1994. Evolution of marsupial and other vertebrate thyroxine-binding plasma proteins. *American Journal of Physiology* 266: R1359–R1370.

Rieppel, O. 1978. The throat musculature of *Sphenodon*, with comments on the primitive character states of the throat muscles in lizards. *Anatomischer Anzeiger* 144: 429–440.

Rieppel, O. 1992. The skull in a hatchling of *Sphenodon punctatus*. *Journal of Herpetology* 26: 80–84.

Rieppel, O., Gauthier, J. and Maisano, J. 2008. Comparative morphology of the dermal palate in squamate reptiles, with comments on phylogenetic implications. *Zoological Journal of the Linnean Society* 152: 131–152.

Robb, J. 1973. Reptiles and amphibia. In: Williams, G. R. (ed.). *The natural history of New Zealand. An ecological survey.* A. H. and A. W. Reed, Wellington, pp. 285–303.

Robb, J. 1977. *The tuatara.* Meadowfield, Durham. 64 p.

Robb, J. 1980. *New Zealand amphibians and reptiles in colour.* Collins, Auckland (revised 1986). 128 p.

Robinson, P. L. 1976. How *Sphenodon* and *Uromastyx* grow their teeth and use them. In: Bellairs, A. d. A. and Cox, C. B. (eds). *Morphology and biology of reptiles.* Linnean Society Symposium Series No. 3. Academic Press, London, pp. 43–64.

Rodda, G. H. and Dean-Bradley, K. D. 2002. Excess density compensation of island herpetofaunal assemblages. *Journal of Biogeography* 29: 623–632.

Roe, W. 2002. Pathology of a cloacal tumour in a tuatara. *Kokako* 9: 5–6.

Roe, W. D., Alley, M. R., Cooper, S. M. and Hazley, L. 2002. Squamous cell carcinoma in a tuatara (*Sphenodon punctatus*). *New Zealand Veterinary Journal* 50: 207–210.

Romer, A. S. 1956. *Osteology of the reptiles*. University of Chicago Press, Chicago. 772 p

Romer, A. S. 1966. *Vertebrate paleontology*. 3rd edn. University of Chicago Press, Chicago. 468 p.

Rosenberg, H. and Gans, C. 1977. Preliminary analysis of mastication in *Sphenodon punctatus*. *American Zoologist* 17: 871 (abstract).

Ross, C. F., Eckhardt, A., Herrel, A., Hylander, W. L., Metzger, K. A., Schaerlaeken, V., Washington, R. L. and Williams, S. H. 2007. Modulation of intra-oral processing in mammals and lepidosaurs. *Integrative and Comparative Biology* 47: 118–136.

Ross, C. F., Baden, A. L., Georgi, J., Herrel, A., Metzger, K. A., Reed, D. A., Schaerlaeken, V. and Wolff, M. S. 2010. Chewing variation in lepidosaurs and primates. *Journal of Experimental Biology* 213: 572–584.

Rout, T. M., McDonald-Madden, E., Martin, T. G., Mitchell, N. J., Possingham, H. P. and Armstrong, D. P. 2013. How to decide whether to move species threatened by climate change. *PLOS ONE* 8 (e75814): 1–7.

Ruffell, J. 2005. *The use of translocation in tuatara (*Sphenodon punctatus punctatus*) conservation and relationships between the tuatara and the tick* Aponomma sphenodonti *(Acari: Ixodidae)*. Unpublished MSc thesis, University of Auckland. 134 p.

Russell, A. P. and Bauer, A. M. 2008. The appendicular locomotor apparatus of *Sphenodon* and normal-limbed squamates. In: Gans, C., Gaunt, A. S. and Adler, K. (eds). *Biology of the Reptilia. Vol. 21. Morphology I. The skull and appendicular locomotor apparatus of Lepidosauria*. Society for the Study of Amphibians and Reptiles, Ithaca, NY. Contributions to Herpetology Vol. 24, pp. 1–465.

Russell, M. 1998. Tuatara, relics of a lost age. *Cold Blooded News* 25: 1–4.

S., N. M. 1966. Tuatara, *Sphenodon punctatus* (Gray, 1842). *International Union for Conservation of Nature and Natural Resources Bulletin* 2: 7–8.

Saint Girons, H. 1980. Thermoregulation in reptiles with special reference to the tuatara and its ecophysiology. *Tuatara* 24: 59–80.

Saint-Girons, H. 1982. Histologie comparée des glandes orbitaires des Lépidosauriens. *Annales des Sciences Naturelles, Zoologie, Paris* 4: 171–191.

Saint Girons, H. 1983. Le *Sphenodon*, particularités écologiques et hypothèses sur son évolution. *Bulletin de la Societe Zoologique de France* 108: 631–634.

Saint Girons, H. 1985. The *Sphenodon*: ecological features and some hypotheses concerning its evolution. *Bulletin of the Chicago Herpetological Society* 20: 48–51.

Saint Girons, H. and Newman, D. G. 1987. The reproductive cycle of the male tuatara, *Sphenodon punctatus*, on Stephens Island, New Zealand. *New Zealand Journal of Zoology* 14: 231–237.

Saint Girons, H., Bell, B. D. and Newman, D. G. 1980. Observations on the activity and thermoregulation of the tuatara, *Sphenodon punctatus* (Reptilia: Rhynchocephalia), on Stephens Island. *New Zealand Journal of Zoology* 7: 551–556.

Sandager, F. 1890. Observations on the Mokohinau Islands and the birds which visit them. *Transactions and Proceedings of the New Zealand Institute* 22: 286–294.

Sanderson, K. 2009. Temperature preference of acclimated tuatara, *Sphenodon punctatus*. *Transactions of the Royal Society of South Australia* 133: 178–180.

Sauerbeck, E. 1906. Eine Gehirnmißbildung bei *Hatteria punctata* (*Sphenodon punctatus*). [Eversio encephali e neuroporo, transgressus persistens laminae nervosa in epidermidem. Anophthalmia duplex partialis (Defectus oculi nervosi et lentis).]. *Nova Acta. Abhandlungen der Kaiserlich Leopoldinisch-Carolinische Deutschen Akademie der Naturforscher, Halle* 85: 1–116 + Tafeln 1–2.

Säve-Söderbergh, G. 1946. On the fossa hypophyseos and the attachment of the retractor bulbi group in *Sphenodon*, *Varanus*, and *Lacerta*. *Arkiv für Zoologi* 38A: 1–24.

Säve-Söderbergh, G. 1947. Notes on the brain-case in *Sphenodon* and certain Lacertilia. *Zoologiska Bidrag fran Uppsala* 25: 489–516.

Schaeffer, B. 1941. The morphological and functional evolution of the tarsus in amphibians and reptiles. *Bulletin of the American Museum of Natural History* 78: 395–472.

Schaerlaeken, V., Herrel, A., Aerts, P. and Ross, C. F. 2008. The functional significance of the lower temporal bar in *Sphenodon punctatus*. *Journal of Experimental Biology* 211: 3908–3914.

Schauinsland, H. 1898a. Zur Entwickelung von *Hatteria*. *Sitzungsberichte der Akademie der Wissenschaften zu Berlin* 44: 629–631.

Schauinsland, H. 1898b. Beiträge zur Biologie der *Hatteria*. *Sitzungsberichte der Akademie der Wissenschaften zu Berlin* 44: 701–704.

Schauinsland, H. 1899. Beiträge zur Biologie und Entwickelung der *Hatteria* nebst Bemerkungen über die Entwickelung der Sauropsiden. *Anatomischer Anzeiger* 15: 309–334 + Tafeln II–III.

Schauinsland, H. 1900. Weitere Beiträge zur Entwicklungsgeschichte der *Hatteria*. Skelettsystem, schallleitender Apparat, Hirnnerven etc. *Archiv fur Mikroskopische Anatomie und Entwicklungsgeschichte* 56: 747–867+ Tafeln XXXII–XXXIV.

Schauinsland, H. 1903a. Beiträge zur Entwickelungsgeschichte und Anatomie der Wirbeltiere I. *Sphenodon*, *Callorhynchus*, *Chamäleo*. *Zoologica, Stuttgart* 16: 1–98 + Tafeln I–XXXI.

Schauinsland, H. 1903b. Beiträge zur Entwickelungsgeschichte und Anatomie der Wirbeltiere II. Studien zur Entwickelungsgeschichte de Sauropsiden. *Zoologica, Stuttgart* 16: 101–143 + Tafeln XXXII–XLVII.

Schauinsland, H. 1903c. Beiträge zur Entwickelungsgeschichte und Anatomie der Wirbeltiere III. Beiträge zur Kenntniss der Eihäute der Sauropsiden. *Zoologica, Stuttgart* 16: 147–168 + Tafeln L–LVI.

Schauinsland, H. 1906a. Die Entwicklung der Eihäute der Reptilien und der Vögel. In: Hertwig, O. (ed.). *Handbuch der vergleichenden und experimentellen Entwicklungslehre der Wirbeltiere. Vol. 1, part 2*. Gustav Fischer, Jena, pp. 177–234.

Schauinsland, H. 1906b. Die Entwicklung der Wirbelsäule nebst Rippen und Brustbein. In: Hertwig, O. (ed.). *Handbuch der vergleichenden und experimentellen Entwicklungslehre der Wirbeltiere. Vol. 3, part 2*. Gustav Fischer, Jena, pp. 339–572.

Schmid, K. L., Howland, H. C. and Howland, M. 1992. Focusing and accommodation in tuatara (*Sphenodon punctatus*). *Journal of Comparative Physiology A* 170: 263–266.

Schmidt, K. P. 1949. To a tuatara alive in my hand. *Tuatara* 2: 90.

Schmidt, K. P. 1952. References to the tuatara in the Stephen Island letter book. *Fieldiana Zoology* 34: 1–10.

Schmidt, K. P. 1953. A visit to Karewa Island, home of the tuatara. *Fieldiana Zoology* 34: 153–164.

Schmidt-Nielsen, B. and Schmidt, D. 1973. Renal function of

Sphenodon punctatum. *Comparative Biochemistry and Physiology* 44A: 121–129.
Schreiber, G., Pettersson, T. M., Southwell, B. R., Aldred, A. R., Harms, P. J., Richardson, S. J., Wettenhall, R. E. H., Duan, W. and Nicol, S. C. 1993. Transthyretin expression evolved more recently in liver than in brain. *Comparative Biochemistry and Physiology* 105B: 317–325.
Schwab, I. R. and O'Connor, G. R. 2004. An enigmatic eye: what can we learn? *Clinical and Experimental Ophthalmology* 32: 559–560.
Schwab, I. R. and O'Connor, G. R. 2005. The lonely eye. *British Journal of Ophthalmology* 89: 256.
Schwenk, K. 1986. Morphology of the tongue in the tuatara, *Sphenodon punctatus* (Reptilia: Lepidosauria), with comments on function and phylogeny. *Journal of Morphology* 188: 129–156.
Schwenk, K. 2000. Feeding in lepidosaurs. In: Schwenk, K. (ed.). *Feeding: form, function, and evolution in tetrapod vertebrates*. Academic Press, San Diego, pp. 175–291.
Sclater, P. L. 1870. Recent additions to the Zoological Society's Gardens. *Nature* 2: 146–148.
Sclater, P. L. 1871. New Zealand animals in the Zoological Society's Gardens. *Nature* 3: 190–192.
Seligmann, H., Beiles, A. and Werner, Y. L. 2003. More injuries in left-footed individual lizards and *Sphenodon*. *Journal of Zoology, London* 260: 129–144.
Seligmann, H., Moravec, J. and Werner, Y. L. 2008. Morphological, functional and evolutionary aspects of tail autotomy and regeneration in the 'living fossil' *Sphenodon* (Reptilia: Rhynchocephalia). *Biological Journal of the Linnean Society* 93: 721–743.
Sewell, F. W. 1931. [On catching tuatara on Mokohinau Island]. *Dominion* (page and date unknown; copy in ANZ M 1 25/611 pt 3).
Sharell, R. 1966. *The tuatara, lizards and frogs of New Zealand*. Collins, London (revised 1975). 94 p.
Shedlock, A. M. 2006. Phylogenomic investigation of CR1 LINE diversity in reptiles. *Systematic Biology* 55: 902–911.
Sherley, G. H., Stringer, I. A. N. and Parrish, G. R. 2010. Summary of native bat, amphibian and terrestrial invertebrate translocations in New Zealand. *Science for Conservation* No. 303. Department of Conservation, Wellington. 39 p.
Siebenrock, F. 1893. Zur Osteologie des *Hatteria* – Kopfes. *Sitzungsberichte der kaiserlichen Akademie der Wissenschaften in Wien, Mathematisch-naturwissenschaftliche Classe* 102: 250–268 + 1 Tafel.
Siebenrock, F. 1894. A contribution to the osteology of the head of *Hatteria*. *Annals and Magazine of Natural History* 13: 297–311.
Simons, J. R. 1965. The heart of the tuatara *Sphenodon punctatus*. *Journal of Zoology* 146: 451–466.
Skegg, P. D. G. 1963. Birds of the Mercury Islands Group. *Notornis* 10: 153–168.
Sladden, B. 1924. Karewa: an island sanctuary. *New Zealand Journal of Science and Technology* 7: 182–187.
Sladden, B. and Falla, R. A. 1928. Alderman Islands. A general description, with notes on the flora and fauna (continued). *New Zealand Journal of Science and Technology* 9: 282–290.
Southey, I. C. 1985. *The ecology of three rare skinks on Middle Island, Mercury Islands*. Unpublished MSc thesis, University of Auckland, Auckland. 99 p.
Spencer, W. B. 1886a. Preliminary communication on the structure and presence in *Sphenodon* and other lizards of the median eye, described by van Graaf in *Anguis fragilis*. *Proceedings of the Royal Society of London* 40: 559–565.
Spencer, W. B. 1886b. The parietal eye of *Hatteria*. *Nature* 34: 33–35.
Spencer, W. B. 1886c. On the presence and structure of the pineal eye in Lacertilia. *Quarterly Journal of Microscopical Science* 27: 165–238 + plates XIV–XX.
Stack, J. W. 1875. On the disappearance of the larger kinds of lizard from North Canterbury. *Transactions and Proceedings of the New Zealand Institute* 7: 295–297.
Stack, J. W. 1898. *South Island Maoris: a sketch of their history and legendary lore*. Whitcombe & Tombs Ltd, Christchurch. Reprinted 1984, Capper Press, Christchurch. 136 p.
Stebbins, R. C. 1958. An experimental study of the 'third eye' of the tuatara. *Copeia* 1958: 183 190.
Stebbins, R. C. and Eakin, R. M. 1958. The role of the 'third eye' in reptilian behavior. *American Museum Novitates* 1870: 1–40.
Subramanian, S., Hay, J. M., Mohandesan, E., Millar, C. D. and Lambert, D. M. 2008. Molecular and morphological evolution in tuatara are decoupled. *Trends in Genetics* 25: 16–18.
Tanaka, Y. 1998. Structure of the reptilian spleen. In: Gans, C. and Gaunt, A. S. (eds). *Biology of the Reptilia. Vol. 19. Morphology G. Visceral organs*. Society for the Study of Amphibians and Reptiles, Ithaca, New York, pp. 533–586.
Tarakawa, T. 1911. Ko tuatara raua ko kumukumu: he korero tara (Tuatara and kumukumu: a fable; translated by the editor). *Journal of the Polynesian Society* 20: 39–41.
Taylor, G. A. 1991. Flora and fauna of Plate (Motunau) Island, Bay of Plenty. *Tane* 33: 113–120.
Taylor, R. 1848. *A leaf from the natural history of New Zealand, or, a vocabulary of its different productions, &c., &c., with their native names*. Robert Stokes, Wellington. 102 p.
Taylor, R. 1855. *Te Ika a Maui, or, New Zealand and its inhabitants: illustrating the origin, manners, customs, mythology, religion, rites, songs, proverbs, fables, and language of the natives ...* Wertheim & MacIntosh, London. 490 p.
Tennyson, A. and Pierce, R. 1995. The presence of Pycroft's Petrel (*Pterodroma pycrofti*) and other petrels on Mauitaha Island, New Zealand. *Notornis* 42: 212–214.
Terezow, M. 2005. *Circadian and ontogenetic changes in activity and anti-predator responses of captive juvenile tuatara (*Sphenodon *spp.)*. Unpublished MSc thesis, Victoria University of Wellington, Wellington. 85 p.
Terezow, M. and Markwell, T. 2005. The circadian locomotor activity of tuatara (*Sphenodon* spp.): ontogenetic changes and proximate causes. In: Abstracts of papers presented at the 11th Biennial Conference of the Society for Research on Amphibians and Reptiles in New Zealand, Springbrook National Park, south-east Queensland, Australia, 7–11 February 2005. *New Zealand Journal of Zoology* 32: 229.
Terezow, M. G., Nelson, N. J. and Markwell, T. J. 2008. Circadian emergence and movement of captive juvenile tuatara (*Sphenodon* spp.). *New Zealand Journal of Zoology* 35: 205–216.
Tetens, V., Brittain, T., Christie, D. L., Robb, J. and Wells, R. M. G. 1984. Characterization and function of isolated hemoglobins from the tuatara, *Sphenodon punctatus* (Reptilia: O. Rhynchocephalia). *Comparative Biochemistry and Physiology* 79B: 119–123.
The Collector. 1913. Ancient reptile: the disappearing tuatara: efforts at preservation. *Evening Post*, Wellington, 11 June, p. 4.
Thilenius, G. 1899. Vorläufiger Bericht über die Eiablage und erste Entwickelung der *Hatteria punctata*. *Sitzungsberichte der Königlich Preussischen Akademie der Wissenschaften zu Berlin* 1: 247–256.
Thomas, A. P. W. 1890. Preliminary note on the development of the tuatara (*Sphenodon punctatum*). *Proceedings of the Royal Society of London B* 48: 152–156.

Thomas, A. P. W. 1891. Preliminary note on the development of the tuatara (*Sphenodon punctatum*). *New Zealand Journal of Science* 1: 27–31.

Thompson, M. and Daugherty, C. 1992. Living a lie: New Zealand's tuatara. *Australian Natural History* 23: 928–935.

Thompson, M. B. 1989. Patterns of metabolism in embryonic reptiles. *Respiration Physiology* 76: 243–256.

Thompson, M. B. 1990. Incubation of eggs of tuatara, *Sphenodon punctatus. Journal of Zoology, London* 222: 303–318.

Thompson, M. B. and Daugherty, C. H. 1997. Metabolism of tuatara, *Sphenodon punctatus*, from Stephens Island. In: Abstracts of papers presented at the 7th Conference of the Society for Research on Amphibians and Reptiles in New Zealand, Kaikoura, New Zealand, 31 January–2 February 1997. *New Zealand Journal of Zoology* 24: 329.

Thompson, M. B. and Daugherty, C. H. 1998. Metabolism of tuatara, *Sphenodon punctatus. Comparative Biochemistry and Physiology* 119A: 519–522.

Thompson, M. B., Newman, D. G. and Watson, P. R. 1991. Use of oxytocin in obtaining eggs from tuatara (*Sphenodon punctatus*). *Journal of Herpetology* 25: 101–104.

Thompson, M. B., Daugherty, C. H., Cree, A., French, D. C., Gillingham, J. C. and Barwick, R. E. 1992. Status and longevity of the tuatara, *Sphenodon guntheri,* and Duvaucel's gecko, *Hoplodactylus duvaucelii,* on North Brother Island, New Zealand. *Journal of the Royal Society of New Zealand* 22: 123–130.

Thompson, M. B., Packard, G. C., Packard, M. J. and Rose, B. 1996. Analysis of the nest environment of tuatara *Sphenodon punctatus. Journal of Zoology, London* 238: 239–251.

Thompson, M. B., Newman, D. G. and Watson, P. R. 1998. Influence of X-rays on incubation in tuatara, *Sphenodon punctatus. New Zealand Journal of Zoology* 25: 295–300.

Thomson, J. A. 1915. The existing state of the tuatara (*Sphenodon punctatus* Gray), with some notes on its habits. Annual Report of the Dominion Museum for the year ending 31st March 1915. *Appendix to the Journals of the House of Representatives* H–33: 22–26.

Thomson, J. A. 1920. Report of the Director, Dominion Museum. *Appendix to the Journals of the House of Representatives* H–22: 12–17.

Thoresen, A. C. 1967. Ecological observations on Stanley and Green Islands Mercury Group. *Notornis* 14: 182–200.

Tintinger, V. 1987. Breeding the tuatara *Sphenodon punctatus* at Auckland Zoo. *International Zoo Yearbook* 26: 183–186.

Toft, R. 1999. Prehistoric treasures. *Pet Reptile* July: 25–28.

Towns, D. R. 1991. Response of lizard assemblages in the Mercury Islands, New Zealand, to removal of an introduced rodent: the kiore (*Rattus exulans*). *Journal of the Royal Society of New Zealand* 21: 119–136.

Towns, D. R. 2004. Sphenodontia: Tuatara (*Sphenodontidae*). In: Hutchins, M., Thoney, D. A. and McDade, M. C. (eds). *Grzimek's Animal Life Encyclopedia. Vol. 7. Reptiles*. Gale, Detroit, pp. 189–193.

Towns, D. R. and Daugherty, C. H. 1994. Patterns of range contractions and extinctions in the New Zealand herpetofauna following human colonisation. *New Zealand Journal of Zoology* 21: 325–339.

Towns, D. R. and Hayward, B. W. 1973. Reptiles of the Aldermen Islands. *Tane* 19: 93–100.

Towns, D. R., Atkinson, I. A. E. and Daugherty, C. H. 1990. The potential for ecological restoration in the Mercury Islands. In: Towns, D. R., Atkinson, I. A. E. and Daugherty, C. H. (eds). *Ecological restoration of New Zealand islands.* Conservation Sciences Publication No. 2. Department of Conservation, Wellington, pp. 91–108.

Towns, D. R., Daugherty, C. H. and Cree, A. 2001. Raising the prospects for a forgotten fauna: a review of 10 years of conservation effort for New Zealand reptiles. *Biological Conservation* 99: 3–16.

Towns, D. R., Parrish, G. R., Tyrrell, C. L., Ussher, G. T., Cree, A., Newman, D. G., Whitaker, A. H. and Westbrooke, I. 2007. Responses of tuatara (*Sphenodon punctatus*) to removal of introduced Pacific rats from islands. *Conservation Biology* 21: 1021–1031.

Towns, D. R., Bellingham, P. J., Mulder, C. P. H. and Lyver, P. O'B. 2012. A research strategy for biodiversity conservation on New Zealand's offshore islands. *New Zealand Journal of Ecology* 36: 1–20.

Townsend, T. M., Larson, A., Louis, E. and Macey, J. R. 2004. Molecular phylogenetics of Squamata: the position of snakes, amphisbaenians, and dibamids, and the root of the squamate tree. *Systematic Biology* 53: 735–757.

Tracy, M. R. 1997. Size variation in tuatara. In: Abstracts of papers presented at the 7th Conference of the Society for Research on Amphibians and Reptiles in New Zealand, Kaikoura, New Zealand, 31 January–2 February 1997. *New Zealand Journal of Zoology* 24: 330.

Tribe, M. and Brambell, F. W. R. 1932. The origin and migration of the primordial germ-cells of *Sphenodon punctatus. Quarterly Journal of Microscopical Science* 75: 251–282 + plates 16–17.

Tribe, M. and Fisk, A. 1940. The development of the hepatic venous system and excretory system in *Sphenodon punctatus. Proceedings of the Zoological Society of London Series B* 110: 153–182 + plates I–II.

Tristem, M., Myles, T. and Hill, F. 1995. A highly divergent retroviral sequence in the tuatara (*Sphenodon*). *Virology* 210: 206–211.

Tsuihiji, T. 2007. Homologies of the longissimus, iliocostalis, and hypaxial muscles in the anterior presacral region of extant Diapsida. *Journal of Morphology* 268: 986–1020.

Twentyman, C. 1999. Diseases in New Zealand reptiles. *Surveillance* 26: 3–5.

Tyrrell, C. 1993. *Corticosterone cycles in wild and captive tuatara (*Sphenodon punctatus*)*. Unpublished MSc thesis, University of Otago, Dunedin. 103 p.

Tyrrell, C. L. 2000. *Reproductive ecology of northern tuatara (*Sphenodon punctatus punctatus*)*. Unpublished PhD thesis, University of Otago, Dunedin. 175 p.

Tyrrell, C. 2001. *Sphenodon punctatus punctatus* (northern tuatara): reproduction. *Herpetological Review* 32: 39–40.

Tyrrell, C. and Cree, A. 1994. Plasma corticosterone concentrations in wild and captive juvenile tuatara (*Sphenodon punctatus*). *New Zealand Journal of Zoology* 21: 407–416.

Tyrrell, C. L. and Cree, A. 1998. Relationships between corticosterone concentration and season, time of day and confinement in a wild reptile (tuatara, *Sphenodon punctatus*). *General and Comparative Endocrinology* 110: 97–108.

Tyrrell, C. and Cree, A. 1999. Hormonal response to short-term confinement stress in northern tuatara (*Sphenodon punctatus punctatus*) on rodent-free and rodent-inhabited islands. In: Abstracts of papers presented at the 8th Conference of the Society for Research on Amphibians and Reptiles in New Zealand, Great Barrier Island, New Zealand, 5–7 February 1999. *New Zealand Journal of Zoology* 26: 261.

Tyrrell, C., Cree, A. and Guillette Jr, L. J. 1993. Low plasma corticosterone concentrations in wild tuatara (*Sphenodon punctatus*). In: Abstracts of papers presented at the 4th Conference of the Society for Research on Amphibians and Reptiles in New Zealand, Invercargill, New Zealand, 27–29 November 1992. *New Zealand Journal of Zoology* 20: 127.

Tyrrell, C. L., Cree, A. and Towns, D. R. 2000. *Variation in reproduction and condition of northern tuatara (*Sphenodon punctatus punctatus*) in the presence and absence of kiore*. Science for Conservation No. 153. Department of Conservation, Wellington. 42 p.

Tytle, T. 1988. The status of New Zealand reptiles exported live to the United States. *Moko*: 9–16.

Underwood, G. 1970. The eye. In: Gans, C. and Parsons, T. S. (eds). *Biology of the Reptilia. Vol. 2. Morphology B*. Academic Press, London, pp. 1–97.

Ung, C. Y.-J. and Molteno, A. C. B. 2004. An enigmatic eye: the histology of the tuatara pineal complex. *Clinical and Experimental Ophthalmology* 32: 614–618.

Unger, L. 1914. Untersuchungen über die Morphologie und Faserung des Reptiliengehirnes. III. Das Vorderhirn der *Hatteria punctata* (*Sphenodon punctatum*). *Sitzungsberichte der mathematisch-naturwissenschaftliche. Klasse der kaiserlichen Akademie der Wissenschaften, Wien* 123: 293–318 + 3 Tafeln.

Unthank, H. W. 1909. [On a skull of *Sphenodon* with abnormal nasal region]. *Proceedings of the Zoological Society of London* 11: 666–667.

Ussher, G. T. 1995. *Feeding ecology and dietary interactions of tuatara and kiore on the Chicken Islands*. Unpublished MSc thesis, University of Auckland, Auckland. 182 p.

Ussher, G. T. 1999a. Dietary competition between kiore (*Rattus exulans*) and tuatara (*Sphenodon punctatus*) on the Chickens Islands. In: Abstracts of papers presented at the 8th Conference of the Society for Research on Amphibians and Reptiles in New Zealand, Great Barrier Island, New Zealand, 5–7 February 1999. *New Zealand Journal of Zoology* 26: 261–262.

Ussher, G. T. 1999b. Method for attaching radiotransmitters to medium-sized reptiles: trials on tuatara (*Sphenodon punctatus*). *Herpetological Review* 30: 151–153.

Ussher, G. 1999c. *Restoration of threatened species populations: tuatara rehabilitations and re-introductions*. Unpublished PhD thesis, University of Auckland, Auckland. 217 p.

Ussher, G. T. 1999d. Tuatara (*Sphenodon punctatus*) feeding ecology in the presence of kiore (*Rattus exulans*). *New Zealand Journal of Zoology* 26: 117–125.

Various authors. 1982. General discussion on the tuatara. In: Newman, D. G. (ed.). *New Zealand herpetology*. New Zealand Wildlife Service Occasional Publication No. 2, Wellington, pp. 289–298.

Vidal, N. and Hedges, S. B. 2005. The phylogeny of squamate reptiles (lizards, snakes, and amphisbaenians) inferred from nine nuclear protein-coding genes. *Comptes Rendus Biologies* 328: 1000–1008.

Vilter, V. 1951a. Valeur morphologique des photorécepteurs rétiniens chez la Hatterie (*Sphenodon punctatus*). *Comptes Rendus Société de Biologie* 145: 20–23.

Vilter, V. 1951b. Organisation générale de la rétine nerveuse chez le *Sphenodon punctatus*. *Comptes Rendus Société de Biologie* 145: 24–26.

Vilter, V. 1951c. Recherches sur les structures fovéales dans la rétine du *Sphenodon punctatus*. *Comptes Rendus Société de Biologie* 145: 26–29.

Virchow, H. 1901. Ueber die Netzhaut von *Hatteria*. *Sitzungsberichte der Gesellschaft Naturforschender Freunde zu Berlin* 1901: 42–62.

von Wettstein, O. 1931. 1. Ordnung der Klasse Reptilia: Rhynchocephalia. In: Kükenthal, W. G. and Krumbach, T. (eds). *Handbuch der Zoologie: eine Naturgeschichte der Stamme des Tierreiches. Vol. Bd 7 Half 1 Lfg 1–2*. W. de Gruyter, Berlin, pp. 1–235.

von Wettstein, O. 1943. *Sphenodon punctatus reischeki* nov. subsp. *Zoologisches Anzeiger* 143: 45–47.

Wade, W. R. 1842. *A journey in the northern island of New Zealand: interspersed with various information relative to the country and people*. George Rolwegan, Hobart Town. 206 p.

Walls, G. L. 1935. A comprehensive morphology of *Sphenodon*. *Science* (New Series) 82: 391.

Walls, G. L. 1942. *The vertebrate eye and its adaptive radiation*. The Cranbrook Institute of Science, New York. Reprinted 1963, Hafner Publishing, New York. 785 p.

Walls, G. Y. 1978. The influence of the tuatara on fairy prion breeding on Stephens Island, Cook Strait. *New Zealand Journal of Ecology* 1: 91–98.

Walls, G. Y. 1981. Feeding ecology of the tuatara (*Sphenodon punctatus*) on Stephens Island, Cook Strait. *New Zealand Journal of Ecology* 4: 89–97.

Walls, G. Y. 1982. Provisional results from a study of the feeding ecology of the tuatara (*Sphenodon punctatus*) on Stephens Island. In: Newman, D. G. (ed.). *New Zealand herpetology*. New Zealand Wildlife Service Occasional Publication No. 2, Wellington, pp. 271–276.

Walls, G. Y. 1983. Activity of the tuatara and its relationships to weather conditions on Stephens Island, Cook Strait, with observations on geckos and invertebrates. *New Zealand Journal of Zoology* 10: 309–318.

Wang, Z., Miyake, T., Edwards, S. V. and Amemiya, C. T. 2006. Tuatara (*Sphenodon*) genomics: BAC library construction, sequence survey, and application to the DMRT gene family. *Journal of Heredity* 97: 541–548.

Ward, H. A. 1882. The *Hatteria* (*Sphenodon*) *punctatus*. *Ward's Natural Science Bulletin* 1: 14–15.

Waters, J. M. and Craw, D. 2006. Goodbye Gondwana? New Zealand biogeography, geology, and the problem of circularity. *Systematic Biology* 55: 351–356.

Weatherhead, B. 1971. Cytology of the neuro-intermediate lobe of the tuatara, *Sphenodon punctatus* Gray. *Zeitschrift für Zellforschung und Mikroskopische Anatomie* 119: 21–42.

Webb, G. J. W., Heatwole, H. and de Bavay, J. 1974. Comparative cardiac anatomy of the Reptilia II. A critique of the literature on the Squamata and Rhynchocephalia. *Journal of Morphology* 142: 1–20.

Webber, G. W. 1953. Tuatara quest: the doctor from Bremen. *Weekly News*, Auckland, 24 June, p. 36.

Weber, R. E., Kleinschmidt, T., Abbassi, A., Wells, R. M. G. and Braunitzer, G. 1989. Allosteric transition in hemoglobin $\alpha_2{}^A\beta_2{}^1$ from the rhynchocephalian reptile relict *Sphenodon punctatus*. *Hemoglobin* 13: 625–636.

Wells, R. M. G. 1991. Respiration and haemoglobin function in the tuatara, *Sphenodon punctatus*. In: Abstracts of papers presented at the 3rd Annual Conference of the Society for Research on Amphibians and Reptiles in New Zealand, Whitianga, Coromandel Peninsula, 28–30 November 1990. *New Zealand Journal of Zoology* 18: 347.

Wells, R. M. G., Tetens, V. and Brittain, T. 1983. Absence of cooperative haemoglobin-oxygen binding in *Sphenodon*, a reptilian relict from the Triassic. *Nature* 306: 500–502.

Wells, R. M. G., Tetens, V., Housley, G. D., Young, A. A., Dawson, N. J. and Johansen, K. 1990. Effect of temperature on control of breathing in the cryophilic rhynchocephalian reptile, *Sphenodon punctatus. Comparative Biochemistry and Physiology* 96A: 333–340.

Werner, F. 1893. Beobachtungen an *Sphenodon* (*Hatteria*) *punctatus. Zoologische Garten* 34: 335–339.

Werner, G. 1962. Das Cranium der Brückenechse, *Sphenodon punctatus* Gray, von 58 mm Gesamtlänge. *Zeitschrift für Anatomie und Entwickelungsgeschichte* 123: 323–368.

Werner, G. 1963. Über das vitalische Organ bei *Sphenodon punctatus* Gray. *Zeitschrift für Anatomie und Entwickelungsgeschichte* 123: 498–504.

Werner, Y. L. and Whitaker, A. H. 1978. Observations and comments on the body temperatures of some New Zealand reptiles. *New Zealand Journal of Zoology* 5: 375–393.

Whitaker, A. H. 1978. The effects of rodents on reptiles and amphibians. *New Zealand Department of Lands and Survey Information Series* 4: 75–86.

White, F. N. 1959. Circulation in the reptilian heart (Squamata). *Anatomical Record* 135: 129–134.

Whiteside, D. I. 1986. The head skeleton of the Rhaetian sphenodontid *Diphydontosaurus avonis* gen. et sp. nov. and the modernizing of a living fossil. *Philosophical Transactions of the Royal Society of London B* 312: 379–430.

Whitworth, E. 2006. *Photothermal orientation and factors associated with egg incubation success in tuatara (*Sphenodon punctatus*).* Unpublished MSc thesis, University of Otago, Dunedin. 95 p.

Wilkinson, M. and Benton, M. J. 1996. Sphenodontid phylogeny and the problems of multiple trees. *Philosophical Transactions of the Royal Society of London B* 351: 1–16.

Willnow, I. and Willnow, R. 1976. Bauplan der Lunge von *Sphenodon punctatus. Acta Anatomica* 94: 504–519.

Wilson, J. 2010. *Population viability and resource competition on North Brother Island: conservation implications for tuatara (*Sphenodon punctatus*) and Duvaucel's gecko (*Hoplodactylus duvaucelii*).* Unpublished MSc thesis, Victoria University of Wellington, Wellington. 99 p.

Wilson, K. J. and Lee, A. K. 1970. Changes in oxygen consumption and heart-rate with activity and body temperature in the tuatara, *Sphenodon punctatum. Comparative Biochemistry and Physiology* 33: 311–322.

Woerner, L. L. B. and Nelson, N. J. 2010. Competition for space and food in tuatara (*Sphenodon* spp.) head-starting facilities. In: Abstracts of papers presented at the Second Meeting of Australasian Societies for Herpetology (Society for Research on Amphibians and Reptiles in New Zealand and the Australia Society for Herpetologists), Massey University, Auckland, New Zealand, 20–22 February 2009. *New Zealand Journal of Zoology* 37: 104–105.

Wojtusiak, R. J. 1973. Some ethological and biological observations on the tuatara in laboratory conditions. *Tuatara* 20: 97–109.

Wojtusiak, R. J. and Majlert, Z. 1973. Bioacoustics of the voice of the tuatara, *Sphenodon punctatus punctatus. New Zealand Journal of Science* 16: 305–313.

Woo, K. L. 2004. *Acquisition of a learned operant and critical flicker-fusion rate in the tuatara (*Sphenodon *spp.).* Unpublished MSc thesis, Victoria University of Wellington, Wellington. 70 p.

Woo, K. L., Hunt, M., Harper, D., Nelson, N. J., Daugherty, C. H. and Bell, B. D. 2005. Acquisition of a learned operant and critical flicker-fusion rate in tuatara (*Sphenodon*). In: Abstracts of papers presented at the 11th Biennial Conference of the Society for Research on Amphibians and Reptiles in New Zealand, Springbrook National Park, south-east Queensland, Australia, 7–11 February 2005. *New Zealand Journal of Zoology* 32: 232.

Woo, K. L., Hunt, M., Harper, D., Nelson, N. J., Daugherty, C. H. and Bell, B. D. 2009. Discrimination of flicker frequency rates in the reptile tuatara (*Sphenodon*). *Naturwissenschaften* 96: 415–419.

Wood, D. 1967. Breeding tuataras *Sphenodon punctatus* at Auckland Zoo. *International Zoo Yearbook* 7: 178–179.

Wood, J. 2006. Subfossil kakapo (*Strigops habroptilus*) remains from near Gibraltar Rock, Cromwell Gorge, Central Otago, New Zealand. *Notornis* 53: 191–193.

Wood, J. R. 2009. Two Late Quaternary avifaunal assemblages from the Dunback district, eastern Otago, South Island, New Zealand. *Notornis* 56: 154–157.

Woodland, W. N. F. 1921. Some observations on caudal autotomy and regeneration in the gecko (*Hemidactylus flaviviridis* Ruppel), with notes on the tails of *Sphenodon* and *Pygopus. Quarterly Journal of Microscopical Science* 65: 63–100.

Wörner, L. L. B. 2009. *Aggression and competition for space and food in captive juvenile tuatara (*Sphenodon punctatus*).* Unpublished MSc thesis, Victoria University of Wellington, Wellington. 103 p.

Worthy, T. H. 1984. Faunal and floral remains from F1, a cave near Waitomo. *Journal of the Royal Society of New Zealand* 14: 367–377.

Worthy, T. H. 1991. Fossil skink bones from Northland, New Zealand, and description of a new species of *Cyclodina*, Scincidae. *Journal of the Royal Society of New Zealand* 21: 329–348.

Worthy, T. H. 1997. Quaternary fossil fauna of South Canterbury, South Island, New Zealand. *Journal of the Royal Society of New Zealand* 27: 67–162.

Worthy, T. H. 1998a. Quaternary fossil faunas of Otago, South Island, New Zealand. *Journal of the Royal Society of New Zealand* 28: 421–521.

Worthy, T. H. 1998b. The Quaternary fossil avifauna of Southland, South Island, New Zealand. *Journal of the Royal Society of New Zealand* 28: 537–589.

Worthy, T. H. 1998c. A remarkable fossil and archaeological avifauna from Marfells Beach, Lake Grassmere, South Island, New Zealand. *Records of the Canterbury Museum* 12: 79–176.

Worthy, T. H. 1998d. *Fossil deposits in Megamania Cave, Gunner River, South Island, New Zealand.* Conservation Advisory Science Notes No. 195. Department of Conservation, Wellington. 14 p.

Worthy, T. H. 2000. Two late-Glacial avifaunas from eastern North Island, New Zealand – Te Aute Swamp and Wheturau Quarry. *Journal of the Royal Society of New Zealand* 30: 1–26.

Worthy, T. H. 2001. A fossil vertebrate fauna accumulated by laughing owls (*Sceloglaux albifacies*) on the Gouland Downs, northwest Nelson, South Island. *Notornis* 48: 225–233.

Worthy, T. H. and Grant-Mackie, J. A. 2003. Late-Pleistocene avifaunas from Cape Wanbrow, Otago, South Island, New Zealand. *Journal of the Royal Society of New Zealand* 33: 427–485.

Worthy, T. H. and Holdaway, R. N. 1993. Quaternary fossil faunas from caves in the Punakaiki area, West Coast, South Island, New Zealand. *Journal of the Royal Society of New Zealand* 23: 147–254.

Worthy, T. H. and Holdaway, R. N. 1994. Quaternary fossil faunas from caves in Takaka Valley and on Takaka Hill, northwest Nelson, South Island, New Zealand. *Journal of the Royal Society of New Zealand* 24: 297–391.

Worthy, T. H. and Holdaway, R. N. 1995. Quaternary fossil faunas from caves on Mt Cookson, North Canterbury, South Island, New Zealand. *Journal of the Royal Society of New Zealand* 25: 333–370.

Worthy, T. H. and Holdaway, R. N. 1996. Quaternary fossil faunas, overlapping taphonomies, and palaeofaunal reconstruction in North Canterbury, South Island, New Zealand. *Journal of the Royal Society of New Zealand* 26: 275–361.

Worthy, T. H. and Holdaway, R. N. 2000. Terrestrial fossil vertebrate faunas from inland Hawke's Bay, North Island, New Zealand. Part 1. *Records of the Canterbury Museum* 14: 89–154.

Worthy, T. H. and Holdaway, R. N. 2002. *The lost world of the moa: prehistoric life of New Zealand.* Canterbury University Press, Christchurch. 718 p.

Worthy, T. H. and Roscoe, D. 2003. Takaka Fossil Cave – a stratified Late Glacial to Late Holocene deposit from Takaka Hill, New Zealand. *Tuhinga* 14: 41–60.

Worthy, T. H., Holdaway, R. N., Alloway, B. V., Jones, J., Winn, J. and Turner, D. 2002a. A rich Pleistocene-Holocene avifaunal sequence from Te Waka #1: terrestrial fossil vertebrate faunas from inland Hawke's Bay, North Island, New Zealand. Part 2. *Tuhinga* 13: 1–38.

Worthy, T. H., Tennyson, A. J. D., Jones, C. and McNamara, J. A. 2002b. A diverse early-Miocene (15–20 MA) terrestrial fauna from New Zealand reveals snakes and mammals. In: Brock, G. A. and Talent, J. A. (eds). *IPC2002 (First International Palaeontological Congress).* Geological Society of Australia, Sydney, pp. 174–175 (abstract).

Worthy, T. H., Miskelly, C. M. and Ching, B. R. A. 2002c. Taxonomy of South Island snipe (Aves: Scolopacidae: Coenocorypha), with analysis of a remarkable collection of snipe bones from Greymouth, New Zealand. *New Zealand Journal of Zoology* 29: 231–244.

Wright, A. 1961. Fairy prion chick attacked by tuatara. *Notornis* 9: 133.

Wright, A. 1963. Predation on fairy prions. *Notornis* 10: 187.

Wu, X.-C. 2003. Functional morphology of the temporal region in the Rhynchocephalia. *Canadian Journal of Earth Science* 40: 589–607.

Wyeth, F. J. 1920. On the development of the auditory apparatus in *Sphenodon punctatus. Proceedings of the Royal Society of London B* 91: 224–228.

Wyeth, F. J. 1924a. The development of the auditory apparatus in *Sphenodon punctatus*; with an account of the visceral pouches, aortic arches, and other accessory structures. *Philosophical Transactions of the Royal Society of London B* 212: 259–368 + plates 11–17.

Wyeth, F. J. 1924b. The development and neuromery of the fore-brain in *Sphenodon punctatus*, with special reference to the presence and neuromeric significance of certain paired metameric diverticula of the central cavity of the fore-brain. *Proceedings of the Zoological Society of London* 1924: 923–959 + plates I–IX.

Wyeth, F. J. 1925. The development and neuromery of the mid-brain and the hind-brain in *Sphenodon punctatus. Proceedings of the Zoological Society of London* 1925: 507–558.

Wyeth, F. J. and Row, R. W. H. 1923. The structure and development of the pituitary body in *Sphenodon punctatus. Acta Zoologica* 4: 1–63.

Wylie, A. P., Veale, A. M. O. and Sands, V. E. 1968. The chromosomes of the tuatara. *Proceedings of the University of Otago Medical School* 46: 22–23.

INDEX

Page numbers in **bold** refer to illustrations.